JOURNAL OF CHROMATOGRAPHY — volume 70

quantitation of amino acids and amines by chromatography

methods and protocols

JOURNAL OF CHROMATOGRAPHY — volume 70

quantitation of amino acids and amines by chromatography

methods and protocols

edited by

Ibolya Molnár-Perl, PhD, DSc
Institute of Inorganic and Analytical Chemistry,
L. Eötvös University, Budapest, Hungary

2005

ELSEVIER

Amsterdam – Boston – Heidelberg – London – New York – Oxford
Paris – San Diego – San Francisco – Singapore – Sydney – Tokyo

ELSEVIER B.V.
Radarweg 29
P.O. Box 211, 1000 AE Amsterdam
The Netherlands

ELSEVIER Inc.
525 B Street Suite 1900
San Diego, CA 92101-4495
USA

ELSEVIER Ltd.
The Boulevard, Langford Lane
Kidlington, Oxford OX5 1GB
UK

ELSEVIER Ltd.
84 Theobalds Road
London WC1X 8RR
UK

First edition 2005

Library of Congress Cataloging in Publication Data
A catalog record is available from the Library of Congress.

British Library Cataloguing in Publication Data
A catalogue record is available from the British Library.

ISBN: 0-444-52050-3 (volume)
ISSN: 0301-4770 (series)

∞ The paper used in this publication meets the requirements of ANSI/NISO Z39.48-1992 (Permanence of Paper).
Printed in The Netherlands.

Dedication and Thanks

To my parents,

Erzsébet and Sándor,

and husband,

Miklós

children

András and Éva,

and grandchildren

AnnMarie, Eszter, Marcel, Daniel and Nóra,

for their love,

and my colleagues and students

for inspiration, encouragement and support of my scientific activity.

Preface

Quantitation of Amino Acids and Amines by Chromatography: Methods and Protocols is intended to serve as a ready-to-use guide for the identification and quantification of amino acids (AAs) and amines (As) in various matrices.

Quantitative determination of AAs and As is in a similar relationship with the chemistry of proteins and peptides as elementary analysis with the chemistry of organic molecules: these methodologies are still relevant, and there is also an urgent need of increasingly sophisticated systems and approaches to clarify the composition and structure of proteomes, genomes, etc.

This book is structured in such a way that the main sections are classified by the chronological order of the development of chromatographic methods: ion exchange chromatography (IEC), (in the case of AAs only); gas chromatography (GC); high-performance liquid chromatography (HPLC); and capillary electrophoresis (CE). Subsections are grouped according to the preliminary preparation/derivatization process(es) as the main ordering concept. This means that first the simple methods will be discussed – those suitable for the analysis of the selected compound(s) in their initial/natural forms – followed by their chromatographic analysis subsequent to various derivatization protocols.

The first approach to the automatic chromatographic analysis of Aas – known today as IEC – was published by Spackman, Stein and Moore in 1958 and who were awarded the Nobel prize in 1972. Now, 46 years later, instead of one day it takes (in special cases) less than five minutes to separate and quantitate the essential protein AAs.

Over the last decades, in addition to the spectacular improvement of the IEC analysis of AAs (in terms of both speed and sensitivity), GC, HPLC and CE have offered unlimited possibilities on both the preparative and analytical scale. The wide choice and sophisticated columns, detectors, derivatization procedures, the development of modern instrumentation and data handling systems have reduced time and costs, and give versatility and automation in Good Laboratory practice (GLP)-controlled conditions for selectivity, sensitivity and reproducibility. It is the responsibility of the researcher to choose the most appropriate method for the given task. The most popular HPLC method for analysis of both AAs and As, (free AAs, present in many natural matrices, biological fluids and tissues, feed- and foodstuffs and of those constituents of protein hydrolyzates), is now reversed-phase (RP) chromatography after pre-column derivatization.

Numerous protocols for derivatization are available. This book will discuss the advantages and drawbacks of the commonly used procedures.

Analysis of AAs in protein hydrolyzates requires a special paragraph dealing with the hydrolysis step itself: since, hydrolysis conditions have a definitive and special influence both on the following anlytical steps and on the reproducibility and recovery of AAs obtained from hydrolyzates. Consequently, hydrolysis of proteins is discussed in detail as a subsection of 1.1.2. , i.e., together with the HPLC analysis of the phenylthiocarbamoyl (PTC) derivatives of AAs: this follows form my multi-facetted experience in the field of protein hydrolysis, and all of this experience is associated with the determination of the PTC-AAs.

Developing new methods and altering/improving old ones is an ongoing process. Despite my more than three decades in the field, and that I follow the literature continuously, readers may have additional, useful proposals or consider some particularly important method to be missing: any questions, comments or suggestions for including other methodologies in a future edition will be gratefully recieved.

I am extremely grateful to the distinguished authors for their time, expertise, and devotion, and for making this book possible.

Contents

Part 2. Amines

2.1. Gas Chromatography

2.2. High Performance Liquid Chromatography

2.3. Capillary Electrophoresis/Capillary Electrochromatography

3. Quantitation of Amino Acids and Amines, Simultaneously

4. Quantitation of Polyamines by Chromatography

Contributors to this Volume

Hans Brückner

Justus-Liebig-Universität Gießen, Interdisziplinä-res Forschungszentrum (IFZ), Heinrich-Buff-Ring 26-32, 35392 Gießen, Germany; e-mail: Hans.Brueckner@ernaehrung.uni-giessen.de

Zilin Chen

Department of Chemical and Biomolecular Engineering, University of Notre Dame, 182 Fitz-patrick Hall, Notre Dame IN 46556-5637, USA; e-mail: zchen@nd.edu

Steven A. Cohen

Waters, Milford, Massachusetts 01757, USA; e-mail: Steven_Cohen@waters.com

Claire Elfakir

Institute de Chimie Organique et Analytique (I.C.O.A.), CNRS UPRES-A, Université d'Orléans, BP 6759, Orléans Cedex 2, France; e-mail: Claire.ELFAKIR@univ-orleans.fr

Charles W. Gehrke

University of Missouri, Columbia, MO., USA; e-mail: UMCHRSSOS@missouri.edu

Petr Hušek

Institute of Endocrinology 11694 Prague 1, Czech Republic; e-mail: phusek@entu.cas.cz

Björn Josefsson

Department of Analytical Chemistry, Stockholm University, Arrhenius Laboratory of Natural Sciences, S-106 91 Stockholm, Sweden; e-mail: bjorn.josefsson@anchem.su.se

Christian W. Klampfl

Department of Analytical Chemistry, Johannes Kepler University, Altenbergerstrasse 69, A-4040 Linz, Austria; e-mail: christian.klampfl@jku.at

Paul Chi Ho

Department of Pharmacy, National University of Singapore, 10 Kent Ridge Crescent, Singapore, 119260, Singapore; e-mail: phahocl@nus.edu.sg

Hiroyuki Kataoka

Faculty of Pharmaceutical Sciences, Okayama University, Tsushima, Okayama 700, Japan; e-mail: www.pharm.okayama-u.ac.jp

Ynze Mengerink

DSM Resolve CT&A Analytics, PO Box 18, 6160 MD Geleen, The Netherlands; e-mail: ynze.mengerink@dsm.com

Ibolya Molnár-Perl

Institute of Inorganic & Analytical Chemistry, L. Eötvös University, Budapest 112, POB 32, Hungary; e-mail: perlne@para.chem.elte.hu

Jacek Namiesnik

Department of Analytical chemistry, Chemical Faculty, Gdańsk University of Technology, 11/12 G. Narutowicza Street, 80-952 Gdańsk, Poland; e-mail: chemanal@pg.gda.pl

Shigeyuki Oguri

Laboratory of Food Sciences 26 Kamikawanari, Hegoshi-cho, Okazaki City, 444-8520, Japan; e-mail: s-oguri@gakusen.ac.jp

Ralf Pätzold

Justus-Liebig-Universität Gießen, Interdisziplinä-res Forschungszentrum (IFZ), Heinrich-Buff-Ring 26-32, 35392 Gießen, Germany

Manuel Silva

Department of Analytical Chemistry Faculty of Sciences, University of Córdoba, E-14004 Córdoba, Spain

Toyohide Takeuchi

Department of Chemistry, Faculty of Engineering, Gifu University, 1-1 Yanagido, Gifu 501-1193, Japan; e-mail: takeuchi@apchem.gifu-u.ac.jp

Thomas Weiss

Department of Surgery, University of Regensburg, Inversitätsstrasse 31, D-93040 Regensburg, Germany; e-mail: thomas.weiss@klinik.uni-regensburg.de

Bogdan Zigmunt

Department of Analytical chemistry, Chemical Faculty, Gdańsk University of Technology, 11/12 G. Narutowicza Street, 80-952 Gdańsk, Poland; e-mail: chemanal@pg.gda.pl

Part 1. Amino Acids

1.1. Gas Chromatography

Ibolya Molnár-Perl (Editor)
Quantitation of Amino Acids and Amines by Chromatography
Journal of Chromatography Library, Vol. 70

1.1.1.

Quantitation of Amino Acids as Chloroformates – A Return to Gas Chromatography

Petr Hušek

Contents

Summary

Alkyl chloroformates (RCF) in aqueous alcohol-pyridine (PYR) media enable amino acids (AAs) to be converted into derivatives amenable to gas chromatography (GC) analysis in seconds. There is no requirement for a dry residue, multiple reaction steps and sample heating. Moreover, the fast conversion of hydrophilic compounds to the organophilic ones has proved to be of general use for most carboxylic acids with the potential to become an integral part of sample work-up. Compatibility of the derivatization with novel sample preparation methods such as solid-phase extraction (SPE) and solid-phase microextraction (SPME), the design of a commercial kit for AA analysis and synthesis of novel fluorinated reagents — all this and more show that RCF-mediated derivatization has become a mature method in GC.

Numerous papers by researchers from various countries and application fields have confirmed the significance and attraction of such a simple and rapid way of changing an analyte's structure prior to GC or even liquid chromatography–mass spectrometric (LC–MS) analysis. Reports published over more than a decade on the analysis of AAs in various materials following treatment with RCF are also summarized in this chapter.

Several methodological lapses and false conclusions in the published papers are discussed throughout the text. In general, the "classic" ECF-ethanol procedure, refined for the whole group of protein AAs, was in most cases simply copied without any optimization of the reaction as were the extraction conditions for targeted applications. It is strange that mainly acidification of the medium, for promoting cyclization of Glu to pyroGlu, was almost always applied in studies where no GLU occurred. It was then clearly useless or even contra-productive.

Further, employment of polar extractants such as chloroform was essential to obtain good yields of polar derivatives, i.e. Ser and Thr (free alcoholic groups) and Gln (free amide) following MCF or ECF treatment. Chloroform also promoted yields of some other analytes a bit. Otherwise, less polar hydrocarbon solvents with a small portion of a polar one were better to handle as an upper and higher-boiling phase. Extraction of, e.g., S- and Se-containing AAs would be feasible in such systems as shown for plasma tHcy and aromatic acids. The more hydrophobic the alkyl of the RCF agent, the less polar solvent was required. Despite this obvious fact, chloroform was used in the follow-up studies almost exclusively. In some papers a molar excess of ECF over PYR was described, but no reason for this was given. In our earlier reports it was shown that by reversing the molar ratio, formation of the mixed (carboxylic-carbonic) anhydrides was promoted, which was undesirable. The importance of clean GC injection port liners was highlighted in one report. In evaluating GC conditions for the derivatives it was found that the choice of the liner used for splitless injections was critical. We can affirm the importance of liner choice from our own studies, especially when MC-ME or EC-EE of the polar and prone to sorption analytes like Ser, Thr and Gln were to be analyzed. The same is true for quality of the GC capillary column.

Regarding the stability of MC-ME and EC-EE, the published reports were somewhere contradictory. Some reported, rather surprisingly, a good stability of EC-EE of GLN (free amide) over 5 days, others reported that IBC-ME were more stable than MC or EC-ME as both the latter tended to decrease upon storage. In our own findings this was especially the case with HIS, the yield of which declined progressively over time if not refrigerated. However, more important for obtaining a good yield of it, and also of SER and GLN, was the quality of the GC column and injection port, as mentioned.

Concerning SPE on exchangers, the possibility to elute directly with a basified reaction medium made the process an integral part of sample pretreatment, simplifying and accelerating it. At the same time, it pointed to robustness of the RCF-derivatization since the reaction proceeded smoothly in presence of the sorbent, sometimes even with improved yields of some analytes.

Contrary to that, SPME did not seem to bring any special advantages in comparison to LLE. Unlike the latter smooth process, requiring seconds and giving high yields, with SPME additional time was required for sorption/desorption of the analytes with more variables. The technique might be beneficial in cases where minute concentrations of compounds of interest occur in large volumes of fluids to be examined.

Introduction of fluorinated alcohols into the process, described first in a study by Wang et al., brought some obvious advantages. It enhanced volatility of the derivatives and lowered their retention in the column, which was in particular desirable in separating AA-enantiomers on the Chirasil-Val column. Furthermore, the strongly acidic fluorinated alcohols were eager to esterify carboxylic groups, leaving thus a minute chance for another alcohol, i.e. to that liberated from the reagent (ECF, IBCF), to compete by forming reaction by-products. Their absence is apparent on Figs. 6 and 7. Eventually, the strongly electron-capturing fluorine atoms allowed substantial enhancement of LOD/LOQ in NICI GC-MS analysis.

As fluorinated RCF (FCF) are commonly not available, they had to be synthesized in the lab. They were made and described for TFECF in AA-enantiomer analysis as TFEC-TFE esters. However, FCF with longer alkyls were synthesized especially by Italian researchers and applied for ultratrace determination (3–30 fmol injected) of highly hydrophilic compounds with multiple carboxylic, hydroxylic, or aminic groups in aqueous solution. Studies on the reactivity differences among four FCF are underway.

Our current projects deal with synthesis of novel, mostly fluorinated agents, too. As a result, novel simplified procedures for treating AAs and other carboxylic acids in biological fluids, beverages and other fluids under optimized conditions are the subject of present tuning studies. Moreover, the reaction mechanism of the processes is being studied in more detail. The employment of FCF is giving interesting insights.

1. Introduction

1.1. Prologue

Let me express a few personal comments. As a result of my earlier studies, a rapid method for derivatization of AAs using dichloro-tetrafluoroacetone (DCTFA) in combination with reactive anhydrides, e.g. trifluoroacetic (TFAA) or heptafluorobutyric (HFBA), was submitted in the early 1980′s. The reaction required an aprotic medium, acetonitrile with PYR; however, both steps could be done in the same medium. Following evaporation of the volatile extraction solvent, an additional brief step also allowed Arg and citruline (Cit) to be determined. The method was reported in the 3-volume book *Amino Acid Analysis by Gas Chromatography*, issued in 1987 [1], and in the *Encyclopedia of Separation Science II* [2].

It was not enough. Like a big bang, a paper on *Amino Acid Derivatization and Analysis in Five Minutes* appeared in 1991 [3]. Only one minute to convert AAs into compounds amenable to gas chromatography (GC), only four minutes for the analysis. A trick? By no means! It was PYR only, a common base in the reaction medium, that caused the miracle. Such a process was discovered by a mere oversight, by omitting DCTFA unwittingly from the reaction medium and admixing methyl chloroformate (MCF) into it to stabilize the His-imidazole [1].

Nevertheless, the first method came at a time when high-performance liquid chromatography (LC) — and later also capillary electrophoresis (CE) — entered the field and became attractive, being *terra incognita*. Therefore, it was not met with much interest. Likewise, the next method was introduced in the 1990's when interest in GC of AAs was clearly fading. *A superior method for a technique leaving the scene*, is what came to my mind.

But the unrivalled way of dealing with polar compounds was slowly recognized as a powerful tool. While in 1996, five years after the discovery, a remark appeared in a paper dealing with LC AA analysis that "the approach using ethyl chloroformate (ECF) for obtaining volatile amino acid derivatives in aqueous samples is not widely used" [4], a decade after one could read that "in relation to GC-based methods for AA analysis the most commonly used one is that of Hušek" [5].

It appears that a decade is usually required to evaluate the potential benefits of a novel finding. During that period more than 100 application papers appeared [6], with citations to the initial report [7] exceeded that hundred. It was concluded, "the rapid and simple sample workup together with the possibility of performing metabolite analysis in complex media is of great industrial and academic interest" [8]

Still, a word about the reagents. They are nasty, poisonous, flammable, corrosive and lacrymatory — without a fume hood they induce tears. But due to the marvelous job they can do, they induce pleasure — in sample preparation for GC analysis at least. The reagents constituted an era — to such a degree that it was also said: BC – before chloroformates, AD – advanced derivatization using chloroformates.

1.2. Retrospection

Until this novel finding, there was no paper published in analytical chemistry, to our knowledge, on an ability of RCF to act as direct and rapid esterification agents, not to say in water-containing media. However, numerous papers by organic chemists dealt with formation of the mixed, i.e. carboxylic–carbonic, anhydrides and with their desirable transformation to the corresponding esters. Some of the attempts were treated in an 1998 review [9], together with the follow-up studies on various classes of carboxylic acids mentioned below.

Unlike the studies of organic chemists done under anhydrous conditions, triethylamine did not prove to function as a suitable esterification catalyst in our experiments. PYR instead appeared to be indispensable for the instantaneous transformation of the carboxylic group into an ester at the analytical microscale. The stunning results obtained with AAs were followed by studies with other carboxylic acids in an attempt to learn more about the reaction mechanism and to optimize reaction conditions for the different classes of compounds.

Esterification of fatty and hydroxycarboxylic acids (FAs, HAs) with ECF performed best in acetonitrile with 1 mol/l PYR and about 4-vol% of ethanol. With MCF, addition of methanol was mostly not required. Should water be part of the reaction medium, alcohol content had to be enhanced. Besides PYR, N-methylpiperidine or 4-dimethylaminopyridine (DMAP) could also be used as catalysts [10, 11]. However, the latter bases failed in derivatizing AAs.

HAs appeared to be tough analytes to derivatize with RCF in aqueous acetonitrile-alcohol media. Along with the main product, the expected O-alkoxycarbonyl alkyl ester, there was also a number of side-products either with shorter or with longer retention times. The former were identified as alkyl esters with a free hydroxyl group, the alkylation of 2-OH-group being partly prevented by alcohol, more by methanol than ethanol. The latter products were shown to be inter-ester oligomers (dimers, trimers) formed by mutual interaction of an activated carboxyl of one molecule with an activated hydroxyl of another one. With acids having the alcohol group not adjacent to the carboxyl, e.g. with 3-OH or 4-OH butyric acids, the formation of inter-esters was not observed, as such remote groups always remained non-alkylated.

However, the main progress in overcoming formation of the side-products ensued from our recent studies on dicarboxylic acids. With acetonitrile prevailing in the reaction medium the derivatization yields of di-C4 and di-C5 acids and their substituents were negligible or low. The acids favoured cyclization to 5- and 6-membered rings of their parent anhydrides unstable enough to sustain injection in the heated block. To promote ester formation, acetonitrile in the reaction medium was replaced by alcohol, the aqueous-alcohol media basified with hydroxide and the reagent added in two portions. Proceeding this way eliminated the aforementioned problems with both the HAs and the dicarboxylic acids [12].

The accomplished studies aided evidently in the search for optimum reaction conditions at some protein AAs, i.e. at Ser and Thr with their β-OH-groups, and at Asp and Glu, being exactly the critical C4 and C5 aminodicarboxylic acids.

2. Derivatization focused on AA analysis

2.1. Using MCF-methanol or ECF-ethanol

RCF with the shortest alkyls and highest reactivity were applied to derivatization studies on protein AAs first [3, 7]. Conversion to the expected products proved to be immediate upon admixing a few microlitres to a solution of AAs in aqueous alcohol with PYR. The volume ratio of the aqueous to the organic portion was 3:2 at the optimum, the latter being ethanol-PYR, 4:1 or methanol-acetonitrile- PYR, 2:2:1. The yield of Glu-diester was low, mainly due to its conversion to pyroGlu. Since a slightly acidified water (50 mM HCl) promoted such a shift it remained a constant part of the medium. The yields of Glu, Asp and Asn were further raised by admixing 1% RCF into chloroform as the extracting solvent. The amide of Asn turned into nitrile while that of Gln remained untouched — the reason for such a different retention behavior. Elution of Arg and Cit failed on any column tested due to insufficiently modified side-chain moieties. The relatively polar chloroform succeeded best in extracting the polar analytes, i.e. Thr and Ser with free hydroxyl group, and Gln with the free amide. Nevertheless, N(O)-methoxycarbonyl methyl esters (MC-ME) of the mentioned analytes were extracted incompletely or slightly (Gln) and were especially prone to sorption in the GC injection port and column. A middle-polar column of 1701 phase-type and a 5–10 m length fit perfectly to separation of both the esters within 4–5 min at a temperature rise of 40 °C/min (Figure1). The N(O)-ethoxy-carbonyl ethyl esters (EC-EE) gave a better separation on the columns tested [7, 13] and an improved GC–FID response of some members than the MC-ME counterparts.

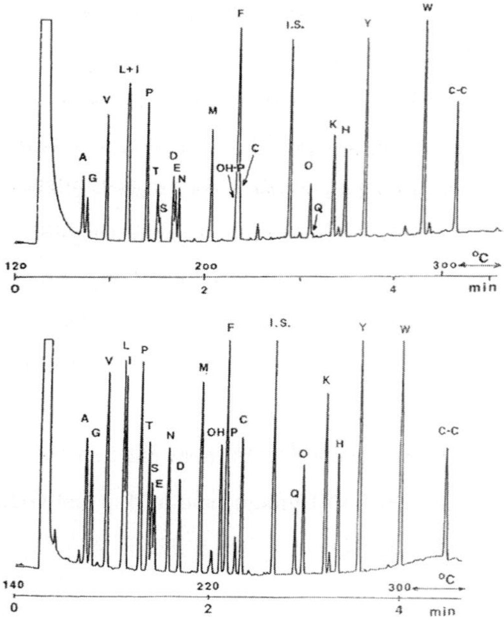

Figure 1 GC-FID analysis of AA standards (initial amount 10 nmol each) as MC-ME (top) and EC-EE on 10 m x 0.25 mm CP-Sil 19 CB (0.2 μm) column in the given temperature range at a rise of 40 °C/min. Hydrogen used as carrier gas at a head pressure of 50 kPa. Reproduced from Reference [7].

Since the procedure was found to be relatively robust and reproducible — CVs did not exceed 5% for most AAs except of Gln and His (8%) — we did not examine the rate of conversion in detail. In a later study, however, an immediate derivatization was put in question. Using methyl laurate as internal standard (I.S.) and exposing the reaction mixture with added ECF to a continuous shaking for several hours, four AAs were said to increase continually within 2–5 h, Val of about 10%, Pro of one third, Ala and Ile of about one half [14].[1]

Anyway, the first reported reaction conditions of EC-EE formation were accepted as a "template" also for other alcohols, reagents and analytes used. The procedure became "classic" in most of the follow-up studies of various research teams, often without giving heed to

[1] Yield increase of some AAs in time cannot be excluded. On the other hand, at the given molar ratio of ECF to PYR a rapid decomposition of the reagent took place in the medium. In our eyes, further additions of ECF were necessary to compensate for the losses and to keep the reaction viable within hours. The rate of conversion was also checked for ECF and heptafluorobutanol (HFB) later [15]. The relative peak areas of some AA-derivatives were determined up to 4 h upon the addition of ECF and before the extraction. No significant alterations were found with the reaction time provided that the proportion of the reagents were correct and in sufficient quantity.

its refinement for the whole class of protein AAs. Should the latter be recognized, simplified treatments could be carried out and improved results mostly achieved.

As mentioned in the preceding section, our studies on dicarboxylic acids [12] aided in understanding the mechanism of side-product formation at Asp and Glu. To promote transfer to the desired diethyl esters the "classic" medium should not be acidified but basified instead. For example, a 50–250 mM solution of sodium hydroxide in the reaction medium was found to be effective in this respect. Addition of MCF or ECF in two portions could be accomplished also via the extractive alkylation, provided that 1–2 % RCF were admixed to chloroform. Profiling of acidic metabolites in body fluids using ECF, while omitting acetonitrile used in the preceding paper[16], will be described in section 6.

Electron-impact (EI) mass spectra together with fragmentation pathways and structures of individual AAs were examined in detail in the follow-up report [17]. GC–MS analysis of EC-EE of AAs using an EI ion source was shown to be convenient and reliable.

2.2. Using RCF and alcohol of the same alkyl

The next commercially available RCF tested were those with larger alkyls, i.e. propyl-, isobutyl- and butyl chloroformates (PCF, IBCF, BCF). Also allyl and isopropyl chloroformate (IPCF) were available but due to instability their benefit was rather limited. Further, secondary alcohols like the IPCF corresponding 2-propanol proved to be very weak esterification agents and should be omitted under the novel conditions in general. But in the two-step reactions, e.g. in treating amino groups with IPCF in the first step and with acidified alcohol in the second one, they were found beneficial in targeted profiling of sulphur AAs [18, 19].

IBCF belonged to the most popular alkylating agent of amino groups as giving a compromise between a higher analyte mass and greater stability of the product. It was found effective e.g. in derivatizing 57 amines in basified water [20]. However, with AAs problems arose with immiscibility of isobutanol (2-methyl-1-propanol), or 1-butanol in case of BCF, with water. Despite the advantages in using mostly net hydrocarbons for liquid-liquid extraction (LLE) of the isobutyloxycarbonyl-isobutyl esters (IBC-IBE), the immiscible medium components resulted in inferior reaction yields especially with basic AAs. In addition, difficulties in eluting high-mass di(iso)butyl esters of disulfidic AAs from GC columns were rather discouraging.

The series of recent papers of Sobolevsky et al. [21–23], dealing with IBCF-mediated derivatization of AAs, is rather a confirmation of such an insufficiency. A mere adoption of the close to "classic" conditions resulted in missing recording of Asp, Glu, His and cystine

(Cys_2), and in a yield lowering for Thr and Ser. Apart from omitting PYR in the text of the first report and unfavorable admixing IBCF in molar excess to PYR in the last one, the chosen 30 m long and thick-coated (0.5 μm) column surely contributed to a partial sorption of Trp and Lys and prevented an effective elution of Cys_2 and His. Nevertheless, comparison of the approach with a silylated one, the latter done under heating in a dry residue, and applying it to a simultaneous determination of 13 AAs, 13 FAs and 6 dicarboxylic acids [22], might be rated a bit more positively. The same concerns the explicit and generally valid remark on an inability of RCF to react with alcoholic groups. It was found "advantageous with biological samples that contain a large amount of sugars. Unlike silylation, selective derivatization of organic and AAs was thus achieved while keeping major components of the biological matrix untouched". IBC-IBE were further exposed to GC-MS detection at positive-ion chemical ionization (PICI) mode using isobutane as the reagent [23]. Estimated detection limits were 6–250 pg in total ion current (TIC) and 3–10 times lower using the selective ion monitoring (SIM) mode.

Replacing isobutanol by 1-propanol led to much improved results in our earlier studies. However, along with propyl esters also the isobutyl ones were formed in amounts of 5–8%. This was a general phenomenon when working in media with different alkyls present, as discussed in the next section.

Based on numerous experiments involving combinations of reagents and alcohols, considering hydrophobicity of the resulting derivatives and their elution behavior on GC columns, the reagent of an intermediate chain length and still sufficient reactivity was chosen. As a result, a unique procedure for AA-pretreatment was developed in cooperation with Phenomenex Inc. (Torrance, CA, USA). It became an integral part of, e.g., body fluid sample workup and obviated the need for serum/plasma protein precipitation. Using a unique SPE there was no need for eluent evaporation following the elution. The procedure comprised 3 rapid steps: a) pre-isolation of AAs on a sorbent placed in pipette filter tips®, b) recovery of the isolated analytes with an extraction–reaction medium, and c) derivatization via phase transfer using an organic phase with the admixed reagent [24, 25].

In comparison to the "classic" method, further improvements were made: the reaction medium was basified, the "reactive" organic phase added in two portions to enhance yields of some AAs, especially Glu, and the effective derivatization was achieved via phase phase-transfer, which aided in augmented yields of some other AAs. The whole process of sample preparation could be accomplished in several minutes and was described in detail in a recent report [5]. GC-MS or GC-FID analysis succeeded also in a few minutes (Figure 2). The

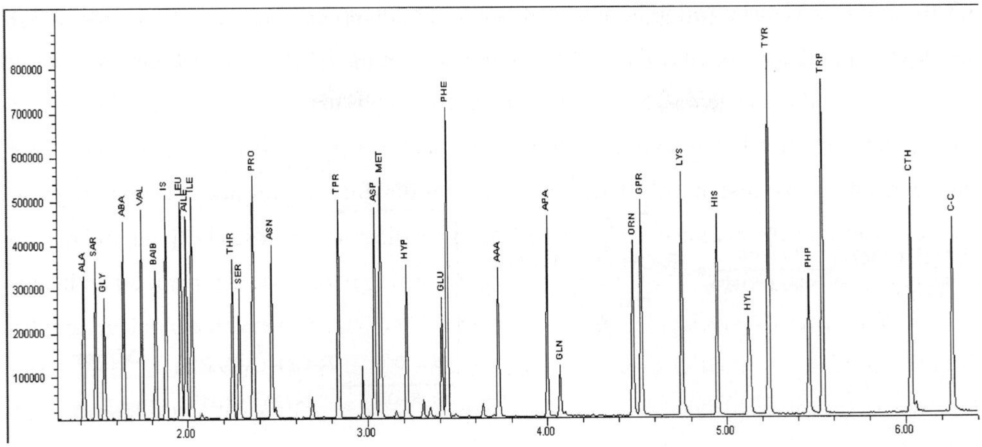

Figure 2. TIC chromatogram of GC-MS analysis of AA standards derivatized with the EZ-Faast Kit and analyzed on 10 m x 0.25 mm ZB-PAAC capillary column under the recommended conditions. Reproduced by courtesy and permission of Phenomenex.

commercial EZ:faast AA-analysis kit (Phenomenex), the first successful kit in the field of AA GC analyses, made compounds ready for both GC–MS and LC–MS analysis [26].

2.3. Using RCF and alcohol of different alkyls

The idea of treating medium containing methanol with ECF and that with ethanol by MCF was tested in our early studies, too. As a matter of fact, EE appeared readily upon treating FAs in chloroform, stabilized with 0.5% ethanol and containing 2% PYR, with MCF [27]. When applied to AAs, ME prevailed upon admixing ECF to the aqueous methanol with PYR. Though no special profit emerged from those trials, it was concluded that a stronger alcohol should be treated with a less reactive reagent to prevent larger formation of side-esters having alkyl of the reagent. By extension, the first useful application of a combined action of RCF and alcohol of different alkyls aimed at separating AA-enantiomers [28; section 3].

However, a detailed and systematic study on a variety of RCF and alcohols used in mutual combinations to obtain a variety of different AA esters was made by Wang et al. [29]. It appeared that the ester moiety of the AA derivative was directly dependent upon the type of alcohol used in the aqueous reaction medium. The alcoholysis, an alcohol exchange reaction with an intermediate mixed anhydride of the carboxyl group, was then responsible for the immediate ester formation. When using an alcohol with an alkyl group different from that in the RCF reagent, the alkoxy group found in the ester derivative was that of the alcohol and not that of the reagent. However, a small amount of side-product always occurred in which the alkyl group in the ester moiety was the same as that in the RCF reagent (**Figure 3**).

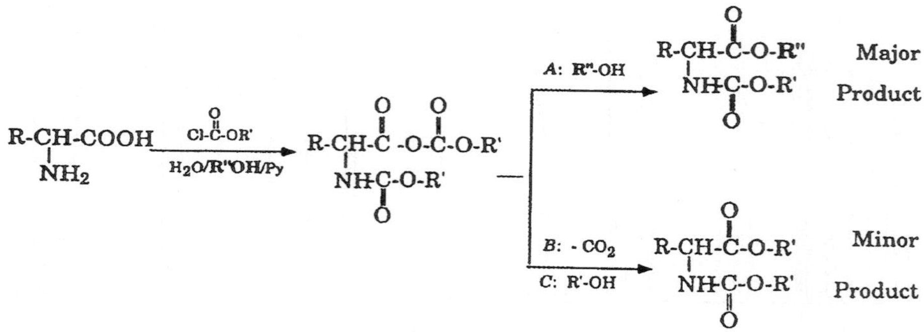

Figure 3. Scheme of AA-derivatization in aqueous-alcohol-PYR media using reagent and alcohol with different alkyls. Reproduced with permission from Reference [29].

Alcohol corresponding to the alkyl of the side-product was produced with a great probability *in situ* from hydrolysis of RCF in the medium. The formation of the minor products was in general less than 10% and more efficiently suppressed when electronegativity of the alcohol was larger than that of the RCF donor group. Also our further finding was confirmed that a molar excess of RCF over PYR contributed to by-product formation to a large extent.

In the study, fluorinated alcohols such as trifluoroethanol (TFE), pentafluoropropanol (PFP) and HFB, together with trimethylsilylmethanol (TMSM), were introduced first into the derivatization process. As corresponding fluorinated RCF reagents were commercially not available, reaction media containing the particular alcohol were treated with ECF, PCF and IBCF. As a result, IBC-HFB (Figure 4), IBC-IB and IBC-TMSM esters produced highest responses regarding peak size. His afforded a larger peak in the first case only, while its response with the two latter combined reagents was very low. The study provided new horizons for preparing esters with different alkoxy groups.

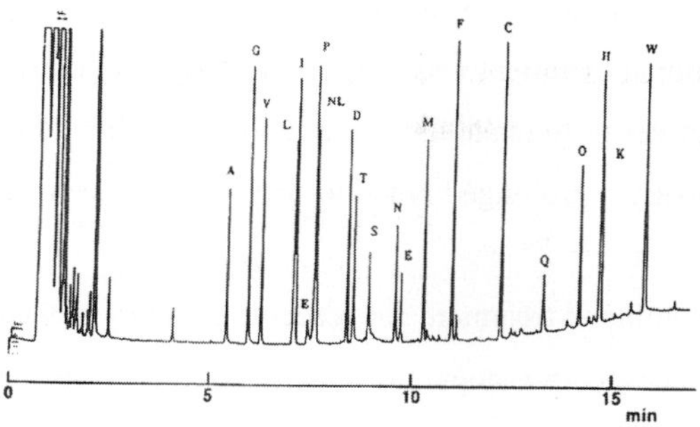

Figure 4 GC-FID record of AA standards (50 ng each) treated with IBCF and HFB and analyzed on 15 m x 0.25 mm/0.25 μm DB-1701 column. Temperature range 100 to 180°C at 10 °C/min, then 15 °C/min to 280 °C. Reproduced with permission from [29].

At the same time, Moini and co-workers modified the "classic" method by replacing ethanol by TFE [30–32]. The intention was to provide for higher sensitivities when using PICI and especially negative-ion chemical ionization (NICI) MS detection. The study [30] indicated that EC-TFE esters of protein AAs possessed higher sensitivity under both detection modes, EI and CI. Twenty-one out of the twenty-three nonprotein AAs studied produced detectable ion chromatograms in both CI modes when methane was used as the CI reagent gas. The detection limits with PICI were in the femtomole range [31]. The derivatization and extraction efficiencies were checked by repeating the treatment following extraction of the derivatives into chloroform and evaporation of the first reaction medium. The former recoveries ranged from 90–99% (except 79% for Asp), the latter being close to 100% [32].

Pentafluorobenzyl chloroformate (PFBCF), the only aryl reagent employed in the novel method, was chosen with the same intention as other fluorinated reagents, i.e. to impart electron affinity in NICI MS detection [33]. Optimization of reaction conditions was first made using Phe and decanol and adopted subsequently for analysis of protein AAs [34]. AAs in aqueous ethanol with PYR plus, i.e. solution of 0.1% DMAP in PYR (Altech Assoc., Deerfield, IL, USA), were treated with 10 μl PFBCF and extracted into toluene. PYR plus as a catalyst gave slightly better yields than PYR alone. Heat treatment at 70°C for 20 min was applied for the decanol, derivatized by PFBCF in the presence of acetonitrile.

Regarding both the fluorinated reagents employed there was no special benefit found except gain in sensitivity. Yields of some AAs were apparently low or nearly none (Ser, basic amino acids). This might further point to the fact that a mere adoption of the "classic" reaction conditions, tailored for ECF and ethanol, might be questionable, as noted in section 2.1.

2.4. Using RCF and additional reagents

Following the rapid derivatization step additional treatment was sometimes chosen to improve the chromatographic properties of the derivatives or to make the compounds accessible to GC analysis at all. The extra step(s) required mostly a prolonged reaction time, sometimes under applied heating.

Huang et al. first employed such a combined treatment for screening urinary acylcarnitines [35]. The procedure was based on 3 consecutive steps: (a) immediate esterification of the isolated analytes with PCF–propanol; (b) ion-pair extraction of the propyl esters with potassium iodide into chloroform; (c) subsequent on-column *N*-demethylation of the resulting acylcarnitine propyl ester iodides. PCF was preferred over the lower-alkyl reagents since it provided more lipophilic esters of especially short-chain acylcarnitines while retaining enough volatility to be easily analyzed by GC–MS with CI.

A sophisticated method for GC determination of S-alk(en)ylcysteine sulfoxides, important secondary metabolites in many plant genera, was developed by Czech researchers [36]. Problems associated with the extremely labile sulfoxide group were solved by reduction of the ECF or MCF derivatized sulfoxides by sodium iodide (1 g/ml aqueous NaI with admixed acetyl chloride, 4:1, standing at room temperature for 24 h; liberated iodine removed with few crystals of $SnCl_2$, extraction with dichloromethane preceded the injection) as shown:

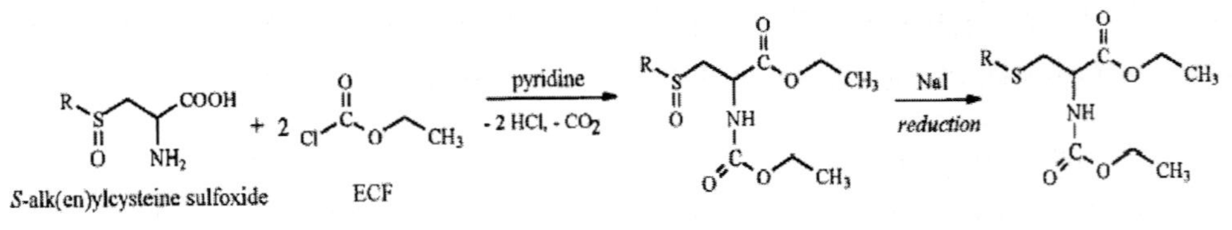

Both MCF and ECF yielded stable derivatives excellently separable by GC-MS. The main advantage of the new method was found in high sensitivity, excellent resolution capability, accuracy and reliability, as well as the possibility to identify unknown compounds.

As discussed previously, the *beta*-hydroxyl groups of Ser and Thr remained non-derivatized, which might cause losses at inferior GC conditions. To prevent it, an additional step was carried in Ser and Thr enantiomer analysis to transform the hydroxyl groups to [37]. After creating IBC-TFE esters the reaction medium (aqueous TFE, PYR and IBCF) was evaporated to dryness. The residue was treated with PYR-IBCF (2:1), sonicated for 5 min, and following extraction with diethyl ether, evaporation and dissolution in dichloromethane, the IBOC-TFE esters of 4 isomers of Thr were then successfully resolved on a Chirasil-D-Val column.

Sequential ethoxycarbonylation, methoximation and silylation were provided for simultaneous determination of amino and carboxylic acids by dual-column GC and GC-MS [38]. The authors abandoned the one-step RCF-based procedure aiming to the same [9], mainly due to "peak tailing and loss of keto acids". The developed 3-step method was based on treating amino groups with ECF in basified water, subsequent oximation of keto groups with methoxy-amine at 60 °C and, following acidification, extraction and evaporation of the extract, the residue with semiderivatized AAs, intact carboxylic acids and methoximated keto acids was silylated with N-methyl,N-*tert*-butyldimethylsilyltrifluoroacet-amide (MTBSTFA) in the presence of toluene at 60 °C for 1 h. A careful study with an awkward derivatization, indeed.

A procedure for a subsequent esterification of RCF-modified amino groups of AAs with a chiral secondary alcohol to get diastereomers of improved separation on GC achiral phases was reported [39]. Likewise, subsequent amidation of the esterified carboxyl was found beneficial in separating AA-enantiomers on a chiral column [40, 41]. For details see the next section.

3. Derivatization focused on analysis of AA enantiomers

Both the field-proven approaches used in enantiomer analyses, i.e. formation of diastereomers and separation on achiral phases or separation of enantiomers on chiral phases, were applied to GC-FID analyses of AA-enantiomers following treatment with RCF.

An optically-active menthyl chloroformate (MenCF, derived from menthol, i.e. 2-isopropyl-5-methylcyclohexanol) was used in combination with aqueous methanol or ethanol for GC-FID analysis of the resulting AA-diastereoisomers [28]. Derivatization succeeded upon admixing 10 μl (-)MenCF to 50 μl of slightly acidified solution of AAs with 40 μl of alcohol-PYR (4:1) within a 10-min standing under occasional shaking and extraction into chloroform. The yields were said to exceed 95%. Favourable separation factors were achieved with most protein AAs including Pro on normal silicone (achiral) phases of DB-210 and OV-1701 type (**Figure 5**).

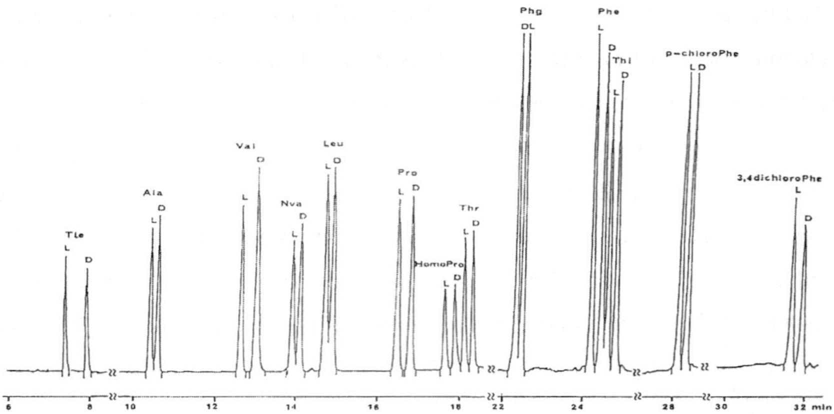

Figure 5. GC-FID analysis of AA standards following treatment with MenCF and ethanol on 30 m x 0.25 mm/ 0.25 μm OV-1701 column. Temperature program, 190 to 223°C at 2 °C/min, to 260 °C at 10 °C/min, to 270 °C at 2 °C/min, to 290 °C at 5 °C/min, then 5 min isothermal. TLe, tert-Leu; Phg, phenylGly; Thi, 3- (2-thienyl)Ala. Reproduced with permission from [28]

Analysis of Arg, His, Trp and cystine failed, GC recording of Ser, Glu and Tyr was not shown and derivatives of Lys and Tyr did not emerge from the OV-1701 column. No menthyl esters as side-products were found.

Abe and coworkers first applied the novel approach to analysis of AA-enantiomers on chiral phases [42]. MC-ME and EC-EE were subjected to analysis on lab-prepared chiral phases of Chirasil-Val and octakis-γ-cyclodextrin/OV-1701 type being lab-coated in pyrex glass tubes (23-26 m x 0.25 mm). With Chirasil-Val, ECF-treated analytes showed mostly better separation factors than MC-ME. Sixteen AAs including His, Tyr and Trp could be separated though with larger size molecules the separation worsened. Higher separation provided the dextrin phase, including a baseline separation of Pro. However, elution of AAs larger than Met was behind the temperature limit of the phase.

In order to improve volatility of the derivatives and to achieve shorter analysis times on the Chirasil-Val column. 2,2,2-Trifluoroethyl chloroformate (TFECF) was synthesized and used in combination with a variety of alcohols [14, 43]. Diminished retention times due to fluorinated alkyls allowed operation at a lower column temperature, which further enhanced the separation. Along with methanol and ethanol also TFE, PFP and the secondary *sym*-hexafluoro-2-propanol were tested as esterifying agents. The latter, however, afforded very low derivatization yields as experienced with secondary alcohols [9]. Among the various AA-derivatives prepared, the TFEC-TFE esters gave fairly stable analytes and attained almost complete separation of all enantiomeric pairs, except for Pro, within 31 min (**Figure 6**). The method was found suitable for enantiomer separation rather than quantitative analysis.

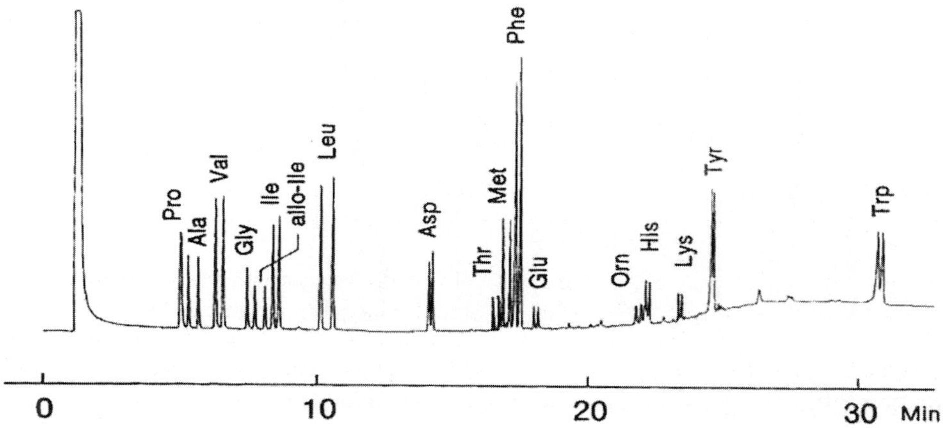

Figure 6. GC-FID analysis of standards of AA enantiomers as TFEC-TFE esters on the chiral 20 m x 0.25 mm Chirasil-Val (0.16 μm) glass capillary. Temperature program: 95 °C, 2 min hold, 3 °C/min to 120 °C, 7°C /min to 225 °C. Carrier gas helium, split ratio 1:40. Reproduced with permission from [14].

Next, the N-alkoxycarbonyl TFE esters were further converted to alkoxycarbonyl alkyl-amides by nucleophylic substitution of the ester group with amines ($R\text{-}COOCH_2CF_3 + H_2NR'$ $\rightarrow R\text{-}CONHR'$)[40, 41]. The ethereal extract of TFEC-TFE esters was evaporated to dryness and isobutyl or n-propylamine was added to the dried residue. Room temperature conversion or that under heating at 100°C in the latter case was applied. Following evaporation of the amines and addition of dichloromethane the final derivatives were separated on Chirasil-Val. All the amidated derivatives showed markedly increased separation factors including Pro that was not separated as its N-perfluoroacyl alkyl ester. The method proved to be suitable for enantiomer separation of relatively simple AAs, while not suited for those with more than 3 functional groups because of their lower volatile properties.

Casal and co-workers chose later a combination of ECF and HFB for enantiomer separation on Chirasil-Val [15]. Several derivatives were prepared using a series of ethyl and isobutyl esters. AA-aqueous solution (125 μl) plus 30 μl of alcohol-pyridine (2:1) were treated with 7 μl ECF and after brief vortexing chloroform extraction was performed. EC-HFB esters showed eventually the best compromise between short retention times, higher responses and good resolution for most AAs tested (Figure 7).

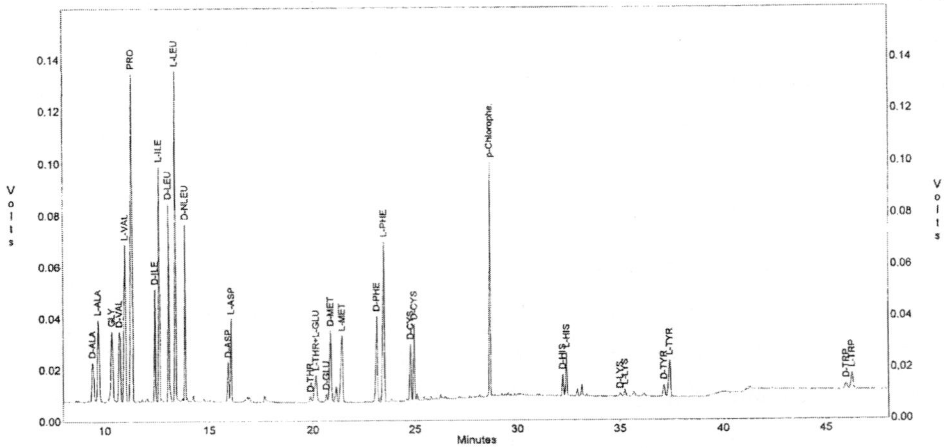

Figure 7. GC-FID analysis of standards of AA enantiomers (400 nmol each) as EC-HFB esters on 25 m x 0.25 mm Chirasil-L-Val (0.12 μm) column. Temperature program: 80°C, 1 min hold, 5 °C/min to 150 °C, 7 min hold, 7 °C/min to 200 °C, 15 min hold. Helium used as carrier gas at 60 kPa initial inlet pressure. Splitless injection used with a time delay of 0.9 min. Reproduced with permission from [15].

However, as is apparent from the record the yields of Thr, Glu and Lys are very low, Ser is fully missing. The stability of EC–HFB esters was tested over a period of 2 weeks, when both the phases of the reaction medium were in contact at 4°C. No significant alterations were observed, and no transformation between the enantiomers occurred. The method was found suitable for routine AA-quantification and was applied to various food samples.

Enantioseparation of 30 racemic AAs in the form of diastereomers on achiral GC columns was reported [39]. Esterification with a chiral secondary alcohol, the (S)-(+)-3-methylbutan-2-ol (at 100°C for 1.5 h in presence of acetyl chloride) followed the preceding step, i.e. phase-transfer derivatization of amino and phenolic hydroxyl groups with ECF in dichloromethane in presence of aqueous phosphate-buffered (pH 8). Among the chiral alcohols tested the one chosen provided the best resolution factors (1.2–8.0). The lengthier two-step procedures, treated in chapter 2 of this book, are always necessary with secondary alcohols, which fail as esterifying agents in the one-step approach.

4. Derivatization and analysis of selected AAs

4.1. Sulphur-containing AAs

Interest in analysis of S-containing AAs was initiated by clinicians in connection with a discovery made in the mid 1980s [44] that plasma homocysteine (Hcy) might be used as a risk factor of cardiovascular disease. As more than 70 % Hcy in plasma is bound to albumin, about 25% oxidized to disulfides (mostly coupled with Cys, much less with another Hcy to form homocystine, Hcy_2), and less than 5% is the free form, a reduction step must precede its determination to get the clinically significant total plasma Hcy (tHcy). Pre-analytical conditions affecting the plasma tHcy levels should be strictly followed. Plasma should be separated from blood cells within 30 min otherwise Hcy levels increase by about 10% per hour. Among numerous stabilizers examined, the 3-deazaadenosine (100 μmol/l) performed best as stabilizing Hcy concentrations for 24 h. However, this stabilizer was restricted to LC methods as not functioning reliably with immunoassays [45].

Clinicians in Norway contributed a great deal to the topic, starting with a radioenzymatic determination of tHcy in body fluids [44], over a fully automated fluorescence assay for plasma tHcy [46] through novel assays based on the modern hyphenated techniques, GC-MS and LC-MS [47, 48]. The ECF method proved to be reliable in GC-MS screening of plasma tHcy, with little bias to the immunological assays [47]. Earlier reviews were focused on plasma levels in health, disease and drug therapy [49], and on methods of the 1990s and clinical applications [50]. Current reviews summarized numerous assay methods elaborated for the measurement of plasma tHcy over the past decade, including the one-step RCF procedures [51–55].

GC procedures based on the one-step RCF treatment not only simplified sample workup but also allowed allied AAs like Cys, Met and possibly also S-dipeptides to be assayed. Unlike hydroxyl groups of Ser and Thr the thiol groups of Cys and Hcy were always alkylated.

The first papers employing the approach for GC-MS determination of plasma tHcy appeared in 1997. In one report combination of ECF–ethanol [56] an in another that of PCF–propanol [57] were used. Following reduction with tributylphosphine or dithiothreitol (DTT), deproteinization with sulfosalicylic (SSA) or trichloroacetic (TCA) acid was performed, and the supernatant treated directly with PCF upon admixing propanol or subjected first to SPE on a cation exchanger [57]. The simple approach enabled analysis of >100 samples within a day.

SPME was engaged to capture RCF-derivatized Met, Cys and Hcy to be assayed by GC-MS [58]. The aqueous portion of the reaction medium contained TCA and DTT, to which propanol-PYR (4:1) were admixed. Various RCF were tested to evaluate affinity of the formed alkoxycarbonyl propyl esters to polydimethylsiloxane (PDMS) or polyacrylate (PA) fiber within 30-min equilibration time. ECF was finally selected as the reagent of choice and optimum conditions for extraction of the EC-propyl esters on the fiber were examined. Some erratic or confusing statements made on reactivity of RCF, yields and responses, and composition of the reaction medium were raised in a critical report [59].

Fasting plasma concentrations of tHcy, Met, Cys, and cystathionine were determined by GC-MS in end-stage renal disease patients receiving daily oral folate and vitamin B6 supplements [60]. Plasma was treated with 1% DTT, deproteinized with SSA, AAs captured on cation ex-change resin and eluted with ammonia. The dried residue was dissolved in water-ethanol-PYR and treated with ECF. Levels of Hcy assayed by GC-MS were compared with those obtained by LC and Abbott Imx immunofluorescence method and a close correlation was found.

Trap and release membrane introduction MS using a removable direct insertion membrane probe was employed to direct quantitation of plasma tHcy alone or together with Cys [61, 62]. DTT for reduction, TCA for protein precipitation and ECF for derivatization were used. There was no need for chromatographic separation. EI GC-MS in SIM mode was applied for quantitation of both Cys and Hcy. The technique proved to be simple and sensitive, linear and reproducible for simultaneous quantitation of selected AAs in plasma and urine after derivatization with ECF. Hcy displayed limit of quantitation (LOQ) of 2 muM.

A combination of IBCF and ethanol to form IBC-EE of Met and Hcy was employed in another study [63]. AAs were captured from the plasma supernatant, containing DTT and TCA, on BondElut SCX cartridge (Varian, Harbor City, CA, USA), and eluted directly by the reaction medium, consisting of water-ethanol-pyridine (15:8:2). Thus, the unique possibility obviating ammonia elution and eluant evaporation [24] was favorably employed. The estimated LOQ was 0.2 μmol/l for plasma Hcy and 0.05 μmol/l for Met.

The same approach, i.e. elution of AAs captured to exchanger by means of reaction medium, was carried out [64]. Moreover, the simplified sample workup also obviated protein precipitation. The reduction of disulfides in plasma with DTT was accomplished within 2–3 min. Following uptake on cation exchanger and elution with reaction medium (150 μl of saline-ethanol-PYR, 15:8:2), AAs were phase-transfer derivatized by shaking with 150 μl of organic phase containing ECF (isooctane-chloroform-ECF, 12:4:1) within seconds. Along with S-AAs, aromatic AAs could also be determined during a 5-min run. GC-FID proved to be sensitive enough to reach plasma Hcy levels, the values being in full agreement with those obtained by LC and close to those obtained by two Hcy immunoassays.

4.2. Selenium-containing AAs

Selenium (Se) belongs to essential metabolic trace elements with intake derived mainly from foods. Naturally abundant and partly identified organoselenium compounds point to a huge complexity of Se-chemistry in the environment and in living organisms. Among Se-AAs, e.g., SeMet, SeCys, $SeCys_2$, Se-methylSeCys, Se-methylSeMet, Se-cysteic acid and some others were identified. The qualitative and quantitative determination of the particular species of this element appeared to be vital in understanding Se-metabolism and its significance in biology, toxicology, clinical chemistry and nutrition. The potential of the RCF-methodology to serve this task was shown in our earlier study [65]. GC-FID analysis of SeMet, selenoethionine (SeEth, I.S.) and $SeCys_2$ in form of EC-EE succeeded on the same 5 m capillary column used for the whole class of protein AAs [13]. Elution of the high-molecular mass $SeCys_2$ was shown to be smooth, certainly in connection to the short column length. Some further studies made use of the rapid one-step treatment but without reflecting a demand for shorter and/or thin-coated capillary columns.

The next two papers might be a clear demonstration of this. Cai and co-workers used derivatization with ECF to identify Se-AAs in vegetables [66]. The analyte of highest mass was Se-allylSeCys in that study so that elution from 30 m long and 1,5 μm thick-coated column was still manageable, though with apparent peak tailings of all the analytes. On the contrary, elution of RCF-derivatized disulfidic AAs posed problems in the recent comprehensive study dealing with influence of different RCF (ECF, MCF and MenCF) on derivatization yields and GC analysis of selected Se-AAs [67]. Since the problems encountered with elution of the high-mass Cys_2 and $SeCys_2$ derivatives, especially the methylated ones, chromatographic parameters and inertness of the different instrumental configurations were intensively studied. It was concluded that "optimization of the inlet temperature and the column flow for split/ splitless injection showed that low inlet temperatures and high column flows led to improved sensitivities... pointing so to a limited thermal stability of the derivatized S- and Se-AAs... and to potential losses in the GC system during the transfer of the analytes from the injector to the detector...". Such conclusions are, however, more or less erratic since they do not hit the heart of the matter[2]. In spite of the limited reproducibility of the method, its repeatability was found to be good enough to allow accurate determination of Se-AAs. This was demonstrated by the analysis of Se-supplementation tablets for human diet that contained SeMet.

Uden and co-workers paid especially great attention to analysis and selective detection of organoselenium compounds, as just reviewed [68]. Such a complex analysis became a big challenge for state-of-the-art hyphenated analytical techniques. Head-space GC-atomic emission detection (AED) proved to be most useful in this respect, possesing the ability to display a number of element-specific chromatograms simultaneously and with a high sensitivity [69]. The AED response was able to flag compounds that contained specific elements in the GC effluent despite their very small abundance or coelution with other components. In this fashion Se–S containing species could be differentiated. In principle, the GC eluent was introduced into a microwave-energized helium plasma coupled to a photodiode array optical emission spectrometer. The energetic plasma was atomizing all the sample elements, exciting then their characteristic atomic emission spectra. Up to 4 elements with adjacent emission wavelengths could be monitored simultaneously [69].

Rapid indirect detection Se-specific nuclear magnetic resonance (NMR) spectroscopic methods were developed to directly analyze aqueous extracts of hydrolyzed Se-yeast without derivatization or separation [70]. If LC-inductively coupled plasma (ICP)-MS, GC-AED and GC-MS analysis was employed, derivatization with diethyl pyrocarbonate (700 μl added to 1.5 ml of water-ethanol-PYR, 60:32:8) or with ECF was performed after proteolytic digestion and aqueous extraction. The former reagent was found equally effective as ECF. A 10-times larger volume of the above medium (15 ml) was treated with 1 ml ECF and applied for analysis of Se-compounds in Se-accumulating *Brassica juncea* (Indian mustard) and in selenized yeast [71]. A key feature of the study was the complementary role of Se- and S-specific detection by LC-ICP-MS and GC-AED. Limits of the LC-detection for such species were in the range 5–50 ng Se/ml in the injected extracts.

Two different derivatization approaches for SeMet and its GC-ICP-MS detection were employed in another study [72]. The one-step ECF treatment was compared with the two-step

[2] Because long columns heavily coated (0.53 μm HP Ultra 2, and 1,8 μm of HP 624) were employed in relation to the study [66], a smooth elution of the heavy analytes was hardly possible. Even more, it was a miracle that they were eluted at all. Less a decomposition but much more analyte sorption led to the problems observed.

approach based on esterification of the carboxylic group with propan-2-ol followed by acylation of the amino group with TFAA. The advantages and disadvantages of both procedures were discussed. The latter method was said to provide cleaner chromatograms and more stable derivatives under the chosen conditions, and was applied to assay of SeMet in a parenteral solution. The faster RCF method was reserved as a screening method for qualitative information on possible presence of Se-AAs in the sample.

SPME was engaged in sample preparation strategy for SeMet, SeEth and $SeCys_2$ GC analysis [73]. IBCF proved to be the most useful RCF reagent since IBC-IBE derivatives were most sensitive to extraction by PDMS-coated fiber, affording a factor two better sensitivity than ECF. Larger volumes of both the reaction medium, i.e. 3.6 ml of water-alcohol-PYR, 6:3:1, and the reagent (0.25 ml) were used for the SPME step, carried out by magnetic stirring for 10 min. GC-ICP-MS was employed for final Se-detection at sub-ppb levels.

Fourteen extraction methods commonly cited in the literature were evaluated for quantitation of Met and SeMet in a yeast candidate certified reference material [74]. Species-specific isotope dilution GC–MS was utilized to compensate effectively for potential errors, such as losses during derivatization and clean-up steps. All the different extraction media were treated with MCF-methanol. Significant differences in measured Met and SeMet levels were obtained in relation to extraction method used. Half of eight tested methods of enzymatic hydrolysis frequently used for extraction of SeMet from yeast submitted very low extraction efficiencies. On the other hand, a 4 M methanesulfonic acid reflux digestion was found to be most efficient for both the analytes. The latter digestion was then applied to analysis of SeMet and Met in yeast [75]. To 1-ml extract 0.48 ml of ammonium hydroxide (concentration not given)[3] and 0.75 mL of methanol-pyridine (3:1) were added and treated with 0.25 mL MCF. After the chloroform extraction the sample was subjected to GC-MS analysis.

---

[3] Ammonium cations were found largely inconvenient in our pilot studies with AAs. They react with RCF and consumpt portion of it, which leads to diminished derivatization yields. This can be partially compensated by admixing unnecessary large amount of RCF as apparently done in this study. Besides of, the authors refer to derivatization procedure of [67] where no ammonium hydroxide was added into the reaction medium.

5. Determination of AAs in various materials

5.1. Clinical material

Body fluids contain numerous products of body metabolism, mostly aliphatic and aromatic carboxylic acids and AAs. Of the wealth of substances some are of diagnostic value but possess structures with functional groups of diverse polarity. A simultaneous transforming of most of those metabolites into derivatives amenable to GC analysis was always difficult to manage. The task became, then, the next major challenge for the RCF methodology.

The employment of ECF enabled marked simplification of body fluid sample processing [16, 76–78] since reducing plasma workup to minutes [16]. First, ethanol provided both plasma de-proteinization and subsequent derivatization. Second, isolation of compounds of interest could be omitted, and neutral lipids extracted off by simple vortex-mixing with hexane. Third, the keto acids did not show a need for preceding oximation. By reversing the mode of reagent and base addition, formation of side-products (see section 1.2.) of the predominant metabolite of serum profile, the lactic acid, was effectively suppressed.

Because of problems experienced with derivatization of C4 and C5-dicarboxylic acids and some of their important substituents later, due to cyclization to parent anhydrides in presence of acetonitrile [12], novel procedures were elaborated for metabolic profiling in body fluids. The details are given in section 7.

Multidimensional enantioselective capillary GC–MS was applied for analysis of the main urinary metabolites excreted in maple syrup urine disease [79]. Without any alcohol added, 1 ml urine was simply mixed with 1 ml PYR followed by careful dropwise addition of 2 ml MCF. The main metabolites of the disorder were monitored in cut-intervals in ethereal extract. Surprisingly, such a simple way of urine workup succeeded in determining the branched-chain carboxylic, oxo, hydroxy and AAs. Even though not optimized for the particular analyte classes it was evaluated as reproducible enough.

A simple procedure was submitted also for urinary AAs [80]. To 300 μl of urine 500 μl of methanol–PYR (4:1) and 1 ml chloroform were admixed and treated with 50 μl ECF. Derivatization and extraction were joined in one step and 1 μl of the organic layer injected directly into a GC–MS instrument. Sample preparation required about 5 min but the method was limited to assay of 15 urinary AAs; Thr, Ser, Asn and Gln did not appear on the chromatogram.

The prolylhydroxyproline dipeptide (PHP), a potential novel osteomarker with a high correlation to urinary total hydroxyproline (HP), was determined by GC-FID following SPE and LLE involving ECF as the derivatizing agent [81].

Derivatization of AAs with PFBCF in aqueous ethanol aimed at gaining sensitivity in GC-NICI (methane)–MS [34]. To demonstrate scope of the approach, AAs from a finger stick were subjected to the derivatization as a 10-µl whole blood sample with no prior extraction or purification (Figure 8).

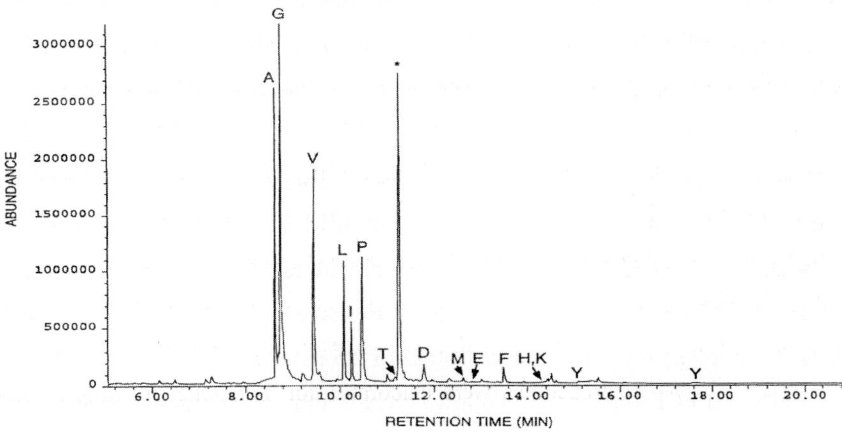

Figure 8. SIM profile from GC-NICI-MS analysis of PFBCF-ethanol treated whole blood (10 µl) sampled from a finger stick. Final volume 100 µl, injected 1µl splitless at 300°C. Column: 15 m x 0.25 mm/0.25 µm DB-17, temperature program was 70 to 150°C at 30 °C/min, 150 to 250 °C at 10 °C/min, and 205 to 300 °C at 30 °C/min with a 3-min hold. Reproduced with permission from [34].

The blood sample was covered with 100 µl of water–ethanol–PYR (containing 0,1% DMAP) treated with 10 µl PFBCF and the derivatives extracted into 100 µl of toluene. His, Lys, and Tyr yielded two derivatives, the yields of basic AAs were apparently low.

Pietzsch and co-workers applied often the one-step derivatizaton to screening AAs and targeted analytes in blood and tissue [56, 82–86]. Stable isotope ratio GC-EIMS analysis of AAs in plasma and protein hydrolysates was done first by using ECF-ethanol [82] and later using ECF-TFE derivatization [83]. In accordance with the earlier findings [30–32], EC-TFE esters were confirmed as derivatives providing a higher sensitivity and specificity when compared with their non-fluorinated counterparts.

The next three papers of the team focused on measurements of specific markers of low-density lipoprotein (LDL) apolipoprotein B-100 oxidation. An alteration of apolipoprotein (apo) B-100 structure by direct oxidative modification is supposed to be an important mechanism involved in atherogenesis. The developed method was based on the oxidation of protein Arg and Pro to γ-glutamyl semialdehyde that was further reduced to 5-OH-2-aminovaleric acid. The latter compound was transferred to EC-EE [84] or EC-TFE ester [85] and analyzed by GC-MS down to femtomole levels. In the latter case, lesion LDL were isolated from the intima of normal and atherosclerotic specimen of human thoracic aortas obtained at necropsy. AAs in residue (max. 50 μg) were covered with 100 μl TFE-PYR (4:1) and mixed with 10 μl ECF by shaking for 30s. The analysis succeeded after extraction into chloroform containing 2% ECF. Eventually, EC-TFE ester of 6-OH-2-aminocaproic acid, a highly specific marker of metal catalyzed protein oxidation, was analyzed in the same way [86].

GC-MS assay of serum Phe and Tyr, the established phenylketonuria (PKU) markers, used ECF and 2-chloroPhe as I.S. [87]. Serum plus aqueous I.S solution (500 μl each) was deproteinized with concentrated HCl and chloroform (75 μl each), followed by heating at 75°C for 5 min. A 200-μl aliquot of supernatant was shaken with 100 μl of hexane to scavange lipids. Then, 60 μl were mixed with 200 μl of 2.5 m*M* aqueous HCl–ethanol–PYR (60:32:8) and 100 μl (!) ECF was added[4]. Following chloroform extraction the analysis was carried out.

ECF-ethanol treatment succeeded in measuring C-13- and N-15-Gln enrichments in plasma samples by GC-MS and GC-combustion/isotope ratio-MS the [88]. The EC-EE derivative of Gln was found to be very stable even after 5 days storage at room temperature. The assay proved to be reproducible and accurate with RSD values below 0.8 and 3.2%, respectively. The method pursued isotopic enrichments in rat plasma after oral force-feeding with [2,5-N-15(2)]-Gln and in human plasma samples after intravenous infusion of [1-C-13]-Gln. Rapid and sensitive method for screening AAs in neonatal blood samples following treatment with IBCF-methanol and SPME was reported [89]. The IBC-ME were headspace extracted by a SPME fiber, desorbed and detected by GC-EIMS. The method was applied to diagnosis of

---

[4] Amount of ECF added is about 10-fold higher than usual. Further, there is a redundant addition of aqueous phase into the reaction medium. Why not add ethanol-PYR only to the acidified aqueous supernatant? For comparison, we point to our study on plasma tHcy assay [64] that allowed a simultaneous screening of the aromatic acids after a more simplified sample workup obviating protein precipitation.

neonatal PKU and maple syrup urine disease (MSUD) by analyzing 5 AAs in blood samples. The approach was said to have a big potential for simultaneous screening of both the diseases.

5.2. Biological and environmental material

The "classic" conditions leading to formation of EC-EE were applied to several analytical tasks soon after being published. Aminomalonic acid (Ama) was detected using GC-MS in alkaline hydrolysates of proteins from *Escherichia coli* ribosomes; about 0.3 Ama per 1000 AAs occurred in the hydrolysates [90]. Aminosalicylates like 5-aminosalicylic acid (5AS) and N-acetyl-5AS and their radical-derived oxidation products were converted to EC-EE and analyzed using GC–MS. Some of the analytes were considered as possible markers for identification of yet unknown metabolites of 5AS in biological material [91]. Further, cycad seeds were investigated for a possible presence of some neurotoxic AAs. Twelve nonprotein AAs, among them a putative neurotoxin β-*N*-methylamino-L-Ala and the known neurotoxin β-*N*-oxalylamino-L-Ala, were identified as EC-EE [92]. The AA sequence of more than 20 low-molecular weight Ser protease inhibitors was determined in insects using ECF in a later study. The authors dealt with the purification and sequence of a small peptide that inhibited chymotrypsin and human leukocyte elastase [93].

The next two studies focused on estimating rate of protein synthesis in vivo. Using MCF and GC–combustion isotope MS, 1-^{13}C values of Val in protein hydrolyzate were determined [94]. For determination of extra/intracellular enrichments of [1-^{13}C]-α-ketoisovalerate using enzymatically converted [1-^{13}C]-Val standardization curves, and the applicability of the method in a [1-^{13}C]-Val tracer infusion in a biological model, the same method was used [95].

Sobolevsky and co-workers compared silylation and esterification/acylation procedures in AA GC-MS analysis and found the one-step treatment using IBCF preferable to analysis of lyophilized *E. coli* microbial culture [22; section 2.2.].

Secreted metabolites from filamentous fungi (extra-cellular media of *Aspergillus terreus* cultivated aerobically on glucose) comprising AAs and nonAAs were analyzed simultaneously as MC-ME [8]. The modified conditions for dicarboxylic acids [12] were adopted, i.e. medium was basified (1% NaOH), methanol content enhanced while excluding acetonitrile, and MCF added in two portions. Lowering GC-inlet temperature to 180°C prevented partial degradation of labile compounds such as phosphoenolpyruvate, 2-oxo-glutarate and oxaloacetate.

An alternative method for quantitation of indole-3-acetic acid and Trp in culture supernatant (*Sinorhizobium meliloti* and *Rhizobium* exudates) avoiding the previous use of diazomethane under anhydrous conditions, employed a combination of ECF and PFP for GC-MS assay [96]. To 1 ml of aqueous supernatant 200 μl PFP, 100 μl PYR and 100 μl ECF were added under vortexing. Following standing for 10 min, extraction with chloroform, evaporation and resuspendation in dichloromethane GC-MS analysis was performed.

As mentioned in 4.2., ECF was applied to screening of anionic and neutral Se-compounds in Se-accumulating plants such as *Brassica juncea* (Indian mustard) by means of GC-AED [71]. Such a plant behavior may provide a cost-effective technology to clean up contaminated soils and waters (phytoremediation). S-(methylSe)Cys was detected in shoots and roots of the plant when grown in the presence of selenate or selenite for the first time.

Concerning the environment, Abe et al. [97] employed IBCF-methanol for treating AAs in hydrolyzates of certain fossils to determine their approximate age. The dating studies were based on changes in enantiomeric ratio of especially Asp, as in living tissues the L-AAs underwent racemization to D-isomers once the life process had ceased. IBC-ME of the enantiomers were separated by capillary GC on a chiral phase. They were said to be more stable than MC or EC-ME as both the latter tended to decrease upon storage.

Three different derivatization methods were tested in order to select and optimize one for the in situ AA analysis in Martian samples [98]. The silylation procedure (MTBSTFA in dimethylformamide, 70°C for 30 min) was easily to automate with a high yield and a large linear response. The alkylation method using tetramethylammonium hydroxide was simple and easily automated, but irreproducible data and by-products were obtained due to reaction in the hot (300°C) GC liner. The MC-ME formation proved to be advantageous regarding sample preparation and short time analysis. The main problem, however, was that the shaking step was difficult to develop in this space application.

5.3. Food and pharmaceuticals

The first application in the food industry was at Nestec Research Centre [99] and was connected with an undesirable process of product browning in food technology. Due to a nonenzymatic glycation of the primary or N-terminal and ε-Lys amino groups in peptides and proteins with aldoses, losses of the essential Lys, e.g. in milk powder, occurred. Periodate oxidation of the glycated products led to formation of *N*-carboxymethyl (i.e. dicarboxylic) AAs, the amount of which directly reflected the extent of glycosylation of the various amino sites in the products. The compounds were effectively derivatized and analyzed as EC-EE as shown:

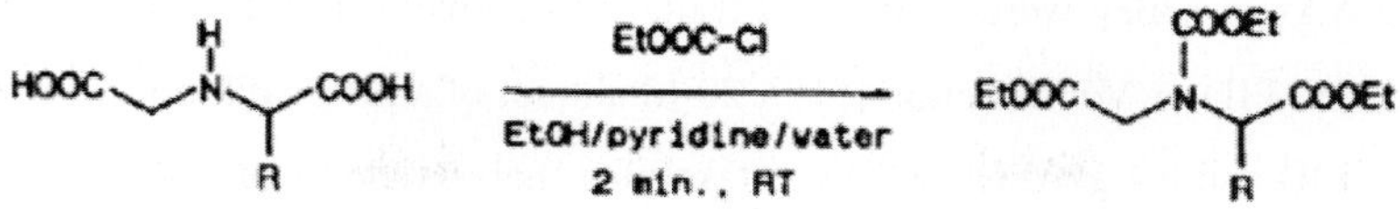

ECF in combination with TFE was shown to enhance sensitivities in analyzing some small peptides down to the femtomole range in GC–MS using EI and NI/PICI modes [32]. The approach was extended to derivatization of twelve dipeptides and one tripeptide, the hydrolytic products of Equal, a sugar substitute containing aspartame (a methyl ester of Asp-Phe) and diprotin B [100].

Se-AAs were identified in ordinary and Se-enriched garlic, onion and broccoli after derivatization with ECF. Lyophilized samples were suspended in 0.1 M HCl, centrifuged and 300 μl of each supernatant was mixed with 200 μl ethanol-PYR (3:1) and 50 μl ECF were added. After chloroform extraction (500 μl) analysis by GC-AED was performed [66].

Kubec et al. developed a method for measurement the content of alliin and other S-alk(en)yl-*Cys* sulfoxides in nine different samples of garlic (Allium sativum L.) and identified a new flavour precursor, the S-ethylCys sulfoxide (ethiin) [101]. In addition, the content of S-methyl-Cys sulfoxide (methiin) was determined in 12 common cruciferous species [102]. AA fraction was isolated by ion-exchange chromatography (IEC), the ammonia eluate (50 ml) evaporated, and the compounds exposed to ECF and NaI treatment [36; section 2.4.]. Excellent sensitivity, accuracy, and reliability was reported, with RSD values (mean of 7%) 4-times lower than by using an LC method. Approximate LOD was 1 ppm of methiin in the fresh tissue.

Portuguese researchers published 4 papers on GC quantification of AAs in food matrices [15, 103–105]. Enantiomeric AAs were converted to the more volatile EC-HFB esters [15, 103], while the non-enantiomeric ones to the EC-EE [104, 105]. AAs were extracted, hydrolyzed and isolated by IEC, transferred to EC-HFB and analyzed by GC-FID on a chiral column. Enantiomeric profile in several food samples (milk, yoghurt, black beer, balsamic vinegar, green coffee) was given [15]. It was also shown that free AAs could be used as a tool for discrimination between coffee species, with a special reference to L-Glu, L-Trp and pipecolic acid [103]. The profile of free AAs in quince fruit (pulp and peel) and jam was reported [104, 105]. Sample workup followed that above using IEC. Due to its rapidity and low cost, this technique was found to be useful in the quality control of quince products.

Free AAs in honey were derivatized using the commercial EZ: faast kit [24–26] and subjected to GC-FID or MS detection [5]. A 20-μl aliquot of a honey dilution with water (0.8 g/ml for GC-FID and 0.4 g/ml GC-MS) was taken for analysis. Following admixing 200 μl of I.S. solution (Nval) the sample was subjected to SPE on a 40 μl resin-packed sorbent tip®. Eluting medium ejected also the sorbent particles, and the slurry was treated with the reagent via phase-transfer derivatization. The method allowed determination of 22 free AAs in honey samples in several minutes. Linearity range, LOQ, reproducibility and accuracy were found suitable for AA quantification in honey.

Concerning pharmaceutical applications, Jegorov and coworkers used MenCF in combination with various alcohols for chiral analysis of some nonprotein AAs occurring in peptide antibiotics [106–109]. The best separation factors were obtained with methyl, 2-fluoroethyl and 2-methylbutyl esters on silicone achiral phases of DB-210 and BPX70 type [106]. The method was employed successfully in chiral analysis of AA residues from new natural and semi-synthetically prepared cyclosporines [107, 108]. Resolution of racemic threo forms of 4-fluoro Glu, an ideal starting material for synthesis of more complicated structures in pharmacy, succeeded by GC after converting the compound into N-(-)-(1R, 2S, 5R)-MenC-ME. The 4 enantiomers were resolved completely on a DB-210 capillary column [109].

Most applications of RCF methodology in the pharmaceutical industry focused on determination of Se-containing AAs, as mentioned in section 4.2. Since common foods have often very low Se content, consumption of Se-yeast supplements has recently become more popular and initiated the analytical interest. Targeted material were Se-containing nutritional supplementation tablets [67, 72], yeast candidate certified reference material [74] and the yeast [70, 75]. In [70] diethyl pyrocarbonate was used instead of ECF and found to be equally effective. SPME [73] and stir bar sorptive extraction (SBSE) in combination with thermal desorption were used as a sample preparation strategy prior to GC analysis. SBSE was tested in GC-MS analysis of pharmaceutical drugs and urine metabolites, and treated favourably also with RCF[110].

5.4. Works of art

A methodology for identification of proteinaceous and oil-binding media used in paintings from collections of art in the Region of Valencia (Spain) was elaborated by Spanish research team and reported in several papers over the years 1997–2004 [111–116]. The approach comprised multicompound GC-FID and GC-MS analysis following mere ECF derivatization

[111–114] or after an additional silylation step [115, 116]. The procedure consisted of 3 steps: 1) Hydrolysis of proteins of the binding media, 2) Extraction of lipophilic compounds into chloroform, 3) Treating both the phases, the aqueous one and the chloroform one, after evaporation, with ECF in water-ethanol-PYR (5:4:1) medium.

Samples (1 mg) were treated with HCl to hydrolyze proteins and glycerides and both the media, the aqueous and the organic, were evaporated and subjected to the ECF treatment. AAs were found in the aqueous phase and C14–C18 saturated fatty acids in the organic one. Moreover, azelaic acid (C9-dicarboxylic) was also detected and found to be distributed equally in the two phases. The ratio of AAs to the fatty acids proved to be extremely useful for identifying the binding media in the paintings in view of planned conservation and restoration work [111]. The next published papers [112–116] used the same methodology for identification of compounds other than AAs in the works of art.

It was concluded that the main advantage of the approach consisted in the possibility of performing simultaneously the analysis of AAs from proteins, fatty acids from drying oils, and diterpenic compounds from natural resins usually found in works of art. The method became of considerable interest due to the required minimum sampling usually involved in the analysis of works of art.

6. ECF in profiling analysis (Protocol)

ECF appeared to be the reagent of choice in the numerous application studies. Though for AAs there was a better alternative (Figure 2), for a simultaneous profiling of AAs with other carboxylic acids, e.g. in body fluids, ECF proved to be the right choice. Mainly due to the employment of ethanol as plasma deproteinizer, which enabled derivatization directly in the supernatant. The novel findings made on dicarboxylic acids and mentioned throughout the text motivated us to develop an innovative procedure unified for both plasma and urine. The derivatization protocols are as follows.

Plasma (Serum)

A) *Deproteinization* (usually done in 1.5-ml plastic tube)

To 150 μl of plasma (or serum) 5–10 μl of I.S. solution [e.g. Nval or 2-phenylbutyric acid (2PB), 15 nmol each] and 250 μl of ethanol are added. After a vigorous mixing and few-minutes standing centrifugation follows. Then 250–300 μl of supernatant are transferred into a glass vial (8 x 40/45), the fluid is basified by addition of 20 μl 2 mol/l NaOH, and the lipids are scavenged by adding 250 µl of hexane and vortexing for few seconds. This step is repeated, the upper phase aspirated and discarded.

B) *Derivatization* (e.g. in culture tubes 6 x 50 mm; Kimble/Kontes, Vineland, NJ, USA)

To 200 µl of the aqueous (bottom) phase from the preceding step, 15 µl PYR (often forgotten!) and 25 µl of isooctane-ECF (4:1) are added and the two-phase system is subjected to a vigorous vortex mixing till clearing the upper phase. The fluid is basified with 20 µl 2 M NaOH, 25 µl of isooctane-ECF again added and shaking continues till clearing. For an effective extraction, 100 µl chloroform (with 1% ECF) are added, the content vortexed for a few seconds, and following addition of 50 µl 2 M HCl vortexed again for a short time. To reach effective separation of the phases a short centrifugation (10-15 s) is recommended. The upper aqueous phase is aspirated and discarded. The step is repeated by adding 50 µl 2 M HCl once more. Following shaking and centrifugation the upper phase can be again aspirated, or aliquot of the bottom organic phase is taken off for the GC analysis.

Desiccation of the chloroform extract by adding a few milligrams of sodium sulfate and/or a volume reduction to about one third followed by refilling with a higher-boiling solvent like isooctane or heptane might be useful in some cases. Note: ECF is very corrosive and the best way to transfer it as a solution or net is to employ adjustable pipettes equipped with glass capillaries.

C) *Derivatization of chromatographic calibrators*

To 5–10 µl of solution of standards, 70–75 µl of saline-1M NaOH (2:1) plus 120 µl ethanol are admixed and after adding 15 µl PYR the sample processing follows that above.

Note: for an exact quantification the medium should be exposed to hexane scavenger, too.

Urine

Urine should not be acidic; if so, few milligrams of solid bicarbonate (or few microliters of saturated solution of $NaHCO_3$) should be added to neutralize it in prior to derivatization.

To 100 µl of urine, 5 µl of I.S. solution, 5 µl of 1 M sodium carbonate, and 100 µl ethanol are added. After admixing 15 µl PYR the sample workup continues as for plasma.

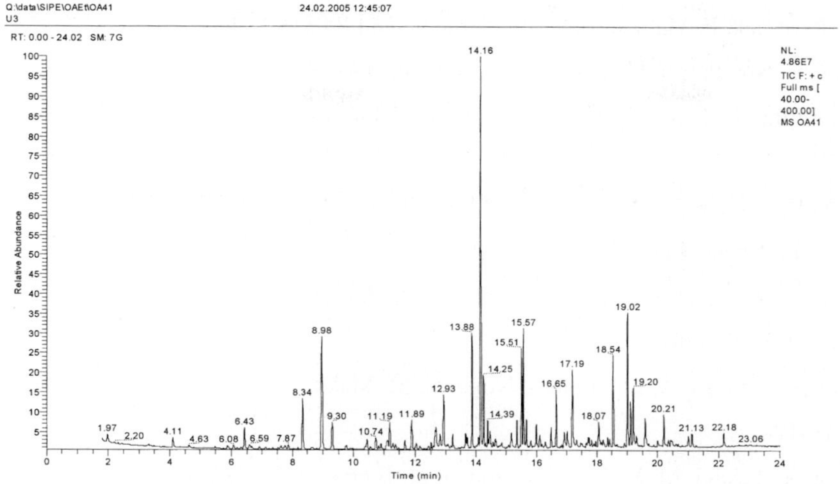

Figure 9. GC-MS analysis of acidic metabolites in urine (and plasma) after treatment with ECF as described in the protocol. TIC (full scan 40-450 u) urine. Column, 15 m x 0.25 mm x 0.15 μm VF50ms, temperature program 60°C (1 min hold), 6 °C/min to 120°C, 12°C/min to 300°C. 1.5μl split injection 8:1, helium 1.2 ml/min.

References

[1] P. Hušek, Cyclic amino acid derivatives in gas chromatography, in: Amino acid analysis by gas chromatography, eds. R.D. Zumwalt, K.C.T. Kuo and C.W. Gehrke, CRC Press, Boca Raton, Florida, U.S.A., Vol. III, p. 93-118, 1987.

[2] P. Hušek, in: Encyclopedia of Separation Science II, Chromatography: Gas-Derivatization, p. 434-443. Academic Press, London, 2000.

[3] P. Hušek, FEBS Lett. 280 (1991) 354-361.

[4] B.M. Polanuer, S.V. Ivanov, J. Chromatogr. A 722 (1996) 311-315.

[5] M.J. Nozal, J.L. Bernal, M.L. Toribio, J.C. Diego, A. Ruiz, J. Chromatogr. A 1047 (2004) 137-146.

[6] P. Hušek, Current Pharmaceut. Anal. 2005, sent to press.

[7] P. Hušek, J. Chromatogr. 552 (1991) 289-299.

[8] S.G. Villas-Boas, D.G. Delicado, M. Akesson, J. Nielsen, Anal. Biochem. 322 (2003) 134-138.

[9] P. Hušek, J. Chromatogr. B 717 (1998) 57-91.

[10] R.B. Vreeken, M.E. Jager, R.T. Ghijsen, U.A.T Brinkman, J. High Resol. Chromatogr. 15 (1992) 785-793.

[11] P. Hušek, P. Šimek, E. Tvrzická, Anal. Chim. Acta, 465 (2002) 433-439.

[12] P. Hušek, P. Šimek, P. Matucha, Chromatographia 58 (2003) 623-630.
[13] P. Hušek, C.C. Sweeley, J. High Resol. Chromatogr. 14 (1991) 751-769.
[14] I. Abe, N. Fujimoto, T. Nishiyama, K. Terada, T. Nakahara, J. Chromatogr. A 722 (1996) 221-227.
[15] S. Casal, M.B. Oliveira, M.A. Ferreira, J. Chromatogr. A 866 (2000) 221-230
[16] P. Hušek, Clin. Chem. 43 (1997) 1999-2001.
[17] Z.H. Huang, J. Wang, D.A. Gage, J.T. Watson, C.C. Sweeley, P. Hušek, J. Chromatogr. 635 (1993) 271-281.
[18] H. Kataoka, H. Tanaka, A. Fujimoto, I. Noguchi, M. Makita, Biomed. Chromatogr. 8 (1994) 119-124.
[19] H. Kataoka, K. Takagi, M. Makita, J. Chromatogr. B 664 (1995) 421-425.
[20] K.R. Kim, M.J. Paik, J.H. Kim, S.W. Dong, D.H. Jeong, J. Pharm. Biomed. Anal. 15 (1997) 1309-1327.
[21] T.G. Sobolevsky, A.I. Revelsky, I.A. Revelsky, B. Miller, V. Oriedo, Eur. J. Mass Spec. 8 (2002) 447-451.
[22] T.G. Sobolevsky, A.I. Revelsky, B. Miller, V. Oriedo, E.S. Chernetsova, I.A. Revelsky, J. Sep. Sci. 26 (2003) 1474-1485.
[23] T.G. Sobolevsky, A.I. Revelsky, I.A. Revelsky, B. Miller, V. Oriedo, J. Chromatogr. B 800 (2004) 101-107.
[24] P. Hušek, P. Šimek, LC-GC North Amer. 19(9) (2001) 986-999.
[25] P. Hušek, T. Farkas, Amer. Biotechnol. Lab. 19(12) (2001) 14-16.
[26] P. Šimek, P. Hušek, Proceedings of 50th ASMS Conf. on Mass Spectrom. and Allied Topics, A021599. pdf. Orlando, FL, USA, June 2002.
[27] P. Hušek, J.A. Rijks, P.A. Leclercq, C.A. Cramers, J. High Resol. Chromatogr. 13 (1990) 633-638.
[28] N. Domergue, M. Pugniere, A. Previero, Anal. Biochem. 214 (1993) 420-424.
[29] J. Wang, Z.H. Huang, D.A. Gage, J.A. Watson, J. Chromatogr. A 663 (1994) 71-78.
[30]M. Vatankhah, M. Moini, Biol. Mass Spec. 23 (1994) 277-286.
[31] P. Cao, M. Moini, J. Chromatogr. A 710 (1995) 303-308.
[32] P. Cao, M. Moini, J. Chromatogr. A 759 (1997) 111-117.
[33]J.T. Simpson, D.S. Torok, S.P. Markey, J. Am. Soc. Mass Spectrom. 6 (1995) 525-541.
[34] J.T. Simpson, D.S. Torok, J.E. Girard, S.P. Markey, Anal. Biochem. 233 (1996) 58-66.

[35] Z.H. Huang, D.A. Gage, L.L. Bieber, C.C. Sweeley, Anal. Biochem 199 (1991) 93-98.

[36] R. Kubec, M. Svobodová, J. Velíšek, J. Chromatogr. A 862 (1999) 85-94.

[37] B. Fransson, U. Ragnarsson, Chem. Lett. (1997) 779-784.

[38] M.J. Paik, K.R. Kim, J. Chromatogr. A 1034 (2004) 13-23.

[39] J. Lee, K.R. Kim, S. Won, J.H. Kim, Analyst 126 (2001) 2128-2143.

[40] I. Abe, M. Kawazuma, N. Fujimoto, T. Nakahara, Chem. Lett. (1995) 329-346.

[41] I. Abe, , T. Nakahara, J. High Resol. Chromatogr. 19 (1996) 511-514.

[42] I. Abe, T. Nishyama, N. Fujimoto, T. Nakahara, Anal. Sci. 10 (1994) 501-509.

[43] I. Abe, N. Fujimoto, T. Nakahara, Chem. Lett. (1995) 113-120.

[44] H. Refsum, S. Helland, P.M. Ueland, Clin. Chem. 31 (1985) 624-628.

[45] M. Nauck, E. Bisse, M. Nauck, H. Wieland, Clin. Chem. Lab. Med. 8 (2001) 675-680.

[46] H. Refsum, P.M. Ueland, A. Svardal, Clin. Chem. 35 (1989) 1921-1937.

[47] E. Nexo, F. Engbaek, P.M. Ueland, C. Westby, P. O'Gorman, C. Johnston, B.F. Kase, A.B. Guttormsen, I. Alfheim, J. McPartlin, D. Smith, J. Moller, K. Rasmussen, R. Clarke, J.M. Scott, H. Refsum, Clin. Chem. 46 (2000) 1150-1156.

[48] H. Refsum, A.W. Grindflek, P.M. Ueland, A. Fredriksen, K. Meyer, A. Ulvik, A.B. Guttormsen, O.E. Iversen, J. Schneede, B.F. Kase, Clin. Chem. 50 (2004) 1769-1784.

[49] P.M. Ueland, H. Refsum, J. Lab. Clin. Med. 144 (1989) 473-501.

[50] P.M. Ueland, H. Refsum, S.P. Stabler, M.R. Malinow, A. Andersson, R.H. Allen, Clin.Chem. 39 (1993) 1764 -1779.

[51] J.B. Ubbink, Semin Thromb. Hemost. 26 (2000) 233-256.

[52] M.I. Amores-Sanchez, M.A. Medina, Clin. Chem. Lab. Med. 38 (2000) 199-204.

[53] K. Rasmussen, J. Moller, Ann. Clin. Biochem. 37 (2000) 627-641.

[54] V. Ducros, K. Demuth, M.P. Sauvant, M. Quillard, E. Causse, M. Candito, M.H. Read, J. Drai, I. Garcia, M.F. Gerhardt, J. Chromatogr. B 781 (2002) 207-267.

[55] O. Nekrassova, N.S. Lawrence, R.G. Compton, Talanta 60 (2003) 1085-1095.

[56] J. Pietzsch, U. Julius, M. Hanefeld, Clin. Chem. 43 (1997) 2001-2004.

[57] J.O. Sass, W. Endres, J. Chromatogr. A 342 (1997) 776-783.

[58] S.W. Myung, M. Kim, H.K. Min, E.A. Yoo, K.R. Kim, J. Chromatogr. B 727 (1999) 1-8.

[59] P. Hušek, J. Chromatogr. 740 (2000) 289-295.

[60] J.I. Sigit, M. Hages, K.A. Brensing, U. Frotscher, K. Pietrzik, K. von Bergmann, D. Lutjohann, Clin. Chem. Lab. Med. 39 (2001) 681-690.

[61] R. Haddad, M.A. Mendes, N.F. Hoehr, M.N. Eberlin, Analyst 126 (2001) 1212-1230.

[62] A.P. Vellasco, R. Haddad, M.N. Eberlin, N.F. Hoehr, Analyst 127 (2002) 1050-1061.

[63] Y. Shinohara, H. Hasegawa, K. Tagoku, T. Hashimoto, J. Chromatogr. B 758 (2001) 283-288.

[64] P. Hušek, P. Matucha, A. Vránková, P. Šimek, J. Chromatogr. B 789 (2003) 311-322.

[65] J. Janák, H.A.H. Billiet, J. Frank, K.C.A.M. Luyben, P. Hušek, J. Chromatogr. A 677 (1994) 192-203.

[66] Cai, X.J., Block, E., Uden, P.C., Zhang, X., Quimby, B.D., Sullivan, J.J., J. Agric. Food Chem. 43 (1995) 1754-1757.

[67] C. Haberhauer-Troyer, G. Alvarez-Llamas, E. Zitting, P. Rodriguez-Gonzales, E. Rosenberg, A. Sanz-Medel, J. Chromatogr. A 1015 (2003) 1-10.

[68] P.C. Uden, H.T. Boakye, C. Kahakachchi, J.F. Tyson, J. Chromatogr. A 1050 (2004) 85-93.

[69] P.C. Uden, Atomic specific chromatographic detection. An overview. In *Element-Specific Chromato-graphic Detection by Atomic Emission Spectroscopy*. ACS Symposium. Series 479, ACS: Washington, DC, 1992, p. 1-24.

[70] E. Block, R.S. Glass, N.E. Jacobsen, S. Johnson, C. Kahakachichi, R. Kaminski, A. Skowronska, H.T. Boakye, J.F. Tyson, P.C. Uden, J. Agric. Food Chem. 52 (2004) 3761- 3794.

[71] C. Kahakachichi, H.T. Boakye, P.C. Uden, J.F. Tyson, J. Chromatogr. 1054 (2004) 303-312.

[72] M.V. Pelaez, M.M. Bayon, J.I.G. Alonso, A. Sanz-Medel, J. Anal. Atomic Spec. 15 (2000) 1217-1222

[73]A.P. Vonderheide, M. Montes-Bayon, J.A. Caruso, Analyst 127 (2002) 49-54.

[74] L. Yang, R.E. Sturgeon, S. McSheehy, Z. Mester, J. Chromatogr. A 1055 (2004) 177-184.

[75] L. Yang, Z. Mester, R.E. Sturgeon, Anal. Chem. 76 (2004) 5149-5173.

[76] P. Hušek, H.M. Liebich, J. Chromatogr. B 656 (1994) 37-43.

[77] P. Hušek, J. Chromatogr.B 669 (1995) 352-357.

[78] P. Hušek, P. Matucha, J. Chromatogr. B 993 (1997) 499-507.

[79] F. Podebrad, M. Heil, S. Leib, B. Geier, T. Beck, A. Mosandl, A.C. Sewell, H. Bohles, J. High Resol. Chromatogr. 20 (1997) 355-362.

[80] A. Namera, M. Yashiki, M. Nishida, T. Kojima, J. Chromatogr. B 776 (2002) 49-55.

[81] P. Hušek, A. Pohlídal, D. Slabík, J. Chromatogr. 767 (2002) 169-174.

[82] J. Pietzsch, S. Nitzsche, B. Wiedemann, U. Julius, W. Leonhardt, M. Hansfeld, J. Mass Spectrom. 30 (1995) S129-S137

[83] J. Pietzsch, U. Julius, M. Hansfeld, Rap. Commun. Mass Spectrom. 11 (1997) 1835-1841.

[84] J. Pietzsch, Biochem. Biophys. Res. Commun. 270 (2000) 852-857.

[85] J. Pietzsch, R. Bergmann, S. Kopprasch, Spectroscopy 18(2) (2004) 177-183

[86] J. Pietzsch, R. Bergmann, Amino Acids 26 (2004) 45-51.

[87] J.B. Laurens, X.Y. Mbianda, J.B. Ubbink, W.J.H. Vermaak, J. Chromatogr. B 762 (2001) 127-136.

[88] F. Montigon, J.J. Boza, L.B. Fay, Rap. Commun. Mass Spectrom 15 (2001) 116-127.

[89] C. Deng, N. Li, X. Zhang, Rap. Commun. Mass Spectrom. 18 (2004) 2558-2561.

[90] D.S. Copley, E. Frank, W.M. Kirsch, T.H. Koch, Anal. Biochem. 201 (1992) 152-161.

[91] C. Fischer, U. Klotz, J. Chromatogr. B 661 (1994) 57-68.

[92] M. Pan, T.J. Mabry, P. Cao, M. Moini, J. Chromatogr. A 787 (1997) 287-294.

[93] R.A. Boigegrain, M. Pugniere, P. Paroutaud, B. Castro, M. Brehelin, Insect Biochem .Molec. Biol. 30 (2000) 145-150.

[94] W. Kulik, J.A.N. Meesterburrie, C. Jakobs, K. DeMeer, J. Chromatogr. B 710 (1998) 37-47.

[95] W. Kulik, C. Jakobs, K. DeMeer, J. Chromatogr. B 729 (1999) 211-216.

[96] F.M. Perrine, B.G. Rolfe, M.F. Hynes, C.H. Hocart, Plant Physiol. Biochem. 42 (2004) 723-729.

[97] I. Abe, H. Yanagi, T. Nakahara, J. High Resol. Chromatogr. 20 (1997) 451-460.

[98] C. Rodier, R. Sternberg, F. Raulin, C. Vidal-Madjar, J. Chromatogr. A 915 (2001) 199-207.

[99] L. Fay, R. Badoud, U. Richli, P. Hušek, J. Chromatogr. 552 (1992) 345-351.

[100] P. Cao, M. Moini, Rapid Comm. Mass Spectrom. 11 (1997) 349-351.

[101] R.Kubec, M. Svobodova, J. Velíšek, J. Agric. Food Chem. 48 (2000) 428-433.

[102] R.Kubec, M. Svobodova, J. Velíšek, J. Eur. Food Res. Technol. 213 (2001 386-388.

[103] S. Casal, M.R. Alves, E. Mendes, M.B. Oliveira, M.A. Ferreira, J. Agric. Food Chem. 51 (2003) 6495-6501.

[104] B.M Silva, S. Casal, P.B. Andrade, R.M. Seabra, M.B. Oliveira, M.A. Ferreira, Anal. Sci. 19 (2003) 1285-1292.

[105] B.M. Silva, S. Casal, P.B. Andrade, R.M. Seabra, M.B. Oliveira, M.A. Ferreira, J. Agric. Food Chem. 52 (2004) 1201-1217.

[106] P. Šimek, A. Heydová, A. Jegorov, Chem. Listy 90 (1996) 90-101.

[107] A. Jegorov, V. Matha, P. Sedmera, V. Havlíček, J. Stuchlík, P. Seidel, P. Šimek, Phytochemistry 38 (1995) 403-407.

[108] A. Jegorov, L. Cvak, A. Husek, P. Šimek, A. Heydová, J. Ondráček, S. Pakhomová, M.Hušák, B. Kratochvíl, P. Sedmera, V. Havlíček, Collect. Czech Chem. Commun. 65 (2000) 1317-1321.

[109] V. Tolman, P. Šimek, J. Fluor. Chem. 101 (2000) 11-19.

[110] B. Tienpont, F. David, T. Benijts, P. Sandra, J. Pharm. Biomed. Anal. 32 (2003) 569-579.

[111] R. Mateo-Castro, M.T. Domenech-Carbo, V. Martinez-Peris, J.V. Adelantado-Gimeno, F. Reig-Bosch, J. Chromatogr. A 778 (1997) 373-381.

[112] R. Mateo-Castro, J.V. Gimeno-Adelantado, F. Bosch-Reig, A. Domenech-Carbo, M.J. Casas-Catalan, L. Osete-Cortina, J. De la Cruz-Canizares, M.T. Domenech-Carbo, Fres. J. Anal. Chem. 369 (2001) 642- 649.

[113] J.V. Gimeno-Adelantado, R. Mateo-Castro, M.T. Domenech-Carbo, F. Bosch-Reiga, A. Domenech-Carbo, J. De la Cruz-Canizares, M.J. Casas-Catalan, Talanta 56 (2002) 71-77.

[114] J. De la Cruz-Canizares, M.T. Domenech-Carbo, J.V. Gimeno-Adelantado, R. Mateo-Castro, F. Bosch-Reig, J. Chromatogr. A 1025 (2004) 277-285.

[115] L. Osete-Cortina, M.T. Domenech-Carbo, R. Mateo-Castro, J.V. Gimeno-Adelantado, F. Bosch-Reig, J. Chromatogr. A 1024 (2004) 187-194.

[116] M.J. Casas-Catalan, M.T. Domenech-Carbo, R. Mateo-Castro, J.V. Gimeno-Adelantado, F. Bosch-Reig, J. Chromatogr. A 1025 (2004) 269-276.

[117] V. Maurino, C. Minero, E. Pelizzetti, S. Angelino, M. Vincenti, J. Am. Soc. Mass Spectrom. 10 (1999)1328-1345

[118] M. Vincenti, N. Ghiglione, M.C. Valsania, P. Davit, S.D. Richardson, Helv. Chim. Acta 87 (2004) 370-377.

Ibolya Molnár-Perl (Editor)
Quantitation of Amino Acids and Amines by Chromatography
Journal of Chromatography Library, Vol. 70

1.1.2.

Quantitation of Amino Acids by Gas-Liquid Chromatography

Charles W. Gehrke

Contents

Summary

The GC analysis of amino acids such as the *N*-trifluoroacetyl (*N*-TFA) *n*-butyl esters – the established method developed principally in our laboratories – provides an effective and reliable means of amino acid determination that is applicable to a very wide range of analytical needs. My research group, graduate students, and colleagues during the period from 1960 to 1975, established the fundamentals of quantitative derivatization, conditions of chromatographic separation, and defined the interactions of the amino acid derivatives with the stationary and support phases. Our studies and continued refinements since 1974 have resulted in a precise and accurate, reliable, straightforward method for amino acid measurement [1–4].

We conducted an extensive array of the applications of GC of amino acid analysis on a wide range of sample matrices, from pine needle extracts to erythrocytes. The *Experimental* section developed (Volume 1, Chapter 1) [2] provides a thorough description of our quantitative analytical procedures, including preparation of ethylene glycol adipate (EGA) and silicone-mixed phase chromatographic columns. The EGA column which is used to separate and quantitate all the protein amino acids, except histidine, arginine, and cystine is composed of 0.65 w/w% stabilized grade EGA on 80/100 mesh acid-washed Chromosorb ® W, 1.5 m x 4 mm ID glass. For quantitation of histidine, arginine, and cystine, the silicone-mixed phase of 1.0 w/w% OV-7 and 0.75 w/w% SP-2401 on 100/120 mesh Gas-Chrom® Q (1.5 m x 4 mm I.D. glass) performs extremely well. We also describe the preparation and use of ion-exchange resins for sample cleanup, and complete sample derivatization to the *N-TFA n*-butyl esters. The amino acids are esterified by reaction with *n*-butanol • 3 *N* HCl for 15 min at 100°C and the excess *n*-butanol • 3 *N* HCl is removed under vacuum at 60°C any remaining moisture is removed azeotropically with dichloromethane; then the amino acid esters are trifluoroacylated by reaction with trifluoroacetic anhydride (TFAA) at 150° C - 5 min in the presence of dichloromethane as solvent. Immediately following the *Experimental* section are valuable comments on various parts of the method [3], which provide guidance to the use of the entire technique, from sample preparation to chromatography to quantitation. Of particular value is a comparison of GLC and IEC results of hydrolysates of diverse matrices. This extensive comparison of an array of sample types showed that the values obtained by the two techniques were generally in close agreement.

The analysis of amino acids as the *N*-TFA *n*-butyl esters is an established technique that offers much to scientists concerned with the determination of amino acids. The method offers excellent precision, accuracy, selectivity, and is an economical complementary technique to the elegant Stein-Moore ion-exchange method.

We also provide both a detailed account and historical perspective on development of GC amino acid analysis and describe the solution of problems encountered as the methods evolved [2]. The *N*-TFA *n*-butyl ester and trimethylsilyl (TMS) derivatives are discussed, including reaction conditions, chromatographic separations, mass spectrometric (MS) identification of both classes of derivatives, interactions of the arginine, histidine, and cystine derivatives with the liquid phase and support materials, and application of the methods in (Volume 1, Chapter 3) [2]. The acylation of arginine posed a problem in early studies; the successful solution of this problem paved the way to a high-temperature acylation procedure, which is now widely used with numerous acylating reagents [5, 6]. Likewise, esterification of the amino acids was investigated in detail, resulting in a direct esterification procedure which quickly and reproducibly converts the amino acids to *n*-butyl esters. This approach has also been widely used to form various amino acid esters [3].

The early development of GLC analysis of iodine- and sulfur-containing amino acids as the TMS derivatives is described, with the finding that bis (trimethylsilyl)-

trifluoroacetamide (BSTFA), a silylating reagent which we invented and patented, is an effective silylating reagent to form amino acid derivatives (Volume 1, Chapter 3) [2, 6]. Our studies on the derivatization of the protein amino acids with our silylation reagent, BSTFA, led the conversion of the amino acids to volatile derivatives in a single reaction step. Although certain amino acids tend to form multiple derivatives which contain varying numbers of TMS groups, high-temperature, long reaction time derivatization permits quantitative analysis of the amino acids as the TMS derivatives. Our studies on the GLC of the TMS amino acids resulted in the development of a 6 m column of 10% OV-11 on Supelcoport® for separation of the TMS derivatives.

The development of a chromatographic column system for the *N*-TFA *n*-butyl esters came about from the realization that the derivatives of arginine, histidine, and cystine were not reproducibly eluted from columns with polyester liquid phases, although this type of column was excellent for analysis of the other protein amino acids. We developed a siloxane mixed phase column specifically for these three amino acids, with the final system being an ethylene glycol adipate (EGA) column for 17 amino acids and the mixed phase column for the remaining 3 [5, 6]. Our summary points out, that the foundation of a successful amino acid analysis by GC is composed of two elements: (a) reproducible and quantitative conversion of amino acids to suitable derivatives, and (b) separation and quantitative elution of the derivatives by the chromatographic column.

A literature review is presented for the period of 1984 to 2005 on topics ranging from N-acyl O-esters of amino acid derivatives to: Enantiomeric composition, racemization, gas chromatographic enantiomer separation, formation of volatile derivatives, mass spectrometric analysis of cyclosporine metabolites, GC of 1- and 3-methyl histidine in biological fluids, rapid analyses, resolution of sulfur-containing amino acids on chiral columns, simultaneous derivatization of functional groups with one-step ethyl chloroformate derivatization, derivatization of chiral amino acids in supercritical CO_2 , separation of diasteromeric esters of α-alkyl- α-amino acids, and capillary GC plus numerous other subjects on GC of amino acids (see titles and authors as follows).

Introduction

In recent years, many investigations have been conducted to develop and refine techniques for quantitatively determining the amino acids in biological materials. These studies have been of intense interest in the fields of biochemistry, nutrition, medical science, bacteriology, and many other areas. The increasingly wide interest in amino acids and proteins has brought with it the need, and indeed, demand, for accurate, sensitive and rapid amino acid analyses.

Investigations by Moore, Stein, Hamilton, Piez, and others have resulted in accurate and precise methods for amino acid analysis by classical ion-exchange chromatographic techniques. However, in the 1960s, gas-liquid chromatographic (GLC) methods have also reached great refinement, with GLC techniques being widely used for the analysis of lipids, carbohydrates, steroids, various metabolites, pesticides, and may drugs. Similar methods for the routine analysis of amino acids have only recently been reported, since the period of 1966 through 1970.

For satisfactory analysis of amino acids by GLC, a complete derivatization of these molecules is essential. Due to the variations in chemical structure and reactivity of the twenty amino acids commonly found in proteins, and other biologically important non-protein amino acids, the quantitative derivatization of all the functional groups under one set of reaction conditions has posed many problems, and solved as described below.

1. Early GLC Studies. Research in the 1970s and 1980s

Earlier reviews of this area by Blau [7] and Weinstein [8] (1960s) discussed in detail various derivatization and chromatographic techniques for the GLC analysis of amino acids. However, prior to 1968, a complete general GLC procedure for the quantitative analysis of the twenty protein amino acids had not been reported.

Extensive research investigations, led by Gehrke, resulted in the development of a GLC technique for quantitatively analyzing the twenty protein amino acids as their N-

DERIVATIZATION OF AMINO ACIDS
TO N-TRIFLUOROACETYL n-BUTYL ESTERS

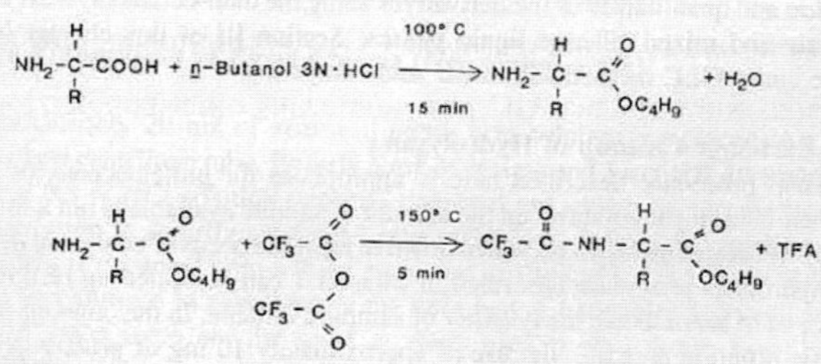

FIGURE 1

The major steps are hydrolysis, cation-exchange cleanup of the hydrolysate, quantitative derivatization of the amino acids to the *N*-TFA *n*-butyl esters (Scheme II), trifluoroacetyl (N-TFA) *n*-butyl ester derivatives. The reaction conditions for quantitatively preparing the amino acid N-TFA *n*-butyl esters of the twenty protein amino acids have been determined [3, 14, 16] (1965-67). (See Figures 1, 2, 3, 4 and 4a for reaction conditions and separations.)

Further, complete GLC resolution on a single column of the protein amino acid derivatives has also been extremely difficult to achieve. Gehrke and Shahrokhi [9] reported in 1966 a mixed polyester liquid phase for the separation of the N-TFA *n*-butyl esters of the twenty amino acids, but reproducible elution of arginine, histidine, and cystine was not achieved using this column. Stefanovic and Walker [10] (1968) investigated the use of ethylene glycol adipate (EGA) as a stationary phase for separation of the twenty amino acid N-TFA *n*-butyl esters derivatives, but these workers also did not achieve quantitative elution of arginine, histidine and cystine.

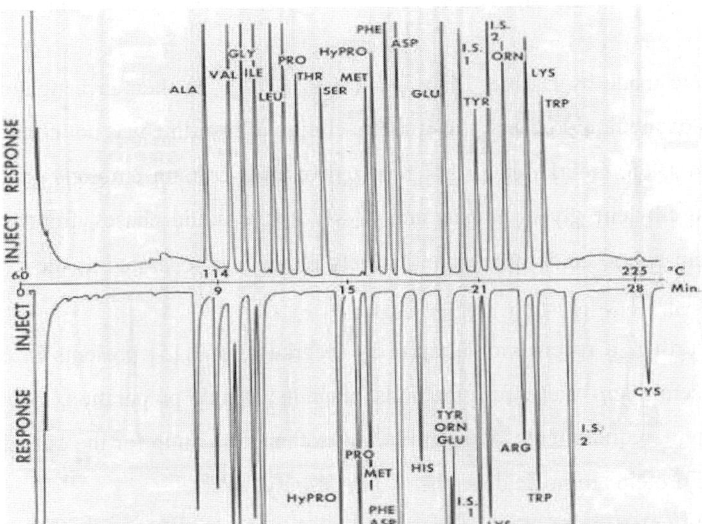

FIGURE 2. Simultaneous GLC separation of the protein amino acids: *N*-TFA *n*-butyl esters, approximately 0.6 µg of each amino acid injected; initial temperature, 60°C; program rate, 6°C/min; final temperature, 225°C; internal standard 1, tranexamic acid; internal standard 2, *n*-butyl stearate. Top chromatogram: EGA column. Bottom chromatogram: mixed phase, 2 w/w% OV-17, 1 w/w% OV-210 on 100/120 mesh Supelcoport®, 1.5 m × 4mm I.D. glass.

Investigations were also conducted by McBride and Klingman [1] (1968) to find a single column which would separate the amino acid N-TFA *n*-butyl esters, and data were reported for all the protein amion acids with the exception of arginine, histidine, and cystine.

Studies on the derivatization and chromatography of the amino acid N-TFA methyl

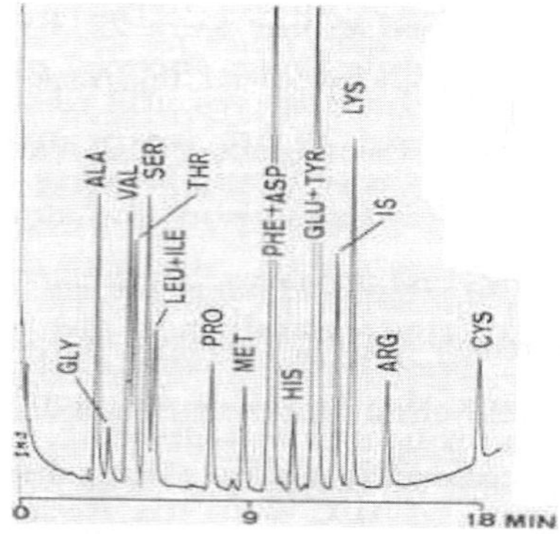

FIGURE 3. GLC analysis of histidine, arginine, and cystine as the *N*-TFA *n*-butyl esters. Sample: ribonuclease hydrolysate; column, mixed phase, 1 w/w% OV-7, 0.75 w/w% SP-2401 on 100/120 mesh Gas-Chrom® Q, 1.8 m × 4 mm I.D. glass.

esters were made by Darbre, Blau and Islam [12,13]. In their investigations using different mixed siloxane liquid phases, quantitative elution of histidine was not obtained.

In 1968, Gehrke *et al.*[14] reported on dual-column chromatographic system, using stabilized ethylene glycol adipate and OV-17 as the liquid phases, from which all twenty of the protein amino acids were quantitatively eluted and separated as their N-TFA *n*-butyl esters.

Further, a recent monograph by Gehrke *et al.* [15] presents in considerable detail macro-, semimicro, and micro methods, reagents, sample preparation, instrumental and chromatographic requirements, and sample ion-exchange cleanup for the quantitative GLC analysis of the protein amino acids as the N-TFA *n*-butyl esters.

Refinements of the GLC method have been reported by Roach et al. [16] (1969) with regard to the quantitative analysis of histidine, and by Roach and Gehrke [17]) on improved performance and reliability of the EGA column of the dual-column system. Conversion of monoacyl histidine to the diacyl derivative on the chromatographic column by injection of trifluoroactic anhydride has obviated the need for the previously reported *n*-butanol injection. These studies have shown that histidine (diacyl) can be completely separated from Asp and

Phe on the siloxane column of the dual-column system. Further, these investigators have reported that columns containing 0.65 w/w% of stabilized EGA coated on 80/100 mesh AW Chromosorb W., dried at 140° C for12 h, is generally superior to 80/100 mesh AW heated-treated Chromosorb G in terms of resolution, reliability, and ease of preparation [17].

Waterfield and Del Favero [18] reported on the use of silica gel column chromatography for purification of amino acid N-TFA *n*-butyl esters. After derivatization of the amino acids to the N-TFA *n*-butyl esters, the samples were applied to a silicic acid column, then the amino acids were eluted with diethyl ether. However, for quantitative analyses, difficulties might be expected to be encountered with regard to hydrolysis of the amino acid derivatives during the clean-up procedure.

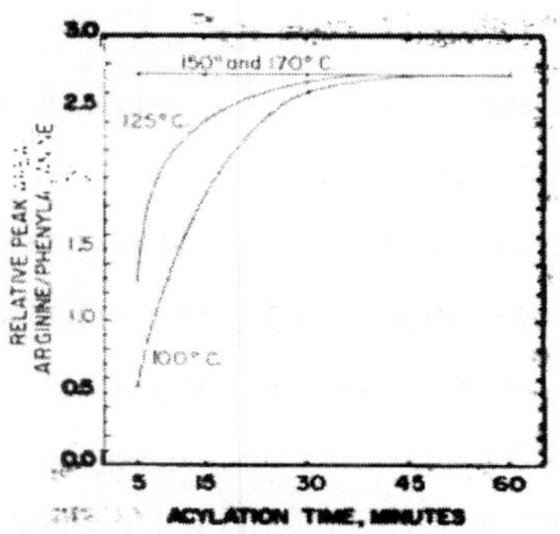

FIGURE 4 **Acylation of arginine as a function of time and temperature.**

In 1969, the N-trimethylsilyl (TMS) esters of the protein amino acids were extensively investigated by Gehrke, Nakamoto et al. [19, 20, 21]. This technique offers certain advantages in that trimethylsilylation of the twenty protein amino acids is completed in a single reaction medium, and can be separated on a single chromatographic column. Although this latest method has not yet reached the level of sophistication that has been attained by N-TFA *n*-butyl ester technique, these researchers have shown that the TMS amino acid derivatives hold great promise as a general and complementary method for routine GLC analysis.

Investigations were also carried out by Gehrke and co-workers on the experimental conditions for silylation and GLC analysis of some biologically important groups of molecules: nucleic acid components [22, 23], iodo-containing amino acids [24], sulfur-containing amino acids [25], and N-acetylneuroaminic acid [26] An extensive series of studies was made on the exact reaction conditions required for quantitative silylation of each organic class. Detailed methods are presented and data reported on the precision, accuracy, recovery, and application of the methods.

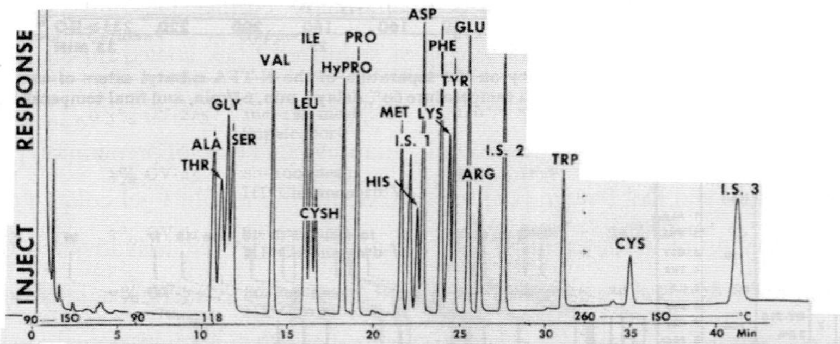

Fig.4a Single-column GLC separation of the N-TFA *n*-butyl esters of the protein amino acids. Column: 10% purified Apiezon M on 80–100 mesh HP Chromosorb W, 2.5 m × 2 mm I.D. glass. Sample: *ca.* 2.4 μg of each. Attenuation: 1×10^{-10} a.f.s. Instrumental conditions: initial temperature 90°, delay 6 min, 6°/min, and final temperature 260°. Internal standards: (1) ornithine, (2) tranexamic acid, and (3) *n*-butyl stearate.

It was the purpose of these investigations to establish the applicability of the developed GLC technique to the quantitative analysis of amino acids as their N-TFA *n*-butyl esters in complex physiological materials, specifically blood plasma and urine. Successful extension of the GLC method into these areas would greatly enhance the utility of the technique. To this end, ion-exchange methods for cleanup of these complex materials prior to derivatization were studied and developed.

Further, experiments were made to determine the quantitation of the GLC procedure over a wide range of amino acid concentrations, with emphasis on the development of techniques for accurately analyzing microgram and submicrogram amounts of acids in physiological substances.

Also, further refinement of the general procedure was studied with regard to the evaluation of various molar excesses of trifluoroacetic anhydride as the acylating reagent. Studies on evaporative losses due to concentration of the N-TFA methyl esters and N-TFA *n*-butyl ester derivatives were also carried out.

It was concluded that quantitative GLC analyses of amino acids can be performed accurately and precisely on the most complex physiological substances if the samples are properly cleaned prior to derivatization and analysis.

These reported GLC methods, developed in the Missouri laboratories, were used by the present authors for the analysis of amino acids in the Apollo 11 and 12 returned lunar samples. The authors served as co-investigator and scientists of the NASA-Ames Consortium of principal and co-investigators under the direction of Dr. Cyril Ponnamperuma, senior scientist, of the Ames Research Center, National Aeronautics and Space Administration, Moffett Field, Calif.

Ion-exchange techniques were found to be well suited for cleanup of biological samples with regard to the removal of substances which interfere with the GLC analysis of the protein amino acids.

Also, quantitative analysis of amino acids by GLC can be made on biological samples containing microgram amounts of amino acids or protein. Good results can be obtained on samples containing 1 to 20 μg of total amino acids or protein, and by concentrating the final 100 μl of acylation solution to 20 μl, and injecting 8 μl, the amount of starting material can be reduced to about 50 to 100 ng of each amino acid and still achieve a semiquantitative analysis. With very careful techniques, ultra pure reagents and exclusion of all moisture, analyses can be made at the 1 to 10 ng level.

The applicability of GLC analysis of biological materials at the semimicro and micro levels was further demonstrated with the analysis of ribonuclease and lysozyme isolated from human milk. These analyses were made in joint investigations with Dr. Shahani, B. Dalaly, R. Eitenmiller of the University of Nebraska, Lincoln.

It was also concluded that the amino acid N-TFA *n*-butyl ester derivative possesses a distinct advantage over the N-TFA methyl esters in that the *n*-butyl derivatives are not lost on any appreciable extent on concentration of the final acylated ample. Sample loss of the methyl esters becomes even more critical when analyses are performed on microgram and submicrogram amounts of amino acids, where concentration of the derivatized sample is necessary.

Manuscripts recently published in the *Journal of Chromatography* by Roach and Gehrke describe in detail a new approach to chromatography [17], and a "direct esterification" method for forming the amino acid *n*-butyl ester derivatives directly from the amino acids (1969) [27], thus eliminating the formation of the amino acid methyl esters and the interesterification reaction steps. These advancements greatly enhance the overall speed and simplicity of sample preparation of GLC analysis of the amino acids.

Research by other investigators in the late 1970s and 1980s – the following paragraphs summarize research conducted by 29 investigators, who are contributors to our 3 volume treatise on Amino Acid Analysis by Gas Chromatography [2]. S. L. McKenzie of the Plant Biotechnology Institute, NRC, Canada, a leader in the successful development of the *N*-heptafluorobutyryl (*N*-HFB) isobutyl ester derivatives, describes the rationale for his extensive work, and presents in detail the derivatization, separation and applications of this derivative in (Volume 1, Chapter 4) [2,28,29,30]. His chapter contains a section entitled *Important Comments*, pointing out that derivatization is the most crucial factor in reproducible analysis of amino acids by GLC. (Figure 5).

Noting that there are some 50 diseases known to be due to anomalies of amino acid metabolism, J. Desgrés and P. Padieu of the National Center for Mass Spectrometry and the Laboratory of Medical Biochemistry at the University of Dijon, France, describe the development of the N-HFB isobutyl derivatives for the clinical analysis of amino acids, and the adaptation of the method to the routine clinical analysis of amino acids in (Volume 1, Chapter 5) [2-31]. The experimental protocol is applied to the analysis of normal and pathologic

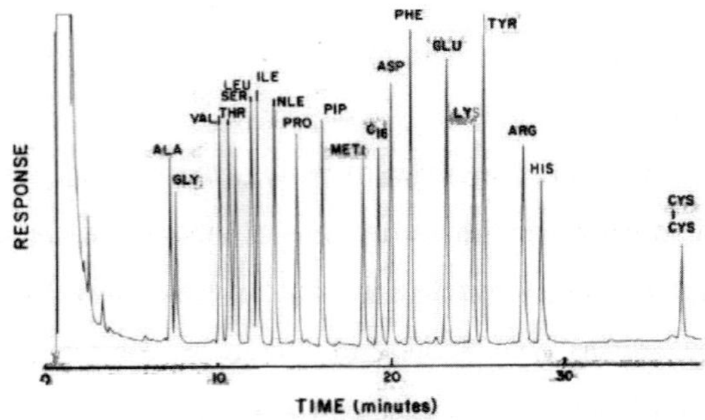

FIGURE 5 Chromatogram illustrating the resolution of *N*-HFB amino acid isobutyl esters on a 2 mm I.D. × 3 m glass column packed with 3% SE-30 on 100/120 mesh HP Chromosorb® W and temperature programmed from 100°C at 4°C/min.

physiological fluids, including phenylketonuria, maple syrup disease, idiopathic glycinemia, cystathionase deficiency, cystathionase synthetase deficiency, and renal absorption disorders. After the use of their method for more than 6 years in a clinical laboratory, they conclude that GLC is perfectly suited for the daily analysis of more than 30 amino acids, and emphasize the importance of GC/MS in elucidating metabolic disorders. (see Figures 6, 7 and Table 1).

I. M. Moodie of the National Research Institute for Nutritional Diseases and Metabolic Unit, Tygerberg Hospital, South Africa, describes in (Volume 1, Chapter 6) [2][32,33] the development of an efficient GLC method specifically to routinely produce accurate determination of protein amino acids in fishery products and presents his choice of suitable derivative and columns, modification of derivative preparation, and sample preparation techniques for packed column analysis of the *N*-HFB isobutyl esters. (Table 2).

Development of the chiral diamide phases for resolution of amino acid enantiomers is the subject of (Volume 2, Chapter 1)[2], by E. Gil-Av, R. Charles and S.-C. Chang of the Weizmann Institute of Science, Israel. Being pioneers in research on the separation of enantiomers, Gil-Av et al. [34,35] discuss the evolution of optically active phases from the α-amino acid derivative phases (e.g., *N*-TFA-L-Ile-lauroyl ester), to dipeptide phases (e.g., *N*-TFA-L-Val-L-Val-*O*-cyclohexyl) to the still more efficient and versatile diamide phases of the formula R1CONHCH(R2)CONHR3. Gil-Av et al., describe the synthesis and purification of the diamide phases, the determination of their optical purity, and the influence of structural features of the diamides on resolution and thermal stability is discussed in detail. In these extensive studies, the structure of R1, R2 and R3 in the above formula were varied with R1 and R3 representing various *n*-alkyl, branched alkyl, and alicyclic groups, and R2 representing various aliphatic and aromatic groups. Chain lengthening of R1 and R3 groups produced the desired increase in thermal stability, yielding phases operable at 200° C and above. The sequel of Gil-Av's research is described in Chapter S-9De [1].

2. Hydrolysis of proteins

The particular method used for the hydrolysis of proteins prior to an amino acid analysis is of considerable importance since some amino acids are preferentially destroyed and the hydrolysis of others is incomplete. In view of the high precision attained in the gas-liquid chromatographic (GLC) analysis of amino acid mixtures, the nature of the hydrolytic conditions plays an increasingly important role and can be easily evaluated. The speed, precision and accuracy of the GLC methods developed by Gehrke *et al.* [36-42] (1965-1969) make possible a thorough investigation of the various parameters involved in the quantitative hydrolysis of different proteins and their compositional characterization. Figure 8.

A hydrolysis reagent of broad specificity is required to break all the possible peptide bonds which are found in natural products of varying complexity. The particular reagent must be capable of cleaving all peptide bonds in a protein. Further, the peptide bonds must be accessible to the hydrolytic agent; however, two features of proteins structure present difficulties in this respect. First, there is steric hindrance due to the bulky side chains of the aliphatic amino acids, and secondly, the macro-molecular structure, *i.e.,* that due to secondary and tertiary bonding of the protein, prevents complete hydrolysis. The degree to which a protein molecule can unfold is limited by its secondary and tertiary structure; therefore, the hydrolysis reagent may react rapidly on one part of a protein molecule and slowly on another. This is evidenced by the number of different hydrolysis methods that are reported.

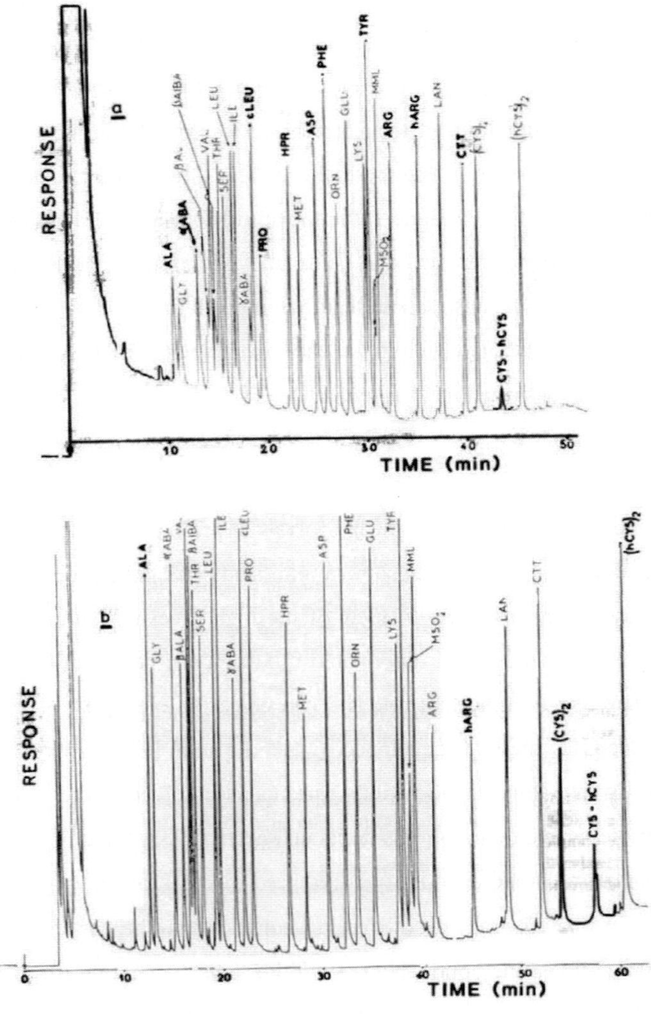

FIGURE 6 Gas chromatograms of the IBU,N(O)-HFB derivatives of 29 amino acids with two primary internal standards, MML and hArg, and one secondary internal standard, cLeu. (a) On a 2.8 m × 3 mm I.D. OV-1 packed glass column. The temperature program was 90°C initial temperature, 5-min isothermal period, and 4°C/min temperature rise to 275°C. (b) On a 25 m × 0.23 mm I.D. OV-101 coated glass capillary column. The temperature program was 90°C initial temperature, 5-min isothermal period, and 3°C/min temperature rise to 275°C. Each amino acid peak represents approximately 1 nmol on the OV-1 packed column and 40 pmol on the OV-101 coated capillary column.

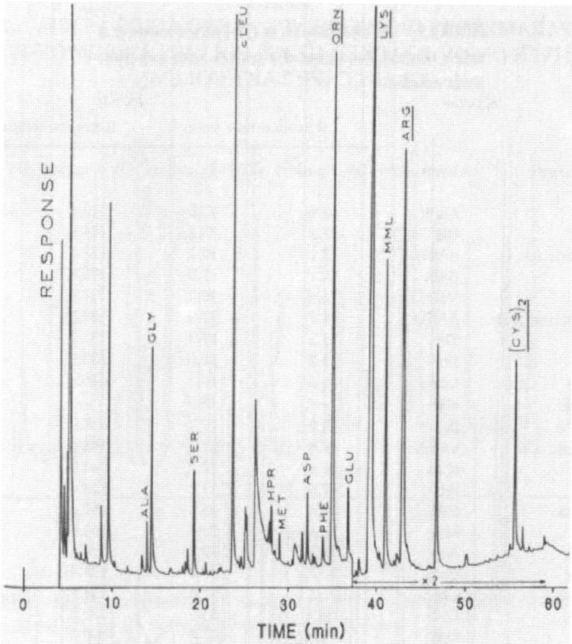

FIGURE 7 Gas chromatogram on OV-101 coated capillary column of urine amino acids from a patient suffering of renal cystinuria with increased excretion of Orn, Lys, Arg, and $(Cys)_2$.

Table 1
GLC PARAMETERS OF REFERENCE AMINO ACIDS AS IBU, *N(O)*-HFB DERIVATIVES ON OV-1 PACKED GLASS COLUMN AND ON OV-101 COATED GLASS CAPILLARY COLUMN

		Retention time (min)		Retention temperature (°C)	
Amino acid	Abbreviation	OV-1 packed	OV-101 capillary	OV-1 packed	OV-101 capillary
Alanine	Ala	10.6	12.6	112.4	112.8
Glycine	Gly	11.2	13.2	114.8	114.6
α-Aminobutyric acid	αABA	13.1	15.2	122.4	120.6
β-Alanine	βAla	14.1	15.0	126.4	123.0
Valine	Val	14.4	16.7	127.6	125.1
β-Aminoisobutyric acid	βAIBA	14.8	17.0	129.2	126.0
Threonine	Thr	15.3	17.3	131.2	126.9
Serine	Ser	15.8	18.0	133.2	129.0
Leucine	Leu	16.6	19.3	136.4	132.9
allo Isoleucine	aIle	—[a]	19.5	—[a]	133.5
Isoleucine	Ile	17.0	19.8	138.0	134.4
γ-Aminobutyric acid	γABA	18.4	21.3	143.6	138.9
Cycloleucine	cLeu	18.8	22.2	145.2	141.6
Proline	Pro	19.6	23.1	148.4	144.3
Hydroxyproline	HPr	22.4	26.8	159.6	155.4
Methionine	Met	23.4	28.4	163.6	160.2
Aspartic acid	Asp	25.1	30.7	170.4	167.1
Phenylalanine	Phe	26.1	32.4	174.4	172.2
Ornithine	Orn	27.2	33.6	178.8	175.8
Glutamic acid	Glu	28.3	35.3	183.2	180.9
Lysine	Lys	30.0	37.8	190.0	188.4
Tyrosine	Tyr	30.4	38.3	191.6	189.9
Methionine sulfone	MSO_2	30.9	39.0	193.6	192.0
ε-*N*-Monomethyllysine	MML	31.2	39.5	194.8	193.5
Arginine	Arg	32.6	41.4	200.4	199.2
Histidine[b]	His	15.2[b]	—	210.8[b]	—
Homoarginine	hArg	35.4	45.4	211.6	211.2
Lanthionine	Lan	37.6	48.8	220.4	221.4
Cystathionine	CTT	40.0	52.2	230.0	231.6
Cystine	$(Cys)_2$	41.3	54.2	235.2	237.6
Cysteinyl-homocysteinyl	Cys-hCys	43.5	57.5	244.0	247.5
Homocystine	$(hCys)_2$	45.7	60.6	252.8	256.8

Note: Retention time and temperature are given for temperature programs: initial temperature, 90°C, 5-min isothermal period, and 4°C/min up to 275°C for OV-1 packed column; initial temperature, 90°C, 5-min isothermal period, and 3°C/min up to 275°C for OV-101 coated capillary column.

[a] This amino acid is not separated from Ile on OV-1 packed columns.
[b] The observed retention time and temperature of His corresponded to its specific chromatographic conditions: 4°C/min from 150 to 250°C.

TABLE 2

AMINO ACID ANALYSIS DATA DERIVED FROM SIX ALIQUOTS OF THE SAME ACID HYDROLYSATE FROM THE SAME COMMERCIAL FISH MEAL (NUMBER 12) EXAMINED AT THE SAME TIME

Amino acid	Mean (% sample)	SD	Relative SD (%)
Ala	4.32	0.04	0.8
Gly	3.74	0.05	1.3
Val	3.33	0.07	2.0
Thr	2.70	0.05	1.8
Ser	2.35	0.06	2.3
Leu	4.73	0.06	1.2
Ile	2.86	0.09	3.0
Pro	2.47	0.02	0.9
HPr	0.33	0.02	5.6
Met	1.96	0.04	2.0
Asp	5.51	0.08	1.5
Phe	2.56	0.06	3.2
Glu	7.85	0.13	1.6
Lys	4.74	0.06	1.4
Tyr	1.70	0.03	1.5
Arg	2.99	0.08	2.8
His	1.44	0.01	4.8

AMINO ACID ANALYSIS DATA DERIVED FROM 17 ALIQUOTS OF THE SAME ACID HYDROLYSATE FROM THE SAME COMMERCIAL FISH MEAL (NUMBER 12) EXAMINED OVER A 2-MONTH PERIOD

Amino acid	Mean (% sample)	SD	Relative SD (%)
Ala	4.34	0.13	3.0
Gly	3.77	0.10	2.7
Val	3.26	0.13	4.0
Thr	2.70	0.06	2.2
Ser	2.38	0.06	2.7
Leu	4.77	0.09	1.9
Ile	2.79	0.13	4.8
Pro	2.52	0.08	3.2
HPr	0.36	0.03	8.2
Met	1.96	0.08	4.3
Asp	5.55	0.12	2.1
Phe	2.58	0.10	3.7
Glu	7.92	0.17	2.1
Lys	4.65	0.17	3.7
Tyr	1.68	0.03	1.8
Arg	3.38	0.33	9.7
His	1.37	0.09	6.6

Differences in the stability of the various functional groups of amino acids necessitate a compromise among several experimental conditions in order to achieve the optimum hydrolysis of the protein. Moore and Stein [43] (1963) reported that the best "all around" hydrolysis can be achieved by reaction for 24 h with 6 *N* HCl at 110° C under conditions rigorously excluding oxygen, non-protein substances, and metals. Oxygen can be excluded by using a sealed tube hydrolysis technique. Generally, acid hydrolysis will yield over 95% recovery for aspartic and glutamic acids, proline, glycine, alanine, methionine, leucine, tyrosine, phenylalanine, lysine, histidine, and arginine. However, tryptophan is completely destroyed, whereas 5–15% of threonine and serine are destroyed. Extrapolation to "zero-time" of hydrolysis can be done, but requires several different times of hydrolysis for each sample. GLC makes studies of "zero-time" hydrolysis practical. The peptide bonds of valine, isoleucine, and leucine are quite stable and thus a longer hydrolysis time is required to obtain maximum yield for these amino acids. Whitfield [4] (1963) has studied this problem and explained it in terms of steric factors. Extending the hydrolysis time to 70 h gives maximum yields [43] for these three amino acids. This, of course, results in lower yields for the other amino acids as compared to a 24 h hydrolysis time. As yet, no satisfactory method has been found for tryptophan, except alkaline, or enzymatic hydrolysis [43].

Table 3
PRECISION OF AMINO ACID ANALYSIS AFTER HYDROLYSIS BY TWO METHODS[a]

Sample	Reference sealed ampule, 110°C-24 hr	Teflon®-lined screw-cap tube, 110°C-24 hr
Soybean meal	2.84	1.39
Poultry feed	1.96	1.42
Fish meal	2.65	1.33
Wheat	1.23	2.11
Pruteen®	1.42	1.16
Orchard leaves	1.11	1.86
Bovine liver	4.12	1.03
Egg white	2.57	1.33
Ribonuclease	1.59	0.94
Mean relative SD, %	2.17	1.40

Note: Beckman® 121M amino acid analyzer.

[a] Each relative SD percent value for each amino acid was calculated from duplicate analyses of three independent hydrolysates, n = 6. It follows then that each relative SD percent in this table represents a summation of the relative SD percent values for all 15 amino acids for each matrix.

The purpose of this research was to study the rates and yields of protein hydrolysis and to determine the optimum reaction conditions which would give maximum yields of all twenty of the protein amino acids in the shortest possible time, using ribonuclease as a representative protein.

Braconnot [45], in 1820, first used sulfuric acid for the hydrolysis of a protein. The use of HCl as a hydrolytic agent was introduced by Bopp [46] in 1849. The hydrolytic agent commonly used today is HCl since the rate of peptide bond cleavage is increased in HCl over what it would be in sulfuric acid of equal concentration. An added advantage of HCl is that it can easily be removed from an amino acid mixture by evaporation. Protein samples are usually hydrolyzed with 2.5 – 5000 times their weight of 6 *N* HCl under reflux for 18–24 h.

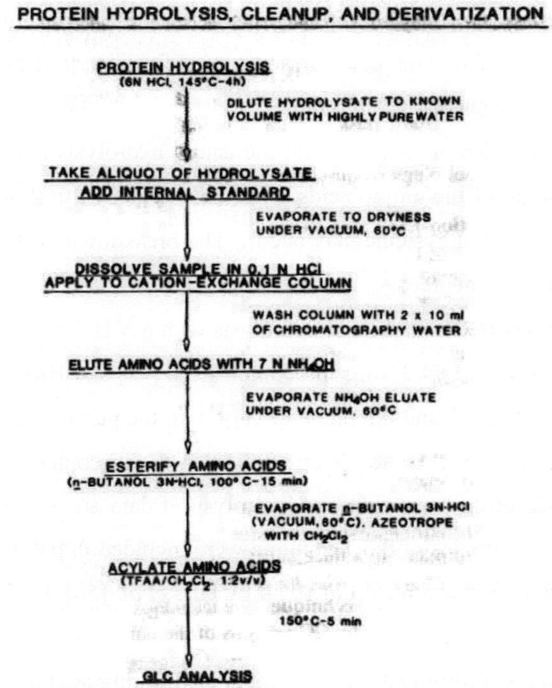

FIGURE 8

The method of MacPherson [47] 1946 is generally recommended for large protein samples (*ca.* 0.2 *g* or larger). A protein sample which has been equilibrated under atmospheric conditions is weighed into a suitable round-bottomed flask which is fitted with a condenser. Concentrated A. R. HCl (36 w/w%) is added (*ca.* 20 ml/g protein), the protein is dissolved on a water bath at 35-40° C, then sufficient hot doubly distilled water is added to bring the concentration of HCl to 20 w/w%. The solution is boiled gently under reflux for 24 h. The excess of HCl is removed under a partial vacuum and the sample is diluted to a suitable volume with 0.1 *N* HCl. An aliquot of this solution is then removed for classical amino acid analysis or GLC amino acid analysis.

The method of Moore and Stein is in common use. A sample of air-dried or lyophilized protein is placed in a 10 x 135 mm heavy-walled Pyrex tube (Corning no. 9860). The protein is suspended in 1 ml of 6 *N* HCl (a 1:1 dilution of concentrated reagent HCl with doubly distilled water). The sample is frozen by placing in a bath of ethanol and solid carbon dioxide. After freezing the sample container is evacuated to below 50 μ, then sealed under vacuum. The hydrolysis is conducted at 110° ± 1° C for 20 h or 70 h, excess HCl is removed under vacuum at 40–45° C, the sample is diluted to a known volume, and aliquots are removed for analysis. This technique or some modification of it is presently the preferred method for the hydrolysis of protein samples.

A serious problem associated with the acidic hydrolysis of proteins is the partial decomposition of some of the amino acids. The destruction of tryptophan is almost complete and a considerable loss of cysteine may occur. The breakdown of the other amino acids generally occurs to a lesser degree.

Rees [48] reported in 1946 that hydrolysis with 6 *N* HCl for 24 h leads to a recovery of only 89.5% for serine and 94.7% for threonine. Rees [48] and Hirs *et al.,* [50] found the rate of decomposition of serine and threonine to vary with the purity of the HCl used in the acidic hydrolysis. However, an accurate determination of the threonine and serine content can be made by extrapolation to "zero-time" of hydrolysis if data are available for several different hydrolysis times. Examples of this technique were included in publications by Harfenist [51] in 1953, by Smith and Stockell [52] in 1954, by Hirs *et al.* [50] in 1954, and by Noltmann *et al.* [53] in 1962.

There is a possibility that proline is degraded during acid hydrolysis. Elliot *et al.*[19] and Zuber and Jaques [3] both suggested an empirical formula of Arg_2 Phe_2 Pro_2 Gly·Ser for the peptide bradykinin from results based on amino acid analyses after acidic hydroysis. Boissonman *et al.* [56], however, synthesized bradykinin and found that the actual structure corresponded to the formula $Arg_2Phe_2Pro_3$Gly·Ser. The variance between the formula determined from amino acid analysis and the actual formula maybe due to the decomposition of proline during the acid hydrolysis prior to analysis.

Lugg [57]) observed that pure tyrosine was not affected by heating it in acid at 100° C for 20 to 30 h. Light and Smith [58] (1962), however, reported that tyrosine was completely destroyed during the acid hydrolysis of the peptide Ala·Val·Gly·Tyr. Shepherd *et al.* [59] also obtained low recoveries of tyrosine from several peptides. This destruction was reduced but not eliminated when the samples were hydrolyzed under a nitrogen atmosphere. The decomposition of tyrosine may involve aspartic acid since tyrosine was quantitatively recovered from the peptideVal·Tyr·Pro but not from Val·Tyr·Pro·Asp. Munier [60] reported that tyrosine may be converted to 3-chlorotyrosine during hydrolysis by reacting with traces of chlorine in the HCl. This reaction could not, however, account entirely for the losses observed by Hirs *et al.* [50].

A large concentration of carbohydrates in the hydrolysis medium may seriously reduce certain amino acids. Tristran [61] noted that arginine was extensively destroyed during acidic hydrolysis in the presence of carbohydrates with the amount of destruction being proportional to the concentration of carbohydrates, and Bailey [62] (1937) reported losses of methionine as high as 20% in samples which were high in carbohydrate content. Osono *et al.*[63] (1955) found that refluxing methionine with 10% HCl resulted in the production of some homocystine, homocysteine, and glycine. Lugg [64] observed only a slight loss of cystine during acidic hydrolysis in the absence of carbohydrate; however, losses of 6 to 7% were noted in the presence of carbohydrates.

Lysine is considered to be the most stable of the diamino acids, but Ishii [65] (1954) reported a loss of 3% when lysine was heated at reflux with 20% HCl. The reported degradation products were aspartic acid, glycine, glutamic acid, and α-aminoadipic acid.

Steric hindrance by bulky side chain residues results in the slow release of some amino acids, particularly valine and isoleucine. Kinetic studies, by Synge [66] in 1945, and by Harris *et al.* [67] in 1956, clearly indicated hindrance by valine, leucine, alanine, and isoleucine and the yields for these amino acids which have been hydrolyzed for varying lengths of time were found to be a function of time. An accurate value for each of these amino acids can be determined by plotting yield as a function of hydrolysis time, and by drawing tangents to the maximums in the curves, then extrapolating to "zero-time".

The rates of decomposition of the amino acids during the acidic hydrolysis are dependent on several factors including: the concentration of the hydrolyzing acid, the purity of the acid used, the time and temperature of hydrolysis, the presence of carbohydrates, aldehydes or metal impurities. Current methods represent a compromise among the several considerations mentioned above. The most common methods for the hydrolysis of proteins are outlined in two excellent review articles by Light and Smith [68] and Moore and Stein [43] (1963).

Hydrolysis of samples in 6 N HCl in water for analysis by both GLC and classical ion exchange

(1) Accurately weigh 25.0 mg of dry protein (ribonuclease) into a large culture tube. (2) Flush tube with filtered N_2. (3) Add 25.0 ml of 6 *N* HCl in water to each tube. (4) Flush each tube again with N_2. (5) Place Teflon-lined cap on each tube and heat at 110° C ± 1°, or 145° C ± 2° in an oil bath for the specified time. (6) Dry the samples at 60° C under a partial vacuum with a rotary evaporator. (7) Accurately pipet 20.0 ml of 0.1 *N* HCl into each of the samples to dissolve the amino acid residue. Mix each sample thoroughly. (8) Draw a 5.0 ml aliquot of each sample and place in a 125 ml flat-bottom boiling flask for GLC analysis, or analyze by classical ion exchange. (9) Dry the samples at 60° C under a partial vacuum with a

rotary evaporator. (10) Add an appropriate quantity of I.S. (0.50 mg butyl stearate, dissolved in 1 ml of BuOH·HCl). (11) Add 1.5 *n*-butanol 3 *N* in HCl per 1.0 mg of total amino acids, ultrasonic mix for 15 sec, esterify at 100° C for 35 min, then dry at 60° C under a partial vacuum, and acylate as described in (37).

Hydrolysis of samples by 6 N HCl in n-butanol

(1) Accurately weight 10 mg of dry protein (ribonuclease) into a large culture tube. (2) Flush tube with filtered nitrogen gas. [3] Add 15.0 ml of *n*-butanol 6 *N* in HCl or *n*-butanol 6 *N* in HCl (1.5 ml of BuOH·HCl per 1.0 mg of protein). (4) Flush reaction vessel again with filtered N_2. [5] Place Teflon-lined cap on each tube and heat at 110° C ± 1°, or 145° C ± 2° in an oil bath for the specified time. [6] Add an appropriate quantity of I.S. (0.50 mg butyl stearate, dissolved in BuOH·HCl.) [7] Dry the samples at 60° C under a partial vacuum with a rotary evaporator [8). Acylate as described by Gehrke et al.[2,37].

The use of *n*-butanol 6 *N* in HCl as a protein hydrolysis reagent would obviate one of the steps in the reported [37] GLC analysis of proteins since the *n*-butyl esters of the amino acids would be formed during the hydrolysis. Thus, these studies were initiated to investigate the yields on hydrolysis of a model protein, ribonuclease, in 6 *N* HCl in *n*-butanol. However, this reagent was found to be unsatisfactory since the rate of hydrolysis was much slower in this medium than it was in 6 *N* HCl in water, and the rates of decomposition of the amino acids were considerably faster.

Since *n*-butanol 6 *N* in HCl was found to be unsuitable for the hydrolysis of proteins, experiments were made to investigate the effect of temperature on the hydrolysis reaction in aqueous 6 *N* HCl with a view to developing a rapid hydrolysis procedure. The maximum yield for all the protein amino acids was obtained at 145° C ± 2° C for the minimal time of 4 h. Essentially equivalent hydrolysis of ribonuclease was achieved at the two different hydrolysis conditions, *i.e.*, 110° C ± 1° C for 26 h, or 145° C ± 2° C for 4 h. The yields obtained were in good agreement.

GLC and IEC analyses of multiple hydrolysates were performed to evaluate the reproducibility of hydrolysate preparation and to compare GLC and IEC analyses of the same hydrolysates. The total amino acids found in the same hydrolysates were essentially identical by both GLC vs. IEC. As the sets of three hydrolysates were prepared at the same time under identical conditions, it might be expected that differences between the GLC and IEC analyses of the same hydrolysates would be greater than the differences between identically prepared hydrolysates. However, the slight differences in the amounts of certain amino acids present in the different hydrolysates can be observed, emphasizing that variations do arise due to the hydrolysis itself, even under preparation conditions most conducive to reproducibility [2, 3].

Table 4
THE EFFECT OF HYDROLYSIS TIME AT 145°C ON PROTEIN AMINO ACIDS[a]

	Amino acids (w/w %) Hydrolysis time (hr)					
	Egg white			Wheat flour		
Amino acid	3	4	5	3	4	5
Asp	8.29	8.47	8.30	1.06	1.05	1.07
Thr	3.53	3.54	3.52	0.38	0.40	0.40
Ser	5.07	4.89	4.94	0.45	0.53	0.50
Glu	10.56	10.78	10.60	2.95	2.98	2.93
Pro	2.96	3.00	3.00	0.94	0.95	0.93
Gly	2.83	2.89	2.85	0.49	0.49	0.48
Ala	4.80	4.90	4.82	0.47	0.48	0.47
Cys	1.47	1.45	1.50	0.06	0.10	0.09
Val	5.51	5.76	5.54	0.61	0.60	0.61
Met	3.28	3.27	3.29	0.16	0.17	0.17
Il	4.14	4.33	4.20	0.53	0.53	0.53
Leu	6.90	7.09	6.95	0.94	0.94	0.93
Tyr	3.44	3.54	3.53	0.39	0.40	0.35
Phe	4.91	5.10	4.96	0.63	0.63	0.61
His	1.92	1.99	1.95	0.31	0.31	0.31
Lys	5.62	5.78	5.64	0.58	0.58	0.59
Arg	4.61	4.72	4.64	0.72	0.72	0.72
Total	79.83	81.50	80.23	11.69	11.84	11.67

[a] Sealed ampule hydrolysis.

As the sulfur-containing amino acids are of particular interest in nutrition, cystine and methionine analyses are discussed in detail. The quantitative determination of amino acids in addition to cystine and methionine in preoxidized hydrolysates by IEC is described [2], and a rapid oxidation-hydrolysis procedure is presented which allows accurate analysis of cystine, methionine, lysine, and nine other amino acids in feedstuffs and other biological matrices. Floyd Kaiser has subsequently used the *N*-TFA *n*-butyl ester method for more than 17 years on a routine basis in our corporate laboratory (Analytical Biochemistry Laboratories, Columbia, Missouri); and his observations on the analysis of an extremely wide range of sample types over this time span are presented as *Experiences of a Commercial Laboratory* and provides valuable practical information into amino acid analysis by GC [2] (Volume 1, Chapter 2, pp. 53-55). Tables 4 and 5.

Table 5
COMPARISON OF INTERLABORATORY HYDROLYSATES OF PROTEIN[a,b]

Amino acid	Soybean meal ESCL[c]	Soybean meal ABC[d]	Relative % difference	Poultry feed ESCL	Poultry feed ABC	Relative % difference	Fish meal ESCL	Fish meal ABC	Relative % difference
Asp	5.46	5.48	0.37	1.98	2.04	2.99	6.13	5.97	2.64
Thr	1.84	1.77	3.89	0.71	0.71	0.00	2.59	2.66	2.65
Ser	2.28	1.99	13.55	0.85	0.76	11.25	2.22	2.35	5.70
Glu	8.65	8.22	5.09	3.60	3.63	0.83	8.68	8.53	1.74
Pro	2.48	2.39	3.69	1.33	1.30	2.27	3.00	3.01	0.33
Gly	2.06	2.02	1.96	0.92	0.94	2.15	4.61	4.60	0.22
Ala	2.08	2.04	1.94	1.07	1.10	2.78	4.08	3.99	2.23
Val	2.46	2.34	5.00	1.04	1.02	1.94	3.26	3.17	2.80
Ile	2.26	2.14	5.45	0.86	0.84	2.35	2.78	2.60	6.69
Leu	3.76	3.67	2.42	1.80	1.81	0.56	4.72	4.56	3.45
Tyr	1.64	1.68	2.41	0.59	0.69	15.62	2.02	2.07	2.44
Phe	2.46	2.41	2.05	1.00	0.99	1.00	2.59	2.57	0.78
His	1.25	1.26	0.79	0.52	0.54	3.77	1.36	1.54	12.41
Lys	3.02	2.92	3.37	1.02	1.02	0.00	4.90	4.71	3.96
Arg	3.41	3.32	2.68	1.24	1.26	1.60	3.86	3.84	0.52
Average			3.64			3.27			3.24

All samples hydrolyzed at 145°C-4 hr.
All analyses conducted at ESCL, University of Missouri at Columbia, by IEC
Hydrolysate prepared at ESCL; values are w/w%.
Hydrolysate prepared at ABC; values are w/w%.

The average recovery of cystine from a wide range of matrices without the use of performic acid was 55.5% as compared to results obtained with performic acid oxidation. Similarly, methionine is preferably analyzed as methionine sulfone. Interlaboratory evaluation of 145° C-4 hr hydrolysis in which one laboratory used sealed ampules, and the other laboratory used Teflon®-lined screw cap tubes, demonstrated excellent agreement of amino acid values. In summary, we found that the hydrolysis of a range of different protein-containing matrices at 145° C-4 h in glass tubes with Teflon®-lined screw caps after vacuum removal of air, nitrogen, purge, and sonication performed as well as sealed glass ampules at both 145°C-4 h and 110°C-24 h hydrolysis conditions. Tables 6 and 7.

With this method a protein can be essentially completely hydrolyzed in 4 h with a minimum of decomposition of the amino acids. Rapid hydrolysis of proteins coupled with quantitative GLC analysis of amino acids provides a powerful tool in protein research, biochemical, and nutritional investigation.

Table 6
COMPARISON OF IEC AND GLC ANALYSES OF THE SAME PROTEIN HYDROLYSATES[a,b]

Amino acid	Soybean meal IEC[c]	Soybean meal GLC[d]	Difference	Relative difference (%)	Poultry feed IEC	Poultry feed GLC	Difference	Relative difference (%)	Fish meal IEC	Fish meal GLC	Difference	Relative difference (%)
Ala	2.04	2.05	+0.01	0.5	1.10	1.08	−0.02	1.8	3.92	3.86	−0.06	1.5
Val	2.34	2.27	−0.07	3.0	1.02	0.97	−0.05	4.9	3.11	2.92	−0.19	6.1
Gly	2.02	2.02	0.00	0.0	0.94	0.92	−0.02	2.1	4.51	4.43	−0.08	1.8
Ile	2.14	2.29	+0.15	7.0	0.84	0.87	+0.03	3.6	2.55	2.64	+0.09	3.5
Leu	3.67	3.59	−0.08	2.2	1.81	1.73	−0.08	4.4	4.48	4.25	−0.23	5.1
Pro	2.39	2.38	−0.01	0.4	1.30	1.27	−0.03	2.3	2.95	2.86	−0.09	3.1
Thr	1.77	1.86	+0.09	5.1	0.71	0.73	+0.02	2.8	2.61	2.54	−0.07	2.7
Ser	—	—	—	—	—	—	—	—	2.31	2.33	+0.02	0.9
Phe	2.41	2.40	−0.01	0.4	0.99	0.95	−0.04	4.0	2.52	2.39	−0.13	5.2
Asp	5.80	5.38	−0.42	7.2	2.04	1.83	−0.21	10.3	5.86	5.59	−0.27	4.6
Glu	8.22	8.28	+0.06	0.7	3.63	3.46	−0.17	4.7	8.37	7.87	−0.50	6.0
Tyr	1.68	1.70	+0.02	1.2	0.69	0.64	−0.05	7.2	2.03	1.88	−0.15	7.4
Lys	2.92	2.91	−0.01	0.3	1.02	0.97	−0.05	4.9	4.62	4.59	−0.03	0.6
His	1.26	1.24	−0.02	1.6	0.54	0.51	−0.03	5.6	1.51	1.38	−0.13	8.6
Arg	3.32	3.16	−0.16	4.8	1.26	1.19	−0.07	5.6	3.77	3.30	−0.47	12.5
Total	41.98	41.55	−0.01	2.5[e]	17.89	17.12	−0.57	4.6[e]	55.12	52.83	−2.29	4.64[e]

[a] Hydrolysis at 145°C-4 hr, sealed ampule.
[b] All hydrolysates prepared at ABC.
[c] IEC analyses performed at ESCL; single analysis of each hydrolysate.
[d] GLC analyses performed at ABC; average of three analyss of one hydrolysate.
[e] Average relative percent difference.

3. Direct esterification of protein amino acids as the N-TFA *n*-butyl esters

The reaction conditions necessary for the "direct esterification" of the protein amino acids to their *n*-butyl esters are described. All of the amino acids were quantitatively esterified in *n*-butanol 3 *N* in hydrochloric acid at 100° C with the exception of isoleucine. This "direct esterification" method with *n*-butanol permits a rapid derivatization and analysis by gas-liquid chromatography of the protein amino acids, thus, one of the major disadvantages of the earlier reported method has been removed.

The amino acids were observed to dissolve very slowly in *n*-butanol 6 *N* in hydrochloric acid even when the samples were subjected to ultrasonic mixing. Fairly rapid dissolution occurred in 1.5 *N* hydrochloric acid but a longer esterification time was noted. The optimum concentration of hydrochloric acid was found to be 3 *N* because the amino acids dissolved quickly in this solution with ultrasonic mixing and short esterification times were obtained. The more insoluble amino acids were broken up by ultrasonic mixing, thus increased rates of solution and esterification to the *n*-butyl esters were achieved. The effect of temperature over the range of 90 to 120° C, on the rate of esterification with *n*-butanol 3 *N* in hydrochloric acid in 15 min at 100°, but 35 min were required for the esterification of isoleucine. However, with the longer esterification time, tryptophan underwent some decomposition (*ca.* 15%).

Table 7
GLC AND IEC ANALYSES OF MULTIPLE HYDROLYSATES

	Bovine liver[a]					
	Hydrolysate number					
	1		2		3	
Amino acid	GLC[b]	IEC[c]	GLC	IEC	GLC	IEC
Ala	3.56	3.42	3.83	3.59	3.65	3.52
Val	3.83	3.79	4.09	3.95	3.87	3.84
Gly	3.42	3.27	3.64	3.50	3.51	3.44
Ile	2.93	2.77	3.18	2.86	2.99	2.80
Leu	5.63	5.52	5.80	5.86	5.74	5.80
Pro	2.92	2.85	3.08	2.92	3.02	2.98
Thr	2.74	2.70	2.81	2.78	2.82	2.78
Ser	2.67	2.62	2.76	2.72	2.76	2.75
Met	1.24	1.52	1.29	1.56	1.29	1.55
Hyp	<0.01	0.36	<0.01	0.35	0.14	0.28
Phe	3.28	3.23	3.40	3.38	3.31	3.35
Asp	5.74	5.71	5.88	5.93	5.80	5.84
Glu	7.70	7.61	7.78	7.97	7.69	7.92
Tyr	2.16	2.36	2.38	2.48	2.21	2.47
Lys	4.12	3.10	3.99	3.30	3.94	3.27
His	1.62	1.74	1.60	1.80	1.73	1.78
Arg	3.12	3.64	3.11	3.80	2.56	3.78
Cys	1.09	0.38	0.98	0.66	0.87	0.57
Total	57.77	56.59	59.60	59.41	57.90	58.72

An esterification time of 35 min is recommended for samples which are to be analyzed for all of the amino acids including isoleucine and tryptophan; however, only 15 min are required if a slight error in the absolute value of isoleucine is permissible

In quantitative work, a reference calibration mixture is analyzed under exactly the same experimental conditions as are the samples, thus correction is made for any breakdown of tryptophan and the response values of all the other amino acids are placed on the same basis.

Using the "direct esterification" method, recoveries of amino acids from a mixture averaged 99.7% with an average-relative standard deviation of 0.92%. This "direct esterification" method allows a rapid, precise, and accurate derivatization of the protein amino acids to their N-TFA-*n*-butyl esters with a minimum of transfers and sample handling. A complete gas-liquid chromatographic analysis of the protein amino acids, including derivatization and chromatography, can be completed in less than 1 h.

In previous studies, gas-liquid chromatography (GLC) methods have proved to be most useful for the analysis of amino acids in biological substances because of their speed and sensitivity. Since the low volatilities of the amino acids has prevented their direct analysis by GLC, suitable volatile derivatives of the amino acids must be prepared. In 1962, Zomzely *et al.* [69] Investigated the n-trifluoroacetyl (N-TFA) *n*-butyl esters as possible derivatives. Lamkin and Gehrke [70] reported in 1965 that the most suitable derivative with respect to volatility and chromatography for the GLC analysis of the natural protein amino acids is the N-TFA *n*-butyl ester. The experimental conditions for quantitative derivatization and chromatographic requirements for separation were detailed by Gehrke and Stalling [71] in 1967. This was followed by a recent monograph by Gehrke et al. [73] which presents macro, semimicro, and micro methods, reagents, sample preparation, instrumental and chromatographic requirements, and sample ion-exchange cleanup for the quantitative GLC analysis of the protein amino acids as their N-TFA *n*-butyl esters.

The chromatographic properties of several derivatives of the protein amino acids, in addition to the N-TFA *n*-butyl esters, have also been evaluated. Among the derivatives are the trimethylsilyl derivative introduced by Rűhlmann and Giesecke [72] in 1961 and studied more recently by Gehrke *et al.* [73,74] ; the *N*-acetyl *n*-amyl esters by Johnson *et al.* [75]; and the N-trifluoroacetyl methyl esters by Saroff and Karmen [76], by Hagen and Black [77], by Cruickshank and Sheehan [78], and by Darbre and Islam [79]. Each of these derivatives has proved to be useful for GLC separations, but none has received the intensive study regarding quantitative analysis as the N-TFA *n* butyl ester. An excellent review was written by Blau [80]) in 1968 on the analysis of amino acids by GLC covering the areas of derivatives, methods and applications.

The N-TFA *n*-butyl esters of the amino acids are less volatile than the N-TFA methyl esters. The N-TFA methyl esters are so volatile that losses can easily occur during the derivatization procedure especially if the samples are concentrated after derivatization. Darbre and Islam [79] report that there is danger of loss of the more volatile N-TFA methyl esters when the samples are dried for longer than 4 min under vacuum at the temperature of an ice bath. Recent investigations [81, 82] resulted in losses of *ca.*10% for the more volatile N-TFA methyl esters when the samples were dried under a stream of nitrogen at room temperature; whereas insignificant losses were obtained for the N-TFA *n*-butyl esters dried under the same conditions. Only *ca.* a 10% loss occurred for the more volatile N-TFA *n*-butyl esters when evaporated at 100%.

During the past eight years, Gehrke and co-workers [70-71, 82] (1965-73), have developed a quantitative single general method for the determination of the twenty protein amino acids and at least forty other non-protein amino acids as their N-TFA *n*-butyl esters.

Mussini and Marcucci [83] (1967) reported the esterification of the amino acids using diazobutane, but the most common method for preparing esters of the amino acids with the higher alcohols is the reaction of the amino acids with the alcohol in the presence of anhydrous hydrogen chloride. Direct formation of the higher alkyl esters of the amino acids has been a problem because of the lack of solubility of cystine and some of the other amino acids in the higher alcohols. Blau and Darbre [84] (1967) suggested dissolving the amino acids in a small amount of anhydrous trifluoroacetic acid before adding the *n*-BuOH·HCl to obviate the solubility problems associated with cystine and some of the others. Stalling *et al.* [85] (1967) reduced the solubility problems by first making the methyl esters by direct esterification with methanolic HCl, then interesterification with *n*-butanolic HCl. These workers found that all of the amino acid methyl ester hydrochlorides were soluble in this reagent.

The quantitative GLC methods developed by Gehrke and co-workers at the University of Missouri have proved to be most satisfactory except that the total derivatization time (methyl ester formation and interesterification) requires about 3.5 h for completion. Thus, in view of the number of laboratories which use this chromatographic method, it is highly desirable to have a faster procedure. From studies directed toward this goal, this research reports a simple, rapid, direct esterification method for preparing the N-TFA *n*-butyl esters of the protein amino acids with *n*-butanol 3 *N* in HCL which can be completed in 15 min. Also, the solubility problems associated with cystine and some of the other amino acids were removed.

The reaction conditions for the "direct quantitative esterification" of the protein amino acids have been established. These experiments conclusively demonstrate that GLC analyses of the amino acids can be made precisely, accurately, and rapidly when the N-TFA *n*-butyl esters of the amino acids are formed directly in *n*-butanol 3 *N* in HCl. A significant advantage of the N-TFA *n*-butyl ester derivative over the N-TFA methyl ester derivative is that no losses occur. Further, the losses of the butyl ester derivative would be minimal if any leaks occurred during derivatization. A complete GLC analysis of the protein amino acids, including derivatization and chromatography, can be completed in less that 1 h. This GLC method should fill an important need of scientists who are faced with the problem of amino acid analyses of biologically important substances.

4. GLC of amino acids – single column separation

The three major difficulties with regard to the single-column separation of the twenty protein amino acid derivatives on a single GLC column are: [1] the on-column loss of arginine, histidine, and cystine, [2] separation of aspartic acid, phenylalanine, and histidine, and [3] separation of leucine and isoleucine. A number of siloxane liquid phases were chosen for evaluation as the use of more polar polyester phases generally results in losses of the three amino acids mentioned above. A careful study of the separation characteristics of the OV series of siloxane liquid phases demonstrated that none of these phases, singly or in combinations, would yield complete and reproducible separation of the protein amino acids on a single column. It was noted, however, that liquid phases of decreasing polarity gave somewhat increased separation of leucine and isoleucine. The non-polar hydrocarbon Apiezon was thus selected for study, and was found to be a promising substrate for the separation of the amino acid derivatives.

Studies were conducted to determine the optimum initial column temperature and hold time for the separation of alanine, threonine, glycine, and serine and of phenylalanine, lysine, and tyrosine by measuring the rates of migration of these derivatives under isothermal conditions.

Analysis of amino acids by liquid-chromatography (GLC) has been extensively studied due to both the importance of the amino acids, and the speed, sensitivity, and resolving power of GLC. As the low volatility of the amino acids has prevented their direct analysis by GLC, the formation of suitable derivatives of increased volatility has been the obvious approach toward their analysis. Several different types of derivatives have been studied by a number of investigators [86], however, continued experiments have shown that the N-trifluoroacetyl (N-TFA) *n*-butyl esters are among the most useful derivatives. The organic reaction conditions for derivatizing the twenty protein amino acids to their N-TFA *n*-butyl esters have been established as the result of earlier investigations by Zomzley *et al.*(87) and Gehrke and co-workers (88-95). Later reports presented improved methods for esterification [91], GLC separation [93], and the detection of extremely small amounts of protein amino acids [94].

The complete GLC separation of twenty protein amino acid derivatives on a single column has been an intractable problem, much more difficult than the quantitative synthesis of the derivatives. In 1971, a chromatographic system was reported by Gehrke *et al.*[93, 94] which employed dual columns; one of 0.65% ethylene glycol adipate (stabilized grade EGA) as liquid phase on 80-100 mesh acid-washed Chromosorb W for separation of seventeen amino acid derivatives, and a mixed phase column containing 1.0% (w/w) OV-210 and 2.0% (w/w) OV-17 on Gas-Chrom Q was used for the separation of histidine, arginine, and cystine. Both columns of the system were operated under the same thermal conditions, thus allowing simultaneous operation.

From the results of investigations toward a single-column separation, it is known that polar liquid phases decompose the N-TFA *n*-butyl ester derivatives of arginine, histidine, and cystine. Therefore, non-polar liquid phases were investigated for the reproducible quantitative elution of the twenty protein amino acids.

A careful study of the separation characteristics of the OV series of siloxane liquid phases demonstrated that none of these phases, singly or in combinations would yield complete and reproducible separation of the protein amino acids on a single column. Studies were then made to determine the optimum initial column temperature and hold time for separation of Ala, Thr, Gly, and Ser, and Phe, Lys, and Tyr by measuring the rates of migration of these derivatives under isothermal conditions on non-polar liquid phases.

The optimimum separation of all twenty amino acid derivatives was achieved with a 2.5 m x 2 mm I.D. glass column of 10% (w/w) Apiezon M. on 80–100 mesh HP Chromosorb W with an initial column temperature of 90° C for 6-9 min, then a 6°/min program rate to 160° C. Analyses of dipeptides containing histidine result in no apparent on-column conversion of diacyl histidine to the monacyl derivative. Apiezon M is a useful liquid phase for the single column separation of the protein amino acid N-TFA *n*-butyl esters.

5. GLC of nanogram amounts using a GLC solvent system

The quantitative gas-liquid chromatographic determination of the protein amino acid content of biological substances has been clearly demonstrated with analyses of corn grain, soybean oil meal, blood plasma, and human urine. The precision and accuracy of the gas-liquid chromatographic technique was found to be equal to that of classical ion-exchange chromatography, and superior in some instances.

Further, an instrumental-chromatographic system has been invented which allows the injection of 100 μl or more on a standard analytical column. This device eliminates the TFA peak during analysis of amino acids on the EGA column; results in a more stable baseline due to the decreased amount of solvent and reagents traversing the column; greatly simplifies analysis for nanogram amounts of amino acids; and should find a wide range of applications in many gas-liquid chromatographic and gas chromatography-mass spectometry investigations.

In our previous publications [94, 95], reference is made to the chromatographic separation of the amino acid derivatives. The packing composed of stabilized grade EGA and acid washed Chromosorb W (heated at 140° C for 12 h) is an excellent one and gives effective separation of 17 amino acids. Column packing, consisting of the mixed phase of OV-17 and OV-210 is a superior packing and shows highly efficient and effective separation of the basic

FLOW DIAGRAM OF PRE-COLUMN VENTING SYSTEM

FIGURE 9

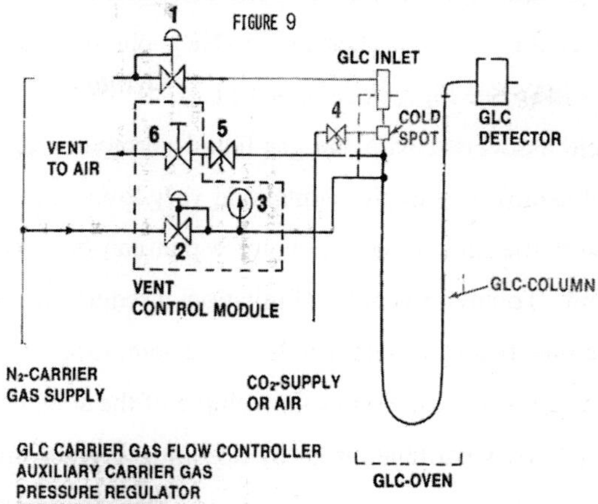

1 GLC CARRIER GAS FLOW CONTROLLER
2 AUXILIARY CARRIER GAS PRESSURE REGULATOR
3 PRESSURE GAUGE
4 CO_2-SOLENOID VALVE
5 VENT SOLENOID VALVE
6 VENT FLOW ADJUST VALVE

GEHRKE ET AL., U. S. PATENT NO. 3, 881, 892 (1975)

amino acids plus cystine over that reported by us earlier. No longer is it necessary to make a computation for histidine.

With these packings (columns I and II),as described above, one can now simultaneously analyze and separate all 20 of the protein amino acids on these two columns in 30 min with automatic electronic integration of all peaks.

A universal gas chromatography system (U.S. Patent no. 3,881,892) is described which has greatly improved the analysis of small samples. A major limitation of analysis at the part per billion (ppb) level and below is the volume of sample solution that can be injected. In order to inject a greater portion of the sample, one must either concentrate the sample solution, thus risking loss of volatile or unstable sample components, or inject larger volumes of sample solution which can produce solvent peak tailing and overlap of early eluting components. The improved capabilities of the described system over conventional GC systems are: (a) the volume of sample solution which can be injected can be increased by 10-fold or greater without adversely affecting the performance of the chromatographic column or detector, thus, significantly increasing the sensitivity of the analytical method, (b) the system eliminates the incompatibility between the various solvents and detectors, (c) it eliminates incompatibility between the sample solvent, reagents, and the analytical column, and (d) long retention time peaks, which are of no interest, may be vented prior to entering the analytical column, thus reducing analytical time. The universal GC system has been used in the analysis of chlorinated hydrocarbons, organophosphates, amino acids, lunar samples, genetic bases, fatty acids, neutral sugars, and biogenic amines. The detection systems used were flame ionization (FID), electron capture (ECD), thermionic (NP), photoionization (PID), and flame photometric detection (FPD). See Figures 9, 10 and 11.

A major obstacle associated with the gas-liquid chromatographic (GLC) analysis of very small quantities of sample or samples containing very low concentrations of the components of interest has been the limited sample volume that can be injected into an analytical chromatographic column. To circumvent this limitation, a reduction of the sample volume to a few microliters or the injection of of larger volumes of sample solution would be necessary. However, in many cases, drastic reduction in the volume of the solvent can result in losses of the components of interest by volatilization, or by decomposition of unstable compounds. In addition, the injection of large amounts of solvent results in excessive solvent peak tailing, thus obscuring the early eluting components of a mixture. Further, solvents and derivatization reagents can be incompatible with selective GLC detectors by causing either greatly enhanced response (e.g. halogen-containing solvents and ECD) or damage to the sensing components of the detectors due to corrosion or the formation of oxide deposits (NP and PID).

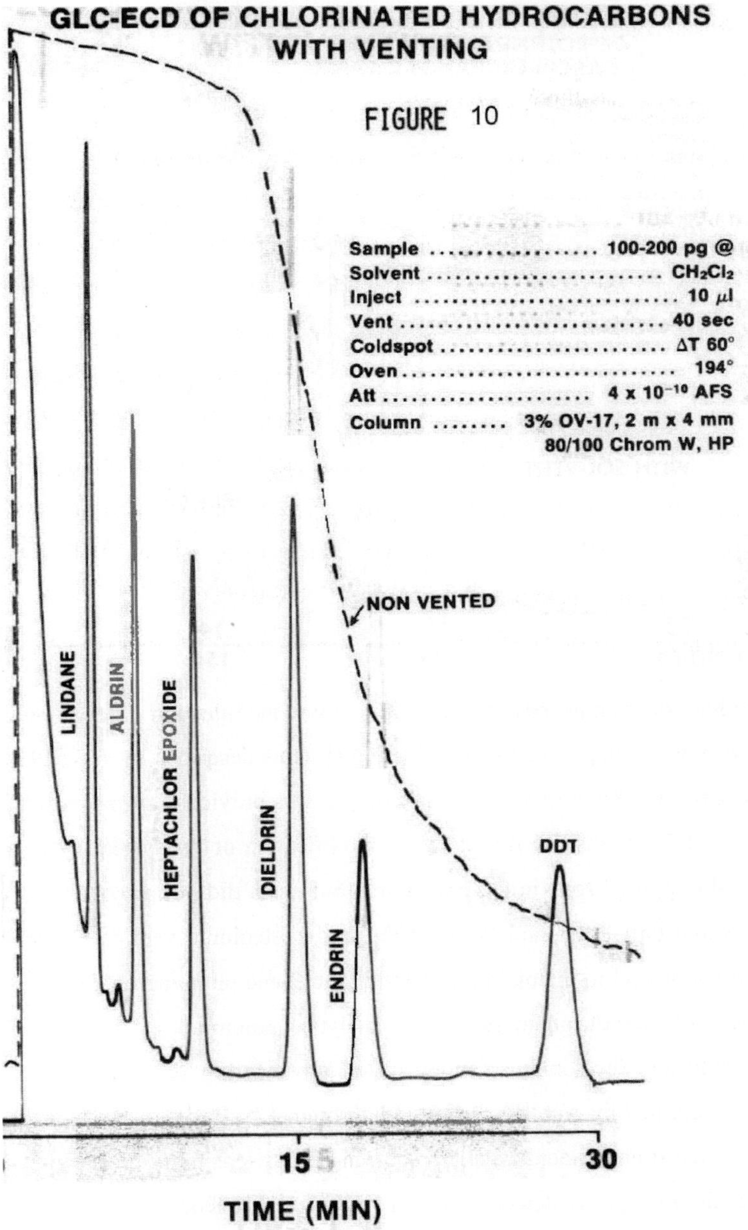
GLC-ECD OF CHLORINATED HYDROCARBONS
WITH VENTING
FIGURE 10
Sample .... 100-200 pg @
Solvent .... CH₂Cl₂
Inject .... 10 μl
Vent .... 40 sec
Coldspot .... ΔT 60°
Oven .... 194°
Att .... 4 x 10⁻¹⁰ AFS
Column .... 3% OV-17, 2 m x 4 mm
80/100 Chrom W, HP
NON VENTED
LINDANE
ALDRIN
HEPTACHLOR EPOXIDE
DIELDRIN
ENDRIN
DDT
15
30
TIME (MIN)

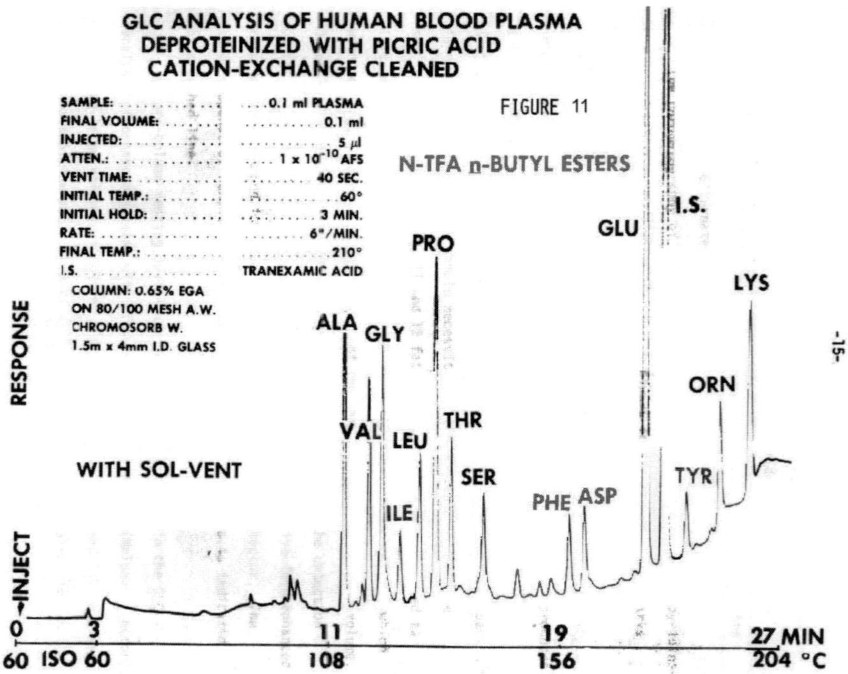

A number of attempts have been made to allow the injection of large volumes of very dilute solutions by venting the solvent. Venting systems described by Abel [96] and Evrard [97] utilized a precolumn, however, no makeup gas was provided during venting, thus eliminating the use of flow sensitive detectors. A modification of the system by Evrard [98g included a "cold trapping" zone in the precolumn, but again did not provide for a makeup gas flow. Claeys and Farr [99]), and Watts [100] used postcolumn venting to protect the ECD from contamination and to maintain the flame in the flame photometric detector, but this approach allowed all components to traverse the analytical column.

Our interest in developing a venting system arose during our search for amino acids in lunar samples returned by Apollo missions 11 through 17. Early analyses of the lunar samples indicated that if amino acids were present, increased sensitivity of the analytical method would be required to clearly detect them. In 1970, Gehrke, Kuo, and Zumwalt [101] developed a manual pre-column venting device to increase the portion of lunar sample extract which could be injected, thus facilitating the analysis for amino acids at the ppb level with FID detection. Our earlier instrumentation was then modified and improved by Gehrke *et al.* and used by Kaiser *et al.* [103]) in the quantitative analysis of amino acids in biological samples. The system was also used in the search for amino acids in Apollo returned lunar soil aqueous extracts [103].

A later electronically controlled version of a universal gas chromatography venting system was developed by Gehrke *et al.* [104, 105] and used in this study. With this system, venting of the solvent from a GC column can be performed under isothermal and temperature programmed conditions. The venting device can be interfaced readily with various models of gas chromatographs and detectors which are affected by carrier flow rate changes or certain solvents and derivatization reagents.

A schematic diagram of the GC solvent venting system is shown in Figure 9. The glass GLC column consists of a short pre-column segment (ca.10 cm), onto which the sample is injected, and an analytical column of various lengths depending on the required column efficiency for a particular analysis.

An important feature in this venting system is a "cold spot". The "cold spot" is accomplished by directing a stream of CO_2 or air on a 10 mm section of the pre-column. The "cold spot" produces a desired temperature differential between a small section of the precolumn and column oven. The chamber is made of brass with holes drilled around the circumference for the coolant (CO_2, air) to escape. The "cold spot" retards the migration of the sample molecules to be analyzed while allowing the solvent to be vented.

A "cold spot" combined with the venting device lends versatility to the chromatography and aids in accomplishing a variety of GLC analyses. Solvents and/or excess reagents interfering with the chromatographic column or detector performance are removed prior to the analytical column and detector. Good reproducibility was achieved on repeated injections (RSD for N =28 of 3.4%).

6. GLC of trimethylsilyl derivatives

The number of chromatographic peaks for the TMS derivatives of glycine and arginine are determined by the polarity of the solvent. With hexane, methylene chloride, chloroform, and 1,2 –dichloroethane, one peak is obtained for glycine (GLY_2) and two peaks (GLY_2 & GLY_3) in six other more polar solvents. Arginine gives no peak in the four less polar solvents studied and one peak in the other six more polar solvents.

We reported on chromatography using a 6 m x 2 mm I.D. glass column of 10 w/w% OV-11 on 110/120 mesh Supelcoport of which good resolution was obtained for the TMS derivatives of nineteen of the protein amino acids.

Since introduction of the trimethylsilyl (TMS) derivatives of the amino acids by Rűhlmann and Giesecke [106], efforts have been made by several groups of researchers in attempts to use these derivatives for the quantitative gas-liquid chromatographic (GLC) analysis of the twenty protein amino acids. The primary interest in the TMS derivative is the one-step derivatization procedure, whereas almost all other derivatives are formed by two or more reaction steps. Rühlmann and Giesecke used hexamethyldisilazane and trimethylchlorosilane (TMCS) to obtain derivatives for most of the protein amino acids. Smith *et al.* [107,108] made a study of the optimum silylation conditions for leucine, serine, and aspartic acid. They concluded that trimethylsilyldiethylamine with some kind of catalyst was the best silylation reagent. Trimethylsilyldimethylamine recently has been claimed to be more volatile and was recommended for these reasons [109]. N-Trimethylsilyl-N-methyl-acetamide has been recommended by Birkofer and Donike [110]. Klebe *et al.* [111] used bis(trimethylsilyl) acetamide (BSA) to obtain sharp single peaks for all the protein amino acids except arginine, which showed indications of decomposition on the column. However, they were unable to separate the derivatives of glycine and alanine from BSA on an SE-30 column. The introduction of bis(trimethylsilyl) trifluoroacetamide (BSTFA) by Stalling *et al.*[19] has solved the problem of separation of the TMS derivatives of glycine and alanine from the reagents and reaction products. Then, Gehrke *et al.* [20,21] published a comprehensive method for the GLC analysis of all twenty protein amino acids as their TMS derivatives using BSTFA as the silylating agent, but they reported that problems existed in the analysis of biological fluids such as urine. Figure 12.

Bergström *et al.* [112] recently reported on the trimethylsilylation of amino acids. These workers used BSTFA with and without solvent at 125° C for 15 min to obtain chromatographic peaks for eighteen amino acids. They also reported on the mass spectra of the two derivatives of glycine and lysine. The results of the mass spectra are in agreement with the structural assignment given by Gehrke and coworkers [20, 21].

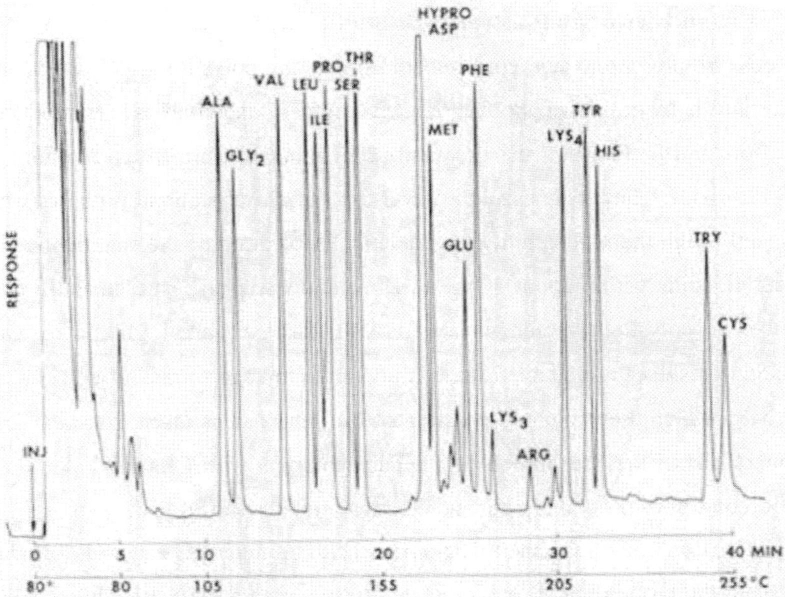

FIGURE 12 GLC chromatogram of TMS amino acids: sample, 0.2 mg of each amino acid in 2 mℓ CH_3CN/BSTFA, 1:1 v/v; closed tube silylation, 135°C-15 min, 5 μℓ injected; column, mixed liquid phase 6 w/w% OV-7 and 3 w/w% OV-22 on 100/120 mesh Supelcoport* 2 m × 4 mm I.D. glass; I. T., 80°C, 5 min, hold, then 5°C/min.

Gehrke and coworkers reported that urine samples containing large amounts of glycine presented difficulties in obtaining a single peak for glycine. Both the di-trimethylsilyl (GLY_2) derivative and the tri-trimethylsilyl (GLY_3, *ca.* 10%) derivatives were obtained when the samples were derivatized at 135° C for both 10 and 15 min. Because of the large quantity of glycine in the urine samples, the GLY_3 peak interfered with the resolution of TMS isoleucine and TMS proline. In preliminary work on the TMS derivative of glycine by the present authors, it was observed that only the first peak for glycine (GLY_2) was obtained when using methylene chloride as a solvent instead of acetonitrile. We investigated the effect of different solvents on silylation and reported on improvements in the chromatographic separation, thus permitting the analysis of biological fluids such as urine.

This chromatogram shows excellent resolution for nineteen amino acids. In the analysis of biological fluids containing large amounts of glycine it may be necessary to derivatize the sample with two different solvents to avoid interference of the GLY peak with isoleucine and proline.

7. Enantiomeric and Diasteromeric Studies

The separation of amino acid enantiomers using chiral polysiloxane liquid phases and quantitative analysis by the novel approach of "enantiomeric labeling" is presented by E. Bayer, G. Nicholson, and H. Frank of the University of Tübingen, Germany in (Volume 2, Chapter 2) [2] [113, 114]. They synthesized a novel type of chiral stationary phase, which possesses sufficiently high thermal stability to enable GC separation of the enantiomers of all the protein amino acids within about 30 min. The stationary phase, subsequently named Chirasil-Val®, is a polysiloxane functionalized with carboxyl alkyl groups to which valine-*t*-butylamide residues are coupled through an amide linkage. Frank et al. [113, 114], describe the criteria which the stationary phase must fulfill, the steps taken to synthesize the required stationary phase, the preparation of capillary columns with Chirasil-Val® stationary phase, and GC conditions needed to optimize the separations. Figure 13 and 14.

The concept of the use of D-amino acids as internal standards for the quantitative analysis of the L-amino acids in biological fluids, a several step procedure, is usually required to separate the amino acids from high molecular weight material, carbohydrate derivatives, fatty acids, oligopeptides, biogenic amines, etc., and that recovery of the amino acids during this process must be either quantitative or of a constant and known percentage. Also, variations can occur during derivatization and in the response of the detector, and discriminatory effects in the injector can arise when using the split technique of sample injection.

FIGURE 13

SYNTHESIS PATHWAY FOR CHIRASIL-VAL

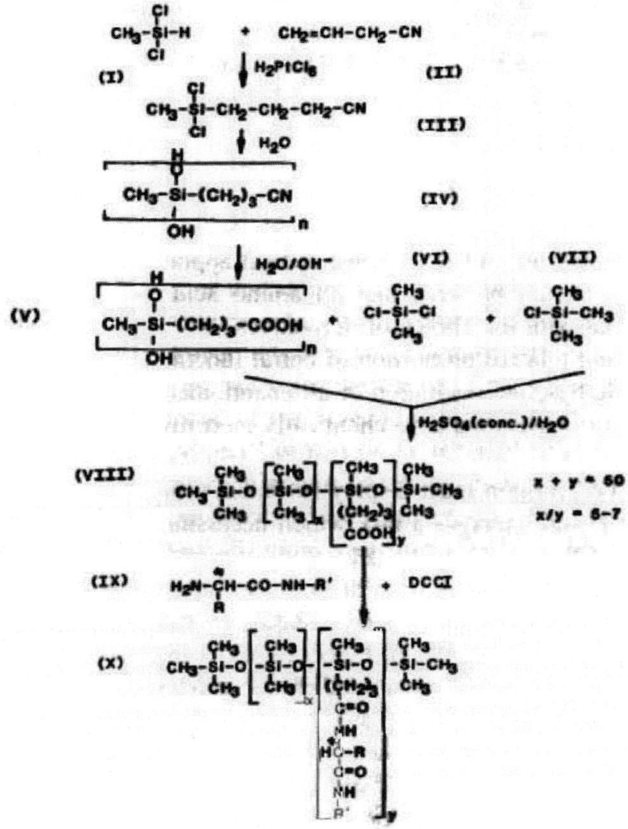

By using the optical antipodes as internal standards, not only are volumetric losses during work-up derivatization and GC compensated, but also differences in chemical behavior, such as recovery from ion-exchange columns during cleanup, yield of derivatization, and stability of the derivatives. With "enantiomeric labeling", the internal standard suffers exactly the same fate as its optical antipode. Frank *et al.* [115], illustrate the use of "enantiomeric labeling" for analyses of blood serum and gelatin, present in the mathematical expressions for determining the concentration of each amino acid in samples. Sensitivity, precision, and accuracy of the enantiomeric labeling technique are discussed, with accuracy of determination of synthetic samples surpassing ion-exchange chromatography (IEC). Thus GC analysis of amino acids by "enantiomeric labeling" is a powerful technique for a wide variety of applications.

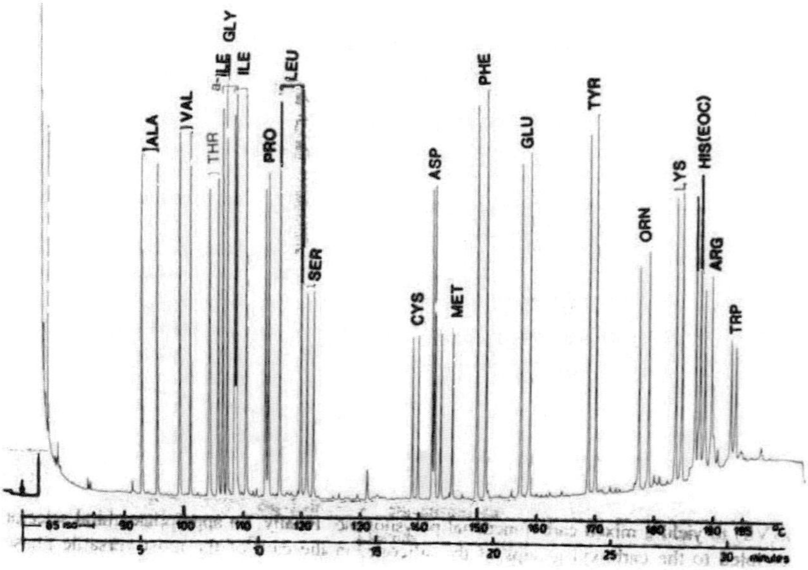

A

FIGURE 14 (A) Racemic amino acid test mixture. Derivatives: *N*-(*O,S*)-pentafluoropropionyl/*n*-propyl ester; column: 20 m × 0.3 mm borosilicate glass, etched with HCl and washed with methanol; stationary phase: Chirasil-Val®, film thickness 0,21 μm; temperature program: 85°C, 3 min isothermal, 4°/min to 195°C; inlet pressure: 45 kPa H_2. (In all cases, the D-enantiomer is eluted prior to the L-enantiomer.) (B) Racemic amino acid test mixture. Derivatives: *N*(*O,S*)-PFP/*n*-isopropyl ester; column as in Figure 2A; temperature program: 50°C, 2 min, 85°C, 3 min, 4°/min to 190°C. (C) Racemic amino acid test mixture. Derivatives: *N*(*O,S*)-TFA/*n*-propyl ester; column: 20 m × 0.25 mm fused silica; stationary phase: Chirasil-Val®, film thickness 0.19 μm; temperature program: 80°C, 3 min isothermal, 4°/min to 200°C; inlet pressure: 50 kPa H_2.

When many analyses are required, Bayer also has introduced a new automated derivatization robot [116].

The analysis of amino acid racemization and the separation of amino acid enantiomers is presented by H. Frank of the University of Tübingen, Germany, in Volume 2, Chapter 3 [2] [116, 117]. The author presents examples of enantiomers of free amino acids and amino acid analogues that exhibit different biological properties. A tragic illustration is the glutamic acid derivative, thalidomide, a sedative drug used in the 1950s which induced malformations of newborns. Later, the L-enantiomer of the metabolite was found to be teratogenic; whereas, the D-antipode is not (see Chapter S-9De) [2]. The stereochemical analysis of amino acids is important for studies on the biochemical, pharmacological, or toxicological properties of enantiomers, as well as its utilization in structure elucidation, peptide synthesis, racemization monitoring, and pharmaceutical research.

For a meaningful determination of lower percentages of D-amino acids in proteins, Frank, (Volume 2, Chapter 3) [2] [116,117] resorted to deuterium/hydrogen exchange of GC/MS which enables the determination of 0.01% of D-enantiomer originally present in a protein, found an even smaller quantity of D-enantiomers by use of an added standard compound as a reference peak. By using a series of dilutions, ideally D-enantiomers could be measured when present at 0.01 to 0.001% of the L-enantiomer. The methods for determination of optical purity of amino acids have been greatly refined, making previously intractable problems now appear relatively straightforward [2].

A. Darbre, of Kings College, London, one of the first investigators to use GC for determination of amino acids, includes an interesting perspective in (volume 2, Chapter 4), [2] [118]. He describes studies of the *N*-TFA amyl and methyl esters, including column evaluations, various methods of esterification, trifluoroacylation, and quantitative studies using a density balance and flame ionization detectors. His investigation includes use of radioactive amino acids to study the breakdown of some derivatives during chromatography by monitoring the column effluent for radioactivity. Darbre describes his methods of solid sample injection, *in situ* precolumn derivatization, and quantitative studies of biological samples using the *N*-TFA amino acid methyl esters. He also discussed GC separation of acylated dipeptide methyl esters, pointing out the recent advances in MS techniques for peptide analysis. Darbre's most recent work has centered on *N*-HFB isobutyl esters, a practical derivatization method and its application to analysis of blood plasma and other biological samples. The use of electron-caputre (ECD), alkali flame ionization (AFID), and flame-ionization detection (FID) are compared for relative molar response (RMR) values of the amino acid derivatives and the analysis of biological samples by capillary columns with FID, AFID, and ECD. Darbre foresees the next stage of development to include work at even higher levels of sensitivity (nanomole to picomole amounts), and perhaps eventually the use of aseptic conditions to eliminate the problems of background contamination.

The *N*-TFA *n*-propyl esters have been the subject of an in-depth study by G. Gamerith, at the University of Graz, Austria, (volume 2, Chapter 5) [2] [119, 120] for medical diagnostic applications. Gamerith thoroughly discusses his effective amino acid analysis method for screening individuals with potential inborn errors of metabolism and demonstrates the application of the method for quantitative screening of urine specimens of patients with mental retardation.

Cyril Ponnamperuma of NASA and the University of Maryland elegantly presents the role of GC in the study of cosmochemistry (Chapter 6, Volume 2) [2] [121]. One of the most fundamental questions of all science is the "origin of life". How did it first begin? Since the laws of physics and chemistry are universal laws, it may be legitimate to extrapolate from the circumstances on earth to elsewhere in the universe, and our present day telescopes reveal that there are 10^{23} stars which like our own sun can provide the photochemical basis for plant and animal life. Ponnamperuma presents the central role of GC and interfaced MS spanning two decades of study on chemical evolution, prebiotic chemistry, and his search for life molecules on the moon and planets, ancient sediments, and the extraterrestrial Murchison meteorite. The detection and quantitation of minute traces of amino acids in complex mixtures was made possibly by highly sensitive and selective GC techniques that were first developed at the University of Missouri by Gehrke, Kuo, and Zumwalt (Volume 2, Chapter 6) [2].

Our experiments in 1984, with the methods of HPLC followed by capillary GC interfaced with MS presented for the first time unambiguous evidence for the existence of extraterrestrial nucleobases in the Murchison meteorite. The nucleobases uracil, thymine, cystosine, adenine, and xanthine were also found in the primordial soup generated by an electric discharge. GC/MS has thus provided the chemist with a powerful research tool for the study of organic cosmochemistry.

W.A. König of the University of Hamburg, Germany, describes the analysis of amino acids and *N*-methylamino acids by capillary GC of diastereomeric derivatives on nonchiral phases and the use of isocyanates as reagents for enantiomer separation on chiral phases in (Volume 3, Chapter 1) [2] [122]. König discusses practical applications of configurational analysis of amino acids, and presents in detail the use of diastereomeric derivatives. Factors that affect the precision of determination of the proportion of a mixture of enantiomers are described, including optical purity of the chiral reagent and reaction kinetics of enantiomers in the formation of diastereomeric derivatives.

I. Abe and S. Kuramoto, Volume 3, Chapter 2 [2] [123,124] address in depth two topics. First, a comparison of the resolution of various D-amino acid derivatives on optically active columns, and second, the "age dating" of sediments by determination of the extent of racemization of amino acids in fossil shells. The influence of the ester group on chromatographic resolution was evaluated by synthesis and analysis of the methyl, ethyl, *n*-propyl, isopropyl, and neopentyl esters, and the effect of the acyl group was investigated by preparation and analysis of the TFA, PFPF, and HFB derivatives. A total of nine different derivatives and the different substituent groups were investigated in terms of retention times, D and L peak-to-peak resolution, and elution order on both glass and fused silica Chirasil-Val® columns. This systematic study provides detailed information on the separation of amino acid enantiomers; and his greatest difficulty was in the separation of the lower molecular weight amino acids (glycine, threonine, isoleucine, *allo*isoleucine) which elute within a very narrow range.

Abe and S. Kuramoto [123] also describe the application of enantiomeric analysis to age dating of sediments of the Osaka Plain. The rate of amino acid racemization was dependent on temperature, pH, and ionic strength. Temperature especially affects the rate of racemization, e.g., racemization half-lives at 0° C are about 100 times longer than at 25° C. The expressions for calculation of the age of fossils from the enantiomer ratios are then presented. Sediments from 50 and 64.5 m in depth were dated at 27,000 to 28,000 and 29,000 to 30,000 years before the present; this illustrates the potential for age dating from D/L ratios of several amino acids.

G. Odham and G. Bengtsson of the University of Lund, Sweden, Volume 3, Chapter 4 [2] [125, 126], present an excellent description of the GC microanalysis of amino acids using electron capture detection (ECD) and MS with emphasis on micro environmental applications, especially in studies on the interactions between microorganisms and other cells. Their research has focused on situations in which the sample size is limited to microliters or micrograms, and the amino acid concentrations are in the ppb or ppt range; this combined limitation of both very low sample amount and very low concentrations within the sample creates special analytical demands, which the authors clearly address. They describe microscale sample preparations including hydrolysis (for proteins and peptides), extraction (for free amino acids), ion-exchange cleanup, and derivatization, providing much practical information on each procedure. For capillary GC separation, their use of splitless and on-column injections is described, as limited sample quantity prohibited use of the split technique. The authors point out that both splitless and on-column injections permit quantitative analysis of amino acids, while split injection is unreliable at least for the higher boiling amino acids.

The oxazolidinones are the subject of (Volume 3, Chapter 5) [2] [127, 128]. P. Hušek of the Endocrinology Institute, Prague, Czechoslovakia, has developed a novel derivative; he considers alternatives to the esterification-acylation approach to amino acid derivatization and presents possibilities for the simultaneous derivatization of β-amino and carboxyl groups. He investigated the use of perhalogenated acetones, i.e., 1,3-dichloro-1, 1, 3, 3 - tetrafluoroacetone (DCTFA), for forming cyclic derivatives thus developing a "single medium, two reagent" reaction approach for amino acid derivatization. Amino acids that do not contain side-chain reactive groups can be analyzed after cyclization only, while those with reactive side-chain groups are analyzed after treatment with an acylating agent. Hušek clearly describes the derivatization procedures and the chromatographic analysis after cyclization and acylation. He also discusses applications of his analytical method to analysis of physiological fluids; recommends a practical procedure for isolation of free amino acids from 50 to 100 μl of serum or urine, and presents a rapid procedure for the determination of thyroid hormonal compounds, diiodothyronine, triiodothyronine, thyroxine, and their deamination and decarboxylation products.

Capillary GC, invented by M.J.E. Golay, was first described in 1957. At the beginning, capillary columns were mainly made of stainless steel or other metals, but with the introduction of fused silica columns in 1979 by Dandeneau and Zerenner, fused silica has largely replaced all other column materials. Our study took advantage of the fact that two different derivative types have been thoroughly investigated and are now well-established for packed column amino acid analysis: the *N*-TFA *n*-butyl esters developed by Gehrke *et al.* (Volume 1,Chapters 1 to 3) [2, 20], and the *N*-HFB isobutyl derivatives primarily developed by MacKenzie and coworkers (Volume 1, Chapter 4) [2] [28, 29].

C.W. Gehrke and R.W. Zumwalt have demonstrated, (Volume 3, Chapter 7) [2, 129] that fused silica bonded phase columns are a very effective means for separating and quantitatively measuring amino acids in protein hydrolysates as both the *N*-TFA *n*-butyl and *N*-HFB isobutyl derivatives. Comparisons are presented of columns with differing liquid phase polarities, providing information on the effect of polarity on separation of particular amino acids, and evaluation of split, splitless, and on-column injection methods revealed the on-column injection technique to be most reproducible. Fused silica capillary columns with chemically bonded phases of 0.25 μm film thickness (J&W Durabond DB-1, methyl silicone, and DB-5, bonded phenyl (5%) methylsilicone) were used to separate and quantitate the amino acid derivatives. All the protein amino acid *N*-TFA *n*-butyl esters were separated on a short (20 m x 0.25 mm I.D.) DB-5 column, while only the derivatives of lysine and tyrosine were not well resolved on a DB-1 column of the same length and diameter. Conversely, the DB-5 column gave better separation of all the protein amino acid *N*-HFB isobutyl esters than did the DB-1 column, with the DB-5 column providing an improved lysine/tyrosine separation. A number of analytical applications on diverse sample types are presented in which data from fused silica capillary GC analyses are compared to results obtained from IEC.

Amino acid analysis of commercial animal feeds, feed ingredients, lysozyme, β-lactoglobulin, and ribonuclease by capillary GLC as both the *N*-TFA *n*-butyl esters and the *N*-HFB isobutyl esters were in agreement with the IEC results [129]. The comparison clearly demonstrates that the capillary GC method is a powerful tool for the measurement of amino acids.

We describe the uniting of established methods of amino acid derivatization which were developed by Kaiser *et al.* in 1974 [102], and MacKenzie in 1974 [28], with the recent development of fused silica immobilized phase capillary columns into a high resolution, precise, and *accurate* method for measurement of amino acids.

8. Amino acids in Apollo returned soil

The lunar samples from Apollo flights 11 through 17 provided the students of chemical evolution with an opportunity of examining extraterrestrial materials for evidence of early prebiological chemistry in the solar system. Our search was directed to water-extractable compounds with emphasis on amino acids. Gas chromatography, ion-exchange chromatography and gas chromatography combined with mass spectrometry were used for the analysis. It is our conclusion that amino acids are not present in the lunar regolith above the background levels of our investigations of 1.0 ng/g.

The characterization of carbon compounds indigenous to the lunar surface is of particular interest as these investigations could result in findings which would advance our knowledge of the processes of chemical evolution. The Apollo missions have provided us with the requisite extraterrestrial material for study.

A number of investigators have studied returned lunar samples for biologically important organic compounds since 1969 (Fox *et al.*, 1973; Hamilton and Nagy, 1972;Gehrke *et al.*,1972a). Our paper presents (138) both our search for water-extractable organic compounds, with emphasis on amino acids, in Apollo 17 fines, and a summary of the analysis for amino acids in samples from Apollo flights 11 through 17.

Apollo 17 sample nos. 72501.62 (4 g) and 70011.37 SECS (4 g) were used in these studies. Sample 72501.62 was a soil sample from material sampled to a depth of 5 cm on a steep uphill slope. The unsieved material was less than 1 mm in size. Sample 70011.37 was especially provided to those interested in organic carbon analysis. This sample was presumed and reported to have been exposed to some rocket exhaust.

The extracts were prepared using the hydrothermal extraction technique described by Cheng and Ponnamperuma (1974). A two gram portion of lunar fines was placed in two separate Pyrex glass extraction tubes (4 g total), ca. 6 ml of water were added to each and the tubes were sealed with a hydrogen-oxygen torch. The sealed tubes were shaken well, then placed in a 165–170° C oven for 1 hr.

The GLC analysis of a standard mixture containing 10 ng of each amino acid is presented in Figure 15. Corresponding to Figure 15 for GLC, standards of 16 amino acids at the

2 ng and 10 ng level were analyzed by CIE. In Figure 16 we show the responses for the 6 amino acids of major concern, aspartic acid, threonine, serine, glutamic acid, glycine, and alanine.

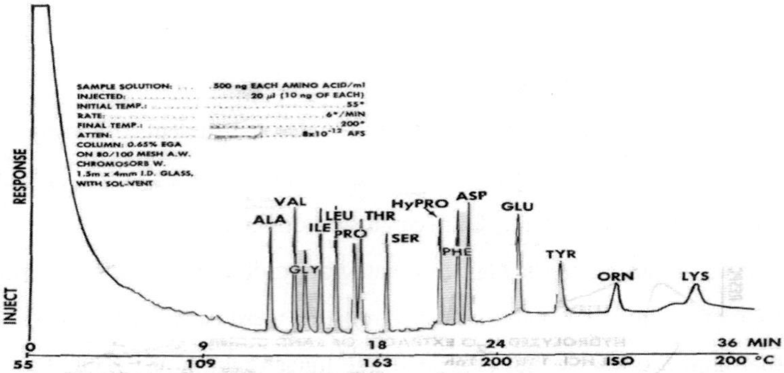

Fig. 15 GLC analysis of a standard mixture of amino acids N-TFA *n*-butyl esters.

Figure 17 presents the chromatograms obtained from the *undydrolyzed and hydrolyzed extracts* of Apollo 17 sample no. 72501.62. Three significant peaks were observed in the unhydrolyzed extract with the largest identified tentatively as oxalic acid (ca. 25 ng), eluted at ca. 145° C. This peak, along with the two smaller peaks eluted near threonine and serine, have been observed in the analyses of lunar samples returned by earlier Apollo flights (Zumwalt *et al.*, 1971: Gehrke *et al.*, 1971, 1973). These peaks were significantly reduced in size after hydrolyzing with 6 N HCL, with the largest also being eluted ca. 1 min earlier than the largest peak in the unhydrolyzed sample. The three large peaks were not observed on the CIE chromatogram and thus were ninhydrin negative.

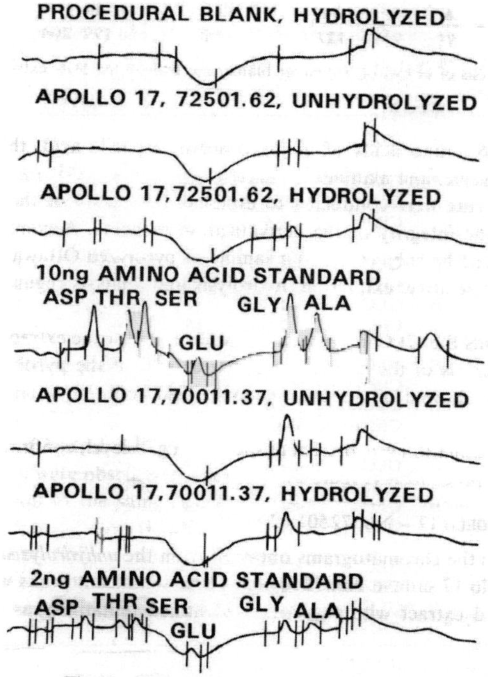

Fig. 16 CIE analysis of Apollo 17 lunar fines.

The GLC data of Apollo 17 – no. 70011.37 SECS for the free and 6 N HCl hydrolyzed aqueous extracts are given in Table 8. Glycine was found at a level of 12 ng g $^{-1}$ (*F*) and 5 ng g $^{-1}$ (*H*), however, the other amino acids were not found in either the free or hydrolyzed extracts at concentrations greater than 1 ng g $^{-1}$.

Throughout these investigations of the returned lunar samples, our primary aim has been the search for, and identification of, water extractable, derivatizable organic compounds by the GLC method and ninhydrin-active organic compounds by the CIE methods with particular emphasis on the amino acids. Analyses by both GLC and CIE on the same sample extracts produced independent yet supporting analytical results.

TABLE 8

Amino acids in Apollo 12, 14, 15, 16, and 17 lunar fines[a]

		Amino acid	Ala	Gly	Thr	Ser	Asp	Glu
12	*H*	GLC	<1	<1		0.5	<1	<1
14	*F*	GLC	<1	<1	[b]	[b]	<1	<1
	H	GLC[c]	<0.5	3	[b]	1	<0.5	<0.5
		CIE[c]	<0.5	3	[d]	1	<0.5	<0.5
	F	GLC	<1	[b]	[b]	<1	–	[b]
15		CIE	–	3	–	<1	–	–
	H	GLC	<1	3	[b]	1	–	[b]
		CIE	2	16	–	<5	3	<1
	F	GLC		[b]	[b]	<1	–	[b]
		CIE	<1	2.5	–	1	–	–
16	*H*	GLC	2.0	7.5	[b]	2.5	2.5	[b]
		CIE	<1	8	–	2	1.5	
17 (72501.62)[e]	*F*[f]	GLC	<1	[b]	<2[b]	<1	<0.5	<1
		CIE	–	1.5	–	<1	<1	–
	H[f]	GLC	<1	<1	<2[b]	<1	<0.5	<1
		CIE	<1	4	–	2		<1
	F	GLC	<3	12	–	n.d.	<1	
17		CIE	1.5	19	4	<1	<1	
(70011.37)	*H*	GLC		≃5				
		CIE	1	11	–	–	1	–

[a] All data as ng g^{-1}.
[b] Elution position obscured by non-amino acid peak.
[c] Gehrke *et al.* (1972a).
[d] Not reported in joint studies (Ibid).
[e] Results from two independent samples.
[f] Average of two separate analyses.
F=Free.
H=Hydrolyzed.
– Indicates none found.

Table 8 presents a compilation of the data obtained from our analyses for amino acids in lunar samples returned by Apollo flights 12, 14, 15, 16 and 17, and is a summary of the results obtained by both GLC and CIE analyses. The total procedural blanks for the analyses which included pyrolyzed sand blanks, showed no amino acid contamination above the 1ng/g level by GLC and CIE. Figures 15, 16, and 17.

The differences in amino acid levels, in particular, glycine, that have been found between the Apollo 17 sample known to have been contaminated by rocket exhaust (70011.37) and the samples taken from a slight depth or distance, could implicate hypergolic rocket fuels as the source of the glycine found.

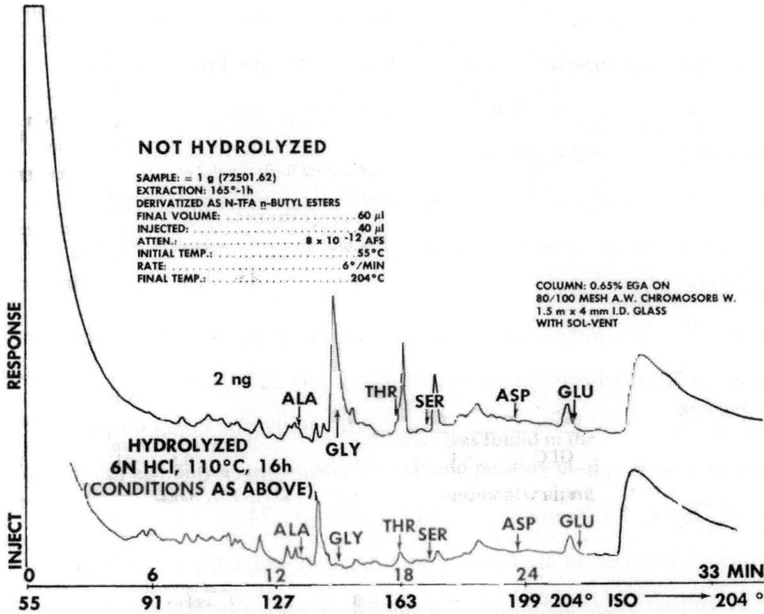

Fig.17 GLC analysis of water extract of Apollo 17 lunar fines.

9. Other GLC methods and reports for amino acid analysis and separation

A literature review follows on the research on amino acids for the period of 1984 to 2005.

Bibliographical information

[1] Hartman, N. R. and Jardine, I., *Mass spectrometric analysis of the cyclosporine metabolites*, Biomed. Env. Mass Spectr. 13 (1986) 362-372.

[2] MacKenzie, S.L., *Gas chromatographic analysis of amino acids as the N-heptafluorobutyryl isobutyl esters*, J. Assoc. Off. Anal. Chem. 70 (1)(1987) 151-160.

[3] Gehrke, C. W. et al., *Analysis of amino acids by gas chromatography as the N-trifluoroacetyl n-butyl esters*, Journal – Assoc. Off. Anal. Chem. 70(1)(1987) 160-70.

[4] Bruckner, H. et al., *Determination of alpha-alkyl-alpha-amino acids and alpha-amino alcohols by chiral-phase capillary gas chromatography and reverse-phase high-performance liquid chromatography*, J. Chromatogr. 395 (1987) 569-90.

[5] Rogoskin, V.A., et al., *Gas chromatographic determination of 1- and 3-methylhistidine in biological fluids*, J. Chromatogr. 423(1987)33-40.

[6] Buser W. and Erbersdobler, H.F., *Gas chromatographic determination of amino acids with nitrogen-selective detection,* Z. Lebensm.-Unters. Forsch. 186 (1988) 509-13.

[7] Borodina, V. L. et al., *Rapid gas liquid chromatographic assay of amino acids*, Laboratornoe Delo (7) (1984) 395-8.

[8] Sumitomo Chemical Co., Ltd., Japan., *Optically active amino acid derivatives as stationary phase for gas chromatographic resolution of amino acids*, Jpn. Kokai Tokyo Koho (1984) 6 pp

[9] Kuesters, E., et al., *Resolution of sulfur-containing amino acids by chiral phase gas chromatography*, Chromatographia 18 (6) (1984) 287-93.

[10] Gamerith, Gernot, *Influence of trifluoroacetic anhydride concentration during derivatization and chromatography of N (O,S)-trifluoroacetyl amino acid n-propyl esters*, J. Chromatogr. 318 (1) (1985) 65-74.

[11] Sakamoto, Munenori et al., *Gas chromatography-mass spectrometry of chemically modified amino acids in silk fibers treated with alkyl glycidyl ethers*, , Japan. Sen'I Gakkaishi 41 (1985) T-156-T163.

[12] Moodie, I.M. et al., *Gas-liquid chromatography of amino acids. Variable arginine response*, J. Chromatogr. 347 (1985) 179-82.

[13] Krylov, A.I. et al., *Gas chromatographic determination of amino acids using micropacked columns,* Zhurnal Analiticheskoi Khimii 40 (1985) 2235-40.

[14] Oi, Takafumi, and Sawada, Yoko, *Stationary phase for gas chromatography of amino acid enantiomers,* (Sumitomo Chemical Co., Ltd., Japan), Jpn. Kokai Tokkyo Koho (1985) 4 pp.

[15] Chen, Gang et al., *Gas chromatographic determination of twenty serum amino acids – preparation of 3-fluoroacetyl butylate derivatives,* Sepu 4 (1986) 56-9.

[16] Facklam, Christine, *An N-acyl-group modified valinamide phase for gas chromatography in the enantiomeric separation of amino-acid derivatives*, Z. Chem. 25 (2) (1985) 438-9.

[17] Palla, Gerardo, et al., *Oligo-oxaalkanoyl tetraamides derived from L-phenylalanine as stationary phases in capillary gas chromatographic resolution of D, L-amino acids*, , J. Chromatogr. 349 (1985) 9-15.

[18] Abe, Iwao, et al., *Study on N(O)-acyl substituents for the resolution of amino acid enantiomers by gas chromatography*, Japan, Chemistry Express 1(6) (1986) 335-8.

[19] Brückner, Hans et al., *Determination of α-alkyl- α amino acids and α-amino alcohols by chiral-phase capillary gas chromatography and reversed-phase high-performance liquid chromatography*, J. Chromatogr. 395 (1987) 569-90.

[20] Wagner, J., et al., *Resolution of enantiomers of various α-substituted ornithine and lysine analogs by high-performance liquid chromatography with chiral eluant and by gas chromatography on Chirasil-Val*, Anal. Biochem. 164 (1987) 102-16.

[21] Brückner, Hans, *Determination of non-protein α-alkyl- α-amino acids by reversed-phase high-performance liquid chromatography in comparison with ion-exchange and capillary gas chromatography*, J. Chromatogr. 386 (1987) 251-72.

[22] Sakamoto, Munenori, *Gas chromatographic and mass spectrometric behavior of O-alkylserines as their N-trifluoroacetyl butyl esters*, J. Chromatogr. 411 (1987) 259-73.

[23] Rogoskin, V., et al., *Gas chromatographic determination of 1- and 3-methylhistidine in biological fluids*, J. Chromatogr. 423 (1987) 33-40.

[24] Bueser, Wofgang, and Erbersdobler, Helmut F., *Determination of amino acids by gas-liquid chromatography with nitrogen-selective detection*, Z. Lebens.-Unters.–Forsch. 186 (1988) 509-13.

[25] Koenig, W.A., et al., *Cyclodextrins as chiral stationary phases in capillary gas chromatography. Part II: heptakis(3-O-acetyl-2,6-di-O-pentyl)-β-cyclodextrin*, HRCC. 11 (1988) 506-9.

[26] Abdalla, S. et al., *Derivatives for separation of amino acid enantiomers*, Chromatographia 23 (2) (1987) 83-5.

[27] Brückner, H. et al., *Chiral separations of α-alkyl- α-amino acids on α-hydroxymethyl α-amino acids*, Colloque INSERM (1989), 174) (Forum Pept., 2nd, 1988) 291-5.

[28] Buyuktimkin, N., et al., *GC-resolution of racemic amino acid methyl esters via diastereomeric amides of (1S)-(-)camphanic acid*, Scientia Pharmaceutica 57 (1) (1989) 21-6.

[29] Domergue, N. et al., *One-step conversion of amino acids into N-methyloxycarbonyl alkyl ester derivatives for chiral gas chromatography*, Anal. Biochem. 214 (1993) 420-5.

[30] Nokihara, K. and Gerhardt, J. *Development of an improved automated gas-chromatographic chiral analysis system: application to non-natural amino acids and natural protein hydrolysates*, Shimadzu Scientific Research Inc., Chirality 13 (2001) 431-4.

[31] Caja Lopez Maria del Mar et al., *Derivatization of chiral amino acids in supercritical carbon dioxide,* Anal.Chem. 76 (2004) 736-41.

[32] Vonderheide, Anne P. et al., *Solid-phase microextraction as a sample preparation strategy for the analysis of seleno amino acids by gas chromatography-inductively coupled plasma mass spectrometry*, Analyst 127(2002) 49-53.

[33] Chunhui, Deng, et al., *Rapid determination of amino acids in neonatal blood samples based on derivatization with isobutyl chloroformate followed by solid-phase microextraction and gas chromatography/mass spectrometry*, Mass Spectrom. RCM 18 (2004) 2558-64.

[34] Kawulka, Karen E., et al., *Structure of subtilosin A, a cyclic antimicrobial peptide from Bacillus subtillis with unusual sulfur to alpha-carbon cross-links:formation and reduction of alpha-thio-alpha-amino acid derivatives*, Biochemistry 43 (2004) 3385-95.

[35] Armstrong, Daniel et al., *Capillary gas chromatographic separation of enantiomers with stable dipentyl- α-, β- and γ-cyclodextrin-derivatized stationary phases,* Anal. Chim. Acta 234 (1990) 365-80.

[36] Pensabene, John W., and Fiddler, Walter, *Determination of ten N-nitrosoamine acids in cured meat products,* Assoc. Off. Anal. Chem. 73 (1990) 226-30.

[37] Seebach, Dieter et al., *C-Alkylation of peptides through polylilthiated and lithium chloride solvated derivatives containing sarcosine lithium enolate units,* Helv. Chim. Acta 74 (1991) 197-224.

[38] Brückner, Hans and Langer, Matthias, *Gas chromatographic separatiaon of diastereomeric esters of α-alkyl- α-amino acids on dimethylpolysiloxane,* J. Chromatogr. 521 (1990) 09-21.

[39] Watabe, Katsunori, *Preparation of N-stearoyl-L-leucine tert-butylamide as a stationary phase for gas chromatography for optical resolution,* Jpn.Kokai Tokkyo Koho (1991) 4pp.

[40] Arbain, Dayar, et al., *The determination of chirality of N-methyl- and N,N-dimethylphenylalanine by GLC,* Austr. J. Chem. 44 (6) (1991) 887-90.

[41] Lou, Xianwen, et al., *Investigation of the effect of acylation on the Enantiomeric separation of amino acid isopropyl esters by gas chromatography*, J. Chromatogr. 626 (1992) 231-8.

[42] Koenig, W. A., *Determination of the Enantiomeric composition of amino acids by enantioselective gas chromatography*, Mod. Methods Protein-Nuc. Acid Res. (1990) 213-30.

[43] Brückner, Hans and Langer, Matthias, *Gas chromatographic separation of stereoisomeric esters of α-amino acids and α-alkyl- α-amino acids on chiral stationary phases,* J. Chromatogr. 542 (1991) 161-72.

[44] Sato, Katyua, et al., *Development of diamide phases of counter-current gas chromatography,* Japan, Kuromatogurafi 13 (1992) 305-6.

[45] Miller, Scott A., et al., *C-alkylation of sarcosine residues in cyclic tetrapeptides via lithium enolates,* Helv. Chim. Acta 76 (1993) 563-95.

[46] Abe, Iwao, et al., *Rapid derivatization of amino acids for separation of enantiomeric isomers by chiral phase capillary gas chromatography,* Japan, Inter. Symp. on Chromatogr., (1995) 22-25, 713-18.

[47] Abe, Iwao, et al., *N-Alkyloxycarbonyl isobutylamides as readily prepared diamide derivatives of amino acids for separation of Enantiomeric isomers by chiral phase capillary gas chromatography*, Chem. Lett. (4) (1995) 329-30.

[48] Liu, Deming, *Gas chromatographic separation of amino acid derivatives on Chirasil-Val stationary phases,* Fenxi Cheshi Xuebao 12 (1993) 37-40.

[49] Benzencon, Oliver, and Seebach, Dieter, *Alkylations of chiral imidazolidinones derived from di- and triglycine and attempts at cyclizations to give cycloisodityrosines*, (8) (1996) 1259-1276.

[50] Benicka, Eva, et al., *Retention behavior of N-TFA-O-alkyl derivatives of selected amino acid enantiomers separated on modified cyclodextrins by HRGC*, J. Microcol. Sep. 8 (1996) 57-65.

[51] Abe, Iwao, et al., *Rapid analysis of amino acid enantiomers by chiral-phase capillary gas chromatography*, J. Chromatogr. A 722 (1996) 221-7.

[52] Cardillo, Giuliana, et al., *Enzymic resolution of α-alkyl β-amino acids using immobilized penicillin G acylase,* J. Org. Chem. 61 (1996) 8651-8654.

[53] Abe, Iwao, and Nakahara, T., *Enantiomer separation of amino acids as their N-alkyloxycarbonyl alkylamide derivatives by chiral phase capillary GC,* HRCC. 19 (1996) 511-514.

[54] Dallakian, P., and Budzikiewicz, H., *Gas chromatography-chemical ionization mass spectrometry in amino acid analysis pyoverdins,* J. Chromatogr. A 787 (1997) 195-203.

[55] Abe, Iwao, et al., *N-pivaloyl methyl esters as novel derivatives of amino acid enantiomers for chiral-phase capillary gas chromatography,* Chem. Letters (1997) 629-630.

[56] Macko, Stephen et al., *Stable nitrogen isotope analysis of amino acid enantiomers by Gas chromatography/combustion/isotope radio mass spectrometry,* Anal. Chem. 69 (1997) 926-929.

[57] Zhou, Liang-Mo et al., *Enthalpy-entropy compensation effect in gas chromatographic enantiomeric separation,* Chinese J. Chem. 17 (1999)363-376.

[58] Dallakian, Paul et al., *Assignment of the absolute configuration of the amino acids of pyoverdins by GC/MS,* Chirality 11(5/6)(1999)381-386.

[59] Pfeiffer, J., Schurig, V., *Enantiomer separation of amino acid derivatives on a new polymeric chiral resorc (4) arene stationary phase by capillary gas chromatography,* J. Chromatogr. A 840 (1999) 145-150.

[60] Colombini, M. P., Modugno, F., Giacomelli, M., Franscesconi, S., *Characterization of proteinaceous binders and drying oils in wall painting samples by gas chromatography-mass spectrometry,* J. Chromatogr. A 846 (1999) 113-124.

[61] Colombini, M. P., Modugno, F., Giacomelli, A., *Two procedures for suppressing interference from inorganic pigments in the analysis by gas chromatography-mass spectrometry of proteinaceous binders in paintings,* J. Chromatogr. A 846 (1999) 101-111.

[62] Katona, Zs.F., Sass, P., Molnár-Perl, I., *Simulatenous determination of sugars, sugar alcohols, acids and amino acids in apricots by gas chromatography-mass spectrometry,* J. Chromatogr. A 847 (1999) 91-102.

[63] Husek, Petr., *Method of preparing sample for amino acid analysis and kit for analyzing the same,* (Phenomenex, Inc., USA), Eur. Pat. Appl. (2000), 12 pp.

[64] Vazquez Pelaez, Marta, et al., *A comparison of different derivatization approaches for the determination of selenomethionine by GC-ICP-MS,* University of Oviedo, Oviedo, Spain, J. Anal. Atomic Spectrom. 15 (9)(2000) 1217-1222.

[65] Colombini, M.P., Modugno, F., Menicagli, E., Fuoco, R., Giacomelli, A., *GC-MS characterization of proteinaceous and lipid binders in UV aged polychrome artifacts,* Microchem. J. 67 (2000) 291-300.

[66] Molnár-Perl, I., Katona, Zs. F., *GC-MS of Amino Acids as their Trimethylsilyl/t-butyldimethylsilyl Derivatives in Model Solutions III*, Chromatographia 51 (2000) 228-236.

[67] Raulin,F., Vidal-Madjar,C., Cabane, M., Israel, G., *Detection of Martian amino acids by chemical derivatization coupled to gas chromatography: in situ and laboratory analysis*, Adv. Space Res. 27 (2001) 195-199.

[68] Dodier, C., Sternberg, R., Raulin, F., Vidal-Madjar, C., *Chemical derivatization of amino acids for in situ analysis of Martian samples by gas chromatography,* J. Chromatogr. A 915 (2001) 199-207.

[69] Chiavari, G., Fabbri, D., Parti, S., *Gas chromatographic-mass spectrometric analysis of products arising from pyrolysis of amino acids in the presence of hexamethyldisilazane,* J. Chromatogr. A 922 (2001) 235-241.

[70] Spanik, Ivan, et al., *Evaluation of non-polar interactions in chiral recognition by alkylated β- and γ-cyclodextrin chiral stationary phases,* Department of Analytical Chemistry, Slovak University of Technology, Bratislava, Slovakia, J. Separation Sc. 25(1/2) (2002) 45/52.

[71] Nokihara, Kiyoshi, and Gerhardt, Jurgen, *Development of an improved automated gas-chromatographic chiral analysis system: application to non-natural amino acids and natural protein hydrolyzates,* Shimadzu Scientific Research, Inc., Tokyo, Japan, Chirality 13 (8) (2001) 431-434.

[72] Deng, C., Shang., C., Hu, Y., Zhang, X., *Rapid diagnosis of phenylketonuria and other aminoacidemias by quantitative analysis of amino acids in neonatal blood spots by gas chromatography-mass spectrometry,* J. Chromatogr. B 775 (2002) 115-120.

[73] Evans, Claire J., et al., *Compound-specific stable isotope analysis of soil mesofauna using thermally assisted hydrolysis and methylation for ecological investigations,* Anal. Chem. 75(22)(2003) 6056-6062.

[74] Abe, Iwao, et al., *N-pivaloyl methyl ester derivatives for amino acids for separation of enantiomers by chiral-phase capillary gas chromatography*, Jpn, J. Sep. Sci. 25 (2002) 661-664.

[75] Fiamegos, Y.C., Nantos, C.G., Stalikas, C.D., *Ultrasonic-assisted derivatization reaction of amino acids prior to their determination in urine by using single-drop microextraction in conjunction with gas chromatography,* J. Chromatogr. B 813 (2004) 89-94.

[76] Sobolesky, Tim G., et al., *Comparison of silylation and esterification/acylation procedures in GC-MS analysis of amino acids*, J. Sep. Sci. 26 (2003) 1474-1478.

[77] Smerdka, Joachim, et al., *Polymer-bound alkyltriazenes for mild racemization-free esterification of amino acid and peptide derivatives,* J. Peptide Sci. 10 (2004) 603-611.

[78] Paik, M. J., Kim, K. R., *Sequential ethoxycarbonylation, methoximation, and tert-butyldimethylsilylation for simultaneous determination of amino acids and carboxylic acids by dual-column gas chromatography*, J. Chromatogr. A 1034 (2004) 13-23.

[79] Hope, J. L., Prazen, B.J., Nilsson, E. J., Lidstrom, M.E., Synovec, R.E., *Comprehensive two-dimensional gas chromatography with time-of-flight mass spectrometry detection: analysis of amino acid and organic acid trimethylsilyl derivatives, with application to the analysis of metabolites in rye grass samples,* Talanta 65 (2005) 380-388.

References

[1] C.W. Gehrke, R.L Wixom, E.Bayer†, A Century of Discovery (1900-2000) – A New Discipline of Science, the Bridge to the Sciences and Technology – 2 Volumes – J. Chrom. Library Series, Elseveir Sci. Amsterdam, 64, (Dec.2001), 1140 pages.

[2] C.W. Gehrke, K.C. Kuo, R.W. Zumwalt, Amino Acid Analysis, Vols., I, II, III, CRC Press, Boca Raton, FL., (1987), 163 pp., 172 pp., 208 pp.

[3] F.E. Kaiser,C.W. Gehrke,., Zumwalt,W., K. Robert, C.Kenneth, J. Chromatogr. 94 (1974) 113-133.

[4] C.W. Gehrke, L.L. Wall, Sr. Absheer, S. Joe, F.E. Kaiser, R.W. Zumwalt, J. Assoc. Off. Anal. Chem. 68 (1985) 811-821.

[5] D.L. Stalling, C.W. Gehrke, Biochem. Biophys. Res. Commun. 22 (1965) 329-335.

[6] C.W. Gehrke, K. Kuo, R.W. Zumwalt, J. Chromatogr. 57 (1971) 209-217.

[7] K. Blau, Biomedical applications of gas chromatography, Vol. 2, Plenum Press, New York, 1968.

[8] B. Weinstein, Methods Biochem. Anal. 16 (1966) 203-211.

[9] C.W. Gehrke, F. Shahrokhi, Anal. Biochem. 15 (1966) 97-108.

[10] M. Stefanovic, Walker, B.L., Anal. Chem. 39 (1967) 710-713.

[11] W.J. McBride, J.D. Klingman, Anal. Biochem. 25 (1968) 109-122.

[12] A. Darbre, K. Blau, J. Chromatogr. 29 (1967) 49-57

[13] A. Darbre, A. Islam, J. Biochem. 106 (1968) 923-928.

[14] C.W. Gehrke, R.W. Zumwalt, Wall, L.L., J. Chromatogr. 37 (1968) 398-413.

[15] C.W. Gehrke, D. Roach, R.W. Zumwalt, D.L. Stalling, L.L. Wall, Quantitative gas-liquid chromatography of amino acids in proteins and biological substances, Analytical Biochemistry Laboratories, Inc., P.O. Box 1097, Columbia, MO 1968.

[16] D. Roach, C.W. Gehrke, R.W. Zumwalt, J. Chromatogr. 43 (1969) 311-321.

[17] D. Roach, C.W. Gehrke, J. Chromatogr. 43 (1969) 303-310.

[18] M.D. Waterfield, A.D. Favero, J. Chromatogr. 40 (1969) 294-296.

[19] C.W. Gehrke, Stalling, D.L., R.W. Zumwalt, Biochem. Biophys. Res. Commun. 31 (1968) 616-622.

[20] C.W. Gehrke, H. Nakamoto, R.W. Zumwalt, J. Chromatogr. 45 (1969) 24-51.

[21] Nakamoto, H., Master's Thesis, University of Missouri, June, 1969.

[22] C.W. Gehrke, D.L. Stalling, C.D. Ruyle, Biochem. Biophys. Res. Commun. 28 (1967) 869874.

[23] C.W. Gehrke, C.D. Ruyle, J. Chromatogr. 38 (1968) 473-491.

[24] F. Shahrokhi, C.W. Gehrke, Anal. Biochem. 24 (1968) 281-291.

[25] F. Shahrokhi, C.W. Gehrke, J. Chromatogr. 36 (1968) 31-41.

[26] D. Craven, C.W. Gehrke, J. Chromatogr. 37 (1968) 414-421.

[27] D. Roach, C.W. Gehrke, J. Chromatogr. 44 (1969) 269-278.

[28] S.L. MacKenzie, D. Tenaschuk, J. Chromatogr. 97 (1974) 19-24.

[29] S.L. MacKenzie, Tenaschuk, J. Chromatogr. 171 (1979) 195-208.

[30] S.L. MacKenzie, Tenaschuk, J. Chromatogr. 173 (1979) 53-63.

[31] J. Desgrés, D. Boisson P. Padieu, Chromatogr. Biomed. Appl. 162 (1979) 133-152.

[32] I.M. Moodie, G.S. Collier, J.A. Burger, B.C. Werb, J. Sci Food Agric. 33 (1982) 345-354.

[33] I.M. Moodie, G.S. Shephard, D. Labadarios, J. Chromatogr. 362 (1986) 407-412.

[34] E. Gil-Av, J. Mol. Evol. 6 (1975) 131-144.

[35] E. Gil-Av ,S. Weinstein, W.S. Hanchock (Ed.), Handbook for the HPLC Separa tion of Amino Acids, Peptides, and Proteins, Volume I, CRC Press, Boca Raton, Fl., USA, (1994), 445 pp.

[36] C.W. Gehrke, W.M. Lamkin, D.L. Stalling, F. Shahrokhi, Biochem. Biophys. Res. Comun. 13 (1965) 328-337.

[37] C.W. Gehrke, D. Roach, R.W. Zumwalt, D.L. Stalling, L.L. Wall, Analytical BioChemistry Laboratories, Inc., P.O. Box 1097, Columbia, MO 65201, 1968.

[38] C.W. Gehrke, F. Shahrokhi, Anal. Biochem. 15 (1966) 97-105.

[39] C.W. Gehrke, D.L. Stalling, Sep. Sci. 2 (1967) 101-138.

[40] C.W. Gehrke, R.W. Zumwalt, L.L. Wall, J. Chromatogr.26 (1968) 398-411.

[41] D. Roach, C.W. Gehrke, J. Chromatogr. 43 (1969) 303-310.

[42] D. Roach., C. W. Gehrke, R.W. Zumwalt, J. Chromatogr. 43 (1969) 311-321

[43] S. Moore, W.H. Stein, Methods Enzymol. 6 (1963) 819-827.

[44] A. Whitfield, Science 142 (1963) 577-585

[45] H. Braconnot, Ann. Chim. Phys.13 (1820) 113-139.

[46] F. Bopp, Ann. 69 (1849) 16-27.

[47] H.T. MacPherson, Biochem. J., 40 (1946) 470-481.

[48] M.W. Rees, Biochem. J. 40 (1946) 632-639.

[49] M.C. Corfield, A. Robson, Biochem. J. 55 (1953) 517-522.

[50] C.H.W. Hirs, W.H. Stein, S. Moore, J. Biol. Chem. 211 (1954) 941-950.

[51] E.L. Harfenist, J. Am. Chem. Soc. 75 (1953) 5528-5541.

[52] E.L. Smith, A. Stockell, J. Boil. Chem. 207 (1954) 501-506.

[53] E.A. Noltmann, T.A. Mahowald, S.A. Kuby,. J. Biol. Chem. 237 (1962) 1146-1173.

[54] D.F. Elliot, G.P. Lewis, E.W. Horton, Biochem. J., 74 (1960) 15-27.

[55] H. Zuber, R. Jaques, Helv. Cheim. Acta, 43 (1960) 1128-1130.

[56] R.A. Boissonmas, S. Guttman, P.A. Jaquenoud, Helv. Chim. Acta, 43 (1960) 1349-1364..

[57] J.W.H. Lugg, Biochem, J. 32 (1938) 775-783.

[58] A. Light, E.L. Smith, J. Biol. Chem. 237 (1962) 2537-2543.

[59] R.G. Shepherd, S.D. Willson, K.S. Howard, D.S. Davies, E.A. Eigner, N.E. Shake speare, J. Am. Chem. Soc. 78 (1958) 5067-5070.

[60] R.J. Munier, J. Chromatogr. I (1958) 524-533.

[61] G.R. Tristran, Biochem. J. 33 (1939) 1271-1305.

[62] K. Bailey, Biochem. J. 31 (1937) 1396-1406

[63] K. Osono, I. Mukai, F. Tominaga, Met. Nagasaki (Igakkai Zasshi, 30 (1955) 156-174.

[64] J.W.H Lugg, Biochem. J. 27 (1933) 1022-1039.

[65] K. Ishii, Met. Nagasaki Igakkai Zasshi, 29 (1954) 218; C.A., 48 (1954) 8837-8841.

[66] R.L.M. Synge, Biochem. J. 39 (1945) 351-373.

[67] J.I. Harris, R.D. Cole, N.G. Pon, Biochem. J. 62 (1956) 154-176.

[68] A. Light, E.L. Smith, H. Neurath (Editor), The Proteins, Vol. I, 2nd ed., Academic Press, New York, 1963, p. 32.

[69] C. Zomzely, G. Marco, E. Emery, Anal. Chem., 34 (1962) 1414-1417.

[70] W.M. Lamkin, C.W. Gehrke, Anal. Chem., 37 (1965) 383-389.

[71] C.W. Gehrke, D.L. Stalling, Separation Sci., 2 (1967) 101-138.

[72] K. Rühlman, W. Giesecke, Angew. Chem., 73 (1961) 113-127.

[73] D.L. Stalling, C.W. Gehrke, C.W. Zumwalt, Biochem. Biophys. Res. Commun. 31 (1968) 616-622.

[74] C.W. Gehrke, H. Nakamoto, R.W. Zumwalt, J. Chromatogr. 45 (1969) 134-136.

[75] D.E. Johnson, S.J. Scott, A. Meister, Anal. Chem. 33 (1961) 669-693.

[76] H.A. Saroff, A. Karmen, Anal. Biochem. J. (1960) 344-350.

[77] P. Hagen, W. Black, Federation Proc. 23 (1964) 371-397.

[78] P.A Cruickshank, J.C. Sheehan, Anal. Chem., 36 (1964) 1191-1207.

[79] A. Darbre, A. Islam Biochem. J. 106 (1968) 923-925.

[80] K. Blau Biomed. Appl. Gas Chromatogr. 2 (1968) 1.

[81] R.W. Zumwalt, D. Roach, C.W. Gehrke, J. Chromatogr. 53 (1970) 171-194.

[82] C.W. Gehrke, R.W. Zumwalt, L.L. Wall, J. Chromatogr. 37 (1968) 398-413.

[83] E. Mussini, F. Marcucci, J. Chromatogr. 26 (1967) 481-494.

[84] K. BlauA. Darbre J. Chromatogr. 26 (1967) 35-40.

[85] D.L: Stalling, G. Gille, C.W. Gehrke, Anal. Biochem. 18 (1967) 118-125.

[86] K. Blau, Biomed Appl. Gas Chromatogr. 2 (1968) 1

[87] C. Zomzely, G. Marco, E. Emery, Anal. Chem. 34 (1962) 1414-1417.

[88] Gehrke, C.W., Lamkin, W. M., D.L. Stalling, F. Shahrokhi, Biochem. Biophys. Res. Commun. 19 (1965) 328-337.

[89] W.M. Lamkin, C.W. Gehrke, Anal. Chem. 37 (1965) 383-389.

[90] D.L. Stalling, C.W. Gehrke, Biochem., Biophys. Res. Commun. 22 (1966) 329-335.

[91] D. Roach, C.W. Gehrke, J. Chromatogr. 44 (1969) 269-278.

[92] C.W. Gehrke, K. Kuo, R.W. Zumwalt, J. Chromatogr. 57 (1971) 209-217.

[93] R.W. Zumwalt, K. Kuo, C.W. Gehrke, J. Chromatogr. 57 (1971) 193-208

[94] C.W. Gehrke, R.W. Zumwalt, L.L. Wall, J. Chromatogr. 37 (1968) 398-413.

[95] D. Roach, C.W. Gehrke, J. Chromatogr. 43 (1969) 303.

[96] K. Abel, J. Chromatogr. 13 (1964) 14-21.

[97] E. Evrard, J. Chromatogr. 27 (1967) 40-46.

[98] E. Evrard, J. Chromatogr. 68 (1972) 47-54.

[99] R.R. Clayes,. T. Farr, Anal. Chem. 40 (1968) 847-848.

[100] R.R. Watts, JAOAC 53 (1970) 787-788.

[101] R.W. Zumwalt, K. Kuo, C.W. Gehrke, J. Chromatogr. 57 (1971) 193-208.

[102] F.E. Kaiser, C.W. Gehrke, R.W. Zumwalt, K.C. Kuo, J. Chromatogr. 94 (1974) 113-133.

[103] C.W. Gehrke, R.W. Zumwalt, K.C. Kuo, Ponnamperuma, C. and Shimoyama, A., Origins of Life 6 (1975) 541-550.

[104] C.W. Gehrke, K.O. Gerhardt, K.C. Kuo, D.R. Younker, Environmental Health Proceedings, XI, 1977, D.D. Hemphill, Ed. University of Missouri, Columbia,

MO.

[105] C.W Gehrke, K.O Gerhardt, K.C. Kuo, D.R. Younker, Environmental Health – Proceedings, XI. 1977. D.D. Hemphill, Ed., University of Missouri, Columbia, MO.

[106] K. Rühlmann, W. Giesecke, Angew. Chem. 73 (1961) 113-127.

[107] E.D. Smith, H. Shepard Jr, Nature 208 (1965) 878-893

[108] P.S. Mason, Smith, E.D., J. Gas Chromatogr. 4 (1966) 398-404

[109] Chromatography/Lipids, Supelco, Inc., Bellefonte, Pa., I (1967) 3-7.

[110] L. Birkofer, M. Donike, J. Chromatogr. 26 (1967) 270-298.

[111] J. Klebe, H. Finkbeiner, D. White, J. Am. Chem. Soc., 88 (1966) 3390-3395

[112] K. Bergström, J. Gürtler, R. Blomstrand, Anal. Biochem. 34 (1970) 74-84.

[113] H. Frank, G.J. Nicholson, E. Bayer, J. Chromatogr. Sci. 15 (1977) 174-180.

[114] H. Frank, G.J. Nicholson, E. Bayer, Angew. Chem. Int. Ed. Eng. 17 (1878) 363-365.

[115] H. Frank, A Rettenmaier, H. Weicker, G.J. Nicholson, E. Bayer, Anal. Chem. 54 (1982) 715-719.

[116] H. Frank, W. Woiode, G.J. Nicholson, E. Bayer, E.R. Klein, P.D. Klein (Eds.) Stable Isotopes, Proc. 3rd Int. Conf., Academic Press, New York, NY (1979)165 pp.

[117] H. Frank, G.J. Nicholson, E. Bayer, J. Chromatogr. 146 (1978) 197-206.

[118] A. Darbre, K. Blau, J.Chromatogr. 26 (1967) 35-40; 17 (1965) 31-49.

[119] G. Gamerith, J. Chromatogr. 256 (1983) 267-281.

[120] G. Gamerith, C.F.X. Zuder, Guiliani, A. and Brantner, J. Chromatogr. 328 (1985) 241-252.

[121] C. Ponnamperuma, Cosmochemistry and the Origin of Life, D. Reidel, Amster dam, 1983 and Ann. NY Acad. Sci. 194 (1972) 56-70.

[122] W.A. König, Benecke, I and Sievers, S., J. Chromatogr. 238 (1981) 427-432 ; J. High Resolut.Chrom. 5 (1982) 588-595.

[123] I. Abe, K. Izumi, S. Kuramoto, S. Musha, HRCC 4 (1981) 549-552.

[124] I. Abe, T. Kohno, Musha, S., Chromatographia 11 (1978) 393-396.

[125] G. Bengstsson, G. Odham, G. Westerdahl, Anal. Biochem. 111 (1981) 163-175.

[126] G, Odham, L. Larsson,. P.A. Mårdh, quantitative mass spectrometry and its ap plication in microbiology, in: Chapter 9, G. Odham, L. Larsson and P.-A.Mårdh (Eds.), Gas Chromatography/Mass Spectrometry Applications in Microbiology,

Plenum Press, New York, USA, 1983, 444 pp.

[127] P. Hušek, J. Chromatogr. 234 (1982) 381-393.

[128] P. Hušek, V. Felt, M. Matucha, J. Chromatrogr. 252 (2982) 217-224.

[129] C.W. Gehrke, L.L. Wall, J.S. Absheer Sr., F.E. Kaiser, R.W. Zumwalt, J. Assoc. Off. Anal. Chem. 68 (1985) 811-821.

Ibolya Molnár-Perl (Editor)
Quantitation of Amino Acids and Amines by Chromatography
Journal of Chromatography Library, Vol. 70

1.1.3.

Chiral Separation of Amino Acids by Gas Chromatography

Ralf Pätzold and Hans Brückner

Contents

Summary

The chirality of amino acids plays an important role in many fields of biosciences including synthetic peptide chemistry and peptide drug design. Analytical methods are required to assure their optical configuration in either their free state or peptide bonded form. Besides established enzymic and a multitude of advanced liquid chromatographic approaches, gas chromatographic procedures are still attractive. This is attributed to their capability to resolve complex mixtures of amino acid enantiomers after suitable derivatization, and to determine enantiomeric ratios with high sensitivity and reliability using GC-MS in the selected ion monitoring mode. Whereas the indirect determination of amino acids by forming diastereomeric using chiral derivatizing reagents has now limited applications (but does not need a special column), direct enantioseparation on chiral stationary phases is widely used. Among the numerous chiral phases described in the literature only Chirasil®-Val, representing dimethylpolysiloxane bonded valine *tert* butylamide (in either the L- or the D-form) and, more recently, functionalized cyclodextrins are commercially available for routine applications. The deuterium labeling method together with GC-MS enables the estimation of the amount of racemization of amino acids caused by total hydrolysis in peptides and proteins. Limitations of the gas chromatographic approaches, however, are the need for derivatization of amino acids in order to generate volatile derivatives, problems encountered with the basic amino acids and amino acid amides, and limited stability of derivatives of hydroxy amino acids.

Abbreviations: common amino acids are abbreviated to the three letter nomenclature; special amino acids are abbreviated as follows: Nle, norleucine; Dap, 2,4-diaminopimelic acid; Hyl, hydroxylysine; Hyp, hydroxyproline; Gaba, γ-aminobutyric acid Iva, isovaline or 2-ethylalanine; TFA, trifluoroacetic acid; TFAA, trifluoroacetic acid anhydride; PFPA, pentafluoropropionic acid; PFPAA, pentafluoropropionic acid anhydride; Me, methyl; Prp, propyl, DCM, dichlormethane; BHT, *tert*. Butylhydroxy toluene; GC, gas chromatography; MS, mass spectrometry; SIM, selected ion monitoring

1. Introduction

1.1. Indirect chiral separation using diastereoisomers

Because diastereomers, contrary to enantiomers, have distinctive chemical and physical properties they can be resolved on achiral stationary phases. Advantages of this approach are that an abundance of established, very robust and thermally stable and relatively inexpensive capillary columns of varying polarity are on the market such as permethylated, cyanolated, or mixed acylarylpolysiloxanes. For the formation of volatile diastereomeric amino acid derivatives either esterification with an optically active alcohol such as 2-methylbutanol (or higher homologues) has been reported to be suitable, followed by acylation using a non-chiral acylating reagent. Alternatively, acylation of non-chiral amino acid esters using a chiral acylating reagent such as 2-L-α-chloroisovaleryl chloride can be used [1,2,3].

An advantage of the method is that a fair selection of suitable chiral derivatizing reagents is on the market, in most cases as (+) and (-) enantiomer.

These approaches, however, have several inherent disadvantages:

(i) chiral derivatizing reagents (CDRs) have to be of high optical purity (at least when relatively low quantities of an enantiomer has to be determined in an excess of the opposite enantiomer), or when the ratio of enantiomers has to be determined very precisely. This also involves that CDRs remain homochiral under storage as well as derivatization conditions.

(ii) CDRs are usually expensive and high throughput analyses rapidly become cost intensive

(iii) Owing to the bulkiness of the CDR kinetic discrimination in the course of derivatization has to be taken into account as well as the lowered volatility of the diastereomers formed.

1.2. Direct chiral separations

As a result of the commercial availability of thermally stable and effective chiral stationary phases (CSPs) the indirect approaches nowadays are much less frequently used.

Advantages of the direct approaches are:

(i) relatively inexpensive non-chiral derivatizing reagents can be used to create d erivatives of high volatility

(ii) kinetic discrimination of enantiomers does not occur

(iii) the low column bleeding of the columns enables coupling with mass spectrometry and GC-SIM-MS makes analyses sensitive and reliable (this also holds for the indirect approach).

1.3. Chiral stationary phases: Chirasil®-Val and modified cyclodextrins

From a pratical point of view the chiral diamide phases and related types, pioneered by E. Gil-Av and coworkers, or the chiral *s*-triazinyl phases of Oi et al., are rather of historical and theoretical interest and for structural details we refer to the literature [1,4]. The most frequently used CSPs for the enantioseparation of amino acids are functionalized polysiloxane phases using valine *tert* butyl amide as chiral selector [5] and, more recently various CSPs based on various functionalized cyclodextrins [6,7,8,9].

Consequently, in the following we will focus on practical aspects of these types of CSPs.

It should be pointed out that the field of chiral amino acid analysis of complex mixtures is still dominated by the Chirasil®-Val columns, whereas the derivatized cyclodextrin columns, owing to their multiple interactions including intermediate formation of inclusion compounds, can be applied to literally all types of chiral compounds. They might be analyzed in their native or derivatized form and cover the whole range of the chiral or stereoisomeric pool of natural or synthetik compounds ranging from flavour compounds and drugs to pesticides.

1.4. Comparison of enantioseparation on Chirasil®-Val and Cyclodextrin columns

It should be noted that the resolution of enantiomers on columns depends to some extent on the manufacturers (providers) and operating conditions, in particular when GC-MS is used. Data of the literature in many instances were obtained from selected and laboratory made glass columns using hydrogen as carrier gas. Commercially available columns are provided with test chromatograms showing separation of few enantiomeric pairs of amino acids, usually as their TFA methyl esters.

In practice, however, in many cases mixtures comprising all protein amino acids and their corresponding D-enantiomers have to be resolved and helium is used as carrier gas. Consequently, columns have to be calibrated and conditions optimized using standard mixtures. Advantageously a mixture containing an odd ratio of enantiomers is employed in order to facilitate calibration (see Figures of standards).

Whereas the Chirasil®-L-Val column is the standard column manufacturers provide also a Chirasil®-D-Val column. The opposite chirality of the selector reverses the elution order of enantiomers. This is of advantage for the confident assignment of enantiomers in complex matrices with background interferences, in particular when an FID is used.

As a result of the different separation mechanisms it was a breakthrough when a series of functionalized cylodextrins suitable for the chiral resolution of enantiomers in general [6,7,8]. This approach still develops [9,10]. In particular octakis(3-*O*-butanoyl-2,6-di-*O*-*n*-pentyl)-γ-cyclodextrin was found to be highly suitable for the enantiomer separation of derivatized amino acid enantiomers [6]. This column is purchased under the trade names Lipodex® E (Macherey-Nagel). Best overall resolution of amino acid derivatives are obtained for TFA methyl esters. As an advantage, DL-Pro that shows lowest resolution on Chirasil®-Val columns, is excellently resolved. However, this effect is somewhat counterbalanced by the close elution of D-Pro and L-Asp. Furthermore, aromatic side chain amino acids such as DL-

Phe and DL-Trp as well as DL-Glu are unsatisfactorily resolved, at least on some of the commercially available columns and in dependence of the derivatives employed. Fine tuning of experimental conditions and columns certainly will overcome this problem. On Chirasil®-L-Valine columns the D-amino acids are eluted prior to the L-amino acids and enantiomeric pairs elute one after the other. On cyclodextrin columns the elution order of some enantiomers is reversed and elution of enantiomeric pairs shows a more complex pattern.

Consequently, care has to be taken to assign the enantiomers carefully in dependence of the column and derivatives used and the column employed. The enantiomer separation of standard DL-amino acids on Chirasil®-Val and Chirasil®-γ-Dex, recently was compared and discussed [11]. Inspection of a chromatogram shows that this cyclodextrin column provides good or satisfactorily resolution of DL-Glu and and DL-Phe, but not of DL-Trp. Further, enantiomers are eluted close together at the beginning of the chromatogram.

1.5. Enantiomer labeling for quantitative amino acid analysis

An inherent disadvantage of quantitative gas chromatographic amino acid analysis using capillary columns are derivatization and the need for calibration using response factors for derivatives. It has been reported that there might be some discrimination and loss of derivatives in the injector thus making results less reliable in comparison to conventional amino acid analysis. An approach named enantiomer labeling was reported [12]. Defined amounts of D-amino acids are added to protein hydrolysates or physiological samples and amino acids determined on Chirasil®-Val. Since D-amino acids chemically behave like L-amino acids all losses in the course of handling procedures of samples should be compensated. Sensitivity important for physiologcal samples, is enhanced 100 two 200 fold employing an alkali flame-ionization detector in the nitrogen mode. In the early eighties however, it was not fully recognized that physiological samples, such as urine or tissues, can be rich in certain D-amino acids and that D-amino acids are abundant in particular in bacterial fermented food. This requires determination of a blank of the analyte and limits the applicability of the method. Moreover, the availability of bench top GC-SIM-MS has drastically lowered the need for element specific detectors. Nevertheless, construction and use of a fully automated amino acid analyzer based on GC and employing enantiomer labeling has been reported [13].

1.6. Deuterium-labeling Method

Total hydrolysis of peptides or proteins is accompanied by the generation of D-amino acids from L-amino acids. Altough under conditions of acidic total hydrolysis the amounts formed are neglectable in most cases, depending on the sequences of proteins or previous treatment among naturally occurring D-amino acids and those formed on hydrolysis have to be distinguished. An elegant approach is performing total hydrolysis with deuterium chloride (DCl) in deuterated water (D_2O) in place of aqueous HCl [14,15]. Acid catalyzed racemization will lead to an exchange of the C^{α}-hydrogen against deuterium. The resulting deuterated D-amino acid can be distinguished by mass spectrometry owing to the mass difference of 1 Da. Thus the degree of acid catalyzed racemization can be calculated. Availablity of such a method is of importance if low amounts of D-amino acids have do be determined in protein hydrolysates, which were generated as result of biological aging processes, diseases, or have to be determined for age dating of fossils based on time-dependent amino acid racemization. A fully automated procedure using this approach has been implemented [16].

1.7. New derivatization approaches of DL-amino acids and applications to Chirasil®-Val

Derivatization of amino acids requires a two-step derivatization procedure, i.e. esterification followed by acylation. It was demonstrated that the well-known chloroformates, under suitable conditions, are capable of derivatizing amino acids in a single step, thus allowing the extremely fast quantitative analysis of the resulting alkoxycarbonal amino acid esters on fused silica capillary columns [17]. Modification of this approach, used for the rapid non-chiral amino acid analysis of amino acids occurring e.g. in physiological fluids, were also applied to chiral amino acid analysis [18,19]. Isolated amino acids were converted at ambient temperature within a few minutes e.g. into *N(O,S)*-ethoxycarbonylamino acid heptafluorobutyl esters using heptafluoro-1-butanol in pyridine followed by treatment with ethylchloroformate (ethyoxycarbonyl chloride). The resulting derivatives were resolved on a Chirasil®-L-Val column. As an disadvantage, resolution of derivatives of DL-Pro was not achieved and the basic amino acids Arg and His cause still problems.

2. Applications of Chirasil®-Val and Lipodex® E columns

Separations of standard mixtures of DL-amino acids on commercially available Chirasil®-L-Val and Chirasil®-D-Val columns are shown in Figure 1 and Figure 2, respectively. Note the reversal of the elution order of enantiomers and that for optimal separations PFP/1- or 2- propyl esters have been used. The odd ratio of D- and L-enantiomers in the standard simplifies calibration.

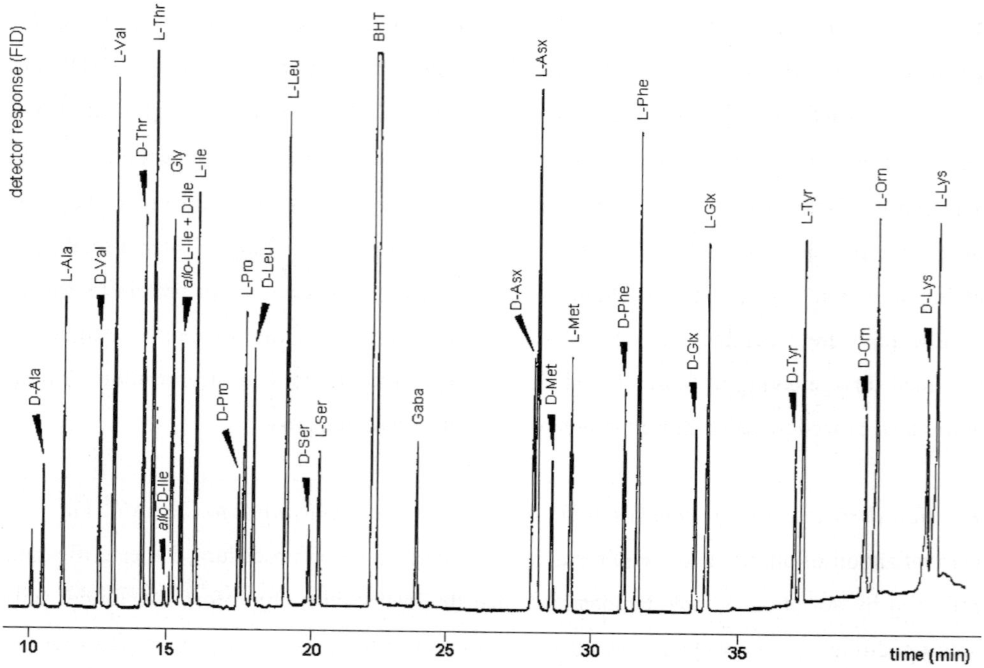

Figure 1Separation (GC-FID) of a standard mixture of DL-amino acids (PFP/1-Prp esters) on Chirasil®-L-Val (Chrompack); CE 5160 instrument (Carlo Erba, Milan); carrier gas, hydrogen; inlet pressure, 45 kPa; temperature program, 80 °C, 5 min isothermal, 3.5 °C/min to 120 °C, 3 min isothermal, 4 °C/min to 152 °C, 3 min isothermal , 5 °C/min to 195 °C, 20 min isothermal; arrows indicate D-amino acids in chromatograms; ratio D: L-enantiomers approx. 1:2; for abbreviations of components in all chromatograms see text; all WCOT fused capillary columns hat 25 m length, 0.25 mm internal diameter and a film thickness of 0.12 μm.

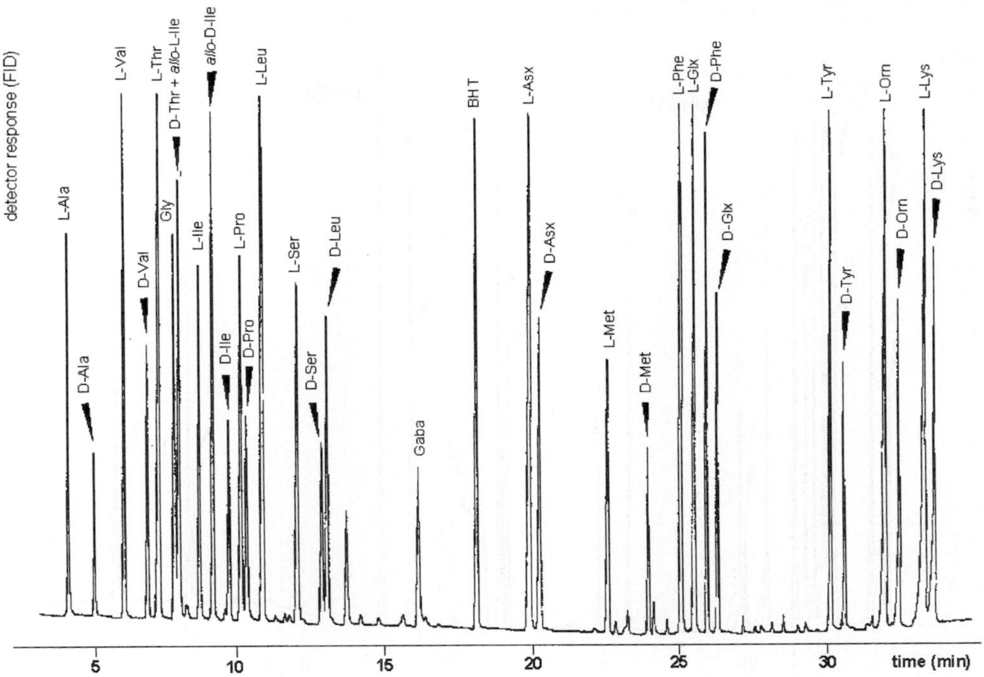

Figure 2 Separation (GC-FID) of a standard mixture of DL-amino acids (PFP/2-Prp esters) on Chirasil®-D-Val (Chrompack); CE 5160 instrument (Carlo Erba, Milan); carrier gas, hydrogen; inlet pressure, 60 kPa; temperature program, 70 to 100 °C at 2.4 °C/min; 2 min isothermal; 3.0°C/min to 125 °C, then 6.0 °C/min to 195 °C, 15 min isothermal; arrows refer to D-amino acids in chromatograms; for abbreviations of components see text. Non-chiral Gaba was added as plant amino acid, serving also as internal marker. Note the reversal of elution of enantiomers in comparison to Figure 1.

Resolution of a standard mixture of DL-amino acids on a cyclodextrin column (Lipodex® E) is shown in Figure 3. Best resolution was obtained for TFA/1-Prp esters. Some enantiomeric pairs (DL-Phe, DL-Glu) are not or unsatisfactorily resolved in comparison to Chirasil®-Val columns. Notably, polysiloxane linked Lipodex® E. is capable of resolving these problematic amino acids [17]. This phase, named Chirasil®-γ-Dex, is currently not commercially available.

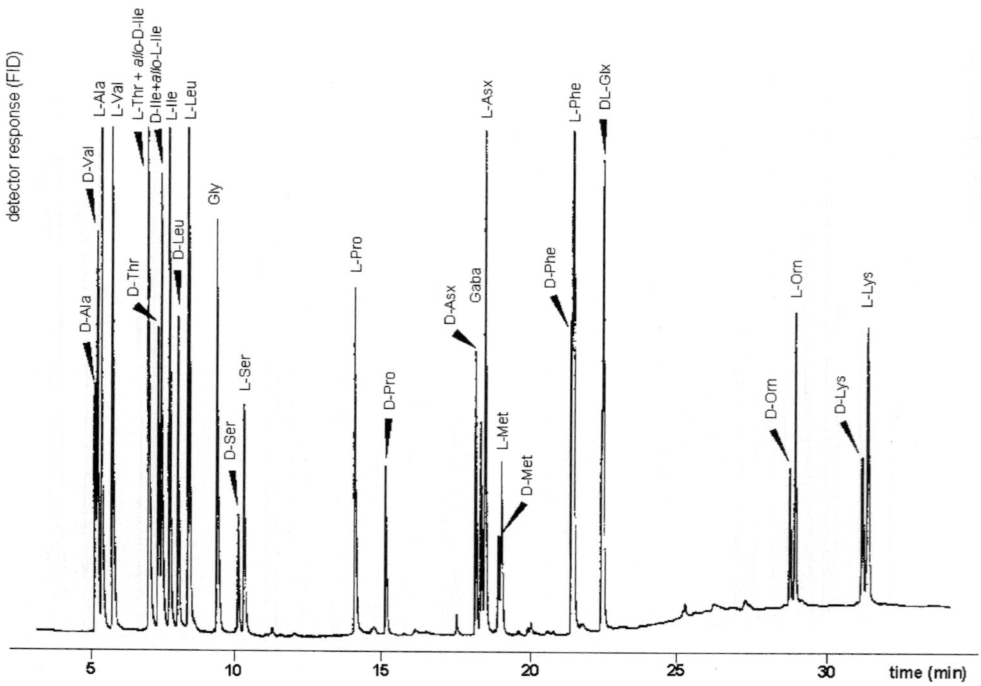

Figure 3 Separation (GC-FID) of a standard mixture of DL-amino acids (TFA/1-Prp esters) on Lipodex® E (Machery-Nagel); CE 5160 instrumen (Carlo Erba, Milan); carrier gas, hydrogen; inlet pressure, 60 kPa; temperature program, 90 °C to 165 °C at 4.0 °C/min; then 7.0 °C/ min to 200 °C at 7.0 °C/ min; 10 min isothermal.

Separation of a standard mixture related to gelatine (collagen) amino acids on Chirasil®-L-Val is shown in Figure 4. Stereoisomers of the special amino acids Hyp and Hyl (amino acids of collagen) have been included (see also Fig. 7c). Optimization of resolution required the use of TFA/2-Prp esters.

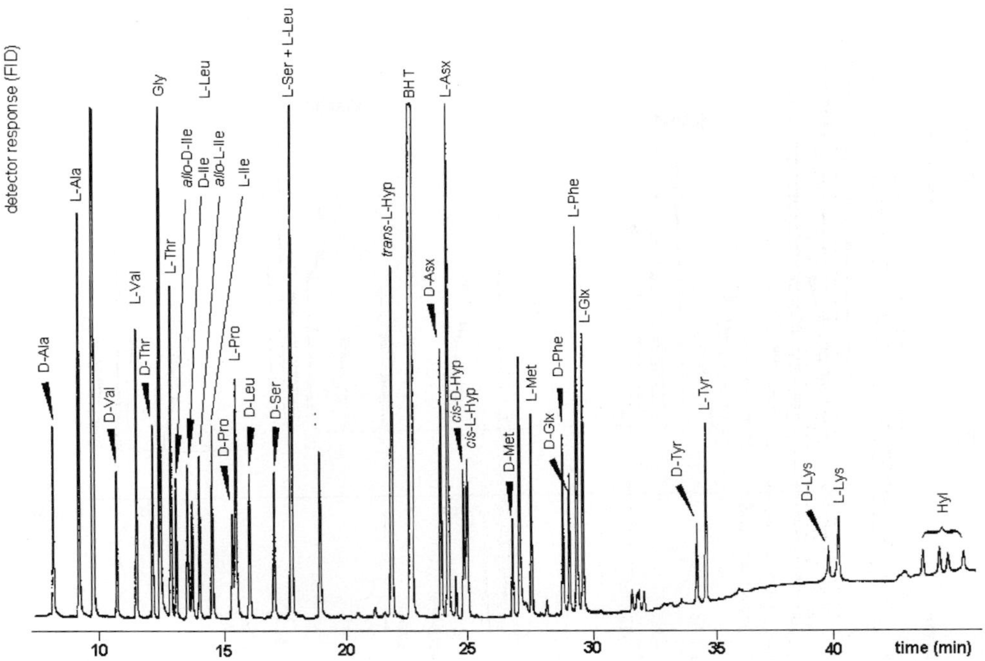

Figure 4 Separation (GC-FID) of a standard mixture of DL-amino acids related to gelatine (TFA/2-Prp esters) on Chirasil®-L-Val (Chrompack); CE 5160 instrument (Carlo Erba, Milan); carrier gas, hydrogen; inlet pressure, 45 kPa; instrument,; temperature program, 80 °C, 5 min isothermal, 3.5 °C/min to 112 °C, 3 min isothermal, then 4 °C/min to 168 °C, 3 min isothermal, then 5.5 °C/min to 195 °C, 20 min isothermal. Note that stereoisomers of Hyl are resolved but not assigned.

A chromatogram of amino acid enantiomers isolated from an alcoholic fermented beverage (Sherry wine) is presented in Figure 5b and compared to a standard resolved on the same Chirasil®-L-Val column (Figure 5a). Very high amounts of some D-amino acids could be detected in many fortified wines [20] and fermented foods [21]. For such samples enantiomer labeling for quantification (see text) is not recommended. Analyses were performed in the SIM mode.

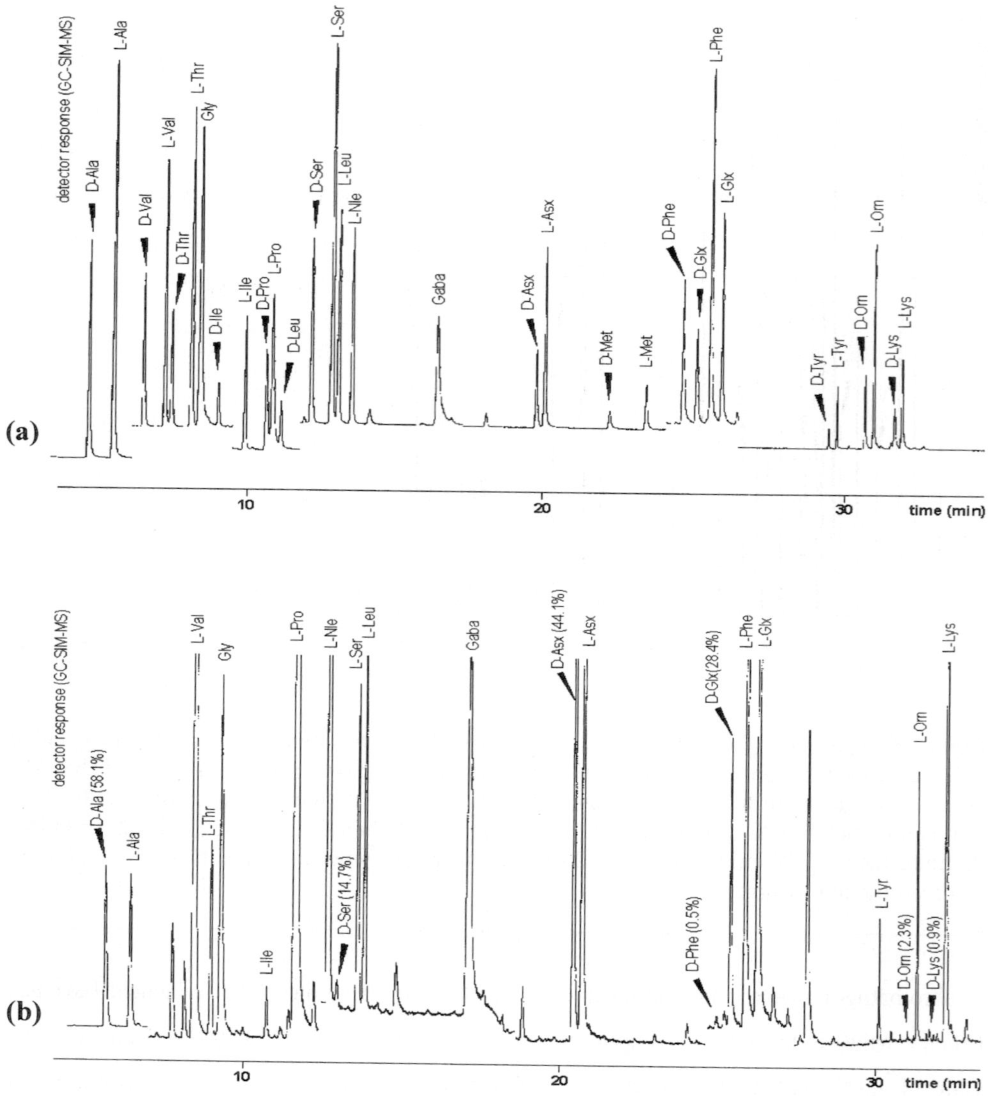

Figure 5a,b GC-SIM- MS (PFP-2-Prp esters) of (a) a standard mixture of D- and L-amino acids (ratio D : L approx. 1 : 2) and (b) of amino acid enantiomers isolated from a Fino Sherry wine; column, Chirasil®-L-Val (Varian, Darmstadt); GC 17A and QP5000 instrument and software (Shimadzu, Kyoto, Japan); carrier gas, helium; injector and interface temperature 250 C;. Note that varying base line levels of SIM-MS result from appropriately selected and amplified ion sets (see table 1).

A chromatogram of the resolution of a standard mixture of stereoisomers of dipeptides on Chirasil®-L-Val is presented in Figure 6. Few reports on attempts so separate isomeric peptides by GC have been published [22]. The TFA/OMe esters were used and prepared at ambient temperature in order to avoid formation of diketopiperazines from dipeptides at elevated temperature (see Experimental). Limitation of the method is the high volatility of most dipeptide derivatives which are not eluted from the Chirasil®-Val column at the upper operating temperature of about 200 °C. Free D-amino acids as well as several dipeptides containg D-amino acids have been detected in plants [23]. As an example we present the GC-SIM-MS detection of the enantiomers of alanylalanine in native tobacco leaves (see insert of Figure 6).

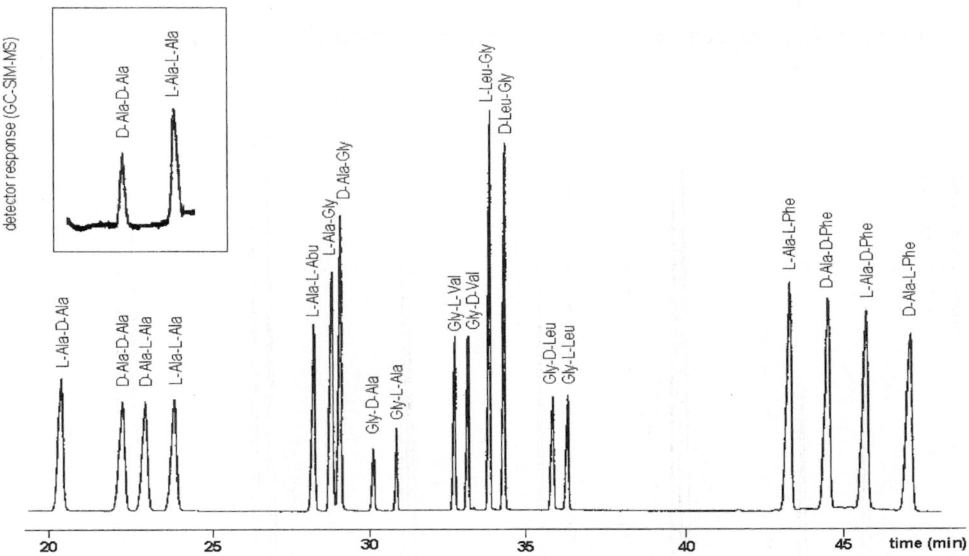

Figure 6 GC-SIM-MS (TFA/OMe esters) of a standard mixture of stereoisomers of synthetic dipeptides separated on Chirasil®-L-Val (Varian); instrument and carrier gas as in Fig. 5; temperature program,140 °C, 10 min isothermal; then 5.0 °C/min to 150 °C, 15 min isothermal; then 3.0 °C mit to 195 °C, 13 min isothermal; pressure program, 5.0 kPa for 10 min; then 1.0 kPa/min to 7.0 kPa, 15 min isobaric; then 0.5 kPa/min to 15.0 kPa, 12 min isobaric. Insert shows appplication of the method to freshly harvested leaves of tobacco (*Nicotina tabacum* L.) and detection of enantiomeric D-Ala-D-Ala and L-Ala-L-Ala dipeptides therein.

Enantiomeric separations of special amino acids are presented in Figure 7. Non-proteinogenic, C^{α}-alkylated amino acids (C^{α}-dialkylted glycines, α-alkylamino acids) are difficult to resolved on Chirasil®-Val columns but resolvable on Lipodex® E (Figure 7a). In particular the enantiomeric ratio of DL-isovaline (α-ethylalanine) and homologs in fungal and meteorites peptide metabolites attracted much attention [24,25,26]. According to the literature enantiomers of *N*-methylamino acids, occurring in peptide antibiotics, are also resolvable on Lipodex® E [1].

The separation of the stereoisomers of hydroxyprolins (occurring in collagen and plants) is presented in Figure 7b and topics related to these compounds have been discussed [27].

A characteristic marker for the unique bacterial peptidoglycan is 2,4 diaminopimelic acid. Presence, quantities and stereochemistry can be used as indicators for bacterial activity and taxonomy [28]. All stereoisomers are resolvabe on Chirasil®-L-Val (Figure 7c).

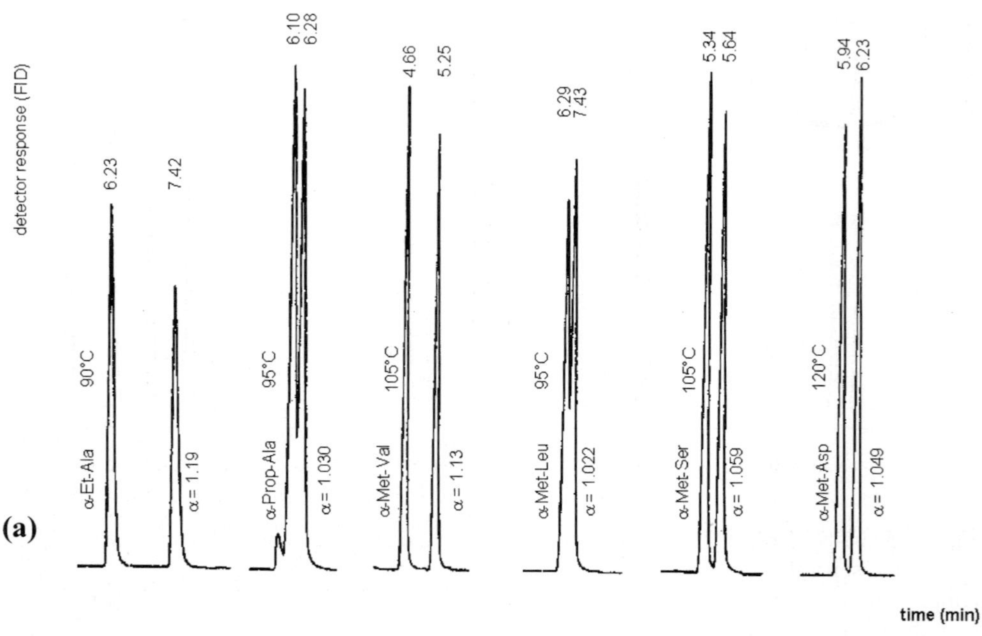

(a)

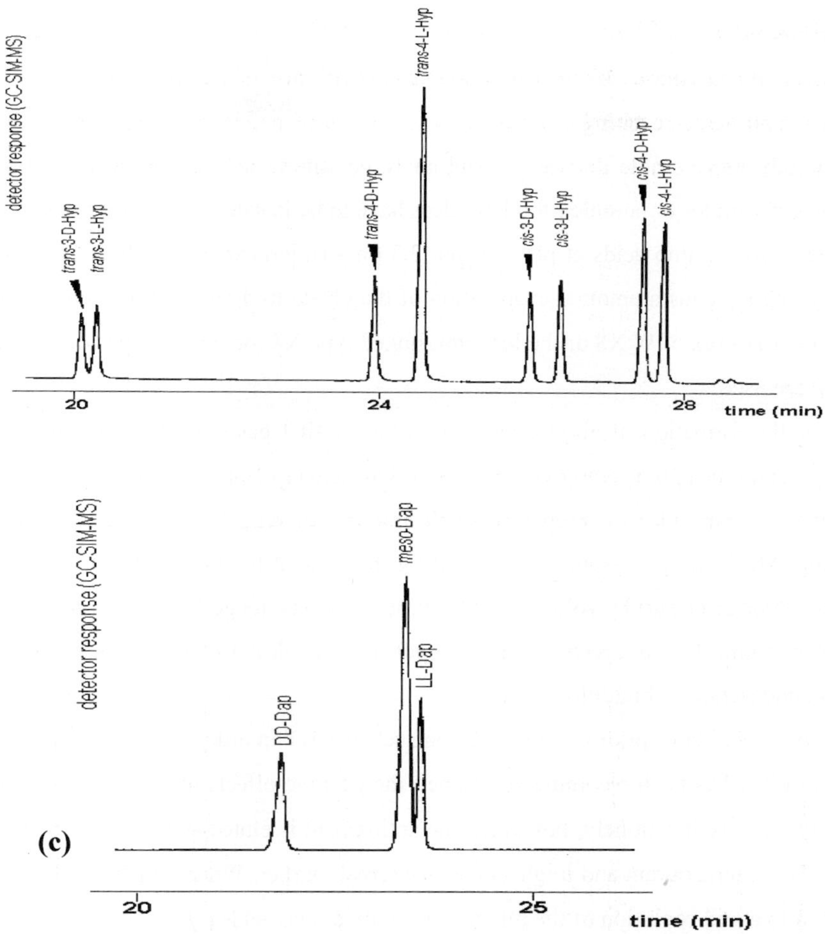

Figure 7a-c Separation of enantiomers and stereoisomers of standards of special amino acids of (a) a series of non-proteinogenic/meteoritic C^{α}-dialkylated a-amino acids (Lipodex® E), (b) collagen related *cis* and *trans* 4-hydroxy-DL-proline and *cis* and *trans* 3-hydroxy-DL-proline, and (c) bacterial LL-, DD- and *meso* diaminopimelic acid. Conditions for (a) GC-FID of TFA/O Me esters on Lipodex® E; isothermal temperatures, net retention times and chiral separation factors, a, of derivatives are given; GC 5160 instrument (Carlo-Erba), helium at 1.0 bar, (b) GC-SIM-MS of TFA/2-Prp esters on Chirasil®-L-Val; details as described in legend of Fig. 5, (b) GC-SIM-MS of TFA/2-Prp esters on Chirasil®-L-Val, instrument as described in Figure 5; temperature program, 165 °C, 20 min isothermal; then 5 °C/min to 190 °C, 10 min isothermal; pressure program 5 kPa, 20 min isobaric; then 2.0 kPa/min to 15 kPa, 10 min isobaric.

3. Experimental

3.1. General considerations for the isolation and derivatizaton of amino acids

Whereas amino acids resulting from total hydrolysates of peptides or proteins can be converted directly into volatile derivatives and analyzed, amino acid enantiomers occurring in physiological samples or complex food matrices have to be isolated. An established procedure is binding of free amino acids at pH of about 2.3 on a cation exchanger (H^+ - form) and desorption using aqueous ammonia, evaporation of the eluate to drynesse followed by derivatization. Either Dowex 50W X8 or the less crosslinked type X4 (or similar types of other manufacturers) are used.

For the formation of alkyl esters a solution of HCl gas (e.g. 4 molar) in the corresponding alcohol usually is employed. However, gas delivery from a lecture bottle, weighting and titration in particular for preparing small amounts of reagent is cumbersome and time consuming. Much more convenient ist preparing the reagent by mixing the alcohol (4 parts) and acetyl chloride (1 part by volume) with chilling [29]. The reagent can be stored at ambient for about a month. It is not recommended to put the reagent bottle in the freezer in order to prevent condensation of humidity on opening.

Addition of antioxidants prior to derivatization such as BHT in order to minimize oxidation of Cys, Met or Tyr has been recommended, albeit the positive effects have been discussed controversally. We use it routinely, however. The antioxidant is eluted as a major and two minor peaks in the chromatogram and might serve as internal marker. When an FID is used care has to be taken to avoid coelution of the minor peaks with amino acids (*cf.* Figures).

Problematic amino acids

The guanidino group of Arg and the imidazol group of His are not derivatized under standard conditions and the derivatives obtained under routine conditions are not eluted from fused silica capillary columns.It has been suggested to acylated Arg at elevated temperature (e.g. 15 min at 150°C) and to derivatize the imidazolyl group of His in a third step using ethylchloroformate. Problems are still encountered using fused silica columns. We have observed that acidic treatment of Arg yields of small quantities of Orn.

The aromatic amino acids Tyr and Trp are prone to oxidation and addition of an anti-oxdans (BHT) has been recommended.

Monomeric cystein is well separated on Chirasil®-Val but is easily oxidized to the sparingly soluble dimeric cystin. Addition of thiols or reducing agents (BHT) has been suggested.

The *allo* forms ofThr, Ile and *cis/trans* Hyp are rarely taken into account since the *allo* forms of amino acids are not proteinogenic. They occur, however, in microbial peptides. The eight stereoisomers of *cis/trans* 3 and 4-DL-hyroxyproline are resolved on Chirasil®-Val, (see Figure 7b).

Derivatives of Ser and Tyr are not stable and deteriorate fast even when stored in the freezer. Although the relative enantiomeric ratio does not change, for quantification analytes should be measured immediately after derivatization.

Asn and Gln are converted almost completely to Asp and Glu, respectively, depending to some extend on the derivatization temperature and time. Notably, non-proteinogenic C^{α} dialkyled amino acids such as DL-isovaline (2-ethylalanine) and higher homologues are sterically hindered and should be checked for complete derivatization.

Notably, most of the chromatograms presented by the inventors and their coworkers of the Chirasil®-Val-columns resulted from single laboratory made, desactivated glass columns (not fused silica columns) using hydrogen as carrier gas thus enabling also elution of DL-Arg, DL-Trp and DL-His (the latter as N^{im} ethylchloroformate derivative) under special derivatization conditions (see above).

3.3. Protocol for the isolation and derivatization of amino acids from samples

Whereas amino acids resulting from total hydrolysates of peptides and proteins can be directly derivatized as described below, free amino acids occurring in physiological samples are foods have to be isolated and proteinaceous compounds to be removed. Proteins are precipitated by addition of picric acid, sulfosalicylic acid, trichloroacetic acid, or organic solvents and removed by centrifugation. It is of importance to adjust the remaining solution to pH 2.3 in order to assure complete adsorption on the cation exchanger. A lower pH will cause losses of acidic amino acids. Amino acids are desorbed using 4 M aqueous ammonia and eluates evaporated to dryness under reduced pressure in a conical flask. The remaining residue is dissolved in 0.01 M HCl, transferred into an vial (e.g. ReactiVial from Wheaton) and evaporated in a stream of nitrogen. The antioxidant BHT (10 μl of a solution of 10 mg /ml of the respecitve alcohol (e.g. 2-PrpOH) and a mixture of 2-PrpOH/AcCl (4:1, v/v, 500 μl) are added and the tighly closed vials heated for 1 h at 100 C. Solvents were removed in a stream of nitrogen and dichloromethane (200 µl) and and the acylating reagent (PFPAA, 50 µl) were added. Excess of reagents were removed in a gentle stream of nitrogen but care has to be taken to avoid losses of volatile amino acid derivatives. The residue is dissolved in dichloromethane or ethyl acetate or acetonitrile (200 µl) and alquots of about 1 µl are subjected to GC-MS in the split mode (splitting ration about 1:30). Relative and absolute quantities of enantiomers are calculated from peak areas of derivatives as described above [20,23,27,28].

3.4. *Protocol for the derivatization of dipeptides at ambient temperature*

To the dipeptide mixture (ca. 0.5 mg) a mixture of MeOH/AcCl (4:1, vv, 500 µl) was added and left ina closed vial at ambient for 17 h. Excess of reagents was removed in a stream of nitrogen, dichloromethane (200 µl) and TFAA were added and the mixture maintained for 2 h. Reagents were removed in a stream of nitrogen, DCM was added and 1 µl- aliquots were subjected to GC-SIM-MS.

3.5. Quantitative enantiomeric amino acid analysis

The relative ratios of D- and L-amino acids can be calculated from the peak areas of the respective enantiomers according to the equation

$\%D = 100 \times A_D/(A_D+A_L)$

where %D is the relative amount the the D-enantiomer and A_D and A_L are the peak areas of the D- and L-enantiomers, respectively, determined by GC-FID or GC-MS.

For the determination of the absolute amounts of amino acid enantiomers using GC the response factors of amino acids have to be determined in relation to the internal standard (IS), usually L-Nle, have to be determined. The IS also compensates for losses in the course of sample work up.

Response factors were calculated according to the equation

$f_R = A_{AA} / A_{IS}$

where f_R is the response factor of the amino acids to be determined, A_{AA} the peak area of the AA to be determined and A_{IS} the peak are of IS.

Amino acids in samples were quantified according to the equation

$c_{AA} = (1/f_R \times A_{AA})/(A_{IS} \times c_{IS})$

where c_{AA} is the amount of amino acid, f_R the response factor, A_{AA} the peak area of amino acid in sample, A_{IS} the peak area of the IS in the sample, and c_{IS} the amount of the IS added to the sample. For typical data we refer to the literature [21].

For GC-MS in the selected ion monitoring mode (SIM) characteristic mass fragments of derivatives have to be selected (for PFP/ 2-Prp esters of standard amino acids see Table 1 .

Table 1 Selected fragment ions (*m/z*) of PFP/2-Prp esters of standard amino acids used for GC-SIM-MS.

	(*m/z*)			
Ala	190	191	235	
Asp	189	234	235	262
Gaba	176	204	232	249
Glu	202	230	248	276
Gly	176	177		
Ile	203	221	232	
Leu	190	232	233	
Lys	176	230		
Met	203	221	263	
Nle	176	190	232	
Phe	91	148	190	266
Pro	216			
Ser	188	189		
Thr	202	203		
Tyr	253	266		
Val	203	218	219	

3.6. Remarks and research needs

Enantioseparation of protein L-amino acids and their corresponding D-enantiomers by GC on chiral stationary phases and their relative and absolute quantification has achieved already a high level with regard to selectivity, sensitivity and reliability. Direct methods are completed by indirect methods using diastereomeric approaches. In particular commercially available Chirasil®-Val columns and cyclodextin based columns (Lipodex® E) can be applied routinely to complex mixtures and physiological fluids [30] and can also be used to separate enantiomers of non-protein and special amino acids. However, manufacturers without exception, provide columns that have been tested with few enantiomeric pairs of DL-amino acids under isothermal conditions. Optimization for resolution of complex mixtures resulting from biomatrices remains with the customer.

Cyclodextrin based stationary phases are thermally more stable and show lower phase bleeding. Overall resolution and capacity, however, are not as good as on Chirasil®-Val columns. For chiral amino acid analysis, derivatives of γ-cyclodextrins are more suitable in comparison to those derived from α- or β-cyclodetrins.

GC-SIM-MS, in comparison to GC-FID , enhances sensitivity and reliability. Co-eluting enantiomersof different amino acids can be distinguished and can also be quantified based on appropriately selected ion fragments.

Whereas relative quantification of enantiomers is very precise, determination of absolute amounts of amino acids by GC suffers to some extent from the need to remove excess of reagents in order to protect columns from destruction. This can lead to a loss in particular of the very volatile amino acid derivatives eluting at the beginning. The injection on capillary columns, usually performed in the split mode, has been reported to discriminate among derivatives if split ratio is changed. Further, quantification relies on complete derivatization of amino acids and accurate determination of response factors. Decomposition in particular of derivatives of Ser and Tyr and problems encountered with enantiomer labeling for quantification have been discussed. The advantage of the use of DCl in D_2O for total hydrolysis of proteins in the field of biosciences should be further demonstrated.

Gas chromatographic quantification of the basic amino acids His and Arg is still a problem and also reliable gas chromatographic differentiation among Gln/Glu and Asn/Asp. Although two step derivatization of amino acids at elevated temperature is common, single step methods at ambient have been described and should be fully tested and further developed for chiral amino acid analysis.

Manufacturers of chiral columns suitable for chiral amino acid analysis

Chirasil®-L-Val and Chirasil®-D-Val are merchandised under the trade names Permabond® L-Chirasil®-Val and Permabond® D-Chirasil®-Val and the Lipodex® E column under the name FS-Lipodex® E by Macherey-Nagel, Düren Germany.

Chirasil®-Val columns are also available under the name CP-Chirasil® Val by Astec (Whippany, NY, USA) and Varian Inc, (Palo Alta, Ca, USA) companies (both antipodes, L and D, are available) and under the name Heliflex® Chirasil®-Val from Alltech Associates, Inc, (Nicholasville, Ky,USA)

The Chiraldex™ G-BP column of Astec is reported to be suitable for amino acid analysis

References

[1] W.A. König, The Practice of Enantiomer Separation by Capillary Gas Chromatography, Hüthig Verlag (1987).

[2] W.A. König, K. Stoelting, K. Kruse, Chromatographia, 10 (1977) 444-448.

[3] H. Brückner, M. Langer, Journal of Chromatography, 521 (1990) 109-121.

[4] R.W. Zumwalt, K.C.T. Kuo, C.W. Gehrke (Editors) Amino acid Analysis by Gas Chromatography Volume II chapter 1 and Volume III chapter3, CRC Press (1987).

[5] H. Frank, G.J. Nicholson, E. Bayer, Journal of Chromatographic Science, 15 (1977) 174-176.

[6] W.A. König, R. Krebber, P. Mischnick, Journal of High Resolution Chromatography, 35 (1989) 732-738.

[7] W.A. König, Gas Chromatographic Enantiomer Separation with Modified Cyclodextrins, Hüthig Verlag (1992).

[8] V. Schurig, H.P. Nowotny, Angewandte Chemie Int. Ed. Engl., 29 (1988) 939-957.

[9] I. Spanik, J. Krupcik, I. Skacáni, P. Sandra, D.W. Armstrong, Journal of Chromatography A, 1071 (2005) 59-66.

[10] V. Schurig, Journal of Chromatography A, 906 (2001) 275-299.

[11] V. Schurig, M. Juza, M. Preschel, G.J. Nicholson, E. Bayer, Enantiomer, 4 (1999) 297-303.

[12] H. Frank, N. Vujtovic-Ockenga, A. Rettenmeier, Journal of Chromatography, 279 (1983) 507-514.

[13] J. Gerhardt, Biotechnology Forum Europe, 8 (1991) 196-200.

[14] H. Frank, W. Woiwode, G.J. Nicholson, E. Bayer, Stable Isotopes: Methodology, instrumentation and Techniques (Proceedings of the 3rd Int. Conference), Academic Press, Inc. (1981).

[15] R. Liardon, S. Ledermann, U. Ott, Journal of Chromatography, 203 (1981) 385-395.

[16] J. Gerhard, G.J. Nicholson, in J. Martinez, J.A. Fehrentz (Editors), Peptides 2000, Editions EDK (2001) 563-564.

[17] P. Husek, Journal of Chromatography B, 717 (1998) 57-91.

[18] I. Abe, N. Fujimoto, T. Nishiyama, K. Terada, T. Nakahara, Journal of Chromatography A, 722 (1996) 221-227.

[19] S. Casal, M.B. Oliveira, M.A. Ferreira, Journal of Chromatography A, 866 (2000) 221-

230.

[20] R. Pätzold, A. Nieto-Rodriguez, H. Brückner, Chromatographia Supplement, 57 (2003) S 207- S 211.

[21] T. Erbe, H. Brückner, Zeitschrift für Lebensmitteluntersuchung und Forschung A, 207 (1998) 400–409.

[22] B. Koppenhöfer, H. Allmendinger, L.P. Chang, L.B. Cheng, Journal of Chromatography, 441 (1988) 89-98.

[23] H. Brückner, T. Westhauser, Amino Acids, 24 (2003) 43–55.

[24] H. Brückner, G. Nicholson, G. Jung, K. Kruse, W.A. König, Chromatographia, 13 (1980) 209-214.

[25] A. Jaworski , H. Brückner, Journal of Peptide Science 6 (2000) 149-167.

[26] J.R. Cronin, S. Pizzarello, C.B. Moore, Science, 275 (1997) 951.

[27] T. Erbe, H. Brückner, Zeitschrift für Lebensmitteluntersuchung und Forschung A, 208 (1999) 424-428.

[28] A. Schieber, H. Brückner, J.R. Ling, Biomedical Chromatography, 13 (1999) 46-50.

[29] H. Frank, D. Bimboes, G.J. Nicholson, Chromatographia, 12 (1979) 168-170.

[30] H. Brückner, A. Schieber Journal of High Resolution Chromatography 23 (2000) 567.

1.2. High Performance Liquid Chromatography

Ibolya Molnár-Perl (Editor)
Quantitation of Amino Acids and Amines by Chromatography
Journal of Chromatography Library, Vol. 70

1.2.1.

HPLC of Amino Acids without Derivatization

Claire Elfakir

Contents

Summary

The recent advances in evaporative light scattering (ELS), chemiluminescent nitrogen (CLN) and mass spectrometry (MS) techniques provide new opportunities in native amino acid determination. Seven chromatographic detectors (low-wavelength UV, refractive index, conductivity detector, nuclear magnetic resonance, ELSD, CLND, MS/MS) are compatible with a direct detection of amino acids. The triple quadrupole ESI-MS/MS was found to be the most promising detector due to its high sensitivity. However, in spite of its high specificity, a high number of amino acids have to be separated due to the intrinsic limitations of MS when the simultaneous analysis of closely related chemical structures is required. By using perfluorinated carboxylic acids of longer n-alkyl chain than trifluoroacetic acid, as ion-pairing reagents in the mobile phase, high amino acid selectivities can be obtained on C8, C18 silica or porous graphitic carbon. They are efficient reagents for the separation of underivatized amino acids and are volatile enough for their use in MS/MS. LC coupled with MS has been shown to be a viable alternative to traditional detection techniques for the separation of underivatized amino acids and their determination in biological extracts without any further treatment. The use of stable isotopically labeled internal standards ensured the accuracy of quantification and eliminated matrix effects. Quantification limits for each molecule were found compatible with their measurement in plasma and urine.

1. Introduction

Since amino acids (AAs) are non volatile compounds and most of them have little UV absorbance, they have been commonly analyzed by liquid chromatography (LC) methods with precolumn or postcolumn derivatization using UV chromophore or fluorophore reagents. Initially, free amino acids were separated by cation-exchange chromatography and treated with ninhydrin in a post-column derivatization step, and the derivatives were detected with a UV detector [1]. Several of these methods have been validated, however interlaboratory studies [2] have shown that one of the sources of variability in the results obtained was variability in derivatization conditions. Therefore, the quantification of amino acids from a biological matrix without derivatization is advantageous in order to reduce the errors introduced by such problems as derivative instability, side reactions and reagent interferences. The problems of assaying underivatized AA consist of both the separation and the detection of these compounds. In the last few years, the remarkable progress in detection technologies such as diode array ultraviolet (UV), evaporative light scattering (ELS), chemiluminescence nitrogen (CLN) and mass spectrometry (MS) detection have provided new opportunities in native amino acid determination and restored new interest in native amino acid analyses.

The aim of this chapter is to highlight the state of the art of the LC analysis of amino acids without derivatization.

2. Detection of underivatized amino acids

2.1. Suitable detection modes

The detector that has been investigated the most for underivatized AA analysis is the amperometric/electrochemical detector however the number of AAs that can be detected by conventional carbon electrodes is limited to only six amino acids (tyrosine, tryptophan, histidine, cysteine, cystine and methionine) [3–13]. In 1999, Clarke et al. [14] described an optimized waveform (changes of potential applied to a gold working electrode vs. time) capable of detecting all primary and secondary amino acids and subsequently, several reports described the use of integrated pulsed amperometric detection in conjunction with high-pH anion exchange chromatographic separation [15–17]. Brazill et al. [18] demonstrated direct AA detection utilising sinusoidal voltametry at a copper electrode. The authors reported that this method is sensitive, selective (allowing three-dimensional data for all analytes detected) with reduced problems of electrode fouling. Pei and Li [19] described an amperometric sensor specifically for amino acids, which employs immobilized amino acid oxidase and was used as detector in LC for the determination of a mixture of AAs with satisfactory results.

Nuclear magnetic resonance (NMR) has also been used for the direct detection of AAs with previous chromatographic separation [20,21] however the application of LC-NMR has been hindered by poor detection sensitivity. Direct detection of amino acids has been further achieved by the ELS detector [22–32]. The detection of these compounds is straightforward thanks to the universal character of this detector for analytes which are less volatile than the chromatographic eluent. The recently developed CLN detector allows the specific and sensitive measurement AAs without derivatization [33,35].

With the development of MS/MS technique, high sensitivity and selectivity can be achieved in the assay of complex samples; this opens the possibility that trace amounts of AAs can be determined without derivatization. Furthermore, Hemmermeister et al. [36] reported that native amino acid actually had better sensitivity than their dansyl chloride derivatized counterparts for MS detection. Concerning mass spectrometry, a variety of different ionisation methods has been used for the ionization and fragmentation investigation of the free amino acids such as chemical ionization [37], field desorption [38], plasma desorption [39], secondary ion mass spectrometry [40], laser mass spectrometry [41], fast atom bombardment [42,43], thermospray (TSP) [44], electrospray ionization (ESI) [45–49]. The first analysis of four underivatized amino acids was obtained by van Leuken *et al.* [44] in 1993 by LC-TSP-MS. However TSP ionisation has been completely outmoded by the even softer and more versatile electrospray ionisation mode. Our group was the first to introduce the LC-ESI-MS of the 20 proteinogenic amino acids [30] and more recently, their LC-ESI tandem MS analysis [46]. At the same time, other interface/ionisation modes have been tested for the LC-MS analysis of underivatized amino acids such as the particle-beam interface [50], the dual oscillating nebulizer atmospheric pressure microwave-induced plasma ionisation [51] and the atmospheric pressure chemical ionization (APCI) [52].

2.2. Comparative study of detection modes

Comparison of different analytical instruments is very useful as it helps scientists to choose among them, in relation to their analytical requirements. Seven commercial LC detectors which can detect amino acids under their underivatized form, were compared in terms of linearity, limit of detection, advantages and disadvantages as well as special features (capacity to provide structural information, specificity, quantification with single calibration curve etc.) [53]. The detectors tested are: CLND, ELSD, NMR, conductivity detector (CD), refractive index detector (RID), UV and electrospray quadrupole mass spectrometry (ESI-MS) in simple and tandem mode. A standard mixture of 12 underivatized AAs (taurine (Tau), hypotaurine (Hpt), aspartic acid (Asp), hydroxyproline (Hyp), asparagine (Asn), serine (Ser), glycine (Gly), glutamine (Gln), cysteine (Cys), glutamic acid (Glu), threonine (Thr) and alanine (Ala)) was analysed with each detector. In order to test the applicability of these detectors in "real world" samples, the amino acid stoichiometry of the tetrapeptide Gly-Gly-Asp-Ala was determined with each detector after acid hydrolysis.

To compare all these detectors the same chromatographic conditions should be used for amino acid separations. This is particularly difficult due to the different chromatographic requirements of the tested detectors. The ELSD, CLND and MS need a volatile mobile phase, the CLND requires the total absence of any nitrogen containing solvents or additives (CH_3CN, NH_3, alkylamines etc.), the UV needs solvents and additives which are transparent at low UV wavelength, the CD needs low conductive mobile phases for maximum solute sensitivity, the ^{1}H-NMR needs the absence of H in the mobile phase and the RID is compatible only with isocratic separations. Finally, the mobile phase should be as simple as possible (with a minimal number of solvents/additives) in order to avoid the system peaks observed with some detectors [54-56]. As ELSD, CLND and MS require a volatile mobile phase, an ion-pair reversed phase liquid chromatographic system was selected, consisting of a Purospher RP-18e octadecyl column and an aqueous mobile phase containing pentadecafluorooctanoic acid (PDFOA) as volatile ion-pairing reagent. Moreover PDFOA was a suitable ion-pairing reagent for NMR experiments due to proton absence in its structure, as well as its low concentration (0.5 mM) required in the mobile phase to obtain satisfactory LC separations. It must be pointed out that compatible conditions for all the detectors tested do not necessarily mean optimum one conditions. For example the ELSD and MS, whose detection limits depend on the spray stability [57-58], are the most penalised by a 100% aqueous phase.

Figure 1 depicts the simultaneous chromatographic analysis of the 12 native amino acid standard mixture by each detection mode and Table 1 summarises the limits of detection (LODs) obtained for each amino acid with the different detectors.

The UV and RI detectors failed to detect all the amino acids investigated either due to the absence of a chromophore group in their structure or due to system interferences. The ELSD was very easy to use with fairly good LOD. The CD was operated in indirect mode and gave good LOD for all the amino acids except for glutamic acid. With the exception of the refractive index detector all the other detectors were compatible with the chromatographic conditions used, gave satisfactory results concerning linearity and good agreement between theoretical and experimental values of the amino acid stoichiometry of the tetrapeptide hydrolysate. The most promising of the detectors tested in this study seems to be tandem MS and the CLND due to their higher sensitivity, and specificity. Furthermore, the CLND equimolarity will be very useful for the amino acid stoichiometry of peptide-protein hydrolysates.

Table 1 Summary of the amino acid limits of detection obtained with the different detectors. From Reference [53].

Amino acids	Limits of detection (mg l^{-1})							
	ELSD	UV	CD	RID	CLND	MS	MS/MS	NMR
Tau[a]	1	NA	5	50	0.6	0.4	0.5	100
Hpt	2	NA	5	50	0.5	0.5	0.4	150
Asp	2.5	3	1	50	0.6	0.4	0.2	250
Hyp	2.5	4.5	4	50	0.6	0.2	0.08	300
Asn	2.5	0.9	3	50	0.33	1	0.8	300
Ser	2.5	3	3	50	0.5	0.5	0.2	300
Gly	5	2	1.5	50	0.4	0.5	0.5	200
Gln	5	2.7	4	50	0.6	0.3	0.1	400
Cys	10	3	5	NA	0.3	1	0.8	500
Glu	5	3	25	NA	0.6	5	0.15	500
Thr	5	3	3	NA	0.7	0.3	0.1	500
Ala	7.5	3	1.5	NA	0.6	1	0.3	500

[a] Detection of the analyte can not be achieved by this detector or was disturbed by system interferences.

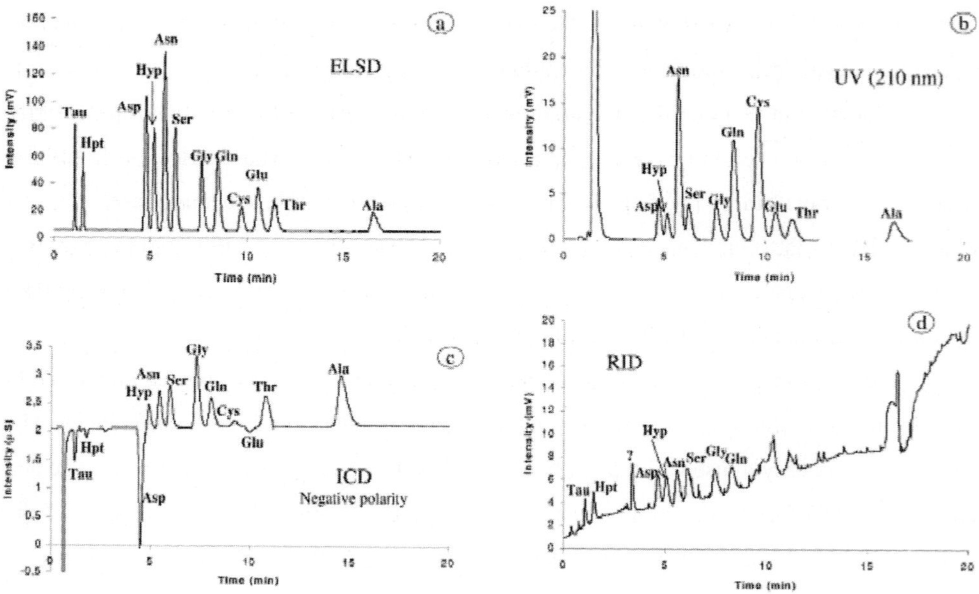

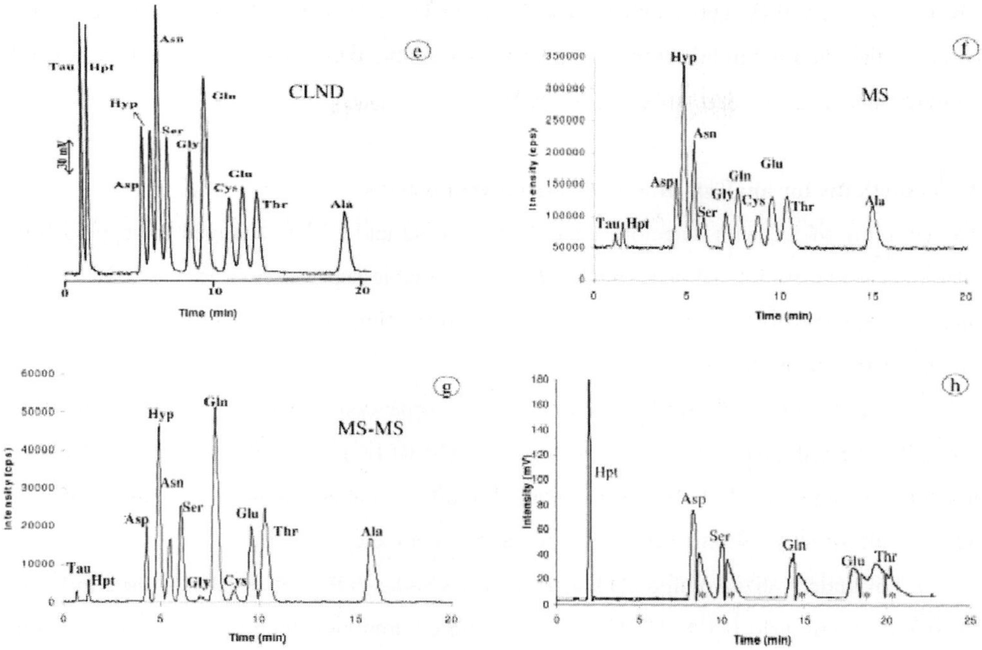

Figure 1 Liquid chromatographic isocratic simultaneous analysis of 12 (6 for NMR) underivatized amino acid mixtures. Column: Merck, Purospher RP-18e (125 x 4 mm). Mobile phase: 0.5 mM pentadecafluroctanoic acid in water (in deuteriumoxide for NMR). Detection by: (a): evaporative light scattering , (b): UV at 210 nm, (c): indirect conductivity, (d): refractive index, (e): chemiluminescence nitrogen, (f): simple MS, (g): tandem MS, (h): ELSD when used as monitor detector for NMR. Asterisks: negative peaks which correspond to the time needed for the mobile phase to reach the ELSD photomultiplier, once the LC restarts. Injected concentrations: 50 mg l^{-1} for Tau and Hpt, 100 mg l^{-1} for the other amino acids; for NMR only: 2000 mg l^{-1}. Injected loop: 20 μl for NMR, 10 μl for all the other detectors. Reproduced with permission from Reference [53].

Each method has its advantages and drawbacks and depending on the matrices investigated as well as their amino acid contents (absolute quantities, existence of low concentrations of amino acids in the presence of high concentrations of others etc.), certain detectors can successfully determine the amino acids while others may fail to do so.

The combination of the above detectors with a more powerful/adequate chromatographic system than the one used in this study could be an interesting alternative methodology to replace amino acid analysis by derivatization methods.

3. LC methods for analysis of underivatized amino acids

The coupling of LC with detectors such as MS, ELSD and CLND changed the approach of analytical scientists for the selection of the most suitable separation conditions. One of the major parameters to consider with the above hyphenations is the volatility of the mobile phase/electrolyte used.

Traditionally, underivatized amino acids are separated by ion-exchange [1,8,59,60] or ion-pair reversed phase liquid chromatography (IP-RPLC) [61–64]. However, the above methods use non-volatile buffers (phosphate, lithium etc.) and/or non-volatile ion-pairing reagents (salts of alkylsulfonic acids, tetrabutylammonium, etc.).

For underivatized amino acid analysis, Chaves-das-Neves et al. [24,25] used mixtures of trifluoroacetic acid (TFA)-H_2O-acetonitrile and column switching or mixed-bed columns (reversed-phase cation exchange) coupled with ELSD. In another interesting approach, Peterson et al. [27] employed a cation exchange column and an aqueous TFA-ammonium acetate buffer with ELSD. However, the selectivity towards high polar AAs remains bad and several coelutions are observed under these chromatographic conditions.

IP-RPLC presents some advantages in comparison with ion-exchange chromatography due to the lower cost, higher viability as well as higher flexibility of reversed phase columns compared to ion exchange supports. Indeed, the dynamic ion-exchange capacity of an ion-pair system can be modulated through the use of different concentrations and/or different nature of the ion-pairing reagent as well as of variable organic solvent content in mobile phase.

3.1. IP-RPLC systems using volatile ion pairing reagents

Several volatile perfluorinated carboxylic acids (trifluoroacetic (TFA), heptafluorobutyric (HFBA), nonafluoropentanoic (NFPA), tridecafluoroheptanoic (TDFHA) and pentadecafluorooctanoic (PDFOA) acid) have been proposed as highly suitable ion-pairing reagents for underivatized amino acids separation by IP-RPLC on C8 and C18 silica [28, 30] or porous graphitic carbon [31] stationary phases. These reagents not only exhibited better selectivity compared with alkylsulfonic acids [64] but were also volatile enough to be used with ELS and MS detection [28-31,46]. The surfactants with the longest side chains (1 mM TDFHA in water or 0.5 mM PDFOA in water) gave the best amino acid selectivities on C18 silica supports and allow the successful isocratic separation with satisfactory baseline resolution of ten polar amino acids (see Figure 2): Asp, Asn, Ser, Gly, Gln, Cys, Glu, Thr, Ala and Pro [28] as well as the possible determination in about 40 min and in gradient elution, of 17 of the 20 proteinogenic AAs with an ELS detection [30]. It appears that the most efficient stationary phase was the most hydrophobic octadecylsilica support.

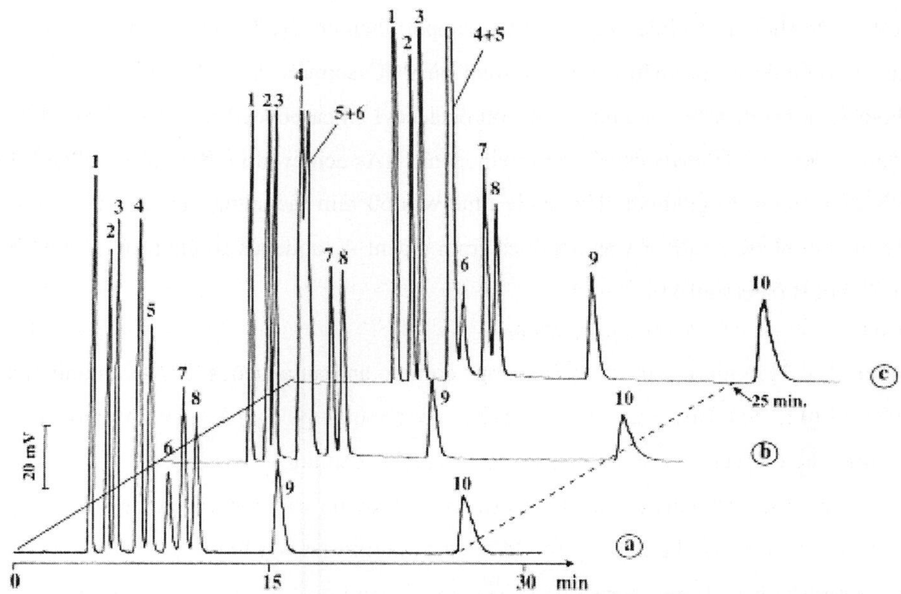

Figure 2 LC-ELSD analysis of a 100 mg l^{-1} standard solution of 10 underivatized amino acids under isocratic conditions on: (a) Purospher RP-18e 125x4 mm I.D. (eluent PDFOA 0.5 mM in water); (b) LiChrospher RP-Select B 125x4 mm I.D. (eluent, PDFOA 0.75 mM in water); and (c) Supelcosil ABZ^{+}Plus 150x4.6 mm I.D. (eluent, PDFOA 0.75 mM in water). Flow rate: 1 ml min^{-1}. Model Sedex 45 evaporative light scattering detector set as follows: drift tube temperature: 60 °C, nebulizer gas pressure: 2.2 bar and photomultiplier: 9. Elution order: 1, Asp; 2, Asn; 3, Ser; 4, Gly; 5, Gln; 6, Cys; 7, Glu; 8, Thr; 9, Ala; 10, Pro. Reproduced with permission from Reference [28].

The elution gradient was based on simultaneously increasing the concentration of acetonitrile and decreasing the concentration of the ion-pairing reagent. The complete determination by LC-ELSD on C18 support was prevented by coelutions or by induced peaks occurring during the gradient elution that mask the elution of some AAs. The use of LC-MS was necessary to identify all the proteinogenic AAs on C18 support. Porous graphitic carbon (PGC) appears to be an interesting alternative to conventional reversed-phase packing material for IP-RPLC analysis of underivatized AAs [31]. On PGC the best selectivities were obtained with 20 mM to 25 mM of NFPA as ion pairing reagent in mobile phase. The use of surfactant with a shorter side chain in mobile phase on PGC support than on C_{18}-bonded silica column involves that the optimized chromatographic system on PGC support has a faster equilibration time (about 15 min) than the one optimized on octadecyl silica bonded material (about 200 min). Figure 2 depicts the analysis of 20 proteinogenic AAs achieved in 40 min on a PGC support with an acetonitrile gradient. The cycle time was 60 min including the reequilibration time. The mobile phase gradient was obtained from eluent A (aqueous solution of 20 mM NFPA) and eluent B (acetonitrile).

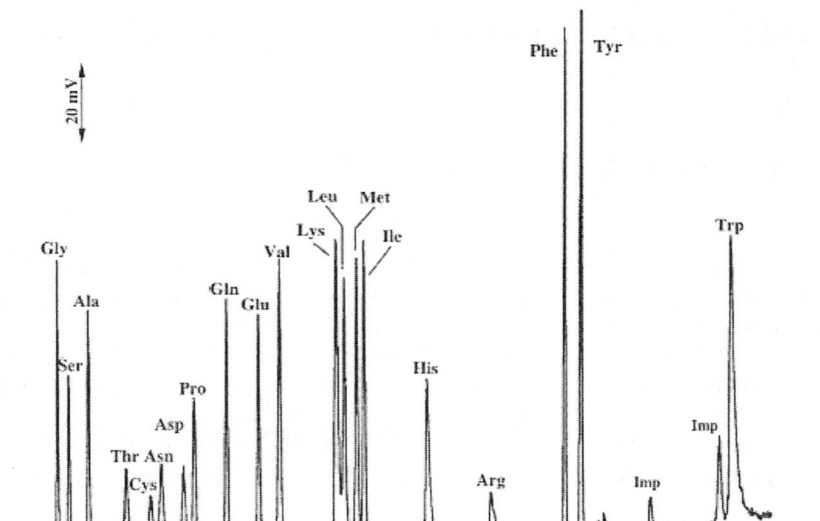

Figure 3 LC-ELSD analysis of a 100 mg l^{-1} standard solution of the 20 proteinogenic underivatized amino acids in gradient elution. Column: Hypercarb (100 x 2.1 mm I.D.). Gradient elution: solvent A: 20 mM NFPA in water, solvent B: acetonitrile; gradient starts from 0 to 15% ACN in 10 min, then 26% ACN in 10 min and finally 50% ACN in a further 10 min, 50% ACN is maintained until the end. Column temperature is 10°C. Flow rate: 200 μl min^{-1}. Detection: ELSD model Sedex 45 set as follows: drift tube temperature: 60 °C, nebulizer gas pressure: 2.2 bar and photomultiplier: 9. Reproduced with permission from Reference [31]

The resolution observed for the Lys-Leu pair and the Met, Ile pair is strongly dependent on the column temperature. The main advantage of the chromatographic system optimized on PGC in comparison with the analysis on Purospher RP 18e support [30] is the absence of a system peak occurring during the gradient elution. Under these LC-ELSD chromatographic conditions the determination of the 20 proteinogenic underivatized amino acids is achieved with detection limits varying from about 1.25 mg l^{-1} (injected amount around 30 pmol) for most of the amino acids to about 2.5 mg l^{-1} (inject amount around 60 pmol) for Asp and Cys. The elution order is different on PGC and on RP packing material: for example, an opposite elution order was observed for the Ile,Leu pair on the two different systems. These two LC-ELSD systems provide complementary informations and can be used favorably as complementary techniques for identification of trace amounts of minor amino acids present in mixture with a major amino acids for example.

4. LC-MS/MS analysis of underivatized amino acids

Without any doubt, the most specific liquid chromatographic detection method nowadays is mass spectrometry. Its high specificity decreases the necessity for powerful chromatographic systems in which all compounds of interest have to be separated before detection. It thus reduces the analysis time.

4.1. Achiral analysis of underivatized amino acids

The first step in the development of a LC-MS/MS analytical methodology was to establish for each molecule its potential to generate $[M+H]^+$ or $[M-H]^-$ ions, and its possibility of fragmentation, giving product ion spectra in positive or negative ionization mode. The ESI-MS/MS fragmentation of 20 proteinogenic underivatized amino acids has been extensively studied in several reports in positive ionization mode [46,49,65] and in negative ionization mode [49]. A more comprehensive ESI-MS/MS analysis of 79 molecules of biological interest in positive and negative ionization mode has been also published [49]. The product ions of each AA are sorted by their maximum intensities (i.e., the intensity of a product ion acquired under its optimal MS parameters, especially the collision energy) from abundant to weak. Except for three molecules (orotic acid, orotidine-markers of urea cycle disorders-and phosphoethanolamine) all the molecules studied and the 20 proteinogenic AAs in particular, could be analyzed in positive ionization mode, with variable sensitivities, but enough for their detection in biological samples.

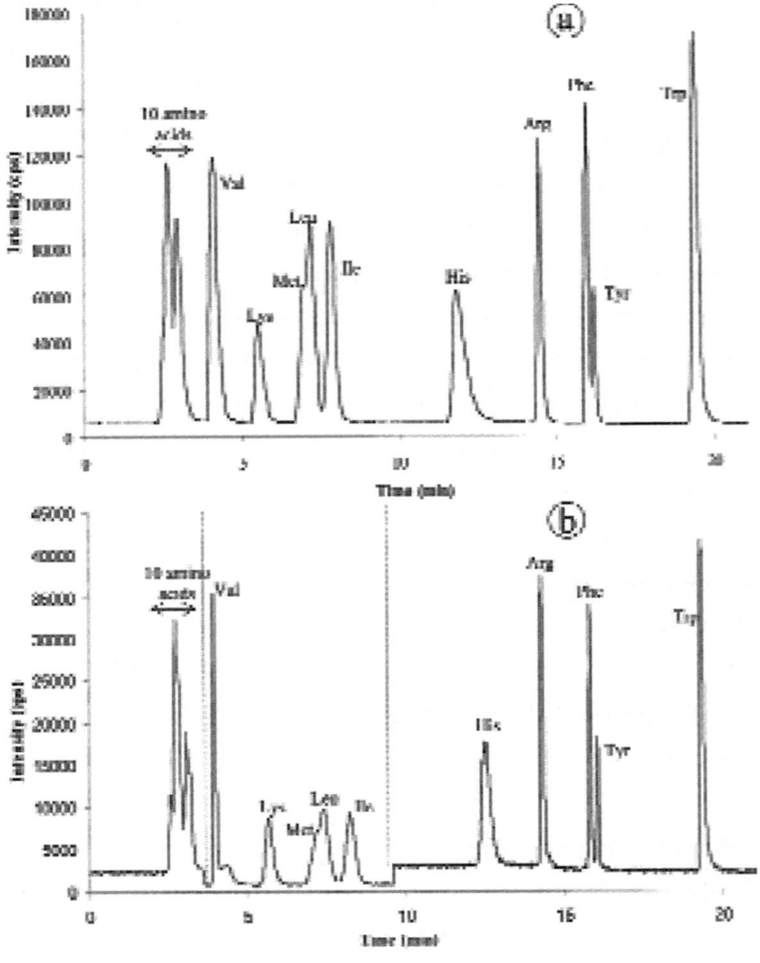

Figure 4 LC-ESI-MS/MS analysis of a standard mixture of the 20 proteinogenic underivatized amino acids in gradient elution. Column: Hypercarb (100 x 2.1 mm I.D.). Gradient elution: solvent A: 20 mM NFPA in water, solvent B: acetonitrile; gradient starts at 10%B for 1 min, from 10 to 72% in 7 min, then from 72% to 90% in 1 min, then 90% is maintained to the end of the analysis. Flow rate: 200 µl min^{-1}; split 1/10. (a) SRM of a 50 mg l^{-1} standard solution (18 ion transitions monitored) (dwell time: 100 ms). (b): SRM of a 10 mg l^{-1} standard solution "cut" in three sections: from 0 to 3 min (10 ion transitions monitored, dwell time: 200 ms) from 3.2 to 9 min (4 ion transitions monitored, dwell time: 400 ms), from 9 min until the end of analysis (5 ion transitions monitored, dwell time: 300 ms). Reproduced with permission from Reference [46]

The specificity of the MS/MS method is linked to the combined analysis of the precursor ion and one of its fragments. Every molecule that can give the same fragmentation constitutes an interference when these two molecules are present in the same mixture. Analysing so many molecules simultaneously highlighted a large number of interferences. These interfering molecules have been indexed [46,49,65]. They are due either to the presence of isobaric compounds (such as Gln and Lys, or Leu and Ile), to natural isotopic distribution of ^{13}C (interference between Asn and Asp or Gln and Glu) and ^{34}S , or to in-source CID fragmentation (interference of Met versus Asn or interference of Asp versus Pro). When the interference is quantitatively important, molecules involved in this phenomenon must be separated by LC before analysis in the tandem mass spectrometer.

The use of perfluorocarboxylic acids as ion-pairing reagents associated to porous graphitic carbon column provided the necessary separation among isomer/isobar amino acids as well as the amino acids that had to be separated due to intrinsic mass spectrometry problems. Figure 4 shows the selective reaction monitoring (SRM) of the 20 underivatized proteinogenic amino acids (18 ion transitions monitoring) obtained by LC-ESI-MS/MS. The specificity of the SRM mode allowed the chromatographic single run analysis of the 20 AAs in less than 20 min. If higher dwell times are needed, it is possible as shown in Fig. 2b, to "cut" the chromatogram in three sections: from 0 to 3 min where 10 ions are monitored, from 3 to 9.5 min (monitor of 4 ions) and lastly from 9.5 min until the end of the analysis (monitor of 5 ions). This operation allows the 1-4 fold increase of the dwell time. Actually, detection limits for tandem MS varied depending on the amino acid from 500 fmol to 40 pmol (using a 10μl loop) [46]. The overall sensitivity of the underivatized proteinogenic AAs was of the same order of magnitude as that of post-column derivatization UV spectrometric methods. Rapid developments in the instrumentation of ion sources and mass spectrometers as well as in the corresponding software will lead to higher sensitivities.

4.2. Chiral separation of underivatized amino acids

A variety of AAs are found in most clinical and environmental samples. Therefore, the ability to separate multiple AAs both chirally and achirally in a simultaneous analysis is also of fundamental interest. The resolution of all the native AAs and their respective enantiomers would require greater separation selectivity than is currently practical, resulting in even longer retention times. The use of MS detection, however, can overcome these problems. MS detection provides a second dimension of discrimination allowing the extraction of specific m/z values from unresolved AAs, thereby significantly reducing the analysis time. In spite of the high specificity of MS, the distinction between the L and D enantiomers of AAs is not possible as they have the same molecular mass and identical fragmentation behaviour.

Fifteen chiral underivatized AAs in standard mixture were analysed simultaneously under isocratic conditions (acetonitrile-water, 75:25) in less than 25 min using a teicoplanin-based chiral stationary phase and ESI-MS/MS detection [66]. Under these chromatographic conditions, positional isomeric amino acids (e.g., L,D-Leu/Ile, or L,D-Val/Iva) were separated. For maximum sensitivity, post-column addition of 500 mM aqueous HCOOH was necessary. Detection limits varied from 2.5 to 50 μg l^{-1} depending on the amino acid.

A simultaneous separation of a standard mixture of all 19 native proteinogenic amino acid enantiomers in less than 20 min on a teicoplanin-based support and APCI-MS detection was also reported [67]. A mobile phase consisting of 1% (w/w) ammonium trifluoroactete in methanol and 0.1% (w/w) formic acid in water was used in step gradient. Unfortunately, the D- and L-forms of leucine and isoleucine could not be fully separated from one another. Asp and Glu enantiomers co-elute with Pro and Leu/Ile enantiomers, respectively. The ionization of Asp and Gln could be suppressed by the presence of those co-eluting species, resulting in poor peak shape of the reconstructed ion chromatograms.

5. Quantitative analysis of underivatized amino acids

Up to now, few direct analytical methods (without a pre or post column derivatization step) have been validated for amino acid analysis. Peterson *et al.* [68] have developed and validated a procedure for the amino acid analysis of peptide hydrolysates based on cation-exchange chromatography coupled with ELSD. This quantitative method detects and separates 18 of the common AAs in a one hour run time. The procedure was tested by analyzing the hydrolysate of human parathyroid hormone 1-34, a synthetic peptide. With the exception of Cys, for the other 17 AAs analyzed, detection limits varied from 250 to 2500 pmol injected. Petritis *et al.* [69] have developed and validated an IP-RPLC procedure coupled with ELSD for the direct quantification of 8 underivatized AAs present simultaneously in a commercial preparation. The chromatographic separation was achieved on a Purospher RP-18e column with an acetonitrile gradient and using HFBA as volatile ion pairing reagent. The accuracy and precision are comparable to other traditional methods. Detection limits varied from 1 mg l^{-1} to 2 mg l^{-1} depending on the AA being analyzed. In spite of the use of an ion-pairing reagent (heptafluorobutyric acid) which needs some time to be adsorbed in the column, the method was very repeatable and reproducible especially in terms of retention time. Candela *et al.* [70] reported the validation of an analysis method for 4-amino-3-hydroxybutyric acid in finished pharmaceutical products based on ion-pair chromatography with heptanesulphonate as ion-pairing reagent and UV detection at 210 nm. Qu *et al.* [71] proposed an LC-MS/MS method for the quantification of free amino acids in a fermentation medium by using pentadecafluorooctanoic acid (PDFOA) as volatile ion-pairing reagent on a C_{18} column. Injections were made directly after centrifugation of the samples, without further preparation. The detection limits range from 0.1 to 5 μg l^{-1}. The analysis time was 12 min per sample, suitable for large-scale analysis. Matrix effects were not observed, as demonstrated by the good recoveries and precisions of the analyses.

A validated quantification of 22 underivatized AAs (20 proteinogenic AAs + ornithine and citrulline) in human serum samples was proposed by Qu *et al.* [65]. The AAs are analysed after a separation on a C18-silica column. The mobile phase consisted of two solvents in gradient elution within 36 min. Solvent A: 0.8 mM TDFHA and 0.05% TFA in water; solvent B: acetonitrile. Some problems associated with the use of ion-pairing reagent were reported: an increase in retention time for several AAs was noticed after nine consecutive assays. To overcome this, after every 6 chromatographic runs, the column is flushed with 100 % acetonitrile for 30 min. The use of stable isotopically labeled internal standards ensured the accuracy of quantification and correct the ion suppression phenomenon. The calibration curves were linear over the range of 0.10–100 mg l^{-1} and the detection limits were 0.03 to 20 pmol injected. The quantitative results obtained with this method on 8 serum samples were in good agreement with those obtained by an established OPA-prederivatization HPLC method.

With such low limits of detection, application of the LC-MS/MS method to the analysis of biological and environmental samples is definitely possible.

References

[1] D.H. Spackman, W.H. Stein, S. Moore, Anal. Chem. 30 (1958) 1190–1206.

[2] J.W. Crabb, L.H. Ericsson, D. Atherton, A.J. Smith, R. Kutny, (1990) A collaborative amino acid analysis study form the association of biomolecular resource facilities, in J.J. Villafranca, ed., Current Research in Protein Chemistry, Academic Press, San Diego (1990) 49-61.

[3] L.A. Allison, R.E. Shoup, Anal. Chem. 55 (1983) 8.

[4] S. Mural, H. Masuda, T. Itoh, J. Neurochem. 50 (1988) 473.

[5] K. Irlyama, M. Yoshiura, T. Iwamoto, J. Liquid Chromatogr. 9 (1986) 2955.

[6] J.A. Cox, T.J. Gray, Anal. Chem. 61 (1989) 2462.

[7] M.K. Halbert, R.P. Baldwin, J. Chromatogr. 345 (1985) 43.

[8] L.E. Welch, W.R. LaCourse, D.A. Mead, D.C. Johnson, Anal. Chem. 61 (1989) 555-559.

[9] L. Fleming, N. Reynolds, J. Chromatogr. 375 (1986) 65-73.

[10] L. Fleming, N. Reynolds, J. Chromatogr. 431 (1988) 65-76.

[11] N.C. Reynolds, B.M. Kissela, L.H. Fleming, Electroanalysis 7 (1995) 1177-1181.

[12] C.D. Paras, R.T. Kennedy, Electroanalysis 9 (1997) 203-208.

[13] D.A. Martens, W.T. Frankenberger, J. Liquid Chromatogr. 15 (1992) 423-439.

[14] A.P. Clarke, P. Jandik, R.D. Rocklin, Y. Liu, N. Avdalovic, Anal. Chem. 71 (1999) 2774.

[15] P. Jandik, C. Pohl, V. Barreto, N. Avdalovic. Anion exchange chromatography and integrated amperometric detection of amino acids, in C. Cooper, N. Packer, K. Williams, ed., Methods in Molecular Biology: Amino acid analysis Protocols, vol. 159, Humana Press, Totowa, (2001) 63-85.

[16] H. Yu, Y. Ding, S. Mou, J. chromatogr. A 997 (2003) 145.

[17] P. Jandik, J. Cheng, N. Avdalovic, J. Biochem. Biophys. Methods 60 (2004) 191-203.

[18] S.A. Brazill, P. Singhal, W.G. Kuhr, Anal. Chem. 72 (2000) 5542-5548.

[19] J. Pei, X.Y. Li, Fresenius J. Anal. Chem. 367 (2000) 707-713.

[20] N. Wu, A. Webb, T.L. Peck, J.V. Sweedler, Anal. Chem. 67 (1995) 3101.

[21] K. Petritis, I. Gillaizeau, C. Elfakir, M. Dreux, A. Petit, N. Bongibault, W. Luijten. J. Sep. Sci. 25 (2002) 593-600.

[22] V. Camel, D. Thiebaut, M. Caude, M. Dreux, J. Chromatogr. 605 (1992) 95.

[23] A.J. Berry, E.D. Ramsey M. Newby, D.E. Games, J. Chromatogr. Sci. 34 (1996) 245.

[24] H.J. Chaves das Neves, Z.B. Morais, Anal. Quim. 93 (1997) 98-101.

[25] H.J. Chaves das Neves, Z.B. Morais, J. High Resolut. Chromatogr. 20 (1997) 115-118.

[26] H.J. Chaves das Neves, Am. Biotechnol. Lab. 17 (1999) 54-55.

[27] J.A. Peterson, L.J. Lorenz, D.S. Risley, B.J. Sandmann, J. Liq. Chromatogr. & Rel. Technol. 22 (1999) 1009-1025.

[28] K. Petritis, P. Chaimbault C. Elfakir, M. Dreux, J. Chromatogr. A 833 (1999) 147-155.

[29] K. Petritis, P. Chaimbault C. Elfakir, M. Dreux, Spectra 2000 Anal. 211 (1999) 31-32.

[30] P. Chaimbault, K. Petritis, C. Elfakir, M. Dreux, J. Chromatogr. A 855 (1999) 191-202.

[31] P. Chaimbault, K. Petritis, C. Elfakir, M. Dreux, J. Chromatogr. A 870 (2000) 245-254.

[32] K. Petritis, M. de Person, C. Elfakir, M. Dreux, Chromatographia 60 (2004) 293-298.

[33] E.M. Fujinari, E. Ribble, M.V. Piserchio, Developments in Food Science 32 (1993) 75-89.

[34] R. Bizanek, J.D. Manes, E.M. Fujinari, Peptide Res. 9 (1996) 40-44.

[35] K. Petritis, C. Elfakir, M. Dreux, LC-GC Europe 14 (2001) 389-395.

[36] D.E. Hemmermeister, J. Serrano, P. Schmeider, D. Kuel, Rap. Commun. Mass Spectrom. 14 (2000) 503.

[37] G.W. Milne, T. Axenrod, H.M. Fales, J. Am. Chem. Soc. 92 (1970) 5170.

[38] H.U. Winkler, H.D. Beckey, Org. Mass Spectrom. 6 (1972) 655.

[39] S. Bouchonnet, J.P. Denhez, Y. Hoppilliard, C. Mauriac, Anal. Chem. 64 (1992) 743.

[40] L.K. Liu, K.L. Busch, R.G. Cooks, Anal. Chem. 53 (1981) 109.

[41] C.D. Parker, D.M. Hercules, Anal. Chem. 58 (1986) 25.

[42] W. Kulik, W. Heerna, Biomed. Mass Spectrom.18 (1988) 419.

[43] N.N. Dookeran, T. Yalcin, A.G. Harrison, J. Mass Spectrom. 31 (1996) 500.

[44] R.G.J. Van Leuken, G.T.C. Kwakkenbos, A.L.L. Duchateau, J. Chromatogr. 647 (1993) 131.

[45] F. Rogalewicz, Y. Hoppilliard, G. Ohanessian, Int. J. Mass Spectrom. 195/196 (2000) 565.

[46] K. Petritis, P. Chaimbault, C. Elfakir, M. Dreux, J. Chromatogr. A 896 (2000) 253-263.

[47] K. Hiraoka, K. Murata, I. Kudaka, J. Mass Spectrom. Soc. Jap. 43 (1995) 127.

[48] B. Mansoori, D. Volmer, R.K. Boyd, Rap. Commun. Mass Spectrom. 11 (1997) 1120.

[49] M. Piraud, C. Vianey-Saban, K. Petritis, C. Elfakir, J.P. Steghens, A. Morla, D. Bouchu, Rap. Commun. Mass Spectrom. 17 (2003) 1297-1311.

[50] E. Magi, C. Ianni, M. Grotti, Anal. Chim. Acta 401 (1999) 55.

[51] J.-Y. Kwon, M. Moini, J. Am. Soc. Mass Spectrom.12 (2001)117.

[52] M.J. Desai, D.W. Armstrong, J. Mass Spectrom. 39 (2004) 177-187.

[53] K. Petritis, C. Elfakir, M. Dreux, J. Chromatogr. A 961 (2002) 9-21.

[54] S. Levin, E. Grushka, Anal. Chem. 58 (1986) 1602.

[55] S. Levin, S. Abu-Lafi, S. Golshan-Shirazi, G. Guiochon, J. Chromatogr. A 679 (1994) 213.

[56] S. Golshan-Shirazi, G. Guiochon, Anal. Chem. 62 (1990) 923.

[57] M. Dreux, M. Lafosse, LC GC Intl. 9 (1996) 148.

[58] K.W.M. Siu, R. Guerremont, J.C.Y. LeBlanc, G.J. Gardner, S.S Berman, J. Chromatogr. 554 (1991) 27.

[59] J.A. Grunau, J.M. Swiader, J. Chromatogr. 594 (1992) 165-171.

[60] P. Luo, F. Zhang, R.P. Baldwin, Anal. Chem. 63 (1991) 1702-1707.

[61] J. Crommen, P. Herné, J. Pharm. Biomed. Anal. 2 (1984) 241-253.

[62] P.G. Rigas, S.J. Arvanitis, D.J. Pietrzyk, J. Liq. Chromatogr. 10 (1987) 2891-2910.

[63] J. Saurina, S. Hernandez-Cassou, J. Chromatogr. A 676 (1994) 311-319.

[64] Y. Yokoyama, T. Amaki, S. Horikoshi, H. Sato, Anal. Sci., 13 (1997) 963-967.

[65] J. Qu, Y. Wang, G. Luo, Z. Wu, C. Yang, Anal. Chem. 74 (2002) 2034-2040.

[66] K. Petritis, A. Valleix, C. Elfakir, M. Dreux, J. Chromatogr. A 913 (2001) 331-340.

[67] M.J. Desai, D.W. Armstrong, J. Mass Spectrom. 39 (2004) 177-187.

[68] J.A. Peterson, L.J. Lorentz, D.S. Risley, B.J. Sandmann, J. Liq. Chrom & Rel. Technol. 22 (1999) 1009-1025.

[69] K. Petritis, M. De Person, C. Elfakir, M. Dreux, Chromatographia 60 (2004) 293-298.

[70] M. Candella, A. Ruiz, F.J. Feo, J. Chromatogr. A, 890 (2000) 273-280.

[71] J. Qu, W. Chen, G. Luo, Y. Wang, S. Xiao, Z. Ling, G. Chen, Analyst, 127 (2002) 66-69.

Ibolya Molnár-Perl (Editor)
Quantitation of Amino Acids and Amines by Chromatography
Journal of Chromatography Library, Vol. 70

1.2.2.

HPLC of Amino Acids as Phenylthiocarbamoyl Derivatives

Ibolya Molnár-Perl

Contents

Summary

Some of the general problems encountered in the analysis of amino acids derivatized with various isothiocyanates, determined both as phenylthiocarbamoyl (PTC) and as phenylthiohydantoin (PTH) derivatives, are presented. Based on our own experience and the literature associated with sample preparation, stability, optimum chromatographic and quantitation conditions are discussed. Twenty-seven PTC-AAs have been separated on seven columns, with the same gradient program, applying various temperatures and eluent flow rates, within 40 min, including equilibration time. Response values and spectral characteristics of the 27 AAs proved to be independent of the chromatographic conditions performed. Particular attention was paid to the issue of sensitive amino acids under hydrolysis (cystine/cysteine, methionine, tryptophan), and derivatization (cystine/cysteine) conditions as well. The special behavior of the PTC-cystine/cysteine has been examined by LC/API/MS measurements showing that, irrespective of the initial compound to be derivatized, the same stereoisomer pairs are formed at m/z = 255, PTC-cysteines (~80 % of the total) and their oxidized version at m/z = 287 (~20 % of the total). Whether there is an advantage of the microwave irradiation procedure is still not decided: however the hydrolysis time of lysozyme has been shortened considerably, but optimum hydrolysis conditions were not uniformly found for all essential amino acids.

1. Introduction

The quantification of amino acids (AAs) such as phenylthiocarbamoyl (PTC) derivatives (along with *o*-phthalaldehyde derivatization) is the second most common process [1] prior to their analysis by high-performance liquid chromatography (HPLC). The reason is simple: (i) phenyl isothiocyanate (PITC) reacts both with the primary and secondary amino groups, at room temperature, within 5 min. (ii) Stability of the PTC–AAs is excellent: in dried conditions they can be stored in the freezer for an unlimited time, and in solutions for several hours, and, (iii) their elution and detection needs a binary gradient pump and a UV detector, only. (iv) The transformation of the PTC-derivatives into the phenylthiohydantoin (PTH) ones serves as the basis in sequencing, identification and quantitification of peptides and proteins. (v) The two disadvantages of the phenylthiocarbamoylation protocol proved to be of secondary importance: the limited sensitivity of derivatives (due to the fact that they do not provide fluorogenic derivatives) and the need for the removal of excess reagent.

The pioneering invention [2], in the last twenty five years has been followed by exhaustive optimization studies of pre-column derivatization and elution conditions, both in the case of PTC- [2–65] and PTH-derivatives [66–82]. From alterations in the reagent's side chain [83–94] improvements have been achieved in terms of chromatographic advantages [83–89] and preparing fluorogenic PTH-derivatives [90–94], equally.

1.1. The phenylthiocarbamoyl derivatives

Recent papers dealing with the HPLC quantitation of PTC-AAs suggested a number of modifications and improvements both at the basic research level [2–10,12,15,18–22,30,31,34,36, 37,39,40,42,43,46–48,52–55,59,60,62,64] and also in application of the protocol; in the determination of AAs (in initially free forms or in protein hydrolyzates), present in various biological matrices such as human plasma [16,26,27,38,45], dried blood [65], human [17,24,28,33,44,61, 63] and rat tissues [25], bacterial cell walls [41], in different vegetables and foods such as apples [11,62], tomato xylem [35] legumes [50,56], must and wine samples

[23,32,57,58] and feed stuffs [40,51]. Representative protocols, selected proposals (Table 1) for the quantitation of AAs as PTC derivatives present in various matrices, as well as those from the leading column manufacturers [34,43] revealed (i) in practice, the fast elutions using the Waters, Picotag column [33,36,53] provide incomplete resolution also for the separation of 18 AAs [36,53], and even more so for 23 AAs [33]. (ii) For the long-lasting elutions (>50 min), neither for the classical 20 [32,34,43], still less for other AAs [24,26,30–32,34,37],did the extended elution time fulfil expectations. (iii) However, excellent separations have been shown recently [34,43] by the manufacturers (promising complete separations within unbelievable short elution times; 4.5 min for 17 [34], 8 min for 18 [43] AAs, without equilibration times), but practical realizations with these protocols have not yet been published. The HPLC of 27 PTC-AAs within 40 min (including equilibration time) is a multi-facetted optimized separation achieved to date [62]: providing acceptable resolution of the classical 20, simultaneously, in addition to asparagine, glutamine, β-alanine, γ-amino butyric acid (GABA), homoserine, homoarginine and 1-aminocyclopropane-1-carboxylic acid, (ACPCA) (Detailed study in Paragraphs 2–4).

1.2. The phenylthiohydantoin derivatives

A major step in peptide/protein identification is their end-group analysis on the basis of the Edman principle, consisting of three main steps: (1) the coupling reaction of the N-terminal AAs with PITC, (2) the cyclization/cleavage reaction for the formation of anilinothiazolinone amino acids (ATZ-AAs) using anhydrous trifluoroacetic acid (TFAA), and, (3) the conversion of the ATZ-AAs into the most stable phenylthiohydantoin amino acids (PTH-AAs) applying aqueous TFAA. This very common and conventional Edman protocol is widely used, applying many different modifications and improvements in the chromatographic separation of the PTH-AAs, both in the reaction procedure and in their detection. Early proposals related to the optimization of HPLC separation of the PTH-AAs [66–73,82], applying mainly UV and DAD detections (Table 2). One of the most spectacular, and still the best chromatographic separation, was achieved by Glajch and Kirkland [68]. Recent proposals are focusing mainly on identifications by mass spectrometry (MS), applying tandem MS [74–76,79,80] with electrospray ionization (ESI) [74–76,] or Fourier transform ion cyclotron resonance (FT-ICR) mass analysis [80] which provides superior mass accuracy to enhance the specificity of database searching.

Studies aiming to clarify/suppress the undesired racemization of AAs during the Edman sequencing are important contributions to the field [77,78,81]. Development of the simultaneous determination of D/L-configuration of derivatives (by HPLC using a combination of reversed phase and a chiral stationary phase of cyclodextrin) made it possible to examine then significantly supress the step responsible for racemization. Boron trifluoride and HCL-methanol (1:10, v/v) were suggested as the cyclization/cleavage/conversion reagents instead of the successive use of anhydrous and aqueous TFAA, excluding the extraction step and resulting in the PTH-AAs with supresssed racemization. Racemization ratios of each cycle were 1.6% (tyrosine[1]), 6.4% (proline[2]), 5% (phenylalanine[3]) and 5.4 % (proline[4]), respectively [81] (Figure 1).

1.3. Derivatives obtained with various isothiocyanates

Out of the continuously introduced substitutes of PITC [83–94] only the fluorogenic derivatives providing ones [83–85,91,93,94] offer adavantages over the classical phenylthiocarbamylation. Alternative derivatisations, with butyl- [87,89] and benzylisothiocyanates [88], do not provide advantages over the classical PITC; moreover, they need to be performed for prolonged reaction time at elevated temperature, (30 min, at 40 °C for butylthiocarbamylation [87,89], 30 min, at 50 °C for benzylthiocarbamylation [88]. Derivatization with 4-nitrophenylisothiocyanate was performed at ambient temperature [86] offering improved sensitivity (1 pM) and selectivity (detection at 340 nm). However, because of difficulties in the extraction of reagent's excess this derivatization did not gain wide acceptance, either.

The (+)-1-(1-naphtyl)ethyl isocyanate was an early proposal for the identification of diastereoisomeric naphtylethylcarbamoyl derivatives [83] applying fluorescence detection. Ferrocenylisothiocyanate was prepared for the ultrasensitive electrochemical detection (0.05 pM) of AAs in biological samples [85]. Dimethylaminobezylisothiocyanate [90] was used for preparing the corresponding PTH derivatives, including optimization of their chromatographic separation. The fluorogenic, 4-(2-cyanoisoindolyl)phenylisothiocyanate reagent [91] ensures the derivatisation of AAs under the reaction condition similar to those for the Edman reagent, (however at 80 °C for 5 min), with the detection sensitivity of two order of magnitude higher than PITC.

Derivatisation with various fluorescent chiral Edman reagents [93,94], such as {4-(3-isothio-cyanatopyrrolidine-1-yl-7-nitro-2,1,3-benzoxadiazole R(-)- and S(+)- variants}[93], (at 55 °C for 10 min) and as 4-(5',6'-dimethoxybenzothiazolyl)phenyliziothiocyanate [94] (at 80 °C for 15 min) have been utilized for the resolution of AA enantiomers as diastereoisomers.

Table 1 HPLC Conditions Suggested for the Determination of PTC-Amino Acids (AAs): Literature Data

Practical conditions			Detection/quantitation			Remarks			Ref
column: cm x mm µm	product of	eluents: composition, pH, temperature	detector	RSD %; cc	analyte, mole	matrix/etc.	AAs not resolved	No: AA/ time:min	
25 4.6 5	Supelcosil LC 18	A: 0.14M Na.ac+0.5ml/l TEA, *pH 6.4* B:60% ACN in water, 38 ^{o}C	DAD	-	250 pM	fresh pig muscle, cured ham	Arg/Thr/Ala/Pro	21/50	[29]
25 4.6 5	Spherisorb[e]	A: 50mM Na.ac,+0.05%TEA+ 0.25%ACN+1 ppm EDTA *pH 6.8*, B: ACN/ H_2O=70/30; C:H_2O/ACN=3/7; 55 ^{o}C	UV 254	<5.0	-	AAs in wines	Ser/Asn	24/71	[32]
10 4.6 2	TSKgel[f]	A: 50 mM Na.ac *(pH 6.0)*/ACN=97/3; B:ACN/H_2O=6/4;	UV 254	-	-	-	Ala/Pro	17/4#	[34]
15 3.9 -	Picotag, W	A: 0.14M Na.ac+0.5ml/l TEA, *pH 6.4* B:60% ACN in water, 38 ^{o}C; (v/v)	UV 254	-	-	AAs of glycoproteins	Thr/Ala/Pro, Isol/Leu	18/12	[36]
15 4.6 5	Hypersil, Shandon	A: 0.05 M Na.ac, *pH 7.2*;B: A/ACN/Met=46/44/10 (amb)	UV,DAD APCIMS	<6.0	25–800 pM	hydr.study,tryptoph, cyst(e)ine,	Isol/Leu	19/32	[37 &]
30 3.9 -	Picotag, W	A: 140 mM Na.Ac, *(pH 6.35)*, cont. 0.05 % TEA; B: ACN/H_2O=6/4; 46°C	UV 254	<5.0	39-1250 pM	plasma AAs	Asn/Ser/Gly; Isol/Leu	26/88	[38]
12.5 4.6 5	Partisphere[a]	A: 50mM Na.ac, TEA 4 ml/l, *pH 6.4*; B:A/ACN/Metanol=50/40/10; 25 ^{o}C;	UV 254	-	>10 pM	AAs in mixed fish meals	Asp/Glu/Hyp, Arg/Thr, Isol/Leu	17/30	[40]
12.5 4 3	GROM-SIL 80 ODS	A: 50 mM Na.ac *(pH 6.5)* ; B: 50 mM Na.ac *(pH 6.5)*/ACN=50/50; 55°C	UV 254	-	-	-	complete resolution	18/8#	[43]
30 3.9 4	Nova-Pak[c]	A: 0.07 M Na.ac*(pH 6.5)*/ACN=975/25; B: ACN/Metanol/H_2O=45/15/40; 46°C	UV 254	<10	-	biological samples	Ser/Asn/Gly,	38/70	[44]
15 4.6 3	Econosphere[b]	A: 0.05 M amm.ac.,*pH 5.1*, /ACN-98/2 B: H_2O/ACN=4/6; 41°C	UV 254	<15.6	>1 pM	model study; plasma AAS	Asn/Gly, Isol/Leu	26/35	[45]
25 4.6 5	Spherisorb[e]	A:0.1Na.ac, 0.68 ml/lTEA, 0.2 mg/ml SPS B: H_2O, C=ACN, 35°C; *pH 5.27*	DAD	<7	>63 pM	AAs, in musts, in wine	Ser/Asn, GABA/Arg, PITC/Tyr	26/50	[49]
25 4.6 5	Alltima[d]	A: 0.1 M amm.ac., *pH 6.4*; 43°C B:ACN/Metanol/H_2O=46/10/44 ,*pH 6.4*	DAD	<3.66	0.02-3nM	protein, toxines	Ser/Asn, Arg/Thr/GABA/Ala	29/65	[50]
15 3.9 5	Picotag, W	A: Na.ac, *pH 6.4*, B:60% ACN (not detailed), 38°C	UV 254	<1.94	>1.25pM	AAs in kelp	Asp/Glu, Isol/Leu	17/18	[53]

25 4.6 5	Spherisorb[e]	A: 140 mM Na.Ac, cont.0.5 ml/ITEA, *(pH 6.2)* B: ACN/H_2O=6/4; 30 °C	UV 254	<3.98	8.1–13.3 pM	AAs in green bean hydr.	Thr/Ala/Pro	17/37	[56]
30 3.9 5	Picotag, W	A: 70 mM Na.Ac, cont.2.5% ACN, *(pH 6.5)* B: ACN/H_2O/Me=45/40/15; 46 °C	UV 254	<6. 8	20–1500 pM	human tissues	GABA/Citr, β-AiBA/Car, OH-Lys/PheAla	34/60	[61]
25 4.6 5	Hypersil, Shandon	A: 0.05 M Na.ac, *pH 7.2;* B: A/ACN/Met=46/44/10 (50 °C)	UV 254 + DAD	<4.0	15–1500 pM	basic elution study; apples	complete resolution	27/40	[62]

Indications: [a]= C18; Whatman; [b]=C18 Alltech; [c]=C18 column; [d]= C18 Alltech; SPS=sodium pentanesulfonate; [e]=C18 ODS-2; TEA=triethylamine; THF=tetrahydrofurane; [f]=Super ODS, TosoHaas product; #=without equilibration time; -=no data available; [37&]=37,39,42,47,48, 51,52,55]

Table 2 HPLC Conditions Suggested for the Determination of PTH-Amino Acids (AAs): Literature Data

Practical conditions			Detetion/ quantitation		Remarks			Ref
column: cm x mm µm	product of	eluents: composition, pH, temperature	de-tector	analyte, mole	aim	S: Splitting conditions; R: AAs not resolved:	No:AA/ time:min	
8 6.2 3	Zorbax (DuPont)	Isocratic, ACN/2mM phos. buffer (pH 3.2) =36/64, (35 °C)	UV 254	-	chromtgr. opt.study	S: -; R: Gly/ Cmcys, Asp/dehydroThr	20/25	[66]
12.5 4.0 5	LiChrocart Supersphere	A: 30mM Na.ac (pH 5,3)/ACN=5/1; B: 60% 2-propanol; 37°C	UV 254	10–100 pM	tryptic digest	S: Beckman 890C sequencer; R: complete resolution	17/17	[68]
25 4.6 5	Nucleosil 5–18	Isocratic, ACN/ H_2O=40/60 contg. 0,1% TEA; 23 °C	UV 254	2–20 pM	column opt.study	S: -; R: complete resolution	18/23	[69]
20 2.1 5	Hypersil, ODS, megabore	A: 50mM amm.phos., pH *3.0*; B: 80% ACN in water; 42°C	DAD 225 NMR, MS	250 pM	identification study	S: -; R: complete resolution	17/15	[71]
3 4.6 -	Non porous C18 silica	A: 10mM amm.phos., pH 4.5; B: 90% ACN in water;	UV 268	-	chromtgr. opt.study	S: -; R: complete resolution	6/1	[72]
25 4.6 5	Supelcosil-LC 18-DB	A: 0.33%TEA+9.5% Met= 10.5%ACN (v/v), pH 3.55; B:Met/IPA/ H_2O=60/20/20;	UV 254	-	chromtgr. opt.study	S: Millipore Prosequencer; R: complete resolution	20/23	[73]
15 4.6 - 5 4.6 -	Polymer grafted silica*	Isocratic, phosphate buffer, pH 6.0 Ionic strength= 0.1M	UV 280	-	column opt.study	S: -; R: complete resolution	6/20	[82]

Indications: as in Table 1, as well as: *=packed into stainless steel conventional columns

2. Sample preparation of the phenylthiocarbamoyl amino acids

2.1. Hydrolysis of protein

The analysis of free AAs in various biological, fruit/vegetable, food and feed matrices is an easier task of the everyday praxis, compared to the overwhelming part of jobs, concerning the separation, identification and quantitation of AAs that needs to be performed in protein hydrolyzates. Why? Because the method of hydrolysis is the limiting parameter also for the reliable preparation of the sample prior to the chromatographic process – a properly carried out hydrolysis is a prerequisite of a successful analysis. According to Pickering and Newton [95], in a paper entitled "Amino acid hydrolysis: old problems, new solutions", published twelve years ago, they stated that "With new instrumentation, amino acid hydrolysis is becoming a fast and automated process. But the problems of degradation, conversion, contamination and incomplete hydrolysis must be solved". Unfortunately, these problems still need to be solved. However, to overcome these difficulties, being intrinsic features of the hydrolysis process, as a result of ongoing efforts on the field, significant developments were achieved in the last fifteen years, i.e., (i) reduction of hydrolysis time by using microwave irradiation energy and by increasing temperature of hydrolyses, (ii) following gas phase hydrolysis instead of the liquid phase one, (iii) improvement in sensitivity and chromatographic resolution, in particular in the separation and quantification of the enantiomeric forms of amino acids, (iv) improvement in the cleavage of the hardly hydrolyzable peptide bonds (valine/valine, alanine/alanine isoleucine/alanine, etc.) and in the inhibition, or in the decrease to the minimum extent the decomposition of sensitive AAs (cysteine, cystine, methionine, tryptophan, tyrosine, serine, threonine).

2.1.1. Tryptophan

For the determination of tryptophan, hydrolysis by alkali [96,97,98], by hydrochloric acid+additives [39] have been proposed. Alkaline hydrolyses (performed with NaOH [96,98], $Ba(OH)_2$ [96,97] or LiOH [97]), being extremely laborius and providing the quantification of tryptophan only, with an 80–90% recovery of the theoretical value, can be regarded as the simplest but most costly solution of the problem. Sulphur containing organic acids (non-volatile compounds such as *p*-toluenesulphonic acid [99], methanesulphonic acid [13,100] and mercaptoethanesulphonic acid [101]) lead to a number of shortcomings in the subsequent derivatization and chromatographic separation of the acids in hydrolyzates. Promising procedures are the classical hydrochloride acid hydrolyses in the presence of additives (thioglycolic acid [102], phenol [103], B-mercaptoethanol [104], ethanedithiol [105], thioglycolic acid+triluoroacetic acid+indole [106] and tryptamine [39,107], providing the quantification of tryptophan simultaneously with all other amino acids; of the suggested additives, only the last two [39,106,107] proved to be satisfactory in vapour-phase hydrolyses.

Applying the combined additives [106], a >75% tryptophan yield from lysozyme, myoglobin and papain, monitored by conventional IEC, has been reported. Performing vapour-phase hydrolysis in the PITC derivatization, a >80% tryptophan recovery could be obtained, measured with a number of proteins and without any disturbing effect on the determination of the other amino acids [37,107]. The same results were achieved in vapour-phase hydrolysis accessory set (160°C), 110 psi, 65% power) [48].

2.1.2. Cystine/Cysteine

The determination of cystine/cystein as their PTC derivatives proved to be an extremely complex problem, and in many papers this determination is not considered [34]. In those cases where it is also necessary to know the amount of cyst(e)ine, mainly the classical performic acid oxidation is applied 108]. As a result of a 16-h reaction time, storing the tubes in an ice-bath at 5 °C in a refrigerator, from cystine and cysteine the same product, *i.e.*, cysteic acid, and from methionine methionine sulphone are formed. The common, but very time-consuming and tedious (involving multiple reaction steps) procedure, suffers from a number of limitations [95,108] . With complex protein matrices, the performic acid oxidation results in substantial destruction of histidine (25%) and tyrosine (87%). Therefore, if cyst(e)ine, histidine and tyrosine determinations are reqiured simultaneously, two separate hydrolyses need to be performed [95], with and without performic acid oxidation. Further possibilities for the determination total of cystine and cysteine, during acid hydrolysis of proteins in the presence of sodium azide [109], after reductive alkylation, in the form of the PTC derivatives of carboxymethylcysteine [19,10], pyridylethylcysteine [110] and sulphopropylcysteine [111] have been reported. Investigations based on the individual determination of all six possible racemates (L-, D- and DL-cystine and –cysteine), partly stoichiometric [34] partly compositional analysis performed by LC/API/MS measurements [52] demonstrated that (i) the interaction of all six racemates and PITC results in two main and two trace PTC derivatives and the ratios of the derivatives are characteristic of the initial compounds; (ii) the total molar responses of the two main PTC derivatives are the same before and after hydrolysis and are independent of the original racemate form of cyst(e)ines. .

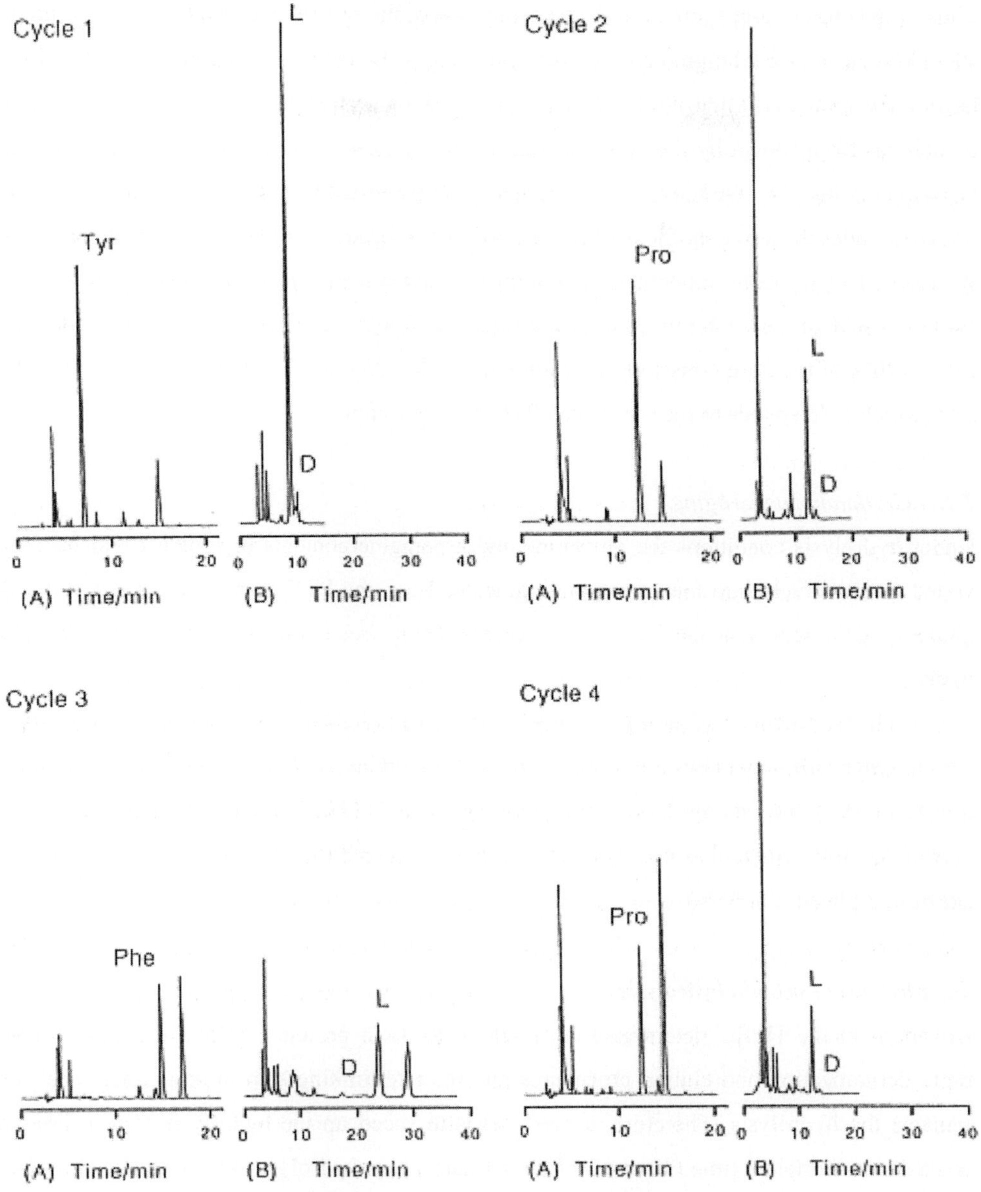

Figure 1 Amino acid sequencing and D/L-configuration determination of β-casomorphine-7 (5nM). (A) Sequence determination of the RP-HPLC (Column: ODS Wakopak WS-PTH, 250 x 4.6 mm, 5μm, Eluent: PTH-amino acid mobile phase, Wako, 1mL/min, Detection: UV, 269 nm). (B) The D/L-configuration determination on the chiral stationary HPLC. (Column: CHIRALCEL OJ-R/ODR, 150 x 4.6 mm, 5μm, Eluent: 2M sodium perchlorate in acetonitrile, 0.5mL/min, Detection: UV, 269 nm) Reproduced with permission from Ref. [81].

Thus, if the elution procedure permits the separation of the two main derivatives (their elution should be ensured by the gradient program generally to be before isoleucine), the determination of the total cyst(e)ine does not need any previous pretreament. The utility of this approach has been shown by the determination of the cyst(e)ine content in the hydrolyzates of various proteins [34]. On the basis of fragmentation patterns of the six PTC-cyst(e)ines [52], obtained under the soft conditions of the LC/API/MS procedure, it has been proved that in all six cases, PTC-cysteine monomers appear in two stereoisomer pairs: the two main stereoisomers, >80% of the total, proved to be the fragment at m/z = 255, while the two trace derivatives, <20% of total are consisting of fragment at m/z =255 and its oxidized version at m/z= 287 (which corresponds to the protonated PTC-cysteine sulphone.

2.1.3. Glutamine, asparagine

Under hydrolysis conditions the glutamine and asparagine contents of proteins will be converted quantitatively into the corresponding acids. Thus, the PTC derivatives of aspartic and glutamic acids represent the total of the corresponding free acids and acid amides, respectively.

The treatment of several proteins with [bis-(trifluoroacetoxy)iodo]benzene, a possible way to determine separately their asparagine and glutamine contents through the determination of their corresponding 2,3-diaminopropionic acids [112]. A rapid and qauntitative conversion (30 min, 100%) has been reported for glutamine, but the reaction of asparagine could not be completed (2h, 65%).

2.2. Methods of protein hydrolysis

Advances in the HPLC determination of amino acids in protein hydrolyzates now ensure rapid derivatization and elution procedures and the rate-limiting step in amino acid analysis remains the hydrolysis. Therefore, a need exists to speed up the hydrolysis step in protein analysis. A shortened time (4h at 160°C) for liquid phase hydrolyses was introduced by Gehrke et al. [113], and a useful vapour-phase technique (4h, 145°C) is offered by Waters (work station) [114]. Both of these possibilities provided the same values for amino acid components as the classical liquid-phase procedure (6 M HCL, 24 h, 110°C). Recently we demonstrated [39] that by means of the work station the same recovery can be obtained for all essential amino acids, also under milder conditions (145°C for 2 h and 125°C for 5h).

2.2.1. Microwave hydrolysis

From our detailed experience with the work station [37,39,42,48,51,55], we intended to start with probably the most advantageous microwave conditions proposed for the CEM hydrolysis system, i.e., with the shortest hydrolysis time, which means also with the highest pressure f0.827 MPA (120 p.s.i.g.)], as advised in the manual [115]. The power applied was 650 W throughout. At the beginning of our experiments with the CEM station, we attempted to work with the shortest possible time and at the highest pressure f0.827 MPa (120 p.s.i.g.)] However, of the four hydrolysis flasks supplied with the apparatus, only one was tight enough to withstand this pressure. The producer states [115] that it is not possible to operate the CEM station at maximum pressure, i.e., above 0.621 and 0.689 MPa (90 and 100 p.s.i.g.) for different hydrolysis times (under the latter conditions the CEM hydrolysis system works excellently).

2.2.2. Recovery study obtained from microwave hydrolyses

Lysozyme was used as a standard protein [37,39,42,48,51,55]. The results obtained were compared with the data provided by the work station applying vapour-phase hydrolysis for 4 h at 145 °C (Table 1, values in braces). The following conclusions can be drawn.

(1) General optimum conditions applicable to all amino acids apparently cannot be established and the laboratory routine will always be some sort of compromise.

(2) In accordance also with literature data [116,117], the yields of threonine, cyst(e)ine, isoleucine, valine and trypthophan were lower than the vlues of the reference methods under all conditions tested.

(3) Nearly the theoretical recovery could be obtained for serine, in order of listing, at 0.621 and 0.698 MPa (90 and 100 p.s.i.g.) for 10 and 5 min, similar to that reported at 180 °C for 4 min (pressure not given) [116]. After longer hydrolysis times (40, 40 and 20 min), even under mild conditions at 0.621, 0689 and 0.379 MPa (90, 100 and 55 p.s.i.g.), considerable losses of serine were observed, 15, 27 and 10% [39], respectively.

(4) The recovery of histidine is usualy high, being 175% at 180 °C after 4 min [116]. According to our data, the histidine recovery increase with increasing pressure and hydrolysis time at 0.621 MPA (90 p.s.i.g.) (55, 86,83 and 100%) and at 0.621 MPa (100 p.s.i.g.) (90, 96, 100 and 104%) respectively (Table 1).

(5) Tyrosine proved to be very sensitive to the hydrolysis conditions: (i) at 180 °C after 4 min only an 80% recovery was reported [10], whereas (ii) according to our results at 0.621 MPa (90 p.s.i.g.) the cleavage of its peptide bond needed at least 10 min.

(6) Regarding the expected amount of lysine, the cleavage of its peptide bond needs stronger conditions: at 0.621 MPa, (90 p.s.i.g.) after only 40 min and at 0.689 MPA (100 p.s.i.g.) after any hydrolysis time the amount of lysine proved to be satisfactory. The reported [10] 119% yield at 180C after 4 min is unlikely, to our knowledge.

(7) The recoveries of the rest of amino acids, are within acceptable experimental error.

2.2.3. Reproducibility of the microwave hydrolyses

Generally, the reproducibility, characterized by the relative standard deviation percentages (RSD %) (Table 3), is higher than in previous experiments. In the case of microwave hydrolyzed lysozyme, the highest error can be attributed mainly to incomplete hydrolysis [tests performed at 0.621 MPa 890 p.s.i.g.) for 5 min and 10 min and at 0.689 MPa (100 p.s.i.g.) for 5 min, and also for all values for phenylalanine]. Regarding the other data, the reproducibility of these measurements proved to be <8.0% (RSD).

2.3. Derivatization reaction

The derivatization reaction consists of two steps. The first, known as the re-drying or coupling step, is performed under alkaline conditions, in order to eliminate the last traces of water and to ensure that the amino groups of acids are in the "free" condition. The mainly used "coupling buffers" [5,9,34,37,39,42,48,62] contain ethanol-water-triethylamine (2:2:1 (v/v) or according to other papers [6,7,30,21] acetonitrile-pyridine-triethylamine-water (10:5:2:3, v/v) . The second step, the real derivatization with phenyl isothiocyanate (PITC), is a very simple, fast and quantitative reaction requiring 5 min at ambient temperature; to be on the safe side, the suggested reaction time is 20 min. The composition of the mainly used reagent, following the coupling step with ethanol-water-triethylamine, is ethanol-triethylamine-water-PITC (7:1:1:1, v/v), and in other proposals [6,7,21,30,34,37,39,42,48,62] PITC is added to the above pyridine-containing coupling agent.

The most important and also the most time consuming step of the derivatization procedure is the removal of the excess of PITC. For this purpose, according to the earlier work [3–6], a high vacuum is needed in supplying the end pressure of <60mTorr (1 Torr=133.322 Pa). This special requirement can be comfortably achieved by applying the Waters Pico Tag work station. This apparatus serves as an evaporator in PTC derivatization and as an oven for protein hydrolysis, simultaneously [113].

Table 3 Recovery of Amino Acids from Microwave Hydrolyzate of Lysozyme Obtained Under Different Conditions

Conditions			Amino Acid (%)[a]																	
Pressure (MPa) (p.s.i.g.)	Time (min)	Reproducibility	Aspartic acid	Glutamin acid	Serine	Glycine	Histidine	Threonine	Alanine	Proline	Arginine	Tyrosine	Valine	Methionine	Cyst(e)ine	Isoleucine	Leucine	Phenylalanine	Tryotophan	Lysine
Values, from the⇒ Work station			(20) {18.0}	(5) {5.0}	(18) {6.6}	(12) {10.1}	(1) {1.0}	(7) {5.2}	(12) {12.6}	(2) {2.0}	(11) {9.0}	(3) {2.9}	(6) {4.9}	(2) {1.9}	(8) {8.9}	(6) {4.7}	(8) {7.5}	(3) {2.5}	(6) {5.1}	(6) {6.6]
0.621	5	A	79	84	84	92	55	53	73	73	60	69	39	83	53	70	82	43	53	66
		B	1.5	7.6	1.8	2.9	5.8	3.7	4.3	8.4	7.2	7.2	9.2	4.8	8.5	3.9	1.6	11.4	4.7	5.9
0.621	10	A	95	82	103	105	86	88	104	105	93	95	70	98	66	86	97	68	83	87
		B	0.8	6.7	1.7	2.9	6.6	2.3	1.0	7.4	1.8	1.2	5.0	10.5	16.4	3.0	1.4	14.4	12.9	16
0.621	20	A	108	75	96	110	83	93	99	98	98	100	68	102	90	92	94	63	92	86
		B	0.5	6.7	1.7	1.7	7.0	2.3	1.9	3.3	1.3	6.0	8.9	2.5	7.0	2.0	2.5	2.0	3.3	1.2
0.621	40	A	80	105	85	105	100	78	97	96	101	94	82	96	81	97	100	76	79	100
		B	3.2	4.7	2.8	1.3	1.1	1.1	0.4	1.1	0.6	4.4	7.5	7.3	7.2	1.5	0.3	6.0	6.1	3.4
0.689	5	A	84	85	106	87	94	86	85	85	86	115	60	105	71	64	82	95	85	100
		B	16.6	7.7	14.0	4.6	1.5	14.2	2.4	6.9	5.9	6.9	0.9	5.4	21.4	0.2	1.5	5.8	14.4	4.8
0.689	10	A	86	90	95	98	96	77	106	100	85	97	75	92	73	78	89	111	80	101
		B	0.45	2.2	2.2	0.7	4.2	1.5	0.8	2.0	1.8	3.9	4.2	5.8	4.0	3.0	1.5	5.3	7.3	2.2
0.689	20	A	92	95	84	98	100	77	108	100	89	93	88	93	95	85	105	110	67	99
		B	2.8	5.9	4.0	1.4	2.0	0.54	4.8	3.7	1.4	6.3	5.4	7.2	5.8	5.7	2.4	4.2	8.0	3.2
0.689	40	A	95	101	73	101	104	76	108	100	96	95	98	92	88	94	104	112	63	102
		B	4.9	3.6	3.1	4.2	4.2	2.2	4.0	7.7	2.9	4.0	4.6	7.5	5.2	4.8	3.2	3.2	7.4	4.6
Literature values																				
n.m.	4	10	92	88	102	94	175	100	100	n.m..	91	80	98	105	92	104	89	104	n.m.	119
0.379 (55)	20	17	100	102	90	93	100	90	100	91	93	100	87	95	93	93	99	97	n.m.	87

[a] Expressed as percentages of data obtained from hydrolyses performed in the work station values, in parentheses, literature values given in ref. [113]; values in brackes, values of lysozyme components expressed in pmol: these serve as the basis of comparison. A=Averages of four separate hydrolyses performed in three vials; B=relative standard deviation percentages (RSD%); n.m.= not measured.

Later, in order to eliminate the excess of PITC [15,20,114] and those of its substitutes (1-naphythyl isothiocyanate [84] and 4-nitroisothiocyanate [86]), extraction procedures were recommended instead of the high-vacuum conditions, with diethyl ether [86] as extractant. The complete automation of the PTC derivatization procedure with the Varian Model 9095 autosampler also involves an extraction step instead of the high vacuum removal of PITC.

2.4. Storage/stability of derivatives

With regard to the stability if derivatized solid samples, in our experience they can be stored in vacuum-sealed flasks, or even in flasks closed with a ground-glass stopper, in a freezer for unlimited time [52].

Regarding the stability of dissolved PTC derivatives, in general, they are considered to be stable compounds. Concerning the exact "lifetime" of the individual PTC amino acids there are contradictory experiences [4,9,8,19,30,37]: the statement that no loss could be detected after 3 days (in the cold) proved to be incorrect [9,19,30]. On keeping the dissolved PTC amino acids at 5–8 °C their stability lasted 16 h [9], whereas others [19,30] reported a ca. 5 % loss after storage for 48 h at 4 °C. As a result of an exhaustive study in the author's laboratory [37], in which 21 PTC amino acids, dissolved in 0.05 M sodium acetate, were kept in a refrigerator at 4°C and tested immediately after preparation and after 0.5, 7, 18, 26, 44, 73, 102 and 190 h, revealed that the maximum time during which the dissolved derivatives retained their initial amounts was <7h. After 18 h losses of serine (4%), histidine (7%), alanine (3%) could be determined. After 190 h the maximum loss for hydroxyproline was demonstrated.

3. Chromatographic conditions

3.1. Columns, flow rate and temperature optimization study

The separation and identification possibilities of 27 PTC-amino acids (AAs), with particular attention to those present in free forms (vegetables/apples), were optimized, in the frame of an exhaustive study, on seven RP columns, subsequently to developing an optimum gradient program (4. Experimental section: 40 min, including equilibration: Table 5) [62].

This optimum gradient program (Table 5) was the basis of the separation of 27 AAs which intended to be further optimized varying the length and particle size of stationary phases, temperature and flow rate of eluents, in accordance/competition with each other, taking into account the responses of the derivatives (the faster the elution the smaller the response) and the column's life time (<55 ^{o}C temperature and <3500 psi column pressure were preferred).

Four Hypersil ODS bonded phase, such as Hypersil 5 μm: 130+(+20 guard) x 4.6 mm, 150+(+20 guard) x 4.6 mm and 200+(+20 guard) x 4.6 mm, Hypersil 3 μm: 150+(+20 guard) x 4.6 mm, two Nucleosil (3 and 5 μm: 150(+20 guard) x 4.6 mm and one Gromsil 3 μm: 150(+10 guard) x 4.0 mm columns were tested (in order of listing T1–T7 columns: Figure 2). A UV range photodiode array (PDA) detection was employed.

In the possession of the optimum gradient program elaborated on T-4 and T-6 columns at 30 ^{o}C, a unified optimization strategy was followed on all seven test columns (T-1–T-7). First, at 30 ^{o}C temperature the flow rates were varied until the best separation could be achieved. This was close to the highest flow rate, on all seven test columns, that could be used without exceeding the 3500 psi limit: it means this limit would be limited for T-1–T-3 and T-5 columns (5 μm) with a flow rate of 2.1 ml/min, while for T-4, T-6 and T-7 columns (3 μm) with 1.3 ml/min. The next task to complete the separation of the 27 AAs was to increase the temperature of elutions (35, 40, 45, 50 and 55 ^{o}C) performing 1.3 ml/min (3 μm) and 2.1 ml/min (5 μm) flow rates, respectively.

Evaluating elution profiles shown at Figure 2, it can be stated that:

(i) acceptable separations could be achieved, performing the optimum gradient program, by all columns tested, in most cases at 45–50 ^{o}C (with the only exception of column T-5: 30 ^{o}C), applying the highest flow rate possible.

(ii) Retention orders of selected PTC-derivatives were changed by increasing the temperature of elutions, however these changes proved to be also independent of columns (T-1–T-7), as follows:

- The retention times of PTC-amines (asparagine, glutamine) by increasing the temperature, decreased relatively faster, comparing them to their neighbours. Thus, this feature of PTC-amines limited the upper temperature level as high as 50 ^{o}C (compare the resolution of asparagine/glycine on T-1–T-4, T-6, T-7 columns, at ≥45 ^{o}C to that on T-5 one, at 30°C).

- The decreased retention time of PTC-glutamine resulted in the retention order's change of β-alanine/homoserine/glutamine (30 ^{o}C) for β-alanine/glutamine/homoserine (50 ^{o}C).

- The retention time of ACPCA was succesively increased, it means that the retention order of GABA/histidine/ACPCA/threonine/alanine (30°C) was altered for GABA/histidine/threonine /alanine/ACPCA (50°C).

- Similarly, the retention orders of proline/arginine and ornithine/tryptophan were also changed by increasing the temperature from 30 ^{o}C to 50°C.

Optimum conditions, at 2.1 ml/min, at 50 ^{o}C, with 40 min run time, including equilibration, have been obtained with the Hypersil, 150(+20 guard) x 4 mm column.

3.2. Quantitation properties/reproducibility study

Response factors of AAs proved to be independent of column length, type of stationary phase, elution temperatures and flow rates (Table 4). As the most advantageous column, in respect of resolution characteristics (Table 4, values in parentheses), cost and column life time (5 μm), the T-2 column was selected (values in Table 4, third vertical column, Figure 1, T-2 chromatogram).

Comparing response values, obtained on various columns, under different elution conditions (Table 4, last vertical column: RSD percentages = < 7.6%) furnished almost as good results as those obtained on the same column (T-2) varying the amount of AAs between 15 and 1500 pM (Table 5, RSD = <4.8%). The exception with its 9.2 RSD% proved to be cyst(e)ine [37].

The utility of the protocol was shown in the quantitation of the free AA content of three apple varieties 62]. Optimum conditions, at 2.1 ml/min, at 50 ^{o}C, with 40 min run time, including equilibration, have been obtained with the Hypersil, 150(+20 guard) x 4 mm column.

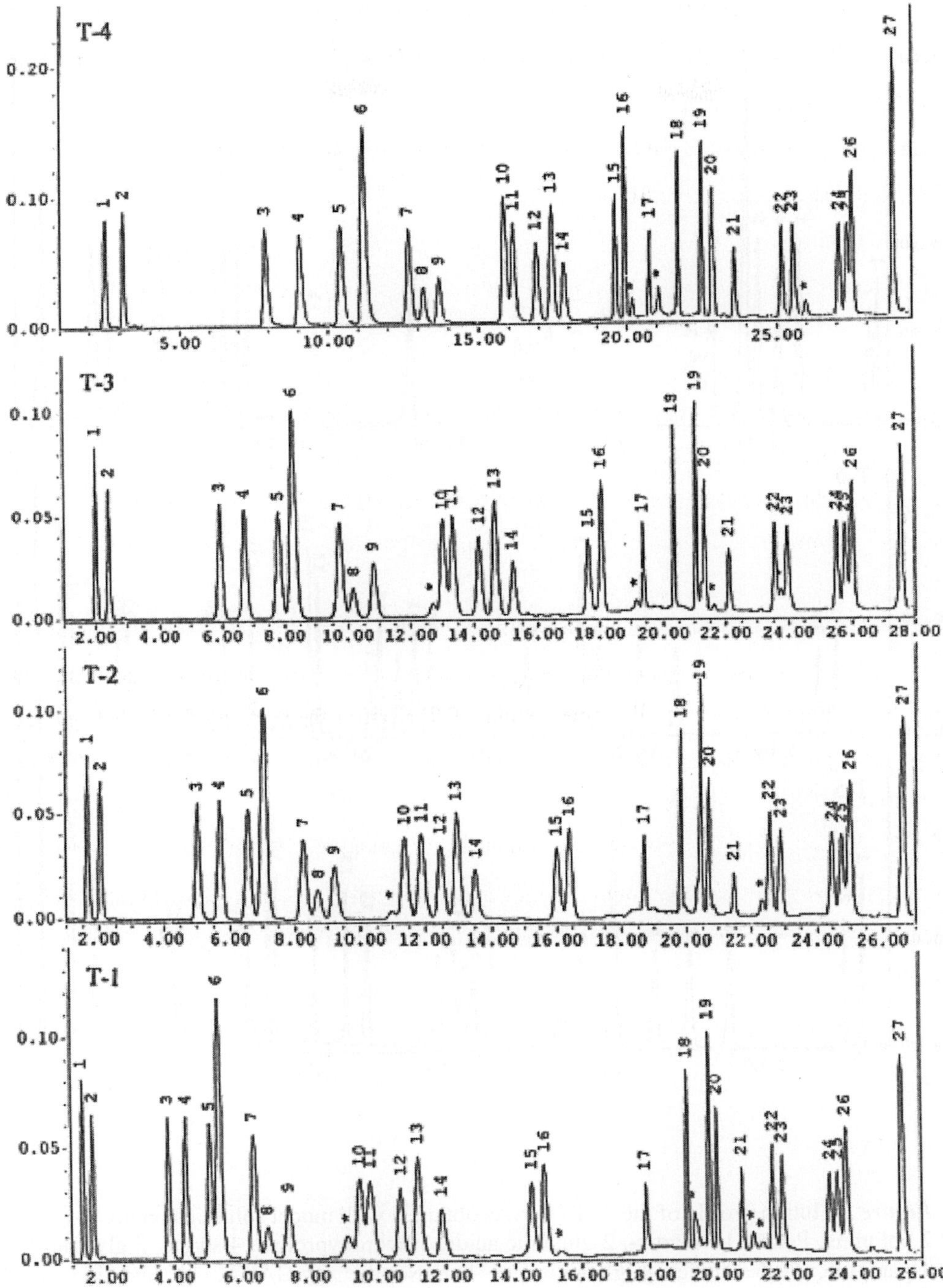
T-4
T-3
T-2
T-1
0.20
0.10
0.05
0.00
5.00
10.00
15.00
20.00
25.00
2.00
4.00
6.00
8.00
10.00
12.00
14.00
16.00
18.00
20.00
22.00
24.00
26.00
28.00

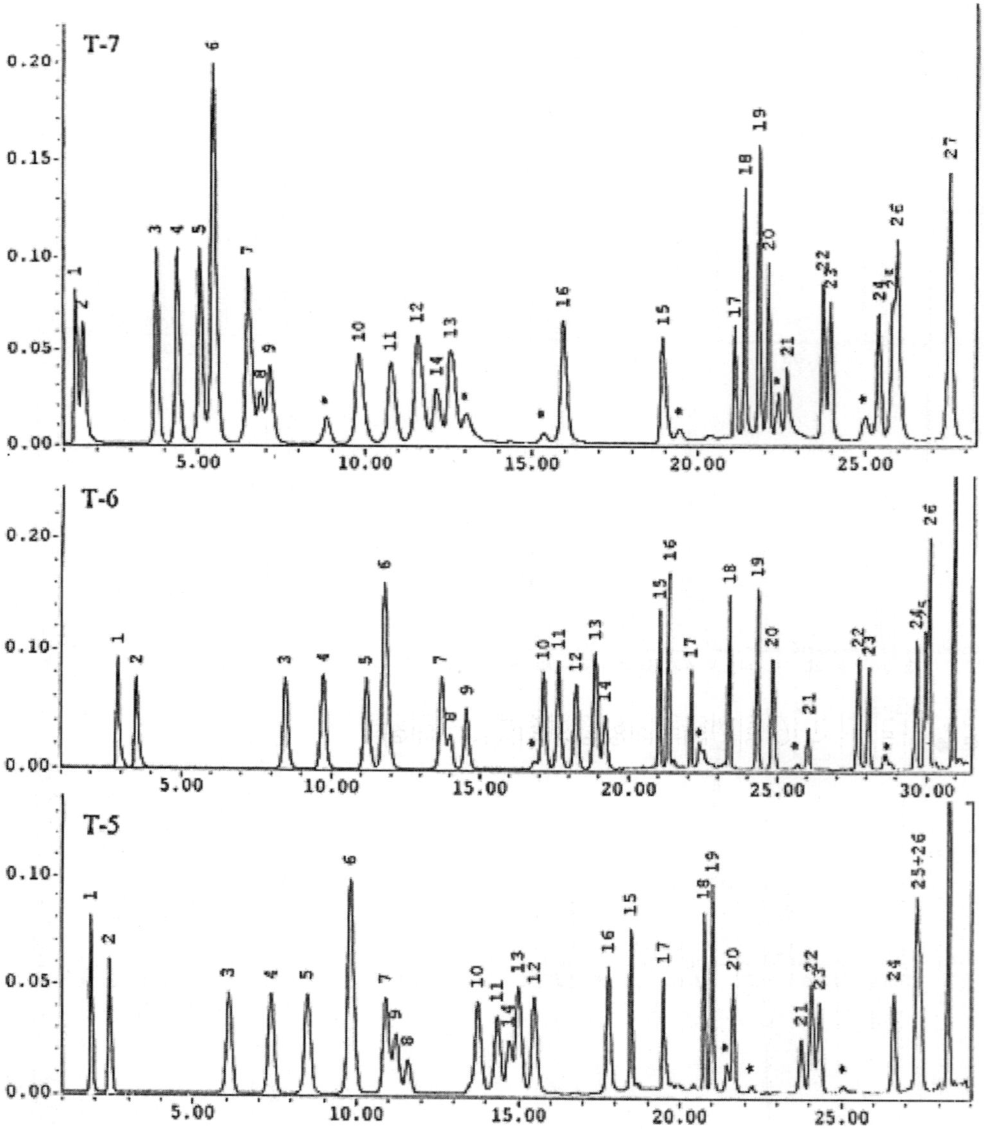

Figure 2 Elution profile of the 27 PTC-AAs obtained with model solutions on the T-1–T-7 columns; Peaks: 1 aspartic-, 2 glutamic acids, 3 hydroxyproline, 4 serine, 5 glycine, 6 asparagine, 7 β-alanine, 8 glutamine, 9 homoserine, 10 GABA, 11 histidine, 12 threonine, 13 alanine, 14 ACPCA, 15 arginine, 16 proline, 17 homoarginine, 18 tyrosine, 19 valine, 20 methionine, 21 cyst(e)ine, 22 isoleucine, 23 n-leucine, 24 phenylalanine, 25 tryptophan, 26 ornithine, 27 lysine; *=system peak. (Reproduced with permission from Ref [62]

Table 4 Reproducibility and Resolution Features in the Quantitation of PTC-AAs Obtained on the Seven Test Columns, Performing Optimum Conditions: at Various Elution Temperatures and Elution Flow Rates*

AAs ⇓		Arbitrary units/ 1pM AA, normalized to 1ml/min flow rate, Resolutions in parentheses								
Peak	Name	Column (T-1–T-7), *ml/min, (^{O}C)							Average#	RSD %
		T-1,2.3(50)	T-2, 2.1, (50)	T-3, 2.1, (50)	T-4,1.5,(45-50)	T-5, 2.3, (30)	T6,1.3,(50)	T-7, 1.2, (45)		
1	Aspartic acid	611	650	561	625	685	647	518	644	4.0
2	Glutamic acid	656 *(2.1)*	675 *(2.4)*	572 *(2.8)*	660 *(3.6)*	710 *(3.5)*	671 *(2.9)*	680 *(0.95)*	661	6.5
3	Hyroxyproline	683	687	704	745	744	672	716	707	4.1
4	Serine	673 *(2.4)*	692 *(2.6)*	665 *(2.9)*	702 *(4.0)*	751 *(4.0)*	679 *(4.0)*	697 *(2.7)*	694	4.1
5	Glycine	661 *(3.0)*	678 *(3.0)*	689 *(3.6)*	730 *(4.5)*	748 *(3.1)*	681 *(4.4)*	728 *(2.8)*	702	4.7
6	Asparagine	703 *(1.0)*	720 *(1.4)*	695 *(1.4)*	737 *(2.3)*	776 *(3.7)*	703 *(1.7)*	726 *(1.4)*	723	3.8
7	β-Alanine	821 *(3.6)*	580 *(3.4)*	676 *(4.4)*	693 *(4.9)*	766 *(3.0)*	665 *(5.5)*	785 *(3.3)*	717	7.6
8	Glutamine	666 *(1.3)*	678 *(1.3)*	655 *(1.3)*	698 *(1.7)*	738	667*(~0.8)*	716 *(~1.1)*	688	4.4
9	Homoserine	497 *(1.7)*	497 *(1.3)*	461 *(1.9)*	359 *(1.9)*	536	491 *(1.7)*	438 *(~1.1)*	487	6.9
10	GABA	580	615	626	698	738	597	607	621	6.6
11	Histidine	658 *(0.72)*	743 *(1.4)*	699 *(0.95)*	778 *(1.1)*	653 *(1.7)*	724 *(1.9)*	569 *(2.2)*	696	5.7
12	Threonine	679 *(2.2)*	706 *(1.7)*	672 *(2.8)*	711 *(2.9)*	924	712 *(2.2)*	876 *(1.8)*	696	2.7
13	Alanine	648 *(1.5)*	671 *(1.4)*	637 *(1.7)*	674 *(1.9)*	717	652 *(2.3)*	624	661	4.6
14	ACPCA	486 *(2.1)*	472 *(1.7)*	458 (2.0)	524 *(1.6)*	528	459 *(1.2)*	466 *(1.8)*	485	6.2
15	Arginine	681	692	630	689	753	669	679	685	5.3
16	Proline	674 *(1.3)*	689 *(1.3)*	737 *(1.8)*	770 *(2.2)*	763 *(3.7)*	698 *(2.32)*	689 *(9.7)*	717	5.4
17	Homoarginine	689	723	702	729	752	709	716	717	2.8
18	Tyrosine	768	789	710	802	790	803	798	780	4.2
19	Valine	626	738	641	659	787	699	712	679	6.5
20	Methionine	573 *(1.9)*	602 *(1.8)*	538 *(2.1)*	621 *(2.1)*	662 *(3.7)*	580 *(3.2)*	596 *(2.0)*	585	4.9
22	Isoleucine	621	750	607	735	805	696	689	698	7.2
23	Leucine	673 *(1.6)*	693 *(1.5)*	606 *(1.8)*	725 *(2.0)*	806 *(1.1)*	683 *(2.1)*	666 *(1.2)*	688	3.3
24	Phenylalanine	672	736	667	770	765	753	745	730	5.9
25	Tryptophan	588 *(0.98)*	608 (1.1)	637 1.2	599 *(0.7)*	699 *(2.6)*	621 (1.9)	587 *(~2.0)*	620	6.3
26	Ornithine	1286 *(1.0)*	1255 *(0.95)*	1304 *(0.98)*	1328 *(0.7)*	1363 *(-)*	1303 *(0.95)*	1471 *(~0.6)*	1307	2.8
27	Lysine	1081	1119	1094	1168	1191	1140	1103	1128	3.6

Indications: # = averages of responses obtained from the seven test columns, without underlined data; Resolutions in parentheses: given by the software, or estimated (~); Resolutions accordingly to changed retention orders: on T-5 column, β-alanine/homoserine=*(~0.8)*; homoserine/glutamine=*(~1.1)*; histidine/ACPCA=*(~1.1)*; ACPCA/alanine=*(~0.9)*; alanine/threonine =*(1.4)*; on T-7 column, threonine/ACPCA=*(1.2)*; ACPCA/alanine =*(0.93)*

4. Experimental

4.1. Derivatization of AAs with PITC

Standards of free AAs, one by one, (such as aspartic acid, glutamic acid, hydroxyproline, serine, asparagine, glycine, glutamine, homoserine, histidine, GABA, threonine, ACP-CA, alanine, proline, arginine, homoarginine, valine, tyrosine, methionine, cyst(e)ine, iso- and n-leucine, phenylalanine, tryptophan and lysine), as well as in a mixture (containing ~1.5×10^{-8} - 1.5×10^{-6} mole of each amino acids, and the corresponding 5–100 fold excess of asparagine, expected in apples), were placed in the 50 x 6 mm I.D. tubes and dried under vacuum.

Apple pulps, 5.0000 g, were filtered on GF/F, Whatman, glass microfibre filter, (particle retention from liquid: 0.7 μm), into a 5 ml volumetric flask acidified with 0.2 ml distilled hydrochloric acid and completed by washing the residue with distilled water up to 5 ml. 1 and 2 ml aliquots of these solutions were subjected to cation exchange clean-up and treated according to our earlier experiences [119]. Samples were dried, by adding 20 μl of ethanol /water/TEA (2:2:1) to each tube. Thereafter to each redried sample 20 μl of derivatization reagent [ethanol/TEA/water/PITC] (7:1:1:1) were added and Vortex mixed. They were prepared according to the Waters Picotag™ Work Station manual.

The derivatization standards were dissolved, in order of listing, in 50 μl acetonitrile, 50 μl water and 900 μl 0.05 M sodium acetate solution (pH 7.2), final volume 1.00 ml. The stock solution of apples were prepared as described above, with the only exception that their solutions were completed by 150 μl 0.05 M sodium acetate solution (pH 7.2), final volume 250 μl. Hence 10 μl of standards containing 15 – 1500 pmole of each amino acid and the corresponding excess of asparagine, as well as 20 μl of the apple stock solution were injected.

4.2. Chromatography

The system was a Waters HPLC instrument (Waters Pharmaceutical Division, Milford, MA, U.S.A.), consisted of Waters^{TM} 996 PDA Detector, Waters^{TM} 600 Controller quaternary pump with thermostattable column area, Waters^{TM} 717 Autosampler, operating with the Millennium Software (version 2010, 1992–95, validated by ISO 9002). Out of the above deateiled T-1–T-7 test columns T-2 was used (150+(+20 guard) x 4.6 mm Hypersil ODS bonded phase, 5μm, Supplier (BST, Budapest, Hungary);

Elutions were followed at an absorbance setting of 254 nm; spectra were taken between 205–400 nm, with an optical resolution of 4.8 nm.

The eluent systems consisted of two components: (A) eluent was 0.0500 M sodium acetate of various pH (6.0, 7.2 and 7.8, optimum pH 7.2), while (B) eluent was prepared with the corresponding pH consisting of 0.100 M sodium acetate/acetonitrile/metanol (46:44:10) (mixed in volume ratios and titrated with glacial acetic acid or 50 % sodium hydroxide to pH 6.0, 7.2 and 7.8). Different gradient programs were followed (given in details with the corresponding sections), the optimum version is shown in Table 4 (further on: optimum gradient program).

Table 5 Optimum gradient program

Step	Time: min	% A	% B
1	0.00	100	0.0
2	6.00	99.0	1.0
3	14.00	91.0	9.0
4	17.00	65.0	35.0
5	20.00	65.0	35.0
6	24.00	55.0	45.0
7	26.00	0.0	100.0
8	29.00	0.0	100.0
9	31.00	100.0	0.0
10	40.00	100.0	0.0

Derivatization procedures were performed at ambient temperature, while elutions were carried out between 30 and 55 ^{o}C.

References

[1] I. Molnár-Perl, (*Review*), J. Chromatogr. 913 (2001) 283–302.

[2] D.R. Koop, E.T. Morgan, G.E. Tarr, M.J. Coon, J. Biol. Chem. 257 (1982) 8472–8480.

[3] T.H. Maugh, II, Science 225 (1984) 42.

[4] B.A. Bidlingmeyer, S.A. Cohen, T.L. Tarvin, J. Chromatogr. 336 (1984) 93–104.

[5] S.A. Cohen, T.L. Tarvin, B.A. Bidlingmeyer, Am. Lab. 8 (1984) 49–51.

[6] R.L. Heinrikson, S.C. Meredith, Anal. Biochem. 136 (1984) 65–74.

[7] R.F. Ebert, Anal. Biochem. 154 (1986) 431–435.

[8] B.A. Bidlingmeyer, T.L. Tarvin, S.A. Cohen, in K.A. Walsh (Editor), Methods in Protein Sequence Analysis, Humana Press (1987). 229–245.

[9] P.S.L. Janssen, J.W. van Nispen, P.A.T.A. Melgers, H.W.M. van der Bogaart, R.L.A.E. Hamelinck, B.C. Goverde, Chromatographia 22 (1986) 345–350.

[10] P.S.L. Janssen, J.W. van Nispen, P.A.T.A. Melgers, H.W.M.van der Bogaart, G.W.M. van Aalst, B.C. Goverde, Chromatographia 22 (1986) 351–357.

[11] D. Lanneluc-Sanson, C.T. Phan, R.L. Granger, Anal. Biochem. 155 (1986) 322–327.

[12] B.A. Bidlingmeyer, S.A. Cohen, T.L. Tarvin, B. Frost, J. Assoc. Off. Anal. Chem. 70 (1987) 241–243.

[13] S.A. Cohen, B.A. Bidlingmeyer, T.L. Tarvin, Nature 320 (1986) 769–770.

[14] M.M.T. O'Hare, O. Tortora, U. Gether, H.V. Nielsen, T.W. Schwartz, J. Chromatogr. 389 (1987) 379–388.

[15] J.L. Tedesco, R. Schafer, J. Chromatogr. 403 (1987) 299–306.

[16] L .Robitaille and L.J. Hoffer, Can. J. Physiol. Pharmacol. 66 (1988) 613–617.

[17] H.P.J. Bennett, S. Solomon, J. Chromatogr., 359 (1986) 221–230.

[18] S.A. Cohen, T.L. Tarvin, J.J. B.A. Bidlingmeyer, (In modern methods in protein chem. istry; J.J. L'Italien Ed.) (1988) 207–213.

[19] S.A. Cohen, D.J. Strydom, Anal. Biochem.174 (1988) 1–16.

[20] A.S. Inglis, N.A. Bartone, J.R. Finlayson, J. Biochem. Biophys. Methods 15 (1988) 249–254.

[21] R. Mora, K.D. Berndt, H. Tsai S.C. Meredith, Anal. Biochem. 172 (1988) 368–376.

[22] B.C. Pramanik, C.R. Moomaw, C.T. Evans, S.A. Cohen, C.A. Slaughter, Anal. Biochem. 176 (1989) 269–277.

[23] R.M. Marcé, M. Callul, J. Guasch, F. Borrul, J. Enol. Vitic. 40 (1989) 194–198.

[24] M. Raghavan, C.K. Smith, C.E. Shutt, Anal. Biochem. 178 (1989) 194–197.

[25] S. Gunawan, N.Y. Walton. D.M. Treiman, J. Chromatogr. 503 (1990) 177–187.

[26] I. Molnár-Perl, J. Chromatogr. 529 (1990) 258–262.

[27] R.A. Sherwood, A.C. Titheradge, D.A. Richards, J. Chromatogr. 528. (1990) 293–303.

[28] E. Tarcsa, L. Fésüs, Anal. Biochem. 186 (1990) 135–140.

[29] M.C. Aristoy, F. Toldrá, J. Agric. Food Chem. 39 (1991) 1792–1795.

[30] A. Guitart, P.H. Orte, J. Cacho, Analyst 116 (1991) 399–403.

[31] D. Fouques, J. Landry, Analyst 116 (1991) 529–531.

[32] M. Calull, J. Fábregas, R.M. Marcé, F. Borrul, Chromatographia 31 (1991) 272–276.

[33] V. Fierabracci, P. Masiello, M. Novelli, E. Bergamini, J. Chromatogr. 570 (1991) 285–291.

[34] Patent Application No. 157, Reversed-Phase Chromatography, TosoHaas.

[35] M.H.M.N. Senden, A.J.G.M. Van der Meer, J. Limborgh, H.TH. Wolterbeek, Plant and Soil 142 (1992) 81–89.

[36] D.E.H. Palladino, R.M. House, K.A. Cohen, J. Chromatogr. 599 (1992) 3–11.

[37] M. Morvai, V. Fábián, I. Molnár-Perl, J. Chromatogr. 600 (1992) 87–92.

[38] A.S. Feste, J. Chromatogr. 574 (1992) 23–34.

[39] I. Molnár-Perl, M. Khalifa, Chromatographia 36 (1993) 43–46.

[40] E.Y. Suzuki, R.J. Early, J. Chromatogr. A 657 (1993) 204–207.

[41] S.R. Hagen, J. Chromatogr. 632 (1993) 63–68.

[42] I. Molnár-Perl, M. Morvai, Chromatographia 34 (1992) 132–136.

[43] Grom Manufacturer Application, Amino Acid Analysis, Grom Analytik +HPLC GmbG, 50 Copyright 1994.

[44] S.R. Hagen, J. Augustin, E. Grings, P. Tassinari, Food Chem. 46 (1993) 319–323.

[45] M. Hariharan, S. Naga, T. Van Noord, J. Chromatogr. 62 (1993) 15–22.

[46] D.J. Strydom, S.A. Cohen, Anal. Biochem. 222 (1994) 19–28.

[47] I. Molnár-Perl, J. Chromatogr. A 661 (1994) 43–50.

[48] E. Tatár, M. Khalifa, Gy. Záray, I. Molnár-Perl, J. Chromatogr. A 672 (1994) 109–115.

[49] M.P. Deu, S. Buxaderas, J. Chromatogr. A 685 (1994) 21–30.

[50] J.K. Khan, Y.H. Kuo, N. Kebede, F. Lambein, J. Chromatogr. A 687 (1994) 113–119.

[51] I. Molnár-Perl, M. Khalifa, LG-GC Int. 7 (1994) 395–398.

[52] K. Schmeer, M. Khalifa, J. Császár, Gy. Farkas, E. Bayer, I. Molnár-Perl, J. Chromatogr. A 691 (1995) 285–299.

[53] S.F. Shang, H. Wang, Chromatographia 43 (1996) 309–312.

[54] A. Guitart, P.H. Orte, J. Cacho, Quim. Analitica 15 (1996) 217–223.

[55] M. Khalifa, I. Molnár-Perl, LG-GC Int. 9 (1996) 143–157.

[56] M.J. G. Casrto, J.L. Hernández, J.S. Lozano, M.J.O. Concha, J. Chromatogr. 35 (1997) 181–185.

[57] A. Guitart, P.H. Orte, J. Cacho, Anal. Chem. 36 (1997) 43–47.

[58] P.H. Ortez, A. Guitart, J. Cacho, Am. J. Enol. Vitic. 48 (1997)

[59] T. Iida, H. Matsunaga, T. Fukushima, T. Santa, H. Homma, K. Imai, Anal. Chem. 69 (1997) 4463–4468.

[60] M. Stark, Y. Wang, O. Danielsson, H. Jörnvall, J. Johansson, Anal. Biochem. 265 (1998) 97–102.

[61] A. Battaglia, A. Bertoluzza, F. Calbucci, V. Eusebi, P. Giorgianni, J. Chromatogr. B 730 (1999) 81–93.

[62] A. Vasanits, I. Molnár-Perl, J. Chromatogr. A 832 (1999) 109–122.

[63] L. Campanelle, G. Crescentini, P. Avino, J. Chromatogr. A 833 (1999) 137–145.

[64] S.Y. Tham, S. Agatonovic-Kustrin. J. Pharm. Biom. Anal. 28 (2002) 581–590.

[65] Y. Dale, V. Mackey, R. Mushi, A. Nyanda, M. Maleque, J. Ike, J. Chromatogr. B 788 (2003) 1–8.

[66] R.W. Stout, J.J. De Stafano, J. Chromatogr. 261 (1983) 189–212.

[67] C. Fonk, S. Frutiger, G.J. Hughes, J. Chromatogr. 370 (1986) 339–343.

[68] J. L. Glajch, J.J. Kirkland, J. Chromatogr. Sci. 25 (1987) 4–11.

[69] K. Hayakawa, J. Oizumi, J. Chromatogr. 487 (1987) 161–166.

[70] T.A. Berger, J.F. Deye, J. Chromatogr. Sci. 27 (1989)

[71] Z. Huang, C.S. Ough, J. Agric. Food Chem. 39 (1991) 2218–2222.

[72] H. Itoh, T. Kinoshita, N. Nimura, J. Liq. Chromatogr. 16 (1993) 809–823.

[73] D.J. Strydom, J. Chromatogr. A 662 (1994) 227–233.

[74] N. Chauvaux, W.V. Dongen, E.L. Esmans, H.AV. Onckelen, J. Chromatogr. A 775 (1997) 143–150.

[75] J. Zhou, S. Hefta, T.D. Lee, J. Am. Soc. Mass Spectrom. 8 (1997) 1165–1174.

[76] T. Yalcin, W. Gabryelski, L. Li, J. Mass Spectrom. 33 (1998) 543–553.

[77] T. Iida, T. Santa, A. Toriba, K. Imai, Analyst 123 (1998) 2829–2834.

[78] T. Iida, H. Matsunaga, T. Santa, T. Fukushima, H. Homma, K. Imai, J. Chromatogr. A 813 (1998) 267–275.

[79] S.G. Summerfield, H. Steen, M. O'Malley, S.J. Gaskell, Int. J. Mass Spectrom. 188 (1999) 95–103.

[80] G.van der Rest, F. He, M.R. Emett, A.G. Marshall, J. Am. Soc. Mass Spectrom. 12 (2001) 288–295.

[81] N. Okiyama, T. Santa, A. Torima, K. Imai, Anal. Chim. Acta 429 (2001) 293–300.

[82] C. Sakamoto, Y. Okada, H. Kanazawa, E. Ayano, T. Nishimura, M. Ando, A. Kikuchi, T. Okana, J. Chromatogr. A 1030 247–253.

[83] D.S. Dunlop, A. Niedle, Anal. Biochem. 165 (1987) 36–44.

[84] A. Neidle, B.S. Miriam, S. Sacks, D.S. Dunlop, Anal. Biochem. 180 (1989) 291–297.

[85] K. Shimada, Y. Kawai, T. Oe, T. Nambara, J. Liq. Chromatogr. 12 (1989) 359–371.

[86] S.A. Cohen, J. Chromatogr. 512 (1990) 283–290.

[87] K.L. Woo, S.H. Lee, J. Chromatogr. A 667 (1994) 105–111.

[88] K L. Woo, Y.K. Ahan, J. Chromatogr. A 740 (1996) 41–50.

[89] K.L. Woo, Q.C. Hwang, H.S. Kim, J. Chromatog. A 740 (1996) 31–40.

[90] J.A. Prieto, C. Collar, C.B. de Barber, J. Chromatogr. Sci. 29 (1991) 205–209.

[91] O. Imakyure, M. Kai, Y. Ohkura, Anal. Chim. Acta 291 (1994) 197–204.

[92] C. Basic, J.M. Bailey, T.D. Lee, J. Am. Soc. Mass Spectrom. 6 (1995) 1211–1220.

[93] T. Toyo'oka, Y.M. Liu, J. Chromatogr. A 689 (1995) 23–30.

[94] T. Iwata, H. Mitoma, M. Yamaguchi, Anal. Chim. Acta 416 (2000) 69–75.

[95] M.V. Pickering, P. Newton, LC-GC Int. 3 (1992) 22–26.

[96] S. Delhaye, J. Landry, Analyst 117 (1992) 1875–1877.

[97] J. Landry, S. Delhaye, J. Sci. Food Agric. 58 (1992) 439–441.

[98] V.P. Hanko, J.S. Rohrer, Anal. Biochem. 308 (2002) 204–209.

[99] T.Y. Liu, Y.H. Chang, J. Biol. Chem. 246 (1971) 2842–2848.

[100] R.J. Simpson, M.R. Neuberger, T.Y. Liu, J. Biol. Chem. 251 (1976) 1936–1940.

[101] B. Penke, R. Ferenczi, K. Kovács, Anal. Biochem. 60 (1974) 45–50.

[102] R.B. Ashworth, J. Assoc. Off. Anal. Chem. 70 (1987) 80–85.
[103] K. Muramoto, H. Kamiya, Anal. Biochem. 189 (1990) 223–230.
[104] L.T. Ng, A. Pascaud, M. Pascuad, Anal. Biochem. 167 (1987) 47–52.
[105] P. Felker, Anal. Biochem. 76 (1976) 192–213.
[106] H. Yano, K. Aso, A. Tsugita, J. Biochem. 108 (1990) 579–582.
[107] I. Molnár-Perl, M. Pintér-Szakács, M. Khalifa, J. Chromatogr. 632 (1993) 57–61.
[108] S.R. Hagen, B. Frost, J.Augutin, J. Assoc. Off. Anal. Chem. 72 (1989) 912–916.
[109] M. Mennenberg, H.W. Lahm, M. Fountoulakis, Anal. Biochem. 224 (1995) 122–127.
[110] K. Okazaki, T. Imoto, H. Yamada, Anal. Biochem. 145 (1985) 87–90.
[111] U.T. Ruegg, J. Rudinger, Methods Enzymol. 47 (1977) 116–122.
[112] D. Fouques, J. Landry, Analyst 116 (1991) 529–531.
[113] C.W. Gehrke. L.L. Wall, J.S. Absheer, F.E. Kaiser, R.W. Zumwalt, J. Assoc. Off. Anal. Chem. 68 (1985) 811–821.
[114] Pico Tag Work Station Operator's Manual, NO 86746, Millipore, Waters Chromatography, Milford, MA, 1988, Revision C.
[115] Installation and Operation Manual: Microwave Sample Preparation System MD-S2100. CEM, Matthews, NC, 1991, p. IV-14–24.
[116] S.H. Chio, K.T. Wang, J. Chromatogr. 491 (1989) 424–431.
[117] S.A. Margolis, L. Jassie, H.M. Kingston, J. Autom. Chem. 13 (1991) 93–95.
[118] A. Pecavar, M. Prosek, D.F. Temeljotov, J. Marsel, Chromatographia 30 (1990) 159–162.
[119] V. Fábián, M. Morvai, M. Pintér-Szakács, I. Molnár-Perl, J. Chromatogr. 553 (1991) 87–92.

Ibolya Molnár-Perl (Editor)
Quantitation of Amino Acids and Amines by Chromatography
Journal of Chromatography Library, Vol. 70

1.2.3.

HPLC of Amino Acids as *o*-Phthalaldehyde Derivatives

Ibolya Molnár-Perl

Contents

Summary

1. Based on stoichiometric and mass spectrometric studies of the OPA derivatives of AAs the nightmare of their instability has solved.
2. It was confirmed, for the first time, that the common characteristic of the believed to be less stable OPA derivatives is associated with their initial molecular structure: they do contain the

-NH_2-CH_2-R moiety and all of them furnish more than one OPA derivative.

3. In the light of the common structure of the multiple OPA derivatives providing compounds a plausible reaction pathway has been described. According to the proposed mechanism hydrogen atoms of the -CH_2- group, neighbor to the primary amino group, play the key role in the reaction with an additional molecule of OPA: resulting in consecutive OPA-derivatives transformed from the initially obtained species.

4. As analytical consequences, the author of this review, based on recent experience, takes the responsibility to state: OPA derivatives of AAs are very much more stable than they are generally believed to be, considering that some of them are providing more than one reaction product (For the time being the successively formed products have not been taken into account). Consequently, in the HPLC analysis of AAs the following requirements are to be followed: (i) the molar ratios of the OPA/SH-group additive/AA should be close to 20/60/1, (ii) the reagents, saved in the refrigerator at ≤4 ^{o}C, should be at least 90 min and maximum 9 days old, (iii) the reaction time between the OPA reagent and the AAs should be at least 7 min, (iv) the OPA derivatives should not be acidified before loading them onto the chromatographic column, (v) the HPLC elution procedure should be proper for the separation and quantitation of all components formed, including the more than one OPA derivative providing AAs, (vi) last but not least, in order to be able to get reliable AA contents, in particular when working in the low pM level, the blank value of the reagent should be carried out at least every day and the impurity values are to be substracted from the coeluting AA derivative.

1. Introduction

All of those analytical chemists who have been worked and/or are working on the field of AA analysis and tried to use various derivatization techniques to their chromatographic determination, know, that for simplicity, selectivity, sensitivity and for minimizing work and time the HPLC of the precolumn derivatized OPA-AAs is the method of choice. The distribution of derivatization methods, proposed in order to quantitate AAs by HPLC proves this statement continuously [1–4]. Since the very welcomed introduction [3] of the OPA-method its popularity is continuously increasing; in the period of 1982–1998 its application represented 30.4 % of all derivatizations performed in HPLC, while later, in the period of 1998–2004 these figures varied between 30.4-39.5% at a constant, high level. Thus, it seemed to be inevitable necessary and useful to compile, examine then possibly clarify all of those advantages/disadvantages, difficulties, chromatographic advancements and application areas that characterize this very common procedure.

The instability of the isoindoles formed from AAs and OPA, in the presence of different SH-group containing compounds, as well as the instability of the reagent itself are known as two of the main disadvantages of this worldwide common principle [3]: suitable for the fast derivatization of AAs in aqueous media prior to their HPLC.

On the basis of all those uncertainties reported in the literature [4–87], including the pioneer work [3], and our own experiences [4–6,67–71] the following questions were intended to be answered:

1. What is the correlation between the extremely different and contradictory stability values of OPA-AAs, i.e., can be this phenomenon associated with the extremely different preparation and storage conditions of the reagents?
2. Do have the OPA/MPA(NAC, MCE) reagents self fluorescence and/or self absorbance which are to be taken into account in the quantitation of the AA derivatives, under HPLC conditions, applying the most common fluorescence (Fl) and photo diode array (DAD) detections, simultaneously?
3. What is the lifetime of the OPA reagents until providing optimum responses?
4. What is the reason (if any) of the believed lower stability of the glycine, β-alanine, γ-aminobut-yric acid (GABA), histidine, ornithine and lysine derivatives, in comparison to the rest of AAs?
5. What are the optimum UV/Fl responses of the isoindoles, also in comparison to each other, performing various molar ratios of the different SH-additive containing reagents to the AAs to be determined?
6. What is the structure of the transformed OPA derivatives and the reaction mechanism they are originating from?

In order to get answer to these questions in editor's laboratory two approaches were followed: stoichiometric investigations were carried out with simultaneous DAD and Fl detections [4–6,67–71] while structure elucidation were obtained by on line HPLC/DAD/mass spectrometry (MS) applying electro spray ionization (ESI) in the positive mode [69,71].

2. Composition, blank value, life time and optimum storage condition of reagents

2.1. Composition of the reagent; Derivatization conditions

The practical conditions proposed in the literature can be regarded as a nightmare. Diversity of the reagent composition, the mole ratios of OPA to the SH group containing additive and to the AAs to be derivatized, the pH, the buffer concentration and time of interactions suggested as optimal in the case of the OPA/MCE, OPA/MPA and OPA/NAC derivatives have been examined and summarized in detail [5,6,70]: they proved to be extremely different (Table 1).

As seen, even in the last decade, little attention has been paid to any of the most important parameters, such as the absolute concentration both of the reactants and the buffer, the pH and time of derivatization reaction, storage condition and life time of the reagent, etc. All these uncertainties result in basically incomparable analytical data and consequences drawn from them – in particular in the cases of the so-called less stable AAs, such as glycine, GABA, β-alanine, histidine, ornithine and lysine.

According to recent experience in our laboratory, the composition of the reagent, i.e., the molar ratio of the OPA to the SH additive, the pH and the reaction time of derivatizations proved to be of primary importance.

Consequently, for all derivatizations of AAs to be comparable (if not otherwise stated), the OPA/SH-additive was applied in the 1/3 molar ratios, while the OPA was used in 20 times molar excess to AAs to be derivatized. The reagent was stored in the refrigerator at 4 °C, and used for nine days only. The pH of the reagent was 9.3±0.05 ensured by 0.2M borate buffer, and the elution temperature was controlled by column thermostat.

2.2. Blank value of the reagent

As for blank values, in general [3,7–16], OPA reagents with SH-group containing additives are regarded as "...nonfluorescent itself, and when present in excess does not break down

Table 1 Derivatization Condition of AAs as OPA/MCE, OPA/MPA and OPA/NAC Derivatives: Literature Data

Related to final volume				Storage		Reac. time: m	Detection		Reference
Buffer		Molar ratios					UV: nm	Fl: Ex/Em	
M	pH	MCE/ OPA	OPA/ AAs	a, r	Life time				
0.075	8.0	0.34	-	r	-	2	338	338/458	17
0.5	9.5	6.7	$2x10^7$	r	1d	1	-	-	18
0.4	8-9.7	2.9	$8x10^3$	r	1m	1-9,3	-	335/425	19
0.2	10.4	1.7	170	r	2w	2.5	-	337/452	20
-	9.5	1.8	5.5	-	-	3.5	268	330/450	21
0.5	9.5	29	20-400	a	1w	3.5-15	-	330/455	22
0.5	10.4	1.8	$1.8x10^4$	r	1w	2	-	330/408	23
0.082	9.5	3.5	$1.4x10^4$	-	-	2	-	330/418	72
0.22	10.4	1.8	588	-	1d	1	-	340/425	73
0.314	10.4	4.2	-	-	1d	2	-	230/440	74
0.1	9.3	0.35	-	r	-	-	electrochemical		75
0.045	9.5	0.11	-	-	-	1	-	360/455	76
0.083	9.4	9.4	-	a	-	-	electrochemical		77
0.53	9.5	8.5	224	-	-	1	-	330/445	78
0.64	10.5	8.8	-	1d	-	-	-	330/445	79
0.09	9.4	2.0	-	r	3d	2	-	350/480	80
0.1	9.5	1.4	-	-	-	1	-	330/450	81
0.4	10.5	9.6	-	-	-	-	-	345/455	82
0.1	10.4	96	-	r	1d	-	-	340/450	83
-	-	5.7	247	-	-	1	-	-	84
0.1	9.3	0.35	42	-	-	-	electrochemical		85
0.1	-	9.5	-	-	1w	-	210	340/430	86
0.1	10.0	5.7	-	-	2d	-	-	300/950	87

Related to final volume				Storage		Reac.	Detection		r
Buffer		Molar ratios					UV:	Fl:	
M	pH	MPA/ OPA	OPA/ AAs	a,r	Life time				
0.45	9.5	1.54	$1.7x10^3$	r	1w	5	334	334/445	44
0.22	10.2	2.56	31	r	1w	2.5	338	230/450	51
0.45	9.5	1.58	42	r	1w	2.5	-	330/450	34
0.45	10.4	1.54	5.9	-	-	2	-	335/440	62
0.08	10.0	1.54	$3.0x10^4$	a	3d	3	-	230/389	61
0.16	9.4	3	19-236	a/r	0-28d	1-240	334	337/454	6,67-69
		NAC/ OPA							
0.05^x	-	1.03	-	r	1w	2	-	360/405	45
0.31	10.0	1.22	<1	a	2d	10	280	344/443	88
0.38^x	10.0	1.03	-	a	1m	1	-	348/450	46
0.03	9.5	1.0	5-200	a	1w	<1	340	340/445	7
0.1	9.4	0.5	74-400	r	3w	15	-	340/450	63
0.1	9.5	1.0	≥10	r	-	1	336	--	89
0.28	9.0	0.61	139	a	1d	2	-	344/443	66
0.25	10.4	1.54	12	-	-	2	-	230/445	52
0.16	9.4	3	19-236	a/r	0-28d	1-240	334	337/442	6,67-69

Abbreviations/indications: M=mole/L; a=ambient temperature; r=refrigerator; d=day; w=week; m=months; in paper [66]: N-tert.butyl-oxycarbonyl-L-cysteine and in paper [52]: N-isobutyryl-L-cysteine were used or react to form fluorescent byproducts." [7]. Although, selected observations [8–16], including the pioneering work [3] are in contradiction with the above general accepted impression and call attention to the fact that in particular with the use of its large excess, the chromatographic elution of the reagent can result in numerous spurious peaks [8–15]. In the pioneering work [3] the author's warning is unambiguous: "the blank fluorescence is the main factor limiting the sensitivity of the method." On the basis of this study, the self-fluorescence of impurities was defined as the half of 1 μL 10^{-3} M alanine solutions applying the OPA/MCE reagent in 10 molar excess. By others [8,9] the fluorescence response of a matured 48 hrs old OPA/MCE reagent "...was found to be approximately 10 times less intense than the response seen from 100 pM of each OPA/MCE AA derivative..." [8]. According to another report [9] "when working with amounts of material less than about 100 pM, appropriate control must be run to compensate for the minute amounts of AAs found in even the best of commercially available reagents." Further experiences [10–15] were mainly qualitative, advising to discard the OPA/MCE reagent solution if contaminant peaks appear, or earlier if necessary, at two months [10], or one week [11]: due to the extremely different, believed lifetime of the same reagent (note of the editor). The impurity peaks proved to be extremely disturbing by using electrochemical detection (ED) [12–15]. In the case of the OPA/MPA reagent substantially more extraneous peaks were detected applying ED compared to fluorescent detection [12]. In order to improve sensitivity either special chromatographic conditions [13] or the elimination of the excess reagent was proposed [14,15]. Scavenging the excess of the OPA/MCE reagent by adding iodoacetamide or N-ethylmaleimide resulted in limited success only [14]. Scavenging the excess of the OPA/tert.-butylthiol (BT) reagent with the excess of glycine and iodoacetamide [15] resulted in the quantitative elimination of interfering reagent peaks and in a detection limit of 50 aM AA. The stabilizing effect of nitrilotriacetic acid acting as a metal chelating agent was found to increase the OPA/MCE reagent stability at least four fold in the sample vial of an automated system [16].

Recently [5,6,70), an exhaustive study was carried out in our laboratory in order to examine the characteristics of the OPA/MPA, OPA/NAC and OPA/MCE reagents. These studies performed with DAD and Fl detections, simultaneously (Table 2) confirmed that all three reagents do have blank values both in the UV- and in the Fl regions.

Table 2 Impurities of the OPA/MCE=1/3, and OPA/MPA=1/3 Reagents

Impurity peaks			Integrator unit**/impurity peaks: as a function of the OPA/MCE reagent's age#										
			Reagent1							Reagent2			
No	Rt: m	UV: nm	90 m	230m	270m	370m	7 d	11 d	27 d	90 m	4 d	21 d	23 d
*1 UV	3.26	216, 253	2.7	4.6	4.4	5.9	3.1	49.0	55.5	3.5	12.7	44.1	2.0
*2 UV	5.24	229, 339	0.7	0.7	1.3	0.6	17.7	3.7	12.7	-	-	-	-
*3 UV	6.13	215, 248	0.3	-	-	--	-	4.1	6.0	-	-	4.2	-
4 Fl	7.58		38.0	26.0	85.0	151.0	56.9	119.1	26.5	93.9	-	-	-
UV		253, 334	1.5	2.1	2.0	4.1	1.3	4.1	0.5	1.9	0.9	-	-
5 Fl	8.21		89.0	-	-	-	2.3	89.2	-	75.0	16.5	95.0	106.0
UV		220, 263	1.7	2.3	1.1	1.4	-	2.4	10.6	1.8	1.4	-	3.3
6 Fl	9.29		-	-	-	-	-	-	-	-	28.7	94.0	82.0
UV		267, 339	3.2	2.3	1.1	1.4	0.3	40.0	20.9	4.8	11.0	14.3	-
*7 UV	13.26	220, 272	7.2	-	-	-	4.7	2.7	4.5	50.9	49.6	172.0	151.0
Fl in total			127 *(30)*	26.0 *(20)*	85.0 *(20)*	151 *(25)*	59.2 *(14)*	208 *(50)*	26.5 *(6.3)*	169 *(40)*	45.2 *(11)*	189 *(473)*	188 *(45)*
UV in total			17.3 *(58)*	12.0 *(40)*	9.9 *(33)*	13.4 *(45)*	27.2 *(91)*	107 *(357)*	111 *(370)*	62.9 *(210)*	75.6 *(252)*	235 *(783)*	156 *(520)*

Impurity peaks			Integrator unit**/impurity peaks: as a function of the OPA/MPA=1/3 reagent's age#							
No	Rt: m	UV nm	50 m	240m	1 d	2 d	3 d	7 d	10 d	22 d
*1UV	3.13	219, 252	1.4	5.5	14	23	37	77	112	160
2 Fl	5.84		53	93	61	112	60	45	20	11
UV		228, 333	2.8	5.2	3.2	6.2	3.2	1.9	0.8	0.4
3 Fl	6.56		38	62	24	62	20	14	7.5	0.5
UV		214, 252	7.1	16	22	23	21	19	15	14
4 Fl	7.14		9.5	21	13	25	11	7.3	4.3	1.4
UV		219, 342	2.5	3.2	0.7	1.3	0.6	<0.3	<0.3	<0.3
5 Fl	7.68		4.9	6.0	5.3	7.8	5.2	5.1	3.4	1.6
UV		224, 266	1.6	1.7	2.1	2.7	2.5	3.5	1.3	2.5
6 Fl	8.78		18	29	46	38	43	31	5.3	<0.3
UV		233, 276	0.3	0.4	0.6	0.4	0.4	<0.3	<0.3	<0.3
7 Fl	9.33		43	47	56	50	46	30	12	3.0
UV		224, 261	23	19	13	17	17	12	9.5	1.3
8 Fl	11.24		13	17	21	24	27	25	14	1.0
UV		233, 276	2.8	3.5	4.5	5.2	6.6	6.9	5.3	<0.3
*9UV	11.53	243, 333	25	29	24	15	12	6.3	2.9	0.4
10 Fl	12.29		72	72	83	76	77	73	26	1.5
UV		228, 333	84	84	92	85	84	78	23	1.0
11 Fl	13.08		50	42	31	22	17	11	21	110
UV		228, 252	6.9	3.6	1.3	0.5	<0.3	<0.3	0.4	3.5
Fl in total			167 (67)	258 (103)	205 (82)	295 (118)	185 (74)	132 (53)	52 (21)	17 (7)
UV in total			39 (234)	51 (306)	55 (320)	75 (450)	82 (442)	115 (690)	139 (834)	181 (1086)

Indications, as in Table 1, as well as: Rt: m=retention time: min; #=averages of four injections; **=reagent's impurity values have been calculated with ~4.2 and ~0.3 (OPA/MCE reagent) and with ~2.5 and ~15 (OPA/MPA reagent) integrator units/1 pM amino acid derivative performing Fl and UV detections: average integrator units obtained from 15 amino acids; Italic printed values in parentheses represent the total of impurities, expressed in pM; detection limit << 0.3 integration unit; *=impurities indicated by asterisks do not have Fl or UV intensities

The quality and quantity of peaks originating from the reagents depend on their age and certainly on their concentrations. It is to be emphasized, even when using the same preparation and storage conditions (detailed description in the Experimental section), that the amounts of the individual impurities are different from reagent to reagent (in spite of the fact that the tendency concerning the increase and decrease of the individual impurities, as a function of reagent age and quality are similar, and characteristic of it).

Consequently, it is obligatory to take into consideration the reagent peaks in order to subtract them from the coeluting AAs, particularly in those cases when a relatively higher reagent concentration is needed, because AAs are measured in extremely different concentrations in the presence of their matrix [6]. Since the quality and quantity of reagent impurity peaks are changing, (the aged reagents contain less than the fresh ones), the blank value of the reagent should be tested at least every day.

As to the expected amount of impurity peaks using the OPA/MPA, OPA/NAC or OPA/MCE reagents, measured with the same reagent between 90 min and 22 days, they are changing: in the elution range of AA derivatives to be determined the amount of the co-eluting impurities can vary between 50 and 300 pM injected in total. Certainly, which of the AAs will be co-eluting with which reagent peaks depends on the column and the gradient applied. Thus, the blank value depends upon the system and conditions.

2.3. Lifetime and optimum storage condition of reagents: depending on the SH group additive

The OPA/MCE and the OPA/*tert*.butylthiol (BT) reagents were reported to be appropriate for derivatization up to 1 day [18], 1 week [22,23], 2 weeks [20] or 1 month [5], usually saved in the refrigerator at 4 ^{o}C [17–20,23], or at room temperature [15,22]. Others [15,17,21] do not attach substantial importance to these phenomena: as a basis of quantitative and reproducible derivatizations [5,6]. Concerning other conditions, the most desirable pH (pH 8–11.5) and/or the applied mole ratios of the reagent to the analyte (Table 1: [OPA]/[AAs] = 5.5-1.8x10^4) proved to be extremely different.

To extend the lifetime of the primarily introduced OPA/MCE reagent it has been reactivated by completing it with the MCE [2–12] continuously (daily [2], in every second [3–5], or third days [6–8] and weekly [9–12]).

The behavior of the OPA/MPA, OPA/NAC and the OPA/MCE reagents – stored in the refrigerator at 4 °C – has been followed with UV/DAD and Fl detections, simultaneously, with storage from 90 min to 28 days (Table 3) [6,70] have been characterized also on the basis of their derivatization capability.

Table 3 Reproducibility of the OPA/MPA-AA Derivatives Obtained with Reagents Saved for Various Period of Time in the Refrigerator

Amino Acid	Arbitrary units/ 1 pM AA, reagent saved at ~ 4 °C, for 1-28 days										
	1	4	7	8	9	10	A**	RSD%	15 (%)&	21 (%)&	28 (%)&
Aspartic Acid	2.81	3.01	2.86	2.86	2.96	2.83	2.90	2.5	1.86 (64)	0.17 (5.9)	0.033 (1.0)
Glutamic Acid	3.28	3.32	3.26	3.26	3.30	3.33	3.29	0.91	2.20 (67)	0.05 (1.5)	0.007 (0.2)
Serine-*2	3.81	3.86	3.79	3.87	4.04	4.07	3.91	3.1	2.68 (69)	0.10 (2.6)	0.104 (2.7)
Gly1-*3+Gly2	4.12	4.16	4.16	4.30	4.37	4.18	4.22	2.3	3.02 (72)	0.04 (0.9)	0.027 (0.6)
Alanine-*4	3.56	3.76	3.74	3.75	3.83	3.61	3.71	2.7	2.63 (71)	0.07 (1.9)	0.022 (0.6)
GABA1+GABA2	4.04	3.98	4.15	4.16	4.24	3.94	4.09	2.9	3.39 (83)	0.04 (1.0)	0.031 (0.8)
Tyrosine-*5	3.59	3.60	3.56	3.63	3.64	3.49	3.59	1.5	2.53 (70)	0.06 (1.7)	0.019 (0.5)
Valine-*6	3.81	3.79	3.80	3.77	3.82	3.75	3.79	0.70	2.64 (70)	0.15 (4.0)	0.057 (1.5)
Phenylalanine-*7	3.60	3.67	3.58	3.82	3.92	3.87	3.74	3.9	2.88 (77)	0.10 (2.7)	0.028 (0.7)
Leucine-Lys1	3.66	3.76	3.79	3.80	3.85	3.76	3.77	1.7	2.80 (74)	0.10 (2.7)	0.013 (0.3)
Lys1+Lys2+Lys3	1.57	1.58	1.74	1.75	1.70	1.59	1.65	5.0	1.28(77)	0.01(0.6)	0.001 (<0.1)

*1 -*7= reagent impurities subtracted from the corresponding OPA derivative;
A** = Average of fluorescence intensities; (%)& = expressed in the average of 1-10 days

It is worthy of mention that in the case of the OPA/MPA, similarly to the OPA/NAC reagents, the older the reagents the smaller the amount of impurities found [5]. Thus, the best solution of the problem might have been the use of aged reagents. Unfortunately, all three reagents can be applied up until 9 days only, remaining on the safe side, in order to obtain a quantitative derivatization [5,70].

3. Stability of the AA derivatives

The stability dilemma and the use of the OPA method are inseparable. Special studies devoted primarily to the HPLC of the OPA-AAs, without exception, became involved in the stability phenomena: providing new SH-group containing alternatives, improved derivatization/chromatographic conditions and/or new application areas.

Before going into details it is worth mentioning that in the literature as model compounds of stability studies, (from the beginning and thereafter continuously), glycine, GABA, ß-alanine, histidine, ornithine and lysine have been selected as the less stable ones: with a lack of knowledge of their intrinsic characteristics, clarified recently [5,6,68–71]. Since, (i) these six AAs provide more than one OPA derivative, (ii) resulting in products of similar high stability as any other AAs: so long as the total of their Fl or UV responses, i.e., the sum of derivatives obtained from the given AA were taken into account in their quantitation (detailed studies in the 3.1.2. sections).

3.1. Comparative studies applying different SH-group containing additives.

3.1.1. Early literature observations

In the 1980s exhaustive investigations were performed to clarify factors affecting the stability of the primary amino group with the OPA/MCE [38], OPA/3-mercapto-1-propanol (MCP) [38] and the OPA/ET [39] reagents. The kinetic stability of isoindole products was evaluated under conditions similar to those for analytical application. As model compounds n-propylamine, GABA, ß-alanine, glycine, ß-aminobutyric acid and alanine have been tested [38,39]. Stability conditions of their derivatives obtained at 40.5 °C and at pH 8.95, applying the mole ratios of OPA/MCE/AA = 1430/149/1, (1 = 1 x 10^{-6}M) were characterized with the 50% decomposition of the product, expressed as the half time of their decomposition (T/2), which proved to be in order of listing 18.2, 21.1, 24.9, 30.8, 57.8 and 98.9 min: i.e., the decomposition of ß-alanine was 4 times faster than that of alanine (98.9/24.9 = 3.97). Decreasing the OPA concentration in OPA/MCE/ß-alanine to 74.8 x1 0^{-6}M and 37.7 x 10^{-6}M, the T/2 values increased for 35.7 and 43.3 min. Increasing the OPA concentration from 149 x 10^{-6}M to 298.9 x 10^{-6}M and to 447.9 x

10^{-6}M the rate of decomposition decreased for 16.5 and 12.0 min, respectively: proving the accelerating effect of the OPA-excess. The superior stability of the OPA/3MCP derivatives was compared by the T/2 of its ß-alanine derivative which proved to be 346.6 min with the 149 x 10^{-6} M OPA concentration (346.6/24.9 = 13.9). Consequently, these investigations resulted in two ways to improve the OPA derivatizations: changin reactant ratios, taking into account that the high excess of OPA accelerates decomposition, and altering the SH-group component.

A convincing approach, regarding the destructive effect of the reagent excess, was presented [8], by holding the OPA/MCE-AAs for various time intervals in the injection loop before injecting them onto the HPLC column (Figure 1A) and comparing them to those 2 min after injection of derivatives onto the HPLC column, the solvent flow rate was stopped for a different time interval to examine the stability of the OPA/MCE AA derivatives on the column (Figure 1B).

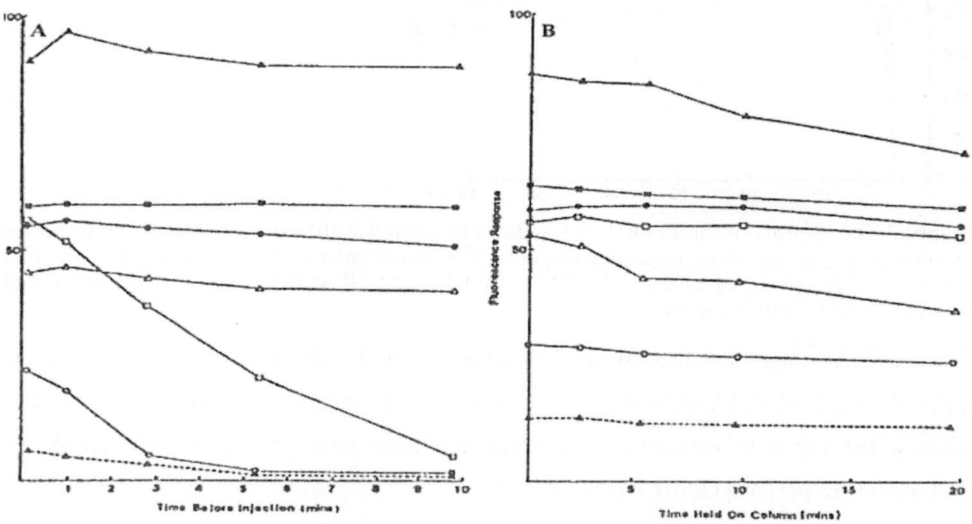

Figures 1A,B Stabilities of the OPA/MCE AAs tested with different reaction times before injections. A, in the presence, B, in the absence of the reagent's excess, the time held on column when the solvent flow was stoppeed. Symbols: ▲, glutamic acid; Δ, aspartic acid; □, glycine; •, serine; ▪, valine; o, lysine; ...▲..., ornithine (Note: the pH of solutions were pH 6.5, the temperature was not given). Reproduced with permission from Reference [4]

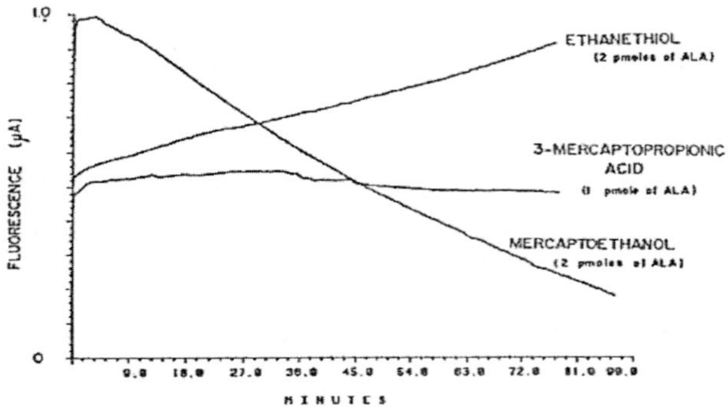

Figure 2 The effect of various mercapto reagents on the stability of OPA-alanine. Reproduced with permission from reference [4]

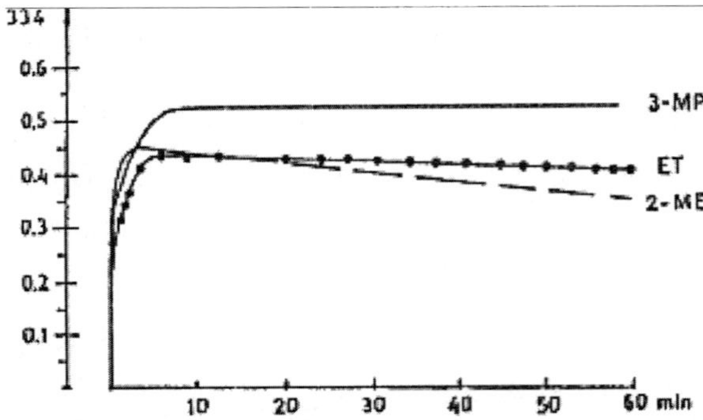

Figure 3 The stability of the 17 OPA-AA adducts illustrated as the UV response (334 nm) versus reaction time for the three mercapto reagents: UV absorbencies of the total of 17 AAs (2,5 nM/ml of each) obtained as the OPA/MCE, OPA/ET and OPA/MPA derivatives. Reproduced with permission from Reference [4]

Special suggestions were made to enhance the stability of the OPA/MCE reagent: using nitrilotriacetic acid (NTAA), in the mass ratio of OPA/NTAA = 5/1 resulted in its four-fold stability [16]. The increased methyl alcohol concentration of the reagent improved the stability of the OPA/MCE- [41] and OPA/ET AAs [42].

The stability of the OPA/MPA-AAs was compared with the corresponding OPA/MCE and the OPA/ET derivatives [43,44]. Evaluating the stabilities and responses of the three derivatives of alanine (Figure 2, [43]) based on Fl intensities and those of the three model solutions, based on the UV responses (334 nm) of the total of 17 AAs (Figure 3, [44]) proved the constant stability of the OPA/MPA derivatives.

The advantage of the OPA/N-alkyl-cysteine reagents over the OPA/MCE and OPA/ET ones, due to their chiral recognition characteristics, attracted only few scientists [7,45,46]. The introduction of NAC as an SH-group containing alternative was explained mainly in a qualitative manner [45].

3.1.2. Recent exhaustive study on the stability of the OPA/MPA, OPA/NAC and OPA/MCE- AAs, with particular attention to the multiple derivatives providing compounds

3.1.2.1. Stability of the single derivatives providing AAs

The comparison of the stability of 26 OPA/MPA, OPA/NAC and OPA/MCE-AA derivatives performed recently [4–6,68–71] (still under investigation as OPA/ET derivatives), proved that all of them are significantly more stable than they are generally believed to be: (i) the single derivatives providing 20 AAs [70], as OPA/MPA or OPA/NAC derivatives exhibited less than 4% degradation, even after 6 h, while the OPA/MCE ones after 75 min reaction time. (ii) The AAs providing more than one derivative (glycine, GABA, β-alanine, histidine, ornithine and lysine), evaluating the total of their peaks, are as stable as the single derivative furnishing OPA-AAs [5,6,67,68].

3.1.2.2. Stability of the multiple derivatives providing AAs

The principle of the multiple derivatives providing AAs proved to be the crucial point of their stability issue [5,6,34]. Earlier observations [18,47–53] did not attach great importance to this experiment. The first reference to the double peaks was related to the OPA/ET derivatives of lysine and ornithine [47]. The lower responses of the OPA/MCE derivatives of lysine and ornithine were assumed (not proved: editor's note) [48] due... "to the presence of two fluorescent isoindole structures." An isotope detection study [49] of the OPA/MCE derivatives of glutamic acid, arginine and ornithine resulted in an additional five, in total of six ornithine derivatives. The double peaks of the OPA/MPA-lysine were reported as the basis of its quantification [50], while in another work [51] the presence of the OPA/MPA derivatives of lysine and ornithine was expected but could not be confirmed. The two GABA derivatives obtained with the OPA/isobutyryl-L/D-cysteine reagents were shown without comment [52]. The formation of the double OPA/MPA derivatives of OH-lysine was shown [53], instead of the expected two OPA/MPA-lysine and ornithine derivatives [47,49,50], which have not been found [53]. The double derivatives of reduced glutathione (GSH) and γ-glutamylcysteine (γ-Glcys) were detected as their OPA/MCE derivatives after 15 min reaction time in extremely high concentrations [18].

Our first approach [6] related to the main contradiction of OPA derivatizations. In order to get a clear answer to the background of the 'believed to be extreme low stability' of glycine, β-alanine, GABA, histidine, ornithine and lysine, these six AAs have been derivatized in two groups as OPA/MPA derivatives (Figures 4A,B). Changes in responses, have been followed by UV and fluorescence detections, simultaneously, as a function of the reaction time, strictly under the same, well defined practical conditions (optimum derivatization conditions in the Experimental section).

On the basis of our results it has been proved (Figures 4AB, Table 4) that:

(i) in the cases of those AA derivatives which are providing single derivatives, and are not listed in Table 4, significant losses (<4%) have not been detected, not even after 6 hrs reaction time. (ii) Maximum Fl intensities and UV absorbances for the OPA/MPA derivatives of aspartic acid, glutamic acid, β-alanine, alanine, GABA, isoleucine, lysine and ornitine, as well as for the OPA/NAC derivatives of threonine were measured after 28 min reaction time.

(iii) Derivatives of serine, threonine, tyrosine, methionine, tryptophan, isoleucine and leucine furnished, not significant, but detectable loss of responses (Table 4) which after 6 hrs reaction times proved to be slightly higher for the OPA/NAC derivatives (88–94%) in comparison to the corresponding OPA/MPA ones (93–97%).

(iv) Experience obtained with glycine, GABA, β-alanine, histidine, lysine and ornithine are of particular importance, furnishing substantially new principles in the understanding of the characteristics of – for the time being believed – unstable derivatives. On the basis of our stoichiometric studies it became evident that the stability of these derivatives is similar to the others in so far as they are eluting in more than one peak. Consequently, evaluating the total of peaks (Table 4)

- Glycine proved to be one of the most stable derivatives providing AA: the total of glycine1 and glycine2 do not change even after 6 hrs reaction time.
- In the cases of β-alanine, GABA, ornithine and lysine, accordingly our earlier experiences [5] the maximum fluorescences were measured after extended reaction times (7–28 min).
- The less stable OPA/MPA-histidine – after 6 hrs – provided 63%, the OPA/NAC-histidine 82% responses. The loss in the cases of OPA/MPA-GABA (90%), OPA/MPA-ornithine (95%) and OPA/NAC-alanine (98%) were as low as manifested any others.

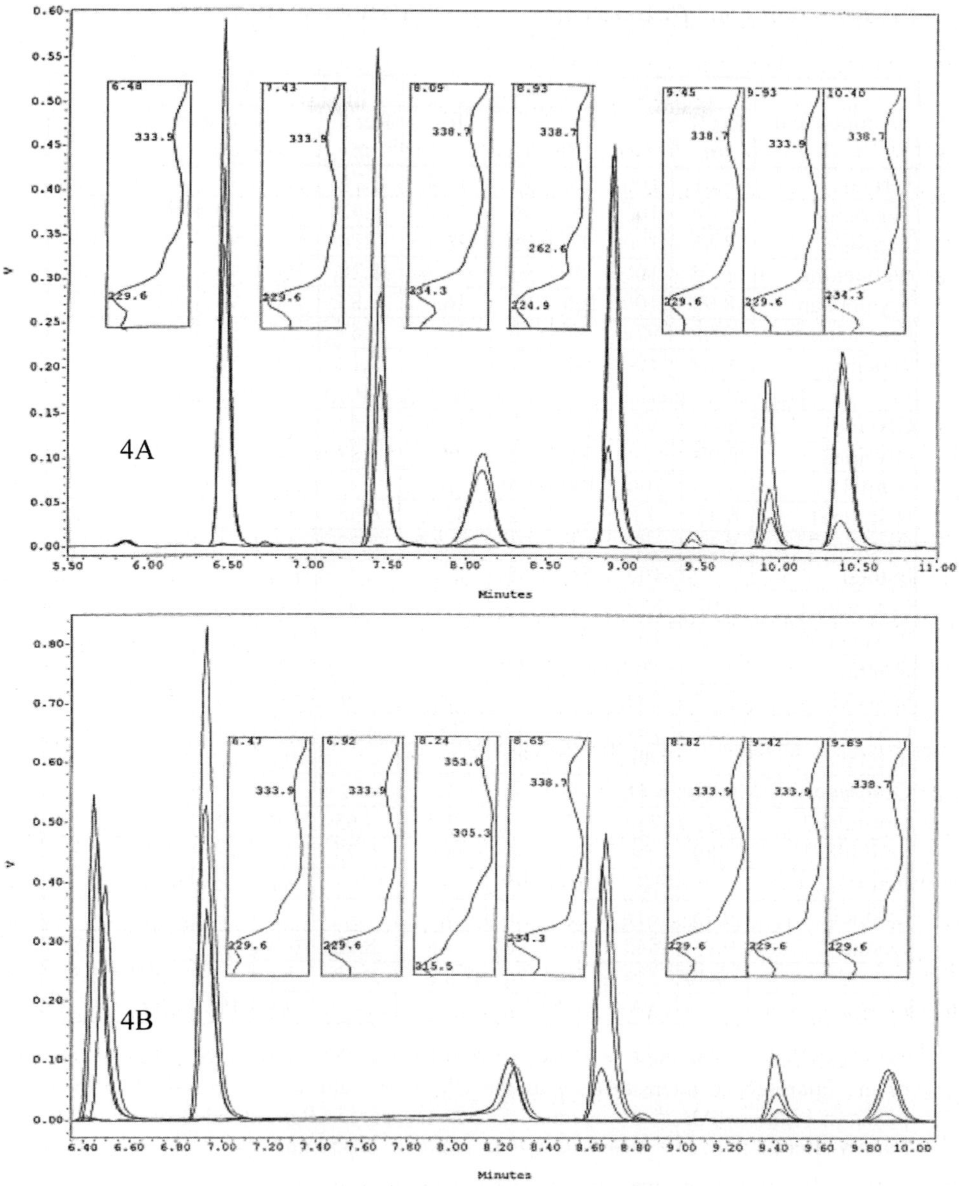

Figures 4A,B Simultaneous Fl chromatogram and DAD spectra of the OPA/MPA derivatives of glycine, GABA and lysine (Figure 4A) and histidine, β-alanine and ornithine, (Figure 4B) as a function of the reaction time (7 min, 3 hrs and 6 hrs; detailed results in Table 4). Reproduced with permission from Reference [5]

Table 4 Stability and Characteristics of the OPA/MPA and OPA/NAC-AA Derivatives as a Function of the Reaction Time

Amino Acid	OPA/MPA-AAs					OPA/NAC-AAs				
	Ret. t: m	response %* 7 m	3h	6 h	RR	Ret. t: m	response % 7 m	3h	6 h	RR
Serine	5.81	100	95	93	16.4	5.74	100	93	92	12.9
Threonine	6.48	100	98	96	14.0	6.42	98#	99	90	13.2
Tyrosine	7.65	100	97	95	16.1	7.74	100	95	89	11.1
Methionine	8.45	100	99	97	19.2	8.28	100	96	94	11.6
Tryptophan	8.98	100	100	97	16.0	8.83	100	95	94	11.9
Isoleucine	9.08	97#	99	96	16.0	8.67	100	95	94	14.7
Leucine	9.63	100	98	97	18.8	9.12	100	88	88	9.0
Derivatives eluting in two or three (OPA/MPA/Lys + Orn) peaks										
Glycine1	6.48	95	67	58	15.9	6.43	97	75	62	15.5
Glycine2	8.09	5	33	42	24.6	7.87	3	25	38	17.5
in total		100	100	100			100	100	100	
Histidine1	6.47	73	63	42	15.8	6.35	57	55	44	10.8
Histidine2	8.24	27	23	21	22.8&	8.33	43	36	38	18.4&
in total		100	86	63			100	91	82	
β-Alanine1	6.92	85#	52	39	17.2	6.63	89#	73	56	15.1
β-Alanine2	8.65	10#	48	61	25.6	8.39	6#	27	42	16.9
in total		95	100	100			95	100	98	
GABA1	7.43	73#	39	26	18.1	6.98	88#	72	59	14.6
GABA2	8.93	16#	61	64	23.5	8.43	5#	28	41	16.6
in total		89	100	90			93	100	100	
Ornithine1	8.82	<1#	3	7	1	-				
Ornithine2	9.42	71#	27	12	2.4	8.65	79#	49#	35	2.2
Ornithine3	9.89	5#	70	66	4.9	9.23	5#	46#	65	3.9
in total		77	100	95			84	95	100	
Lysine1	9.45	<1#	3	6	5.0	-				
Lysine2	9.93	54#	19	10	3.8	8.92	67#	48	31	2.8
Lysine3	10.40	13#	78	84	8.0	9.48	6#	52	69	5.5
in total		68	100	100			73	100	100	

%* = expressed in the total; m = min; h = hrs; RR=Fl/UV represent the response ratios of the fluorescence intensities versus the UV ones; values are corresponding to the peaks in Figures 4AB; # = maximum Fl was measured after extended reaction time, mainly after 28 min; & = measured at their maxima, at 353 nm;

Already, in these introductory investigations we assumed that the molar ratio of the OPA/SH-group additive plays a special and key role in the transformation of the initially formed OPA derivative: therefore we varied these ratios from 1/0.5 through 1/3 to 1/10 (Figures 5ABC)

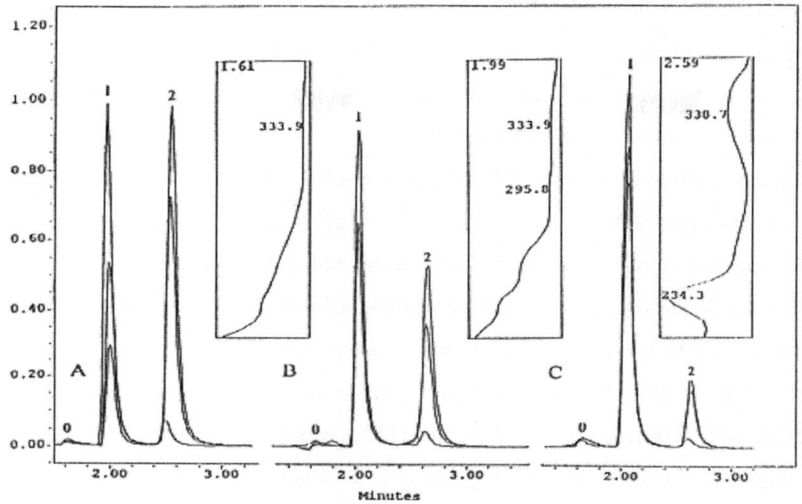

Figures 5A,B,C Chromatograms (A,B,C), obtained by Fl/DAD detections, applying the molar ratios of OPA/MPA/AA=20/10/1 (A), OPA/MPA/AA=20/60/1 (B), OPA/MPA/ AA=20/200/1 (C) (1=1x10^{-9}M) as a function of the reaction time (7 min, 3h, 6h). Peaks: 1, gly0; 2, gly1; 3, gly2; Reproduced with permission from Reference [68]

As seen (Figure 5A,B,C), proven also by exact data [68], decreasing the molar ratios of OPA/MPA in the reagent from 1/0.5 (Figure 5A) through 1/3 (Figure 5B) to 1/10 (Figure 5C), led to a spectacular rate reduction regarding the transformation of the first products to the forth-coming ones. For example, in all six cases, for both derivatives (OPA/MPA, OPA/NAC), the peak ratios of glycine (Figures 5A,B,C) similarly to those of GABA, β-alanine, histidine, or-nithine and lysine (not shown) reveal that the total of products does not decrease considerably, not even after 6 hrs reaction time (data in Table 4, detailed data in paper [68]).

These phenomena proved to be a very important indication, pointing ahead to solve the is-sue of the formation/transformation of the multiple OPA-derivatives.

Evaluating these experiences from an analytical point of view, they proved to be really promising. Namely: taking into account the total of responses of these more than one derivative providing AAs, we could prove that these amino acids manifest the same stability more than all others. Consequently, the only requirement was to develop a proper chromatographic technique that ensured the quantitation of total of derivatives formed [67] (section 7)

3.2. The role of pH in the stability of the OPA derivatives of AAs

The two selected factors affecting the transformation of the primarily formed products to their transformed ones are (i) the mole ratios of OPA to the SH-group additive (detailed above) and (ii) the pH of solution containing the OPA-AAs [4,68].

Acidifying the solution of the OPA/MPA-AAs from pH 9.3 to pH 7.2, for 7 min, before injecting them – (subsequently to 7 min reaction time at pH 9.3) – resulted in considerable changes in their responses. In particular, in the cases of the multiple derivatives providing AAs: the transformation rate of the first forming derivative upon acidification became highly accelerated, the decompostion of the total of derivatives increased from 0 to 16%.

Responses of the single OPA-AAs, acidifying them under the same condition as detailed above are also decreased, in order of listing: threonine (11%), serine (10%), leucine (4%), isoleucine (3%), all others ≤2%. This experience, regarding its tendency, proved to be in accordance with that of the OPA/MCE-AAs [22].

Although, under a very similar pH condition the OPA/MCE-AAs showed extremely dramatic drop of responses (Figure 6A,B) even in comparison to our experience (likely due the initially lower stability of the OPA/MCE-AAs). These observations are important because several papers suggest the acidification of the OPA-AAs containing solutions before injecting them.

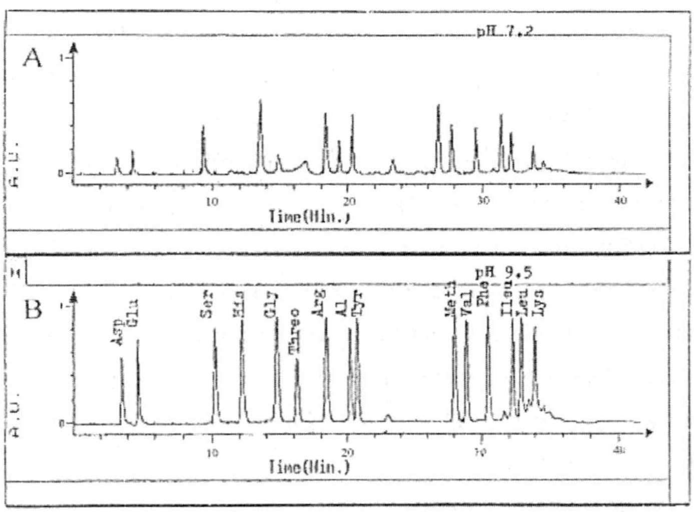

Figure 6A,B Responses of OPA/MCE-AAs at pH 7.2 and 9.5 in phosphate buffers. Reproduced with permission from Reference [4]

4. Reaction mechanism studies

4.1. Earlier literature data

The formation and composition of the fluorescent adduct, the 'main' reaction product, proved to be 1-alkylthio-2-alkyl-substituted isoindoles [24–33]: obtained from 1-1-1 moles of OPA, NH_2- and SH-group containing compounds with the elimination of 2 moles of H_2O: based on spectral and NMR [24–28], kinetic [29,30], as well as on those of mass spectrometric (MS) evidence [31–33,35.36]. Although, in addition to the 'main' product, very likely as a result of its decomposition (disproportionation/autooxidation), the presence of the 1,3 dithio-substituted-2-alkylisoindoles was also identified, based on NMR evidence [27], supported also by kinetic data [29,30].

Recently, on line HPLC-MS determinations [17,35,36] have been reported. The thermospray (TSP)-HPLC-MS was applied in order to determine S-nitrosoglutathione in the presence of an enormous excess of reduced glutathione as their OPA/MCE derivatives [17].

The diastereomeric isoindoles of the OPA/NAC derivatives of alanine, valine, leucine phenylalanine, phenylglycine and their corresponding amides have been separated and identified by TSP-HPLC-MS [35], all on the basis of the protonated molecular ions of the classical isoindoles ($[M+1]^+$. The isotope enrichment of citrulline, arginine, tyrosine, valine and leucine has been measured by HPLC-MS(ESI) as their OPA/MPA derivatives [36] working in the negative ion mode.

4.2. Recent on line HPLC-MS(ESI) studies

All of our experience detailed in section 3.1.2.2. raised the questions:

(i) What is the composition of the transformed derivatives; consequently, what is the unknown reaction mechanism they are originating from?

(ii) What is the common and determinative peculiarity of all those compounds that are providing more than one OPA derivative?

(iii) Is it possible, and in what way, to inhibit or at least to decrease the transformation of the initially formed derivatives?

The answers to these questions have been given one by one in the following manner.

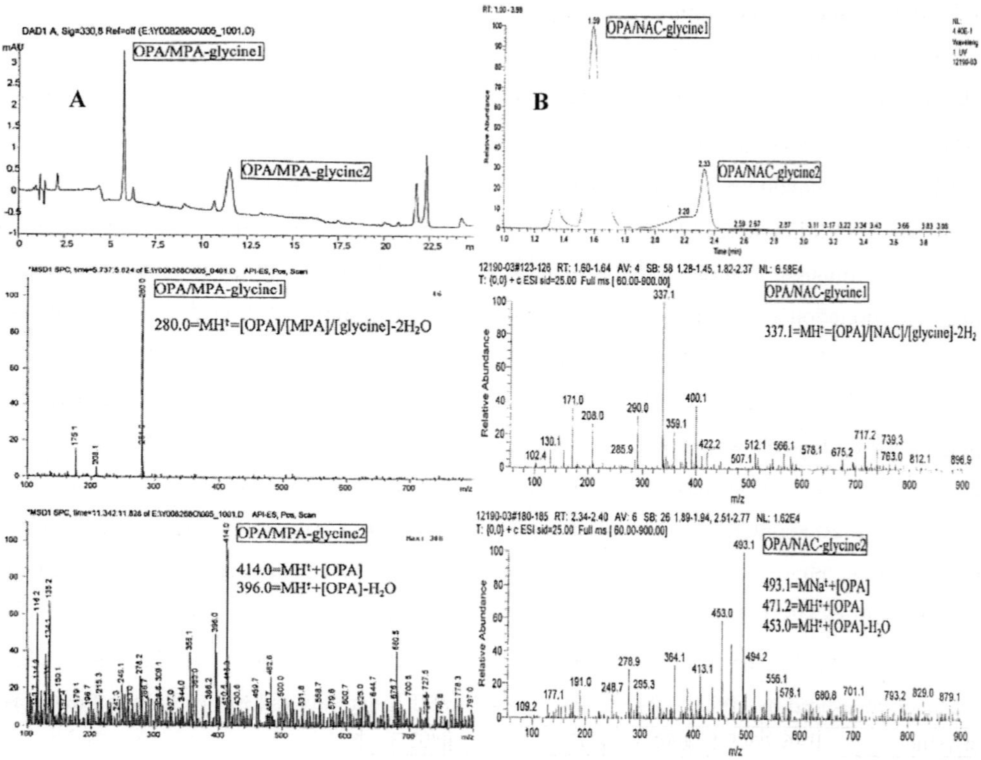

Figures 7A,B DAD/UV chromatograms (first lines: glycine1, glycine2), and MS spectra of the initially obtained (glycine1: spectra in second lines) and transformed (glycine2: spectra in third lines) OPA/MPA- (Figure 7A) and OPA/NAC-glycine (Figure 7B) derivatives; (Detailed composition of fragments in Table 5). With permission from Reference [69]

Composition of the OPA/MPA and OPA/NAC derivatives of glycine, β-alanine, GABA, histidine, ornithine, lysine and ε-amino caproic acid, applying on line HPLC-DAD-MS(ESI), were measured in the positive mode. (Table 5, Figures 7A,B). Data revealed unambiguously, that the initially formed products of all of these AAs were the well-known isoindoles while their transformed versions, without exception, contain an additional molecule of OPA.

Spectra of the individual derivatives proved to be particularly informative (Table 5, Figures 7A,B). They served as unambiguous evidence both to earlier results with OPA derivatives of -NH-CH_2-bonding (Table 6), and, to the suggested reaction scheme (Figure 8). Evaluating the MS spectra the following conclusions can be drown.

Table 5 Fragmentation Patterns of the OPA/MPA and OPA/NAC Derivatives of Amino Acids

(MW_{OPA}=134.1; MW_{MPA}=106.1; MW_{NAC}=163.2)

Amino Acids, ⇓Ions ⇒	MW	***First OPA/MPA Deriva-***				***Transformed OPA/MPA Deriva-***				
		MH^+ calc.	MH^+ obt.	MH^+ - COO	MH^+ - m/z =105	MH^+ + OPA	MH^+ + OPA - H_2O	MH^+	2OPA +MPA -$2H_2O$	$OPAH^+$
Glycine	75.1	280.3	**280.0**	-	175.0	**414.0**	396.0	280.0		-
β-Alanine	89.1	294.3	**294.0**	-	189.1	**428.0**	410.0	294.0	338.0	-
Lysine	146.2	555.6	**555.2**	511.0	405.1#	**689.0**	671.0	555.0	338.0	
ε-NH_2-caproic	131.2	336.4	**336.1**	292.0	231.1	470.1	**452.2**	336.1		135.1
Amino Acids, ⇓Ions ⇒	MW	***First OPA/NAC Deriva-***				***Transformed OPA/NAC Deriva-***				
		MH^+ calc.	MH^+ obt.	MNa^+	MH^+ - m/z= 129	MH^+ + OPA	MNa^+ + OPA	MH^+ + OPA - H_2O	MNa^+ + OPA +Na	MNa^+ + OPA +2Na
Glycine	75.1	337.4	**337.1**	359.1	208.0	471.2	**493.1**	453.0	-	-
β-Alanine	89.1	351.4	**351.1**	373.0	222.0	485.1	**507.2**	467.2	-	-
GABA	103.1	365.4	**365.1**	387.1	236.1	528.6	**521.2**	481.2	543.1	564.8
Histidine	155.2	417.5	417.3	**461.4**	310.3	-	-	-	-	-
Ornithine	132.1	655.7	**655.3**	677.2	-	789.2	-	-	-	-
Lysine	146.2	669.8	**669.3**	691.4	-	803.3	**825.3**	785.3	847.3	869.3

Indications: Concentrations: [OPA]/[MPA]([NAC])/[AA]([A])=20/60/1, 1=~$1x10^{-9}$; bold printed values = abundant ions;

4.1.1. Fragments of the initially formed derivatives (Table 5: initially formed OPA/MPA and OPA/NAC derivatives (Figures 7A,B), compounds indicated by number1, as well as ornithine2 and lysine2 (the tiny amounts of ornithine1 and lysine1 could not be ionized).

(i) The protonated molecular ions (MH^+) and/or their cationized versions (MNa^+) were detected, as abundant fragments, in spectra of all initially formed derivatives, with the exception of OPA/NAC derivatives of histidine

(ii) Further selective fragment ions (SFI) were obtained by partial loss of the aliphatic chain of both SH-group additives (by losses of m/z = 72 instead of m/z = 73 in case of OPA/MPA-glycine1: Figure 7A, m/z = (280-72) = 208 and those of m/z = 129 instead of m/z =1 30 in eight spectra of the initially formed OPA/NAC-derivatives: m/z values in Table 5, seventh vertical column). The elimination of the MPA anion with the mass of m/z = 105 was obtained in the spectra of the initially formed glycine1, β-alanine1 and ε-aminocaproic acid1 while from lysine2 the fragment of m/z = 405.1 was formed by the simultaneous loss of one molecule CO_2 and MPA (Figure 7A) spectra of glycine1, β-alanine1, ε-aminocaproic acid1 and lysine2).

4.1.2. Fragments of the transformed derivatives

(Table 5: transformed OPA/MPA and OPA/NAC derivatives, Figures 7A,B, compounds indicated by number2, {with exceptions of ornithine2 and lysine2 which are initially formed OPA species}, as well as ornithine3, lysine3)

(i) The protonated molecular ions (MH^{+}+OPA) and/or their cationized version(s) (MNa^{+}+OPA) were detected in all transformed species (the only exception was histidine: its transformed derivative could not be identified except under specified conditions [71]: section 6).

(ii) Dehydrated species, obtained from the transformed molecular ions by elimination of one molecule of water (MH^{+}+OPA-H_2O), were masses of high intensity. The presence of these fragments is of primary importance corresponding to the stabilized end-product of our proposed mechanism scheme (Figure 8: compounds VII and VIII).

(iii) Fragments with the same masses, i.e., those of m/z = 278.2–279.1 are of particular importance. The existence of these ions (MW = $C_{17}H_{12}NOS$ = 278, its protonated form: m/z = 279) proved to be independent of the reagent's SH-group containing component and that of the NH_2-group containing compound. Consequently, it served as strong evidence for our proposed mechanism, being formed by the complete elimination of the aliphatic-chains of the transformed end-product, belonging both to the SH-group and to the NH_2 containing parts [69] (detailed examination of the mechanism in chapter 2.2.1. HPLC of amines as OPA-derivatives)

4.2. Basic molecular structure of the multiple derivatives furnishing primary amino-group containing compounds

As to the initial molecular structure of all of those derivatives examined in our laboratory [4–6, 67–71] we distinguished them according to their common structural characteristics, i.e., to the more than one and to the single OPA derivative providing species (Table 6). According to our data and its explanations, it is clear that the substituents of the neighboring C atom to the primary amino group play a key role. Referring to Figure 8, the mobilizable H atom of the CH_2-group next to the isoindole Nitrogen initializes the reaction with an additional OPA molecule (Figure 8: compound III), i.e., this is the first step leading to the transformation of the initially formed OPA species. Consequently, the nucleophile reaction of OPA with the isoindole must start at the CH_2 group only, following the pathway shown in Figure 8. This assumed mechanism was supported by MS data (Figures 7A,B, Table 5) and further completed by analytical results.

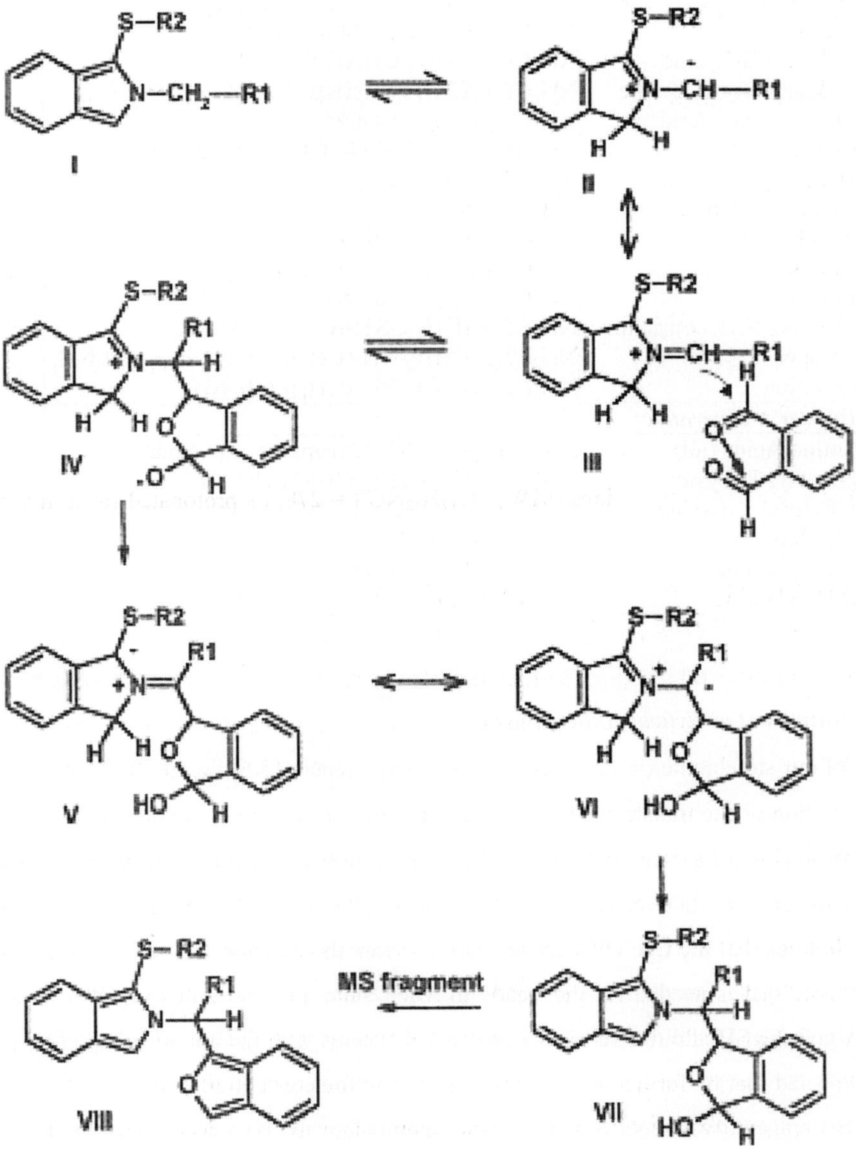

Figure 8 Reaction pathway of the formation/transformation of the multiple derivative providing amino acids and amines (list of compounds in Table 5). With permission from Reference [69]

Table 6 Structural Properties of the Multiple/Single OPA Derivatives Providing Amino Acids and Amines

More than One OPA Derivative	
Glycine	$NH_2\mathit{CH_2}\text{-}COOH$
β-Alanine	$NH_2\mathit{CH_2}\text{-}CH_2)\text{-}COOH$
γ-Aminobutyric Acid	$NH_2\mathit{CH_2}\text{-}(CH_2)_2\text{-}COOH$
Ornithine	$NH_2\mathit{CH_2}\text{-}(CH_2)_2\text{-}CHNH_2\text{-}COOH$
Lysine	$NH_2\mathit{CH_2}\text{-}(CH_2)_3\text{-}CHNH_2\text{-}COOH$
N-α-Acetyl-L-Lysine	$NH_2\mathit{CH_2}\text{-}(CH_2)_3\text{-}CHNH(CH_3CO)\text{-}COOH$
N-α-Benzyloxy-L-Lysine	$NH_2\mathit{CH_2}\text{-}(CH_2)_3\text{-}CHNH(C_6H_5CH_2O)\text{-}COOH$
ε-Aminocaproic Acid	$NH_2\mathit{CH_2}\text{-}(CH_2)_4\text{-}COOH$
Histidine	$C_3N_2H_3\text{-}CH_2\text{-}\mathit{CH}NH_2\text{-}COOH$ *(irregular)*
Ethanolamine	$NH_2\mathit{CH_2}\text{-}CH_2\text{-}OH$
n-C_1-$C_{5,}$ i-C_4 Amines	$NH_2\mathit{CH_3}$, $NH_2\text{-}(\mathit{CH_2})_{2\text{-}5}$
C_2-C_5 Diamines	$NH_2\mathit{CH_2}\text{-}(CH_2)_{1\text{-}4}\text{-}NH_2$
Agmatine	$NH_2\mathit{CH_2}\text{-}(CH_2)_3\text{-}NHC(=NH)NH_2$
Tyramine	$NH_2\mathit{CH_2}\text{-}CH_2\text{-}C_6H_4\text{-}OH$
Bis(hexamethylene)triamine	$NH_2\mathit{CH_2}\text{--}(CH_2)_5\text{-}NH\text{-}(CH_2)_5\text{-}\ \mathit{CH_2}NH_2$
Spermine	$NH_2\mathit{CH_2}\text{-}(CH_2)_2\text{-}NH\text{-}(CH_2)_4\text{-}NH)\text{-}(CH_2)_2\text{-}\mathit{CH_2}NH_2$
Spermidine	$NH_2\mathit{CH_2}\text{-}(CH_2)_3\text{-}NH\text{-}(CH_2)_2\text{-}\mathit{CH_2}NH_2$
Single OPA Derivative	
α-Amino Acids (all)*	$R\text{-}\mathit{CH}NH_2\text{-}COOH$, *except the above ones
N-ε-Acetyl-L-Lysine	$NH(CH_3CO)\text{-}(CH_2)_2\text{-}\mathit{CH}NH_2COOH$
N-ε-Formyl-L-Lysine	$NH(CHCO)\text{-}(CH2)_2\text{-}\mathit{CH}NH_2COOH$
i-Propylamine	$CH_3\text{-}\mathit{CH}NH_2\text{-}CH_3$
sec.-Butylamine	$CH_3\text{-}\mathit{CH}NH_2\text{-}CH_2CH_3$
tert.-Butylamine	$CH_3\text{-}CNH_2CH_3\text{-}CH_3$

4.3. Considerations of analytical consequences.

In the light of our stoichiometric mass spectrometric experiences [5,6] i.e., in the knowledge of the composition of the transformed derivatives, and that of the reaction pathway they are originating from (Figure 8), we decided to change the composition of the reagent. Molar ratios of OPA to the SH additive were altered for OPA/MPA and for OPA/NAC=1/50. This observation implies that the free OPA concentration favors the transformation of the initially formed isoindole that is needed for the 'ready to react' state, i.e., those of free OPA molecules. (OPA and the SH additive react as a preformed species with the amino group [7–11]). Thus, we expected that by further considerably increasing the concentration of the SH-group additive in the reagent, two benefits should result simultaneously: (i) a decrease of the transformation rate of the initially formed derivative, and, (ii) an increase of the overall stability of the total derivatives.

The realization of this assumption leaded to the expected results in the case of C_1-C_4 aliphatic monoamines (chapter 2.2.1. of this book). However in the case of the multiple derivative providing AAs this resulted in increasing number of transformed species. Consequently, for analytical purposes, the use of the OPA/MPA or the OPA/NAC reagents are to be preferred, applying strictly the same optimum analytical conditions (OPA/MPA(NAC) = 1/3; sections 7 and 8 of this chapter).

5. The special behavior of histidine

The irregular behavior of histidine in its reaction with the OPA reagents has been clarified recently [71]. Histidine provides more than one OPA derivative, similarly to all those primary amino group-containing compounds that do have the -CH_2-NH_2 moiety in their initial structure (Table 6). The ratio of histidines initially formed and transformed OPA derivatives depends on the temperature: very likely due to the fact that elevated temperature favors the intra-molecular rearrangement of histidine resulting in the formation of the -CH_2-NH_2 moiety containing tautomer(s). The higher the temperature the higher the amount of the transformed species. The composition of the initially and transformed OPA derivatives of histidine was identified on the basis of their on-line HPLC-MS(ESI) spectra, applying selective ion monitoring and computations [71] The initially formed species has been identified as the classical isoindole, while the transformed one contains an additional OPA molecule. Stoichiometric and MS studies confirmed

(i) that the transformation of the initially formed OPA derivatives of histidine was associated with its virtual tautomerization leading to the species containing the -CH_2-NH_2 moiety predestinated to react with an additional OPA molecule.

(ii) The ratio of its initially formed and transformed OPA derivatives depends on the temperature, but is independent on the composition of the OPA reagents: likely due to the fact that elevated temperature favors the formation of the -CH_2-NH_2 moiety containing HIS-tautomers.

(iii) The behavior of 3-methylhistidine (MHIS) – in its initial structure unable to tautomerize – furnishing an unambiguously single and stable OPA derivative, serves as completing evidence to the mechanism of the OPA derivatization of HIS.

(iv) The composition of its initially obtained and transformed derivatives was proved on the basis of their MS spectra, confirmed by the possible reaction pathway they are originating from, applying computation.

(v) From an analytical point of view it proved to be unambiguous that the consequences of virtual tautomerization can not be avoided: consequently, in order to obtain reliable and reproducible histidine values, its content must be calculated on the basis of the total of their derivatives.

6. Trials in order to derivatize the secondary amino group

The second main disadvantage of the OPA reagents is their inability to derivatize the secondary amino group (after the first one which is the uncertainty in the stability of derivatives). To overcome this drawback one possibility, without any analytical data, was reported even in the pioneering work [3]: the oxidation of the secondary amino group with sodium hypochlorite or chloraminT, prior to the reaction of AAs with the OPA reagents. Japanese researchers [54,55] elaborated the practical details of this two-step derivatization technique. Later on, an other derivatization method requiring two steps was patented [56] and introduced in HPLC practice [57], based on the the reaction of the secondary amino group with FMOC subsequently to the reaction of the primary AAs with the OPA/MPA reagent. Recently, a special OPA/clornitrobenzofurazan (NBF) reagent was suggested to derivatize proline and OH-proline for their HPLC separation [21]. Unfortunately, the OPA/NBF reagent does not react with the primary amino group; consequently two separate reactions (OPA/MCE and OPA/NBF) and elutions had to be performed [21].

6.1. The two-step method: oxidation of the secondary amino group by hypochlorite prior to the reaction of AAs with the OPA reagent

The two step-method has been elaborated [45], improved [46,54,55] and patented also [58] in the postcolumn mode. Involving all of disadvantages inherent to the postcolumn techniques, i.e., the need of extra pumps, temperature controllable reaction coils resulting in high cost, increased time and decreased sensitivity. Instead of OPA/MCE-oxidized proline [54,55,58], providing low fluorescence, the introduction of the OPA/NAC derivative was an advancement being as stable and as sensitive as all other OPA/NAC-AAs [45,46]. In spite of selected cases which inevitably need simultaneous quantitation of the primary and secondary AAs, because of the above detailed disadvantages the hypochlorite oxidation has not gained wide acceptance.

6.2. Double precolumn derivatisation with the OPA/MPA/FMOC reagent

The first introduction of the OPA/MPA/FMOC derivatization was patented [56] and practically described [57] in the automated version. This pioneering work [57] contained several chromatographic proposals: the quantitation of 17 primary AAs+proline within 12 min, 28 primary AAs +OH-proline and proline witine 20 min, as well as 36 primary AAs+OH-proline and proline within 40 min. All separations were performed on the same column (Hypersil ODS, 20 cm x 4.6 mm, 5μ) in order to determine the AAs of various special matrices under optimum conditions: from bovine serum albumin, wort, beer and plants with 12 min, from plasma deprotonized by acetonitrile with 20 min and by sulfosalycylic acid with 40 min run times.

The AA contents of β-lactoglobulin and peptides, including proline, have been determined by HPLC and identified by MS, in the 10–100 pM range within 12 min [58]. Further improvements have been published [51,53,59] for the quantitation AAs in plasma [51], in potato tuber (20 AAs/8 min) [59] and 42 AAs in urine [53].

7. Advances in the HPLC of AAs Quantitated as OPA/MPA or OPA/NAC Derivatives

The separation/identification possibilities of 25 AAs as their OPA/MPA and OPA/NAC derivatives have been optimized (with particular attention to those AAs which are eluting with more than one derivative (glycine, histidine, GABA, β-alanine, ornithine, lysine), and expected to be present in natural matrices, such as apples, in free forms, etc). Optimum separation conditions have been reported on six RP test column (T1–T6: detailed in section 8.5. of this chapter)

Elutions have been followed, simultaneously, with DAD and Fl detectors, connected in line. Optimization studies performed in model solutions, depending on temperature (30–55 ^{o}C) and flow rate (0.8–2.5 mL/min), of eluents proved that optimum resolutions are associated with the highest flow rate applicable, (remaining on the safe side with a column pressure of <<3500 psi), in the temperature range of 30–50 ^{o}C. 25 AAs, eluting in 31 separate, characteristic derivatives have been determined on all six columns, (the main component asparagine, present in overwhelming excess, together with the minor constituents glutamine,

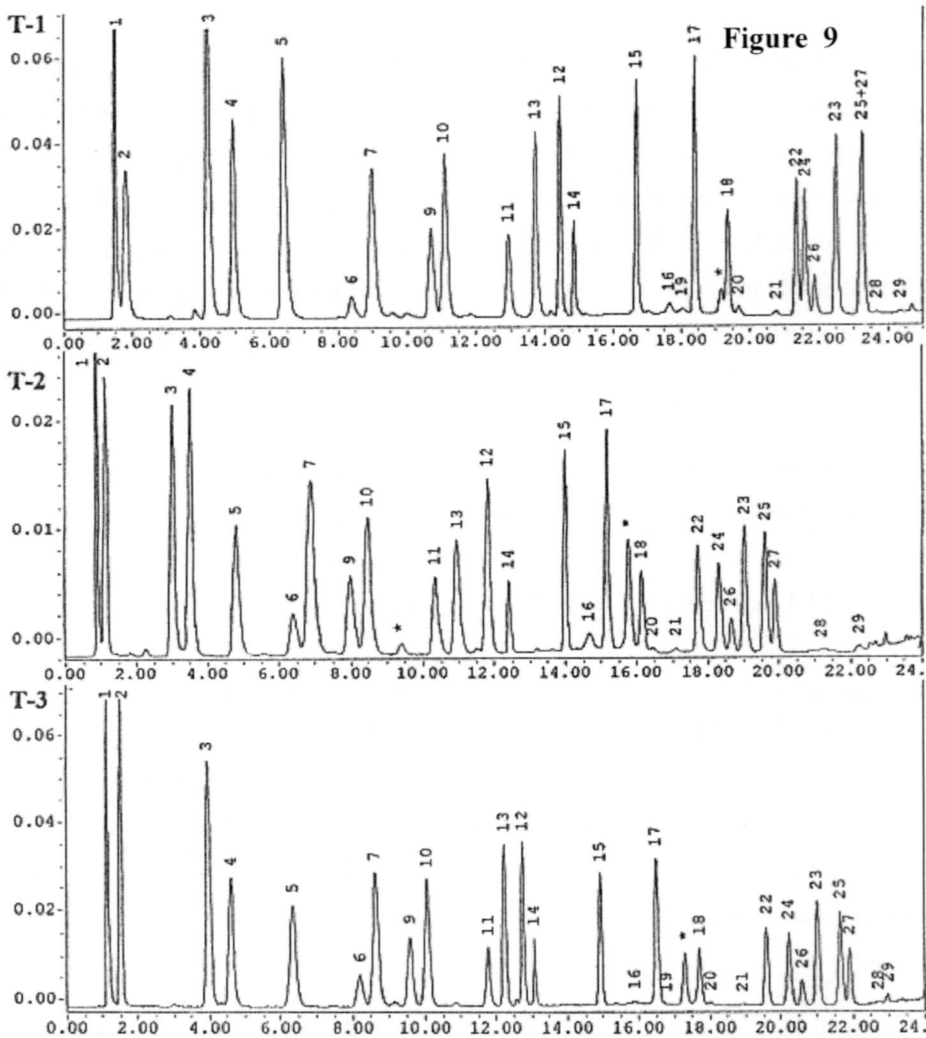

Figure 9 (above: this page)**,** ***Figure 10*** (below: next page) Elution profile of 25+4 OPA/MPA-AAs (Figure 9: T-1–T-3, columns) obtained with model solutions, as well as with the OPA/MPA derivatives of Jonagored apple on column (Figure 10A-D, T3); A: derivatized according to section 8.4.4.; B: blank test belongs to the A elution; C: derivatized in the eluate of a cation-exchanged pulp [90] and D: blank, obtained from the cation exchange process, belongs to the C elution); Peaks: 1 aspartic-, 2 glutamic acids, 3 asparagine, 4 serine, 5 glutamine, 6 histidine, 7 glycine, 8 homoserine, 9 threonine, 10 β-alanine, 11 arginine, 12 alanine, 13 GABA, 14 homoarginine, 15 tyrosine, 16 glycine2, 17 valine, 18, methionine, 19 cyst(e)ine, 20 β-alanine2, 21 GABA2, 22 tryptophan, 23 phenylalanine, 24 i-leucine, 25 n-leucine, 26 ornithine2, 27 lysine2, 28 ornithine3, 29 lysine3; *=system peaks, **not AA type impurities in apple; ***unknown apple constituents providing AA type spectra. With permission from Reference [6]

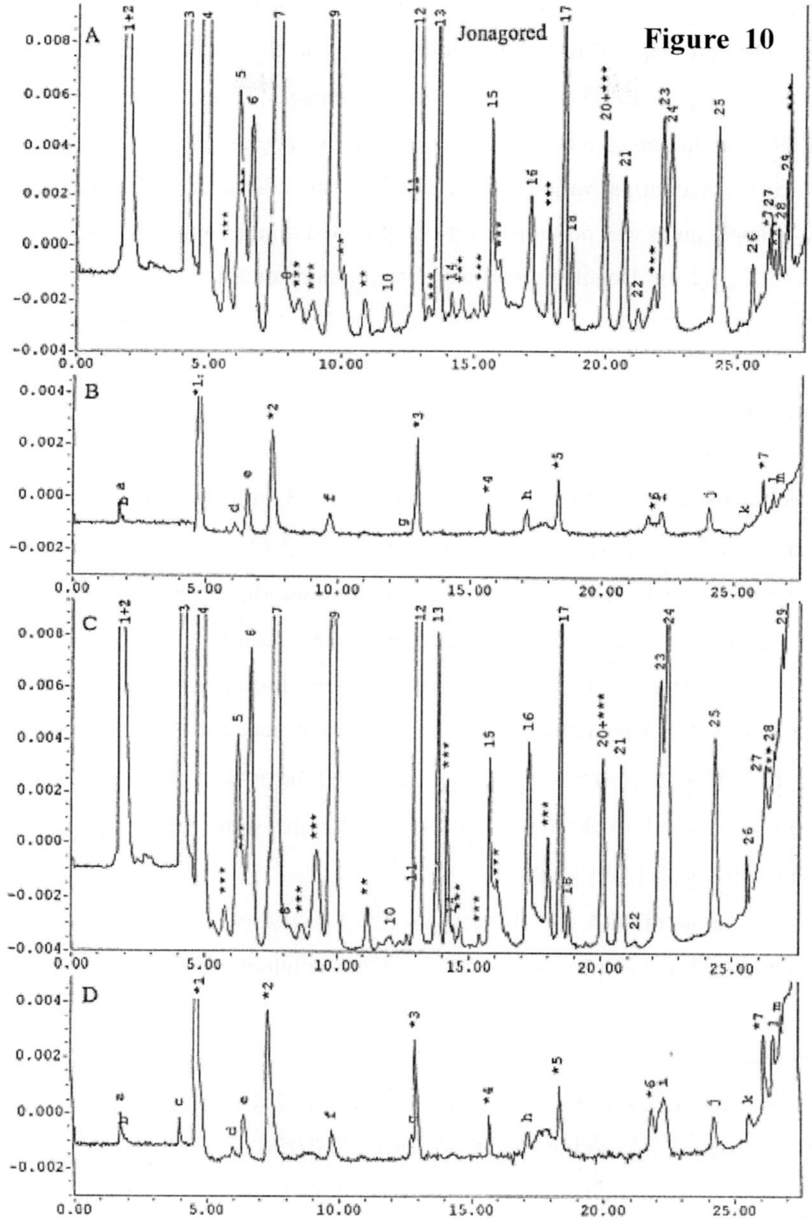

β-alanine, GABA, homoserine, and homoarginine. Optimum conditions, in the case of both derivatives, have been obtained on the same type of column (Hypersil, 5 μm), as follows: for the OPA/MPA AAs with programmed flow rate (1.3–2.3 ml/min, column: 200(+20 guard) x 4 mm), at 50 °C, while, for the OPA/NAC AAs at 2.1 ml/ min flow rate, at 30 °C, (column: 150(+20 guard) x 4 mm), with 40 and 37 min run times, including equilibration. Responses of the corresponding AAs proved to be independent of the column used [67]; Reproducibility in the concentration range of 6–12000 pmole, relating to the injected amount of AAs was <3.4 RSD % (average relative standard deviation percentages). The utility of the protocol was shown in the quantitation of the free AA content of five apple varieties (Jonagored, Idared, Jonica, Florina, Freedom), at various harvesting dates and after different storage times. Derivatization of apple pulps was performed with the filtrated sample, applying a special isolation process [67] (Figure 9). Detailed description of optimum conditions is given in section 8 of this chapter.

8. Experimental

8.1. Standard solutions

Standard solutions (ST) of free AAs have been prepared with distilled water [16], in two stock solutions: ST-1 and ST-2. ST-1 contained 2.5×10^{-4} M/L of each (L-aspartic and L-glutamic acids, L-serine, L-histidine, glycine, DL-homoserine, L-threonine, β-alanine, L-arginine, L-alanine, GABA, L-homoarginine, L-tyrosine, L-valine, L-methionine, L-cysteine, L-tryptophan, L-phenyl-alanine, L-isoleucine, L-leucine, L-ornithine and L-lysine;), except of L-glutamine and L-asparagine (ST-2). ST-2 solution contained various amounts of L-asparagine (2.5×10^{-2}–2.5×10^{-4}M/L and L-glutamine (2.5×10^{-3}–2.5×10^{-4}M/L), and has been prepared in all second week: ST-1 and ST-2 solutions were further diluted to obtain $[OPA]/[MPA]$-$([NAC]):[AA]^T \cong 6.89:1 - 120:1$.

Stock solution of OPA contained 0.256–0.512 g OPA (weighed with analytical precision) in 25 mL methanol (further on: methanolic OPA solution).

8.2. Buffer solution

Borate buffer was mixed in 50/50 volume ratios from 0.2M boric acid (dissolved in 0.2M potassium chloride) /0.2 M sodium hydroxide (pH 9.9±0.05).

8.3 Akaline solution for neutralization of apple pulps (further on: neutralization solution)

10 g sodium hydroxide, 7.4 g potassium chloride and 6.1 g boric acid have been dissolved in 100 mL distilled water (pH ≅ 12.4)

8.3 Reagent solutions

OPA/MPA reagent was obtained by mixing, in order of listing, 5.0 mL methanolic OPA, 20.0 mL borate buffer and 50–200 μL MPA solutions: final pH 9.3±0.05.

OPA/NAC reagent was prepared from 5 mL methanolic OPA solution, 20.0 mL borate buffer containing 0.1–0.4 g NAC: final pH 9.3±0.05. The mole ratios of OPA to the MPA and NAC were, without exception, OPA/MPA(NAC) = 1/3.

8.4. Derivatization

8.4.1. Characterization of the reagent solutions. Blank measurements were performed with freshly prepared (reagent's age ≥90 min [6]) reagent solutions, saved in the refrigerator, at ~4 ^{o}C, no longer as ≤9 days and injected by the robotic autosampler, every day at least two times (Waters 717, thermostated for ~4 ^{o}C).

8.4.2. Method development study with the OPA/MPA(NAC)/AA solutions.

Derivatizations were performed with reagents prepared at least 90 min earlier before use, and saved no longer as ≤9 days [2]. The same amount of reagent solutions (120 μL) were mixed with 380 μL ST-1+ST-2 of AAs, and, let them react for 7 min before injection.

8.4.3. Reproducibility study of the isoindole derivatives

Derivatizations were carried out with reagent solutions prepared at least 90 min earlier before use, and retained no longer as ≤9 days. The same amount of reagent solutions (120 μL) were mixed up with variously diluted (380 μL) ST-1 and ST-2 of AAs and injected as detailed above.

8.4.4. Preparation of the apple pulps prior to their derivatization

Peeled apples were homogenized and sieved. Pulps were used immediately, or saved in the refrigerator at -20 ^{o}C. Prior to derivatization 2–5 g pulp were weighed and filtrated as follows: a 10 mL volumetric flask, in its neck with a funnel (diameter, 5 cm), covered with a GF/F glass microfibre paper, all – i.e. flask+funnel with the paper – were weighed with 0.01 g precision (the microfibre paper was weighed also separately, with analytical precision). The 2–5 g apple pulp was pipetted into the funnel, weighed again with 0.01 g precision. The amount of the apple pulp was obtained from the difference. After the filtration was complete the undissolved residue was washed with 4 x 1 mL distilled water, and the pH of the homogenized filtrate was alkalinized, by means of a pH meter with the neutralization solution (0.35 – 0.40 mL) up to pH 9.3–10.2. Thereafter its volume was completed up to 10.0 mL (apple's stock solution). Derivatization was performed with 100 μL apple's stock solution according to section 8.4.3. (The funnel with the paper, containing the undissolved residue of apple pulps, was put into the oven, dried at 50 ^{o}C, and the dried filter + residue was weighed with analytical precision.)

8.5. Chromatography

The system was a Waters HPLC instrument (Waters Pharmaceutical Division, Milford, MA, U.S.A.), consisted of Waters^{TM} 996 PDA, and Waters 274 Fluorescence Detectors, Waters^{TM} 600 Controller quaternary pump with thermostattable column area, Waters^{TM} 717 Autosampler, operating with the Millennium Software (version 2010, 1992-95, validated by ISO 9002). Separations were optimized on six columns Test columns (T1-T6) T-1–T-3 test columns were Hypersil ODS bonded phase. T-1: 150mm x 4mm, 3μm; T-2: 150mm x 4mm, 5μm; T-3: 200mm x 4mm, 5μm. (Shandon; Supplier: BST, Budapest, Hungary). T-4 was 125mm x 4mm, 3μm, Gromsil, (Grom Analytic+HPLC; Supplier: LAB-COMP, Budapest, Hungary). T-5 and T-6: 50mm x 4mm Nucleosil 120, C_{18}, 3μm and 5μm columns (Macherey-Nagel; Supplier: BST). T-1–T-3, T-5, T-6 columns were used with 20mm x 4 mm, while T-4 with 10 mm x 4mm guard columns. (henceforth: T-1–T6 columns).

Detections have been performed simultaneously, applying the PDA (Waters 996) and Fluorescence (Fl) (Waters 274) detectors, connected in order of listing. Blank tests, concentration dependence and the apple pulps have been detected between 190 and 400 nm (PDA), evaluated at 334 nm (OPA/MPA(NAC)/AAs), as well as at the optimum Ex/Em wavelengths: at Ex/Em= 337/454 nm.

The eluent system consisted of two components: (A) eluent was 0.05 M sodium acetate of pH 7.2, (in its final version containing 1% THF, v/v) while (B) eluent was prepared from 0.1 M sodium acetate/acetonitril/metanol (46:44:10) (mixed in volume ratios and titrated with glacial acetic acid or 1 M sodium hydroxide to pH 7.2): separate gradient programs were followed: Optimum gradient programs for the OPA/MPA (Prg-1, T-3 column, 50 ^{o}C) and for the OPA/NAC AAs (Prg-2, T/2 column, 30 ^{o}C)

OPA/MPA program (Prg-1)

Step	Time: min	Flow mL	% A	% B
1	0.00	1.3	99.5	0.5
2	6.00	1.3	99.0	1.0
3	8.00	1.3	97.0	3.0
4	9.00	1.3	93.0	7.0
5	11.00	2.3	85.0	15.0
6	12.00	2.3	81.0	19.0
7	16.00	2.3	72.0	28.0
8	19.00	2.3	72.0	28.0
9	22.00	2.3	65.0	35.0
10	26.00	2.3	0.0	100.0
11	28.00	2.3	0.0	100.0
12	29.00	2.3	99.5	0.5
13	40.00	2.3	99.5	0.5

OPA/NAC program (Prg-2)

Step	Time: min	Flow mL	% A	% B
1	0.00	2.1	99.5	0.5
2	4.00	2.1	99.0	1.0
3	8.00	2.1	90.0	10.0
5	10.0	2.1	75.0	25.0
6	15.0	2.1	65.0	35.0
7	19.0	2.1	55.0	45.0
8	21.0	2.1	0.0	100.0
9	24.0	2.1	0.0	100.0
10	26.0	2.1	99.5	0.5
11	37.0	2.1	99.5	0.5

References

[1] Handbook of Derivatization Reactions for HPLC, G. Lunn, L.C. Hellwig, Editors, Amino Acids, Wiley (1998), pp 625-821.

[2] I. Molnár-Perl, Amino Acids, in Advanced Chromatographic and Electromigration Meth ods in Biosciences, Z. Deyl et al, Editors, Elsevier Science, (1998), pp 416-464.

[3] M. Roth, Anal. Chem. 43 (1971) 880-882.

[4] I. Molnár-Perl, (Review) J. Chromatogr. A 913 (2001) 283-302.

[5] I. Molnár-Perl, A. Vasanits, J. Chromatogr.A 835 (1999) 73-91.

[6] A. Vasanits, D. Kutlán, P. Sass, I. Molnár-Perl, J. Chromatogr. A 870 (2000) 271-287.

[7] M.C.G. Avarez-Coque, M.J.M. Hernández, R.M.V. Camañas, C.M. Fernández, Anal. Biochem. 180 (1989) 172-176.

[8] J.D.H. Cooper, G. Ogden, J. McIntosh, D.C. Turnell, Anal. Biochem. 142 (1984) 98-102.

[9] G.L. Lookhart, B.L. Jones, Cereal Chem. 62 (1985) 97-102.

[10] H.W. Jarret, K.D. Cooksy, B. Ellis, J.M. Anderson, Anal. Biochem. 153 (1986) 189-198.

[11] H.S. Sista, J. Chromatogr. 359 (1986) 231-240.

[12] T.A. Durkin, G.M. Anderson, D.J. Cohen, J. Chromatogr. 428 (1988) 9-15.

[13] J. Kehr, U. Ungerstedt, J. Neurochem. 51 (1988) 1308-1310.

[14] O.Orwar, S. Folestad, S. Einarsson, P. Andiné, M. Sandberg, J. Chromatogr. 566 (1991) 39-55.

[15] B.W. Boyd, R.T. Kennedy, Analyst 123 (1998) 2119-2124.

[16] A.M. Uhe, G.R. Collier, E.A. McLennan, D.J. Tucker, K. O'Dea, J. Chromatogr. 564 (1991) 81-91.

[17] D. Tsikas, J. Sandmann, D. Holzberg, P. Pantazis, M. Raida, J.C. Fröhlich, Anal. Biochem. 273 (1999) 32-40.

[18] G. Noctor, C.H. Foyer, Anal. Biochem. 264 (1998) 98-110.

[19] R.C. Dorresteijn, L.G. Berwald, G. Zomer, C.D. de Gooijer, G. Wieten, E.C. Beuvery, J. Chromatogr. A 724 (1996) 159-167.

[20] D. Fekkes, A. vanDalen, M. Edelman, A. Voskuilen, J. Chromatogr. A 669 (1995) 177-186.

[21] B.H. Klein, J.W. Dudenhausen, J. Liqu. Chromatogr. Rel. Techn. 18 (1995) 4007-4028.

[22] G. Michael, G. Henrion, Git. Fachz. Lab. 9 (1995) 769-773.

[23] G. Georgi, C. Pietsch, G. Sawatzki, J. Chromatogr. 613 (1993) 35-42.

[24] S.S. Simons, Jr., D. F. Johnson, J. Am. Chem. Soc. 98 (1976) 7098-7099.

[25] S.S. Simons, Jr., D. F. Johnson, J.C.S. Chem.Comm. (1977) 374-375.

[26] S.S. Simons, Jr., D. F. Johnson, Anal. Biochem., 82 (1977) 250-254.

[27] S.S. Simons, Jr., D. F. Johnson, J. Org. Chem. 43 (1978) 2886-2891.

[28] S.S. Simons, Jr., D. F. Johnson, J. Chromatogr. 261 (1983) 407-414.

[29] J.F. Stobaugh, A.J. Repta, L.A. Sternson, J.Org Chem. 49 (1984) 4306-4309.

[30] O.S. Wong, L.A. Sternson, R.L. Schowen, J. Am. Chem. Soc. 1077 (1985) 6421-6422.

[31] R.C. Simpson, J. Spriggle, H. Veening, J. Chromatogr. 261 (1983) 407-414.

[32] P.A. Metz, J. Gehas, J. Spriggle, H. Veening, J. Chromatogr. 330 (1985) 307-313.

[33] P.A. Tippet, B.E. Clayton, A.I. Mallet, Biom. Env. Mass Spectr. 14 (1987) 737-738.

[34] S.Palermo, M. deMarchis, M. prati, E. Fugassa, Anal. Biochem.. 202 (1992) 152-158.

[35] R.G.J. van Leuken, A.L.L. Duchateau, G.T.C. Kwakkenbos, J. Pharm. Biomed. Anal. 13 (1995) 1459-1464.

[36] H.M.H. van Eijk, D.R. Rooyakkers, P.B. Soeters, E.P. Deutz, Anal Biochem. 271 (1999) 8-17.

[37] V.J.K. Svedas, I.J. Galajev, I.L. Borisov, I.V. Berezin, Anal. Biochem. 101 (1980) 188-195.

[38] J.F. Stobaugh, A.J. Repta, L.A. Sternson, K.W. Garren, Anal Biochem. 135 (1983) 495-504.

[39] J.F. Stobaugh, A.J. Repta, L.A. Sternson, J. Pharmaceut. Biomed. Anal. 4 (1986) 341-351

[40] J.R. Cronin, S. Pizarello, W.E. Gandi, Anal. Biochem. 93, (1979) 174-179.

[41] A.P. Halfpenny, Ph.R. Brown, HRC, 8 (1985) 243-248.

[42] D.W. Hill, F.H. Walters T.D. Wilson, J.D. Stuart, Anal. Chem. 51 (1979)1338-1341.

[43] P. Kucera, H. Umgat, J. Chromatogr. 255 (1983) 563-579.

[44] H. Godel, T. Graser, P. Földi, P. Pfaender, P. Fürst, J. Chromatogr. 297 (1984) 49-61.

[45] N. Nimura, T. Kinoshita, J. Chromatogr. 352 (1986) 169-177.

[46] M. Fujiwara, Y. Ishida, N. Nimura, A. Toyama, T. Kinoshita, Anal. Biochem. 166 (1987) 72-78.

[47] P. Lindroth, K. Mopper, Anal.Chem. 51 (1979) 1667-1674.

[48] G.A. Qureshi, L. Fohlin, J. Bergström, J. Chromatogr. 297 (1984) 91-100.

[49] B.J. Mixcallef, B.J. Shelp, R.O. Ball, J. Liqu. Chromatogr. 12 (1989) 1281-1300.

[50] U. Bütikofer, D. Fuchs, J.O. Bosset, W. Gmür, Chromatographia 31 (1991) 441-447.

[51] H.G. Worthen & H. Liu, J. Liqu. Chromatogr. 15 (1991) 3323-3341.

[52] H. Brückner, T. Westhauser, Chromatographia 39 (1994) 419-426.

[53] C. Carducci, M. Birarelli, V. Leuzzi, G. Santagata, P. Serafini, I. Antonozzi, J. Chromatogr. A 729 (1996) 173-180.

[54] Y. Ishida, T. Fujita, K. Akai, J. Chromatogr. 204 (1981) 143-148.

[55] J. Haginaka, J. Wakai, Anal Biochem. 171 (1988) 398-403.

[56] R. Schuster, A. Apffel, Appl. Note Pub.No. 12-5954-6257.

[57] R. Schuster, J. Chromatogr. 431 (1988) 271-284.

[58] D.T. Blankenship, M.A. Krivanek, B.L. Ackermann, A.D. Cardin, Anal. Biochem. 178 (1989) 227-232.

[59] T. Bartók, G. Szalai, Zs. Lõrincz, G. Börcsök, F. Sági, J. Liqu, Chromatogr. 17 (1994) 4391-4403.

[60] V Rizzo, A. Anesi, L. Montalbetti, G. Bellantoni, R. Trotti, G.V. Melzi d'Eril, J. Chromatogr. A 729 (1996) 181-188.

[61] T. Teerlink, P. A.M. vanLeeuwen, A Houdijk, Clin. Chem. 40 (1994) 245-249.

[62] H.M.H. van Eijk, D.R. Rooyakkers, N.E.P. Deutz, J. Chromatogr. 620 (1993) 143-148.

[63] M. Zhao & J.L. Bada, J. Chromatogr.A 690 (1995) 55-63.

[64] H. Brückner, M. Langer, M. Lüpke, T. Westhauser & H. Godel, J. Chromatogr.A 697 (1995) 229-245.

[65] H.P. Fitznar, J.M. Lobbes, G. Kattner, J. Chromatogr.A 832 (1999) 123-132.

[66] A. Hashimoto, T. Nishikawa, T. Oka, K. Takahashi, T. Hayashi, J. Chromatogr. 582 (1992) 41-48.

[67] A. Vasanits, D. Kutlán, P. Sass, I. Molnár-Perl, J. Chromatogr. A 870 (2000) 271-287.

[68] D. Kutlán & I. Molnár-Perl, Chromatographia 53 (2001) 1-11.

[69] Y. Mengerink, F. Tóth, D. Kutlán & I. Molnár-Perl, J. Chromatogr. A 949 (2002) 99-124

[70] R. Hanczkó & I. Molnár-Perl, Chromatographia 57 (2003) S-103–S-113.

[71] A. Csámpai, D. Kutlán, F. Tóth & I. Molnár-Perl, J. Chromatogr. A 1031 (2004) 67-78.

[72] F.R. Antoine, C.I. Wei, R.C. Littel, M.R. Marshall, J. Agric. Food Chem. 47 (1999) 5100-5107.

[73] M.R. Alberto, M.E. Arena, M.C. Manca de Nadra, Food Control 13 (2002) 125-129.

[74] P. Lehtonen, M. Saarinen, M. Vesanto, M.L. Riekkola, Z. Lebensm. Unters. Forsch. 194 (1992) 434-437.

[75] P. Gamache, E. Ryan, C. Svendsen, K. Murayama, I.N. Acworth, J. Chromatogr. 614 (1993) 213-220.

[76] D. Egger, G. Reisbach, L. Hültner, J. Chromatogr. B 662 (1994) 103-107.

[77] T.B. Jensen, P.D. Marley, J. Chromatogr. B 670 (1995) 199-207.

[78] O. Busto, M. Guasch, J. F. Borrull, J. Chromatogr. A 718 (1995) 309-317.

[79] O. Busto, M. Mestres, J. Guasch, F. Borrull, Chromatographia 40 (1995) 404-410

[80] Y. Feng, A.E. Halaris, J.E. Piletz, J. Chromatogr. B 691 (1997) 277-286.

[81] E.M. Siaghy, Y. Devaux, H. Schroeder, N. Sfaksi, D. Ungureanu-Longrois, F. Zannad, J.P. Villemot, P. Nabet, P.M. Mertes, J. Chromatogr. B 745 (2000) 279-286.

[82] A. Sass-Kiss, E. Szerdahelyi, , G. Hajós Chromatographia 51 (2000) S-316-S-320.

[83] E. Soufleros, M.L. Barrios, A. Bertrand, Am. J. Enol. Vitic. 49 (1998) 266-278.

[84] T.P. Piepponen, A. Skujins, J. Chromatogr. B 757 (2001) 277-283.

[85] G. Achilli, , G.P. Cellerino, G. Melzi d'Eril, J. Chromatogr. A 661 (1994) 201-205.

[86] Y. Yokoyama, O. Ozaki, H. Sato, J. Chromatogr. A 739 (1996) 333-342.

[87] I.M. Mackie, L. Pirie, A.H. Ritchie, H. Yamanaka, Food Chem. 60 (1997) 291-295.

[88] R.H. Buck, K. Krummen, J. Chromatogr. 387 (1987) 255-265.

[89] M.Catalá-Icardo, M.J.M. Hernandez, M.C.G. Alvarez-Coque, J. Liq. Chromatogr. 18 (1995) 2827-2841.

[90] V. Fábián, M. Morvai, M. Pintér-Szakács, I. Molnár-Perl, J. Chromatogr. 553 (1991) 87 92.

Ibolya Molnár-Perl (Editor)
Quantitation of Amino Acids and Amines by Chromatography
Journal of Chromatography Library, Vol. 70

1.2.4.

HPLC of Amino Acids as Chloroformate Derivatives

Björn Josefsson

Contents

Summary

The chloroformate chemistry of AAs is reviewed taking different aspects into account such as reactivity, method constructions, automation, sensitivity, detector arrangements, sample matrix problems. The exceptional vigourous derivatization reaction is sensitive to pH and buffer capacities. These influence the formation of FMOC-derivatives with tyrosine and sometimes histidine. During stronger basic conditions coupling as well as deblocking of AAs occurs.

The FMOC-Cl, a classical reagent within peptide synthesis, is proved to yield derivatives without racemization. The analogues (+)- and (-)- FLEC chiral reagents are useful for trace analysis of enantiomeric AAs. The optical purity problems in ascertaining accuracy in the enantiomer determinations with chiral chloroformate reagents are discussed.

Anthracene is proposed as a chromogenic label instead of fluorine since the sensitivity is considerably raised. For instance as a consequence of the attained sensitivities the matrix problems in complex constituted samples can be solved just by dilution. High resolution separation systems with capillary columns and laser induced fluorescence detection are exemplified.

1. Introduction

1.1 *General chemistry of chloroformates*

The first chloroformates were prepared already 1834 by Dumas [1]. The chloroformates acquired great attention in connection with the development of new polymers in the plastic industry. The wide range of reactions with different functional groups is presented in a comprehensive review by Matzner et al. [2].

In 1974 Carpino [3,4] proposed a successful method to protect the amino groups during the peptide synthesis by using 9-fluorenylmethyl chloroformate (FMOC) chloride reagent (FMOC=9-fluorenylmethyloxycarbonyl). The reaction is catalyzed by a base and it proceeds very rapidly (seconds) in water at room temperature. The chloroformates react in buffered solutions with the amino functions to form carbamates (urethanes) leaving hydrogen chloride under so-called Schotten-Baumann conditions. The mechanism is of S_n2 type, where the amines are the nucleophilic components. The reaction claim appropriate buffered basic interval to be selective for primary and secondary amino groups. The chloroformate reacts also with water yielding hydrogen chloride, carbon dioxide and the hydroxy compound from which it is derived from. Besides, a symmetrical carbonate is formed presumably by the reaction of the starting chloroformate with the hydroxy side-compound. The reagent excess as well as the side-products can be observed in the subsequent chromatography in case they are not removed. The by-products amounts are dependent on the excess conditions, pH intervals and period of time. The presence of ionizable acidic groups in the amino acids makes the removal of excess hydrophobic reagent relatively simple, either by extraction with nonpolar solvents or to let the reagent excess react with hydrophobic amine thereby consuming the reagent. In the latter case the added amine derivative will appear after the amino acids in the chromatogram. The rapid derivatization of primary and secondary amino acids, the formed stable derivatives gave the foundations of the first fully automated amino acid HPLC method [5,6].

Chloroformates appear under other names as chlorocarbonates and oxycarbonyl chlorides, however, in this presentation chloroformates designation is consequent upon CA current practice of indexing.

1.2 *Protein amino acids*

The precolumn HPLC reagent FMOC-Cl was found to possess certain attractive properties suitable for sensitive, rapid determination of AAs. Thus the sample can be stabilized for subsequent manual or automated performed analysis. The reagent itself is fluorogenic, which utilized in a proper way leads to a highly sensitive method with minimal sample preparation. Thereby the excess problems have to be solved adequately.

The introduction of FMOC-Cl reagent for AA determinations was developed by Einarsson et al. [5]. The reaction was carried out in a borate buffer pH 7.7 within 30 s. at room temperature. The excess of reagent and the side-product the hydrolysis product (FMOC-OH) were extracted away twice with pentane. The left solution was ready for HPLC determination with fluorescence detection. In such a way 16 AAs in physiological fluids as blood serum diluted (1:50), cerebrospinal liquid (1:10) and urine (1:200) were determined after HPLC-separation within 20 min. In the range 0.1 – 50 μM the R.S.D were 2.4 - 6.4 %. The method was evaluated by comparison studies with classical IEC-Ninhydrin method on protein hydrolyzates and the OPA method on cerebrospinal fluid. Cunico et al. [6] automated the method where all operations were made in the same sample container (Autosampler equipment). The on-line performance includes even sample hydrolysis since the matrix interference was low. They modified the derivatization and extraction conditions somewhat compared to Einarsson et al. [5]. The derivatization is then performed with a bicarbonate buffer at pH 8 in 10 min. The on-line extraction of the reagent, and the FMOC-OH was carried out with a pentane/ethyl acetate (1/1) mixture once. The method Amino Tag (Varian TM) was applied to AA composition determinations of different peptide acid hydrolyzates. The results obtained were in good agreement with the ninhydrin method and the expected composition derived from known sequence data. Näsholm et al. [7] applied the Einarsson et al. [5]. method to study extracted plant materials.

Free AA compositions of leaf exudates and pheom sap were determined by Weibull et al. [8]. They measured 19 AAs, however, Asn and Gln were overlapping in the chromatogram. Quantitation of AAs in collagen hydrolysis products was carried out by Miller et al. [9]. The secondary AAs 3-hydroxyproline, 4-hydroxyproline and proline were completely resolved from all the other protein-derived AAs within 18 min. The AA composition of the collagen chains were well in agreement with data obtained by ion-exchange chromatography (IEC). They used the slightly modified Amino Tag method and found that it was possible to quantitate the secondary AAs together with the other 18 AAs in a wide concentration range 1 – 800 pM. The derivatization reaction time was 10 min with approximately 20-fold excess of reagent. The reagent excess and FMOC-OH was removed by extraction twice with pentane/ethyl acetate (90/10). The separation was completed at 34 min. followed by an 11 min. equilibrium interval with starting buffer mixture. The total running and recycling time required 45 min.

After methanesulfonic acid hydrolysis of leaves, larvae and their faeces, the AAs compositions were determined by Malmer et al. [10]. They found that derivatization at pH

8.0 gave reproducible results for measured AAs. Sodium borate buffer was easier to use and gave better precision of recovery than sodium bicarbonate buffer. The reagent access was extracted twice with pentane. Studies on reagent excess 1.5 to 200 (molar ratio to total AAs) showed no derivative formation difference. The formed derivates were stable for a 24 h runs after loading the autosampler. Complete HPLC separation was achieved during a 40 min. run time. The AAs in the biological samples were determined with a mean C.V. under 5 % for all of them on daily routine basis.

As an alternative to solvent extraction of reagent excess Betnér et al. [11] introduced a scavenging method. The reagent surplus is consumed by a hydrophobic amine, 1-aminoadamantane (ADAM) where the derivative appeared after Lys at the end of the chromatogram. Excess of ADAM is added shortly after the ordinary derivatization to suppress the FMOC-OH formation. With this modification Gustavsson et al. [12] derivatized the AAs in a borate buffer pH 7.7 and an incubation for 45 s before autoinjection. Instead of ADAM Kirschbaum et al. [13] used heptylamine in a similar way. Arnold et al. [14] applied the FMOC-ADAM method to measure 27 AAs in coffee beans. Carratù et al. [15] used the FMOC-ADAM method to determine AAs in parenteral solutions. Fabiani et al. [16] could quantitatively in the mM range determine AAs in different juices. The samples were directly derivatized after centrifugation using the ADAM scavenger method.

Haynes et al. [17, 18] worked out a method to obtain only monosubstituted FMOC derivatives of His and Tyr with no requirement of extraction procedure. The derivatization was proceeded in borate buffer pH 8.5 for 90s. Thereafter an alkaline cleavage reagent solution consisting of sodium hydroxide, hydroxylamine hydrochloride and 2-(methylthio)ethanol was added for 3.5 min. The reaction was then stopped by the addition of a quenching reagent of acetonitrile/water/acetic acid (20/3/2). These reagents convert the disubstituted His and Tyr into stable mono derivatives. The mixture was injected into the HPLC column. The FMOC-hydroxylamine product remains soluble in the reaction mixture and does not interfere with the chromatography of the AA derivatives. The quenching reagent also consume the excess FMOC-Cl reagent to the FMOC-hydroxylamine products thus the removal of the excess unnecessary. The addition of the thioether, 2-(methylthio)ethanol was to act as an antioxidant while Met otherwise is oxidized to sulphone and sulphoxide artefacts. This method automated was applied to determine AAs in protein hydrolyzates as well at free AAs in physiological fluids. The method was involved in a screening programme for detection of AA metabolism disorders. Ou et al. [19] found that the Haynes [18] method gave high reproducibility of the chromatography. The lifetime of the column covered 800 runs of samples from blots of two-dimensional gels. The method was also routinely used for determination of mucin glycoprotein AAs [20].

Bank et al. [21] drastically changed the derivatization conditions to 40 min in the borate buffer pH 11.4 at ambient temperature. They claimed that they exclusively obtained mono His and di Tyr eluted as free peaks in the chromatogram. They practised 75-fold reagent and three times extraction with pentane. The appeared FMOC-OH peak was 30 % lower than during the ordinary reaction conditions.

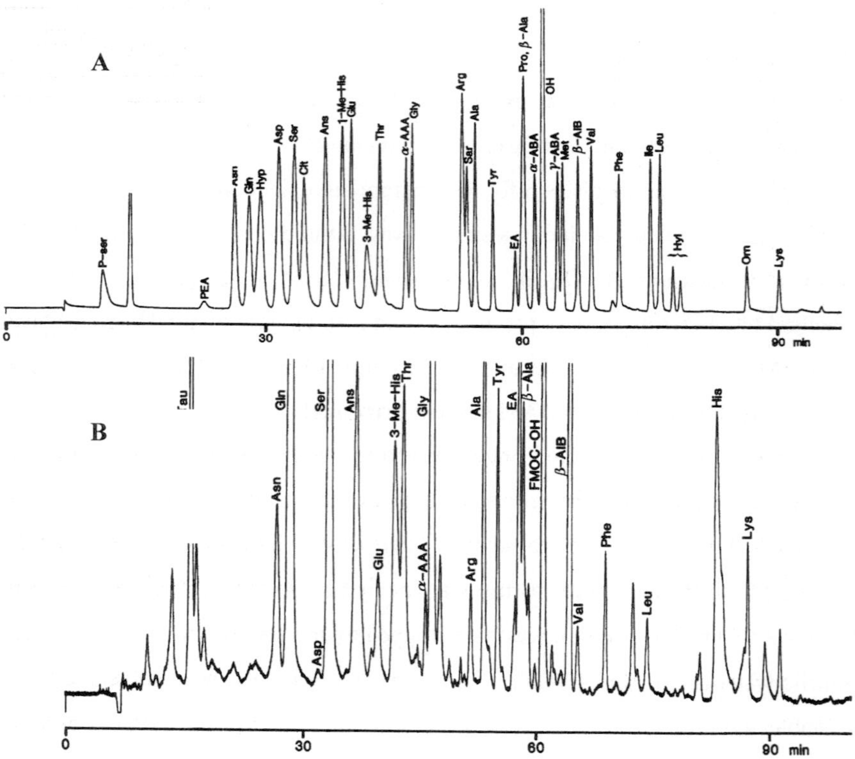

Figures 1 A,B Gradient separation of a standard AAs mixture (A). Column: 500 x 0.226 mm fused silica, Spherisorb octyl, 5μm. Gradient: 0-15 min. 25 % ACN / 60 % buffer (pH 4.08) / 15 % H_2O (26/60/15) 15-90 min. 25/60/15-65/28/7. Flow rate: 2 μl /min. Separation of an urine sample (B). The sample was diluted with distilled water (1:19) before derivatization. Separation conditions are as above. (Reproduced with permission from Reference [94].)

Packed fused silica capillaries (50 cm x 0.226 mm ID) for high resolution RP-HPLC separation was applied to separate and determine free AAs in urine [22] Figure 1. The derivatized AAs were protonized to utilize the peak compression effect in the injection. In contrast, they showed that a short column 75 x 4.6 mm ID with a mobile phase ACN/buffer (pH 4.3) gradient could resolve 18 protein AAs in about 6 min. Figure 2.

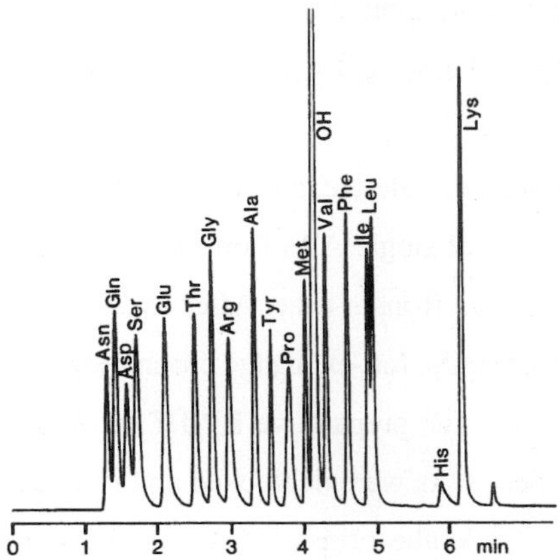

Figure 2. Separation of a standard mixture of protein AAs. Column: 50 x 4.6 mm Spherisorb ODS-2, 3μm. Moile phase: ACN / buffer pH 4.31, linear gradient 22-60 % ACN from 0.7-6.0 min. (Reproduced with permission from Reference [94])

1.3.Selected amino acids

Lysine was determined in biological samples (e.g. rumen fluid and hydrolyzate of mixed microorganisms) with isocratic HPLC within 6 min. by Or-Rashid et al. [23]. The isolated samples including also human plasma and human urine were derivatized at pH 10 and the reagent excess was extracted twice with heptane. In a subsequent work [24] they extended the study to include Thr, Met, Cys Hcys, MSD, MSO_2, Hser, Gly, 2-aminobutyric acid and cystathionine with a ternary gradient and UV detection. They observed that no degradation of the derivatives occurred up to 20 h with the exception of Hcys, Met and cystathionine. However, these tolerated 4 h storage time. Sultana et al. [25] determined Arg, Cit, Orn, Pro, 4-aminobutyric acid, 5-aminovaleric acid and Lys in rumen fluids with some changes. The FMOC-Cl reagent was solved in acetonitrile instead of acetone to prevent surface adsorption of AAs in Eppendorf tubes. They found that optimum pH for the derivatization was 9.2.

Cubedo [26] carried out quantitative analysis of Met, Cys and Lys in feedstuffs. The samples were subjected performic acid pretreatment to protect sulfur AAs before acid hydrolysis thereby Met and Cys were analyzed as methionine sulfone and cysteic acid respectively. Thomas et al. [27] quantitated alk(en)yt-L-cysteine sulfoxides and some related AAs in extracts of selected plant tissues from the genus *Allium*.

Bauza et al. [28] determined amines and those of their precursor AAs in urine Phe, Tyr and His. They used the method by Haynes et al. [18] with cleavage and quenching reagents.

Hydroxyproline was studied to estimate the collagen content in different tissues by Schilb et al. [29]. The borate buffer for the derivatization was changed to pH 9.5 and the reagent was dissolved in THF. They injected the sample after centrifugation directly into the HPLC column without any extraction. Bank et al. [30] applied their earlier published method [21] (reaction time 40 min. pH 11.4) to HOLys, HOPro and glycosylated HOLys in collagenous proteins.

O´Connell et al. [31] developed a preparative scale technique to isolate and measure both carbon and nitrogen isotopic values (^{13}C, ^{15}N) of single AAs from protein hydrolyzates. The archaeological samples were bone collagen from Roman time. After hydrolysis the AAs were derivatized using the ADAM scavenger technique. Ion-exchange chromatography could not be used since it leads to isotopic fractionation. The preparative RP-HPLC was here carried out on 250 x 25 mm ID column. Ten of twenty AAs were deblocked and isolated by base hydrolysis. The FMOC-AA is then cleaved to yield the original AA and the molecule 9-methylene-fluorene.

Isocratic HPLC method to determine 3-methylhistidine in urine was developed by Teerlink et al. [32]. The excess of FMOC-Cl was removed by extraction with diethyl ether twice.

The neurotoxic AA β-N-methyl-L-alanine (BMAA) is found in free form in seeds of *Cycas* species, a palmtree. The AA is suggested as a trigger of e.g. Parkinsonism-dementia. Kisby et al. [33] reported a reproducible and highly sensitive procedure for the subpicomole level determination of BMAA in seeds, monkey serum, cerebrospinal fluid and brain tissue. The FMOC-Cl reaction time was increased to 10 min. to ensure complete derivatization of all free AAs. The BMAA eluted last or 19 min. in a binary gradient HPLC system. Shah et al. [34] studied nonprotein AAs in *Acacia* seed for taxonomic classification. The optimum conditions for the derivatization of the AAs were found to be the same as Einarsson et al. [5], Niedbalski et al. [35] and Ringer [36] reported a sensitive HPLC method for the determination of phosphoserine, phosphothreonin and phosphotyrosine as FMOC derivatives isocratically on an ion-exchange column.

S-carboxymethyl-L-cysteine and some of its metabolites were determined in urine and serum with the FMOC-Cl reagent [37,38]. The sample need no clean-up and could directly be derivatized in boric acid buffer pH 7.7 during 50 s. The derivatization was stopped by the addition of 0.5 M ammonium acetate before injection. The excitotoxic AAs, kainic acid and domoic acid which cause selective neurodegeneration as e.g. Alzheimer's disease were studied by measurements of the release of Glu and Asp from rat brain synaptosomes [39,40]. Gilbert et al. [41] used HPLC ion-exchange chromatography as a clean-up procedure to measure the basic AAs. His, Arg, Orn and Lys in biological samples. L-tryptophan together with neutral AAs was determined in plasma of children with insulin-dependent diabetes [42].

The haemoglobin-sulphur mustard adducts were used as biomarkers for exposure to sulphur mustard [43]. The alkylation could be studied with the globin adducts of His, Glu, Asp, Cys and N-terminal Val. The adducts were tentatively quantitated and finally identified mass-spectrometrically as X-(2-hydroxyethylthioethyl)-AAs. The AA-adducts were derivatized according to ref [5] and analyzed with LC-tandem MS. The fumigant methyl bromide is a methylating agent with mutagenic and probably carcinogenic properties. The exposure to that chemical could be monitored from produced blood protein adducts [44]. With the FMOC-Cl method on hydrolyzed albumin and globin samples the S-methyl Cys was determined in the fmol range.

Van Eigk et al. [45] isolated and quantitated isotopically labelled AAs from biological samples. The FMOC-Cl method was used since the label could be removed (deblocked) after preparative LC separation. In the derivatization procedure acetone was change to ACN, thus enabling the preparation of more concentrated FMOC-Cl solutions. They increased the reaction pH to 8.75 thus increasing the capacity of the borate buffer to ensure better reproducibility. Gartenmann et al. [46] studied 17 AAs with RP-HPLC/ESI-MS. Underivatized AAs at nmol concentration resulted in that only Met, Tyr and Phe could be identified. The relative abundance of the ions was low suggesting poor ionization of the analyte. However, FMOC-derivatized AAs at pmol concentrations resulted in the identification of all AA residues.

2. Derivatization

2.1 *Derivatization conditions*

The reaction FMOC-Cl with the aminfunction is very rapid and completes in about a minute at pH 7.8-8.0. At higher pH the phenolic function of Tyr participate to give a mixture of mono- and disubstituted Tyr. His with an imidazole moiety is double derivatized. Dependent on pH and period of time the diform is gradually converted to the monoderivative. The higher pH the faster is the return to the monosubstituent. The kinetic is reported with different figures since the time interval change between derivatization and injection. The pH of the derivatization reaction mixture is crucial.

The optimal conditions should be maintained independent of the sample compositions. The buffers used, boric acid pK_a=9.2 and carbonate pK_1 =6.4 pK_2=10.3 and combinations of them, are not ideal since they have poor capacity in the range up to pH 8.0. Phosphate buffer pH range 6.2 – 8.2 or Tris 7.3 - 9.3 are not feasible since they react with the FMOC-Cl. Other buffers have been discussed as e.g. sterically hindered phenolic buffers e.g. sodium-2,6-dichlorophenol-6-sulphonate pK_a7.7 might be an interesting alternative [47,48].

The hydrolysis rate of the FMOC-Cl to FMOC-OH increases with higher pH. It is desirable to keep the FMOC-OH peak small in the middle of chromatogram. The FMOC-OH formation is dependent on excess of reagent, pH and reaction time. In ideal case a reagent excess up to 5 molar ratio, pH 7.8 and about 1-2 min. reaction time results in that the sample can be directly run on HPLC.

The variability of residual HCl in protein hydrolyzates leads to that the pH should be checked and adjusted if necessary to 7.5–8.0. Hydrolyzates may contain components e.g. antioxidants as phenols which consume the reagent. Commonly, a reagent excess of 10–100 times (molar ratio) is used. Then there is a need to terminate the reactions by extraction or quenching.

Banks et al. [21] recommended quite another reaction scheme in 0.1 M borate buffer (pH 11.4, 75 times reagent excess and 40 min.) During these conditions His appeared only as monosubstituted and Tyr as disubstituted derivatives. The FMOC-OH peak was broad, however, free in the chromatogram. At this pH of 11.4 the AA derivatives are not stable. Therefore the derivatized samples were diluted with boric acid buffer to lower the pH to 9.4. Acetonitrile (25 % $^V/_V$) was also added to prevent adsorption on surfaces.

The derivatization at pH 11.4 leads to that other functional groups than amines are accessible to reactions that diminish the selectivity [2].

The derivatization conditions are dependent on the nature of the sample. There is no one common recipe. The high detection sensitivity of the FMOC groups makes the possibility to dilute the sample and thus suppress the matrix influence. Physiological fluid can be subjected to derivatization after 100-fold dilution [5].

2.2 *Solvent extraction of excess reagent*

The formed AA derivatives are ionic thus easy to isolate from reagent, reagent by-products and extractable hydrophobic components. Different solvents have been proposed. Pentane extraction two or three times works well without loss of AA derivatives. One extraction with pentane/ethyl acetate (1/1) was introduced by Cunico et al. in the Amino Tag automated method [6]. The solubility of the FMOC-Cl reagent in water is low necessitating the presence of acetone in the reaction mixture. The pentane extractions reduce the acetone content by a factor of four [5]. This allows larger volumes to be injected into the RP-HPLC column.

2.3 *Adding of scavenging hydrophobic amines*

After the derivatization of the AAs the excess FMOC-Cl is reacted with a hydrophobic amine, 1-aminoadamantane (ADAM) to form a hydrophobic amine-FMOC complex [11,12]. This derivative has a longer retention time than the AAs and appears at the end of the chromatogram. The ADAM scavenger reagent 40mM in water/acetone (1/3) is added 45 s and/or not exceeding 60 s after the FMOC-Cl introduction. In that way the FMOC-OH concentration is kept low. Then after at least 30 s the mixture is automatically injected. The FMOC-Cl/ADAM ratio should be approximately 1/40 in order to minimize the FMOC-OH peak.

Attention should be drawn to factors for the reproduction of the method. The HPLC must be completely re-equilibrated between runs thus at least a ten minutes extra is needed.

2.4 *Cleavage and quenching reagents*

The Haynes et al. [17] method leads to that the Tyr and His are appearing as mono substituted derivatives and the excess FMOC-Cl reagent is removed without the need of solvent extraction. After derivatizing (note pH 8.5 which yields di Tyr derivative) a cleavage reagent solution is added for 3.5 min. followed by a quenching reagent to stop the reaction. The mixture is then ready for HPLC.

The cleavage reagent consist of a mixture of 680 µl of 850 mM NaOH solution with 300 µl of 500 mM hydroxylamine (H_2NOH) and 20 µl of 2-(methylthio)ethanol. The NaOH and H_2NOH components are proved to effectively convert the di FMOC-Tyr and di FMOC-His to the monoderivatives. The cleavage reagent has little effect on the other FMOC derivatives. Furthermore, the reagent converts the excess FMOC-Cl to unreactive FMOC-hydroxylamine product which appear late in the chromatogram. Methionine was oxidized by hydroxylamine to sulphone and sulphoxide derivatives. Then the added antioxidant 2-(methylthio)ethanol in the cleavage solution promoted the normal FMOC-Met derivative formation. The quenching reagent consists of ACN/H_2O/CH_3COOH (20/3/2). This reagent stops the quenching reactions. The prepared solution is ready for injection without extraction. The FMOC-derivatives here are stable for at least 24 h. Note that the rest-products appear in an area of the chromatogram without interference of the AA derivatives.

3. Chromatographic conditions of the FMOC amino acids

3.1. Column

Studies on protein hydrolyzates with a limited number of AAs are generally performed on 150 x 4.6 mm columns (Table 1). Separation of 18 protein AAs could be achieved within about 6 min. on a short 50 x 4.6 mm column Spherisorb ODS-2, (3 μm) with Ile, Leu partly overlapped [22] Figure 2. Complex mixtures of AAs as in e.g. physiological fluids may call on more efficient columns especially when trace amounts of exclusive AAs are intended to be measured. Haynes et al. [18] showed that a 30 AA standard at 100 pmol level could be separated in 28 min. on a 150 x 4.6 mm Spherisorb ODS-2 (3 μm), protected by a guard column. Two Amino Tag columns 150 x 4.6 mm in series have capacity to resolve about 45 species. (author comm.) Otherwise, longer packed fused silica capillaries exhibit high separation efficiency and low dispersion (Figure 1). Nowadays such columns are commercially available.

3.2. Mobile phases

The optimizations of the mobile phase systems are based on varying the pH, ionic strength, organic modifiers and temperature. In the literature a lot of information is available, see Table 1. A chemometric approach as experimental design will be of great help for optimization [49]. Commonly known is that acidic AAs are separated in the beginning while the hydrophobic derivatives are separated with organic modifiers typically THF, ACN and MeOH. The retentions of all AA-derivatives are reduced as pH varies from 2.8 to 5 while the retention of FMOC-OH, the hydrolysis product, is unaffected. Lower pH favour the resolution of earlier eluting derivatives, while higher pH enhance separation of the more strongly retained AA-derivatives. Acetonitrile as modifier provides best resolution for early eluting compounds while MeOH influences separation of late peaks. Cunico et al. [6] proposed a ternary solvent elution scheme employing a pH gradient from 2.8 to 4.5 combined with a gradient in which proportions of ACN and MeOH were varied throughout the elution program. Using an incorrect ionic strength of the mobile phase, may cause co-elution. Arginine is slightly affected by the buffer concentration. Cystine is favoured by an increase in ionic strength resulting in a sharper peak, since the interaction with the matrix diminish. Acetonitrile gradient was not sufficient to separate some hydrophobic AA-derivatives. The introduction of THF solved this problem. The high salt concentration results in an accelerated hydrolysis of the bonded phase which leads to an undesirable retention shift and column life. Decrease of sodium acetate and

Table 1 HPLC Conditions Suggested for the Determination of FMOC-Amino Acids (AAs): Literature Data

Practical conditions			Detection/quantitation			Remarks			Ref
Column cm×mm µm	product of	eluents: composition, pH, temperature	detector man/aut	RSD % cc	analyt, mole	matrix/ etc.	AAs not resolved	No: AA/time min	
12.5×4.6 3	Hypersil ODS Shandon	A: ACN/MeOH/0.7% Na.ac. +0.1% TEA pH=4.2 (10:40:50) B: ACN/No.ac. (50:50) amb	fluorescence man	<6.4	100 pM	Protein hydrolysates Physiolog., fluid		19/20	[5]
15×4.6 5	TSK ODS 80TM Varian	A: 0.02 M Na.cit + 0.005 M TEA; H_3PO_4 PH = 2.85 B: A + H_3PO_4 pH 4.5/MeOH (80:20) C: ACN 30°	fl aut	<4.3	125 pM	Protein hydrolysates		16/25	[6]
50×0.226 3	Spherisorb octyl	A: ACN/Na.ac. pH =4.06 (25:75) B: ACN/Na.ac. pH =4.08/H_2O (25:60:15) amb	fl man	average 6.6		Urine		33/90 (18/6)	[22]
12.5×4 4	LiChroCart Superspher CH8 Merck	A: ACN/50mM Na.ac. pH=4.2 (20:80) B: ACN/50mM Na.ac pH=4.2 (70:30) 35°	fl aut			Protein hydrolysates	Asp / HO-Pro	22/43	[11]
25×4.6 5	Hypersil ODS	A: MeOH/0.7% Na,ac, +0.1% TEA pH=4.2 (50:50) B: MeOH amb	fl man	<10	µM	Conifer extracts		18+GABA/40	[7]
15×4.6 5	Amino Tag Varian	A: 0.02M Na.ac + 0.005M TEA; H_3PO_4 pH 2.85 B: Buffer A+H_3PO_4 pH=4.5/MeOH (80:20) C. ACN 32°	fl + UV man			Plant materials and sap of them	Asn + Gln	19/28	[8]
25×4 5	Kromasil C_8 Aczo Nobel	A: 100 mM Na.ac pH=4.4/THF/ACN (75:15:10) B: ACN/THF (85:15) 45°	fl + UV man aut	man <8.4 aut<4.8		Protein Hydrolysates		19/30	[12]
15×4.6 5	Micro Pak ODS Varian	A: 0.02M Na.cit.+0.0005M TEA pH=2.85 B: BufferA+H_3PO_4 pH=4.5/MeOH (80:20) C: ACN 30°	fl man		low pM	Collagen hydrolysates		22/32	[9]
15×4.6 5	Amino Tag	A: 15mM Na.cit. +10mM TEA/ACN (73:37) pH=3.25 B: BufferA/ACN/THF (60:35:5) pH=5.3 C: BufferA/ACN/THF (25:62:13) pH=6.2 30°	fl + UV aut	pentane etr.<3.6 ADAM<8.9 biol.m.<9.4	nM	Biological materials CytochromeC	Asn/Gln Asp/Glu	19/40	[10]
15×4.6 3	Sphersorb ODS-2	A: 20mM Ammonium dihydroregenorthophosphate/MeOH (85:15) pH=6.5 B: ACN/H_2O (90:10) 35°	fl aut	<5	100pM	Physiological samples Plasma etc	Asp/Asn Glu/Gln	30/29	[17,18]
15×4.6 5	Hypersil ODS	A: 30mM Ammonium dihydrogenorthophosphate/MeOH (85:15) pH=6.5 B: BufferA / MeOH (85:15) C: ACN / H_2O (90:10) 38°	fl aut	<5	100pM	Protein hydrolysates		16/35	[19,20]
15×4.6 5	Micro Pak ODS	A: 20mM Na.cit.+5mM TEA +0.01% Na.azide pH=2.85 B: 20mM + 5mM TEA +0.01% Na.azide pH=4.5 C: ACN 40°	fl aut	<4.4		Collagen hydrolysates Protein hydrolysates		19/29	[21]

the adding of triethyl-ammonium acetate resulted in same resolution while reproducibility was better and column life extended. In that way regeneration was carried out after 300 injections and the column could be used for 800 samples [19].

3.3. Detection

The 9-fluorenylmethyl chloroformate derivatives are suitable for both UV and fluorescence detection. The wavelengths maxima are absorption around 260 mm and uncorrected fluorescence emission 315 mm. The molar absorption coefficient of the fluorene chromophore is 19400 L mol $^{-1}$ cm^{-1} at 262 nm, which results in UV sensitivity in the picomolar range. With a quantum yield of 0.5 the fluorescence limit of detection is in the order of three magnitudes lower. The sensitivity is dependent on the way to suppress the straylight. The Stoke shift is small and in the UV area which demands special optical arrangements. The introduction of liquid filters is shown to give best results since these are free of UV straylights in contrast to glass gratings [5].

The fluorescence detection with the fluorene label is little affected with changes of the pH or the mobile phase compositions. The sensitivity can be further enhanced by diminishing the column diameter to microbore (1 à 2 mm I.D.) [11] or packed fused silica capillary (0.25 mm I.D.) [22] columns. On column laser induced fluorescence (LIF) excitation on capillaries make attomole sensitivity attainable with an other chromophore [50]. However, continuous wave lasers working in UV around 250-260 nm region matching the fluorene chromophore has not been available until recently. Now a solid state continuous wave laser wavelength 266 nm (power up to 500 mW) is commercialized by Spectra Physics. This laser (Millennia UV) is of semiconductor type and very simple to handle.

The relative response of peak area of protein AAs related to FMOC alanine was investigated by Gustavsson et al. [12]. Regarding internal fluorescence quenching following were stated: the four AAs possessing own aromatic moieties Phe, Trp Tyr and His as mono FMOC derivatives gave 1.39, 0.05, 0.89 and 0.44 respective. This shows that only Trp (imidazole moiety) was affected to a larger extent. With doubled derivatized Lys, Tyr and His the responses were 2.15, not shown and 0.19. The di-FMOC His is quenched. Sörensen [51] found that this concern even di-FMOC Tyr. Tryptophan is seldom reported while it is often destroyed already in the hydrolysis step. After methanesulfonic acid hydrolysis Trp has been determined in intravenous solution [52] biological materials [10] and as free AA by Weibull et al. [8]. Tyrosine is problematic.

3.3.1 Alternative chromophores

The FMOC-CL reagent was not developed for either UV or fluorescence optimal detection. The search for alternative labels lead to the anthracene molecule [50,51,53]. Anthracene possess absorption maxima 256 (186000), 351 (6600), 366 (11000) and 386 (11000) nm, (L mol^{-1} cm^{-1}) respective in cyclohexane. The large molar absorptivity at 256 nm results in very high sensitivity compared to the fluorene chromophore which at 262 nm (19400 L mol^{-1} cm^{-1}). The analogous 2-(9-anthryl)ethyl chloroformate (AEOC) (see Figure 3) reagent derivatives exhibit limit of detection in the range 0.5 pmol for UV determinations. The anthracene alternative was shown to be about 17 times more sensitive than the FMOC [54]. The different absorption maxima at 351, 366 and 386 nm besides 256 nm can also be used for confirmation and identification purposes.

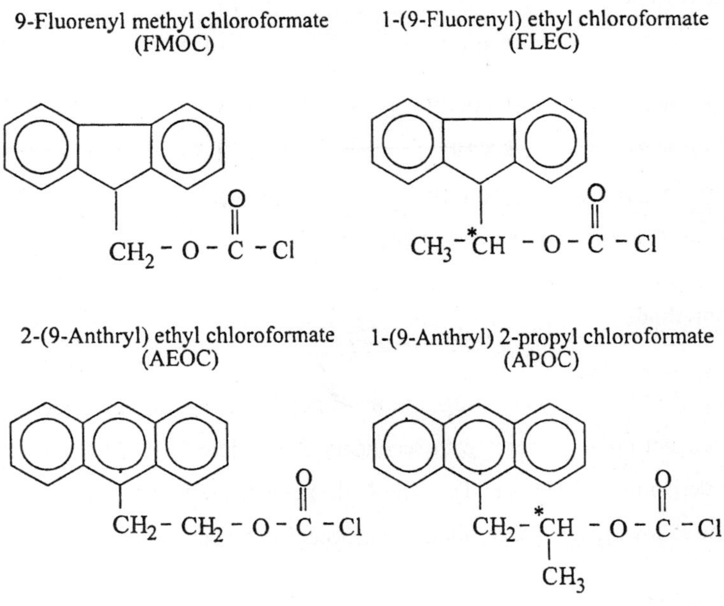

Figure 3. Structures of the different reagents. Star marks the asymmetric carbon atoms

The fluorescence quantum yield is 0.54 for FMOC derivatives which is larger than for anthracene analogue which is about half of that value. The far larger molar absorptivity of AEOC still results in quite much higher fluorescence emission and thus limit of detection about 0.06 pmol [54]. The bifunctional AAs $(Cys)^2$, His, Lys and Tyr when double derivatized have a higher limit of detection presumably due to intramolecular quenching (forming excimers) [51].

The absorption maxima at 351 nm match the commercially available CW Ar ion laser with the wavelength 351 nm which facilitates LIF detection.
With LIF detection on column the limit of detection of monoderivatized AAs was reported to be a few fmoles [50] (Table 2).

Other chromophore analogues have been reported such as 2-(9-carbazole)ethyl and 4-phenylazobenzyl chloroformates [55,56] and [57,58] respective.

Table 2 Detection limits for AEOC with the different systems.

Separation mode	Detection mode	Conc. LOD (M)	Mass LOD (mol)
HPLC	UV	20×10^{-9}	500×10^{-15}
HPLC	fluorescence	3×10^{-9}	60×10^{-15}
HPLC*	LIF	300×10^{-12}	300×10^{-18}
CE	UV	150×10^{-9}	1×10^{-15}
CE	fluorescence	4×10^{-9}	10×10^{-18}
CE	LIF	100×10^{-12}	100×10^{-21}

* *Packed capillary HPLC*

4. OPA-FMOC hybrid methods

4.1 Introduction

The combination of OPA and FMOC reagents bridges some of the deficiency of the respective methods. The OPA reagent does not react with secondary AAs while FMOC-CL can yield problematic double derivatives of His and Tyr. The both methods are particularly described at other parts. In the following just the combination aspects will be considered.

4.2 Masking method for facilitating secondary amino acids determinations.

Einarsson [59] introduced the way to apply two different reagents on the same sample. The primary AAs are first concealed by the reaction with OPA reagent then the FMOC-Cl reagent reacts with the left unaffected secondary AAs. This masking procedure work as far as the OPA reagent is not influencing the followed FMOC chemistry. However, the mercaptoethanol in the OPA reagent reacts with FMOC-Cl to give late eluting peak that interferes in subsequent runs. Furthermore, the mercaptoethanol is consuming FMOC-Cl. Einarsson found that iodoacetamide could be used as a blocking reagent to the mercaptoethanol. The unreacted secondary AAs are then available to form stable FMOC derivatives. The whole procedure is completed within two minutes in a borate buffer pH 10.4. The fluorescence emission of the two different reaction products isoindoles (OPA) and carbamates (FMOC) is in the range 400-500 nm respective 300-350 nm.

The method was applied on sea water, urine and hydrolyzates. At pmol level the R.S.D. was 2.1–3.6 for Hyp, Sar and Pro while the limit of detection was about 50 fmol. The secondary AAs could isocratically be separated in about one minute. Hydroxyproline has been determined in urine [60,62] collagen [63,64] and rat lung tissue [65].

Notice that the masking techniques with OPA reagent can be practised to clean-up from contaminating AAs in ultra trace determinations.

4.3 Combination method

Schuster [66,67] carried out a modification of the OPA reagent by changing the mercaptoethanol to 3-mercaptopropionic acid (3-MPA) which made it possible to combine and measure the formed OPA and FMOC derivatives in the same run at shifted wavelengths. The two reagents are mixed in buffered (pH 10.4) sample stepwise with first OPA-3-MPA to derivatize all primary AAs and then FMOC-Cl for the secondary AAs. All steps are automated and precisely time-controlled and finished within 90 s. The incorporation of 3-MPA into the isoindols decreases their hydrophobicity resulting in a elution time of less than 15 min. and in front of the FMOC derivatives. Hydrolyzate samples can be automated run including column conditioning time within 20 min. cycle (nowadays 14 min.). Physiological AAs separations take about 60 min.

The detection is carried out with either UV or fluorescence with ability to change wavelengths during run. The UV absorption is recorded with diode array or multiwave-length detector at respective 266 nm and 338 nm. The fluorescence emission is facilitated with programmed wavelengths switching operating at level down to 150 fmol.

Reproducibility of 24 AAs regarding peak areas was found to a mean of 2.0%. The method is known as Amino Quant (Agilent TM). The automated system was tested on different AA determinations as hydrolyzates of proteins and variety of samples in different matrices such as physiological fluids, plants and beer fermentations. Most of the samples could be directly injected in a single procedure without interference from matrix compounds.

Blankenship et al. [68] used the system for protein structure determinations. Godel et al. [69] tested narrow-bore columns (2.1 mm I.D.) which increased the sensitivity approximately five times compared to standard columns (4.6 mm I.D.). They also treated the cysteine, cystine problem as to convert the cysteine to cys-MPA. The separation of 40 AAs in 92 min. was achieved by Carducci et al. [70]. Liu et al. [71–73] measured the free AAs in ultrafiltrates of blood plasma. Patients suffering from several types of AA metabolic disorders were detected.

Hydrolyzates were automatically processed for AA content in cheese [74] and plant materials of potato tubers [75]. The AA content in must and port wine could be measured after filtration without clean-up and interference of high sugar content [76,77]. AA profiles in honey were used to control the authenticity [79]. The different AA patterns were statistically processed by principal component analysis (PCA) on 19 different AAs.

With separation cycle of 10 min. Hpr and Pro were automatically determined in collagen by Nathans et al. [79]. Heems et al. [80] practised an alternative system to Amino Quant consistent of the ASTED Gilson automated sampling processor. The samples were subjected to clean-up by on-line dialysis before injection into column. The samples were red wine and orange juice as well as hydrolyzates of food.

5. Literature/Comparison studies of FMOC to other AA methods

In the literature there are some articles devoted to comparison studies of different AA methods. The examples given are presented in perspective of time since the different methods are subjected to continuous developments.

Bruton [81] 1986.

Methods: FMOC (Amino Tag) IEC-Ninhydrin, IEC-OPA, OPA, PITC, DNS.

Samples: Protein and peptide hydrolyzates (16 protein AAs).

Conclusive remarks: The precision and accuracy are comparable to or superior to those achieved with IEC and furthermore the sensitivity is higher.

Smith et al. [82] 1989.

Methods: Amino Tag IEC-Ninhydrin, PTC.

Samples: Protein standard hydrolyzated range 5–500 picomoles (16 protein AAs)

Conclusive remarks: At lower concentrations the IEC method was not adequate. The derivatization pH 8.5 favoured the formation of di-FMOC derivatives of His and Tyr. The reagent excess is important since the present antioxidant phenol consume FMOC-CL. The recovery of internal standard must be carefully monitored to ensure a proper excess of reagent. The FMOC method compared favourably with the PTC except for Gly.

Fürst et al. [83] 1990.

Methods: FMOC manual and automated, OPA-3-MPA, PITC. DNS.

Samples: Human plasma, protein and peptide hydrolyzates. Major AAs (number of AAs varied between 21 and 26) based on experiences of together 13–14000 analyses pro year.
Conclusive remarks: The automated OPA-3-MPA method was found to be favourably for routine analyses of primary free AAs except cystine. When secondary AAs are included the FMOC or OPA-FMOC method is recommended. Application of the PITC method might be beneficial if sufficient sample material is available.

Bütikofer et al. [74] 1991.
Methods: IEC-Ninhydrin, OPA-FMOC (Amino Quant).
Samples: Hydrolyzed cheese (16 AAs).
Conclusive remarks: The HPLC method gave better resolved peaks, more stable baseline and higher sensitivity.

Lai et al. [84,85] 1993.
Methods: Amino Tag, PITC, NDA modified OPA.
Samples: AAs standard with added matrices such as Tris-HCl, phenol, sulfosalicylic, acid (SSA), sodium dodecyl sulfate (SDS), sodium citrate.
Conclusive remarks: There are significant differences in matrix effects among the three different methods. The reagent excess with FMOC and NDA methods has an impact on reagent depletion. SSA decreases the reaction pH for FMOC and NDA methods. The PITC method has a more stable buffered reaction medium with large excess of TEA. The detection limit for FMOC method is low thus the sample can be diluted to diminish the matrix effects.

Kroll et al. [86] 1997.
Methods: Amino Tag, FMOC after Haynes et al. [17], Amino Quant, OPA, PITC, DNS/DBS.
Samples: AA standard.
Conclusive remarks: The Haynes method gave one stable derivative of His and Tyr thus better precision. The solvent extraction procedure is not necessary. Simpler linear eluent gradient, pH 6.5 which also prolong the column life.

6. Chiral separations

6.1. Introduction

The importance of stereochemistry has resulted in very exciting development of new methods to determine the enantiomeric composition and purity of e.g. biologically active peptides, food and therapeutic drugs. The methods can be divided into two different categories called direct and indirect separations. Direct methods means separation by chiral recognition in the separation system either through chiral stationary bonded phases or by adding asymmetric substances in the mobile phase.

Indirect methods are based on precolumn derivatization with an asymmetric reagent to form diastereomers which are then separable by conventional achiral separation systems. Chloroformate reagents have been employed in both type of methods. In contradistinction to the racemization, sometimes observed with acylating agents, the use of FMOC-Cl induces no detectable racemization when optically active AAs are employed. This is important fact in the peptide synthesis [3,4]. The close related chloroformate reagents: (±)-1-(9-fluorenyl)ethyl chloroformate (±) FLEC and (±)-1-(9-anthryl)2propyl chloroformate (Figure 3) are not racemizing during the derivatization reaction as well.

6.1.1. Direct separation methods

Most protein AAs lack strong UV absorbance except for the three aromatic AAs Phe, Trp and Tyr. Therefore derivatization becomes necessary for sensitive detection. Furthermore derivatization is expected to improve selectivity because more bulky molecules have to be separated.

Zukowski et al. [87] used a chiral bonded phase R-(-)-1-(1-naphtyl)ethyl carbamoylated-β-cyclodextrin (RN-β-CD) selector in conjunction with nonaqueous polar mobile phase for highly selective enantioseparation of secondary AAs as FMOC derivatives. The derivatization was made in accordance with Einarsson et al. [5]. The FMOC-Cl reagent served to lock the imino acids into their conformation avoiding racemization. The detection limits for FMOC derivatives of Pro and others were in low femtomole range. Enantiomeric impurities as low as 1 ppm could be determined.

Other chiral bonded phases such as Chiralcel-OD [88], β-and γ-cyclodextrin [89], quinine-derived chiral anion-exchange phase [90] and γ-cyclodextrin phase [92] have been employed with the separation of FMOC-AAs.

Spöndlin et al. [88] described the determination of enantiomeric purity of 17 FMOC AAs on Chiralcel-OD. The fluorescence sensitivity and large separation factors (α 1.5–2.2) allowed detection limit below 0.05 %. In contrary to general belief the elution order of Trp enantiomers with a straight mobile phase were opposite to that obtained with the other AAs.

Ekberg-Ott et al. [92] determined the concentration and enantiomeric purity of selected free AAs in fermented beers with cyclodextrin type bonded phase RN-β-CD and ACN/acetic acid 1000/7 ($^V/_V$) mobile phase.

6.1.2. Indirect separations

The prerequisite for successful application of the indirect separation method is that the purity of the chiral reagents applied must be ascertained or known; this is in particular necessary for trace enantiomeric impurity determinations. The demand on the reagents as to react without racemization or discrimination is important. The reagent products which as diastereomers possess different physico-chemical properties are readily separated in a achiral, conventional preferably reversed phase system. Furthermore the formed diastereomers must be stable. Amino acids are ideal for resolution by diastereomer-formation with chiral reagents because the amine-functional group is connected directly to the asymmetric carbon atom.

If the optical purity of the chiral reagent is known and considered to be "pure" >99.9 % the precision in the determination of L and D forms will solve the most problems. If a still higher demand, the enantiomeric impurity has to be specified with high accuracy. This can be done with direct chiral separation with glycine (achiral AA) with a chromatographic (HPLC or MEKC) system where typically cyclodextrins act as chiral selectors [93]. If the purity of the chiral reagent is known the enantiomeric purity of the compound of interest can be calculated by the general formula:

$$y = \frac{1}{2}\left[\frac{M_A - I_A}{(2x\text{-}1)(M_A + I_A)} + 1\right] \qquad (1)$$

Where y is defined as the molar fraction of one enantiomer y = R / (R + S) of the compound to be investigated, and x is defined in the same way as the molar fraction of one form of the reagent x = R / (R + S). M_A and I_A are the peak areas of the main and the impurity peak respective in the chromatogram. If the chiral reagent is 100 % enantiomeric excess (ee) x = 1 , Eq. 1 will be

$$y = \frac{1}{2}\left[\frac{M_A - I_A}{M_A + I_A} + 1\right] = \frac{M_A}{M_A + I_A} \qquad (2)$$

Einarsson et al. [94] synthetisized the chiral reagent (+)-1-(9-fluorenyl)ethyl chloroformate (+) FLEC (Figure 3) by incorporating a methyl group on the methylene adjacent to the chloroformate group. In that way an asymmetric carbon atom was created. The chiral (+) FLEC reagent is an analogue of the FMOC reagent and it behaves in the same way except that the (+) FLEC reacts slower by an average factor of 2.8. Under identical derivatization conditions besides a prolonged reaction time to 4 min. the diastereomers of 17 AAs were formed ready for separation on a conventional RP-HPLC system (Figure 4). The D-AAs appeared in front of the corresponding common L-AAs which is an advantage. The reagent is commercially available in (+) and (-) form with high optical purity nearly 100 % [93,95].

Okuma et al. [96] determined 19 different AAs in D and L form in nervous tissues of crustaceous with the (+) FLEC after derivatization essentially according to Ref [94]. The RP-HPLC separation was carried out on a (250 x 4.6 mm I.D.) ODS column with a ternary gradient with an ion-pair reagent tetramethyl ammonium present. Todoroki et al. [97] determined N-methyl-D-aspartate (NMDA) a neuroexitatory compound in bivalves.

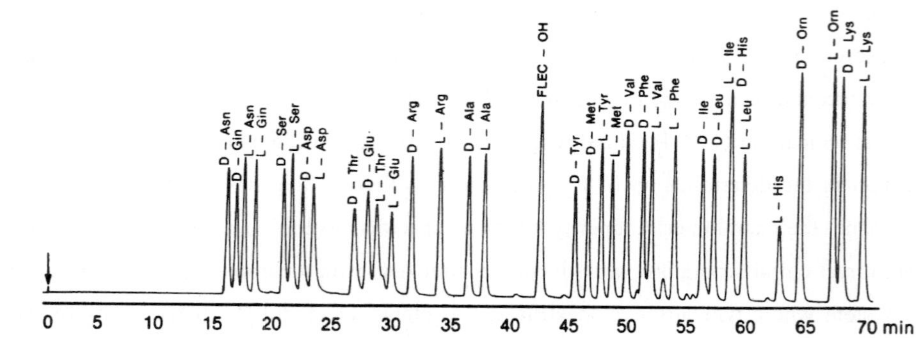

Figure 4. RP-HPLC separation of L- and D- AAs. Column: 150 x 4.6 mm 3μm Spherisorb octyl. Mobile phase: A: acetic acid pH 4.35, B: ACN, C: THF. Gradient, 0-8 min. 8 % ACN / 17 % THF / 75 % buffer (8/17/75); 8-22 min, 8/17/75 to 0/30/70; 22-70 min. 0/30/70 to 0/50/50. Flow rate was 0.8 ml/min.

In the clean-up method they used OPA for removal of acidic primary AAs. The NMDA was derivatized with (+) FLEC and thereafter isocratically eluted on a HPLC system.

L-Carnitine is a vitamin-like AA derivative which plays an important role in fatty acid metabolism. It has a highly therapeutic effect and, therefore, it is used in various nutritional and pharmaceutical applications. However, the D-carnitine is quite toxic and it must be carefully monitored in formulations. The D-enantiomer has been determined in promille range by (+) FLEC derivatization and HPLC or CZE [98-100]. Enantioselective determination of glufosinate, a phosphorous-containing AA-type of herbicide, was performed by Hori et al. [101] with (+) FLEC derivatization and RP-HPLC.

A chiral analogue to AEOC and FLEC, (±)-1-(9-anthryl)-2-propyl chloroformate APOC (see Figure 3) was proposed by Thorsén et al. [102]. The detection limit for the AA derivatives of APOC with UV detection is in the femtomole range while in the attomole range with laser induced fluorescence (LIF) detection after MEKC separation. The high detection sensitivity with LIF makes enantiomeric determinations of ppm levels feasible. Stereoselective determinations of AAs in β-amyloid peptides and senile plaques from deceased Alzheimer's disease patients was accomplished by Thorsén et al. [103]. The elution order of the L- and D-AAs depends on which reagent is used (+) or (-) (Figure 5). The reversal order, the peak height and the reagent optical purity all together result in confirmative information to secure identity and moreover better quantity measure. Varying purities of the (+) and (-) reagents give valuable additional contribution.

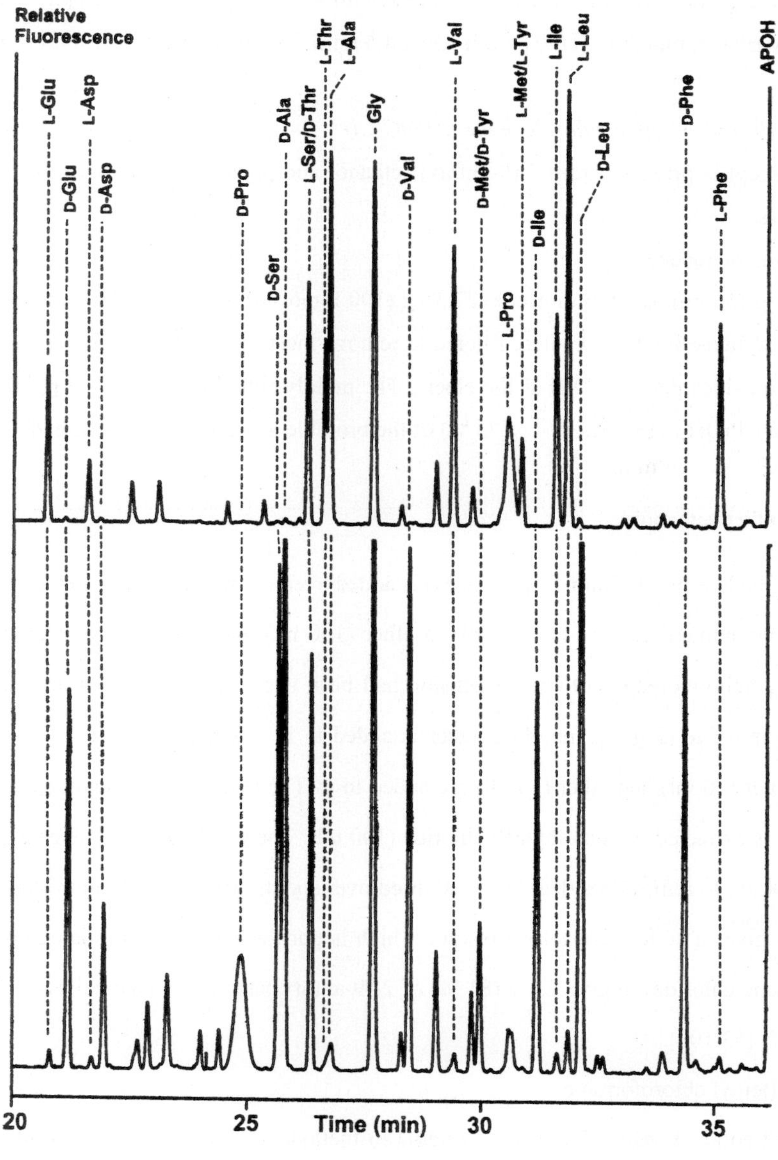

Figure 5 Chiral separation of AAs from hydrolyzed senile plaque samples: upper trace derivatized with (+)-APOC and lower trace derivatized with (-)-APOC.

7. Experimental

7.1. Equipment

The automated AA determination is performed with a ternary gradient delivery system Varian 9012 Q with Varian 9100 autosampler. Detectors: Varian 9050 UV detector, Agilent diode array detector DAD. Shimadzu RF-530 fluorescence spectromonitor (could be coupled to the outlet of the UV-detector). Integration with the peaks can be made with Varian Star Chromatography data system. Column: Varian Amino Tag (ODS, 5 μm 150 x 4.6 mm). Mobile phase flow 1.8 ml/min.

7.2. Reagent 2-(9-Anthryl)ethyl chloroformate (AEOC-Cl)

The AEOC-Cl reagent is prepared from 2-(9-anthryl) ethanol and phosgene. The three step synthesis as follows:

i) 9-Bromoanthracene

A mixture of 8.91 g (50 mmol) of anthracene, 22.34 g (100 mmol) of a hydrous $CuBr_2$, and 300 ml of CCl_4 is refluxed for 16 h, cooled, filtered to remove most of CuBr. The mixture is chromatographed on silica gel with PhH as the eluent. The rapidly eluted material (11.9 g) is recrystallized from i-PrOH to separate 9.4 g (73 %) of the bromide as yellow needles (mp 98-99°) [104].

ii) 2-(9-Anthryl)ethanol

Commercially available n-butyllithium (9 ml, 1.6 M) is added over a 5 min. period to an ice-cold solution of 9-bromoanthracene (3 g, 11.7 mmol) in ether (100 ml) under nitrogen blanketing. After the resulting yellow-orange colored suspension had been stirred for 30 min, an ice-cold solution of ethylene oxide (3 g, 68 mmol) in ether is added to it. After the mixture had been stirred for 1 h, water (70 ml) and ether (50 ml) are added to it. The ether layer is separated and the aqueous layer is extracted with methylene chloride (100 ml). The combined organic extracts are washed with water to remove residual base and dried over sodium sulphate. Vacuum evaporation of solvents gives a yellow crystalline residue which is purified by flash chromatography (silica gel/methylene chloride) to give 2.1 g (81 %) of 2-(9-anthryl)ethanol as pale yellow crystals (mp 106-108 °) [53,105].

iii) 2-(9-Anthryl)ethyl chloroformate

A solution of 2-(9-anthryl)ethanol 3.5 g (15.7 mmol) in methylene chloride (20 ml) is added slowly to an ice-cold solution of phosgene (20 ml, 38.4 mmol, 20 % in toluene, commercially available). The reaction mixture is kept for 3 h in an ice bath, and then overnight at room temperature. Workup by vacuum evaporation of solvents gives crystalline residue which is recrystallized from methylene chloride solution by precipitation with hexane which gives about 3.4 g (75 %) of AEOC-Cl as virtually colorless crystals, mp 85-87 ° [51,53].

7.3. Derivatization

The autosampler is programmed to mix 25 µl of the AEOC (10 mM in ACN) with 25 µl of the sample buffered with 250 mM $NaBO_4$ (prepared by mixing the sample with 1M $NaBO_4$ pH 9.2 and H_2O to the desired concentration). After a waiting period of 3 min. excess reagent is removed from the aqueous phase by addition of n-hexane to the mixture in five consecutive portions of 30 µl each, with mixing in between. 10 µl of the aqueous phase is injected after a 2 min. wait to ensure phase separation. The derivatization performs at ambient temperature.

7.4. Chromatography (AEOC or FMOC)

The separation of the derivatives is carried out with a ternary gradient elution scheme. The eluents are two buffer solutions consisting of sodium acetate (15 mM) and tetramethylammonium hydroxide (10 mM), with pH adjusted by addition of phosphoric acid to (A) 3.0, and (B) 5.5, respectively, and (C) acetonitrile with 10 % methanol.

Table 3 Gradient Program

Step	Time/min.	%A	%B	%C
1	0	66	0	34
2	15	12	48	40
3	37	25	5	70
4	40	20	0	80

The column is equilibrated during the automated derivatization procedure.

7.5. Detection alternatives

Table 4 Wavelengths

	AEOC	FMOC
UV	254 nm (ε = 150 000 in water)	264 nm (ε = 17 000 in water)
	351 nm	
	366 nm	
	388 nm	
Fluorescence	420 nm	320 nm

References

[1] J. Dumas, Ann Chim. (Paris) 56 (1834) 113.

[2] M. Matzner, R.P. Kurkju, R.J. Cotter, Chem. Rev. 64 (1964) 645-687.

[3] L.A. Carpino, G.Y. Han, J. Am. Chem. Soc. 37 (1972) 3404-3409.

[4] L.A. Carpino, Acc. Chem. Res. 6 (1973) 191-199.

[5] S. Einarsson, B. Josefsson, S. Lagerkvist, J. Chromatogr. 282 (1983) 609-618.

[6] R. Cunico, A.G. Mayer, C.T. Wehr, T.L. Sheehan, Bio Chrom. 1 (1986) 7-14.

[7] T. Näsholm, G. Sandberg, A. Ericsson, J. Chromatogr. 396 (1987) 225-236.

[8] J. Weibull, F. Ronquist, S. Brishammar, Plant Physiol. 92 (1990) 222-226.

[9] E.J. Miller, A.J. Narkates, M.A. Niemann, Anal. Biochem. 190 (1990) 92-97.

[10] M.F. Malmer, L.A. Schroeder, J. Chromatogr. 514 (1990) 227-239.

[11] J. Betnér, P. Földi, Chromatographia 22 (1986) 381-387.

[12] B. Gustavsson, J. Betnér, J. Chromatogr. 507 (1990) 67-77.

[13] J. Kirschbaum, B. Luckas, W-D. Beinert, J. Chromatogr. A 661 (1994) 193-199.

[14] U. Arnold, E. Ludwig, R. Kühn, U. Möschwitzer, Z. Lebensm. Unters. Forsch. 199 (1994) 22-25.

[15] B. Carratù, C. Boniglia, G. Bellamonte, J. Chromatogr. A 708 (1995) 203-208.

[16] A. Fabiani, A. Versari, G.P. Parpinello, M. Castellari, S. Galassi, J. Chromatogr. Sci. 40 (2002) 14-18.

[17] P.A. Haynes, D. Sheumack, J. Kibby, J.W. Redmond, J. Chromatogr. 540 (1991) 177-

185.
[18] P.A. Haynes, D. Sheumack, L.G. Greig, J. Kibby, J.W. Redmond, J. Chromatogr. 588 (1991) 107-114.
[19] K. Ou, M.R. Wilkins, J.X. Yan, A.A. Gooley, Y. Fung, D. Sheumack, K.L. Williams, J. Chromatogr. A 723 (1996) 219-225.
[20] J.X. Yan, N.H. Packer, Methods in Molecular Biology, 125 (2000) 113-119.
[21] R.A. Bank, E.J. Jansen, B. Beekman, J.M. te Koppele, Anal. Biochem. 240 (1996) 167-176.
[22] S. Einarsson, S. Folestad, B. Josefsson, S. Lagerkvist, Anal. Chem. 58 (1986) 1638-1643.
[23] M.M. Or-Rashid, R. Onodera, S. Wadud, M-E. A. Nasser, M.R. Amin, J. Assoc. Off. Anal. Chem. 82 (1999) 809-813.
[24] M.M. Or-Rashid, R. Onodera, S. Wadud, N. Mohammed, J. Chromotogr. B 741 (2000) 279-287.
[25] H. Sultana, R. Onodera, M.M. Or-Rashid, S. Wadud, J. Chromatogr. B 755 (2001) 321-329.
[26] A. Cubedo Fernandez-Trapiella, J. Assoc. Off. Anal. Chem. 73 (1990) 935-939.
[27] D.J. Thomas, K.L. Parkin, J. Agric. Food Chem. 42 (1994) 1632-1638.
[28] T. Bauza, A. Blaise, F. Daumas, J.C. Cabanis, J. Chromatogr. A 707 (1995) 373-379.
[29] L.A. Schilb, V.D. Fiegel, D.R. Knighton, J. Liq. Chromatogr. 13 (1990) 557-567.
[30] R.A. Bank, B. Beekman, R. Tenni, J.M. te Koppele, J. Chromatogr. B 703 (1997) 267-272.
[31] T.C. O'Connel, R.E.M. Hedges, Archaeometry 43 (2001) 421-438.
[32] T. Teerlink, E. de Boer, J. Chromatogr. 491 (1989) 418-423.
[33] G.E. Kisby, D.N. Roy, P.S. Spencer, J. Neurosci. Meth. 26 (1988) 45-54.
[34] A.J. Shah, D.A. Younic, M.W. Adlard, C.S. Evans, Phytochem. Anal. 3 (1992) 20-25.
[35] J.S. Niedbalski, D.R. Ringer, Anal. Biochem. 158 (1986) 138-145.
[36] D.P. Ringer, Methods in Enzymology 201 (1991) 3-10.
[37] B. Staffeldt, J. Brockmöller, I. Roots, J. Chromatogr. 571 (1991) 133-147.
[38] J. Brockmöller, B. Staffeldt, I. Roots, Eur. J. Clin. Pharmacol. 40 (1991) 387-392.
[39] J.A. Brown, M.S. Nijjar, Mol. Cellular. Biochem. 151 (1995) 49-54.
[40] R. Pocklington, S.S. Bates, J.E. Milley, C.J. Bird, A.S.W. De Freitas, M.A. Quilliam, Int. J. Environ. Anal. Chem. 38 (1990) 351-368.

[41] R.S. Gilbert, G.G. Gonzalez, L. Hawel III, C.V. Byus, Anal. Biochem. 199 (1991) 86-92.

[42] R. Herrera, G. Manjarrez, E. Nishimura, J. Hernandez, Pediatric Neurology 28 (2003) 20-23.

[43] D. Noort, A.G. Hulst, H.C. Trap, L.P.A. de Jong, H.P. Benschop, Arch. Toxicol. 71 (1997) 171-178.

[44] A.M.F. Müller, E. Hallier, G. Westphal, K.R. Schröder, H.M. Bolt, Fresenius J. Anal. Chem. 350 (1994) 712-715.

[45] H.M.H. van Eijk, D.R. Rooyakkers, A.J.M. Wagenmakers, P.B. Soeters, N.E.P. Deutz. J. Chromatogr. B 691 (1997) 287-296.

[46] K. Gartenmann, S. Kochhar, J. Agric. Food. Chem. 47 (1999) 5068-5071.

[47] S. Einarsson personal communication.

[48] H.K. Hall jr, J. Am. Chem. Soc. 79 (1957) 5439-5441.

[49] R. Carlson, Design and Optimization in Organic Synthesis. Elsevier, Amsterdam 1992.

[50] A. Engström, P.E. Andersson, B. Josefsson, W.D. Pfeffer, Anal. Chem. 67 (1995) 3018-3022.

[51] H. Sörensen, Ph.D. Thesis, Gothenburg University, Gothenburg, Sweden 1988.

[52] A.G. Meyer, T.L. Sheehan, 17th Int. Symp. Chromatography Vienna poster 1988.

[53] A.J. Faulkner, H. Veening, H-D. Becker, Anal. Chem. 63 (1991) 292-296.

[54] J. Björklund, S. Einarsson, A. Engström, A. Grzegorczyk, H-D. Becker, B. Josefsson, J. Chromatogr. A 798 (1998) 1-8.

[55] X. Fan, J. You, J. Kang, Q. Ou, Q. Zhu, Anal. Chem. Acta 367 (1998) 81-91.

[56] J. You, Y. Shan, L. Zhen, L. Zhang, Y. Zhang, Anal. Biochem. 313 (2003) 17-27.

[57] H. Brückner, M. Lüpke, J. Chromatogr. A 697 (1995) 295-307.

[58] J. Kirschbaum, H. Brückner, Chromatographia 43 (1996) 275-278.

[59] S. Einarsson, J. Chromatogr. 348 (1985) 213-220.

[60] T. Teerlink, P. Tavenier, J. Coen Netelenbos, Clin. Chim. Acta 183 (1989) 309-325.

[61] S. Castelain, S. Kamel, C. Picard, G. Desmet, J.L. Sebert, M. Brazier, Clin. Chim. Acta 235 (1995) 81-90.

[62] G. Mazzi, F. Fioravanzo, E. Burti, J. Chromatogr. B. 678 (1996) 165-172.

[63] V. Monboisse, J.C. Monboisse, J.P. Bord, A. Rondoux, Anal. Biochem. 176 (1989) 395-399.

[64] A. Kondo, O. Ishikawa, K. Okada, Y. Miyachi, S. Abe, Y. Kuboki, Anal. Biochem. 252 (1997) 255-259.

[65] P.R. Hutson, M.E. Crawford, R.L. Sorkness, J. Chromatogr. B 791 (2003) 427-430.
[66] R. Schuster, J. Chromatogr. 431 (1988) 271-284.
[67] R. Schuster, A. Apfel, Publ. No 12-5954-6257 Hewlett-Packard Int. Palo Alto Ca 1986.
[68] D.T. Blankenship, M.A. Krivanek, B.L. Ackerman, A.D. Cardin, Anal. Biochem. 178 (1989) 227-232.
[69] H. Godel, P. Seitz, M. Verhoef, LC-GC Int. 5 no2 (1991) 44-49.
[70] C. Carducci, M. Birarelli, V. Leuzzi, G. Santagata, P. Serafini, J. Antonozzi, J. Chromatogr. A 729 (1996) 173-180.
[71] H. Liu, H.G. Worthen, J. Chromatogr. 579 (1992) 215-224.
[72] H.G. Worthen, H. Liu, J. Liq. Chromatogr. 15 (1992) 3323-3341.
[73] H. Liu, Methods in Molecular Biology 159 (2001) 123-140.
[74] U. Bütikofer, D. Fuchs, J.O. Bosset, W. Gmür, Chromatographia 31 (1991) 441-447.
[75] T. Bartok, G. Szalai, ZS. Lövincz, G. Börcsök, F. Sagi, J. Liq. Chromatogr. 17 (1994) 4391-4403.
[76] P. Herbert, P. Barros, N. Ratola, A. Alves, J. Food Sci. 65 (2000) 1130-1133.
[77] P. Herbert, L. Santos, A. Alves, J. Food. Sci. 66 (2001) 1319-1325.
[78] J.F. Cotte, H. Casabianca, B. Giroud, M. Albert, J. Lheritier, M.F. Grenier-Loustalot, Anal. Bioanal. Chem. 378 (2004) 1342-1350.
[79] G.R. Nathans, D.R. Gere, Anal. Biochem. 202 (1992) 262-267.
[80] D. Heems, G. Luck, C. Fraudeau, E. Verette, J. Chromatogr. A 798 (1998) 9-17.
[81] C. Bruton, Int. Lab. June (1986) 30-38.
[82] A.J. Smith, J.M. Presley, W. Mc Intire, Tech. Protein Chem. (1989) 255-265.
[83] P. Fürst, L. Pollack, T.A. Graser, H. Godel, P. Stehle, J. Chromatogr. 499 (1990) 557-569.
[84] F.Lai, A. Mayer, T. Sheehan, BioTechniques 11 (1991) 236-243.
[85] F. Lai, T. Sheehan BioTechniques 14 (1993) 642-648.
[86] C. Kroll, J. Hartmann Labor Praxis 21 (2) (1997) 62-66.
[87] J. Zukowski, M. Pawlowska, D.W. Armstrong, J. Chromatogr. 623 (1992) 33-41.
[88] Ch. Spöndlin, E. Küsters, Chromatographia 35 (1994) 325-329.
[89] Y. Tang, J. Zukowski, D.W. Armstrong, J. Chromatogr. A 743 (1996) 261-271.
[90] A. Peter, G. Török, G. Tóth, W. Lindner, J. High Resol. Chromatogr. 23 (2000) 628-636.
[91] T-Y. Kim, H-J. Kim, J. Chromatogr. A 933 (2001) 99-106.
[92] K.H. Ekborg-Ott, D.W. Armstrong, Chirality 8 (1996) 49-57.

[93] A. Engström, H. Wan, P.E. Andersson, B. Josefsson, J. Chromotogr. A 715 (1995) 151-158.

[94] S. Einarsson, B. Josefsson, P. Möller, D. Sanchez, Anal. Chem. 59 (1987) 1191-1195.

[95] D. Sanchez, P. Möller, S. Einarsson, B. Josefsson, Janssen Chim. Acta 6 (1) (1988) 10-14.

[96] E. Okuma, H. Abe, J. Chromatogr. B 660 (1994) 243-250.

[97] N. Todoroki, K. Shibata, T. Yamada, Y. Kera, R-H Yamada, J. Chromatogr. B 728 (1999) 41-47.

[98] P. De Witt, R. Deias, S. Muck, B. Galletti, D. Meloni, P. Celletti, A. Marzo, J. Chromatogr. B 657 (1994) 67-73.

[99] C. Vogt, A. Georgi, G. Werner, Chromatographia 40 (1995) 287-295.

[100] S. Freimüller, H. Altorfer, J. Pharm. Biomed. Anal. 30 (2002) 209-218.

[101] Y. Hori, M. Fujisawa, K. Shimada, M. Sato, M. Honda, Y. Hirose, J. Chromatogr. B 776 (2002) 191-198.

[102] G. Thorsén, A. Engström, B. Josefsson, J. Chromatogr. A 786 (1997) 347-354.

[103] G. Thorsén, J. Bergquist, A. Westlind-Danielsson, B. Josefsson, Anal. Chem. 73 (2001) 2625-2631.

[104] H.O. House, N.I. Gali, J.L. Haack, D. Van Derveer, J. Org. Chem. 45 (1980) 1807-1817

[105] H-D. Becker, L. Hansen, K. Andersson, J. Org. Chem. 51 (1986) 2956-2961.

Ibolya Molnár-Perl (Editor)
Quantitation of Amino Acids and Amines by Chromatography
Journal of Chromatography Library, Vol. 70

1.2.5.

HPLC of Amino Acids as Dansyl and Dabsyl Derivatives

Toyohide Takeuchi

Contents

Summary

Pre-column derivatization of amino acids with 5-dimethylaminonaphthalene-1-sulfonyl chloride (dansyl chloride) or 4-dimethylaminoazobenzene-4'-sulfonyl chloride (dabsyl chloride) is described. These pre-column derivatization methods enable a sensitive HPLC analysis of amino acids.

Dansyl derivatization has originally been applied to the sequential analysis of peptides and proteins. It finds another use in biochemistry for the fluorogenic labeling of proteins and enzymes. Dansyl chloride is the most widely used for the derivatization of amino acids. Dansyl chloride readily reacts with primary and secondary amino groups of amino acids. The reaction medium is usually an aqueous-organic mixture (e.g., 1:1 acetone-water) adjusted to a pH of 9.5–10. Dansylation reaction is usually carried out at elevated temperatures. Various dansylation reaction conditions are reported, involving 60 °C for 60 min or at 38 °C for 90–120 min. Dansyl amino acids absorb light in the UV region. For example, absorption maxima are observed at 214, 246 and 325 nm for dansyl glycine, and the absorption at 214 nm is the strongest. Dansyl group is also fluorescent and dansyl amino acids can therefore be detected by a fluorimetric detector. The excitation and emission wavelengths for dansyl glycine are 324 and 559 nm, respectively. Dansyl amino acids are mostly separated in the reversed-phase mode on a C8 or C18 column with a linear gradient.

Dabsyl chloride has a number of advantages over other derivatization methods, including a simple derivatization procedure, very good stability, good reproducibility and a good limit of detection for the method, complete HPLC separation of all the amino acids, and specific detection at a wavelength in the visible region. Dabsyl chloride also reacts with primary and secondary amino groups of amino acids as does dansyl chloride. Dabsylation reagent solution can be prepared by dissolving dabsyl chloride in acetone or in acetonitrile, followed by mixing with buffer (e.g., carbonate, pH 8.5-9.5). Derivatization of amino acid with dabsyl chloride is also carried out at elevated temperatures as for dansylation. For example, samples were incubated at 70 °C for 15-30 min.

There have been a number of literatures dealing with application of dansylation to the determination of amino acids contained in various sample sources, involving biological fluids, tissues, foods, peptide or protein hydrolyzates, etc. Dabsylation also covers nearly the same application areas as for dansylation. Enantiomeric separation of dansyl and dabsyl amino acids can be carried out by ligand-exchange chromatography. The use of native, and derivatized β- and γ-cyclodextrin stationary phases as well as the use of β-cyclodextrin mobile phase additive are other options for the chiral separation of these derivatized amino acids.

1. Introduction

1.1. Dansyl derivatives

Pre-column derivatization with 5-dimethylaminonaphthalene-1-sulfonyl chloride (dansyl chloride) is known to enable a sensitive HPLC analysis of amino acids, as do derivatization with *o*-phthalaldehyde, phenylisothiocyanate, fluorescamine or 2,4-dinitrofluorobenzene. Dansyl derivatization has originally been developed by Gray and co-workers [1–3] and applied to the sequential analysis of peptides and proteins. It finds another use in biochemistry for the fluorogenic labeling of proteins and enzymes.

Dansyl chloride is the most widely used for the derivatization of amino acids. Dansyl chloride readily reacts with primary and secondary amino groups of amino acids. It can also react with phenols, thiols and imidazoles less rapidly, and very slowly with alcohols. Therefore, some amino acids can form a few different dansylated derivatives depending on the derivatization conditions. For example, tyrosine forms *N*-dansylated, *O*-dansylated and didansylated derivatives [4]. Dansyl amino acids have usually been separated in the reversed-phase mode, followed by fluorimetric or UV spectrophotometric detection. They have mostly been separated by gradient elution, but isocratic separations have also been reported.

1.2. Dabsyl derivatives

After introduction of 4-dimethylaminoazobenzene-4'-sulfonyl chloride (dabsyl chloride) as a precolumn derivatization agent for amino acids [5], numerous applications have been published, and most of them related to the analysis of amino acids in biological fluids or protein hydrolyzates by HPLC [6, 7]. Dabsyl chloride has a number of advantages over other derivatization methods, including a simple derivatization procedure, very good stability, good reproducibility and a good limit of detection for the method, complete HPLC separation of all the amino acids, and specific detection at a wavelength in the visible region [5–8]. Dabsyl chloride also reacts with primary and secondary amino groups of amino acids as does dansyl chloride, and it forms mono-dabsyl derivatives as well as bis-dabsyl derivatives with amino acids such as lysine, tyrosine and histidine [9].

2. Derivatization conditions

2.1. Dansyl derivatization

The reaction medium is usually an aqueous-organic mixture (e.g., 1:1 acetone-water) adjusted to a pH of 9.5–10. Figure 1 shows the reaction of a primary or secondary amino acid (HNR_1R_2) with dansyl chloride. Dansylation reaction is usually carried out at elevated temperatures. Various dansylation reaction conditions are reported, involving 60 °C for 60 min [10] or at 38 °C for 90-120 min [11].

The dansylation reaction provides high yield derivatization while the yield appears to be independent of the ratio of dansyl chloride to amino acid over 1000-fold range [12]. In addition, dansyl chloride stock solution is kept in a freezer to avoid degradation.

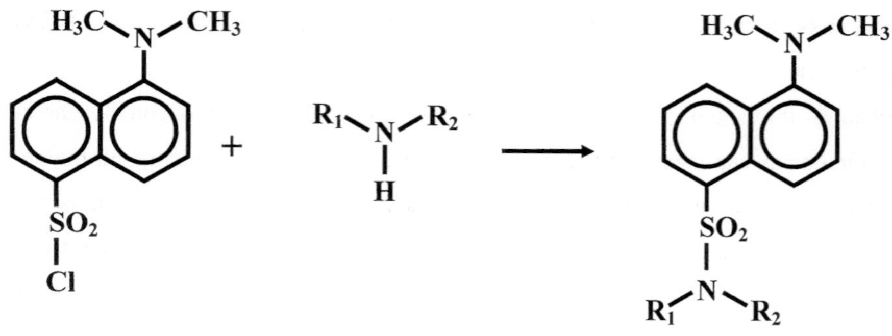

Figure 1 Dansyl derivatization of amino acids

2.2. Dabsyl derivatization

Dabsylation reagent solution can be prepared by dissolving dabsyl chloride in acetone or in acetonitrile, followed by mixing with buffer (e.g., carbonate, pH 8.5–9.5). Derivatization of amino acid with dabsyl chloride is also carried out at elevated temperatures as for dansylation. For example, samples were incubated at 70 °C for 15–30 min [13, 14]. It should be noted that dabsyl chloride stock solution is also kept in a freezer.

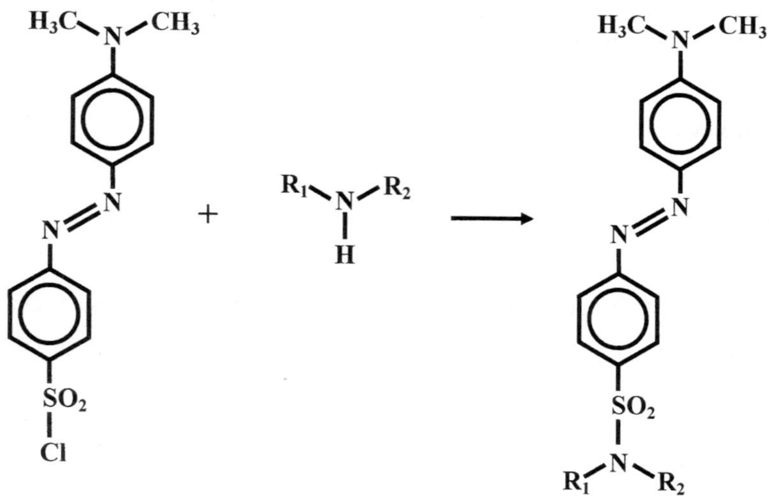

Figure 2 Dabsyl derivatization of amino acids

3. Chromatographic conditions for dansyl and dabsyl amino acids

3.1. Dansyl amino acids

3.1.1. Detection

Dansyl amino acids absorb light in the UV region. Figure 3 shows UV absorption spectra of 0.01 and 0.1 mM dansyl glycine aqueous solution. It can be seen that absorption maxima are observed at 214, 246 and 325 nm, and the absorption at 214 nm is the strongest. The molar extinction coefficients of dansyl glycine at 214, 246 and 325 nm are 3.5×10^4, 1.3×10^4 and 0.39×10^4, respectively (see Table 1).

The dansyl group is fluorescent and dansyl amino acids can therefore be detected by a fluorimetric detector [15]. The excitation and emission wavelengths for dansyl glycine are 324 and 559 nm, respectively [16]. The excitation wavelength is close to the wavelength of a He/Cd laser, which leads to highly sensitive detection of dansyl amino acid [17]. Fluorimetric detection with a packed flow cell was applied to dansyl amino acids in LC using conventional size columns [18, 19] and microcolumns [20]. Both peak area and peak height of analytes increased with increasing retention owing to focusing and environmental effects of the stationary phase [18-20]. Chromatograms obtained in the presence of the stationary phase are demonstrated in Figure 4 [18].

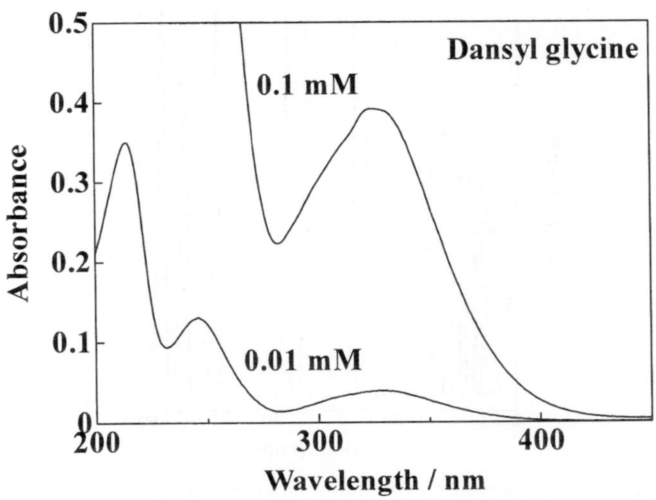

Figure3 UV spectra of dansyl glycine

Table 1. Molar extinction coefficients of dansyl and dabsyl glycine.

Amino acid	Wavelength	ε
	nm	M^{-1} cm^{-1}
Dan-Gly	325	0.39×10^4
	246	1.3×10^4
	214	3.5×10^4
Dab-Gly	473	1.9×10^4
	277	0.65×10^4

When dansyl amino acids eluted from the column are combined at a mixing tee with tris(2,2'-bipyridyl)ruthenium(II) [$Ru(bpy)_3^{2+}$] and $Ru(bpy)_3^{3+}$ is electrochemically oxidized to $Ru(bpy)_3^{3+}$ at the electrode, dansyl amino acids and $Ru(bpy)_3^{3+}$react to emit light. This chemiluminescence detection achieves the detection limit of 2 pmol injected. [21].

Dansyl amino acids were separated by reversed-phase gradient HPLC and detected by optical activity and UV absorbance. This method allowed the determination of enantiomeric ratios of dansyl amino acids without physical separation of the isomers [22].

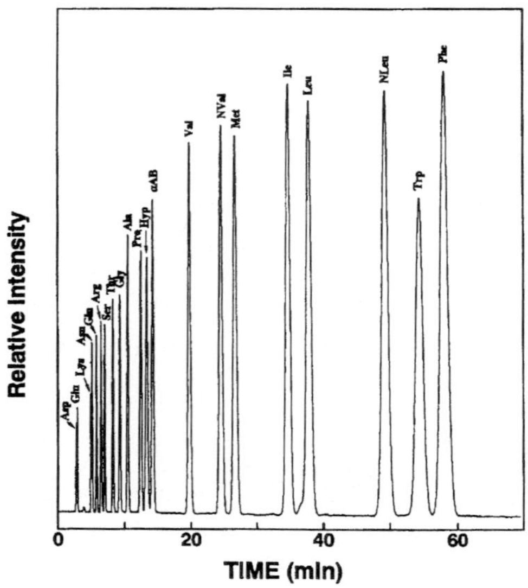

Figure 4 Fluorimetric detection of dansyl amino acids in the presence of the stationary phase. Column, L-column ODS, 150×4.6 mm I.D. Mobile phase, aqueous acetonitrile (22%, v/v) in 40 mM ammonium acetate (pH 7.20). Flow rate, 1.0 mL/min. Detection, 335 nm for excitation and 522 nm for emission. Samples, 21 μL solution containing 0.21 pmol each dansyl amino acids. (Reproduced from Ref. 18 with the permission from the publisher.)

3.1.2. Separation

Amino acids from protein hydrolyzates can be separated in the reversed-phase mode on a C8 or C18 column with a linear gradient. The mobile phase is composed of a mixture of buffer and an organic modifier such as methanol or acetonitrile. The buffer can be selected from Tris-HCl (pH 7.55) [23], ammonium acetate (pH 7.20) [18], sodium acetate (pH 7.5) [24], phosphate (pH 7.4) [25], etc.

Enantiomeric separation of dansyl amino acids can be carried out by ligand-exchange chromatography with (*S*)- and (*R*)-phenylalaninamide-modified silica gel as the stationary phase and copper (II) as complexing metal ion [26]. Human serum albumin [27] as well as bovine serum albumin stationary phases [28] allowed the resolution of dansyl amino acids. Enantiomeric separation of dansyl amino acids were also performed on native, and derivatized β- and γ-cyclodextrin stationary phases [29, 30] as well as on a chiral β-CD-based polymer [31]. Norvancomycin [32], vancomycin [33] and teicoplanin [34] are another stationary phases for the enantiomeric separation of dansyl amino acids. Enantiomeric separation of dansyl amino acids by reversed-phase micro HPLC with a β-cyclodextrin mobile phase additive was reported, in which twelve pairs of dansyl amino acids were separated in a single chromatographic run [35].

3.2. Dabsyl amino acids

3.2.1. Detection

Dabsyl amino acids absorb light in the UV region as well as in the visible region. This feature leads to an improvement of the selectivity when dabsyl amino acids are detected at longer wavelengths. Figure 5 shows a UV/visible absorption spectrum of dabsyl glycine. It can be seen that the absorption maxima are observed at 277 and 473 nm, and the latter absorption is much larger than the former. The molar extinction coefficients of dabsyl glycine at 277 and 473 nm are 0.65×10^4 and 1.9×10^4, respectively (see Table 1). Dabsyl amino acids are normally determined by using visible absorption detector. A linear relation between peak area and concentration was observed from 1.25 to 1,250 pmol and the detection limit was between 0.12 and 0.52 pmol [13].

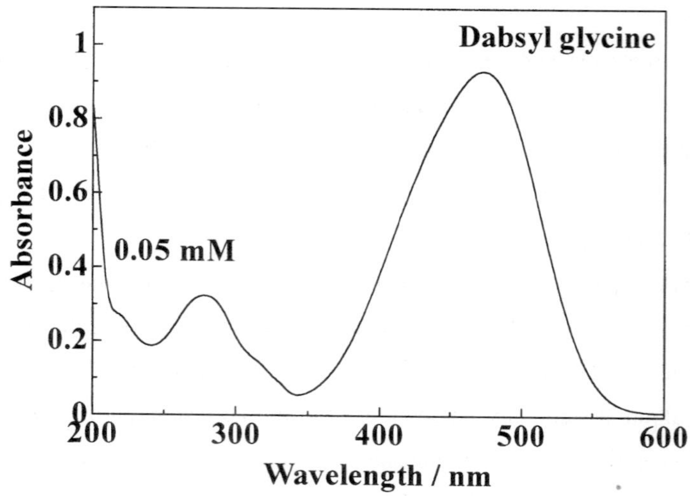

Figure 5. UV-visible spectra of dabsyl glycine.

Detection limits of 0.75 fmol of dabsyl amino acids were achieved by using a 0.25 mm I.D. column and a laser-induced crossed-beam thermal lens detector [36]. Absorbance detection of dabsyl amino acids by laser wave mixing was demonstrated in microbore LC, where an argon ion laser was operated at 488 nm [37]. Tandem mass (MS/MS) detector was used for the determination of dabsyl amino acids, in which common amino acid derivatives had good linear relationship between the peak area and the concentration between 10-250 μM [38].

3.2.2. Separation

Dabsyl amino acids can also be separated in the reversed-phase mode with a gradient consisting of buffer and an organic modifier. Acetonitrile, methanol and ethanol can be employed as the organic modifier, whereas sodium phosphate (pH 6.55) [13] and sodium acetate (pH 4.0) was used as the buffer [39]. It is reported that triethylamine was a very effective additive to optimize separation efficiency [13].

Enantiomeric separation of dabsyl amino acids can be carried out by ligand-exchange chromatography with (*S*)- and (*R*)-phenylalaninamide-modified silica gel as the stationary phase and copper (II) as complexing metal ion [26]. Addition of L-benzoylproline-copper (II) chelate in the mobile phase allowed enantiomeric separation of dabsyl amino acids using a reversed-phase column (C18) [40]. Native and derivatized β-cyclodextrin-bonded stationary phased could also be used for the chiral separation of dabsyl amino acids. Dabsyl amino acid enantiomers were resolved best on a (*R*)- and (*S*)-naphthyl-ethylcarbamate-β-cyclodextrin-bonded phase [41].

4. Application

4.1. Dansyl amino acids

There have been a number of literatures dealing with application of dansylation to the determination of amino acids contained in various sample sources, involving plasma [15, 42], monkey saliva [24], rat cortical tissue [43], bacteria broths [12], forest tree tissues and cell cultures [10], soy sauce [11], peptide or protein hydrolyzates [25, 44–52], and NH_2-terminal analysis of lysozyme [9].

4.2. Dabsyl amino acids

Amino acids contained in various sources have also been derivatized with dabsyl chloride for the determination. Amino acids from protein hydrolyzates [13], plasma [13], urine [39, 53], various tissues [53], food [13, 54] are typical applications.

Figure 6 demonstrates reversed-phase HPLC separation of dabsyl derivatives from human plasma using gradient elution [13], where amino acids and amines are separated in 70 min. The repeatability was reported to be between 0.2 and 3.3%.

Selective determination of secondary amino acids such as proline and hydroxylproline are reported [55, 56]. The method involves an HPLC method with two derivatizations, the first with *o*-phthalaldehyde in order to eliminate interferences due to some primary amino acids eluting with retention times similar to those of L-proline and *trans*-4-hydroxy-L-proline and the second with dabsyl chloride, was developed and evaluated. The method was applied to the analysis of rat tail collagen, and the contents of hydroxyproline and proline were 1.55 ± 0.04 and 2.03 ±0.04 nmol/μg, respectively [56].

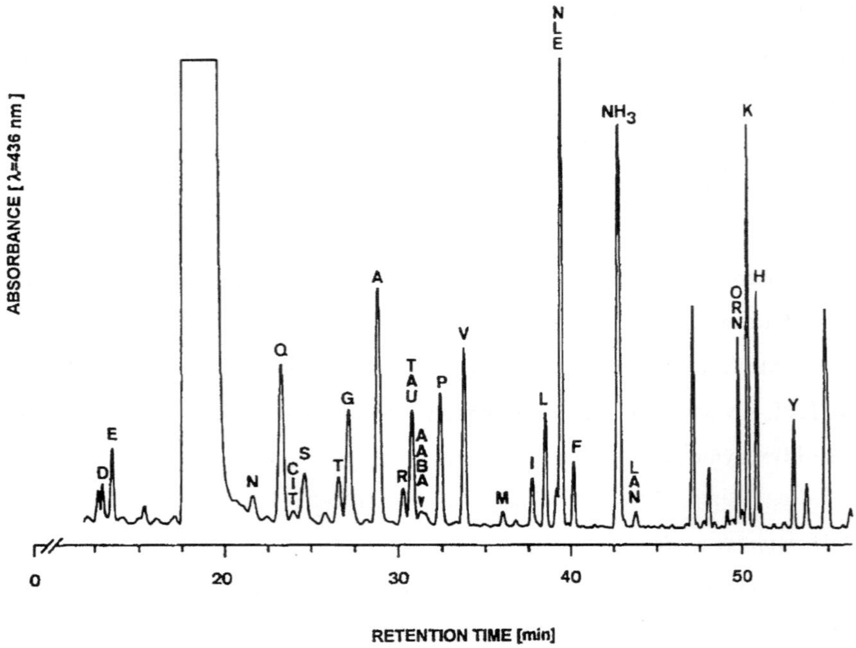

Figure 6 Reversed-phase HPLC separation of dabsyl derivatives from human plasma. Column, Spherisorb ODS-2, 150×4.6 mm I.D. Mobile phase A, 9 mM sodium dihydrogenphosphate, 4% dimethylformamide and 0.16% triethylamine (pH adjusted to 6.55 with phosphoric acid). Mobile phase B, 80% (v/v) aqueous acetonitrile. Gradient, 8% B (0min) to 100% (66 min), almost linear, see Reference 13 for details. Flow rate, 1 mL/min. Wavelength of detection, 436 nm. (Reproduced from Reference 13 with the permission from the publisher.)

Experimental

5.1. Derivatization of amino acids with dansyl chloride

Standard free amino acids were dissolved in a carbonated buffer (pH 9.7) and mixed with acetonitrile-acetone solution of dansyl chloride, followed by heating at 38°C for 90 min for dansylation. The amino acids could be derivatized with addition of dansyl chloride about 5 fold excess of total amino acids. In addition, some dansyl amino acids are commercially available.

Derivatization of amino acids in sake with dansyl chloride can be carried out by the following procedures [11]. The pH of sake was adjusted to 9.7 with 0.2 M sodium bicarbonate and sodium hydroxide. The prepared solution (1 mL) was then placed in a 5-mL vial and mixed with 0.13 mL of 1% (w/v) dansyl chloride dissolved in acetonitrile-acetone (90:10). The vial was placed in an oven and heated at 38°C for 90 min for dansylation.

5.2. Chromatography

The liquid chromatograph was assembled from two PU-980 LC pumps (Jasco, Tokyo, Japan) with an HG-980-30 mixer, a Model 7000 six-port valve (Rheodyne, Cotati, Ca, USA) with a laboratory-made loop (ca. 20 μL) as injector, a 150×4.6 mm I.D. An L-column ODS separation column (5 μm; Chemical Inspection & Testing Institute, Tokyo, Japan), an FP-920 fluorimetric detector (335 nm for excitation and 522 nm for emission; Jasco), and a Chromatopac C-R4AX data processor (Shimadzu, Kyoto, Japan) [19]. The flow rate of the pump was 1 mL/min.

The solutions delivered by the two pumps were 40 mM ammonium acetate and 50% (v/v) acetonitrile in 40 mM ammonium acetate. The composition of acetonitrile was linearly programmed in the high-pressure gradient mode. The acetonitrile concentration was changed from 18% at a slope of 0.35%/min for 21 dansyl amino acids [19]. It should be noted that the dansyl amino acids can also be sensitively detected by a UV detector at 214 nm.

References

[1] W. R. Gray, B. S. Hartley, Biochem. J.89 (1963) 379-380.

[2] W. R. Gray, J. F. Smith, Anal. Biochem. 33 (1970) 36-42.

[3] B. S. Hartley, Biochem. J. 119 (1970) 805-822.

[4] P. L. Felgner, J. E. Wilson, Anal. Biochem. 80 (1977) 601-611.

[5] J. K. Lin, J. Y. Chang, Anal. Chem. 47 (1975) 1634-1638.

[6] J.-Y. Chang, R. Knecht, D. Braun, Methods Enzymol. 91 (1983) 41-48.

[7] H. J. Schneider, Chromatographia, 28 (1989) 45-48.

[8] E. H. J. M. Jansen, R. H. Van den Berg, R. Both-Miedema, L. Doorn, J. Chromatogr. 553 (1991) 123-133.

[9] N. Kaneda, M. Sato, K.Yagi, Anal. Biochem. 127 (1982) 49-54.

[10] R. Minocha, S. Long, J. Chromatogr. A 1035 (2004) 63-73.

[11] T. Takeuchi, M. Yamazaki, D. Ishii, J. Chromatogr. 295 (1984) 333-339.

[12] Y. Tapuhi, D. E. Schmidt, W. Lindner, B. L. Karger, Anal. Biochem. 115 (1981) 123-129.

[13] I. Krause, A. Bockhardt, H. Neckermann, T. Henle, H. Klostermeyer, J. Chromatogr. A 715 (1995) 67-79.

[14] M. A. Castillo, R. C. Castells, J. Chromatogr. A 921 (2001) 121-133.

[15] J. T. Tyopponen, J. Chromatogr. 413 (1987) 25-31.

[16] Y. Kitamaki, T. Takeuchi, unpublished data.

[17] B. G. Belen'kii, V. A. Mostovnikov, S. V. Nechaev, A. F. Lobazov, I. V. Nazimov, Zhurnal Prikladnoi Spektroskopii 44 (1986) 511-517.

[18] T. Takeuchi, T. Miwa, Chromatographia 40 (1995) 545-549.

[19] T. Takeuchi, T. Miwa, Chromatographia 41 (1995) 148-152.

[20] T. Takeuchi, T. Miwa, Anal. Chim. Acta 311 (1995) 231-236.

[21] W.-Y. Lee, T. A. Nieman, J. Chromatogr. A 659 (1994) 111-118.

[22] B. H. Reitsma, E. S. Yeung, Anal. Chem. 59 (1987) 1059-1061.

[23] N. Kaneda, M. Sato, K.Yagi, Anal. Biochem. 127 (1982) 49-54.

[24] R. A. Miller, N. E. Bussell, J. R. Wynkoop, II, R. Bongiovanni, T. M. Boehm, Chromatogr. Sci. Series 18 (1981) 235-246.

[25] A. R. Martins, A. P. Padovan, J. Liq. Chromatogr. Relat. Techno. 19 (1996) 467-476.

[26] B. Galli, F. Gasparrini, D. Misiti, C. Villani, J. Chromatogr. A 666 (1994) 77-89.

[27] E. Peyrin, Y. C. Guillaume, N. Morin, C. Guinchard, J. Chromatogr. A 808 (1998) 113-120.

[28] Y. Abe, S. Fukui, Y. Koshiji, M. Kobayashi, T. Shoji, S. Sugata, H. Nishizawa, H. Suzuki, K. Iwata, Biochim. Biophys. Acta 1433 (1999) 188-197.

[29] T. Araki, Y. Kashiwamoto, S. Tsunoi, M. Tanaka, J. Chromatogr. A 845 (1999) 455-462.

[30] T.-Y. Kim, H.-J. Kim, J. Chromatogr. A 933 (2001) 99-106.

[31] B. Mayr, F. Sinner, M. R. Buchmeiser, J. Chromatogr. A 907 (2001) 47-56.

[32] G.-S. Ding, Y. Liu, R.-Z. Cong, J.-D. Wang, Talanta 62 (2004) 997-1003.

[33] I. Slama, E. Jourdan, A. Villet, C. Grosset, A. Ravel, E. Peyrin, Chromatographia 58 (2003) 399-404.

[34] E. Peyrin, C. Ravelet, E. Nicolle, A. Villet, C. Grosset, A. Ravel, J. Alary, J. Chromatogr. A 923 (2001) 37-43.

[35] T. Takeuchi, H. Asai, D. Ishii, J. Chromatogr. 357 (1986) 409-415.

[36] T. G. Nolan, N. J. Dovichi, Anal. Chem. 59 (1987) 2803-2805.

[37] Z. Wu, W. G. Tong, J. Chromatogr. A 805 (1998) 63-69.

[38] Y.-H. Chen, L.-L. Shih, S.-E. Liou, C.-C. Chen, Food Sci. Technol. Res. 9 (2003) 276-282.

[39] J.-K. Lin, C.-H. Wang, Clin. Chem. 26 (1980) 579-583.

[40] S. H. Lee, T. S. Oh, Y. C. Lee, Bull. Kor. Chem. Soc. 11 (1990) 411-414.

[41] S. H. Leel, A. Berthod, D. W. Armstrong, J. Chromatogr. 603 (1992) 83-93.

[42] F. J. Márquex, A. R. Quesada, F. Sánchez-Jiménez, I. Núñez De Castro, J Chromatogr. B 380 (1986) 275-283.

[43] M. H. Joseph, J. Halliday, Anal. Biochem. 64 (1975) 389-402.

[44] G. J. Schmidt, D. C. Olson, W. Slavin, J. Liq. Chromatogr. 2 (1979) 1031-1045.

[45] S. Weiner, A. Tishbee, J. Chromatogr. 213 (1981) 501-506.

[46] L. N. Mackey, T. A. Beck, J. Chromatogr. 240 (1982) 455-461.

[47] L. F. Congote, J. Chromatogr. 253 (1982) 276-282.

[48] S. Weinstein, S. Weiner, J. Chromatogr. 303 (1984) 244-250.

[49] R. Badoud, G. Pratz, Chromatographia 19 (1984) 155-164.

[50] A. Negro, S. Garbisa, L. Gotte, M. Spina, Anal. Biochem. 160 (1987) 39-46.

[51] T. Takeuchi, T. Niwa, D. Ishii, J. Chromatogr. 407 (1987) 141-150.

[52] A. M. Rizzi, P. Briza, M. Breitenbach, J. Chromatogr. 582 (1992) 35-40.

[53] T. Abe, Y. Kurozumi, W.-B. Yao, T. Ubuka, J. Chromatogr. 712 (1998) 43-49.

[54] O. Pinho, I. M. P. L. V. O. Ferreira, E. Mendes, M. O. Bruno, F. Margarida, Food Chem. 75 (2001) 287-291.

[55] M. Ikeda, K. Sorimachi, K. Akimoto, Y. Yasumura, J. Chromatogr. 621 (1993) 133-138.

[56] V. Bianchi, L. Mazza, J. Chromatogr. B 665 (1995) 295-302.

Ibolya Molnár-Perl (Editor)
Quantitation of Amino Acids and Amines by Chromatography
Journal of Chromatography Library, Vol. 70

1.3.6.

Quantitation of Amino Acids as 6-Aminoquinolyl-N-hydroxysuccinimidyl Carbamate Derivatives

Steven A. Cohen

Content

Summary

Amino acid analysis following derivatization with 6-aminoquinoly-N-hydroxysuccinimidyl carbamate (AQC) has been applied to a wide variety of samples in the past decade. Derivative stability is sufficient for automated analysis and storage. Protein compositional analysis shows excellent agreement with expected results. Fluorescence detection provides fmol sensitivity and permits accurate analysis of samples down to 50 ng. Application to a number of unusual amino acids has been demonstrated, as well as the analysis of chiral amino acids, peptides and porteins following derivatization. Matrices as diverse as sea water, cell culture fluids, biological fluids and food samples have proven amenable to the procedure.

1. Introduction

Improvements in amino acid analysis have been an important driving force and have often paralleled advances in analytical methods. For the past twenty years amino acid analysis has been split between methods using pre- and post-column derivatization prior to chromatographic separation. My lab was involved with the development of the phenylisothiocyanate (PITC) method that uses the coupling step for Edman degradation as the labeling step. This yields reasonably stable phenylthiocarbamyl derivatives suitable for analysis by liquid chromatography (LC) with UV detection [1, 2]. In comparison to post-column ninhydrin based derivatization that relies on ion-exchange separation of the underivatized amino acids, the PTC method provided faster analysis times, more sensitive detection, and could be performed on standard HPLC equipment rather than dedicated amino acid analyzers.

Even with these improvements, it quickly became clear that scientists required even greater sensitivity, and an easier to perform, more robust derivatization chemistry. The PTC method was shown to work best with highly purified salt-free samples, not always practical with real samples, and the derivatives, while sufficiently stable for automated HPLC analysis, do degrade slowly, even at 4° C, with the losses up to 10% in a few days under refrigeration.

Many other pre-column derivatization reagents, such as ortho-phthalaldehyde (OPA), 9-fluorenylmethyl chloroformate (FMOC-Cl) and 4-dimethylamino azobenzenesulfonyl chloride (Dabsyl- Cl), have been shown suitable for amino acid analysis [3-5], yet in each case one or more key performance attributes have been found lacking, such as OPA derivative stability [6], multiple derivative formation with FMOC-Cl and Dabsyl-Cl [7, 8], narrow optimal pH range for derivatization with Dabsyl-Cl, [8], and salt interference with derivatization by FMOC-Cl [9]. AQC, a more recently introduced reagents for pre-column derivatization, is a highly reactive activated carbamate specifically developed for amino acid analysis [10], and is the focus of this chapter.

2. Basic Chemistry

2.1 Reagent Design

N-hydroxysuccinimide (NHS) activated esters have long been a widely employed moiety for amine derivatization and have been used for both biochemical reactions and in organic synthesis. Anderson et al. first reported the use of NHS activated esters for peptide synthesis [11] and the well-known Bolton-Hunter's reagent [3-(4-hydroxyphenyl)propionic acid, succinimidyl ester [12] is used to provide a facile method for iodination of peptides and proteins.

At mildly alkaline pH (e.g. < 11), the acylation of amines and amino acids by NHS esters is a much faster reaction than hydrolysis, thus permitting derivatization in aqueous solution.

Less well-known are the NHS activated carbamates, which were introduced by the Japanese group led by Ogura [13, 14]. Similar to the activated esters, NHS activated carbamates are highly reactive electrophilic reagents suitable for reaction with nucleophilic targets such as amines and amino acids. Nucleophiles attack the carbonyl center adjacent to the NHS group, which provides an excellent leaving group for enhancing the reactivity. The resulting amides (in the case of the activated esters) and asymmetric ureas (for the carabamates) are very stable at room temperature. Ogura and co-workers synthesized several NHS activated carbamates they employed for amino acid analysis, but these reagents are not commercially available and have been virtually ignored by analytical laboratories.

There are numerous fluorescent groups that are potentially suitable for an activated carbamate derivatizing reagent. The choice of the 6-aminoquinolyl tag was suggested by a somewhat obscure paper from 1981 by Brynes et al., who synthesized a chymotrypsin substrate analog with the 6-aminoquinolyl moiety linked to the carboxyl terminus of phenylalanine via an amide bond [15]. Enzymatic hydrolysis of the bond released 6-aminoquinoline (AMQ), whose fluorescent emission maximum was red shifted over 100 nm compared to that of the substrate (emission maximum ca. 520 nm vs. 390 nm), thus enabling sensitive detection of the product without separation from the starting material. Our design intention was to exploit this red shift to selectively detect derivatized amino acids in the presence of a large excess of the carbamate reagent with minimal interference, both simplifying sample preparation and chromatographic separation development.

Synthesis of AQC was analogous to the method developed by Ogura using a one step reaction combining disuccinimidyl carbonate with AMQ in acetonitrile, and heating at reflux for one hour. Alternate routes starting from 6-isocyanato quinoline were also successful, but ultimately provided no significant advantage over the Ogura procedure. A simple recrystallization from acetonitrile afforded very pure AQC (>99%) suitable for amino acid derivatization [10].

2.2 Derivatization Chemistry

Coupling of amino acids (and other amines) to the quinoline tag of AQC is readily accomplished in alkaline solution in the pH range from 8.0 – 10.5. Although precise kinetics have not been reported, as the reaction is very rapid, attempts to quench the reaction with an excess of ammonia added within five seconds reagent addition had no discernible effect on labeled amino acid yields (S. Cohen, unpublished results), suggesting that the reaction half-life is much less than one second. Simultaneously, although at a much slower rate, excess reagent is hydrolyzed in a pH dependent reaction to produce 6-AMQ, NHS and CO_2 (Figure 1). The estimated reaction half-life for hydrolysis at pH 8.8 is 15s [10].

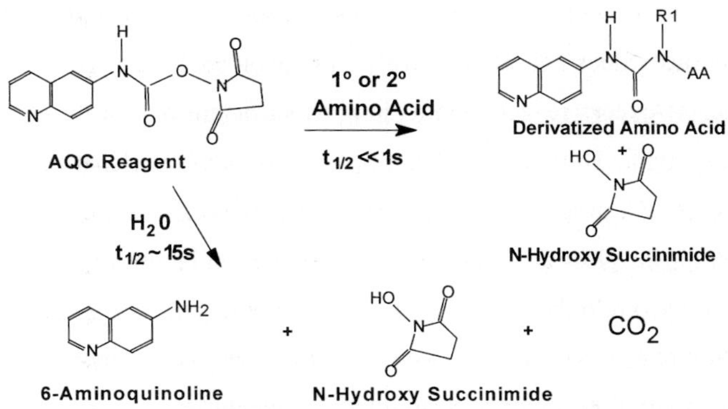

Figure 1. Derivatization scheme for the reaction of AQC with amines, also showing the primary route of excess reagent hydrolysis. Reproduced with permission from Reference [10].

A number of buffers for derivatization were initially studied including borate, bicarbonate, carbonate and phosphate, but ultimately sodium borate proved to provide the best results with excellent yields and the lowest background signal from buffer components. Within the pH range specified above, good reaction yields were observed. Below pH 8.0 there was an increasing amount of singly derivatized Lys products (see below for details), and the yields of slow reacting amino acids (e.g. Asp, Glu, Val, and Ala) were not quantitative. Above pH 10.5, hydrolysis was too rapid to permit effective mixing and complete derivatization before the reagent was destroyed.

Because of the competing hydrolysis reaction, a sufficient molar excess of AQC versus the amount of available derivatizable groups is necessary to effect quantitative conversion to the desired products. The early work by Cohen and Michaud demonstrated that a five-fold molar excess was sufficient as long as the pH was maintained in the optimal range of 8.0 – 10.0 [10]. If sufficient excess is not provided, lower yields are observed for the same slow reacting amino acids seen in the pH studies discussed above.

Side reactions of AQC are limited in nature. Alcohols react very slowly, as can be seen from the reaction of AQC dissolved in methanol. However, there is no evidence that Ser or Thr are labeled on the side chain, and almost undoubtedly the hydrolysis reaction rapidly depletes the reagent before any reaction with the hydroxyl groups.

Because of rapid hydrolysis and the potential for solvolysis in protic solvents, dry, aprotic solvents such as acetonitrile are essential for preparing AQC solutions prior to its addition to a buffered sample. Even minute amounts of water can be deleterious for storing AQC in solution as AQC will scavenge any water present in the solution. This produces AMQ, which fortuitously does not interfere with subsequent analysis as it is a normal product from the hydrolysis of excess reagent and as long as sufficient reagent remains in solution, derivative yields are high. However, AMQ, albeit a very weak nucleophile, will react with another molecule of AQC, producing a di-aminoquinolyl urea. Because of limited solubility of this urea in acetonitrile, slow generation of a precipitate can be observed in long term storage if moisture is allowed into the reagent solution. Fortunately the small concentration of the urea has no affect on chromatographic analysis, having longer retention than any amino acid derivative, and very weak fluorescence due to internal quenching.

In a pH dependent side reaction, the Tyr phenolic group reacts with AQC to produce a di-derivatized product. Yields increase with increasing pH, but at no pH was the conversion to a doubly derivatized product complete, with maximum conversion of approximately 35%. Fortunately, the phenolic adduct is not very stable, with a half life (pH 8.8, 25° C) around an hour. Singly derivatized Tyr, solely tagged on the amino group, can be readily generated by the addition of base (e.g. triethylamine) or simply by heating at 55° C for 10 minutes. Because a high pH solution is preferably avoided so as to minimize introducing a high pH sample onto base-sensitive HPLC columns, the heating route has been universally adopted.

Several amino acids, notably Lys, Orn and cystine, provide multiple derivatization sites for electrophilic reagents like AQC. With Lys, the reactivity of the two sites are dissimilar due to their difference in pKa, the alpha amino group having a pK of 8.95, the side chain epsilon amino group having a pK of 10.53. When the AQC concentration is kept below a five-fold excess over available derivatizable amine groups, or if the reaction pH is below 8.0, the yield of the di-derivatized Lys product is not quantitative. In many instances one or both of the mono-derivatized forms are observed in separations with retention times much less than the di-derivatized form. Typically, these elute in the region of other polar amino acids such as Gly and His [10]. Their disappearance and conversion to the di-derivatized moiety can be readily accomplished by the addition of an additional aliquot of reagent. In contrast, cystine being a symmetrical molecule shows no discrimination between the reactivities of its two amino groups, and gives excellent yields even at limiting reagent concentrations.

A novel method for derivatization was described by Krull and co-workers where a styrene-divinylbenzene polymeric reagent was used to tag amines and amino acids [16]. Thus, the polymer reagent contained an activated aminoquinoline carbamate attached to the polymer backbone via a benzotriazole linker. Alkaline primary or secondary amine containing solutions were passed through a column in a pipette tip containing the reagent and reacted with the aminoquinoline to form AQC derivatized analytes. Analytes were washed from the column with acetonitrile and an aliquot injected for HPLC analysis. While not as straightforward as the solution phase derivatization, it did show some potential as the basis for an automated derivatization procedure, but this has never been reduced to practice.

2.3 Derivative Properties

Derivative stability is far superior to many other amino acid reagents such as PITC or OPA, with little discernible degradation observed at room temperature in a week. Under refrigeration, derivatized standards can be stored up to a month [10, 17].

The quinoline tag provides some of the more interesting properties for the derivatives. They absorb strongly ($\varepsilon \sim 22{,}000$) in the UV with an absorption maximum approximately 248 nm and a smaller local maximum at 325 nm. Thus UV detection is readily accomplished with low pmol sensitivity [18], but the unusual fluorescence properties form the basis of highly sensitive, simplified fluorescence detection. With excitation at 248 nm, derivative emission maximum is ca. 395 nm, but the major fluorescent hydrolysis product, AMQ, has only weak emission at 395 nm with a maximum emission intensity at ca. 520 nm. Thus, direct injection of a derivatization mixture without any AMQ removal is possible with only a small AMQ peak resulting even if an enormous excess of reagent is employed (cf. Fig 2). This unusual fluorescence property can be exploited by using a dual monochromator detector where both excitation and emission wavelengths are selected with narrow bandwidths [10].

Fluorescence response is strong function of solvent pH. Maximum response is observed at pH > 7.0, with response falling to approximately 80% of maximum at pH 5.0, and virtually no fluorescence at pH below 3.0 [10]. Although not supported by independent evidence, it is likely that this is a result of protonation of the quinoline ring on the aminoquinoline, pK approximately 4.5, and poor fluorescence for the protonated form. Thus, chromatographic conditions under mildly acidic conditions (e.g. pH 5.05 as shown in Figure 2), are only modestly impacted by operating at less than optimal pH for fluorescence yield.

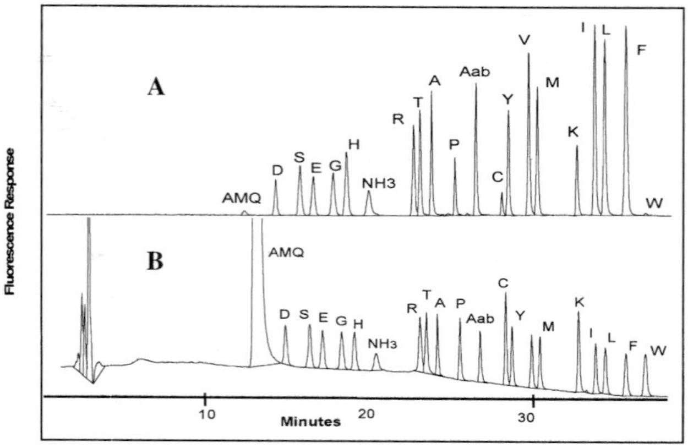

Figure 2 Separation of 50 pmol of a hydrolyzate standard mixture plus Trp and Aab using fluorescence (A) and UV (B) detection.

Fluorescence response for derivatized amino acids varies widely due to two separate, but important quenching phenomena. As seen in the following section, retention of the AQC-derivatized amino acids varies with the hydrophobicity of the amino acid side chain. Predictably the more polar amino acids such as Asp, Glu, Ser and Thr elute in the highly aqueous portion of the gradient separation, which causes a significant reduction in their response due to aqueous quenching. Typically these polar amino acids show half the response of more highly retained amino acids. The aqueous quenching effect was confirmed by Nugent et al. [19] who showed that post-column addition of buffered acetonitrile provided more uniform fluorescence response across the entire gradient range. The high pH addition also allowed the use of an acidic (pH 4.0) mobile phase for the separation without compromising fluorescence detection, with the caveat that band broadening must be minimized in such a system.

Exceptions to the usually greater response of the more highly retained amino acids are observed with disubstituted amino acids such as Lys and cystine. Trp, which contains the fluorescent indole as well as the quinoline group, also exhibits much lower fluorescence response. This is due to significant internal fluorescence quenching in these derviatives, where the energy is transferred from one fluorescent group to another, and the subsequent non-radiative emission from a lower excited state, typically in the infrared. Thus, these derivatives with two fluorescent moieties have much lower response than other hydrophobic amino acids such as Leu and Phe. As the magnitude of this effect is dependent on the spatial arrangement of the two fluorescing groups, fluorescence quenching varies widely, the effect being moderate with Lys, more significant for cystine, and severe for Trp, which has only 1% the fluorescent response of Phe.

The AQC derivatives predictably have very similar UV molar extinction coefficients at their maximum ca. 248 nm [10], typically exceeding 22,000, which is similar to other quinoline derivatives [20]. Liu also showed that this provided very sensitive analysis with UV detection, with the caveat that the separation conditions must be modified to increase the resolution of AMQ and Asp [18]. This can be accomplished in several ways, notably by decreasing the initial gradient slope, or as in the cited paper, lowering the aqueous eluent pH from 5.05, as in the original work, to 4.95. Retention time and response reproducibility, response linearity and compositional accuracy were comparable to that reported for fluorescence detection, with somewhat higher detection limits of 0.5 – 1.0 pmol.

2.4 Chromatographic Separation of Amino Acid Derivatives

As with virtually all amino acid derivative separations, gradient chromatography is required due to the wide hydrophobicity range of the amino acid side chains. Optimized separations require careful manipulation of a number of operating parameters such as eluent ionic strength and pH, ion pairing reagent concentration, gradient shape and column temperature.

The normal conditions for the chromatographic separation of a hydrolyzate standard (50 pmol injected) use fluorescence detection with excitation at 248 nm and emission at 395 nm. A typical analysis of a standard is shown in Figure 2. Features in the separation worth noting include the elution position of the AMQ peak prior to any of the amino acid derivatives. This elution order is highly pH dependent, with low pH eluent favoring AMQ elution prior to the polar amino acids, but at higher pH, the AMQ peak elutes in the polar amino acid region. Low pH elution simplifies separation development as the peak is rendered non-interfering. Fortunately, the fluorescence of AMQ is weak and by setting the fluorescence detector for emission at 395 nm, the peak is very small for the amount of AMQ present. As the typical reaction contains an excess of at least five-fold of reagent over derivatizable amines, a minimum of 80% of the AQC is converted to AMQ. Also evident are the effects of aqueous and internal fluorescence quenching.

By optimizing the eluent pH, the elution positions of the acidic amino acids Asp and Glu can also be carefully controlled, as their retention increases relative to other amino acids with increasing pH. The di-derivatized amino acids are also more sensitive to eluent pH compared to other nearby components, also exhibiting greater retention with increasing pH.

One of the few papers describing alternate conditions for the hydrolyzate amino acids was published by Shindo et al. [21]. By using a hydrophobic ion pairing reagent, tetrabutylammonium, retention of the amino acids, especially acidic amino acids, was greatly increased, thus minimizing the aqueous quenching effect. This improved the detection limits for polar amino acids so that LOD's ranged from 12 – 52 fmol (except for Trp, LOD 1.8 pmol).

2.5 Analytical Performance

Excellent analytical performance is one of the major reasons for the wide spread use of the AQC derivatization method [10]. The highly fluorescent derivatives allow for detection in the fmol range, with detection limits ranging from 50 – 400 fmol on-column being determined. The higher detection limits were for polar and di-derivatized amino acids and the lower limits for hydrophobic amino acids. Although there have been a number of technical improvements in detectors since that publication in 1993, there are no recent publications exploiting the potential for lower detection limits that these may offer. Data from the initial work also demonstrated good linearity over a wide concentration range (2 – 200 μM, $r^2 \geq 0.9992$) for all hydrolyzate amino acids. Replicate analyses of ten standards showed excellent reproducibility for both area and retention time (Figure 3). Samples spiked with salt or detergent showed no discernible effect on derivatization yields (Figure 4).

One of the practical limitations that the AQC derivatization procedure imposes on analysis is dictated by the derivatization solution, which typically contains 20% acetonitrile and is at half saturation for AQC. Injection of more than 10-15 ul of this solution invariably causes excessive peak broadening for early eluting amino acids. This translates to only 10-15% of the total sample being injected if the final solution volume is 100 ul. The lowest practical volume for derivatization is ca. 50 ul, which increases the percentage injected allowed to 30%. However, decreasing the volume ratio of AQC solution is not acceptable as derivative yields are compromised. To solve this problem, Crimmons and Cherian [22] showed that the final solution could be dried, reconstituted with 5% acetonitrile and up to 80% of the total derivatized amount (80/100 ul reconstituted sample) injected with no effect on chromatographic peak shape or resolution. Excellent recoveries of all derivatized amino acids was also demonstrated. A similar result was achieved by Nugent et al. [19] who increased the loading volume capacity of 1 mm ID microbore column by using a polymeric trap column before the separation column. Up to 50 ul of the initial 100 ul derivatization solution was injected without loss of chromatographic performance.

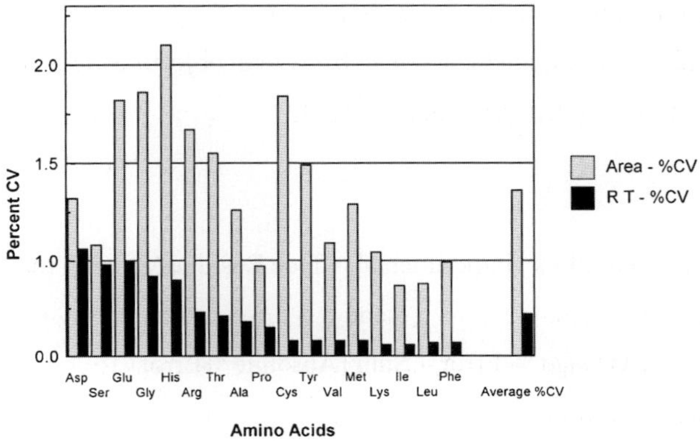

Figure 3 Reproducibility plot for retention time and peak area. A total of 10 samples containing 2 nmol of each amino acid (1 nmol cystine) were derivatized and 5% of the total was analyzed using the hydrolyzate amino acid conditions.

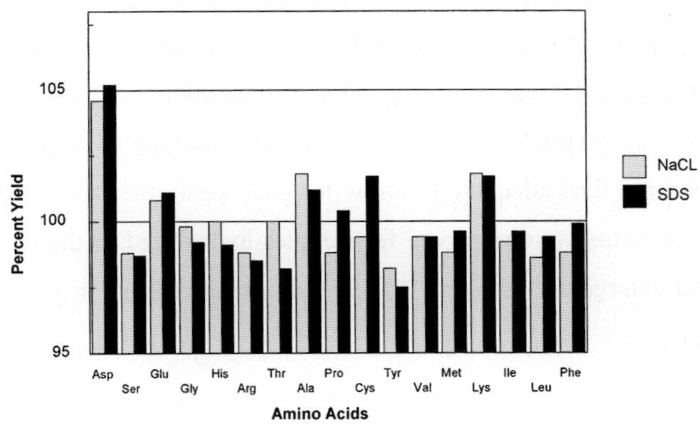

Figure 4 Salt and detergent effects on derivatization efficiency. Standard amino acid solutions (10 μL of a 0.1 mM solution) were spiked with either 50 μL of 0.1 M NaCl or 0.2% sodium dodecyl sulfate solution.

3. Applications

3.1 Compositional Analysis

Accurate amino acid analysis of hydrolyzed proteins is one of the more challenging bioanalytical methods. Issues with the hydrolysis process are covered elsewhere in this volume, but even with good hydrolysis procedures, accurate and reproducible compositional analysis requires a sound methodology for the analysis and careful attention to a variety of operational details, particularly for high sensitivity analysis.

Compositional accuracy can be assessed by several objective methods, with a common approach being that described by Strydom et al. [23].

Equation 1

%Error – 100 x (Experimental – True Residue Value)/True

and

Average % Error – Sum (Absolute %Error)/16

In this calculation, errors for Trp and Cys are ignored, as they are poorly recovered during standard acid hydrolysis procedures. The average errors include hydrolytic losses of acid sensitive amino acids including Met, Ser and Thr.

In a series of papers, Cohen and coworkers [10, 17, 24] provided ample evidence for accurate amino acid analysis using AQC derivatization. Typical average errors were 6% or less if the amount hydrolyzed exceeded 1 ug, and even with submicrogram amounts, average errors less than 10% were routinely obtained (Table I). As can seen in Figure 5, the response for these samples is more than adequate to allow for good peak integration. Only minor increases in the average errors were observed for samples hydrolyzed in the presence of common buffer salts and detergents. Others [18, 25] have obtained similar accuracy results for compositional analysis.

Table I. Compositional Errors in the Analysis B-Lactoglobulin Hydrolyzates: the Effect of Sample Amount on Compositional Accuracy

Sample Amount (ng)	140 - 389	587 - 789	1038 - 1915	9400 - 9770
Number of Analyses	9	5	11	2
Average of average %errors*	9.7	8.6	6.2	4.5
Range of average %errors	5.5 - 16.3	4.9 - 14.6	4.1 - 12.2	4.3 - 4.7
Average sample amount (ng)	231	684	1583	9587

* Errors for each individual hydrolysis are calculated according to the formula in Equation 1. The average errors for all analyses are then calculated for the sample amount range indicated.

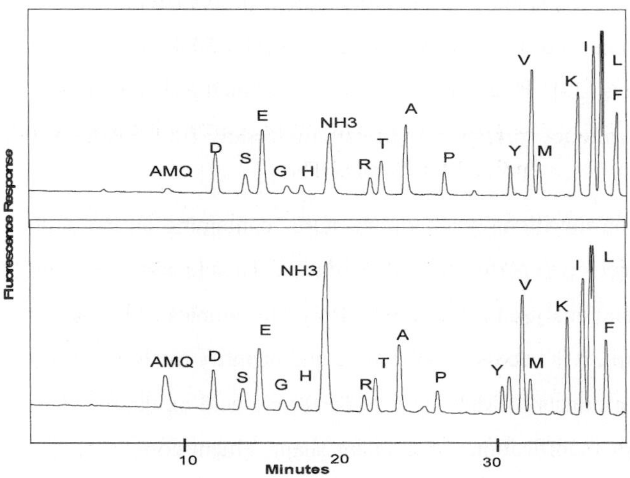

Figure 5 Amino acid analyses of B-lactoglobulin hydrolyzates. Aliquots of protein solutions containing 160 (A) and 1500 (B) ng were dried and hydrolyzed using a vapor phase procedure (24). Reproduced with permission from Reference [24]

Highly sensitive analysis of proteins analyzed by polyacrylamide gel electrophoresis was recently described by Cohen [26]. Samples were electrophoretically transferred from the gel to the PVDF membranes, and the immobilized proteins hydrolyzed by vapor phase hydrolysis. Amino acids were extracted with methanolic HCl solution (30% MeOH in 0.1 M HCl), dried, reconstituted, derivatized and analyzed. Shindo et al. [27] also demonstrated the ability to derive amino acid compostion data from PVDF blotted samples. By using clean techniques, contamination was reduced to the low fmol level.

3.2 Other Amino Acid Applications

Since its introduction in 1993, there have been a number of papers expanded the scope of sample types and applications amenable to analysis using AQC derivatization beyond that of the originally reported protein compositional analysis. The first of these to be reported [28] was the analysis of hydrolyzed crude animal feed samples, a sample matrix well-known for its challenges due to the presence of many extraneous components, especially the high carbohydrate content, that can cause interference with derivatization reactions.

Sample preparation for feed analysis poses issues not encountered with highly purified proteins. The first problem is how to obtain a sample that is representative of the whole. There are no suitable solvents for feed samples, and thus it is not possible to dissolve a sample and aliquot an appropriate amount for hydrolysis and derivatization. Instead, samples are milled to a uniform size, and then a large amount, typically 50-100 mg, of the solid sample is weighed for hydrolysis [29]. Hydrolyzed samples are diluted and filtered to remove insoluble material, and then an aliquot corresponding to approximately 1-5 μg is used for derivatization and analysis.

To analyze the nutritionally important sulfur containing amino acids Met and Cys, feed samples are subjected to performic acid oxidation. Thus in addition to the 17 amino acids normally separated and analyzed in protein hydrolyzate samples (20 protein amino acids minus Asn, Gln and Trp), it is necessary to resolve and quantify methionine sulfone and cysteic acid, the respective products of Met and Cys performic acid oxidation. Cohen and De Antonis [28] used minor modifications of gradient shape, eluent composition and column temperature to better resolve Ala and methionine sulfone, which were poorly resolved using conditions described in Figure 2. Because the analysis is carried out with a relatively large amount of sample, the sensitivity of fluorescence detection is not required, and UV analysis, as described by Liu et al., is sufficient. As a key requirement for these types of samples, several papers have shown excellent agreement with classical ion-exchange analysis for amino acid concentrations in a number of feed samples [28, 30].

Collagen samples also challenge the analysis procedure due to the presence of hydroxyproline and hydroxylysine, as well as the high concentration of Gly, which occurs at every third residue. Successful analysis required conditions to accommodate the two additional components, and improve the separation of Gly and His because of the high ratio of Gly to His. With initial eluent pH near 5.0, the hydroxyproline peak elutes between AMQ and Asp, while the hydroxylysine peaks (due to the presence of diastereomers) elute near Met. Conditions for the oxidized sulfur amino acids were shown to be suitable for complete resolution. An improved collagen hydrolyzate separation in which the hydroxylysine peaks elute after Met has also been reported [31]. This was accomplished by changing the eluent pH midway through the analysis. Despite the added complication of the wide dynamic range in concentration, collagen compositional analysis was also in good agreement with published data.

Glycoproteins contain no unusual amino acids, but most have acetylated forms of glucosamine (GlcNH2), derived from o-linked oligosaccharides, or galactosamine (GalNH2), from N-linked oligosaccharides, that are converted to the free amine during acid hydrolysis. Although yields of the hexosamines are not quantitative, the remaining surviving amino sugars react with AQC, forming stable derivatives. Derivatization of GalNH2 and GlcNH2 produces one major and one minor component. The major components elute prior to the AMQ peak, and thus are easily resolved from the derivatized amino acids. The minor component of GlcNH2 elutes between the Glu and Ser peaks and also does not interfere. However, the GalNH2 minor component elutes near the Ser peak and necessitated a change in the standard hydrolyzate gradient to effectively resolve from Ser [28]. It is hypothesized that the multiple components stem from resolution of the anomeric forms of the amino sugars.

Diaz et al. exploited the method of Cohen and De Antonis to develop a quantitative method for the amino sugars by using a 6 h hydrolysis of glycoprotein containing samples such as soya protein, sludge-containing soils, and the polysaccharide chitin [32]. These researchers noted that because the chromatographic method separated the anomeric forms, no prior reduction was required for the analysis, thus saving several hours of sample preparation.

Other researchers have also reported on the analysis of amino sugars by AQC derivatization. Palace and Phoebe successfully analyzed low levels in neutral polysaccharide polymers such as laminarin, inulin and pullulan, where the amino sugars were in great excess over the amino acids present [33, 34]. This complicated the chromatography and required an eluent pH change to 5.8 to effectively resolve the amino sugars. This study was also notable for the successful analysis of the amine derivatives in the presence of a huge excess of neutral sugars and the complete validation of the quantitative method for pharmaceutical analysis.

Ahmed et al. also analyzed glycation products associated with disease processes such as cataracts, diabetes and Alzheimer's disease [35]. Glycation, the process by which glucose or other saccharides non-enzymatically react with proteins, typically the amino group on Lys or the N-terminus, includes a group complex reactions which are collectively also termed "Maillard" reactions. The glycation products are most often in low concentration their detection in aided by a highly sensitive fluorescent assay. Due to the hydrolytically unstable nature of the glycation products, analysis of glycated proteins was carried out on enzymatically hydrolyzed samples. To effect complete hydrolysis, a cocktail of proteases including pepsin (separately carried out at pH 1.7), pronase, amino peptidase and prolidase, was employed. An antioxidant, diethylene-triaminepenta-acetic acid, was added to the hydrolysis solution to improve the recovery of sensitive glycation products, and did not interfere with the subsequent derivatization. The chromatographic conditions, with modifications to extend the analysis and allow for separation of a number of the glycation adducts with similar retention characteristics, used eluents similar to those previously described [31]. In a 220 minute analysis, more than a dozen glycation products, as well as the protein amino acids were resolved. Epimers of several glycation products were identified from the derivatization of standards and MALDI MS analysis of their derivatives. Although not sufficiently sensitive to analyze endogenous glycation products in plasma in most samples, minimally modified standard proteins were successfully analyzed.

An extreme example of the robustness of the derivatization reaction in the presence of high salt concentrations is provided in the paper by Jorgenson and Jensen [36] describing the analysis of amino acids in hydrolyzed sea water samples. The sample matrix is approximately 3M in NaCl, with numerous other salts (e.g. Mg, Ca etc.) present as well. Results showed that even with these salts, derivatization yields were near quantitative, although an amine contaminant present in seawater apparently caused a falsely high value for Met. Measured concentrations in seawater ranged from 30 – 500 nM. The authors concluded that the comparison of the results obtained with the AQC method with an OPA derivatization procedure favored analysis with AQC.

Van Wandelen and Cohen also described the separation of 26 amino acids commonly found in samples with free amino acids in solutions [31]. A quaternary gradient system was employed, with a key to the optimized separation involving a change in the eluent pH midway through the analysis (Figure 6). While these conditions are commonly applied to the analysis of cell culture fluid and media, they are also useful for a wide variety of samples types containing free amino acids such as foods.

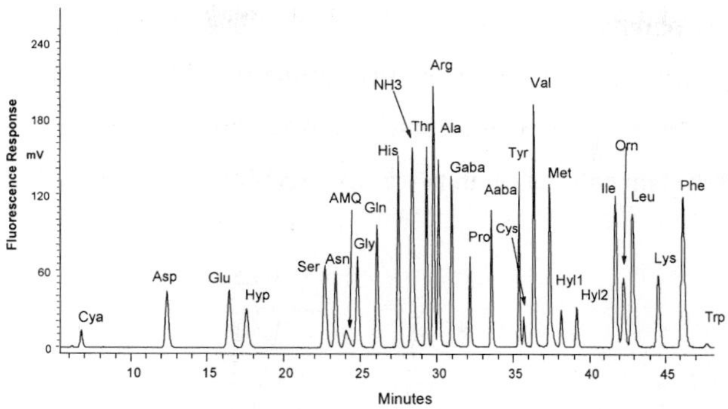

Figure 6 Separation of the standard mixture for free amino acid analysis. Reproduced with permission from Reference [31].

Only a few publications have described the use of AQC for the analysis of amino in physiologic fluids. This is somewhat surprising as deproteinized serum or plasma should, in principle, pose no difficulties for the derivatization reaction. Reverter et al. [37] developed a method for the analysis of 23 amino acids in porcine plasma. Samples were deproteinized by sulfosalicylic acid precipitation, and the pH of the supernatant was adjusted by the addition of borate derivatization buffer. The separation used a sodium phosphate eluent system at pH 6.3 with the gradient conditions described previously [10]. This provided for improved resolution of the additional amino acids. Although the separation of several amino acids was not baseline, it was sufficient to permit analysis of hydroxyproline, Asn, citrulline, and Gln, not analyzed in the original method. Trp was also analyzed using UV, rather than the fluorescence detection used for the other amino acids. Recovery studies using a standard addition approach were excellent with most amino acids recovered in excess of 90%, with several others (His, Tyr and citrulline), being recovered at 80 – 90%. It was noted in the study that improved chromatographic resolution was needed to increase the accuracy of analyzing several of the amino acids.

Selected groups of physiologic amino acids pose a less challenging chromatographic problem. For example, a method targeted at screening newborn infants for several important amino acidopathies, was developed by Cohen and van Wandelen [38] in which Met, Leu and Phe could be analyzed in 10 min (Figure 7). These amino acids are indicators for the important metabolic disorders homocystinuria, maple syrup urine disease and phenylketonuria. The method was applied to blood spot samples dried on filter paper, the common method for collecting samples from heel sticks of newborn infants. Amino acids were extracted with an ethanolic solution containing dibromotyrosine as an internal standard, and an aliquot could be derivatized without prior removal of the extraction solvent. Dhondt and co-workers used the method as a secondary screening method to confirm results from primary newborn screening [39], and a group at the Centers for Disease Control verified the amino acid content of control blood samples used for system calibration in newborn screening laboratories [40, 41].

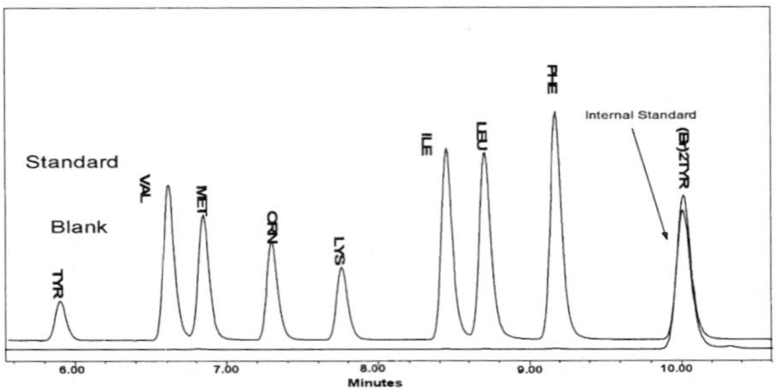

Figure 7 Fast analysis of standards for newborn screening.

Other researchers have focused on the analysis of specific amino acid groups, with the analysis of methylated Arg [42] and various Tyr derivatives [43] in biological fluids being described. The methylated Arg compounds, implicated in neurotransmitter pathways related to nitric oxide function, were analyzed in plasma samples, while 3-nitrotyrosine and the *o*-, *p*-, and *m*-isomers of Tyr, considered biomarkers of a number of diseases states such as lung injury, atherosclerosis, and various cancers, were detected in urine samples.

3.3 Analysis of Non-Protein Amino Acids

Uncommon amino acids have been the subject of several papers. Janak et al. have described the analysis of selenium analogs of sulfur amino acids [44]. Modified gradient conditions were developed to provide resolution of non-hydrolyzate amino acids including Asn, Gln, Orn and Trp. Selenium analogs of Met, Cys and ethionine were all resolved in the system. One interesting observation was the near baseline resolution of the selenocystine diastereomers, and that the achiral separation system could only resolve into two peaks the four possible diastereomers present.

The sensitivity of the AQC method was exploited in a study of actin amino acid composition. Actin contains a very low concentration of 3-methyl histidine (3 MHis), with only

0.65 residues per mole present. Strydom et al. analyzed a hydrolyzed actin sample and were able to quantify the 3 MHis peak eluting between NH_3 and Arg [45]. Due to its rare occurrence in proteins, this was a clear indicator of the presence of actin in the samples. Another unusual His derivative, 2-oxohistidine, was detected in oxidatively inactivated glutamine synthesis samples [46]. Similar to the Tyr oxidation products, this oxidation is also thought to be an indicator of protein oxidation and to play a potential role in aging, atherosclerosis and other chronic diseases. Improved recovery of the unstable amino acid was accomplished by adding dithiothreitol to the hydrolysis solution. The AQC derivative is well resolved and elutes just after His and before the NH_3 peak.

Quantifying proteins present in low concentrations can pose problems for a number of common assays. Because amino acid analysis does not require prior knowledge of a protein's structure, and makes no assumptions regarding its binding properties (as do various dye binding methods), the technique is often employed when there is insufficient sample for less sensitive methods, or basic information, such as its UV extinction coefficient, are not known. So-called quantitative amino acid analysis has applications in determining peptide content in crude or purified synthetic peptide samples, the determination of protein extinction coefficients, or verification of results from alternate methods, including the dye-based ones. Although few reports on using AQC for this application have appeared in the literature, two early papers employed it with highly purified proteins [47, 48].

Analysis of neurotransmitter amino acids has seen significant growth in recent years, as the role of several amino acids in nervous system function has been elucidated. However, the assay is very challenging due to the very low amounts of several of the important transmitter amino acids, such as Asp and GABA. Also studied are taurine, Gly, Glu and Gln. The first reported use of AQC for these samples was a paper by Grabb et al. [49] who used the conditions developed by Cohen. While suitable for the analysis of Gln, Glu, Asp and Tau in brain microdialysates, the analysis was slow (over 2 hours injection-to-injection), and detection limits were insufficient for the analysis of rare amino acids in these samples such as μl) and the maximum injection volume permitted could be increased by lowering the Eluent A pH and using a shallower gradient profile. Detection limits were reported to be in the 3-8 nM range for Asp, Glu, Gly and GABA. Separation of the standard mixture used for neurotransmitter analysis is shown in Figure 8.

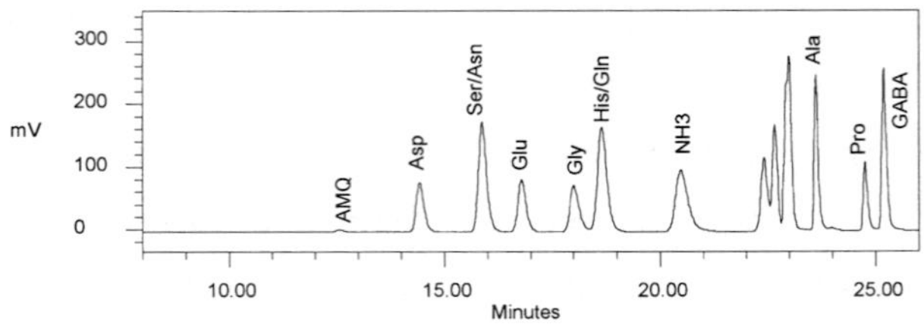

Figure 8 Analysis of neurotransmitter amino acids. Reproduced with permission from Reference [50].

3.4 Chiral Amino Acid Analysis

Sensitive methods for the analysis of enantiomeric amino acids has most commonly employed chemical derivatization. The labeling process both increases analyte detectability and provides improved chromatographic selectivity for separating the enantiomeric pairs. Armstrong and co-workers first reported on the use of AQC for enantiomeric amino acids, with a number of favorable attributes. A variety of bonded cyclodextrin columns, both native and modified, were successfully employed [51]. Either a non-aqueous polar organic eluent or a highly aqueous reversed phase eluent could be used. Enantiomeric selectivity with the reversed phase system, however, could only be achieved with derivatized cyclodextrin columns. Elution order of a number of D,L pairs could be reversed by changing the column type, which allowed either enantiomer to be analyzed in the presence of a large excess of the other enantiomer. In the polar organic system, detection limits were as low as 10 fmol, as aqueous quenching was eliminated. Enantiomeric content as low as 0.0075% could be determined for Phe, and content levels less than 1% were routinely determined with good precision. Enantiomeric stability was also excellent, with no change in enantiomeric ratios observed after seven days.

Armstrong has compared the results to those obtained after FMOC-Cl labeling, and found the AQC method comparable or superior. Rizzi et al. have also compared the separation of AQC labeled amino acid enantiomers with a number of derivatizing reagents [52]. Selectivity was similar to several isothiocyanate labels, but these researchers only used a reversed phase eluent system, rather than the polar organic system of Armstrong.

Swartz et al. have analyzed AQC derivatized amino acid enantiomers by capillary electrophoresis [53]. They employed a chiral surfactant, (S)- or (R)-N-dodecoxycarbonylvaline, to provide enantiomeric selectivity. Elution order reversal could be accomplished by changing from the R to the S form of the surfactant. As many as 8 chiral pairs could be separated in a single 20 min analysis, with excellent peak shape and resolution. Detection was by UV at 254 nm, and typical detection limits were 0.1 μg/ml, which allowed the detection of enantiomeric content as low as 0.1%.

Vespalec et al. also used CE for separating enantiomeric sulfur and selenium containing amino acids. The chiral selector for their work was vancomycin, a macrocyclic antibiotic that has strong chiral separation influence. Extremely high efficiencies, up to 250,000 theoretical plates/m, were observed, which along with high selectivity conferred by the chiral selector, allowed for baseline separation of enantiomers of Cys, Met and their selenium analogs. [54, 55].

3.5 Derivatization of Peptides and Proteins

A logical extension of amino acid derivatization have been the numerous reports on AQC derivatization applied to the analysis of peptides and proteins. The first such paper [56] appeared shortly after the first amino acid publication and described derivatization of synthetic peptides. Efficient derivatization did not require any modification of the amino acid derivatization protocols. Separations were carried out on reversed-phase columns using a sodium acetate mobile phase at pH 5.05. Derivatization of intact peptides has been problematic with a number of other reagents, but evidence for complete derivatization with AQC was obtained from several experiments: (1) loss of > 98% of the underivatized peptide assayed by HPLC, (2) MALDI MS of the derivatized products showing lack of the underivatized forms (3) and no increase in derivative yields with the addition of more reagent to the reaction mixture. Response was linear over three orders of magnitude (8– 8000 nM) with a correlation coefficient of 0.999. Peptides containing a Lys residue were shown to contain two tags, presumably at the N-terminus and on the side chain of Lys. Derivatized peptides are more highly retained than amino acids, and do not suffer the limitation for injection volumes noted with amino acid analysis. The concentration detection limits for a series of similar peptides were approximately 100 pM and the mass detection limit was as little as 20 fmol.

Cohen et al. extended peptide applications to the analysis of standard peptides by CE [57] and to high sensitivity tryptic mapping [58]. In the latter paper a study with methionine enkephalin as a model compound showed that common acidic eluents used for peptide mapping, such as trifluoroacetic acid mobile phases, were unsuitable for fluorescence detection due to the lack of response at acidic pH. Derivatization experiments showed evidence for complete conversion to derivatized species, although this cannot be assumed for all complex mixtures of peptides. Analysis of fmol levels of a cytochrome c digest were demonstrated. More recently this work has been extended to incorporate separations of a derivatized cytochrome c digest by capillary LC with laser induced fluorescence detection, where low fmol detection limits were shown [59]. The capillary system also was interfaced to a mass spectrometer, which confirmed the nature of the derivatized amino acids, demonstrating homogeneous, fully derivatized peptides.

Derivatization of larger peptides, such as insulin, and proteins were the subject of several papers by Krull et al. [60–62] Large peptides and proteins pose problems not encountered with smaller peptides as they potentially contain numerous Lys residues as possible derivatization sites. Quantitative derivatization of all available sites can be difficult to assess, and the occurrence of multiple partially derivatized species with varying numbers of labels is a serious issue. Derivatization may also be sterically hindered by the three dimensional structure of polypeptides. Strong et al. [60] also noted that as the number of labels increases, analyte solubility can be reduced, with precipitation interfering with completion of the derivatization and subsequent analysis. Using the detergent SDS to help solubilize proteins such as bovine serum albumin, which contains 60 potential derivatization sites (59 Lys residues plus the N-terminus), helps maintain the derivatized protein in solution. Derivatization was also facilitated by reduction of the disulfide bonds and alkylation of the Cys residues, which in conjunction with the detergent improves the accessibility of the derivatization sites.

Separation of larger peptides and proteins has been accomplished by HPLC and CE, either by zone electrophoreses [61] or capillary isoelectric focusing [60]. With HPLC and fluorescence detection, detection limits up 6000x better than for the underivatized analytes using UV at 280 nm were demonstrated. MALDI MS analyses of derivatized BSA and α-chymotrypsinogen showed that complete or nearly complete derivatization of these proteins was accomplished.

4. Experimental

4.1 Derivatization

For the optimized reaction conditions [10], 20 μl of 10 mM AQC (3 mg/mL) in acetonitrile are added to 80 μl of buffered sample. Protein hydrolyzates are reconsitituted in 10 μl of 0.1M HCl, which gives slight, but significant improvement to the yield of basic amino acids Arg, His and Lys, and are buffered with the addition of 70 μl of 0.2M sodium borate, pH 8.8.

Inclusion of 5 mM ethylenediamine tetraacetic acid in the buffer also increases His yields. Reactions are complete within 10 s, and Tyr is converted to a single product by heating at 55° C for 10 min.

4.2 Chromatography

For all separations, the column used is a reversed phase C18 AccQ-Tag™ column. Fluorescence detection uses a dual monochromator detector employing excitation at 248 nm and emission at 395 nm, and UV detection is at 248 nm. For hydrolyzate amino acids, the gradient conditions are given in Table II. The aqueous eluent (A) is 140 mm sodium acetate, 7mM TEA, pH titrated to 5.05 with phosphoric acid, the organic eluent (B) is acetonitrile and Eluent C is water. The flow rate is 1.0 mL/min and the column is thermostatted at 37° C.

Table II. Gradient used for Hydrolyzate Amino Acid Separation.

Time	%A	%B	%C	Curve[1]
Initial	100	0	0	*
0.5	99	1	0	11
18.0	95	5	0	6
19.0	91	9	0	6
29.5	83	17	0	6
33.0	0	40	60	11
36.0	100	0	0	11

1. Curve 11 is a step gradient; curve 6 is a linear step.

The separation of the free amino acid mixture for cell culture and cell media analysis used the gradient conditions given in Table III. Eluent A was 140 mM sodium acetate, 17 mM TEA pH 5.70, Eluent B was the same as A except the pH was 6.80, titrated with NaOH from a stock solution of Eluent A. Eluent C was acetonitrile and Eluent D was water. Note how eluent pH is varied by changing the ratio of A:B. The column was thermostatted at 34° C and detection was by fluorescence.

Table III. Gradient used for Free Amino Acid Separation.

Time	Flow Rate	%A	%B	%C	%D	Curve
Initial	1.0	90	10	0	0	*
0.5	1.0	89	10	1	0	11
17	1.0	88	10	2	0	6
24	1.0	86	9	5	0	6
32	1.0	63	25	12	0	6
33.5	1.0	0	87.5	12.5	0	6
33.8	1.3	22	65.5	12.5	0	6
37	1.3	22	65	13	0	6
48	1.3	22	63	15	0	6
48.1	1.3	0	0	60	40	6

Fast analysis of hydrophobic amino acids for newborn screening used the gradient conditions shown in Table IV. The eluents are the same as those used for the hydrolyzate separation. The column was thermostatted at 37° C and detection was by fluorescence.

Table IV. Gradient used Rapid Analysis of Hydrophobic Amino Acids.

Time	Flow	%A	%B	%C	%D	Curve
0.0	1.5	100	0	0	0	0
0.1	1.5	89	11	0	0	6
3.0	1.5	86	14	0	0	6
7.0	1.5	78	22	0	0	6
9.0	1.5	0	60	40	0	11
12.0	1.5	100	0	0	0	11
45.0	1.5	0	60	40	0	11
100.0	0.0	0	60	40	0	6

Analysis of neurotransmitter amino acids used a ternary system consisting of Eluent A as in the hydrolyzate separation at pH 5.05, and Eluent B = Eluent A titrated to pH 4.60 with phosphoric acid. Eluent C was acetonitrile and Eluent D was water. The flow rate ws 1.0 mL/min and the column was thermostatted at 40° C.

Table V. Gradient Conditions for Neurotransmitter Amino Acids

Time	%A	%B	%C	%D	Curve
0.0	90	10	0	0	*
0.5	90	9	1	0	11
18	86	9	5	0	6
19	0	91	9	0	6
25	0	86	14	0	6
25.1	0	0	80	20	6

References

[1] D. R. Koop, E. T. Morgan, G. E. Tarr, M. J. Coon, J. Biol. Chem. 257 (1982) 8472-80

[2] B. A. Bidlingmeyer, S. A. Cohen, T. L. Tarvin, J. Chromatogr. 336, (1984) 93 – 104.

[3] D. W. Hill, F. H. Walters, T. D. Wilson, J. D. Stuart, Anal. Chem. 51, (1979) 1338-341.

[4] S. Einarsson, B. Josefsson and S. Lagerkvist, J. Chromatogr., 282 (1983) 609-618.

[5] R. Knecht and J.-Y. Chang, Anal. Chem., 58 (1986) 2375-2379.

[6] M. C. Garcia Alvarez-Coque, M. J. Medina Hernandez, R. M. Villanueva Camanas, C. Mongay Fernandez, Anal. Biochem. 178 (1989) 1-7.

[7] D. M. Garside, P. M. S. Monteiro, M. J. Orren, S. Afr. Mar. Sci. 6 (1988) 47-53.

[8] J. Vendrell, F. X. Aviles, J. Chromatogr. 358 (1986) 410-413.

[9] F. Lai, A. Mayer, T. Sheehan, BioTechniques 11 (1991) 236-244.

[10] S. A. Cohen, D. P. Michaud, Anal. Biochem. 211 (1993) 279-287.

[11] G.W. Anderson, J.E. Zimmerman, F.M. Callahan, J. Am. Chem. Soc. 86 (1964) 1839.

[12] A.E. Bolton,W.M. Hunter, Biochem. J. 133 (1973) 529.

[13] N. Nimura, K. Iwaki, T. Kinoshita, K. Takeda, H. Ogura, Anal. Chem. 58 (1987) 2372-2375.

[14] K. Iwaki, N. Nimura, Y. Hiraga, T. Kinoshita, K. Takeda, H. Ogura, J. Chromatogr. 407 (1987) 273-279.

[15] P. J. Brynes, P. Bevilacqua, A. Green, Anal. Biochm. 116 (1981) 408-413.

[16] J. H. Yu, G. D. Li, I. S. Krull, S. Cohen, J. Chromatogr. B 658 (1994) 249-260.

[17] D. J. Strydom, S. A. Cohen, Anal. Biochem 222 (1994) 19-28.)

[18] H. J. Liu, J Chromatogr. A 670 (1994) 59-66.

[19] K. Nugent, S. Baldwin, K. Schegg, D. Schooley, Poster from the 9th Annual Symposium of the Protein Society, 1995.

[20] U. T. Kreibich, U. P. Wild, J. Mol. Spectrosc. 47 (1973) 189-93.

[21] N. Shindo, S. Nojima, T. Fujimura, H. Taka, R. Mineki, K. Murayama, Anal. Biochem. 249 (1997) 79-82.

[22] D. L. Crimmins, R. Cherian, Anal. Biochem. 244 (1997) 407-410.

[23] D. J. Strydom, G. E. Tarr, Y.-C. E. Pan, R. J. Paxton, in R. H. Angeletti (Editor), Techniques in Protein Chemistry III, Academic Press, San Diego, CA, 1992, 261-274.

[24] S. A. Cohen, K. M. De Antonis, D. P. Michaud, in R. H. Angeletti (Editor), Techniques in Protein Chemistry, Academic Press, San Diego, CA, 1993, 289-298.

[25] H. Fujita, H. Mohri, H. Kanamori, A. Iwamatsu, T. Okubo, Exp. Cell Res. 217 (1995) 484-489

[26] S. A. Cohen, Meth. Mol.Biol. 211 (2003) 143-154.

[27] N. Shindo, T. Fujimura, S. Nojima-Kazuno, R. Mineki, S. Furusawa, K. Sasaki K., Murayama, Anal. Biochem. 264 (1998) 251-258.

[28] S. A. Cohen, K. M. De Antonis, J. Chromatogr. A 661 (1994) 25-34.

[29] C. W. Gehrke, P. R. Rexroad, R. M. Schisla, J. S. Absheer, R. W. Zumwalt, J. Assoc. Off. Anal. Chem 70 (1987) 171-174.

[30] H. J. Liu, B. Y. Chang, H. W. Yan, F. H. Yu, X. X. Liu, J. AOAC Int. 78 (1995) 736-744.

[31] C. van Wandelen, S. A. Cohen, J. Chromatogr. A 763 (1997) 11-22.

[32] J. Diaz, J. Ll. Lliberia, L. Comellas, F. Broto-Puig J. Chromatogr. A 719 (1996) 171-179.

[33] G. P. Palace, C. H. Phoebe, Jr. Amer. Biotech. Lab August (1997) 74-76.

[34] G. P. Palace, C. H. Phoebe, Jr., Anal. Biochem. 244 (1997) 393-403.

[35] N Ahmed, O. K. Argirov, H. S. Minhas, C. A. A. Cordeiro, P. J. Thornally Biochem J. 364 (2002) 1-14.

[36] N. O. G. Jorgenson and R.E. Jensen, Marine Chem. 57 (1997) 287-297.

[37] M. Reverter, T. Lundh, J. E. Lindberg, J. Chromatogr. B. 696 (1997) 1-8.

[38] S. A. Cohen , C. Van Wandelen, in B. L. Therrell, B. G. Aldis Jr. (Editors.), Proceedings of the 11th National Neonatal Screening Symposium, Corpus Christi, TX (1995) 215-216.

[39] J. L. Dhondt, J. Loeber, L. H. Elvers, E. Paux,. J Med Screen 1998;5:63-66.

[40] B. W. Adam, J. R. Alexander, S. Jay Smith, C. H. Chace, J. G. Loeber, L. H. Elvers, W. H. Hannon, Clin. Chem. 4 (2002) 126-128.

[41] D. H. Chace, B. W. Adam, S. J. Smith, J. R. Alexander, S. L. Hillman' W. H. Hannon, Clin. Chem. 45 (1999) 1269-1277.

[42] T. Heresztyn, M. I. Worthley, J. D. Horowitz., J. Chromatogr. B, 805 (2004) 325–329.

[43] M. Du, W. Wu, N. Ercal, Y. Ma, ., J. Chromatogr. B, 803 (2004) 321–329.
[44] J. Janak, H. A. H. Billiet, J. Frank, K. Ch. A. AM. Luhben, P. Husek, J. Chromatogr. A. 677 (1994) 192-196.
[45] D. J. Strydom, K. Kandror, Biochem. Biophys. Res. Commun. 205 (1994) 195-201.
[46] S. A. Lewisch, R. L. Levine Anal. Biochem. 231 (1995) 440-446
[47] D. Soler, T. Nomizu, W. E. Brown, M. Chen, Q.-Z. Ye, H. E. van Wart, D. S. Auld, , Biochem. Biophys. Res. Commun. 201 (1994) 917-923.
[48] N. Russo, R. Shapiro, K. R. Acharya, J. F. Riorday, B. L. Vallee, Proc. Natl. Acad. Sci USA 91 (1994) 2920-292.
[49] M. C. Grabb, V. M. Sciotti, J. M. Gidday, S. A. Cohen, D. G. van Wylen Neurosci Methods. 82 (1998) 25-34.
[50] H. Liu, M. C. Sanuda-Pena, J. D. Harvey-White, S. Kalra, S. A. Cohen, J Chromatogr A. 828 (1998) 383-95.
[51] M. Paslowska, S. Chen, D. W. Armstrong, J. Chromatogr. 641 (1993) 257-265.
[52] A. M. Rizzi, S. Cladrowa-Runge, H. Jonsson, S. Osla, J. Chromatogr. A, 710 (1995) 287-295.
[53] M. E. Swartz, J. R. Mazzeo, E. R. Grover, P. R. Brown, Anal. Biochem. 231 (1995) 65-71.
[54] R. Vespalec, H Corstjens, H. A. H. Billiet, J. Frank, K. Ch. A. M. Luyben, Anal. Chem 67 (1995) 3223-3228.
[55] R. Vespalec, H. A. H. Billiet, J. Frank, K. Ch. A. M. Luyben, High Resol. Chromatogr. 19 (1996) 137-142.
[56] K. M. De Antonis, P. R. Brown, S. A. Cohen, Anal. Biochem 223 (1994) 191-197
[57] K. M. De Antonis, P. R. Brown, Y.-F. Cheng, S. A. Cohen, J. Chromatogr. A 661 (1994) 279-285.
[58] S. A. Cohen, I. Mechnikov, P. Young, in J. W. Crabb (Editor), Techniques in Protein Chemistry VI, Academic Press, San Diego, CA,1995, 251-258.
[59] H. Liu, I. S. Krull, S. A. Cohen, Anal. Biochem. 294 (2001) 7-18.
[60] R. A. Strong, H. Liu, I. S. Krull, B.-Y. Cho, S. A. Cohen, J. Liq Chromatogr. 23 (2000) 1775-1807
[61] H. Liu, B.-Y. Cho, I. S. Krull, S. A.Cohen, J. Chromatogr. A 927 (2001) 77-89.
[62] H.-J. Liu, R. E. Strong, I. S. Krull, S. A. Cohen, Anal. Biochem 298 (2001) 103-111.

1.3. Capillary Electrophoresis/Capillary Electrochromatography

Ibolya Molnár-Perl (Editor)
Quantitation of Amino Acids and Amines by Chromatography
Journal of Chromatography Library, Vol. 70

1.3.1.

Determination of Underivatized Amino Acids by Capillary Electrophoresis and Capillary Electrochromatography

Christian W. Klampfl

Contents

Summary

In principle electroseparation methods are well suited for the analysis of relatively polar compounds such as underivatized amino acids. These systems provide several benefits such as high performance, the possibility of miniaturization and integration and low cost of analysis. The absence of a derivatization step implicates additional advantages like the avoidance of all types of pitfalls related to this procedure. Unfortunately the analysis of amino acids in their underivatized form also generates some new problems such as the necessity to separate compounds with very similar structures and the need to find detection systems providing sufficient sensitivity for these compounds. The capillary electroseparation method most commonly employed for underivatized amino acids is capillary zone electrophoresis and up to now a substantial number of reports describing the application of this technique for a wide range of different samples can be found. Optimized procedures allowing the separation of the whole set of protein amino acids in a single run have been developed. Detection is mostly performed using either spectrophotometric detectors (mainly because of their presence in most commercially available instruments) or electrochemical detection. Although substantial efforts to improve its sensitivity for underivatized amino acids have been made, spectrophototmetric detection can not be regarded as the best choice for this type of analytes. Electrochemical detection in particular the more recently presented contactless conductivity detector provides better performance. Optimum choice for the detection of underivatized amino acids after their separation using capillary electrophoretic or electrochromatographic methods is the use of a mass spectrometric detector. This can also be seen from the increasing number of publications dealing with this issue.

1. Introduction

From a biochemical point of view, amino acids can be regarded as one of the most important classes of chemical substances. For this reason numerous publications deal with their analysis and almost all types of analytical methods have at some stage been tested with respect to their suitability regarding the determination of these compounds. Whereas their high polarity and their low volatility can be seen as a major problem with respect to their separation, the lack of a strong chromophore group complicates their detection [1]. For these reasons the majority of analytical methods for amino acids include a derivatization step, to transform these analytes into derivatives facilitating both their separation as well as their detection. A wide range of derivatization reagents for amino acid analysis can be found in the literature and research in this field is still going on [2–4]. Unfortunately derivatization is also affected with some disadvantages including derivative instability, insufficient reproducibility of derivative yield, interferences caused by the reagent or time consuming derivatization procedures. To avoid these problems and to speed up and facilitate analysis methods allowing the analysis of underivatized amino acids are of significant interest. For this reason a series of such separation and detection methods have been developed and published [4,5].

2. Electroseparation methods for the analysis of underivatized amino acids

Since the introduction of the first commercially available instruments for capillary electrophoresis (CE), almost 20 years ago, electroseparation methods have gained more and more importance in the analysis of diverse solutes ranging from small ions to large biomolecules in a variety of different matrices. Often these methods can be regarded as an interesting complement or sometimes even as an alternative to the chromatographic techniques commonly employed for separation problems. Major capillary electroseparation techniques can be divided into two principal groups: purely electrophoretic methods and electrochromatographic methods involving both electrophoretic and chromatographic separation principles. The first group comprises capillary zone electrophoresis (CZE) the most frequently used electrophoretic method and capillary isotachophoresis (cITP). Important representatives of the second group are micellar elektrokinetic chromatography (MEKC), microemulsion electrokinetic chromatography (MEEKC) and finally capillary electrochromatography (CEC), a hybrid method between common liquid chromatography and CZE. All of these different methods have (to some degree) been applied to the analysis of amino acids.

2.1. Electrophoretic methods

2.1.1. Capillary zone electrophoresis (CZE)

A principal advantage of CZE is its simplicity with respect to both the instrumentation required as well as the basic mechanisms of separation involved. In addition to that CZE offers some promising features such as short analysis times, high efficiency expressed in high numbers of theoretical plates and separation selectivities that are often completely different from those achieved in chromatography. Especially the latter feature can be exploited to increase the certainty of peak assignment in complex samples. Another benefit of CZE over chromatographic methods is the absence of a stationary phase avoiding the contamination and deterioration of expensive packing materials e.g. by aggressive and dirty matrix components. So in a series of cases, CZE allows the direct analysis of samples which would require extensive sample pretreatment using chromatographic methods. The basic principle of CZE is the migration of charged compounds in an electric field. This so called electrophoretic mobility is determined by the charge/ionic radius ratio of the analytes. The movement of molecules in the electric field is superimposed by a bulk flow, the so called electroosmotic flow (EOF) generated by the charges of the capillary surface. Due to their negatively charged surface caused by the deprotonation of the silanol groups, fused silica capillaries (the most widely employed species) show an EOF directed towards the cathodic end of the capillary. Its magnitude is mainly governed by parameters like the pH of the carrier electrolyte (influencing the degree of deprotonation of the silanol groups), the presence of an organic solvent in the electrolyte (influencing the degree of protonation of the silanol groups as well as the viscosity of the electrolyte) and the ionic strength of the electrolyte solution. To meet the needs of different analytical applications the EOF can be suppressed or even reversed (employing either capillaries made from uncharged polymers or coating fused silica with surfactants leading to a neutral or positively charged surface) allowing the faster co-electroosmotic separation of anions.

Coming to amino acids, CZE has been widely used for the analysis of this class of compounds [9]. Whereas the fact that amino acids in their native form are highly polar compounds makes them ideal candidates for CZE, their similar structures represent a substantial challenge. Due to the zwitterionic character of these substances, amino acid analysis by electrophoresis can be performed in two principally different modes: firstly a cationic mode with protonated species (if highly acidic carrier electrolytes are employed) and secondly an anionic mode (in the case of a strongly basic electrolyte). Comparing these two separation modes with respect to the separation selectivity an almost reversal of the migration order can be observed, as can be seen from Figure 1. An advantage of the analysis in the cationic mode is the possibility to separate the two isomeric amino acids leucine and isoleucine, which is not the case if strongly basic electrolytes are employed. Additionally, arginine tends to migrate with the EOF if the analysis is performed in the anionic mode, a fact that does not allow a proper identification and quantitation of this compound [9]. A principal drawback low pH electrolytes (which are necessary to provide sufficient protonation of all amino acids) are the relatively long analysis times caused by the almost suppressed EOF under these conditions.

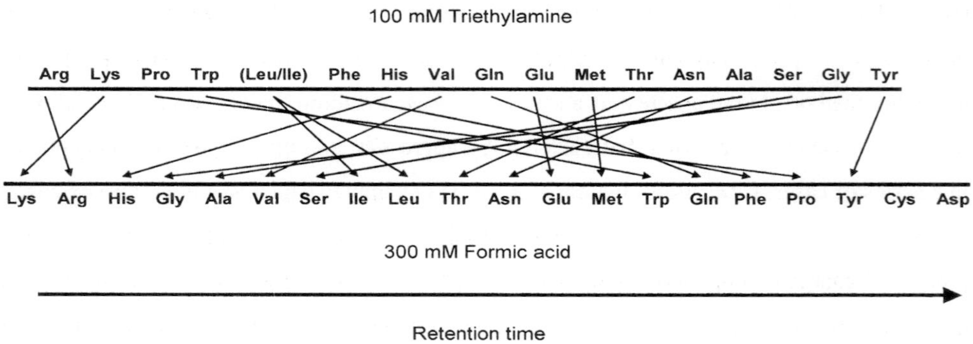

Figure 1 Migration order of amino acids in CZE employing strongly acidic and strongly basic carrier electrolytes

2.1.2. Capillary Isotachophoresis (cITP)

The basic principles of ITP are somewhat different from those in CZE. In ITP the separation system consists not only of a single carrier electrolyte, but of a leading (i.e. a high mobility) and a terminating (i.e. a low mobility) electrolyte with the sample located in between. Whereas the EOF is an important parameter in CZE, it is commonly avoided in ITP analysis. The EOF suppression is achieved either by the addition of water soluble polymers or neutral surfactants to the electrolyte. ITP has got a large potential as a sample pretreatment and a preconcentration process in combination with other techniques in particular with CZE. Krivankova et al. published a paper comparing the sensitivity enhancement achieved by transient ITP induced by the sample (sample self stacking), transient ITP induced by the composition of the electrolyte system and on-line coupling of ITP and CZE. The latter technique was found to be by far superior [1]. ITP in the chip format has been successfully employed for the determination of underivatized amino acids [1]. More common (as already mentioned above) is the use of ITP as a sample pretreatment instrument combined with CZE for the final separation of the amino acids [2,3]. The ITP step allows the injection of large sample volumes as well as sample clean-up and pre-concentration.

2.2. Electrochromatographic methods

Electrochromatographic methods combine electrophoretic principles (i.e. the migration of charged compounds in an electric field) with chromatographic ones (i.e. interactions between a solute and a stationary phase). Depending on the type of stationary phase employed, methods with so called pseudo stationary phases like micellar electrokinetic chromatography (MEKC) and microemulsion electrokinetic chromatography (MEEKC) and methods with true stationary phases like capillary electrochromatography (CEC) can be distinguished.

2.2.1. Micellar electrokinetic chromatography and microemulsion electrokinetic chromatography

MEKC was introduced in 1984 by Terabe [14]. The principal difference to CZE is the presence of micelles as a pseudo stationary phase, caused by the addition of micelle forming agents. Those can either be anionic e.g. sodiumdodecylsulfate (SDS), cationic e.g. quaternary alkylammonium salts or neutral like Brij 35 with the resulting micelles moving towards the anode, cathode or being transported with the EOF. Therefore charged analytes can be separated according to a mixed mechanism (i.e. migration in the electric field and partitioning between the electrolyte and the micelles) whereas neutrals are separated according to differences in their distribution between the two phases. MEEKC, firstly reported by Watarai in 1991, is based on a similar principle with microemulsion droplets instead of micelles. A typical microemulsion consists of an oil phase (usually an alkane or another water insoluble solvent), a surfactant (commonly the same classes of compounds are employed as in MEKC) and a co-surfactant (mostly an alcohol). Advantages of microemulsions over micellar systems are the reduced rigidity of the oil droplets, facilitating the penetration of the analyte, and their better performance in the separation of highly hydrophobic substances [16]. MEKC has been successfully applied to the analysis of derivatized amino acids whereas, due to their more polar nature, this separation method is less suitable for the underivatized species [7–9].

2.2.2 Capillary electrochromatography

Capillary electrochromatography (CEC) can be regarded as a hybrid technique combining the attributes of CZE and liquid chromatography. In principle two different kinds of CEC can be distinguished, according to the separation columns employed. Both, applications of packed or monolithic columns as well as the open tubular (OT) approach can be found in the literature. Despite the majority of CEC separations of amino acids refers to derivatized species a few papers describing the use of this technique for underivatized amino acids have been published [17,18].

3. Detection methods

Detection is an essential point in chemical analysis, co-determining essential parameters like sensitivity or selectivity of the method employed. High sensitivity is generally desired whereas the wish for a universal detection method suitable for a wide range of solutes even with completely different chemical properties is often surpassed by the necessity for high selectivity, allowing the determination of a single analyte in a complex sample matrix. Focusing on the detection of underivatized amino acids a series of different detection methods have been investigated with respect to their performance for this type of analytes. Those include spectrophotometric techniques (mainly direct and indirect UV-detection), electrochemical detection, mass spectrometric detection and some other detection methods that are less common in combination with CE. A comparative study, investigating the suitability of different types of detectors for the analysis of underivatized amino acids has been published recently [1]. Table 1 gives an overview of the performance obtained with some detection methods commonly used in the CZE analysis of underivatized amino acids.

Table 1 Limits of detection (LOD) for some amino acids using CZE with different detection methods

	LOD ($\mu M\ L^{-1}$)				
Amino acid	**ESI-MS SIM positive mode [1]**	**ESI-MS SIM negative mode [2]**	**Direct UV at 185/195 nm [3]**	**Indirect UV at 254 nm [4]**	**Contactless conductivity [5]**
Gly	10	270		9	
Ala	6	61		5	
Ser	6	50		9	
Pro	10	260		7	
Val	3	50	424/1500	3	
Thr	1	96		13	
Ile/leu	1	11	378/1800	9	
Asn	4	135		21	
Asp	11		750/3500	2	1
Lys	1	4		5	2
Glu	6	340		2	1
Met	3	20		5	
His	0.3	38		9	2
Phe	1	14	3/-	4	
Arg	0.3		6/120	9	

[1] electrolyte: 1 M formic acid; sheath liquid: 5 mM ammoniumacetat in 50% methanol at 10 μl min^{-1};data taken from [75].
[2] electrolyte: 100 mM triethylamine; sheath liquid: 0.1% triethylamine in i-propanol/water 8/2 at 4 μl min^{-1}; C.W. Klampfl, unpublished data.
[3] electrolyte: 50 mM phosphoric acid, 35 mM octanesulfonate, pH 2.2; data taken from [20].
[4] electrolyte: 8 mM 4-aminosalicylic acid, 2m M Na_2CO_3, pH 10.15; data taken from [32].
[5] electrolyte: 50 mM 2-amino-2-methyl-1-propanol; data taken from [59].

3.1. Spectrophotometric detection

3.1.1. Direct and indirect UV-detection

As the majority of commercially available CE instruments still are equipped with spectrophotometric units this type of detection plays an outstanding role in all electroseparation methods. Unfortunately UV-vis detection is not the perfect choice for underivatized amino acids. Whereas aromatic amino acids and those showing appropriate functional groups (e.g. arginine) can be detected in the wavelength range commonly employed in UV-vis detection, detection wavelengths as low as 185 nm are needed to achieve acceptable sensitivity for many simple aliphatic amino acids like gylcine or alanine [20], as can be seen from Table 1. This fact is connected to a series of disadvantages. Firstly, due to the requirement of sufficient UV transparency at the detection wavelength only a limited number of buffer components and electrolyte additives can be used [21] and secondly special lamps are needed for UV detection at wavelength below 195 nm. Whereas the Deuterium lamps commonly employed in tune able or diode array detectors show insufficient intensity in this range, mercury lamps providing high radiation intensities at 185 nm only allow single wavelength detection and are just available in some outmoded instrumentation. Despite these facts a number of reports describing the use of direct UV detection for underivatized amino acids can be found in the literature. Indirect UV detection is a very common technique in electromigration method, allowing the detection of non UV-absorbing species. For this reason it is an interesting approach for the detection of underivatized amino acids [21,22]. Indirect detection is based on the use of a carrier electrolyte with an UV-absorbing co-ion. The co-ion is the part of the electrolyte showing the same charge as the analyte molecules. To maintain electroneutrality, the electrolyte co-ions are displaced by analyte molecules in the sample zone, leading to a decrease in UV absorbency. The resulting negative peaks can be reversed using appropriate software [23]. Looking at the data depicted in Table 1 it can be seen that indirect UV detection provides several advantages over the direct detection mode with respect to the analysis of underivatized amino acids. Firstly the detector response is not directly depending on the specific absorptivity of the analyte, so a more leveled sensitivity is achieved for the different amino acids; secondly in particular for the aliphatic amino acids indirect UV detection provides increased sensitivity compared to the direct detection mode and thirdly no special instrumentation like low wavelength UV lamps are required. Disadvantages of this detection method are the frequent occurrence of system peaks as well as interferences in the case of analytes absorbing at the detection wavelength like aromatic amino acids.

Applications

Due to the fact the CE instrumentation is mostly equipped with a spectrophotometric detector this type of detection has been widely used also for the determination of underivatized amino acids, although these compounds often lack suitable chromophores. Starting with direct UV detection (i.e. monitoring the absorbency of the analytes) a series of papers have been published either employing very low detection wavelengths [18,20,25-28] or restricting the selection of analytes to aromatic amino acids or those with appropriate functional groups allowing the use of wavelengths above 210 nm [28,29]. A method to determine underivatized amino acids in different fodders and raw materials was published by Komarova et al. [25]. Employing a basic pH, borate based electrolyte and direct UV detection at 190 nm (the only combination of a basic electrolyte and direct UV detection at low wavelengths that can be found in the literature) four amino acids could be analyzed in real samples down to a minimum concentration in the range of 5 mg L^{-1} (cystine) to 20 mg L^{-1} (lysine). The group of Klampfl published several papers describing the use of highly acidic phosphate based electrolytes together with long chain alkyl sulfonates as additives for the separation of underivatized amino acids in different matrices including beer and fruit juice samples [26-28]. A separation of more than 20 amino acids using such an electrolyte system can be seen in Figure 2. Klampfl and Vo compared direct UV detection at two different wavelengths, indirect UV detection and MS detection with respect to their suitability for the quantification of some amino acids in an infusion solution [20]. Whereas MS detection was considered as too complex for this task, indirect UV detection together with a basic electrolyte did not allow the determination of all analytes of interest. Finally comparing direct UV detection at 185 nm and at 195 nm, the latter wavelength was chosen although improved sensitivity was achieved at 185 nm. The reason for this selection was that due to the significant differences in concentration between the amino acids, only direct UV detection at 195 nm allowed to quantify all solutes in a single run. An OT-CEC approach was employed for the separation of this class of analytes by Charvatova et al. [18]. Separation selectivities were modified by either coating the capillary wall with a porphyrin derivative as well as by changing the electrolyte system employed (Tris-phosphate or Na-phosphate). Protein and non-protein amino acids with aromatic side chains have been separated by Lee et al. employing an acidic, phosphate based electrolyte [29]. Salimi-Moosavi et al. investigated the suitability of non-aqueous carrier electrolytes for the separation of underivatized and dansylated amino acids [30]. Employing a methanol based running buffer, separation selectivities that were significantly different from those obtained in aqueous media were obtained. The principle of indirect UV detection is the displacement of an UV absorbing probe (included in the carrier electrolyte) by the non-absorbing analyte ions. This leads to a reduction in absorbency in the sample zone that can be measured. As the majority of protein amino acids shows only weak absorbency in the wavelength range commonly monitored with UV detectors, indirect detection can be regarded as a well suited tool for this type of compounds. A further advantage over direct detection is the fact, that almost identical sensitivities are achieved for the majority of amino acids whereas dramatic differences in response are observed at low wavelengths such as 185 nm [20]. In two early works Lee and al. describe the use of basic electrolytes with several electrolyte co-ions for the determination of underivatized amino acids [22,23]. Different strategies were tested with respect to modify separation selectivities and thereby increase the number of separable analytes in a single run. Those include the addition of metal ions for complexation, cyclodextrines as well as the modification of the capillary surface with reagents influencing

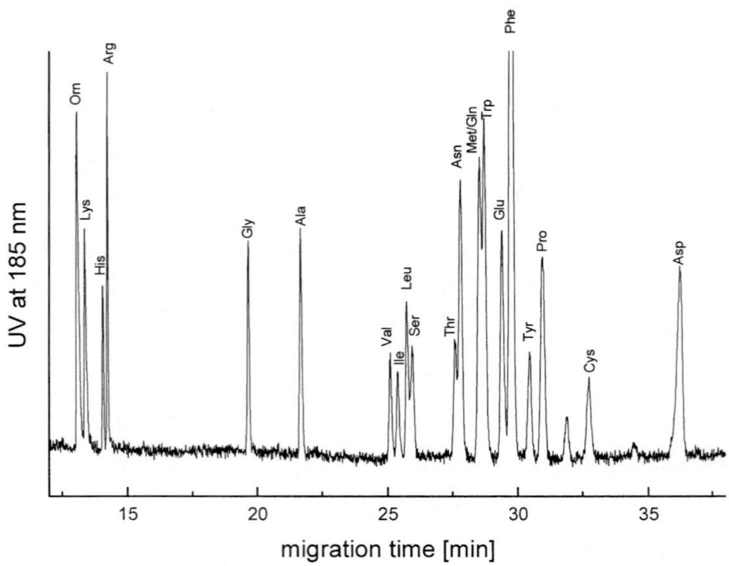

Figure 3 Separation of 21 common amino acids by CZE. Conditions: electrolyte, 10 mM Na-dihydrogenphosphate, 35 mM octanesulfonate (pH 2.36); separation voltage 30 kV; capillary, 100 cm x 50 μm i.d.; detection, direct UV at 185 nm

---

the EOF. Using this approach a series of papers has been published demonstrating the suitability of CZE with alkaline carrier electrolytes and indirect UV detection for the analysis of underivatized amino acids in a wide range of real samples. Fujiya et al. determined the amino acid patterns of protein hydrolyzates employing dinitrobenzoate and phthalate as the electrolyte probe [30]. Up to 30 different amino acids in physiological fluids were separated and detected by Zunic et al. [31]. Using their optimized method based on a pH 10.15, 4-aminosalicylcate electrolyte quantitative analysis of the selected solutes was possible with excellent within run and between run precision. In a single run, Soga et al. separated inorganic anions, organic acids, sugars and amino acids using a pyridine dicarboxylate electrolyte [32. The group of Adams published two papers on the analysis of plant tissues with respect to their amino acid composition [33,34]. A principal disadvantage of indirect UV detection is the increased probability of system peaks, especially when it comes to the analysis of real samples. This can clearly be seen on the example of the determination of underivatized amino acids in urine, published by Chen et al. [35].

3.1.2. Fluorescence detection

Fluorescence detection is an important tool for the sensitive and selective monitoring of compounds that are either showing native fluorescence or that have been labeled using an appropriate tag. Regarding underivatized amino acids, no native fluorescence is observed for most of these compounds. Exceptions are tyrosine and tryptophane, where excitation of fluorescence is possible employing lasers in the range of 255-270 nm This fact allows the determination of these two analytes in biological samples like cerebrospinal fluid [36] or single cells [37]. For the other amino acids similar as in the case of UV detection indirect fluorescence detection can be seen as a possible solution [38].

3.1.3. Chemiluminescence detection

Chemiluminescence detection is an extremely sensitive and selective detection method that can also be used in combination with CE [39]. In principle, like in the case of UV detection, chemiluminescence detection can be employed in a direct and an indirect detection mode. A common experimental setup using the latter type of detection mode is based on the reaction of luminol or luminol derivatives with H_2O_2 in the presence of a catalyst [40]. The formed exited state 3-aminophthalate molecules emit light in the range of 425 nm to 435 nm that can be detected. Quantitative analysis is possible as the intensity observed is proportional to the luminol concentration, the oxidant and the catalyst, commonly a metal ion like Cu^{2+} or Co^{2+}. The number of available catalyst ions can be decreased by partial complexation of these species. This is the case in the analysis of underivatized amino acids, which tend to form complexes with Cu^{2+} ions thereby reducing the intensity of the emitted light [41,42]. Similar results can also be obtained by the reaction of luminol with BrO^- in an alkaline medium, a reaction that is also suppressed in the presence of underivatized amino acids [42]. Direct chemiluminescence detection requires reagents, like tris(2,2′-bipyridyl) ruthenium(III) chloride, providing spontaneous light emission if they are reacted with underivatized amino acids [43,44]. In a first step the reagent tris(2,2′-bipyridyl)ruthenium(II)chloride has to be oxidized to the activated species containing Ru(III). A completely different approach leading to chemiluminescence from N-containing species is their high temperature oxidation using a chemiluminescence nitrogen detector, a device that has been used for years in gas chromatography and HPLC in particular for the determination of pesticides, pharmaceutical, petroleum, nitrosamines and food flavors [45]. As amino acids are N-containing compounds this device is also suitable for the detection of these analytes. The basic principle of this process is the formation of nitric oxide that is reacted with ozone leading to light emission in the range of 600 nm [46].

3.2. Electrochemical detection

Electrochemical detection methods are widely employed in the determination of small ions and carboxylic acids, as they do not rely on the presence of chromophores, unlike spectrophotometric methods. An additional benefit, in particular if they are combined with separation methods in the capillary or even chip format, is the fact that electrochemical detectors are well suited for miniaturization because, unlike in the case of UV detection, sensitivity does not depend on the size of the detection cell. Unfortunately almost no commercially available instrumentation for CZE provides the possibility of electrochemical detection. Exceptions are a (no longer available) contact conductivity detector distributed by Thermo instruments in the 90ies and a just recently commercialized detector for contactless conductivity detection that can be combined with chromatographic as well as electrophoretic instrumentation [47]. Although this implies the necessity to design and fabricate lab-made devices, a number of applications describing the use of these detection methods in combination with capillary electroseparation methods can be found in the literature. Comprehensive reviews on this topic have been published recently [48–52].

3.2.1. Conductometric detection

The most commonly used type of electrochemical detector employed with CE is the conductivity detector. Its basic principle of operation is the monitoring of the conductivity, a so called bulk property of the system inherent to both, the background (i.e. the electrolyte) and the sample. The detector response is therefore depending on the conductivity difference between the analyte molecule and the appropriate co-ion of the electrolyte. If the mobility or conductivity of the analyte ion matches that of the electrolyte co-ion, no signal is observed [52]. Commonly low conductivity electrolytes are used leading to positive sample peaks due to the higher conductance of the sample zone. Naturally this setup can also be reversed, if low conductivity analytes have to be detected. A common way to increase the sensitivity of conductivity detection is the use of a so called suppressor [53]. The suppressor can be understood as an ion-exchanger or an electrochemical device that transforms the ions present in the mobile phase into a less conductive species by protonation or deprotonation respectively. Prerequisite is the use of a suppressable electrolyte such as tetraborate or carbonate in the case of anion analysis. The suppressor converts these highly conducting species into boric acid or carbonic acid respectively, resulting in a drop of the background conductivity. Unfortunately the efforts to combine this technology with CE only led to a single publication describing a prototype with promising features [54]. Due to the expected costs no further development or commercialization was carried out. Focusing on non-suppressed conductivity detection contact and contactless devices can be distinguished. Although a device for contactless conductivity detection has already been described by Gas in 1980 [55] this type of detector has faced increased interest since the publications by Zemann [56] and Fracassi da Silva [57]. The contactless approach provides several benefits over contact conductivity detectors, including the unprecedented ease of cell alignment as well as the inherent prevention of electrode fouling. Due to its universality and its simplicity conductometric detection is well suited for the determination of underivatized amino acids after their separation with electroseparation methods. This can be seen from the substantial number of papers describing the application of this approach to a variety of different samples [1,5,6,9].

Applications

The group of Hauser published several papers on the use of their more recently developed contactless conductivity detector for amino acid analysis. This device was used in combination with conventional CZE instruments [59] as well as with miniaturized instrumentation in the form of integrated chips [60,61]. Employing the first approach amino acids were separated using either highly acidic electrolytes or strongly basic ones. Finally the pH 10.8 carrier electrolyte was employed for the analysis of real samples namely urine and beer [59]. Limits of detection were in the range of 0.2 10^{-6} M for arginine and 10.5 10^{-6} M for aspartate employing the lactic acid electrolyte. Using the same type of detector in combination with an on-chip electrophoresis system, fast separations of standard mixtures containing 3-4 amino acids could be obtained [60,61]. Detection limits achieved were comparable with those mentioned above for the conventional instrument. Vcelkova et al. employed CZE with contactless conductivity detection for the determination of cationic mobilities and pKa values of 20 amino acids [61] as well as for the analysis of these solutes in several different samples [62,63]. In a recently published study, Kaml et al. investigated the amino acid patterns of several proteinaceous binding media by this allowing the differentiation of collagen, egg white and casein, employing this type of instrumentation [63]. Using an acetic acid background electrolyte and hydroxyethylcellulose as an EOF suppressor 20 amino acids could be separated in a run of about 30 minutes as can be seen from Figure 2 [64]. The developed method was applied to the analysis of these compounds in such different matrices as beer samples, alcoholic extracts and biological fluids. In addition to these works employing contactless conductivity detectors a number of papers describing the use of a conventional contact conductivity detector for the determination of amino acids after their separation with capillary electromigration methods can be found in the literature [12,13,64,65. Graß et al. presented a combined cITP and CZE system on a single chip, using the ITP step for sensitivity enhancement. They tested this approach for the analysis of seleno amino acids achieving detection limits in the low μM L^{-1} range [13]. In a more recently published work, Prest et al. describe the analysis of underivatized amino acids using a miniaturized ITP system [12]. The combination of a contact conductivity detector with a poly(methacrylate) based microfluidic device for CZE was presented by Galloway et al. [65]. An interesting approach for the design of a conductivity detector decoupled from the high voltage separation system has been used for the analysis of hydrolyzed hair samples by Mo et al. [66]. Employing this type of instrument amino acid's mixture could be separated/ detected.

electrolyte, 2.3 M acetic acid (pH 2.1); separation voltage 30 kV; capillary, 80 cm x 50 μm

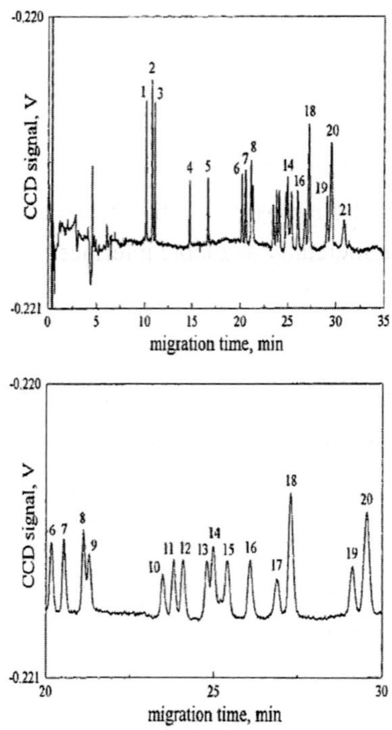

Figure 2 Electrophoretic separation of 20 underivatized essential amino acids. Conditions: i.d.; detection, contactless conductivity; peaks: 1, Lsy; 2, Arg; 3, His; 4, Gly; 5, Ala; 6, Val; 7, Ile; 8, Leu; 9, Ser; 10, Thr; 11, Asn; 12, Met; 13, Gln; 14, Trp; 15, Glu; 16, Phe; 17, Pro; 18, Tyr; 19, Cys-Cys; 20, Cys; 21, Asp. From [64] with permission

---

3.2.2. Other electrochemical detection methods

Other frequently used electrochemical detectors in CZE are potentiometric and amperometric detectors [52]. Underivatized amino acids separated by electromigration methods have been detected using pulsed [67] and constant potential amperometric detection [68,69] and potentiometric detection [70] Microchip electrophoresis was used together with pulsed amperometric detection for the separation and detection of several underivatized amino acids by Garcia et al. [67]. The pulsed mode allowed to apply different voltages for cleaning, reactivation and detection cycles employed for carbohydrate, amino acid and antibiotic analysis. Two somewhat older papers report the use of copper microelectrodes for the amperometric detection of underivatized amino acids [68,69]. The detection of these analytes by this type of electrodes can be performed in two different ways; firstly, under less alkaline conditions complexation with copper can occur and secondly, if highly basic electrolytes are used, amino acids are catalytically oxidized. A copper disk electrode operated in the potentiometric detection mode has been presented by Kappes and Hauser [70].

3.3. Mass spectrometric detection

Over the last 15 years mass spectrometric (MS) detection has become more and more important in combination with electroseparation methods. Particularly the development of electrospray ionization (ESI) greatly boosted the hyphenation of MS with liquid phase separation methods. The development of commercially available interfaces for CZE-MS can be seen as a further step towards an increased use of this hyphenated method also in routine analysis. Examples for comprehensive review papers on this topic published within the last few years can be found in the literature [70–74]. Coming to underivatized amino acids, MS detection can be seen as the best choice compared to the detectors discussed above. Amino acids can easily be protonated or deprotonated (the most common ionization process employed in ESI-MS) and sensitivity is at least equal or even better than in the case of indirect UV detection or electrochemical detection, as can be seen from Table 1. As MS detection provides far more spectral information than UV detectors certainty of peak assignment can also be improved. An additional benefit is the possibility to (spectroscopically) separate co-migrating analytes. This fact does not only allow the determination of an increased number of analytes in a single CZE run, it also reduces the effort in electrolyte optimization. This can be seen if for example the electrolyte systems used for the separation of all protein amino acids employing direct [28] or indirect UV detection [22,23] is compared with those using MS detection [75]. Whereas in the first case only a narrow pH range and the addition of several additives allowed the satisfactory separation of an acceptable number of amino acids, this problem did not occur with the latter system as MS detection provides sufficient selectivity to evaluate overlapping or even co migrating signals. Nevertheless despite these benefits suppression effects should not be overlooked if quantitative analysis is required.

Applications

Already in 1995, Lu et al. described the applicability of CZE-ESI-MS for the selective and also sensitive analysis of some amino acids [76]. Operating the spectrometer in the selected ion monitoring (SIM) mode, amino acids could be detected down to a concentration of 5μg/ml. Further studies showed the suitability of this approach for the analysis of real samples. He et al. employed CZE-ESI-MS for the monitoring of diagnostic metabolites, including some amino acids, in urine [77]. Two papers dealing with the use of CZE-ESI-MS for the determination of amino acids in foods can be found in the literature [10,75]. The whole range of proteinogenic amino acids has been determined in food products by Soga et al. [75]. Klampfl et al. used this methodology for the analysis of infant food samples with respect to their differences in amino acid patterns [10]. A principal advantage of the proposed method was the lack of any sample pretreatment except filtration. The same set of solutes was analyzed in children's blood by Martin-Girardeau et al. as depicted in Figure 4 [77]. A comparison of the results obtained with CZE-ESI-MS with those from a usual LC method involving derivatization with ninhydrine showed excellent agreement. A recent report by Soga et al. described the use of CZE-ESI-MS/MS for the determination of physiological amino acids. Improved certainty in peak assignment was enabled by the characteristic MS/MS spectra obtained for the different amino acids [78].

3.4. Other detection methods

Browsing the literature some applications describing the determination of underivatized amino acids by the combination of electrophoretic separation with of less common detection systems can be found. He et al. constructed a CZE device with a surface enhanced raman scattering (SERS) detector [79. The system was finally evaluated for the analysis of a tryptophane/tyrosine mixture. A CZE system with a refractive index detector was employed for the determination of amino acids and sugars by Ivanov et al. [80].

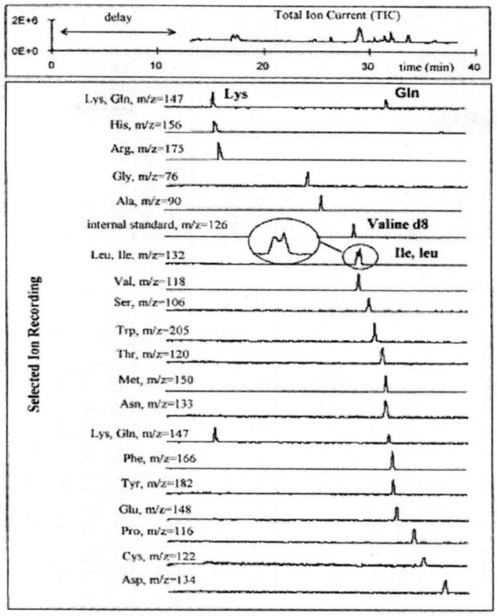

Figure 4 Total ion current and mass electropherograms of a 0.5 mM L^{-1} mixture of 20 amino acids (tyrosine at 0.1 mM L^{-1}). Conditions: electrolyte 50 mM formic acid (pH 2.5); separation voltage 30 kV; capillary, 105 cm x 50 µm i.d.; detection, MS. From [78] with permission.

4. Chiral separations

An important topic in amino acid analysis is the separation of the different enantiomers. Due to the very simple separation mechanism involved, namely the differences in the charge to ionic radius ratio, CZE itself is not suitable for the separation of enantiomers. For this reason this type of analysis is restricted to methods involving additional separation criteria (i.e. chromatographic interactions) or selective pre- or in-column complexation of the analytes. The latter approach has been taken by Moini et al. [81,82]. They separated D and L amino acids by CZE after complexation with 18-crown-6-tetracarboxylic acid. Detection was performed with an ion-trap MS, whereby the complexation of the solutes did not only allow their separation but also significantly improve their detection limits. Ligand exchange CE employing either Lys-Cu^{2+} or 3-aminopyrrolidine-Cu^{2+} as a chiral selector was employed by Lu et al. [84] and Zhao et al. [85] respectively. CEC with a continuous bed column equipped with a L-4-hydroxyproline ligand as a selector was used for the separation of amino acid enantiomers by Schmid et al. [19]. On the example of DL-phenylalanine pressure driven, electro driven and mixed separation modes were compared, whereby the best results with respect to chiral resolution and acceptable analysis times were obtained using pressure assisted CEC.

Table 2 Determination of underivatized amino acids with electroseparation methods (since 1995)

Sample	Analytes	Mode	Electrolyte/Separation column	Detection	Ref.
Human urine	Ala, Arg, Asn, Asp, Cys, Glu, Gln, Gly, His, Ile, Leu, Lys, Met, Phe, Pro, Ser, Thr, Trp, Tyr, Val, ß-Ala, GABA, Orn, α-AAA, MeHis, Cysthi;	CZE	1 M Formic acid;	MS-MS Triple Quad	79
Standard mixture	Trp, Asp, Glu;	CZE on chip	50 mM AMP, 10 mM CAPS; pH 10.8;	Contactless conductivity	60
Standard mixture	Ala, Arg, Asn, Asp, Cys, Glu, Gln, Gly, His, Ile, Leu, Lys, Met, Phe, Pro, Ser, Thr, Trp, Tyr, Val, OH-Pro, ß-Ala;	CZE	Chloroacetic acid, LiOH (10 mM total); pH 2.0-3.2;	Contactless conductivity	62
Standard mixture	Ala, Arg, Asn, Asp, Cys, Glu, Gln, Gly, His, Ile, Leu, Lys, Met, Phe, Pro, Ser, Thr, Trp, Tyr, Val, ß-Ala;	ITP on chip	LE: 10 mM Glycolic acid (2 mM Mg^{2+}), 1 mg/ml HEC; pH 9 with benzylamine ; TE: 20 m ß-Ala; pH 9.7 with $Ba(OH)_2$;	Conductivity	12
Proteinaceous binding media	Ala, Asn, Glu, Gly, His, Ile, Leu, Lys, Phe, Pro, Ser, Thr, Tyr, Val, OH-Pro;	CZE	51.9 mM Chloroacetic acid, 3.05 mM LiOH; pH 2.26;	Contactless conductivity	63
Fodder and raw materials	Ala, Arg, Asn, Asp, Cys, Glu, Gln, Gly, His, Ile, Leu, Lys, Met, Phe, Pro, Ser, Thr, Trp, Tyr, Val;	CZE	10 mM Na-Tetraborate, 10 mM ß-CD; pH 9.18;	Direct UV at 190 nm	25
Standard mixture	Arg, His, Cys;	CZE on chip	10 mM Tetraborate ; pH 9.45 ;	Pulsed amperometry	67
Standard mixture	Ala, Arg, Cys, Gln, Gly, His, Leu, Lys, Met, Phe, Pro, Ser, Thr, Trp, Tyr, Val;	CZE and OT-CEC	a) 50 mM Tris, 100 mM phosphate; pH 2.25 ; b) 100 mM Tris, 200 mM phosphate; pH 2.25 ; c) 50 mM Na^+, 100 mM phosphate ; pH 2.25 ; OT-CEC : Rh(III)tetrakis(phenoxyphenyl)-porphyrinate as wall modifier;	Direct UV at 200 nm	18
Infusion solution	Arg, Asp, Ile, Leu, Val, Orn;	CZE	a) 5 mM 4-Aminosalicylic acid; pH 11; b) 50 mM Phosphoric acid, 35 mM octanesulfonate; pH 2.2 with LiOH; c) 300 mM Formic acid;	a) Indirect UV at 266 nm b) Direct UV at 185 nm c) MS	20

Standard mixture	Trp, Phe, Thr, Tyr;	CZE on chip	50 mM AMP, 10 mM CAPS; pH 10.8;	Contactless conductivity	61
Lysed blood cells	Ala, Arg, Asn, Asp, Cys, Glu, Gln, Gly, His, Ile, Leu, Lys, Met, Phe, Pro, Ser, Thr, Trp, Tyr, Val;	CZE	30 mM 18-C-6-TCA;	MS	82
Urine, beer	Ala, Arg, Asn, Asp, Cys, Glu, Gln, Gly, His, Ile, Leu, Lys, Met, Phe, Pro, Ser, Thr, Trp, Tyr, Val;	CZE	a) 6 mM α-Hydroxybutyric acid; pH 2.8; b) 6 mM Glycolic acid; pH 2.7; c) 6 mM Lactic acid; pH 2.9-2.4; d) 50 mM AMPD; pH 10.8; e) 50 mM AMP (50 μM Mg^{2+}); pH 10.8-11.4;	Contactless conductivity	59
Protein hydrolyzates	Ala, Asn, Asp, Cys, Glu, Gln, Gly, His, Ile, Leu, Lys, Met, Phe, Pro, Ser, Thr, Trp, Tyr, Val;	CZE	a) 10 mM 3,5-Dinitrobenzoate, 0.2 mM CTAB; pH 11; b) 7.5 mM 3,5-Dinitrobenzoate, 15 mM α-CD, 10% MeOH; pH 11; c) 7.5 mM Phthalate, 0.2 mM CTAB; pH 11;	Indirect UV at a,b) 254 nm c) 214 nm	31
Injection solution	Arg, Cys, Met, Pro, Ser, Thr, OH-Pro;	CZE	20 mM Na-Borate, 0.5 mM luminol; pH 9.2;	Chemiluminescence with BrO^-	43
Cerebrospinal fluid	Trp, Tyr;	CZE	10 mM CAPS, 15 mM Na-tetraborate; pH 9.2;	LIF 266 nm (ex)	37
Beer, urine, saliva, herb extracts	Ala, Arg, Asn, Asp, Cys, Glu, Gln, Gly, His, Ile, Leu, Lys, Met, Phe, Pro, Ser, Thr, Trp, Tyr, Val, Cys-Cys;	CZE	2.3 M Acetic acid, 0.1% HEC;	Contactless conductivity	64
Standard mixture, blood	a) Ala, Arg, Asn, Asp, Cys, Glu, Gln, Gly, His, Ile, Leu, Lys, Met, Phe, Pro, Ser, Thr, Trp, Tyr, Val; b) D/L Arg, D/L Thr, D/L Val, D/L Trp, D/L Leu, D/L Phe, D/L Tyr, D/L Met, D/L Lys, D/L Gln, D/L Asp;	CZE	a) 1 M Formic acid; b) 30 mM 18-C-6-TCA;	MS	83
Physiological fluids	Ala, Arg, Asn, Asp, Cys, Glu, Gln, Gly, His, Ile, Leu, Lys, Met, Phe, Pro, Ser, Thr, Trp, Tyr, Val, Orn, GABA, HyLys, ß-Ala, carnosine, Cit, hCyss, hCys, Cys-	CZE	8 mM 4-Aminobenzoic acid, 2 mM Na_2CO_3; pH 10.2;	Indirect UV at 254 nm	32

	Cys;				
Standard mixture	Se-Cys, Se-Met, Se-hCys ;	ITP-CZE on chip	LE : 4mM 5-Bromo-2,4-dihydroxybenzoate, ethanol-amine, 2mM BTP, 0.05 % Mowiol ; pH 9.4 ; TE : 10 mM ß-Ala, $Ba(OH)_2$; pH 10.5 ; CZE: 10 mM ß-Ala, $Ba(OH)_2$, 1 mM hyroquinone; pH 10.2 ;	Conductivity	14
Standard mixture	Ala, Val, Gln, Trp;	CZE on chip	10 mM Triethylammonniumacetate; pH 7,0;	Conductivity	65
Standard mixture	D/L Tyr, D/L Ser, D/L Phe, D/L Trp;	CZE	10 mM NH_4Ac, 50 mM SDS, 3.33 mM Cu^{2+}, 6.67 mM Lys; pH 6.0;	Direct UV at 214 nm	84
Infant food	Ala, Arg, Asn, Asp, Cys, Glu, Gln, Gly, His, Ile, Leu, Lys, Met, Phe, Pro, Ser, Thr, Trp, Tyr, Val;	CZE	a) 100 mM TEA b) 300 mM Formic acid	MS	10
Single cells	Tyr, Trp;	CZE	Borate; pH 8.7;	Fluorescence, 257 nm (ex) 260 - 710 nm (em)	38
Standard mixture	D/L Phe, D/L Trp;	CZE	20mM NH_4Ac, 5 mM $CuCl_2$, 6mM (S)-3-aminopyrrolidine, 4 mM L-His, 0.2% PVA; pH 6.5;	Direct UV at 254 nm	85
Foods and beverages	Ala, Arg, Asn, Asp, Cys, Glu, Gln, Gly, His, Ile, Leu, Lys, Met, Phe, Pro, Ser, Thr, Trp, Tyr, Val;	CZE	20 mM Pyridinedicarboxylic acid, 0.5 mM CTAH; pH 12.1;	Indirect UV at 230 nm	33
Standard mixture	hPhe, Phe, Trp, Tyr, d-OHPhe, other aromatic amino acids;	CZE	100 mM Na-Phosphate; pH 2.0;	Direct UV at 200 nm	29
Standard mixture	D/L Phe;	CEC	50 mM NaH_2PO_4, 0.1 mM Cu^{2+}; pH 4.6 /acrylate based monolith with N-(2-hydroxy-3-allyloxypropyl)-L-4-hydroxy-proline as ligand	Direct UV at 223 nm	19
Standard mixture	Phe;	CZE	0.4% Boric acid / 0.3% tetraborate;	CLND	47
Standard mixture	Tyr, Trp;	CZE	50 mM Phosphate; pH 6.8;	SERS	80

Blood	Ala, Arg, Asn, Asp, Cys, Glu, Gln, Gly, His, Ile, Leu, Lys, Met, Phe, Pro, Ser, Thr, Trp, Tyr, Val;	CZE	50 mM Formic acid;	MS	78
Standard mixture	Arg, His, Lys, Met, Phe, Trp;	CZE	75 mM Phosphate; pH 11.0;	Refractive Index	81
Soya sauce	Ala, Arg, Asn, Asp, Cys, Glu, Gln, Gly, His, Ile, Leu, Lys, Met, Phe, Pro, Ser, Thr, Trp, Tyr, Val;	CZE	1 M Formic acid	MS	75
Plants	Ala, Arg, Asn, Asp, Cys, Glu, Gln, Gly, His, Ile, Leu, Lys, Met, Phe, Pro, Ser, Thr, Trp, Tyr, Val, GABA, Orn, Cit;	CZE	20 mM Benzoate, 0.5 mM MTAB; pH 12.0;	Indirect UV at 225 nm	34
Peptide hydrolysates	Ala, Gly, Phe, Leu, Pro, Val;	CZE	15 mM Na-Borate, 10 % ACN; pH 9.0;	Chemiluminescence with $Ru(bpy)_3^{3+}$	44
Plant tissues	Ala, Asn, Asp, Cys, Glu, Gln, His, Leu, Lys, Met, Phe, Pro, Ser, Thr, Trp, Val, OH-Pro;	CZE	a) 20 mM Salicylate; b) 20 mM Benzoate; c) 20 mM Phthalate; d) 20 mM Trimellitate all pH 11.2;	Indirect UV at 235 nm (a,d), 225 nm (b), 215 nm (c)	35
Urine	Orn;	CZE	300 mM Borate; pH 8.5;	MS	77
Standard mixture	D/L Trp, D/L Tyr, D/L Phe;	CZE	100 mM H_3PO_4, 10-40 mM α-CD; pH 2.5;	Direct UV at 214 nm	
Urine	D/L Trp;	ITP-CZE	LE : 10 mM BTP chloride, 0.2 % MHEC ; pH 9.3 ; TE : 10 mM BTP borate ; pH 9.0 ; CZE: 50 mM BTP borate, 0.2% MHEC, 80 mM α-CD; pH 9.0;	Direct UV at 220 nm	13
Beer	Ala, Arg, Asp, Gly, His, Lys, Phe, Pro, Trp, Tyr;	CZE	10 mM NaH_2PO_4, 30 mM octanesulfonate; pH 2.36	Direct UV at 185 nm	26
Standard mixture	Pro, Val, Ser;	CZE	15 mM Na-Borate; pH 9.5;	Chemiluminescence with $Ru(bpy)_3^{2+}$	45
Orange juice, beer	Ala, Arg, Asp, Gly, His, Lys, Phe, Pro, Trp, Tyr;	CZE	10 mM NaH_2PO_4, 30 mM octanesulfonate, 5% ACN; pH 2.36	Direct UV at 185 nm	27

Standard mixtures	Arg, Asp, Cys, Glu, Gly, Leu, Lys, Met, Pro, Ser, Thr, Tyr, Val;	CZE	6.5 mM Phthalate; pH 11.6;	Indirect UV at 225 nm	
Standard mixtures	Tyr, Phe, His, Tyr, Dph;	CZE	0.01 M TEA in methanol	Indirect UV at 214 nm	30
Standard mixture	Ala, Asn, Asp, Cys, Glu, Gln, Gly, His, Ile, Leu, Lys, Met, Phe, Pro, Ser, Thr, Trp, Tyr, Val ;	CZE	a) 50 mM Ethansulfonate; pH 2.3; b) 10 mM Phosphate, 35mM octansulfonate; pH 2.4; c) $HClO_4$; pH 2.8;(PEO coated capillary)	Direct UV at 185 nm	28
Hydrolyzed hair	Ala, Arg, Asp, Cys, Glu, Gly, His, Ile, Leu, Lys, Met, Pro, Ser, Thr, Tyr, Val;	CZE	10 mM K_2HPO_4;	Conductivity	66
Urine sample	Ala, Asn, Asp, Cys, Glu, Gln, Gly, His, Ile, Leu, Lys, Met, Phe, Pro, Ser, Thr, Trp, Tyr, Val, OH-Pro, Orn, Cit;	CZE	a) 2 mM Salicylate, 5mM carbonate, 0.15 mM MTAB; pH 10.7; b) 2 mM Benzoate, 5mM carbonate, 0.15 mM MTAB; pH 10.7; c) 2mM Salicylate, 5mM carbonate; pH 10.7;	Indirect UV at 230 nm	36
Standard mixture	Arg, Lys, Pro, Phe, Thr, Ser, Tr, Glu, Asp;	CZE	a) 10 mM Carbonate; pH 10.5; b) 3 mM Tetraborate; pH 10.5;	Potentiometry	70
Health drink	Arg, Asn, Asp, Cys, Glu, Gln, Gly, Leu, Ser, Val;	CZE	15 mM CO_3^{2-}, 5 mM luminol, 25 mM H_2O_2; pH 10.0	Indirect chemiluminescence	42
Standard mixture	His, Trp, Phe, Asp;	CZE	10% Acetic acid	MS	76
Infusion fluid	Ala, Arg, Asn, Asp, Cys, Glu, Gln, Gly, His, Ile, Leu, Lys, Met, Phe, Pro, Ser, Thr, Trp, Tyr, Val;	CZE	a) 10 mM 4-Aminosalicylate, 20 mM α-CD; pH 11.0; b) 10 mM Dimethylaminobenzoate, 20 mM α-CD; pH 11.0;	Indirect UV at 266 nm (a), 288 nm (b)	23
Urine, brain microdialysate	Ala, Arg, Asp, Cys, Glu, Gly, His, Leu, Lys, Met, Phe, Pro, Ser, Thr, Trp, Tyr;	CZE	a) 100 mM CAPS; pH 9.8; b) 20 mM Carbonate; pH 9.7;	Amperometry	68
Protein hydrolyzate	Ala, Arg, Asn, Asp, Cys, Glu, Gln, Gly, His, Ile, Leu, Lys, Met, Phe, Pro, Ser, Thr, Trp, Tyr, Val;	CZE	10-20 mM NaOH/10 – 23 mM Na-acetate;	Amperometry	69
Standard mixture	Arg, Leu, Ser, Cys, Asp ;	CZE	15 mM CO_3^{2-}, 5 mM luminol, 25 mM H_2O_2; pH 10.0 ;	Indirect chemiluminescence	41

Abbreviations:

GABA:	γ-Aminobutyric acid
α-AAA:	α-Aminoadipic acid
Me-His:	3-Methylhistidine
Cysthi:	L-Cystathionine
AMP:	2-Amino-2-methyl-1-propanol
CAPS :	3-(Cyclohexylamino)-1-propanesulfonic acid
OH-Pro :	Hyxroxyproline
LE :	Leading electrolyte
TE :	Terminating electrolyte
β-CD:	β-Cyclodextrin
Tris:	Trishydroxymethylaminomethane
PEO:	Polyethyleneoxide
SDS:	Sodiumdodecylsulfate
TEA:	Triethylamine
PVA:	Polyvinylalcohol
CTAH:	Cetyltrimethylammonium hydroxide
hPhe :	Homophenylalanine
d-HPhe :	Dihydroxyphenylalanine
MTAB:	Myristyltrimethylammonium bromide
ACN :	Acetonitrile
α-CD:	α-Cyclodextrin
Dph :	Diphenylalanine

18-C-6-TCA:	18-Crown-6-tetracarboxylic acid
AMPD:	2-Amino-2-methyl-1,3-propanediol
CTAB: HyLys:	Cetyltrimethylammonium bromide
HEC:	Hydroxylysine
Cit :	Hydroxyethylcellulose
hCyss:	Citrulline
hCys:	Hydroxycystine
Se-Cys :	Hydroxycysteine
Se-Meth :	Seleoncysteine
Se-hCys:	Selenomethionine
	Selenohomocysteine
BTP:	1,3-Bis[tris(hydroxymethyl)methylamino] propane

5. Experimental

The optimum experimental parameters employed for CZE separation and detection of underivatized amino acids strongly depend on the actual analytical problem. If the analysis of a maximum number of amino acids in a single run is required, acidic carrier electrolytes are better suited as they allow the proper identification and quantitation of leucine and isoleucine as well as arginine. Regarding detection (employing commercially available instrumentation) direct UV detection and MS detection (preferable if available) are good choices.

5.1. CZE analysis of underivatized amino acids using direct UV detection

Instrumental parameters:	capillary: 80-100 cm x 50 or 75 μm i.d.; (smaller i.d. leads to improved resolution but also to a decrease in sensitivity); eventually 50 mm i.d. capillaries with extended light path e.g. from Agilent can be used. UV detection at 185-195 nm separation voltage: +30 kV
Electrolyte:	50 mM NaH_2PO_4; pH adjusted to 2.3-2.4 with H_3PO_4; up to 35 mM octanesulfonate if improved separation of the group valine/isoleucine/leucine/serine is required;

This system allows the separation of up to 19 proteinogenic amino acids in less than 35 minutes without any sample pretreatment. Sensitivity is good (LOD better than 0.5 mg L^{-1}) for highly absorbing analytes (phenylalanine, tyrosine) but only moderate (LOD around 50 mg L^{-1}) for simple aliphatic amino acids (glycine, alanine…).

5.2. CZE analysis of underivatized amino acids using MS detection

Instrumental parameters:	capillary: 70-100 cm (minimum length depending on the type of CE-MS interface used) x 50 μm i.d.; MS detection in SIM or scan mode separation voltage: +30 kV
Electrolyte:	300-1000 mM formic acid;
Sheath liquid:	either 0.6% formic acid in i-propanol/water (8/2) or 10 mM NH_4formate in methanol/water (1/1) at 2-4 μl min^{-1}

This system allows the separation of all proteinogenic amino acids even in problematic matrices like food samples or biological fluids in less than 35 minutes without any sample pretreatment. Sensitivity is good with LOD values between 0.1 and 2 mg L^{-1}) in the SIM mode using a simple single quadrupol instrument.

References

[1] K. Petritis, C. Elfakir , M. Dreux, J. Chromatogr. A 961 (2002) 9-21.

[2] T. Fukushima, N. Usui, T. Santa ,K. Imai, J. Pharm. Biomed. Anal. 30 (2003) 1655-1687.

[3] I. Molnar-Perl, J. Chromatogr. A, 913 (2001) 283-302.

[4] J.C.M. Waterval, H. Lingeman, A. Bult ,W.J.M. Underberg, Electrophoresis 21 (2000) 4029-4045.

[5] V. Poinsot, C. Bayle F. Couderc, Electrophoresis 24 (2003) 4047-4062.

[6] P. Husek P. Simek, LC-GC Europe 19 (2001) 986-999.

[7] J.T. Smith, Electrophoresis 20 (1999) 3078-3083.

[8] S. Oguri, J. Chromatogr. B 747 (2000) 1-19.

[9] C. Prata, P. Bonnafous, N. Fraysse, M. Treilhou, V. Poinsot F. Couderc, Electrophoresis 22 (2001) 4129-4138.

[10] C.W. Klampfl W. Ahrer, Electrophoresis 22 (2001) 1579-1584.

[11] L. Krivankova, P. Pantuckova P. Bocek, J. Chromatogr. A 883 (1999) 55-70.

[12] . E. Prest, S.J. Baldock, P.R. Fielden, N.J. Goddard B.J. Treves Brown, J. Chromatogr. A 1051 (2004) 221-225.

[13] M. Dankova, D. Kaniansky, S. Fanali F. Ivanyi, J. Chromatogr. A 838 (1999) 31-43.

[14] B. Graß, R. Hergenröder, A. Neyer D. Siepe, J. Sep. Sci. 25 (2002) 135-140.

[15] S. Terabe, K. Otsuka, K. Ichikawa, A. Tsuchiya T. Ando. Anal. Chem. 56 (1984) 111-113.

[16] H. Watarai, Chem. Lett. 231 (1991) 391-394.

[17] C.W. Klampfl, Electrophoresis 24 (2003) 1537-1543.

[18] J. Charvatova, Z. Deyl, V. Kasicka V. Kral, J. Chromatogr. A 990 (2003) 159-167.

[19] M. G. Schmid, N. Grobuschek, C. Tuscher, G. Gübitz, A. Vegvari, E. Machtejevas, A. Maruska S. Hjerten, Electrophoresis 21 (2000) 3141-3144.

[20] C.W. Klampfl T.D.T. Vo, J. Liq. Chrom. & Rel. Technol. 26 (2003) 2783-2796.

[21] C.W. Klampfl, W. Buchberger P.R. Haddad, J. Chromatogr. A 881 (2000) 357-364.

[22] Y.-H- Lee T.-I. Lin, J. Chromatogr. A 680 (1994) 287-297.

[23] Y.H. Lee T. -I. Lin, J. Chromatogr. A 716 (1995) 335-346.

[24] P. Doble, M. Macka P. R. Haddad, TrAC Trends Anal. Chem. 19 (2000) 10-17.

[25] N.V. Komarova, J.S. Kamentsev, A.P. Solomonova R.M. Anufrieva, J. Chromatogr. B 800 (2004) 135-143.

[26] C.W. Klampfl, J. Agric. Food Chem. 47 (1999) 987-990.

[27] C.W. Klampfl, W. Buchberger, M. Turner J.S. Fritz, J. Chromatogr. A 804 (1998) 349-355.

[28] M.J. Thornton, J.S. Fritz C.W. Klampfl, J. High Resol. Chromatogr. 20 (1997) 647-652.

[29] H. Lee, O.-K. Choi, H.S. Jung, K.-R. Kim D.S. Chung, Electrophoresis 21 (2000) 930-934.

[30] H. Salimi-Moosavi R.M. Cassidy, J. Chromatogr. A 790 (1997) 185-193.

[31] N.M. Fujiya and M.F.M. Tavares, J. Sep. Sci. 26 (2003) 562-568.

[32] G. Zunic, Z. Jelic-Ivanovic, M. Colic S. Spasic, J. Chromatogr. B 772 (2002) 19-33.

[33] T. Soga M. Imaizumi, Electrophoresis 2 (2001) 3418-3425.

[34] C.R. Warren M.A. Adams, J. Exp. Botan. 51 (2000) 1147-1157.

[35] Z.L. Chen, C.R. Warren M.A. Adams, Chromatographia 51 (2000) 180-186.

[36] H. Chen, Y. Xu M.P.C. Ip, J. Liq. Chromatogr. & Rel. Technol., 20 (1997) 2475-2493.

[37] C. Bayle, N. Siri, V. Poinsot, M. Treilhou, E. Causse F. Couderc, J. Chromatogr. A 1013 (2003) 123-130.

[38] X. Zhang, R.R. Fuller, R.L. Dahlgren, K. Potgieter, R. Gilette J.V. Sweedler, Fresenius J. Anal. Chem. 369 (2001) 206-211.

[39] W.G. Kuhr E.S. Yeung, Anal. Chem. 60 (1988) 1832-1834.

[40] Y.-M. Liu J.K. Cheng, J. Chromatogr. A 959 (2002) 1-13.

[41] S.-Y. Liao, Y.-C. Chao C.-W. Whang, J. High Resol. Chromatogr. 18 (1995) 667-669.

[42] S.-Y. Liao C.-W. Whang, J. Chromatogr. A 736 (1996) 247-254.

[43] W. Yang, Z. Zhang W. Deng, Talanta 59 (2003) 951-958.

[44] H.P. Hendrickson, P. Anderson, X. Wang, Z. Pittman D. R. Bobbitt, Microchem. J. 65 (2000) 189-195.

[45] D. R. Bobbitt, W.A. Jackson H.P. Hendrickson, Talanta 46 (1998) 565-572.

[46] L.O. Courthaudon E.M. Fujinari, LC-GC, 9 (1992) 732-734.

[47] A.A. Mohammad, E. Fujinari, A.O. Okorodudu J.R. Petersen, J. Chromatogr. A 868 (2000) 121-125.

[48] www.istech.at

[49] L.A. Holland A.M. Leigh, Electrophoresis, 23 (2002) 3649-3658.

[50] J. Tanyaniwa, S. Leuthardt and P.C. Hauser, Electrophoresis, 23 (2002) 3659-3666.

[51] A.J. Zemann, Trends Anal. Chem. 20 (2001) 346-354.

[52] S. Polesello S.M. Valsecchi, J. Chromatogr. A 834 (1999) 103-116.

[53] M.U. Katzmayr, C.W. Klampfl W. Buchberger, J. Chromatogr.A 850 (1999) 355-362.

[54] J. Weiss, “Handbook of Ion Chromatography” 3rd Ed., Wiley-VCH, Weinheim, Germany (2004).

[55] M. Harrold, J. Stillian, L. Bao, R. Rocklin N. Avdalovic, J. Chromatogr. A 717 (1995) 371-383.

[56] B. Gas, M. Demjanenko J. Vacik, J. Chromatogr. 192 (1980) 253-257.

[57] A.J. Zemann, E. Schnell, D. Volgger G.K. Bonn, Anal. Chem., 70 (1998) 563-567.

[58] J.A. Fracassi da Silva C. do Lago, Anal. Chem. 70 (1998) 4339-4343.

[59] J. Tanyaniwa, K. Schweitzer P.C. Hauser, Electrophoresis, 24 (2003) 2119-2124.

[60] J. Tanyaniwa, E.M. Abad-Villar P.C. Hauser, Electrophoresis 25 (2004) 903-908.

[61] J. Tanyaniwa, E.M. Abad-Villar, M.T. Fernandez-Abedul, A. Costa-Garcia, W. Hoffmann, A.E. Gruber, D. Herrmann, A. Gerlach, N. Gottschlich P.C. Hauser, Analyst 128 (2003) 1019-1022.

[62] K. Vcelakova, I. Zuskova, E. Kenndler B. Gas, Electrophoresis, 25 (2004) 309-317.

[63] I. Kaml, K. Vcelakova E. Kenndler, J. Sep. Sci., 27 (2004) 161-166.

[64] P. Coufal, J. Zuka, T. van den Goor, V. Smith B. Gas, Electrophoresis 24 (2003) 671-677.

[65] M. Galloway, W. Stryjewski, A. Henry, S.M. Ford, S. Llopis, R.L. McCarley S. A. Soper, Anal. Chem. 74 (2002) 2407-2415.

[66] J. Mo, W. Liu S. Mo, Anal. Commun. 35 (1998) 365-367.

[67] C.D. Garcia C.S. Henry, Anal. Chem. 75 (2003) 4778-4783.

[68] J. Zhou S.M. Lunte, Electrophoresis 16 (1995) 498-503.

[69] Y. Guo, L.A. Colon, R. Dadoo R.N. Zare, Electrophoresis 16 (1995) 493-497.

[70] T. Kappes P.C. Hauser, Anal. Chim. Acta 354 (1997) 129-134.

[71] C.W. Klampfl, J. Chromatogr. A 1044 (2004) 131-144.

[72] P. Schmitt-Kopplin M. Frommberger, Electrophoresis, 24 (2003) 3837-3867.

[73] C.G. Huber and G. Hölzl in Z. Deyl F. Svec (Ed.), Capillary Electrochromatography, Elsevier, Amsterdam 2001.

[74] A. V. Brocke, G. Nicholson E. Bayer, Electrophoresis 22 (2001) 1251-1266.

[75] T. Soga D.N. Heiger, Anal. Chem. 72 (2000) 1236-1241.

[76] W. Lu, G. Yang R.B. Cole, Electrophoresis 16 (1995) 487-492.

[77] T. He, D. Quinn, E. Fu Y.K. Wang, J. Chromatogr. B 727 (1999) 43-52.

[78] A. Martin-Girardeau M.-F. Renou-Gonnord, J. Chromatogr. B 742 (2000) 163-171.

[79] T. Soga, Y. Kakazu, M. Robert, M. Tomita T. Nishioka, Electrophoresis, 25 (2004) 1964-1972.

[80] L. He, J. Natan C.D. Keating, Anal. Chem. 72 (2000) 5348-5355.

[81] A.R. Ivanov, I.V. Nazimov, A.P. Lobazov G.B. Popkovich, J. Chromatogr. A 894 (2000) 253-257.

[82] M. Moini, C.L. Schultz H. Mahmood, Anal. Chem. 75 (2003) 6282-6287.

[83] C.L. Schultz M. Moini, Anal. Chem. 75 (2003) 1508-1513.

[84] X. Lu, Y. Chen, L. Guo Y. Yang, J. Chromatogr. A 945 (2002) 249-255.

[85] S. Zhao, Y.-M. Liu, Anal. Chim. Acta, 426 (2001)65-70.

Ibolya Molnár-Perl (Editor)
Quantitation of Amino Acids and Amines by Chromatography
Journal of Chromatography Library, Vol. 70

1.3.2.

Quantitation of Amino Acids as *o*-Phthalaldehyde Derivatives by CE

Shigeyuki Oguri

Contents

Summary

In 1971, Roth [3] found OPA reagent could be used as the fluorescence-tagging reagent for primary amnion compounds instead of traditional reagents such as ninhydrine. Subsequently, OPA has played an important part in the science of amino acid and biogenic amines analysis. Because of its versatility and ease of handling, it is possible that other, latent advantages of OPA reagent will be discovered in the near future I believe that OPA reagent will play an important role in future developments in the field of amino-acid analysis using CE.

1. Introduction

In order to make accurate determinations of naturally-occurring amino acids, a considerable number of the chromatographically-based methods have been developed over the past 50 years. As almost all amino acids lack chromophore or fluorophore in their molecules, derivatization has generally been done before or after chromatographic separation in order to improve the chromatography or delectability of a compound. Consequently, many derivatization reagents have been employed [1, 2]. O-phthalaldehyde (OPA) /thiol, which is a well-established derivatization reagent, reacts with primary amino groups in an alkaline medium in the presence of a reducing reagent such as a thiol compound, at room temperature [3]. In most instances, the reaction is completed in less than 5 min., and the resulting derivative fluoresces at a wavelength of 445 nm when irradiated with light having a wavelength at 340 nm. For these reasons, OPA/thiol reagent is frequently used with CE, as well as with other chromatographic methods for the determination of amino acids.

Capillary electrophoresis (CE) as well as capillary electrochromatography (CEC) were developed based on electrophoretic theory [4, 5]. Electrophoresis, one of the most widely-authorized separation techniques, is currently practiced primarily in a slab gel rather than the column format of chromatography. In recent years, CE and CEC have been growing in popularity as a separation technique complementary to HPLC because of remarkable advances in the fabrication techniques of both capillaries and capillary columns. Their main advantages are low solvent consumption, speed and a high-degree of separation efficiency in comparison with traditional chromatographic methods. When a high voltage is applied across a capillary tube filled with a run buffer of pH more than 4 or 5, the run buffer moves from the anodic site to the cathodic site; this flow is called electro osmotic flow (EOF). The image of an EOF profile appears like "plug flow". By contrast, the flow generated by means of a mechanically-pressurized pump appears like "parabolic flow". As the result, much sharper peaks are obtained using CE which is driven with EOF than with LC driven with a mechanically-pressurized pump. Further, because EOF acts just as if it were a mechanically-pressurized pump or HPLC pump, thereby causing the mobile phase or run buffer to flow, it is unnecessary to install any additional mechanical pump onto CE equipment. As the results of these developments, these methods have been applied to almost every kind of naturally-occurring substances, including organic and inorganic compounds. Because of the popularity of the method which combines OPA/thiol and CE, I shall briefly review the variations of this method, e.g., pre-, in- and post-capillary derivatization of amino acids, in this chapter.

2. Pre- and post-capillary derivatization with OPA in CE

OPA is widely utilized for pre-capillary derivatization CE, as well as LC, because, basically, there are no additional hardware requirements. In addition, OPA has the following merits compared with other tagging reagents. OPA itself does not fluoresce. This allows one to save time by not having to remove the excess amount of OPA from the sample solution just before CE or CEC analyses, because OPA itself does not interfere with the fluorescence detection of OPA-amine derivatives.

Novotny and his group [6] made the first attempts at separation of pre-capillary OPA-derivatized amino acids using CE with normal and micellar solutions in the presence of organic modifiers. Derivatization of the amino acids was performed by mixing a standard amino acid solution (ca. 0.6 mg/ ml each) in a phosphate buffer (pH 7.00) and OPA/2-mercaptoethanol (2ME) solution in 0.4 M borate for 1 to 2 min at room temperature. After optimizing the CE conditions, 18 kinds of isoindole derivatives (or OPA/2ME-amino acids derivatives) were separated with a fused silica capillary (100 mm x 50 μm i.d.; 53 cm to the detector) and a run buffer which consisted of

0.05 M sodium dodecyl sulfate (SDS)-0.05 M sodium acetate buffer (pH 9.5)-15 % methanol-1 % tetrahydrofuran (THF). Mank [7] also demonstrated the separation of 18 kinds of amino acids derivatives with the same method. A typical electropherogram of 18 OPA/2ME-amino acid derivatives is shown in Figure 1.

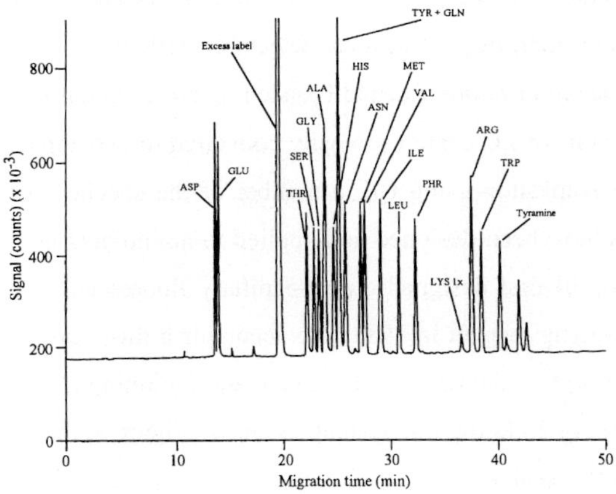

Figure 1. Electromigration profile of OPA/2ME-amino acids in pre-capillary derivatization CE. CE conditions: separation capillary, 33 □m (i.d.) x 75 cm (length); run buffer, borate (pH 9.0, 20 mM)-methanol (72.5:27.5, v/v) and 17.5 mM SDS. Reproduced from Reference [7] with permission.

One finds several articles in the literature of OPA pre-capillary derivatization CE applied for the separation of D, L amino acid diasteroisomers. To produce fluorescence of OPA-amino acid derivatives, as mentioned above, a reducing reagent such as thiol compound is required. When a chiral thiol compound, such as N-acetylcysteine (NAC) [8] or N-tert-butoxycarbonyl-L-cysteine (BOC) [9] was used together with OPA for tagging D, L-amino acid enantiomers, they were converted into their diastereoisomers, which were then separated and detected by means of an ordinary CE apparatus.

OPA, on the other hand, has some demerits. OPA-amine derivatives are less stable than those of Dansyl (Das), 9-fluorenylmethylchloroformate (Fmoc) or 6-aminoquinolyl-N-hydroxysuccinimidyl carbamate (AQC). Because OPA-amine derivatives gradually decompose from handling during the derivatization procedure when the pre-capillary method is employed, it is difficult to obtain assay results with a high degree of reproducibility. For this reason, use of "post-capillary derivatization method" is preferable to that of "pre-capillary derivatization method". In the case of LC, post-column derivatization is a commonly-utilized technique for determination of various compounds in highly-selective and sensitive by using various kinds of derivatizing reagents. Generally, the volume of the reactor in LC has little influence on peak broadening, and the reaction chamber can be heated to shorten the reaction time. Post-capillary derivatization CE, however, cannot be done in the same way as LC, because it requires a special reactor in which the separated analyte meets the derivatization reagent to form detectable derivatives. Fast reaction kinetics, efficient mixing and minimal contribution to band broadening are essential in a post-capillary scheme. This means that the reactor for LC cannot be easily converted to one applicable to CE. As OPA is very convenient for derivatization using CE, a number of the special post-capillary reactor s for OPA labeling reactions have been designed and applied to amino-acid determinations. For example, Jorgenson's group [10] first designed a post-capillary fluorescence detection scheme for CE using OPA/2ME as a tagging reagent in 1988. To accomplish this work, they fabricated a coaxial capillary reactor in order to mix the OPA/2ME reagent and migrating zones without excessive zone broadening, as shown in Fig 2. Using this system, as shown Figure 3, the detection limits for amino acid reached at the attomol level.

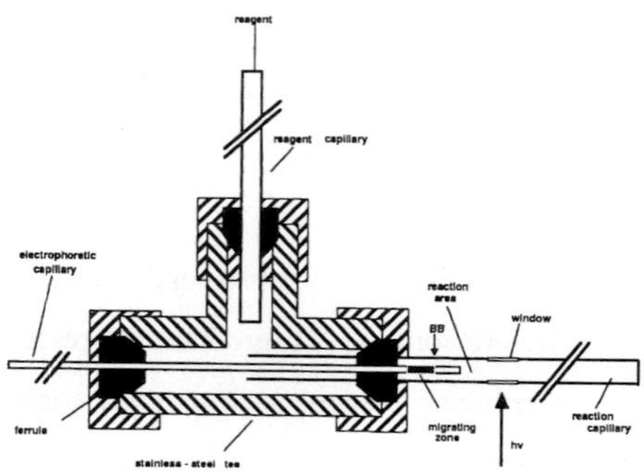

Figure 2 Cross-schematic of post-capillary reactor in stainless-steel tee. Reproduced from Reference [10] with permission

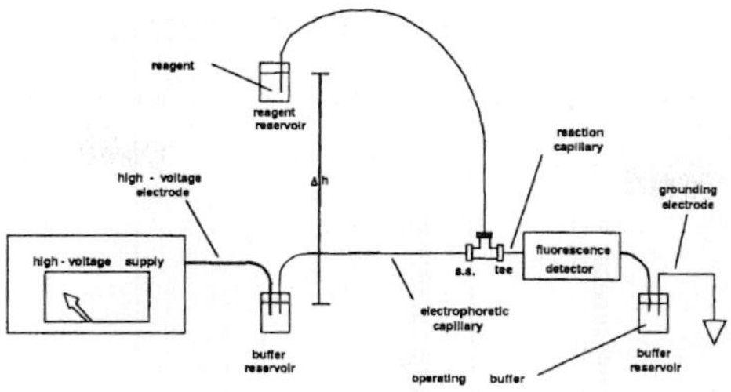

Figure 3. Overall CZE experimental set-up for post-capillary fluorescence detection using the post-capillary reactor as shown in Figure 2. Reproduced from Reference [10] with permission.

Around the same time, in 1988, Zare and his group [11] constructed an on-line connector for post-capillary derivatization CE. The connector, in the form of a cross or tee, is fabricated from the fused silica capillary tubing, itself, and has sufficiently low dead volume to be compatible with the tubing dimensions normally associated with micro-column separation techniques. Their CE system is illustrated in Figure 4 and typical electropherogram separations of OPA/2ME–amino-acid derivatives are shown in Figure 5. Regarding these separations, they commented that zone broadening by the connector was determined to be approximately 10 % for CZE separations performed in a 75 μm (i.d.)-capillary.

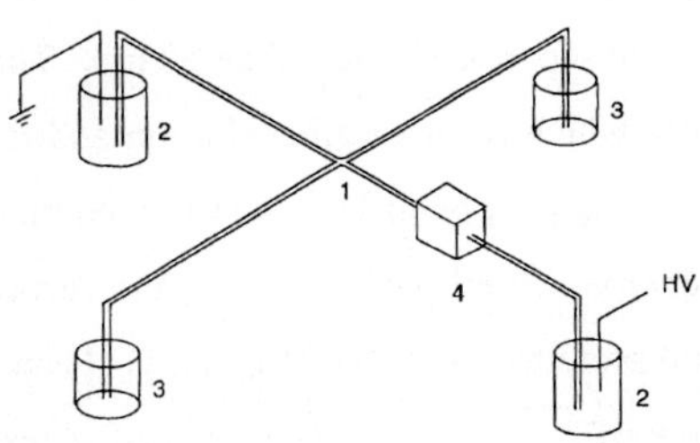

Figure 4. Experimental set-up of the CZE/LIF detector system: 1, on-column connector; 2, buffer reservoirs; 3, derivatization reagent reservoir; 4, LIF detector housing. Reproduced from Reference [11] with permission

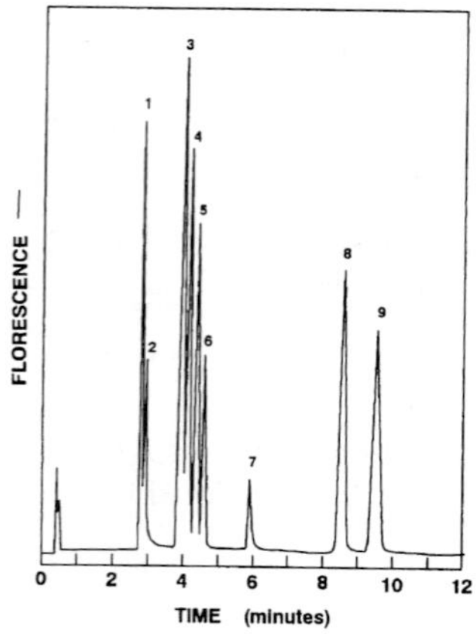

Figure 5 Electromigration profile of OPA/2ME-amino acids in post-capillary derivatization CE. Peak identification: 1, L-arginine; 2, L-lysine; 3, L-isoleucine; 4, L-histidine; 5, L-serine; 6, L-asparagine; 7, L-cysteine; 8, L-glutamic acid; 9, L-aspartic acid. Capillary: 75 μm (i.d.) x (length), 40 cm to connector, 48 cm to LIF detector, 74 cm (total length). Run buffer: 200 mM borate buffer (pH 7.8). Operating voltage: -25 kV. Detection: fluorescence (ex. 325 nm, em 450 nm). Reproduced from Reference [11] with permission

---

Furthermore, several types of post-capillary reactor have been constructed and applied to amino-acid separations using OPA/2ME. For example, Albin et al. [12] reported a gap injection reactor in which two capillaries are separated by less than 50 μm. The principle on which the fluid-gap reactor is based is diffusion, which results in the addition and mixing of the derivatization reagent and analyte. Zhang and Yeung [13] published an improved model of a coaxial capillary reactor in 1996. Using this reactor, mass limits of detection were recorded in the attomole range for various amines, amino acids and proteins. In 1991, Rose Jr. [14] constructed a free-solution reactor for post-capillary derivatization CE. The reactor consists of a reservoir of OPA/2ME reagent, which acts as the cathodic reservoir in which the electrophoretic capillary terminates. Coble [15] reported on a method where the separation capillary was mounted in a sheath flow cell as the post-capillary reactor, and which used OPA/2ME derivatization for the determination of proteins and amino acids, as shown in Figure 6.

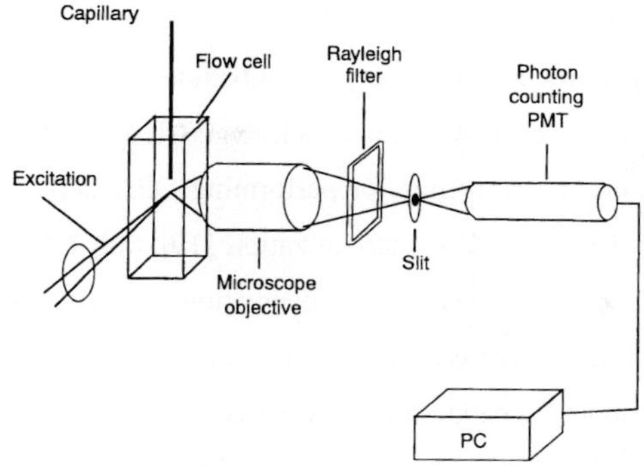

Figure 6 A schematic of the sheath flow post-capillary derivatization reactor. The sheath fluid, which contains the OPA/2ME solution, enters from the top of the flow cell and surrounds the capillary eluent as it exits the capillary. Reproduced from Reference [15] with permission.

Zhu and Kok [16] made a reactor in which a "bubble-cell", made by a porous tube connecting at the end of the separation capillary. The reagent solution is driven by regulated air pressure applied to the separation capillary inlet and the reactor vessel simultaneously. They reported that a ten-fold increase in sensitivity could be obtained by this method.

Albin et al. [17] evaluated the fluorescence detection of amino acids in CE using several kinds of reagents and techniques. In their paper, they compared the performances of pre- and post-capillary derivatization CE with OPA/2ME, as shown in Figure 7. In their studies, post-capillary derivatization was accomplished with the use of the gap injection method.

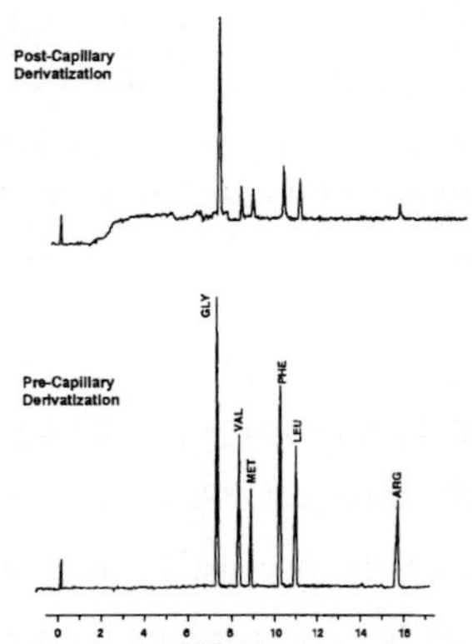

Figure 7 Comparison of electromigration profiles created by pre-capillary method (bottom) and post-capillary method (top) using OPA/2ME-amino acids. Reproduced from Reference [17] with permission

3. μ-chip CE for OPA derivatization

Since the fabrication of a reaction chamber on a glass substrate for use as the CE platform is much easer than installing the reaction chamber onto a traditional CE apparatus, μ-chip CE has recently come to be one of the preferred strategies for performing OPA derivatizations, Ramsey's group demonstrated pre- [18] and post-capillary derivatization [19] with OPA on a μ-chip fabricated by using a standard photolithographic etching, and deposition techniques as shown in Fig 8-A and 8-B, respectively. In each of these figures, the reservoirs on the μ-chip, which are filled with OPA solution, amino acid solution and separation buffer, are connected by channels. The flow of solutions in the channels is generated and controlled by EOF by means of switching the applied voltage potential between each channel. The derivatization reaction occurs at the intersection of the reagent channel and the analyte channel located before or after the separation channel. Although this approach has not yet been perfected, and there are still many problems yet to be solved, I believe that μ-chip CE will become a well-established approach for the determination of amino acids as well as other natural biomolecules using OPA reagent.

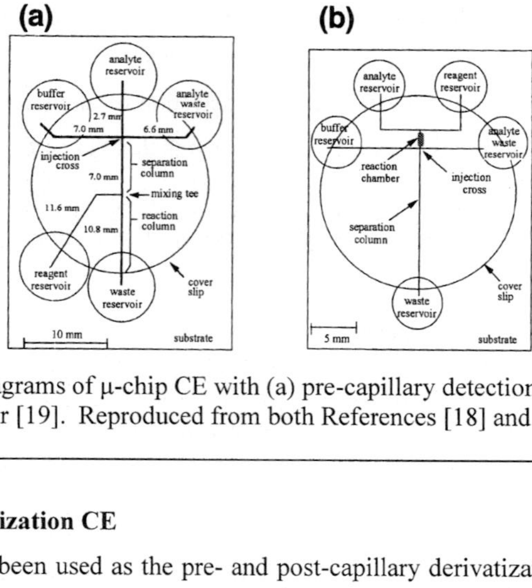

Figure 8 Schematic diagrams of μ-chip CE with (a) pre-capillary detection mixer [18] and (b) post-capillary detection mixer [19]. Reproduced from both References [18] and [19] with permissions.

4. In-capillary derivatization CE

Although OPA has been used as the pre- and post-capillary derivatization reagent as described above, such an assay system, which employs OPA for CE has not become the standard method among scientists, perhaps because, on the one hand, the post-capillary derivatization reactor is mechanically hard to install and, also, because a special reactor is needed. On the other hand, the pre-capillary derivatization method requires very delicate handling, as described in the previous section. To solve both these sets of problems, the Zare group [20] came up with the new idea of using the front-end of a separation capillary tube as the reaction chamber. This idea comes out from the on-line pre-capillary derivatization technique. The sample and labeling reagent plugs are injected separately into an inlet of the capillary tube, then the both plugs are mixed simply by keeping the capillary untouched for several minutes. After completing the reaction, an adequately high voltage is applied across the capillary tube for doing CE. The derivative begins traveling towards a detection window and is detected in the capillary tube. This method is sometime called "in-capillary derivatization". Further, until now, in-capillary derivatization CE had been done almost entirely with OPA reagent. During this same period, Honda's group [21] developed and improved upon a method called "tandem mode" (i.e., reagent solution/sample solution) or "sandwich mode" (i.e., reagent solution/sample solution/reagent solution) as shown in Figure 9.

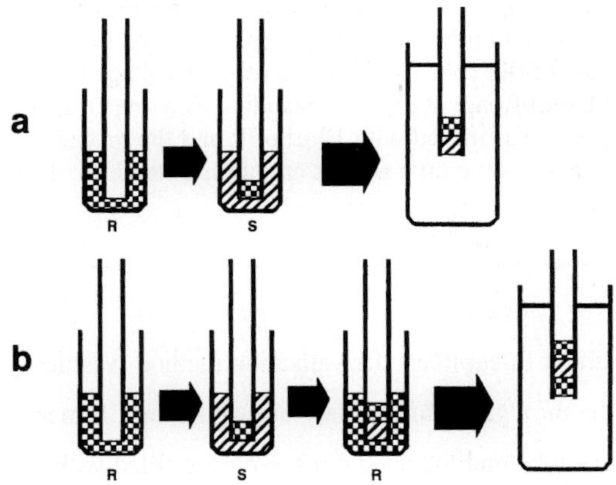

Figure 9 Schematic diagram of the introduction of sample and reagent solutions to the inlet of a capillary by tandem (a) and sndwich (b) modes. sample solution (S), reagent solution (R), electrode solution. Reproduced from Reference [21] with permission

They showed the migration profiles of several kinds of amino-acid derivatives obtained from this method and the pre-capillary method. For comparison's sake, the electropherogram of the in-capillary derivatization CE was obtained form that of a traditional pre-capillary method. (Figure 10) To perform this in-capillary derivatization, before doing CE, the reservoir should be switched several times from being a sample solution reservoir to being a run buffer reservoir, and should be positioned at the injection site of the separation capillary tube.

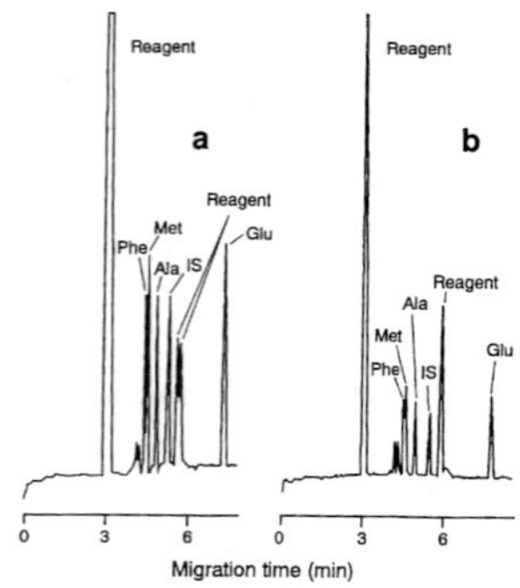

Figure 10 Analysis of OPA derivatives of selected amino acids formed by at-inlet (a) and pre-capillary (b) derivatizations. Capillary: 50 μm (i.d.) x 72 cm (length). Run buffer: 50 mM borate buffer (pH 10.0). Operating voltage: 30 kV. Detection: absorption (230 nm). Sample solution (S): 5 x 10^{-3} % (w/v) amino acids in 50 mM borate buffer (pH 10). Reagent solution (R): 1 x 10^{-2} % (w/v) OPA in 50 mM borate buffer (pH 10). Sample/reagent voltage injection (30 kV) in (a): Reagent (3 sec)-Sample (1 sec)-Reagent (3 sec). Standing time in (a): 20 min. Pre-capillary derivatization in (b): 2 μl of S was mixed with 12 μl of R and the mixture was allowed to stand for 20 min at room temperature. IS: cinnamic acid as an internal standard. Reproduced from Reference [21] with permission

In 1996, a throughout in-capillary derivatization method was designed by Oguri et al. [22, 23] The mechanism of the throughout in-capillary derivatization CE method is illustrated in Figure 11. First, both the capillary tube and the anodic reservoir are filled with a run-buffer containing the derivatization reagent. (Step-1) As for the cathodic reservoir, it does not matter if the derivatization reagent is present in the run-buffer or not. After injection of a free amino-acid solution directly into the capillary by means of either hydrostatic or electrokinetic voltage injection, a high-voltage potential is applied across the capillary. The free amines then begin to travel toward the cathodic site due to driving force of both EOF and electrophoresis. During the migration of free amines, they are converted to their derivatives (Step-2) by the derivatization reagent in the run-buffer. Subsequently, in-line separation happens (Step-3) and detection (Step-4) is performed. This means that the free amines are automatically tagged, separated, and detected while traveling through the capillary tube. 15 kinds of amino acids were determined by using this technique as shown in Figure 12.

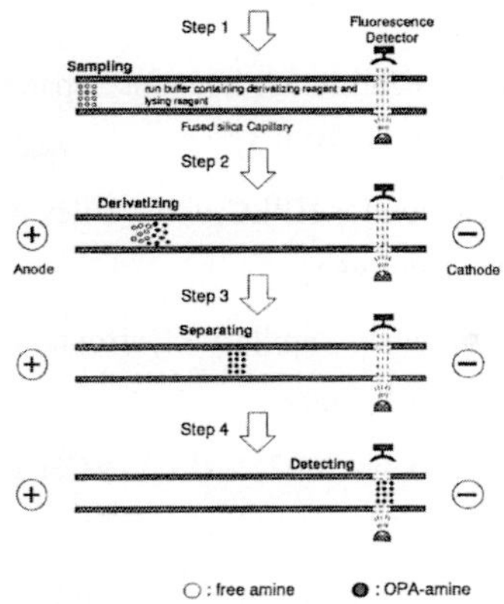

Figure 11 Schematic representation of throughout in-capillary derivatization CE for the determination of amino acid using OPA/n-acetylcysteine. In step 1, the separation capillary tube was filled with run buffer containing OPA/ n-acetylcysteine. After that, in step 2, a free amino acid solution was injected at the anodic side of the capillary by using either a hydrostatic or a voltage injection method. Then, high-potential voltage was applied across the capillary, subsequently, the free amino acids began traveling toward cathodic site. During this traveling, the free amino acids were automatically derivatized (in step 2), separated (in step 3) and detected (in step 4) when OPA-derivatives reached the detection window.

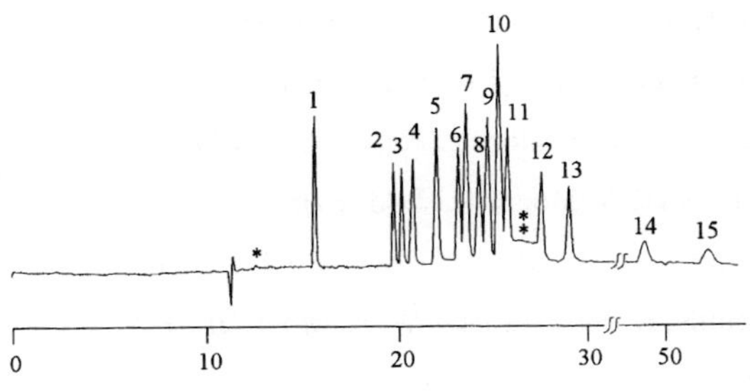

Figure 12 Electromigration profile of OPA/n-acetylcysteine-amino acids in throughout in-capillary derivatization CE. Peak identification: 1. arginine; 2, phenylalanine; 3, tyrosine; 4, leu-cine; 5, isoleucine; 6, methionine; 7, valine; 8, lysine; 9, alanine; 10, glycine; 11, histidine; 12, serine; 13, threonine; 14, glutamic acid; 15, aspartic acid. The symbols * and ** represent ammo-nia and cystine, respectively. Capillary: 50 μ m (i.d.) x 100 cm (length). Run buffer: 20 mM β-cyclodextrin-2 mM OPA/NAC-100 mM phosphate-borate buffer (pH 10). Operating voltage: 22 kV. Detection: fluorescence (ex. 340 nm, em 455 nm). Reproduced from Reference [23] with permission.

References

[1] T. Toyo'oka (Ed.), Modern Derivatization Methods for Separation Science, Wiley & Sons Ltd, England, 1999. G. Lumn, LO. C. Hellwig (Eds.).

[2] Handbook of Derivatization Reaction for HPLC, John Wiley & Sons Ltd, England, 1999.

[3] M. Roth, Anal. Chem., 43 (1971) 880-882.

[4] M. G. Khaledi (Ed.), High performance Capillary Electrophoresis Theory, Techniques, and Applications, John Wiley & Sons, INC., 1998.

[5] A. E. Rassi (Ed), CE and CEC Reviews 2002, WILEY-VCH Verlag GmbH, Weinheim (Federal Republic of Germany), 2002.

[6] J. Liu. K. A. Cobb, M. Novotny, J. Chromatogr. 468 (1988) 55-65.

[7] A. J. G. Mank, E. S. Yeung, J. Chromatogr., A, 708 (1995)309-321.

[8] R. J. H. Houben, H. Gielen, S. van der Wal, J. Chromatogr., 634 (1993) 317-322.

[9] L. Kang, R. H. Buck, Amino Acids, 2 (1992) 103-109.

[10] D. J. Rose, Jr. and J. W. Jorgenson, J. Chromatogr., 447 (1988) 117-131.

[11] S. L. Pentoney, Jr., X. Huang, D. S. Burgi, R. N. Zare, Anal. Chem., 60 (1988) 2625-2629.

[12] M. Albin, R. Weinberger, E. Sapp, S. Moring, Anal. Chem., 63 (1991) 417-422.

[13] L. Zhang, E. S. Yeung, J. Chromatogr. A, 734 (1996) 331-337.

[14] D. J. Rose, Jr., J. Chromatogr., 540 (1991) 343-353.

[15] P. G. Coble, A. T. Timperman, J. Chromatogr. A, 829 (1998) 309-315.

[16] R. Zhu, W. T. Kok, J. Chromatogr., 716 (1995) 123-133.

[17] M. Albin, R. Weinberger, E. Sapp, S. Moring, Anal. Chem. 63 (1991) 417-422.

[18] S. C. Jacobson, R. Hergenröder, A. Moore Jr., J. M. Ramsey, Anal. Chem., 66 (1994) 4127-4132.

[19] S. C. Jacobson, L. B. Koutny, R. Hergenröder, W. A. Moore Jr., J. M. Ramsey, Anal. Chem., 66 (1994) 3472-3478.

[20] H. A. Fischman, J. B. Shear, L. A. Colon, R. N. Zare, US Patent 5 318 680, 1994.

[21] A. Taga, S. Honda, J. Chromatogr. A, 742 (1996) 243-250.

[22] S. Oguri, T. Fujiyoshi, Y. Miki, Analyst, 121 (1996) 1683-1688.

[23] S. Oguri, K. Yokoi, Y. Motohase, J. Chromatogr. A, 787 (1997) 253-260.

Ibolya Molnár-Perl (Editor)
Quantitation of Amino Acids and Amines by Chromatography
Journal of Chromatography Library, Vol. 70

1.3.3.

Capillary Electrophoresis and Capillary Electrochromatography of Amino Acids as Dansyl Derivatives

Zilin Chen

Contents

Summary

This chapter summarizes the advances in the enantioseparation of Dns-DL-AAs by CE and CEC. The basic principle of chiral recognition with the different chiral selectors is discussed by combining the typical examples for resolving the Dns-AAs. Cu(II) complexes with L-amino acids, L-amino acylamides or other complex-forming agents, CDs or CD derivatives, chiral surfactants and other chiral selectors have been well used as the chiral selectors added in the CE background electrolytes for chirally resolving Dns-DL-AAs. With the introduction of micellar phase of anionic surfactants in the CE electrolytes, two separation modes, LE-

MEKC and CD-MEKC, have been developed for improving the separation selectivity. In addition, the combination of different chiral selectors, the addition of organic modifiers, and the adjustment of separation conditions like pH, ionic strength, buffer composition were also used to manipulate the selectivity. In CEC, with the advent of monolithic column technology, chiral stationary phases have been immobilized on the monolithic matrix by *in situ* copolymerization and/or post-modifications or grafting of chiral selectors. Monolithic column technology makes CEC a very powerful technique for chiral separation with high separation selectivity and efficiency. New chiral selectors and column technology are expected to develop for the chiral separation of Dns-AAs in the future

1. Introduction

Amino acids are important biological compounds building peptides and proteins, and exist in food, feeds, body fluids and tissues. It has been recognized that amino acids are chiral compounds and that their D- and L-enantiomers show different biological or physiological activities. Therefore, the separation and detection of chiral amino acid has become an interesting and important research topic in the life sciences, separation science and pharmaceutical, medical and biotechnological fields. The nature of free amino acids usually does not suit many separation and detection techniques due to weak chromatographic retention and poor selectivity for the separations and low sensitivity for the detections. Therefore, derivatization is an important approach for improving the selectivity of separation and the sensitivity of detection in amino acid analysis. Dansyl amino acids (Dns-AA) are one important category of amino acid derivatives. They are obtained by the following derivatization reaction with dansyl chloride.

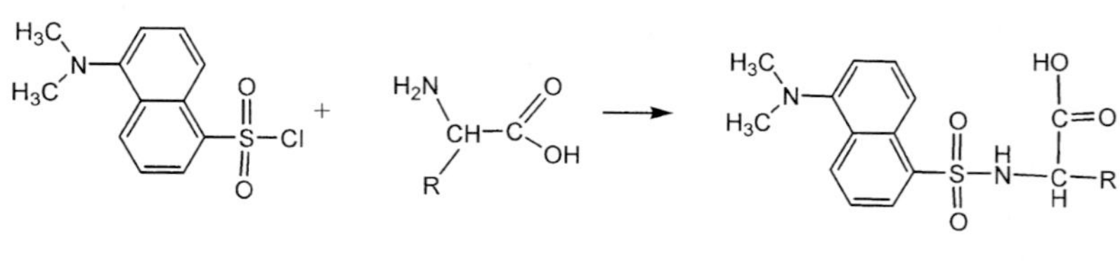

Dansyl chloride Amino acid Dansyl amino acid

Dansylated derivatization of amino acids not only increases the detection selectivity and sensitivity by introducing UV-absorbing and fluorescing dansyl groups into free amino acid molecules, but also improves the chromatographic retention and separation selectivity by enhancing the intermolecular interactions between the dansyl group and the stationary phase in chromatography or by changing the electrophoretic mobility and hydrophobicity to enable the use of suitable electrokintic separation modes. Since the dansyl group is UV absorbing and fluorescing group, Dns-AAs can be detected by UV at the typical absorbance 254 nm, and/or by laser-induced fluorescence (LIF) at 360 nm/570 nm (Ex/Em).

As described in the other chapters, gas chromatography (GC) and high performance liquid chromatography (HPLC) have been employed as conventional techniques for the separation of Dns-AAs with UV absorbance and LIF detections. In recent decades capillary electrophoresis (CE) and capillary electrochromatography (CEC) have been developed as very useful separation techniques of amino acids with high efficiency and selectivity. Very few papers reported achiral separation of several dansyl amino acids by CE [1-3] and CEC [4]. As amino acids are chiral compounds, this chapter will discuss the basic principle and recent advance in the chiral separation of Dns-AAs by CE and CEC.

2. Capillary electrophoresis

Capillary electrophoresis is an attractive and effective tool for the resolution of chiral amino acids. The advantages of CE include extremely high separation efficiency, excellent feasibility in selectivity manipulation, short analysis time, simplicity, versatility and low sample volume and reagent consumption. Chiral separation can be accomplished by CE adding a chiral selector capable of resolving the enantiomers in the background electrolyte. A large number of chiral selectors have been used in the resolution of amino acid enantiomers. Typical chiral selectors for resolving amino acid enantiomers are ligand exchange complexes, cyclodextrins, chiral crown ethers, chiral surfactants, and linear polysaccharides [5]. The chiral separation of Dns-AAs by CE will be discussed in the term of chiral selectors used as the additive in the electrolyte solutions.

2.1 Chiral ligand exchange complexes as chiral selector

Chiral ligand-exchange (LE) was first suggested by Helfferich in 1961 [6] and introduced by Davankov et al. into the LC in the early 1970s [7]. These original LE stationary phases were based on L-proline residues bonded to chloromethylated polystyrene, using Cu(II) or other metal ions as complexing agents for resolving amino acid enantiomers with remarkable enantioselectivity. Gübitz et al. developed the chiral LE stationary phases on silica gel [8, 9]. Subsequently, a number of papers reported chiral separations by LE-LC, as summarized in the review papers [10, 11]. In recent decade, LE has also been well used in CE and CEC for chiral separations. Recent review papers described the advances in the chiral separation based on ligand exchange. [12-15].

2.1.1 Ligand exchange-capillary electrophoresis

Ligand exchange was introduced to CE for chiral separation since 1985 [16]. The basic mechanism in LE-CE is the formation of ternary mixed metal complexes (in most case Cu(II) was used as center ion) by the interchange of ligands between the selector and the analyte. Therefore, this technique is suitable only for resolving analytes like amino acids, hydroxy acids, amino alcohols, diamines, and some peptides, which can serve as ligands for a given central ion. In the case of dansyl amino acids, the analyte (D,L-Dns-AA) replaces one ligand in the bidentate chiral selector of [Cu(II)(L-Sel)$_2$] (Sel: chiral selector) and forms two ternary complexes, (L-Sel)Cu(II)(D-Dns-AA) and (L-Sel)Cu(II)(L-Dns-AA). The following equilibrium represents the LE reaction in the electrolyte solution:

Cu(II)(L-Sel)$_2$+D,L-Dns-AAs → (L-Sel)Cu(II)(D-Dns-AA) + (L-Sel)Cu(II)(L-Dns-AA)

The chiral discrimination is based on the difference in the complex stability constants of the two ternary Cu(II) complexes with D- and L-enantiomers. Figure1 shows the difference in the stereoselectivity of two ternary complexes, when L-Prolinamide as chiral selector [17]. The substitutent group R orientates toward up and down and interacts with the dansyl group, which makes two complexes different stability.

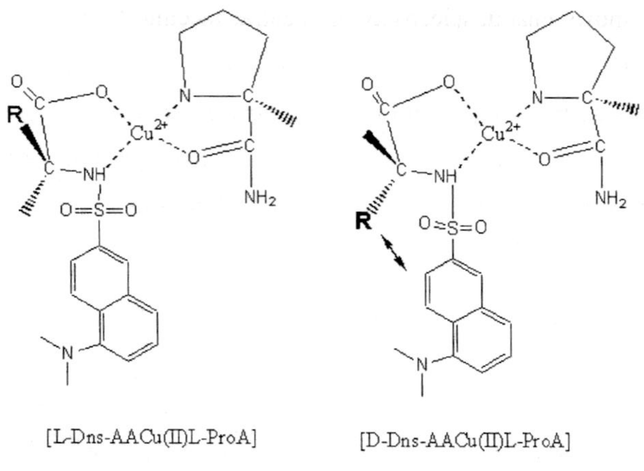

Figure 1 Proposed structures of ternary Cu(II) complexes with mixed L-Prolinamide and Dns-D-AA or Dns-L-AA.

As early as 1985, Zare and co-workers reported the enantioseparation of 12 Dns-DL-AAs by using Cu(II) complexes with L- and D-histidine (His) as a chiral selector [16]. Interestingly, it is the first enantioseparation achieved by CE. LIF was used as the detection method. The chiral recognition of ligand exchange is based on the difference in the thermodynamic stability of diastereomeric ternary complexes: Cu(II)(L-His)(D-Dns-AA) and

Cu(II)(L-His)(L-Dns-AA) formed by Dns-AAs interchanging the ligand of chiral selector Cu(II)(L-His)$_2$. Although the complexation process offers low enantioselectivity, the Dns-DL-AA enantiomers were completely resolved due to the extremely high efficiency of CE. Amino acids bound to Cu(II)(L-His) migrate faster than free amino acids because the Cu(II)(L-His) carries positive charge under neutral pH 7-8. However, amino acids bound more strongly to Cu(II)(His) show a weaker fluorescence signal, which is the result of quenching from association with the copper ion. Enantiomer migration order (EMO) depends on the amino acid structures. The EMO for aromatic amino acids possessing aromatic and polar side-chains was the same; however, the EMO was the opposite in the case of aliphatic amino acids. Cu(II) was replaced by Co(II), no enantioselectivity was observed.

In a subsequent study by the same group, a more effective chiral selector, Cu(II)-aspartame, was employed for resolving 14 racematic Dns-DL-AAs in a single run in less than 12 min [18]. At pH 7.0, ternary Cu(II) complexes are positively charged and migrate faster than neutral free Dns-DL-AAs. D-enantiomers migrate faster than L-enantiomers. Figure 2A shows the enantioseparation of a mixture of 4 Dns-DL-AAs obtained with an electrolyte solution of 2.5 mM Cu(II), 5.0 mM aspartame, and 10 mM NH_4Ac at pH 7.4. Although Cu(II)-Aspartame shows enantioselectivity for neutral racemic Dns-DL-AAs, the separation selectivity among the different Dns-AAs is not sufficient. Sodium tetradecyl sulfate (STS) was incorporated as an additional micellar phase to improve the selectivity among different neutral Dns-AAs. Figure 2B Shows the simultaneous separation of three pairs of neutral Dns-DL-AAs with electrolyte containing 2.5 mM Cu(II), 5.0 mM aspartame, 10 mM NH_4Ac and 20 mM STS at pH 7.81.

Yuan et al. reported the simultaneous separation of 6 Dns-DL-AAs using Cu(II) complex of L-arginine as chiral selector [19]. The authors also investigated the enantiomeric resolution by using other metal ions like Co(II), Zn(II) and Ni(II) and other ligands like L-Asp, L-Glu, and L-Ala. The results showed a loss of chiral recognition in most cases, which suggests that both central ion and ligand play important role in the LE recognition.

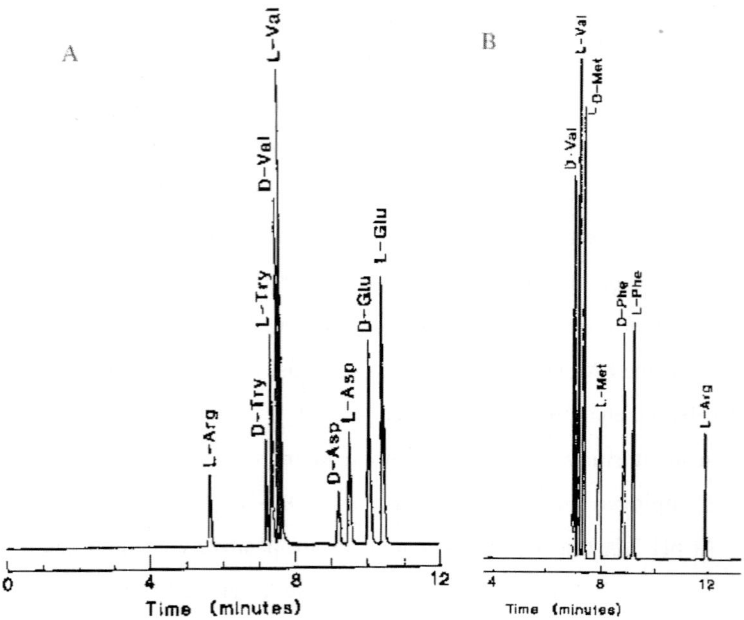

Figure 2 Electropherograms of 2 mixtures of Dns-DL-AAs. Dns-DL-Arg is used as an internal standard. Background electrolytes, (A) 2.5 mM $CuSO_4$, 5.0 mM aspartame and 10 mM NH_4Ac, pH 7.4; (B) 2.5 mM $CuSO_4$, 5.0 mM aspartame, 10 mM NH_4Ac and 20 mM STS, pH 7.8; applied voltage, -30 kV, current 38 μA. (From Reference [18] with permission)

We succeeded in the enantioseparation of Dns-DL-AAs by using Cu(II) complexes with L-amino acylamides namely L-phenylalaninamide (L-PheA), L-prolinamide (L-ProA) and L-alaninamide(L-Ala) as chiral selectors [20,21]. Figure 3 shows the enantiomeric resolution of Dns-DL-AAs using Cu(II)-L-PheA complex as chiral selector. A typical characteristic of Cu(II) complexes with L-amino acylamides is that they are present in several species in an electrolyte and their distribution depends on the pH. These complexes are present in the forms of $[CuL]^{2+}$, $[CuL_2]^{2+}$ at acidic conditions, in $[CuLH_{-1}]^+$ at neutral condition, and in $[CuL_2H_{-1}]^+$ and $[CuL_2H_{-2}]$ at alkaline condition, where L donates the L-amino acylamides. It suggests that the forms of complexes are changed and responsible for the chiral recognition with different enantioselectivity with the change of pH.

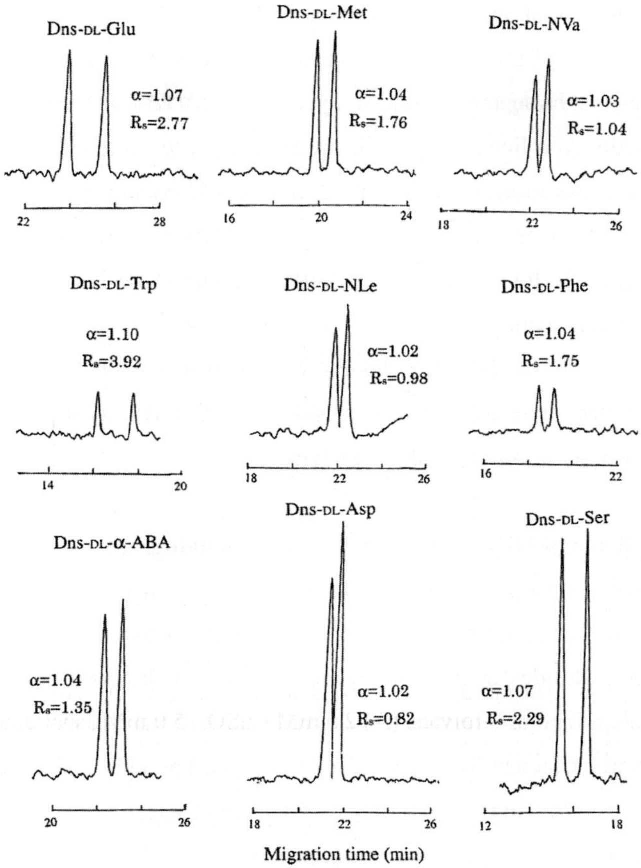

Figure 3 Electropherograms of Dns-DL-AAs resolved by CE using Cu(II)-L-PheA as chiral selector. Running electrolyte: pH 5.0, 10 mM Cu(II)$(Ac)_2$, 20 mM L-phenylalaninamide, and 20 mM NH_4Ac. Capillary total length: 55 cm, effective length: 40 cm, 50 μm i.d. and 375 μm o.d.. Applied voltage: 15 kV. UV detection: 254 nm. L-enantiomers migrate faster than L-enantiomers. (From Reference [20] with permission)

As compared the separation factors among three Cu(II) complexes with L-PheA, L-ProA and L-AlaA, Cu(II)-L-ProA complex shows almost the highest enantioselectivity for all tested Dns-DL-AAs, Cu(II)-L-PheA complexes shows enantioselectivity for most Dns-DL-AAs, whereas, Cu(II)-L-Ala complex only shows enantioselectivity for a very limited number of Dns-DL-AAs. The different enantioselectivity among three complexes of chiral selectors probably results from the influence of the rigid pyrrolidine ring in Cu(II)-L-ProA and the π-π interaction offered by the phenyl group in phenylalaninamide [17]. It was interestingly observed that that D-enantiomers migrate faster than L-enantiomers when Cu(II)-L-ProA complex is used as the chiral selector, in contrast to Cu(II)-L-PheA and Cu(II)-L-Ala systems [20]. The introduction of micellar phase of anionic surfactants, such as sodium dodecyl sulfate (SDS), improves the separation selectivity, but does not reverse the EMO [21].

Cu(II)-L-hydroxyproline and Cu(II)-L-proline complexes were well used as the chiral selectors for enantioseparations of underivatized amino acid enantiomers [22]. Author and coworkers investigated the enantioseparation of Dns-DL-AAs by using Cu(II)-L-hydroxyproline complex as chiral selector. Author's unpublished results show that Cu(II)-hydroxyproline complex shows weak enantioselectivity for several Dns-DL-AAs namely Dns-DL-Glu, Dns-DL-Asp, Dns-DL-Ser, Dns-DL-Thr, Dns-DL-Trp., however, most of which were partially resolved when a electrolyte containing 10 mM Cu(II), 20 mM L-hydroxyproline and 20 mM NH_4Ac at pH 6 was used as a running electrolyte.

2.1.2 Ligand Exchange-Micellar Electrokinetic Chromatography

It has been recognized that LE-CE using Cu(II) complexes offers good enantioselectivity for resolving amino acid enantiomers. However, the selectivity among the analytes is not sufficient in most cases. Besides, some chiral ligands with high hydrophobicity such as didecy-L-alanine has problem for chiral ligands to dissolve in the aqueous electrolyte solutions [23]. Micelles of anionic surfactant like SDS and STS were used as additional additive in the electrolyte solution for improving the solubility and separation selectivity [18, 23], later, further developed into a powerful separation technique called as ligand exchange-micellar electrokinetic chromatography (LE-MEKC) by Chen et al. [24,25].

Cohen et al. [23] reported enantioseparation of a mixture of Dns-DL-Tyr, Dns-DL-Met, Dns-DL-Leu by using Cu(II)-N,N-didecyl-L-alanine (DDAla) complex as chiral selectors in the presence of SDS. Solubilization of DDAla in an aqueous solution was possible only in the presence of SDS. As a result, mixed micellar complexes, Cu(II)-DDAla-SDS-Dns-D-AA and Cu(II)-DDAla-SDS-Dns-L-AA, are formed in the solution. DDALa is incorporated into the micelle in such a way that the hydrophilic chiral centers of L-Ala are on the surface, while the hydrophobic tail is in the interior. Later, Sundin et al. also [26] reported enantiomeric separation of several Dns-DL-AAs by using the similar electrolyte system as Cohen et al. used. In these early reports, micelles were only regarded as an additive. Based on the difference in the mechanism of chiral ligand exchange in the absence and presence of micelles, author and coworkers proposed the separation mode of LE-MEKC, the mechanism of which is considered as a hybrid of LE-CE and MEKC [24,25].

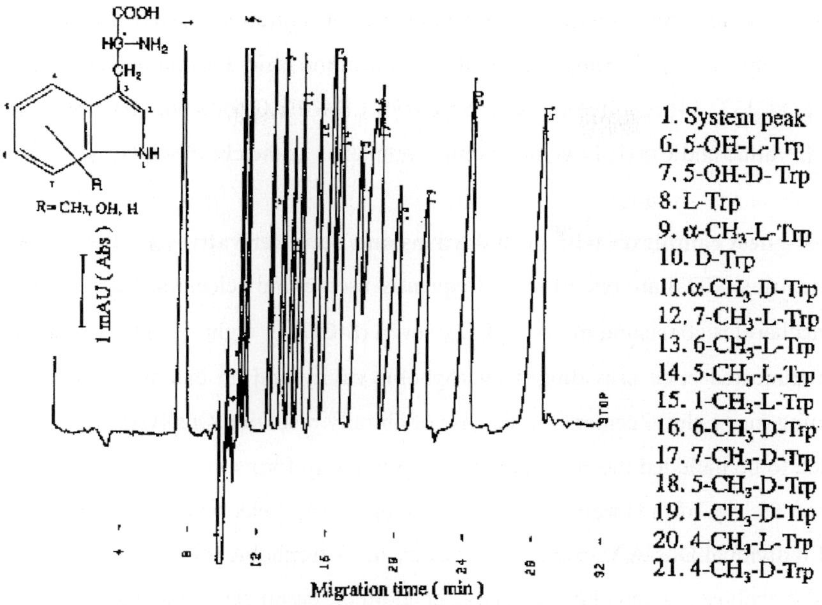

Figure 4 Simultaneous separation of 16 positional enantiomers of tryptophan derivatives. Electrolyte: 25 mM Cu(II), 50 mM L-hydroxyproline and 10 mM SDS at pH 5.2. Applied voltage: 10 kV. Fused silica capillary: 40 cm effective length, 55 cm total length, 50 μm i.d. and 375 μm o.d., UV detection: 208 nm. (From Reference [28] with permission).

The LE-MEKC mode shows a very powerful ability for both enantioselectivity between enantiomers and selectivity among the analytes. A typical application of LE-MEKC is the simultaneous separations of optical and positional isomers [27,28]. Figure 4 Shows the simultaneous of 16 positional enantiomers of tryptophan derivatives by using 25 mM Cu(II), 50 mM L-hydroxyproline and 10 mM SDS at pH 5.2. In LE-MEKC, micelles work as the pseudo-stationary phase and improve the separation selectivity especially of positional isomers, and Cu(II) complexes with L-hydroxyproline as the chiral selectors for chiral recognition. A baseline simultaneous separation can be achieved because the analytes interact concurrently with both the micellar phase of SDS and the chiral selector of the Cu(II)-L-hydroxyproline complex. The migration time of analytes in LE-MEKC is correlated with the hydrophobicity, log P_{ow} (the logarithm of the octanol-water partition coefficient). It is very interesting to examine the EMO in LE-MEKC. It was found that the EMOs of amino acid enantiomers in LE-MEKC were reversed not only by the introduction of the micellar phase of anionic surfactants [22], but also by the change in the chirality of the selector ligand and/or the *cis/trans* confirmation of the 4-hydroxy group in 4-hydroxyproline [29,30]. LE-MEKC was also used for estimation of the critical micelle concentration (CMC) and the stability constants of ternary Cu(II) complexes with mixed amino acid enantiomers [31,32]. Recently Zheng et al. [33, 34] reported the applications of LE-MEKC for resolving Dns-AAs by using Cu(II)-L-valine and Cu(II)-D-penicillamine complexes as the chiral selectors.

2.2 Host-guest complexes with cyclodextrins and their derivatives as chiral selector

Cyclodextrins (CDs) are one of most frequently used chiral selectors. Native CDs are cyclic oligosaccharides consisting of six (α-CD), seven (β-CD) or eight (γ-CD) glucopyranose units with a truncated cone providing a hydrophobic cavity and an external hydrophilic surface containing many chiral centers. In the case of chirally resolving Dns-AAs, the dansyl group is regarded to be included the cavity, and the amino group forms hydrogen bonds with hydroxyl groups at the rim of the toroid, as shown in Figure5 [35]. Selectivity or differential complexation of individual Dns-AA results from the size of hydrophobic group with respect to the ability of the analyte to penetrate the cavity. In addition, chiral selectivity results from hydrogen bonding at the entrance of the cavity with the chiral glucose moiety [35]. Further, CDs show excellent stability at wide pH range (2-12) and the UV absorbance is minimal. These unique characters make CDs the most widely used chiral selectors in chiral CE.

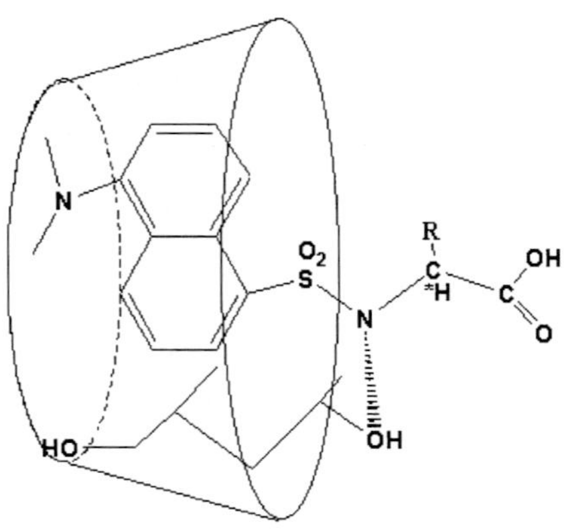

Figure 5 Schematic diagram of CD-Dns-AA complex structure

i. Cyclodextrins

The CD was first used as the chiral selector by Snopek et al. [36]. Guttman et al. reported the enantioseparation of 12 Dns-DL-AAs by using polyacrylamide gel filled capillary with incorporated β-CD as chiral selector [35]. CD type and structure have great influence on the enantioseparation, which also depends on the structure and nature of analytes. Tanaka et al. [37] studied the enantioseparation of Dns-DL-AAs using native α-, β-, and γ-CDs. Among seven pairs of Dns-DL-AAs, α-CD exhibited enantioselectivity only for Dns-DL-Asp, whereas all Dns-DL-AAs could be resolved by β-CD. γ-CD also showed a high enantioselectivity for six Dns-DL-AAs studied. Werner et al. [38] reported the enantioseparations of dansylated drugs and Dns-DL-AAs with α-, β-, and γ-CDs at pH values of 2.6 and 9.5. Figure 6 shows the separation of Dns-DL-AAs with 20 mM γ-CD as chiral selector added in pH 2.6, 0.1 M phosphate and pH 9.5, 0.1 M borate buffers. It was shown that 20 mM γ-CD is the best chiral selector for Dns-DL-AAs among α-, β-, and γ-CDs. The EMO shows L-enantiomers migrates faster than corresponding D-ones with exception of Dns-DL-His when an electrolyte containing 20 mM γ-CDs 0.1 M phosphate buffer at pH 2.6 was used. However, the D-enantiomers migrate faster than L-forms when an electrolyte containing 20 mM γ-CDs 0.1 M borate buffer at pH 9.5 was used. Yoshinaga et al. [39] investigated the effect of urea addition on chiral separation of Dns-DL-AAs by CE with α-, β-, and γ-CDs and di-, trimethylated CDs. The addition of urea increases the β-CD solution concentration above its water solubility. Moreover, the addition of urea and urea derivatives (methyl-, ethyl-, and 1,3-dimethylureas) dramatically enhanced the enantioselectivity of unmodified β-CD, but also reduced the enantioselectivities of the other CDs.

2.2.2. CD derivatives

Terabe [40] reported the enantioseparation of Dns-DL-AAs using the charged CD derivative, cationic 6-β-aminoethyl-amino-6-deoxy-β-CD as chiral selector by electrokinetic chromatography (EKC). The simultaneous separation of six racemic Dns-AAs was achieved. Tanaka and Terabe [41] reported the enantioseparation of Dns-DL-AAs using amphoteric β-CD (AM-β-CD) as chiral selector. Ten Dns-DL-AAs were resolved by AM-β-CD, furthermore, the simultaneous enantioseparation of 4 Dns-DL-AAs was also achieved with 10 mM AM-β-CD in 50 mM phosphate buffer, pH 6.0. Compared with neutral CDs, enantioseparations can be achieved at a relatively low concentration of chiral selector.

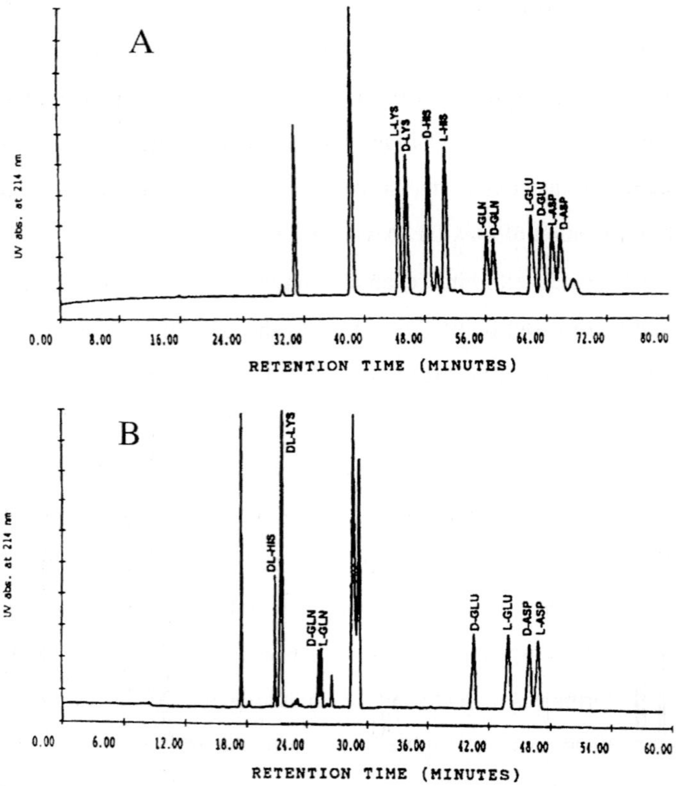

Figure 6 Separation of Dns-DL-AAs with buffers of (A) 20 mM γ-CD and 0.1 M phosphate, pH 2.6, and (B) 20 mM γ-CD and 0.1 M borate, pH 9.5 at 10 kV. Capillary 60 cm × 75 μm ID. (From Reference [38] with permission)

2.2.3. CD-MEKC

To improve the separation selectivity, CD-modified MEKC (CD-MEKC) was introduced by Terabe et al. [42]. Nishi et al. were the first to use CD-MEKC for enantioseparation [43]. Terabe et al. [44] reported the enantioseparation of 9 Dns-DL-AAs by using CD-MEKC. Either β- or γ-CD was used as chiral selector, and SDS as a micelle-forming agent for improving selectivity. All Dns-DL-AAs except Dns-DL-Ser were resolved at least partially with either β- or γ-CD. The addition of 20% methanol significantly improved the chiral resolution in the case of β-CD, but some of Dns-DL-AAs were only partially resolved. The enantioseparation was improved further by using a mixture of β- and γ-CDs. In addition to using CDs as chiral selector, MEKC with chiral surfactants as micellar phase offers high enantioselectivity. In this same study, chiral surfactant of sodium taurodeoxycholate (STDC) was incorporated with CDs for enantioseparation of Dns-DL-AAs by CD-MEKC; the result is shown in Figure 7B. Compared with the results obtained by MEKC using STDC in the absence of CD [45] shown in Figure 7A, the addition of γ-CD significantly improves the resolution and reduced the migration times.

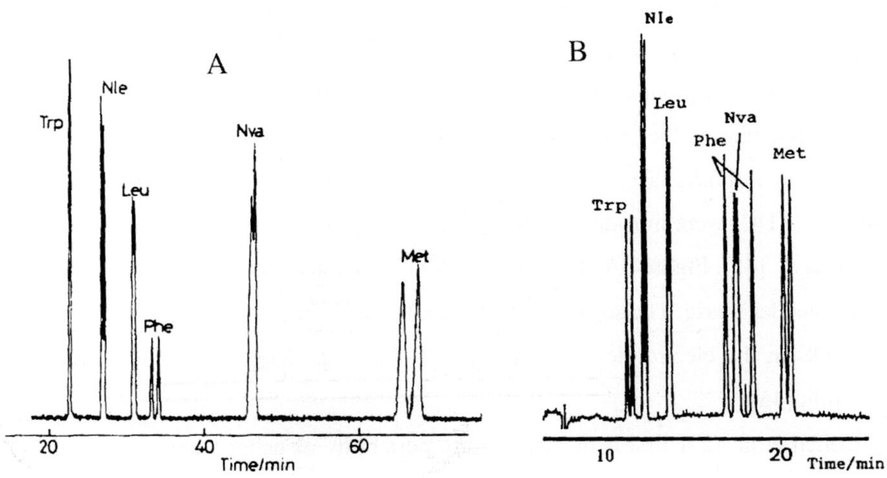

Figure 7 Electropherograms of Dns-DL-AAs in (a) MEKC with 50 mM STDC and (b) CD-MEKC with 10 mM γ-CD and 50 mM STDC in 50 mM phosphate buffer (pH 3.0). (From Reference [44, 45] with permission)

Ozaki et al. [46] reported the separation of Dns-DL-AAs by CD-MEKC using high molecular mass surfactants, butyl acrylate-butyl methacrylic acid copolymer sodium salt (BBMA) and butyl methacrylate-methacryloyloxyethyltrimethyl ammonium chloride co-polymer (BMAC). The former is anionic and the latter a cationic surfactant. These polymers do not form real micelles but show surface-active properties as a whole. Nine pairs from ten Dns-DL-AAs were resolved by CD-MEKC with 10 mM β-CD and 2 % BBMA. The separation factors of 8 pairs of these enantiomers were superior to those in the system containing a higher concentration of β-CD in combination with SDS. Therefore, these results show that BBMA is preferable to SDS for the enantioseparation of Dns-DL-AAs.

2.3. Chiral surfactants and linear polysaccharides as chiral selector

Enantioseparation with chiral surfactants is based on the interactions of a chiral analyte with the chiral micellar phase. There are three possible solute-micellar interactions [47] (i) the solute is adsorbed on the surface of the micelle by electrostatic or dipole interactions; (ii) the solute behaves as a co-surfactant by participating in the formation of the micelle; and (iii) the solute is incorporated into the core of the micelle.

Bile salts are well known natural surfactants with some unique properties. They consist of a relatively flat-shaped steroid portion, in which the A ring is *cis* with respect to the B ring, and a side-chain having a carboxyl group, which may be conjugated by taurine or glycine. All the hydroxyl groups bonded to the steroid portion are oriented in the same direction nearly perpendicular to the steroidal frame, and consequently the bile salts posses a hydrophobic and a hydrophilic face. Bile salts like sodium taurocholate (STC) and sodium taurodeoxycholate (STDC) were introduced as chiral micellar phases for enantioseparation in MEKC by Terabe et al. [45]. Figure 7A shows the electropherogram of Dns-DL-AAs by MEKC with STDC as micellar phase. The separation time is relatively long. However, with the addition of 10 mM SDS in the electrolyte containing STDC, the separation time was shortened to be about 30 min [45].

Cohen et al. [23] first demonstrated the possibility of using a chiral synthetic surfactant, N,N-didecyl-L-alaninate (DDAla), in combination with an achiral synthetic micelle-forming agent, SDS, for enantioseparation of Dns-AAs. The enantioseparation is based on the LE-MEKC as discussed in Section 2.1.2.

Ticke et al. [48] reported that glucopyranoside-based surfactants like dodecyl-β-D-glucopyranoside-4,6-hydrogenphospate (DGHP) and dodecyl-β- D-glucopyranoside-6-hydrogensulfate (DGHS) anionic surfactants as pseudo-stationary phases for chiral separations by MEKC. Figure8 shows the structure of DGHP and enantioseparation of Dns-AAs by using DGHP chiral surfactant. These surfactants have advantages like low CMC, high solubility in aqueous solution, and low absorbance at wavelengths above 200 nm. Besides, unlike natural surfactants, in particular CDs and bile salts, which are available in one stereoisomeric form, these surfactants can be synthesized in both D- or L- forms. This is necessary for one wants to reverse the EMO. Recently, Tran and Kang reported chiral separation of Dns-DL-AAs with octyl-β-thioglucopyranoside (OTG) as chiral selector [49].

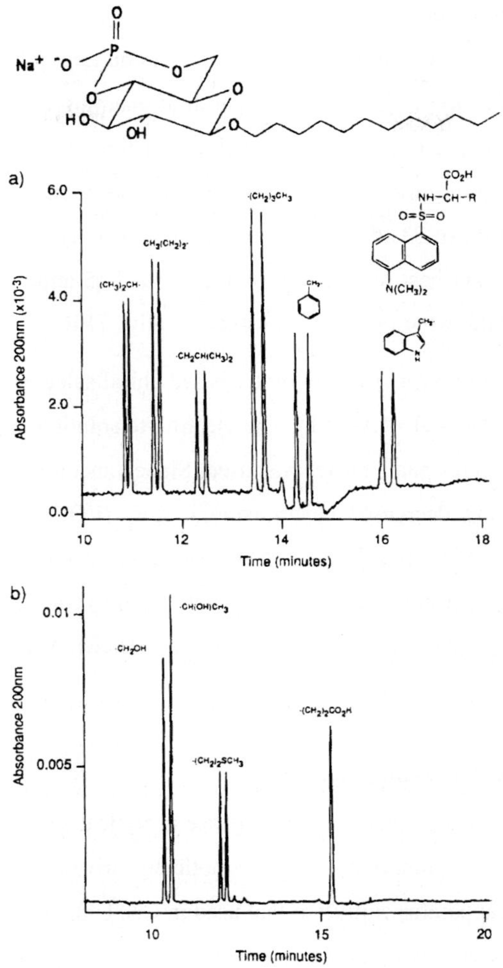

Figure 8 Structure of the chiral surfactant sodium n-dodecyl-β-D-glucopyranoside -4,6-hydrogenphosphate and electropherograms of racemic mixtures of Dns-DL-AAs. Electrophoretic conditions: running electrolyte, pH 8.0 phosphate-borate containing 45 mM chiral surfactant; capillary 50 cm/57 cm × 50 μm ID; applied voltage, 20 kV. (From Reference [48] with permission)

Enantioseparation of amino acids can be accomplished by judiciously adjusting the pH of the electrolyte and the concentration of OTG. Better separation can be achieved when OTG is used with combination of SDS and CD. Interestingly, not only can this OTG-SDS-CD system provides better separation than OTG alone, but also it can separate compounds which cannot be separated when only CD or CD-SDS are used.

2.3. Other chiral selectors

Honzatko et al. reported enantioseparation of Dns-AAs on terguride (TER)-based chiral selectors, allyl- and aminoalkyl-tergurides [50]. The best enantioselectivity was achieved for hydroxyl group-containing amino acids like serine and threonine. Reversal of the EMO was achieved using allyl-terguride with the opposite stereochemical configuration. N(1)-aminoalkyl-substitution of TER increased both the selectivity and the resolution of the discriminative process, and best results were obtained on 1-(5-aminopentyl)-TER, with the exception of Dns-DL-Glu, which was best resolved on allyl-TER.

Armstrong's group reported antibiotic-based chiral selectors, such as vancomycin, for enantioseparation of Dns-DL-AAs [51]. The electrolyte solution containing 2 mM vancomycin in 50 mM pH 7.0 phosphate buffer was used. Since vancomycin tends to interact with the capillary wall, causing decreased electroosmotic flow (EOF) velocity, increased analysis times and less than optimum efficiency, SDS surfactant at the concentration up to 50 mM were added to the background electrolyte so that vancomycin binds strongly to the micelle and acts in conjunction with it, which benefits on the efficiency, analysis times, and the resolution of enantiomers.

2. Capillary electrochromatography

Capillary electrochromatography is a liquid phase analytical separation technique that is carried out with capillary columns by electroosmotically driven mobile phase at high electric field strengths in an apparatus similar to that used in CE. Unlike CE, which only uses an open capillary filled with electrolytes as separation media, CEC uses capillary column with stationary phases as separation media. The separation mechanism of CEC is based on the combination of chromatographic interactions between the stationary phase and the analytes like in LC and the difference in electrophoretic mobility of analytes like in CE [52].

Three types of columns namely open tubular, packed and monolithic columns are used in CEC. In the open tubular column, the stationary phase is bonded or coated on the inner surface of a capillary column. In the packed columns, the particles bonded with chromatographic stationary phases are packed inside the capillary. Frits at the ends of capillary are used for preventing the movement of particles; sol-gel silicate or organic polymer is also used to embed the particles so that they do not move during the separation. The monolithic columns, also known as continuous bed, include polymer-based and silica-based monoliths. Polymeric monolith columns are prepared by *in situ* polymerization of organic monomers in the presence of a porogen reagent using various initiator techniques such as free radical polymerization or photopolymerization. Silica monoliths are prepared by sol-gel process including hydrolysis and polycondensation of precursors in a defined solvent *in-situ* to produce a hydrogel, which then converted to a xerogel upon drying. In the monolithic columns, the stationary phases are bonded on the monolithic matrix by *in situ* co-polymerization or post-modifications of bare monolith. Both packed and monolithic column have been reported for the resolution of Dns-DL-AAs by CEC. However, in recent years monolithic columns have received much increasing attraction, because monolithic columns offer many advantages over conventional open tubular and packed columns, including in-capillary preparation, fritless design, high sample capacity and the absence of bubble during CEC operation. The advance and typical application examples of using these types of columns for the resolution of Dns-AAs will be discussed in the following.

a. Open tubular CEC

Liu et al. [53] reported open tubular CEC (OT-CEC) with adsorbed avidin as a chiral stationary phase (CSP) for chiral separations including several Dns-DL-AAs. Avidin is a basic protein with an isoelectric point of 10-10.5, which has been used as CSP in LC and chiral selector in CE. OT-CEC is only suitable for enantiomers having strong interactions with the stationary phase due to low phase ratio. Later, the same group developed an avidin-physically absorbed monolithic column. As phase ratio is increased, the monolithic column showed better separation capacity as compared with the open tubular column [54].

b. Packed CEC

Li and Lloyd studied [55] the CEC separation of Dns-DL-Thr and DNP-amino acids by using packed β-cyclodextrin stationary phase with a 5μm particle diameter inside 50 μm i.d. capillary, and a mobile phase of methanol-triethylammonium acetate (TEAA). Since TEAA background electrolyte works in some ways similar to cationic surfactant additives such as cetyltrimethylammonium salts, EOF is reversed and flows towards anode due to adsorption of cations onto the silica surface. The efficiency of β-cyclodextrin stationary phase column can be substantially increased through the use of TEAA buffer.

Schurig's group [56] reported enantioseparation of Dns-DL-AAs by nonaqueous and aqueous CEC with capillary packed with β-, γ-CD modified silica (Chiral Dex-beta and ChiralDex-gamma). A 10 mM 2-(N-morpholino)ethanesulfonic acid (MES) in methanol at pH 6.0 was used as nonaqueous mobile phase. Negatively charged Dns-DL-AAs were resolved into their enantiomers by nonaqueous CEC. The addition of a small amount of water to the nonaqueous mobile phase improved the enantioselectivity but increased the retention time. When MES and TEAA used as the background electrolyte, a reversed anodic EOF was observed. The nature of mobile phase and the concentration of background electrolyte strongly influenced the separation behavior.

Chirica and Remcho [57] reported a silicate-entrapped molecular imprinted polymeric (MIP) phase for resolving Dns-DL-Phe by CEC. Lin and Hobo et al. reported the enantioseparation of Dns-DL-Phe by entrapping the MIP particle in acrylamide [58].

Zhang and El Rassi [59] reported the CEC separation of Dns-DL-AAs and organochlorine pesticides on a packed diol-silica dynamically coated with hydroxypropyl-β-cyclodextrin (HP-β-CD). The commercially available achiral diol-silica stationary phase was converted *in situ* to a chiral stationary phase (CSP) by dynamically coating with HP-β-CD. The neutral HP-β-CD in the presence in mobile phase acts simultaneously as a chiral selector and lowers the actual charge-to-mass ratio of the anionic solutes, which speeds up their transport by the cathodic EOF across the packed column. The mobile phase was composed of 5 mM ammonium acetate and 3 mM HP-β-CD in methanol/aqueous ammonium acetate (8:2, v/v), pH 5.5. The same group [60] reported a packed CSP having surface-bound HP-β-CD for CEC separation of Dns-DL-AAs and phenoxy acid herbicides. The HP-β-CD-sulfonated silica (CDSS) CSP was prepared by coating Nucleosil silica with a primary layer of 3-glycidoxypropyldimethylethoxysilane (sublayer) and sandwiching a sulfonated layer through covalent bonding between the hydrophilic sublayer and the HP-β-CD top layer. The sulfonated sublayer generates a relatively strong EOF. The CDSS-CSP provided about 20 % stronger EOF than that generated by the diol-silica stationary phase dynamically coated with HP-β-CD. The mobile phase was composed of acetonitrile and aqueous phosphate buffer containing 1.6 mM sodium phosphate (2:8, v/v), pH 5.5. Figure9 shows the structure of CDSS-CSP and the electrochromatogram of Dns-AAs on the CDSS-CSP.

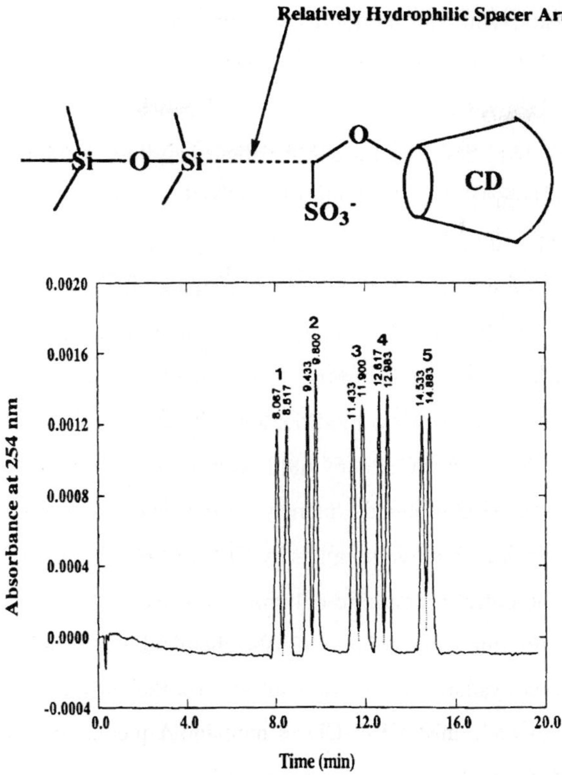

Figure 9 Structure of CDSS-CSP and electrochromatogram of Dns-AAs. Capillary column, 20.5 cm/27 cm ×100 μm ID, whole packed with 5 μm CDSS stationary phase; mobile phase, hydro-organic eluent containing 1.6 mM sodium phosphate and composed of 20 % v/v acetonitrile and 80 % v/v aqueous phosphate buffer, pH 5.5; running applied, 15 kV; injection, electrokinetic 1 kV for 1 s. Solutes: 1, Dns-Leu; 2, Dns-Trp; 3, Dns-Phe, 4, Dns-Met; 5, Dns-Glu. (From Reference [60] with permission)

3.4. Monolithic CEC

Bedair and El Rassi reported [4] a preparation of sulfonated stearyl acrylate monolith and their CEC behavior of Dns-AAs. The monolithic stationary phase was prepared by the *in situ* copolymerization of pentaerythritol diacrylate monostearate (PEDAS) and 2-acrylamido-2-methyl-1-propanesulfonic acid (AMPS) in a ternary porogenic solvent consisting of cyclohexanol/ethylene-glycol/water. AMPS was used to generate the EOF, and PEDAS was introduced to provide the nonpolar sites for chromatographic retention. The monolithic stationary phase exhibited reversed-phase chromatographic behavior toward neutral solutes and generated a relatively strong EOF. For charged solutes like Dns-AAs, nonpolar as well as electrostatic interaction/repulsion with the monoliths was observed in addition to electrophoretic migration. Since achiral selectors are involved in the stationary phase, it cannot be used for resolving enantiomeric Dns-AAs.

Koide and Ueno [61] developed a chiral fritless capillary column filled with a positively charged polyacrylamide gel, to which allyl carbamoylated β-CD (AC-β-CD) derivatives covalently bind. This column was used for the resolution of 12 single Dns-DL-AAs and the simultaneous enantioseparation of 4 Dns-DL-AAs (Dns-DL-Ser, Dns-DL-Val, Dns-DL-Trp, Dns-DL-Leu) by using 200 mM Tris-300 mM Boric acid buffer (pH8.1) as mobile phase.

Author and coworkers succeeded in the development of chiral monolithic columns by using the sol-gel process and chemical modifications. Two categories of chiral CSPs, LE-CSPs [17, 62-66] and CD-modified CSPs [67,68] have been successfully developed to resolve a variety of important chiral compounds, namely amino acids, hydroxyl acids, dipeptides, positional isomers and natural chiral compounds by CEC [62-63, 65-68] and μ-LC [64].

We fabricated the chiral monolithic columns modified with LE-CSPs in 3 steps: (1) the preparation of a monolithic silica column, (2) the chemical modification of a spacer of (3-glycidoxypropyl)-trimethoxysilane and a chiral selector on the surface of the monolithic matrix, and (3) the loading of LE-CSP with Cu(II) ions. L-PheA [62], L-ProA [63], L-Ala [62], L-lysinamide (L-LysA) [66] and L-hydroxyproline [65] has been employed as the CSPs. The chemical structures of the LE-CSPs are shown in Figure10. The scanning electron micrographs (SEM) in Figure11 show that the monolithic bed was well bonded with the capillary inner wall and had morphology with a continuous skeleton and micrometer-scaled through-pore. These LE-CSPs with L-amino acylamides generated a reversed EOF with the direction from the cathode to the anode under an external electric field. The EOF was strong enough to perform the separation by CEC, and dependent on the pH and the compositions of the mobile phases. CSPs modified with L-amino acylamides generated a relatively stronger EOF, because positively charged Cu(II) complexes with L-amino acylamides were formed on the surface of the monolithic columns. The EOF on L-lysA CSP was about twice as strong as that on L-PheA CSP, possibly because there are 3 amino groups in a molecule of lysinamide [66]. However, the EOF generated on L-hydroxyproline CSP was too weak to be determined by acetone as a neutral marker [65], probably because the Cu(II) complexes with L-hydroxyproline were neutral.

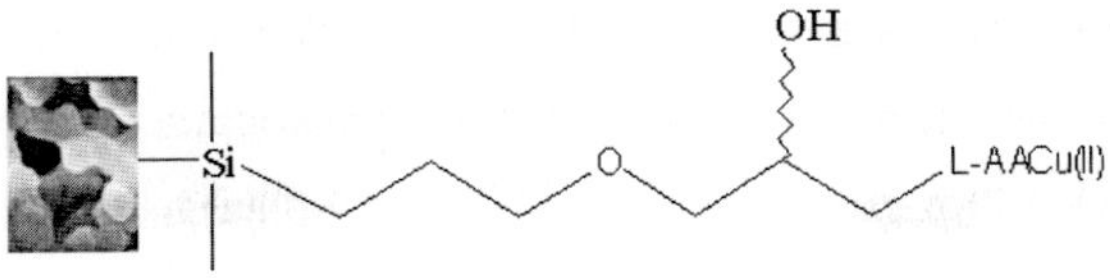

AA: amino acylamide or amino acid

LE-CSP1: AA=L-phenylalaninamide
LE-CSP2: AA=L-alaninamide
LE-CSP3: AA=L-prolinamide
LE-CSP4: AA =L-lysinamide
LE-CSP5: AA=L-hydroxyproline

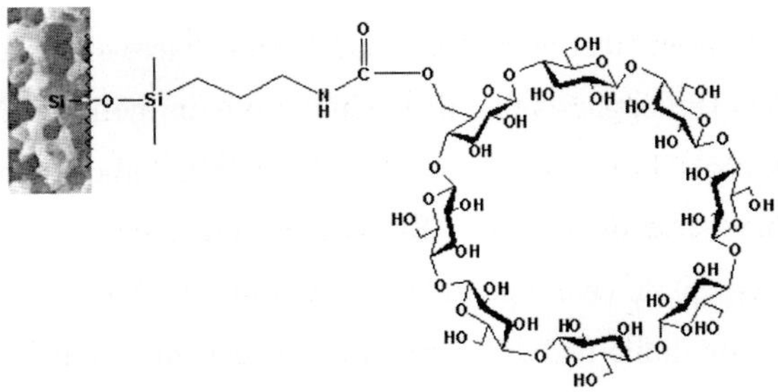

Figure 10 Chemical structures of monolithic LE-CSPs (up) and γ-CD-CSP (down).

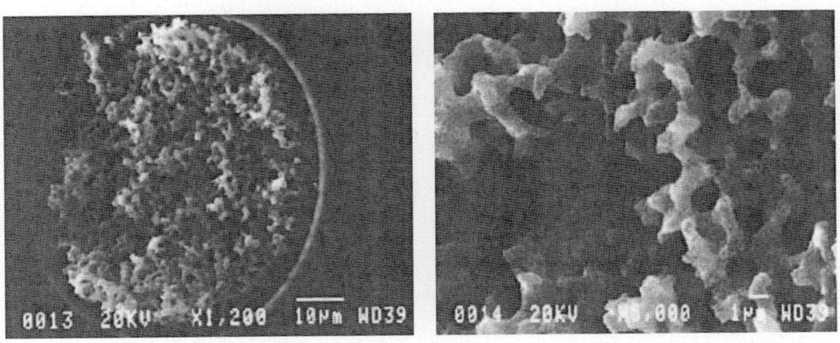

A B

Figure 11 SEM photographs of sol-gel monolithic column. 20 kV, WD 39, and magnification (A) 1200×, (B) 5000×. (From Reference [67] with permission)

Chiral separation has been achieved for a variety of compounds on these monolithic LE-CSPs by CEC and μ-LC with different enantioselectivities. An L-PheA-modified monolithic column was used for resolving 12 dansyl amino acid enantiomers by CEC [62]. Figure12 shows the electrochromatograms obtained when resolving a mixture three pairs of Dns-

DL-AAs. The maximum theoretical plate number up to 90,000 plates/m was obtained for Dns-D-Leu. Figure 13 shows the representative electrochromatogram of Dns-DL-AAs resolved on monolithic L-PheA CSP. It can be seen that L-PheA CSP has excellent enantioselectivity for resolving Dns-DL-AAs. Both the Dns-DL-AAs and several hydroxy acid enantiomers were resolved on an L-ProA-modified monolithic column by CEC and μ-LC [64]. Figure 14 shows the representative electrochromatograms of Dns-DL-AAs and hydroxy acid enantiomers. It was shown that L-ProA CSP exhibited high enantioselectivity, especially for resolving analytes with large substitutent groups such as a phenyl and indole rings at the chiral α-carbon of Dns-AAs and hydroxyl acids [17,63]. When we compared the enantioselectivities of L-AlaA and L-PheA CSPs, we found that higher separation factors were obtained on L-PheA CSP than on L-AlaA CSP. This fact suggests that the phenyl group in the L-PheA played an important role in chiral recognition [17,62], as found with CE [20,21]. Although the EOF was extremely weak on monolithic L-hydroxyproline CSP, the enantioseparation for Dns-DL-AAs, DL-hydroxy acids and dipeptide enantiomers was achieved by CEC [65], since the analytes migrated through the monolithic bed as a result of their electrophoretic mobility. The retention times on monolithic L-hydroxyproline CSP were longer than on monolithic L-ProA CSP, because the EOF on L-hydroxyproline CSP was weaker than that on L-ProA CSP.

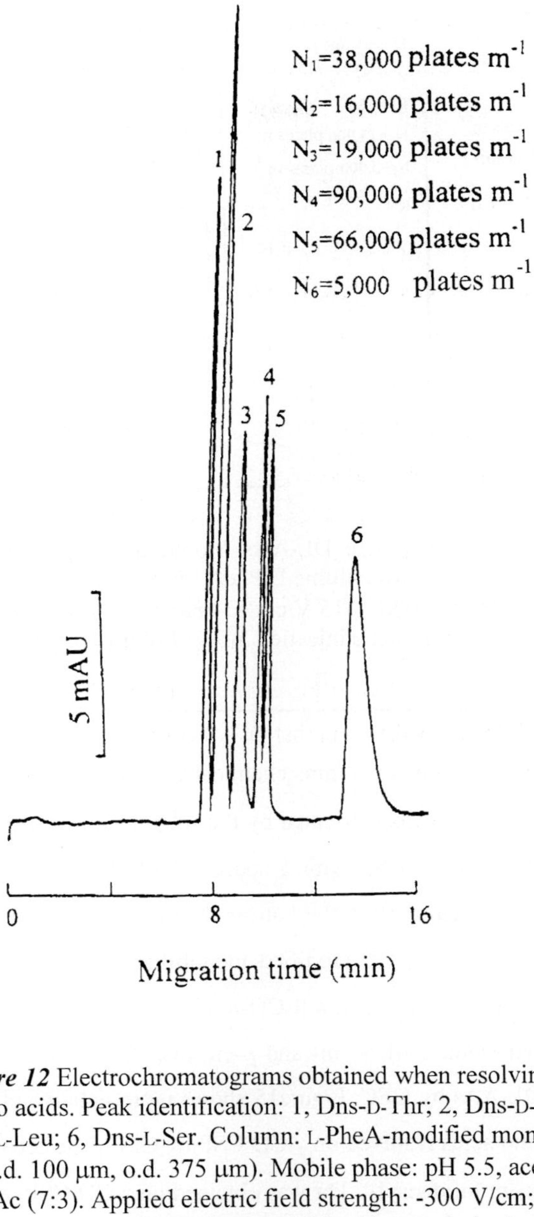

Figure 12 Electrochromatograms obtained when resolving a mixture of three pairs of dansyl amino acids. Peak identification: 1, Dns-D-Thr; 2, Dns-D-Ser; 3, Dns-L-Thr; 4, Dns-D-Leu; 5, Dns-L-Leu; 6, Dns-L-Ser. Column: L-PheA-modified monolithic column (TL 35 cm, EL 26.5 cm; i.d. 100 μm, o.d. 375 μm). Mobile phase: pH 5.5, acetonitrile/0.50 mM $Cu(Ac)_2$-50 mM NH_4Ac (7:3). Applied electric field strength: -300 V/cm; UV detection 254 nm; electrokinetic injection 3-5 s. (From Reference [62] with permission)

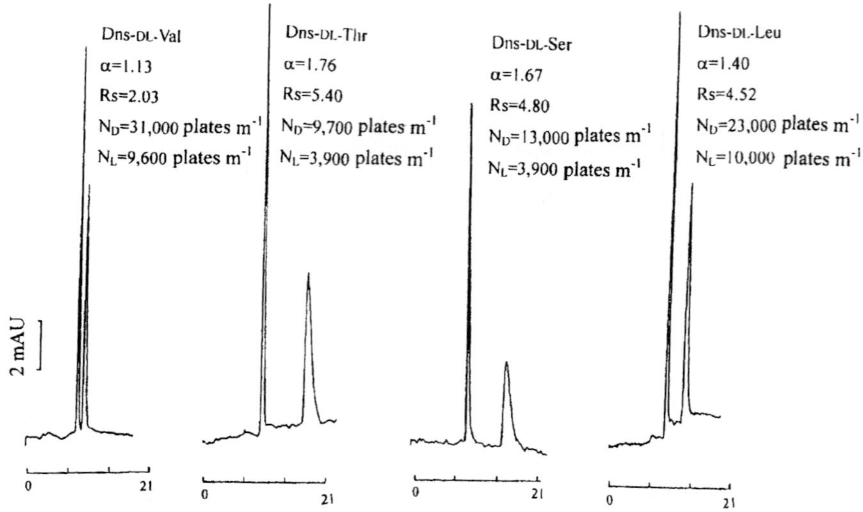

Figure 13 Electrochromatograms of representative Dns-DL-AAs. Mobile phase: pH 7.6, acetonitrile/0.1 M NH4Ac-0.25 mM Cu(Ac)2 (7:3). Monolithic L-PheA CSP: EL 28.5 cm, TL 38 cm, i.d. 100 μm, o.d. 375 μm; Applied electric field: -315 V/cm. D-enantiomers were eluted as the first peaks. UV detection 254 nm; electrokinetic injection 3-5 s. (From Reference [62] with permission)

Recently we developed monolithic silica columns chemically modified with β- and γ-CDs [67, 68]. The monolithic silica column was fabricated by the sol-gel process, and then chemically immobilized using β- or γ-CDs as CSP with a spacer of (3-glycidoxypropyl)-trimethoxysilane by in-column reactions. Figure 10 at the bottom shows the chemical structure of γ-CD-modified monolithic column. A γ-CD-modified monolithic column has been successfully employed for resolving Dns-DL-AAs, and a β-CD-modified monolithic column has been used for resolving positional isomers of *o-*, *m-* and *p*-cresols, the racemic enantiomers of benzoin, and several Dns-DL-AAs by CEC. Figure15 shows representative electrochromatograms of several dansyl amino acids resolved on γ-CD CSP by CEC. The separation efficiency of 5.0×10^4 theoretical plates/m for dansyl-L-Thr was obtained on γ-CD CSP at electric field strength of -300 V/cm in the mobile phase of 50 mM MES-Tris/methanol (70/30) buffer at pH 7.0. L-Enantiomers were eluted as the first peak.

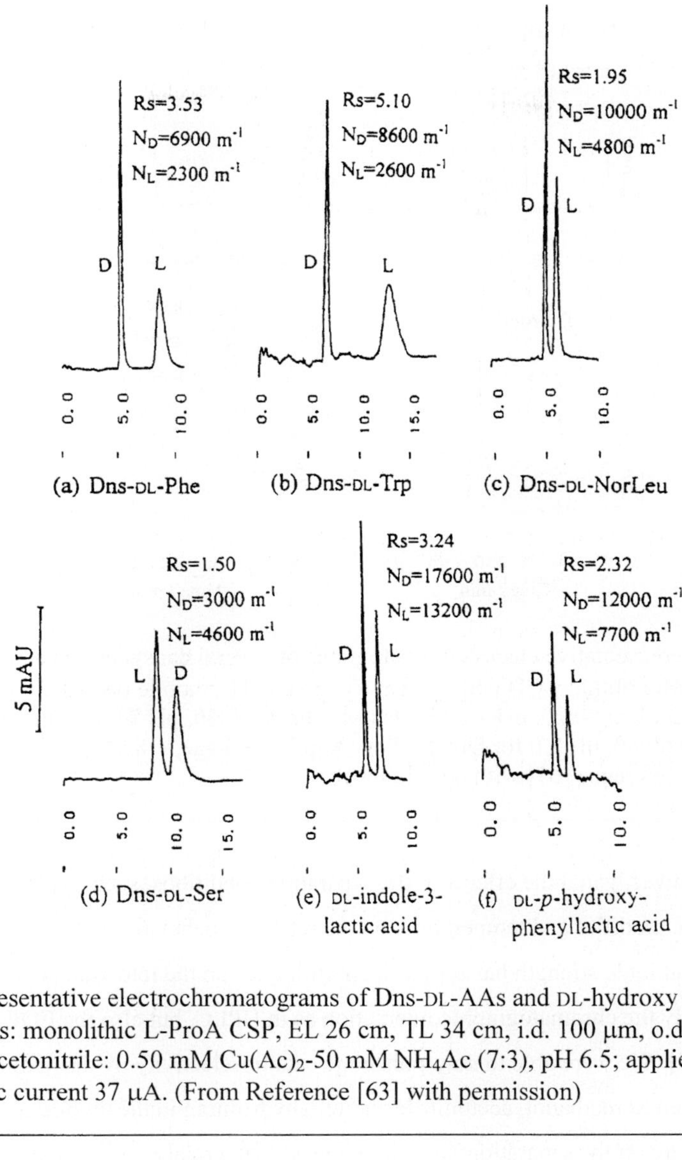

Figure 14 Representative electrochromatograms of Dns-DL-AAs and DL-hydroxy acids. Separation conditions: monolithic L-ProA CSP, EL 26 cm, TL 34 cm, i.d. 100 μm, o.d. 375 μm; mobile phase: acetonitrile: 0.50 mM $Cu(Ac)_2$-50 mM NH_4Ac (7:3), pH 6.5; applied voltage: -13.6 kV, electric current 37 μA. (From Reference [63] with permission)

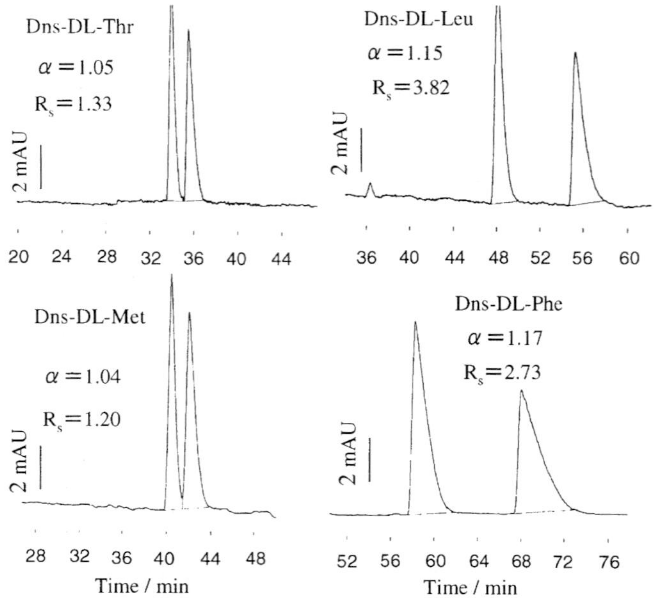

Figure15 Representative electrochromatograms of several dansyl amino acids on monolithic γ-CD CSP. Monolithic γ-CD CSP, 45 cm EL, 55 cm TL; mobile phases, 50 mM MES-Tris buffer/methanol, at 80:20, pH 7.0 for Dns-DL-Thr, at 70:30, pH 7.0 for Dns-DL-Leu and Dns-DL-Met, and 60:40, pH 8.0 for Dns-DL-Phe. Applied voltage, -18 kV, but -15 kV for Dns-DL-Phe. (From Reference [67] with permission)

We investigated the effects of the separation conditions, such as pH, buffer composition, organic modifier and temperature, on the retention behavior of Dns-DL-AAs in CEC. It is known that ionic strength has a significant influence on the retention behavior, since it affects not only the chromatographic interaction as in HPLC, but also the EOF and mobility of analytes [68]. We were able to manipulate the enantioselectivity by the addition of organic solvents, such as methanol, acetonitrile and tetrahydrofuran in the mobile phases. The effect of temperature on the separation behavior showed that a relatively low temperature under enthalpic control benefited the enantioselectivity. Furthermore, we investigated the correlation between the separation factors and the hydrophobicity expressed by log $P_{o/w}$. This investigation indicated that the hydrophobic interaction between CD and the analytes contributed to the retention on the CD-CSP. In addition, the stereoselectivity of the analyte molecules in the CD cavity was considered to be another main factor as regards the chiral recognition on CD-CSPs in CEC [68].

4.Experimental

4.1 Instrumentation

Many commercial instruments for capillary electrophoresis can be easily used for both CE and CEC operations. In our group, the CE and CEC were carried out on home-built setup including an HCZE-30PNO25-LD high voltage power supply (Matsusada Precision Devices, Tokyo, Japan), a CE-1570 intelligent UV-Vis detector (Jasco, Tokyo, Japan) and a C-R7A plus Chromatopac integrator (Shimadzu, Kyoto, Japan). CE-1570 intelligent UV-Vis detector provides on-capillary detection. For UV-Vis detection of dansyl amino acids, the detection wavelength were generally set at 254 nm.

4.2 Capillary electrophoresis and capillary electrochromatography

CE was carried out by using 50 μm i.d. capillary. Before use, the capillary was washed by using 1 M sodium hydroxide, water, and running electrolytes. The capillary was filled with running electrolytes containing chiral selectors like Cu(II) complexes with amino acid enantiomer, and cyclodextrins. CEC was carried out by using capillary columns in the formats of open-tubular, packed or monolithic stationary phases. Both electrokinetic and hydrodynamic injections can be used as the injection methods. The running electrolytes and the mobile phase were filtered through a 0.45 μm membrane and degassed by vacuum and ultrosonication.

References

[1] M. Hashimoto, K. Tsukagoshi, R. Nakajima, K. Kondo, A. Arai, J. Chromatogr. A 867 (2000) 271-279.

[2] K. Tsukagoshi, Y. Obata, R. Nakajima, J. Chromatogr. A, 971(2002)255-260.

[3] M. Stastna, K Slais, J. Chromatogr. A, 832 (1999) 265-271.

[4] M. Bedair, Z. El Rassi, Electrophoresis 23 (2002) 2938-2948.

[5] B. Chankvetadze, Capillary Electrophoresis in Chiral Analysis, John Wiley & Sons, 1997.

[6] F. Helfferich, Nature 189(1961) 1001-1002.

[7] V. A. Davankov, S. V. Rogoxhin, J. Chromatogr. 60 (1970) 280-283.

[8] G. Gübitz, W. Jellenz, G. Laffler, W. Santi, J. High Resol. Chromatogr. Commun 2 (1979)145-146.

[9] G. Gübitz, W. Jellenz, W. Santi, J. Chromatogr. 203 (1981) 377-384.

[10] V. A., Davankov, J. Chromatogr. A 666 (1994) 55-76.

[11] V. A. Davankov, Enantiomer 5 (2000) 209-223.

[12] M. G. Schimid, N. Grobuschek, O. Lecnik, G. Gübitz, J. Biochem. Biophys. Methods 48(2001) 143-154.

[13] V. A. Davankov, J. Chromatogr. A 1000 (2003) 891-915.

[14] Z. Chen, T. Hobo, Recent Res. Devel. Chem. Pharm. Sci. 2 (2002) 15-30, Transworld Research Network, Trivandrum, India.

[15] Z. Chen, Chromatography 25 (2004) 9-23.

[16] E. Gassmann, J. E. Kuo, R. N. Zare, Science 230 (1985) 813-814.

[17] Z. Chen, M. Niitsuma, K. Uchiyama, T. Hobo, J. Chromatogr. A 990 (2003) 75-82.

[18] P. Gozel, E. Gassmann, H. Michelsen, R. N. Zare, Anal. Chem. 59(1998) 44-49.

[19] Z. Yuan, L. Yang, S. S. Zhang, Electrophoresis 20 (1999) 1842-1845.

[20] Z. Chen, M. Niitsuma, T. Nakagama, K. Uchiyama, T. Hobo, J. Sep. Sci. 25 (2002) 1197-1201.

[21] Z. Chen, K. Uchiyama, T. Hobo, Anal. Chim. Acta 523 (2004) 1-7.

[22] M. G. Schmid, G. Gübitz, Enantiomer 1 (1996) 23-27.

[23] A. S. Cohen, A. Paulus, B. L. Karger, Chromatographia 24 (1987) 15-24.

[24] Z. Chen, Bunseki Kagaku (Ph.D Dissertation Digest) 49(2002)713-714.

[25] Z. Chen, Ligand Exchange-Micellar Electrokinetic Chromatography and Its Applications, Ph. D. Dissertation, Tokyo Metropolitan University, Tokyo, March, 2002.

[26] N. G. Sundin, T. M. Dowling, N. Grinberg, G. Bicker, J. Microcol. Sep. 8 (1996)323-329.

[27] Z. Chen, J. Lin, K. Uchiyama, T. Hobo, J. Chromatogr. A 813 (1998) 369-378.

[28] Z. Chen, J. Lin, K. Uchiyama, T. Hobo, Chromatographia, 49 (1999) 436-443.

[29] Z. Chen, J. Lin, K. Uchiyama, T. Hobo, J. Microcol. Sep. 11(1999) 534-540.

[30] Z. Chen, J. Lin, K. Uchiyama, T. Hobo, Anal. Sci. 16 (2000) 131-137.

[31] Z. Chen, J. Lin, K. Uchiyama, T. Hobo, Anal. Chim. Acta 403 (2000) 173-178.

[32] Z. Chen, K. Uchiyama, T. Hobo, Anal. Sci. 16 (2000) 837-841.

[33] Z. X. Zheng, J. M. Lin, F. Qu, J. Chromatogr. A, 1007 (2003) 189-196.

[34] Z. X. Zheng, J. M. Lin, F. Qu, T. Hobo, Electrophoresis 24 (2003) 4221-4226.

[35] A. Guttman, A. Paulus, A. S. Cohen, N. Grinberg, B. L. Karger, J. Chromatogr. 448 (1988) 41-53.

[36] J. Snopek, I. Jelinek, E. Smolkova-Keulemansova, J. Chromatogr. 438 (1988) 211-218.

[37] M. Tanaka, S. Asano, M. Yoshinagha, Y. Kawaguchi, T. Tetsumi, T. Shono, Fresenius J. Anal. Chem. 339 (1991) 63-64.

[38] A. Werner, T. Nassaueer, P. Kiechle, F. Erni, J. Chromatogr. A, 666 (1994) 375-379.

[39] M. Yoshinaga, M. Tanaka, J. Chromatogr. A 710 (1995) 331-337.

[40] S. Terabe, Trends Anal. Chem. 8 (1989) 129-134.

[41] Y. Tanaka, S. Terabe, J. Chromatogr. A 781 (1997) 151-160.

[42] S. Terabe, Y. Miyashita, O. Shibata, E.R. Barnhart, L. R. Alexander, D. G. Patterson, B.L. Karger, K. Hosoya, N. Tanak, J. Chromatogr. 516 (1990) 23-31.

[43] H. Nishi, T. Fukuyama, Y. Miyashita, J. Chromatogr. 553 (1991) 503-516.

[44] S. Terabe, Y. Miyashita, Y. Ishihama, O. Shibata, J. Chromatogr. 636 (1993) 47-55.

[45] S. Terabe, M. Shibata,Y. Miyashita, J. Chromatogr. 480 (1989) 403-411.

[46] H. Ozaki, A. Ichihara, S. Terabe, J. Chromatogr. A 709 (1995) 3-10.

[47] S. Terabe, J. Pharm. Biomed. Anal. 10 (1992) 705-715.

[48] D. C. Tickle, G. N. Okafo, P. Camilleri, R. F. D. Jones, A. J. Kirby, Anal. Chem. 66(1994) 4121-4126.

[49] C. D. Tran, J. Kang, J. Chromatogr. A 978 (2002) 221-230.

[50] A. Honzatko, J. Votruba, J. Olsovska, M. Flieger, F. Bachechi, P. Ferrantelli, M. Sinibaldi, L. Cvak, J. Sep. Sci. 26 (2003) 851-856.

[51] K. L. Rundlett, D. W. Armstrong, Anal. Chem. 67 (1995) 2088-2095.

[52] Z. Deyl and F. Svec (Eds),. Capillary Electrochromatography, Elsevier, Amsterdam 2001.

[53] Z. Liu, K. Otsuka, S. Terabe, J. Sep. Sci. 24 (2001) 17-26.

[54] Z. Liu, K. Otsuka, S. Terabe, M. motokawa, N. Tanaka, Electrophoresis 23 (2002) 2973-2981.

[55] S. Li, D. K. Lloyd, J. Chromatogr. A 666 (1994) 321-335.

[56] D. Wistuba, K. Cabrera, V. Schurig, Electrophoresis 22 (2001) 2600-2605.

[57] G. Chirica and V. T. Remacho, Electrophoresis 20 (1999) 50-56.

[58] J. Lin, K. Uchiyama, T. Hobo, Chromatographia 47 (1998) 625-629.

[59] M. Zhang, Z. El Rassi, Electrophoresis 21 (2000) 3126-3134.

[60] M. Zhang, Z. El Rassi, Electrophoresis 21 (2000) 3135-3140.

[61] T. Koide, K. Ueno, J. High Resol. Chromatogr. 23 (2000) 59-66.

[62] Z. Chen, T. Hobo, Anal. Chem. 73 (2001) 3348-3357.

[63] Z. Chen, T. Hobo, Electrophoresis 22 (2001) 3339-3346.

[64] Z. Chen, K. Uchiyama, T. Hobo, J. Chromatogr. A 942 (2002) 83-91.

[65] Z. Chen, T. Nishiyama, K. Uchiyama, T. Hobo, Anal. Chim. Acta 501 (2004) 17-23.

[66] T. Nishiyama, Z. Chen, T. Nakagama, K. Uchiyama, T. Hobo, Bunseki Kagaku 52 (2003) 1005-1010.

[67] Z. Chen, H. Ozawa, K. Uchiyama, T. Hobo Electrophoresis 24 (2003) 2550-2558.

[68] H. Ozawa, Z. Chen, T. Nakagama, K. Uchiyama, T. Hobo, Bunseki Kagaku 52 (2003) 1105-1112.

Part 2 Chromatography of Amines

2.1. Gas Chromatography

Ibolya Molnár-Perl (Editor)
Quantitation of Amino Acids and Amines by Chromatography
Journal of Chromatography Library, Vol. 70

2.1.1.

Gas Chromatographic Determination of Volatile Aliphatic and Selected Aromatic Amines, without Derivatization: Solid Phase Microextraction

Jacek Namieśnik, Bogdan Zygmunt

Contents

Summary

Amines can be found in many matrices, from environmental samples to industrial raw materials, wastewater and wastes. Due to detrimental effects and other reasons, they are monitored in many media at low levels. Determination of amines is a challenging task and chromatographic techniques, preceded by sample treatment are generally applied. For volatile amines gas chromatography is a method of choice. For separation and detection amines are often converted to derivatives though determination of free amines seems more attractive. Many packed and open tubular columns with specially treated stationary phases have been proposed for separation of free amines directly in air, organic solvents, aqueous samples and headspace of aqueous solutions. Solid phase microextraction has been finding increasingly wide use in sample preparation for determination of free volatile amines. In the procedures commercial SPME fibres and also new custom-made coatings are used. An example of determination of volatile aliphatic amines in air from different sites by means of SPME-GC-FID is described in detail and the results are presented.

1. Introduction

Amines are organic compounds, which can be derived from ammonia in the same sense that alcohols and ethers are derived from water. Lower aliphatic amines are gases at room temperature. while lower aromatic amines are liquids of relatively high vapour pressure. Aliphatic amines have odour similar to ammonia but more pungent and more "fishy" [1-2].

Low molecular mass amines and amino alcohols naturally occur in the environment [1–4]. Many amines were isolated from different plant and animal tissues [5-7]. Methylamine, and ethanolamine occur in distillates of biological plant materials. Amines can be formed in the process of decomposition of larger molecules, e.g. amino acids.

Aliphatic amines are formed in metabolism in animals and microorganisms, and they are present in urine and animal faeces [8-9]. They are formed in decomposition of meat, mainly fish and, after animal death, they are still produced but are no longer converted to other substances and fish, seafood, meat and meat products release unpleasant odour. In decay of animals so-called "cadaveric poisons" are formed: pentamethyldiamine, commonly named cadaverine and tetramethyldiamine called putrescine from the Latin word "*putresco*" which means "to decay" [2].

Short chain aliphatic amines are emitted to the atmosphere from waste incinerators, wastewater treatment plants, stock farming, cigarette smoking, exhaust gases. They are substrates and semi-products in manufacturing pharmacological agents, polymers, pesticides, rubber, dyes, glues, solvents, corrosion and polymerisation inhibitors, surfactants, emulsifying agents, complexion cream components, depilatories, etc.

Amines can penetrate living organisms through skin, by breathing and with food. Low molecular mass amines such as methylamine are quickly and easily assimilated by animals. Volatile amines have unpleasant odour and various health hazards [10-19].

Due to the fact that they are present in many common matrices (air, body fluids, beverages, mineral oils, food, algae and plant samples, wastewater, etc.), are malodorous and have harmful health effects the maximum allowable concentrations for volatile amines are rather low and they must be monitored even when present at low concentration. The determination of amines is a very important and difficult task. All the aspects of amines formation, occurrence and determination have been reviewed elsewhere [20].

2. Gas chromatographic determination of amines – general remarks

In the process of determination of amines, starting from sampling, through transport, storage, sample preparation, and analysis, the composition of samples can undergo some changes due to different physical, chemical and biological processes. This results in increased uncertainty of analysis, which can also be enhanced by possible interference of accompanying components when low concentrations are to be determined. Problems with validation of analytical procedures result from the lack of appropriate Reference materials, including standard gas mixtures, which are typical matrix-free Reference materials. The problems of trace organic analysis as well as validation of analytical procedures, and the use of Reference materials in quality assurance and quality control have been widely discussed in a number of review and monographic papers [21-24].

Gas chromatography (GC) seems a method of choice for determination of volatile amines due to its simplicity, high resolving power, high sensitivity, possible selectivity, when proper detection system is applied, short time of analysis and quite low cost. However, rarely amines can be detected upon direct injection of the real air sample into GC. Some relatively "clean" aqueous samples can be directly analyzed after pH adjustment to convert amines into non-ionic form. In majority of cases, however, real samples are too complex and concentrations are too low for direct GC analysis. Hence, samples must be subjected to more or less complicated preparation, generally based on selective isolation and enrichment of amines. Since amines are polar substances, can adsorb on and decompose in different parts of the gas chromatographic system; tailing analyte peaks, some ghost peaks and lowered detector sensitivity can be observed.

These problems can be partly solved by conversion of analyte amines to less polar and more stable derivatives. Determination of amines with the use of derivatisation is widely discussed in section 2.1.2 of this book.

Also the selective detection has been comprehensively described elsewhere and will not be dealt with in this chapter [25]. For detection with the electron capture detector (ECD), flame photometric detector (FPD) and chemiluminescence detector (CLD) amines are generally converted to derivatives. The universal flame ionization detector (FID), mass spectrometric detector (MS), which can be selective when operated in selective ion monitoring (SIM) mode, and especially selective nitrogen-phosphorous detector (NPD) are usually applied for detection of both free amines as well as their derivatives.

Analysis of free amines has such advantages that it is increasingly often applied. However, it needs a special care at the steps of sample preparation as well as GC separation (columns must be specially prepared to protect chromatographic bands from tailing).

3. GC columns for separation of volatile free amines

Non-derivatized volatile aliphatic amines are polar basic substances and deactivated basic stationary phases are recommended for separation. As molecular mass of the amine decreases the relative effect of amine group increases. The amine group imparts high dipole moment on amine molecules and has the capability to form hydrogen bonds, so it is responsible for strong interaction with silanol groups and siloxane bridges leading to strong sorption on column surface, silica gel supports and, to some extent, on polysiloxane stationary phases. Primary amines have the strongest sorption tendency, followed by secondary and then tertiary amines. So the most difficult to chromatograph are the first members of primary amine homologous series.

Corcia and Samperi [26] used graphitized carbon blacks (GCB) with suitable fraction of KOH (0.3, 0.2 or 0.8%) and polyethylene glycol (PEG-20M or PEG-1500) (1.3, 0.5 or 4%) as stationary phases. They obtained symmetrical quite well resolved peaks of 11 C_1-C_4 aliphatic amines in water solution.

Good separation of free C_1-C_4 amines directly in water solution was also achieved on Chromosorb 102 coated with 20% KOH and Chromosorb 102 treated with TMCS (trimethylchlorosilane) coated with 5% KOH [27].

Pennwalt 223, Triton X-100 and Dowax 9N9 stationary phases were often recommended for amine separation, though Pennwalt was considered the best [28]. All are polyoxyethers. Pennwalt 223 is a mixture of two components; one with a polyether chain terminated by two phenyl rings (hydrocarbon on each ring), and the second containing a methyl group instead of one of the phenyl rings.

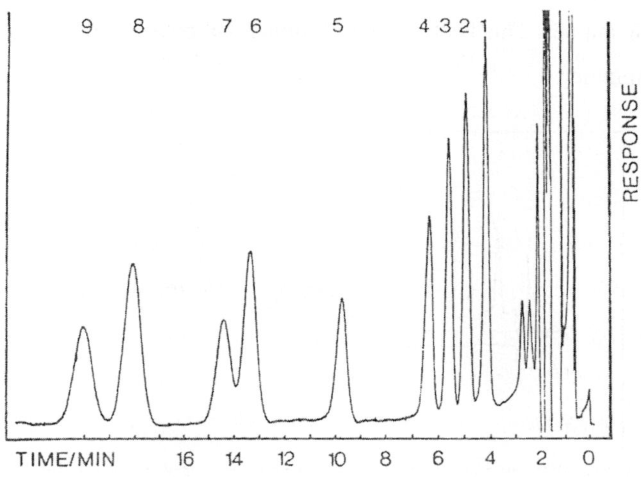

Figure 1 Chromatogram of a mixture of amines in hexane with 500 ppm ammonia added. Injection 2 μL. Solutes (each 0.5 ppm): 1=hexylamine; 2=tripropylamine; 3=cyclohexylamine; 4=heptylamine; 5=octylamine; 6=aniline; 7=tributylamine; 8=*N,N*-dimethylaniline; 9=*o*-toluidine. Column: 28% Pennwalt 223 with 4% KOH on Gas-Chrom R (80-100 mesh) (temperature 190°C) (Reproduced with permission from Reference [29])

Dalene et al. [29] used Pennwalt 223 to separate amines with chain length between one and fifteen. A prerequisite to successful separation was addition of ammonia to all solutions prior to analysis. The chromatogram of quite complex mixture of 8 amines in hexane with ammonia added is presented in Figure 1.

Mathiasson and Lovkvist [30] compared packings used in those days for amine separation with respect to detection limits, peak shape, separation efficiency, and possibility of using water as a solvent. The last factor was very important as many samples are originally aqueous or are produced as a result of absorption of amines from air samples. Chromosorb 103 was found not suitable for amines with more than 5-6 carbon atoms. Pennwalt had higher capacity and hence lower detection limits could be obtained. Long chain amines give symmetrical peaks, while methylamine and ethylamine showed some tailing. Pennwalt 223 could have been used with water as a solvent though some problems existed.

Also Audunsson and Mathiasson [31] showed applicability of 28% Pennwalt 223 with 4% KOH for trace analysis of amines in salt solutions. The chromatogram of selected amines is given in Figure 2. The above stationary phase proved effective in separation of different free amines in organic solvents and alkaline aqueous solutions with varying salt concentrations [32]. It was once again confirmed that addition of ammonia not only minimizes adsorption of amines in syringes and glassware but also reduces adsorption in the column, especially for volatile amines. This leads to more symmetrical peaks, better separation and repeatability of determination

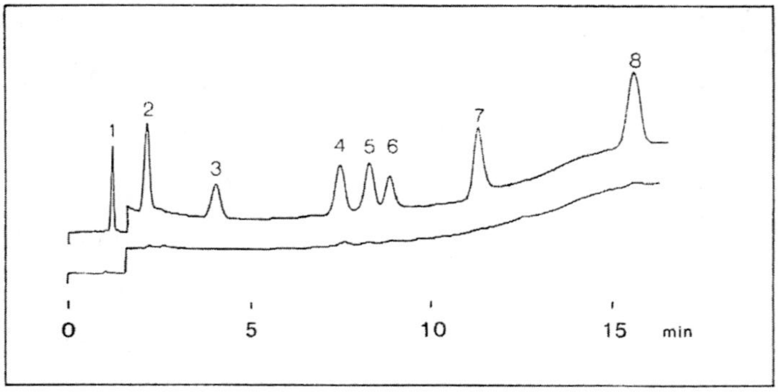

Figure 2 Chromatogram of a mixture of selected amines. Peaks (all amines at 0.3 ppm, except for piperazine and DABCO): 1=dimethylamine; 2=*n*-propylamine; 3=triethylamine; 4=piperidine; 5=methylmorpholine; 6=*n*-hexylamine; 7= piperazine; 8=DABCO. Sample – alkali-treated 0.5 M H_2SO_4; blank injection is shown below. Glass column (2 m x 3 mm I.D.): 28% Pennwalt 223 with 4% KOH on Gas-Chrom R (80-100 mesh). Oven temperature: 130°C (3 min) to 220°C at 10°C (Reproduced with permission from Reference [31]).

Audunsson [33] applied Pennwalt 223 to determine a number of amines in urine. Again an injector was impregnated with hydroxide to prevent adsorption of amines. Urine is a complex matrix and a liquid membrane technique was used for cleaning up samples.

For determination of aliphatic amines in air at the sub-ppt level Gronberg et al. [16] employed impinge sampling in dilute H_2SO_4 and selective enrichment across a PTFE membrane and quantification by GC. The following packings were used: Chromosorb 103, 28% Pennwalt 223 with 4% KOH on Gas-Chrom R, 28% Alltech 223 with 4% KOH on Gas-Chrom R, 10% Carbowax 20M with 2% KOH on Chromosorb W-AW, and 4% Carbowax 20M with 0.8% KOH on Carbopack B.

Carbopack B coated with 4% Carbowax 20M and 0.8% KOH was successfully used to separate ammonia and 23 amines including C_1-C_5 alkylamines [34]. However, after each sample run, 20 μL 15% ammonia solution must have been injected into the hot (150-200°C) column to reduce appearance of ghost peaks.

For separation of ammonia, monomethylamine (MMA), dimethylamine (DMA) and trimethylamine (TMA) a 3 m x 3 mm glass column packed with 5% squalane + 2% KOH either on Chromosorb 104 or on Chromosorb 103 was successfully applied [35].

Using GHP-1 (Mitsubishi Chemical) - spherically shaped styrene-divinylbenzene copolymer) - coated with KOH, Kuwata et al. [36] obtained good separation of all C_1-C_4 alkyl amines without water interference. Amines were isolated and enriched by passing air through handy cartridges packed with Sep-PAK C_{18} treated with phosphoric acid and then eluted with the methanol-water (1:1) mixture. Injecting 2-4 μL aliquots of the alkalized eluate to GC-NPD, the authors achieved the detection limits of 0.1 ppb for air samples.

Using a 3 m x 2 mm glass column packed with Gaschropak 54 + 5% KOH (Gaschro Industry Co., Tokyo), Fujii and Kitai [37] obtained almost symmetrical peaks of volatile amines by direct injection of air samples of a volume of 1 mL. Before measurements two 10 μL portions of 10% KOH in water were injected to coat an injector liner so as to reduce amines adsorption.

With the advent of capillary columns characterized by much higher efficiencies than packed columns, analytical chemists started to use the former for separation of amines in complex mixtures. At the beginning separation was generally preceded by derivatisation. Attempts were made to separate free amines, as well.

Audunsson et al. [32], used a 20 m x 0.32 mm Duran 50 glass capillary column coated with 0.4 μm film of OV-73 stationary phase to separate some aromatic and tertiary amines which are of interest in rubber industry.

Tsukioka et al. [38] separated lower aliphatic tertiary amines (TMA - trimethylamine, TEA - triethylamine, TAA – triallylamine, TPA - tripropylamine, TBA - tributylamine, and also TEA-d_{15}) on a megabore column (30 m x 0.53 mm) coated with a quite thick film (5 μm) of TC-1 stationary phase. The temperature program was 40°C for 2 min, increased to 180°C at 20°C/min, and held at 180°C for 5 min. For detection MS (SIM mode) was coupled with GC. Out of 500 mL water sample or 50 g sediment mixed with 300 mL purified water, 40 mL portions were distilled into 5 mL sulphuric acid solution. After alkalization, the headspace of the mixture was injected into GC. The detection limits were 50, 10, 5, 10 and 5 ng for TMA, TEA, TAA, TPA, and TBA, respectively. The water analysed (Suwa Lake) contained 0.25 μg/L TEA and 0.23 μg/L TBA; while the bottom sediment from Tokyo Bay 0.53 μg/g TMA, 4 μg/kg TEA, 0.26 μg/kg TPA, 1.0 μg/kg TBA and no TAA.

Airborne, volatile tertiary amines from polyurethane foams (pentamethyldiethyltriamine, diazabicyclooctane, *N*-methylmorpholine, *N*-ethylmorpholine, 1,4-dimethylpiperazine, *N,N*-dimethylethanolamine) were well separated in capillary columns coated with 5% phenyl-95% dimethylpolysiloxane: DB-5 (30 m x 0.25 mm, 0.25 μm film thickness) and HP-5 (60 m x 0.25 mm, 0.25 μm film thickness) [39]. GC was coupled with MS and FT-IR for final analysis of aqueous concentrates obtained by solid phase extraction using a high capacity cation-exchange resin based on poly(succinic) acid.

Maris et al. [40] applied a packed precolumn to extend a direct aqueous injection (DAI) technique of volatile aliphatic amines to capillary columns. However, they found that only the PoraPLOT Amines capillary column commercialised by Chromapack was suitable. They achieved baseline separation for 11 C_1-C_4 amines within 15 min; retention factors ranged from 1.3 to 13.1. Repeatability was greatly improved by addition of ammonia: at 50 μM concentration of amines RSD dropped down from 7-16% without ammonia to 3.2-4.9% when ammonia was present at 200 ppm concentration. Symmetrical peaks of amines were obtained but the PoraPLOT Amines column required a regeneration time of 20 min at 220°C to remove water. With GC-NPD the limit of quantification was in the range of 12-40 μg/L. For good separation on PoraPLOT Amines a crucial role was played by the packed precolumn, which trapped acidic vapour. The composition of an aqueous sample was important. Amine chloride salts only sublimate while amine sulphates can decompose at high temperature and hydrochloric acid at 0.12 M concentration was recommended as a matrix. The pre-column trapping efficiency depended strongly on temperature; below 80°C amines adsorption caused peak tailing and decreased the response, above 130°C trapping efficiency was reduced. The temperature rate can be selected that all volatile amines of interest are eluted. The lifetime of the precolumn was evaluated at least at 500 injections. If repeatability decreased, injection of methanol saturated with KOH regenerated the injector coating. The packed pre-column – PoraPLOT Amines column system was found suitable for aromatic (aniline) and heteroaromatic (pyridine) amines.

The packed precolumn is not necessary if aqueous sample headspace is injected into the PoraPLOT Amines capillary column, however ammonia addition is indispensable at a minimal concentration of 0.052 M to have good repeatability (RSD = 0.6-6.4%) [41]. With the PoraPLOT Amines column (27.5 m (+2.5 m particle trap) x 0.32 mm, 10 μm film thickness) the 14 lowest aliphatic amines were baseline separated within 15 min (Figure 3). Applying HS for sample preparation (liquid-gas ratio of 1, incubation temperature of 80°C, pH 13.7, a mixture of NaCl and Na_2SO_4) and NPD for detection the limits of detection were 0.2 μg/L for secondary and tertiary amines and 10 μg/L to 3000 μg/L for primary amines in aqueous samples.

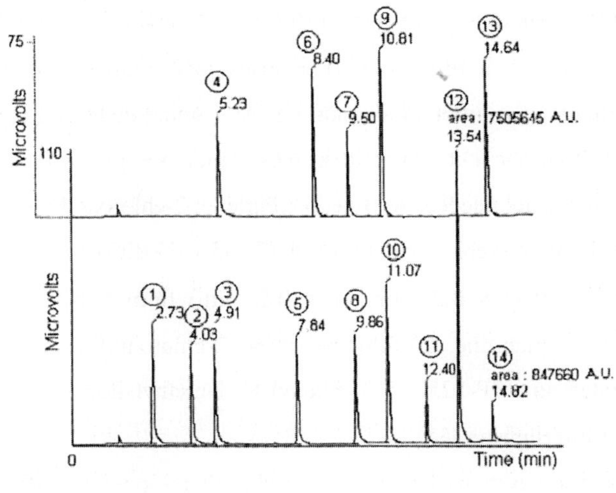

Figure 3 Chromatogram of volatile aliphatic amines (conditions in the text). Peaks: 1=methylamine; 2=dimethylamine; 3=trimethylamine; 4=ethylamine; 5=ethylmethylamine; 6=isopropylamine; 7=dimethylethylamine; 8=propylamine; 9=*tert.*-butylamine; 10=diethylamine; 11=diethylmethylamine; 12=butylamine; 13=diisopropylamine; 14=triethylamine (Reproduced with permission from Reference [41]).

For determination of trimethylamine in fibreglass extracts, Bir and Tutin [42] injected HS samples into a HP Basic Wax column (30 m x 0.25 mm, d_f = 0.5 μm). The initial oven temperature was 40°C held for 4 min, then it was ramped at 35°C to 150°C, and at 10°C/min to 190°C with a final hold time of 2 min. TMA was eluted during the first part of temperature program (t_R < 4 min). Application of the base-modified polyethylene glycol column eliminated tailing associated with the standard polyethylene glycol columns and gave an increase in signal-to-noise ratio greater than expected from peak shape improvement. However, the variation in the peak area and shifts in t_R were observed with aqueous sample injection. HS sampling eliminated the problems. When applying the HS equilibration temperature of 95°C and minimal matrix volume, TMA could be considered fully evaporated, which could make quantitation easier. With MS in SIM mode the detection limit of the procedure was close to 3 ppm.

The columns applied by other authors included a PoraPLOT Amines (30 m x 0.32 mm, 10 μm film thickness) [43], a Carbowax fused silica capillary GC column (30 m x 0.32 mm, 0.25 μm film thickness) [44], a Carbowax Amine column (5 m x 0.53 mm, 1.0 μm film thickness) [45], an SPB1-sulphur (30 m x 0.25 mm, d_f = 5 μm) from Supelco [46], a DB-1 column (60 m x 0.25 mm, d_f = 1.0 μm) from J&W [47] and a Stabilwax-DB/KOH megabore column (60 m x 0.53 mm, d_f = 1.5 μm) from Restek [48], and a CP-Cil 5™ capillary column (10 m x 0.25 mm, d_f = 0.12 μm) from Chrompack [49]. Figure 4 presents an exemplary chromatogram of aqueous solution of alkylamines on a amine-dedicated capillary column.

Separation of monocyclic aromatic amines, which are quite volatile, is much simpler and no special columns are needed. Aniline, *o*-toluidine, 2-chloroaniline, 2,6-dimethylaniline and 2,4,6-trimethylaniline were separated on a PTE-5 (5% diphenyl 95% dimethylsiloxane copolymer) column (30 m x 0.25 mm, d_f = 0.25 μm) from Supelco [50]. Aniline, *N,N*-dimethylaniline, *m*-metylaniline, 2,4-dimethylamine, 2-chloroaniline and 3,4,-dichloroaniline were easily separated on a SP-2250 (50% phenyl 50% methylsiloxane) column (30 m x 0.32 mm, d_f = 20 μm) from Supelco [51].

These separation systems will be discussed in detail together with sample preparation by solid phase microextraction.

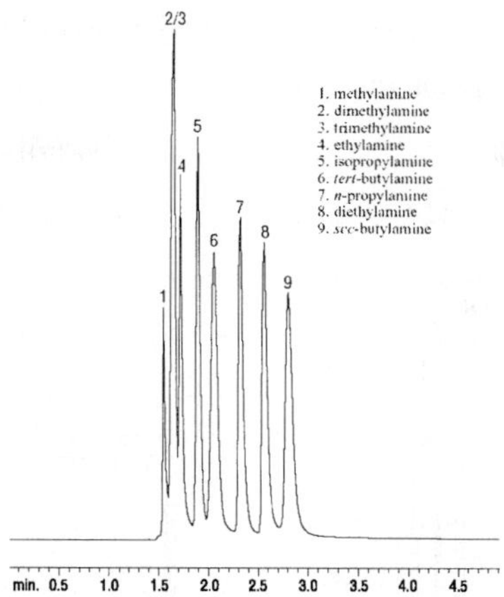

Figure 4 Chromatogram of aqueous solution of alkylamines on a Restek Rtx-35 Amine GC column (30 m x 0.53 mm ID, 1 μm) at 35°C. Peaks: 1=methylamine; 2=dimethylamine; 3=trimethylamine; 4=ethylamine; 5=isopropylamine; 6=*tert.*-butylamine; 7=n-propylamine; 8=diethylamine; 9=*sec.*-butylamine (Reproduced from Restek Catalog, 2005/6).

4. Solid phase microextraction in sampling free amines

Volatile amines are generally looked for in air, water and biological matrices. Very often original matrices must be changed or at least modified with simultaneous amines enrichment to make them separable and detectable in non-derivatised form by gas chromatography. Preconcentration of amines from air is usually performed by absorption in water [52] or sorption on solid sorbents and elution with water or other liquid solvent [27,36,39]. In the case of solid samples amines can also be extracted with water or organic solvent. Original aqueous samples or aqueous concentrates (sampling from air, extraction from solids) can be either directly injected into GC, provided that separation system compatible with aqueous samples can be selected and contents of amines in concentrates are above the corresponding detection limits, or analysed by headspace (HS)–GC. Both approaches have been widely applied (see sub-chapter 3).

Since solid phase microextraction (SPME) was developed by Pawliszyn and his co-workers 15 years ago [53,54] it has become an attractive alternative to common sampling techniques and gained widespread acceptance in many areas [55]; it has been most widely used in trace organic analysis [56,57].

The principles of SPME have been comprehensively described in a number of papers and books [54,58-61].

SPME uses a modified syringe assembly which houses a fused silica fibre externally coated with a thin layer of extraction material (Figure 5).

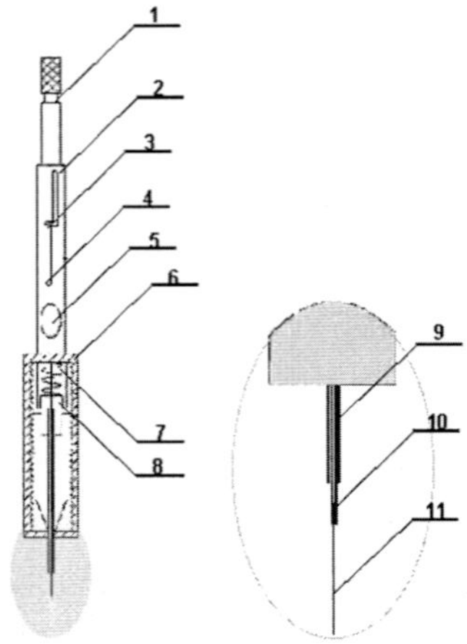

Figure 5 Manual for solid phase microextraction **1**=plunger; **2**=barrel; **3**=plunger retaining screw; **4**=z-slot; **5**=hub-viewing window; **6**=adjustable needle guide/depth gauge; **7**=tensioning spring; **8**=sealing septum; **9**=septum piercing needle; **10**=fibre attachment tubing; **11**=coated fused silica fibre. The drawing is based on the commercial SPME device made by Supelco.

In SPME the fibre is withdrawn from the needle and exposed to the sample. After partitioning between a fibre coating and sample matrix (sampling), analytes are desorbed in an injection port of GC or into HPLC. Both equilibrium and non-equilibrium sampling modes can be employed. Apart from very few situations extraction is not exhaustive and calibration needs a special care, mainly with the use of standard solutions of the same matrix as real samples.

SPME can be used by direct immersing the fibre in a gas or liquid sample or in the headspace (HS) of an aqueous or solid sample. Another approach is placing the SPME extraction fibre in a "dirty" sample protected by a semi-permeable membrane. These situations are presented in Figure 6.

HS and protective membrane SPME introduce additional selectivity and the fibre becomes protected from many impurities, which can deteriorate fibre coating. In determination of volatile amines in different matrices direct and HS approaches were used. It should be mentioned that the number of commercial fibre coatings is limited and it is easier to select effective coating for aromatic amines than for volatile aliphatic amines; generally tertiary amines are more effectively concentrated than secondary and primary amines.

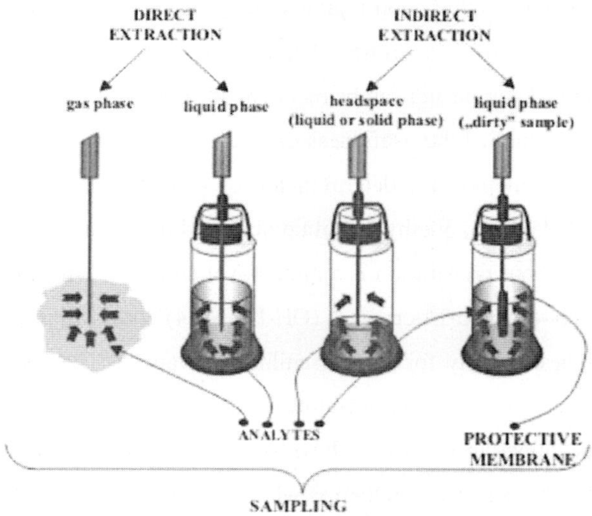

Figure 6 Different modes of isolation and enrichment of analytes by means of solid phase microextraction. The drawing prepared for student classes.

DeBruin et al. [50] applied HS-SPME for determination of monocyclic aromatic amines in biological fluids such as urine, blood, and milk. In the case of HS-SPME three phase system equilibria must be considered. Analytes partition between aqueous sample and HS and between HS and the fibre. The optimisation of extraction conditions can be started from decreasing solubility of amines in the aqueous medium. Monocyclic aromatic amines are weak organic bases and should be converted to non-ionic form for better gas extraction. They are polar substances even in non-ionic form and addition of inorganic salt should decrease their water solubility and, in this way, lower liquid-headspace partition coefficients and increase effectiveness of HS extraction. Finally the authors proposed adding KOH (2.0 M) and NaCl (4.0 M). The temperature of sampling is also very important. Increase in temperature improves extraction to the gas phase but decreases extraction to the fibre coating. Cooling fibres was proposed [62] but this approach is not very popular. The authors [50] tested all commercially available fibres (polydimethylsiloxane - PDMS, polyacrylate - PA, PDMS/divinylbenzene - DVB, Carbowax/DVB and Carboxen), and selected PDMS/DVB basing on extraction and desorption effectiveness. The value of optimum temperature depends on amine; as a compromise the temperature of 45°C was selected for sampling. The kinetics of extraction is not infinitely fast process and stirring the solution was applied to speed up reaching equilibrium. With MS operated in SIM mode and optimum sampling conditions, the authors obtained quite good reproducibility (RSD < 11%) and low detection limits ranging from 0.40 ppb for 2,4,6-trimethylamine in urine to 7.7 ppb for aniline in blood. The procedure is fast and simple and can be recommended for determination of amines.

Zeng et al. [63] studied the usefulness of three SPME fibre coatings of crown ethers prepared by the sol-gel method for determination of aromatic amines such as aniline, *m*-toluidine, *N,N*-dimethylaniline, *N*-ethyl-*m*-toluidine 3,4-dimethylaniline. Application of sol-gel methodology in sorbent preparation is a quite novel and promising approach [64]. It was found that the hydroxy-dibenzo-14-crown-4 (OH-DB14C4) coated fibre produced by sol-gel technology had the best affinity for several aniline derivatives. The coating has very high thermal stability and can withstand temperature up to 340°C without loss of extraction efficiency. It can also be dipped in water for 5 h with only minor loss of efficiency. In the case of all the above amines the sensitivity obtained with the OH-DB14C4 coated fibre was much higher than with commercial fibres coated with PDMS (100 μm) and CW/DVB (65 μm). The chromatographic column used for amines separation was also coated with OH-DB14C4 as a stationary phase. It was relatively short (10 m) and contained relatively small amount of stationary phase (0.25 mm I.D., 0.25 μm film thickness) and yet separation was quite good at isothermal conditions (140°C). When amines were sampled from HS and FID was used as a detector, the system allowed achieving detection ranging from 0.17 – 0.98 ng/mL. Repeatability of determinations was very good – relative standard deviation (n = 5) was 3.23-6.20%.

The broadening of application has been achieved by coating sorbents on metal wires by an electrochemical method [65,66]. Basing on the principle of "like dissolves like", Minjia et al. [51] prepared a novel polyaniline (PANI) coating on a stainless-steel wire for SPME of monoaromatic amines (aniline, *N,N*-dimethylaniline, *m*-metylaniline, 2,4-dimethylaniline, 2-chloroaniline and 3,4,-dichloroaniline) from aqueous sample HS. The highest recovery was when amines were sampled from HS at 65°C. The desorption from the fibre was exhaustive at 250°C. The good separation was achieved for all six target amines on a SP-2250 (50% phenyl 50% methylsiloxane) column (30 m x 0.32 mm, d_f = 20 μm) from Supelco. Oven temperature was initially held at 50°C, programmed at 50°C/min to 100°C, held for 3 min and then increased at 5°C/min to a final temperature of 150°C (1 min). Under optimized conditions the detection limits were from 0.19 to 1.06 μg/L; they were better than those obtained with the commercial CW/DVB fibre by the order of magnitude [67]. Relative standard deviations were found to be 2.02-6%.

SPME sampling of volatile aliphatic amines is much less effective. Mills et al. [44] applied HS-SPME-GC-MS to determine trimethylamine in urine. The sample was alkalized and salted out and deuterated TMA as an internal standard was added. SPME sampling was performed at 50°C. Three fibres were used: polyacrylate (75 μm film thickness), PDMS (100 μm film thickness) and Carboxen/PDMS (75 μm film thickness). It was found that polyacrylate SPME fibre did not extract TMA efficiently as compared to PDMS whereby peak areas of TMA and deuterated TMA were ca. 200 times larger and this fibre was used. TMA was completely desorbed from the fibre within 2 min and no carry over between samples was observed. The authors also tried to apply the Carboxen/PDMS coated fibre which appeared ca. 12 times more sensitive for TMA. Using a Carbowax capillary GC column (30 m x 0.32 mm, 0.25 μm film thickness), which gave sharper peaks than a non-polar BPX 5 column, and a quadrupole MS in SIM mode, the authors obtained within-batch imprecision of 6.1 to 9.9% (RSD) for aqueous solutions at amines concentrations from 10.7 to 95.2 μmol/L and 12.2% for non-spiked urine containing TMA at 2.4 μmol/L. The number of measurements was 7 for aqueous solutions and 8 for non-spiked urine.

Airborne trimethylamine was also determined using SPME-GC [45]. TMA was first sampled on acidified filter. After trapping the filter was immersed in water, alkalized and TMA from HS was extracted at 30°C to SPME fibre. The amines were desorbed from the fibre in an injection port of GC at 250°C, separated on a Carbowax column (5 m x 0.53 mm, 1.0 μm film thickness) at 65°C (isothermally), and detected by FID. Two fibres were tried: PDMS (100 μm) and PDMS/DVB (65 μm). The latter has higher affinity for TMA and shorter equilibration time (2 min). However, it is characterised by fast saturation and poor linearity and is recommended for low concentrations. Linearity for the non-polar PDMS fibre is good though equilibration time is much longer (14 min) and this fibre was selected by Chien et al. [45] for TMA sampling. They determined the reliable quantitation limit based on a 16 L air sample to be 2.5 mg/m^3.

The Carboxen/PDMS fibre (75 μm) was used to sample TMA together with carbon disulphide, dimethylsulphide and dimethyldisulphide from air in wastewater, composting and animal agriculture industries [47]. The extracted analytes including TMA were desorbed at 250°C (the needle with the fibre exposed was held in a heated injector until the end of a chromatographic run to eliminate any possible carry over) and separated on a DB-1 column

(60 m x 0.25 mm, d_f= 1.0 μm) from J&W (32°C for 5 min, 3.5°C/min to 118°C) and detected with MS coupled with GC. The calibration curve for TMA was linear over the range of ca. 3.2-210 ppbv though R^2 was a bit lower than for other compounds due to substantial tailing. The detection limit was 2.38 ppbv. The error between fibres was 7.6% (n = 8) and 5.8% for the same fibre (n = 4).

For sampling C_1-C_6 volatile amines (MMA, DMA, TMA, DEA, TEA) from HS of wastewater, primary and secondary effluents, and sewage-polluted river water, Abalos et al. [43] studied the applicability of four SPME fibres: PDMS (100 μm), CW (Carbowax)/DVB (65 μm), PA (85 μm), Carboxen/PDMS (75 μm), all obtained from Supelco. Kinetics of sorption was quite different for these fibres. With PDMS, equilibrium was reached within 10 min for most of the amines studied, while with PA equilibrium was not reached even within 30 min. Poor sampling kinetics probably results from the fact that PA is low density solid polymer and hence diffusion coefficients of solutes are lower than in the case of liquid PDMS. CW/DVB showed the shortest equilibration time of 2 min which can be explained by the fact that adsorption controls extraction process. Much longer equilibration times for Carboxen/PDMS probably result from the fact that Carboxen is characterised by small pore size and hence slower mass transfer. With CW/DVB and PA much lower sensitivities were obtained than with PDMS and Carboxen/PDMS for all the amines studied. Carboxen/PDMS was the most effective for extraction of medium- to low-volatility amines but PDMS appeared more useful due to the lack of interferences in the determination of target amines. Baseline chromatographic separation of all five target amines (MMA, DMA, TMA, DEA, TEA) was obtained using a PoraPLOT Amines column (30 m x 0.32 mm, 10 μm film thickness) with helium as a carrier gas and temperature programming (60°C for 5 min, 10°C/min to 220°C at which held for 15 min). With the PDMS coated fibre for HS sampling and GC-NPD for final analysis, the detection limits were compound dependent and ranged from 3 to 56 μg/L, being close to or lower than the odour threshold concentrations, and reproducibility was ca. 15% (n = 5). The results proved applicability of SPME for gas phase sampling of volatile amines.

The review presented above shows that SPME is a promising technique for fast and low cost analysis of volatile amines. However, Lestramau et al. [46] indicated that the technique has some limitations when using the PDMS/Carboxen fibre for sampling volatile aliphatic amines from air. These authors made experiments with propylamine, diethyamine, triethylamine and dipropylamine and three fibre coatings: PDMS (100 μm), PDMS/DVB (65 μm) and PDMS/Carboxen (85 μm). Desorption was performed for 2 min at 250°C for the first two fibres and at 300°C for PDMS/Carboxen. Under these conditions production of artefacts was observed in the case of the PDMS/Carboxen fibre. The artefacts were unsaturated amines formed from the studied amines. Reactions were found to occur in a GC injection port, enhanced by temperature and catalysed by particles coming from an inox needle of the SPME device. According to this observation the PDMS/Carboxen, which is a very interesting coating for many VOCs, can not be recommended for aliphatic amines. This can explain why rather other coatings were used in procedures of GC amine determination with the use of SPME sampling.

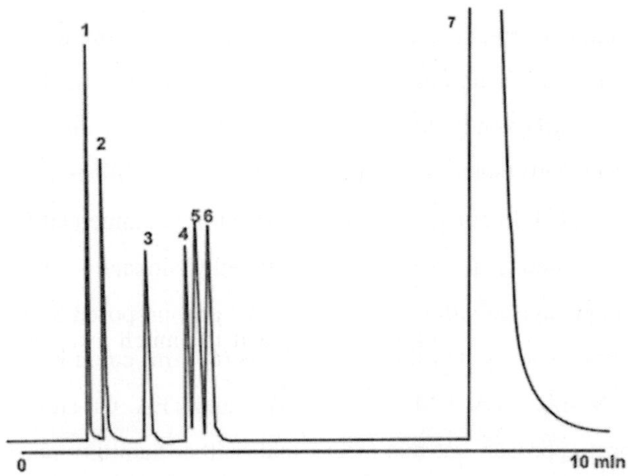

***Figure** 7* Chromatogram of standard solution of aliphatic amines (conditions given in the text). Peaks: 1=ethanamine; 2=2-propanamine; 3=1-propanamine; 4=*N*-ethylethanamine; 5=2-butanone (internal standard); 1-butanamine; 7=benzyl alcohol (solvent). (Reproduced with the permission from Reference [49])

Artefacts formation originating from catalysed temperature decomposition of analyte amines can be avoided using HS microextraction to a single drop for determination of aliphatic amines in aqueous solution. The procedure is also very simple and rapid. Kaykhaii et al. [49] showed applicability of such approach for analysis of aqueous solutions of ethanamine, 2-propanamine, 1-propanamine, *N*-ethylethanamine and butanamine. With 1 μL drop of benzyl alcohol as an extracting medium, separation on a CP-Cil 5^{TM} capillary column (10 m x 0.25 mm, d_f = 0.12 μm) from Chrompack and detection with FID, the detection limits ranged from 25 μg/L for ethanamine to 2.5 μg/L for 1-propanamine with RSDs ranging from

12 to 6% (n = 3). The corresponding chromatogram of the standard solution of aliphatic amines is given in Figure 7.

This approach is quite promising and probably more papers will soon appear on the topic.

5. Experimental; SPME-GC-FID determination of aliphatic amines in different atmospheres in Gdansk area

As discussed above the procedures of trace components determination based on SPME as a technique of analyte isolative sampling, enrichment and introduction into a GC column are relatively simple and convenient and can be applied also to determine amines in air samples. In the case of volatile aliphatic amines generally PDMS (100 μm), PDMS/DVB (65 μm), CW/DVB (65 μm), Carboxen/PDMS (75 μm) have been taken into consideration. As has been shown quite recently some artefacts are formed from amines on Carboxen/PDMS at elevated temperature [49]. In our studies [48] we selected a commercial PDMS/DVB (65 μm) fibre and three custom-made fibres (ethoxy polydimethylsiloxane - 10 μm, polyurethane acrylate - 40 μm, and phenyl polydimethylsiloxane - 10 μm) proposed by Ligor et al. [68]. The preliminary experiments suggested that PDMS/DVB (65 μm) could be the best for sampling target amines, i.e. MMA, ETA, DMA, DEA, TMA, and TEA. On-site SPME sampling was regarded as the best approach but this means that neither humidity nor temperature of real samples would be controlled. Apart from optimisation of time and temperature of extraction, the effects of temperature and air humidity on sampling were investigated. Standard gaseous mixtures of different analyte concentrations, humidity and temperature were generated using a modified version (Figure 8) of a custom generator designed and constructed at the Analytical Chemistry Department, Chemical Faculty, Gdansk University of Technology [69].

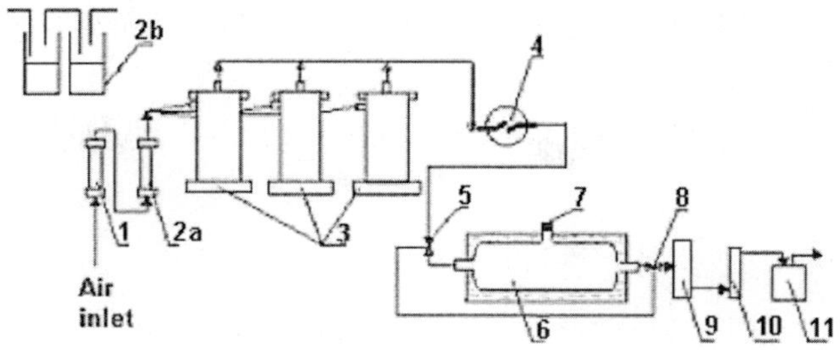

Figure 8 Scheme of apparatus for generating standard mixtures and SPME sampling: 1=air purifier, 2a=drier, 2b=humidifires 3=analytes generators, 4=preliminary mixing chamber, 5=Teflon 3-way valve, 6=thermostated mixing glass chamber, 7=septum, 8=glass 3-way valve, 9=flowmeter, 10=rotameter, 11=suction pump. The drawing prepared for this book; the device was used in our earlier studies [48]

For mixture generation commercial permeation tubes were used (Vici Metronics, Inc.). They were kept at constant temperature; changing the flow rate of diluting gas changed concentration of amines. The concentration ranges of particular amines in the mixtures generated are presented in Table 1.

Table 1 Concentration ranges of amines in standard gas mixtures (flow rates: 15, 35, 75, 100, 150 cm^3/min)

Amine	Concentration range [mg/m^3]
Methylamnine (MMA)	0.86 – 8.67
Ethylamnine (ETA)	0.93 – 9.33
Dimethylamine (DMA)	0.88 – 8.80
Diethylamnine (DEA)	0.54 – 5.47
Trimethyamine (TMA)	0.52-5.00
Triethyamine (TEA)	0.05 – 0.53

With this apparatus standard mixtures of relative humidity of 0, 20, 47, 79, and 92% could be produced.

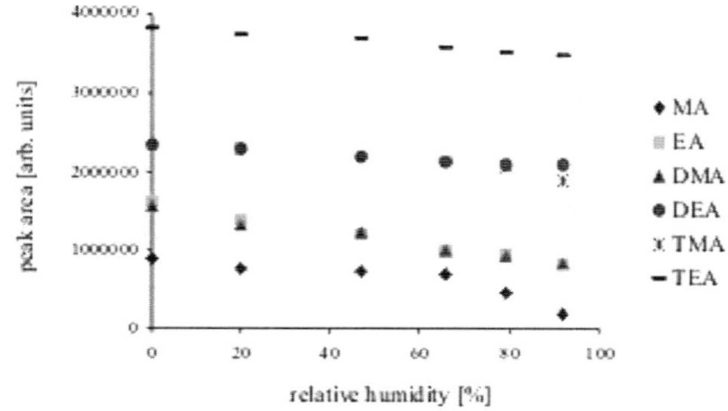

Figure 9 The effect of relative humidity of a gas mixture on amount of amines extracted by the polyurethane acrylate coated SPME fibre (Reproduced with permission from Reference [48])

As can be seen from Figure 9, the increase in relative humidity decreases fibre sorption effectiveness. For the polyurethane acrylate coated SPME fibre the increase in humidity from 0 to 92% at 35°C results in the high drop in the amount adsorbed. This shows that humidity should be taken into consideration in calibration process. This can be done in the following way: the humidity of a real sample is measured, the SPME-GC system is calibrated with the same standard mixture at two bracketing humidities and the results interpolated into the humidity of a real sample.

The sampling chamber temperature can be controlled within some limits. Experiments with standard mixtures were made at 15, 20, 25, 30, and 35°C. For SPME-GC calibration the temperature of standard gas mixture was set equal to the temperature of the real sample.

The fibre can be easily exposed to the mixture under static and dynamic conditions though linear velocities were rather small (up to ca. 0.09 cm/s). It was shown that at the 95% probability level sampled amounts under static and dynamic conditions did not differ significantly.

On the basis of experiments with standard gaseous mixtures of amines, the following parameters were selected:

Sampling temperature - 15 min

Desorption temperature - 200°C

Desorption time - 2 min

Separation conditions: a Stabilwax-DB/KOH megabore column (60 m x 0.53 mm, 1.5 μm film thickness); hydrogen as a carrier gas; oven temperature - 50°C.

Detection and quantitation limits calculated from calibration curves are given in Table 2 [48].

Table 2 Detection and quantitation limits for 'air amines' determined from calibration curves

Compound	Quantitation limits [mg/m^3]	Detection limits [mg/m^3]
Triethlamine (TEA)	0.86	0.30
Diethylamine (DEA)	0.57	0.21
Ethylamine (ETA)	1.06	0.38
Methylamine (MMA)	1.03	0.31
Dimethylamine (DMA)	1.88	0.67
Trimethylamine (TMA)	0.55	0.19

The procedure developed was used to study air pollution with amines in different sites in the Gdansk area [48]: pharmaceutical plant - A, chemical storeroom - B, car paint shop (freshly renovated staff lounge) - C, city market (samples collected on a market day, ca. 2 m away from fish stands) - D, city market (samples collected within a 20 m radius of fish stands) - E.

SPME fibres were cleaned by thermal desorption at the temperature recommended by the manufacturer (250°C). The fibres were exposed to the air analysed for 20 min. They were placed at a height of breathing zone. The fibre was withdrawn into the needle, sealed with a silicone rubber septum and placed on dry ice for transport. The temperature and humidity of the sampled air were measured.
The GC analysis of amines extracted into the fibre coating was conducted under the conditions given above.

The contents of particular amines determined and also some criteria are given in Table 3 [48].

Table 3 Concentrations of selected aliphatic amines [mg/m^3] in the air from different sampling sites in Gdansk area

Analyte	Sampling site (see text)					Ctiteria	
	A	B	C	D	E	STEL*)	TLV**)
MMA	n.d.	n.d.	n.d.	112.4	21.1	15	5
ETA	16.3	5.1	4.3	n.d.	n.d.	15	5
DMA	n.d.	n.d.	n.d.	140.1	79.2	18	9
DEA	83.2	6.3	n.d.	n.d.	n.d.	75	30
TMA	n.d.	n.d.	n.d.	24.3	2.2	24	12
TEA	148.2	2.5	n.d.	n.d.	n.d.	n.s.	n.s.

n.d. = not detected
n.s. = not specified
*) STEL = Mean short term exposure or maximum concentration of a substances which a worker may be exposed to for a continuous 30 minute period, with a low probability of adverse health effects for the worker and his/her future generations [70].
**) TLV = Maximum permissible 8-h time-weighted average concentration of a material which a worker may be exposed to with a low probability of adverse health effects for the worker and his/her future generations when working 42 h per week [70].

References

[1] M.L. Richardson, The Dictionary of Substances and Their Effects, Royal Society of Chemistry, England, 1992.

[2] J. Chodakowski, Słownik Chemiczny (Dictionary of Chemistry), WP, Warsaw 1995.

[3] P.H. Howard: Handbook of Environmental Fate and Exposure Data for Organic Chemicals, Lewis Publishers, Michigan, USA, 1990.

[4] J.W. Vincoli, Risk Management for Hazardous Chemicals, CRC Press, New York, USA, 1997.

[5] A. Bouchereau, P. Guénot, F. Larher, J. Chromatogr. A 747 (2000) 49-67.

[6] O. Busto, M. Miracle, J. Guash, F. Borrull, J. Chromatogr. A 757 (1997) 311-318.

[7] O. Busto, J. Guash, F. Borrull, J. Chromatogr. A 718 (1995) 309-317.

[8] F. Sacher, S. Lenz, H-J. Brauch, J. Chromatogr. A 764 (1997), 85-93.

[9] T. Kawasaki, M. Maeda, A. Tsuji, J. Chromatogr. A 328 (1985), 121-126.

[10] Y. Hwang, T. Matsuo, K. Hanaki, N. Suzuki, Water Res. 29 (1995) 711-718.

[11] S.E. Manahan, Fundamentals of Environmental Chemistry, Lewis Publishers, Michigan, USA, 1993.

[12] F. Sacher, S. Lenz, J.H. Brauch, J. Chromatogr. A 764 (1997) 85-93.

[13] G.W. Schade, P.J. Crutzen, J. Atmos. Chem. 22 (1995) 319-346.

[14] P.L. Hanst, J.W. Spence, M. Miller, Environ. Sci. Techn. 11 (1977) 403-405.

[15] B. Sahasrabuddhey, A. Jain, K.K. Verma, Analyst 124 (1999) 1017-1021.

[16] L. Gronberg, P. Lovkvist, J.A. Jonsson, Chromatographia 33 (1992) 77-82.

[17] J-K. Lin, C-C. Lai, Anal Chem. 52 (1980) 630-635.

[18] W. Seńczuk, Toxicology, PZWL, Warsaw, Poland, 1990.

[19] E.Vo, S.P. Berardinelli, J Environ Monit., 1 (1999) 545-548.

[20] A. Jastrzębska, B. Zygmunt, B. Żukowska, W. Murawski, J. Namieśnik, Chem. Inż. Ekol. 10 (2003) 609-626.

[21] J. Namieśnik, T. Górecki, Pol. J. Environ. Stud. 10 (2001) 77-84.

[22] J. Namieśnik, Crit. Rev. Anal. Chem. 32 (2002) 271-300.

[23] M. Rawa-Adkonis, L. Wolska, J. Namieśnik, Crit. Rev. Anal. Chem. 33 (2003) 199-248.

[24] B. Zygmunt, J. Namieśnik, Accred. Qual. Assur. 5 (2000) 191-197.

[25] H. Kataoka, S. Yamamoto, S. Narimatsu, Gas chromatographic analysis of environmental amines with selective detectors in W. Kleibohmer (ed.), Environmental analysis, Handbook of Analytical Separations, Vol. 3, Elsevier, 2001.

[26] A. Di Corcia, R. Samperi, Anal. Chem. 46 (1974) 977–981.

[27] K. Kuwata, Y. Yomazaki, M. Uebori, Anal. Chem. 52 (1980) 1980–1982.

[28] Technical Bulletein No. 21, Applied Science Labs., State College, PA.

[29] M. Dalene, M. Mathiasson, J.A. Jonsson, J. Chromatogr. 207 (1981) 37-46.

[30] L. Mathiasson, P. Lovkvist, J. Chromatogr. 217 (1981) 177-181.

[31] G. Audunsson, L. Mathiasson, J. Chromatogr. 315 (1984) 299-312.

[32] G. Audunsson, M. Dalene, J.A. Jonsson, P. Lovkvist, L. Mathiasson, Int. J. Environ. Anal. Chem. 20 (1985) 85-100.

[33] P G.Audunsson, Anal. Chem. 60 (1988) 1340-1347.

[34] X-H. Yang, C. Lee, M.I. Scranton, Anal. Chem. 65 (1993) 572-576.

[35] N. Kashihira, K. Makino, K.Kirita, Y. Watanabe, J. Chromatogr. 239 (1982) 617-624.

[36] K. Kuwata, E. Akiyama, Y. Yomazaki, H. Yomasaki, Y. Kuge, Anal. Chem. 55 (1983) 2199–2201.

[37] T. Fujii, T. Kitai, Anal. Chem. 59 (1987) 379-382.

[38]. T. Tsukioka, H. Ozawa, T. Murakami, J. Chromatogr. 642 (1993) 395-400.

[39] G. Seeber, M.R. Buchmeiser, G.K. Bonn, T. Bertsch, J. Chromatogr. 809 (1998) 121-129.

[40] C. Maris, A. Laplanche, J. Morvan, M. Bloquel, Wat. Sci. Tech. 40 (1999) 141-148.

[41] C. Maris, A. Laplanche, J. Morvan, M. Bloquel. J. Chromatogr. 846 (1999) 331-339.

[42] D. Bir, K. Tutin, J. Chromatogr. Sci. 40 (2002) 337-342.

[43] M. Abalos, M.J. Bayona, F. Ventura, Anal. Chem. 71 (1999) 3531-3537.

[44] G.A. Mills, V. Walker. H. Mughal, J. Chromatogr. 723 (1999) 281-285.

[45] Y-C. Chien, S-N. Uang, C-T. Kuo, T-S. Shih, J-F. Jen, Anal. Chim. Acta 419 (2000) 73-79.

[46] F. Lestremau, V. Desauziers, J.L. Fanlo, Analyst 126 (2001) 1969-1973.

[47] H. Kim, C. Nochetto, L.L. McConnell, Anal. Chem. 74 (2002) 1054-1060.

[48] J. Namieśnik, A. Jastrzębska, B. Zygmunt, J. Chromatogr. 1016 (2003) 1-9.

[49] M. Kaykhaii, S. Nazari, M. Chamsaz, Talanta 65 (2005) 223-228.

[50] L.S. De Bruin, P.D. Josephy, J.B. Pawliszyn, Anal. Chem. 70 (1998) 1986-1992.

[51] H. Minjia, T. Chao, Z. Qunfang, J. Guibin, J. Chromatogr. A 1048 (2004) 257-262.

[52] P G. Audunsson, Anal. Chem. 58 (1986) 2714-2723.

[53] R.P. Belardi, J.Pawliszyn, Water Pollut Res. J. Can. 24 (1989) 179-191.

[54] C.L. Arthur, J. Pawliszyn, Anal. Chem. 62 (1990) 2145-2148.

[55] J. Pawliszyn, Application of Solid Phase Microextraction, Royal Society of Chemistry, Cambridge, UK, 1999

[56] J. Namieśnik, B. Zygmunt, A. Jastrzębska, J. Chromatogr. A 885 (2000) 405-418.

[57] B. Zygmunt, A. Jastrzębska, J. Namieśnik, Crit. Rev. Anal. Chem. 31 (2001) 1-18.

[58] Z. Zhang, J. Pawliszyn, Anal. Chem. 66 (1994) 844 A-853.

[59] R. Eisert, K. Lavsen, J. Chromatogr. 733 (1996) 143-157.

[60] H. Lord, J. Pawliszyn, J. Chromatogr. A 885 (2000) 153-193.

[61] J. Pawliszyn, Solid Phase Microextraction. Theory and Practice, Wiley-VCH, Canada 1997.

[62] Z. Zhang, J. Pawliszyn, Anal. Chem. 67 (1995) 34-43.

[63] Z. Zeng, W. Qiu, M. Yang, X. Wei, Z. Huang, F. Li, J. Chromatogr. A 934 (2001) 51-57.

[64] A. Malik, New polymeric extraction materials, pp. 1023-1080 in Comprehensive Analytical Chemistry (B. Barcelo, ed.); Vol. 37: Sampling and Sample Preparation for Field and Laboratory (J. Pawliszyn, ed.), Elsevier Science B.V., Amsterdam 2002.

[65] J. Wu, J. Pawliszyn, Anal. Chem. 73 (2001) 55-63.

[66] J. Wu, J. Pawliszyn, J. Chromatogr. A 909 (2001) 37-52.

[67] L. Muller, E. Fattore, E. Benfenati, J. Chromatogr. 791 (1997) 221-230.

[68] M. Ligor, M. Ściborek, B. Buszewski, J. Microcol. Sep. 11 (1999) 377-383.

[69] D. Gorlo, L. Wolska, B. Zygmunt, J. Namieśnik, Talanta 44 (1997) 1543-1550.

[70] Decree of the Ministry of Employment and Social Policy from June 27, 1998, on the maximum allowable levels and intensities of substances and physical factors detrimental to health in workplace, Dziennik Ustaw, Vol. 79, Item 513, Warsaw, Poland.

Ibolya Molnár-Perl (Editor)
Quantitation of Amino Acids and Amines by Chromatography
Journal of Chromatography Library, Vol. 70

2.1.2.

Gas Chromatography of Amines as Various Derivatives

Hiroyuki Kataoka

Contents

Summary

In this chapter, a wide variety of derivatization reactions used for the GC analysis of amines has been surveyed. Some potential problems with derivatization procedures include the formation of unwanted derivatives, the presence of unchanged derivatization reagents, and the requirement for non-aqueous reaction conditions. Amines in environmental and biological samples are usually found in aqueous solutions except for air samples, and are often at trace levels among complex matrices containing a number of coexisting substances. It is therefore desirable that these amines be derivatized in aqueous solutions and be sensitively detected. Of the derivatization reactions presented in this chapter, silylation and acylation usually require anhydrous reaction conditions because their respective reagents are very sensitive to moisture and hydrolyze. In contrast, alkylation, halogenation and the formation of Schiff base, carbamates, sulfonamide and phosphonamide derivatives can be generally performed in aqueous solutions. GC detectors must be not only sensitive to the minute amounts of analytes, but selective enough to discriminate among coexisting substances. The nitrogen selective detector NPD and GC-MS have been used to obtain results with high sensitivity and selectivity. In addition, the conversions of amines into fluorinated derivatives to introduce ECD responses and into sulfur- and phosphorus-derivatives to introduce FPD responses have been devised for these purposes. The GC methods with selective detectors, described in this chapter, have advantages and disadvantages and to date there is no all-powerful analytical methods. The choice of an analytical method depends on the presence of amines at low parts per billion or less and the variety and complexity of the sample. To achieve the efficient isolation and preconcentration of amines in environmental and biological samples, several methods of sample preparation have been developed using a number of different purification techniques. Among these, SPME is simple and rapid, and may be applied directly to the sample in combination with derivatization. Finally, it is the author's hope that this review will serve as a guide for choosing the most effective techniques for the GC analysis of amines.

Derivatization for the protection of amino groups is widely employed to increase volatility, to form diastereomers, and to improve the chromatographic and, if possible, the mass spectral properties of derivatives. In this chapter, we summarize derivatization methods for the determination of amines by gas chromatography (GC), especially with regard to the reactivity, selectivity and sensitivity of these reactions. Among the derivatives used for GC analysis are the silyl, acyl, alkyl, carbamate, sulfonamide, phosphonamide, Shiff base and thiourea derivatives of amines. Some derivatives can be selectively and sensitively detected using a variety of methods, including thermionic, flame photometric, electron capture and mass selective detectors. The recent applications of these derivatization methods to the analysis of individual aliphatic and aromatic amines in food, environmental, clinical chemistry and other areas are also described.

1. Introduction

1.1. Derivatization of amines for gas chromatography

GC has been widely used for amine analysis because of its inherent advantages, including its simplicity, high resolving power, high sensitivity, short analysis time and low cost. In general, amines are analyzed in their free forms or as derivatives. GC analysis of free amines is described in detail in a previous review [1] and in chapter 2.1.1 in this book. Free amines can be separated on strongly basic stationary phases. Following additional of alkali, the samples are analyzed by direct injection or by the headspace analysis technique, or they are extracted into an organic solvent before analysis. Solid-phase microextraction (SPME) is another simple and convenient technique for the extraction and concentration of amines and can be combined with GC analysis. The SPME technique minimizes sample preparation, thereby reducing sample contamination. However, GC analysis of free amines generally has several inherent drawbacks, which are related to the difficulty in handling low-molecular mass amines, due to their high solubility in water and their high volatility. Free amines are therefore, difficult to extract from water, and are not easily chromatographed due to their polarity. Furthermore, free amines tend to be adsorbed by and decompose on columns, resulting in tailed elution peaks, ghosting phenomena and low detector sensitivity. The adsorption capacity of amines in analytical systems, (*i.e.*, sample vessels, injection system, glass wool and GC columns), is in the order primary > secondary > tertiary, and it is generally more difficult to chromatograph aliphatic than aromatic amines.

A common method of overcoming these problems is to convert polar compounds to relatively non-polar derivatives more suitable for GC analysis. Derivatization methods have been employed to reduce the polarity of the amino group and to improve their ability to be chromatographed on GC columns. Among these derivatives are silyl, acyl, alkyl, carbamate, sulfonamide, phosphonamide, Shiff base and thiourea derivatives of amines. The formations of amide, carbamate, urea and isourea derivatives have also been used to separate chiral amines. Derivatization reactions, which are often selective for amine type (*i.e.*, primary, secondary, or tertiary amines), have been used to improve their detection and separation. For example, introducing sulfur-, phosphorus- and halogen-containing groups into a molecule has been shown to enhance the response of flame photometric detector (FPD) and electron capture detector (ECD). In addition to FPD and ECD, other amine detectors including hydrogen flame ionization detector (FID), thermionic detector (FTD), each of these has increased selectivity and sensitivity for specific amines. Furthermore, the combination of GC-mass spectrometry (MS), together with mass selective detector (MSD), which use a method based on selected ion monitoring (SIM), can provide structure information allowing for the unequivocal identification of specific amines. By using these detectors, sub-nanogram detection limits can be achieved. Although most detectors respond directly to amines, several, such as FPD and ECD, require derivatization prior to detection, as described above. These chemical derivatization reactions and selective detection methods have been described in detail in previous reviews [2–10] and books [11–13].

1.2. Objective and scope

This chapter is concerned with utilizing chemical derivatization and GC analysis for the determination of amines. I, Section 2, general aspects of amino group derivatization for GC analysis are surveyed according to type of reaction, and derivatization combined with SPME is also described. In Section 3, general aspects of the selective and sensitive detection of amines by GC are surveyed according to type of detector. In Section 4, we describe several applications of amine group derivatization and GC analysis to food, environmental, and clinical chemistry according to the type of amine. This chapter covers not only aliphatic and aromatic primary-, secondary- and tertiary-amines but also biogenic and heterocyclic amines. Of particular note are references published over the past two decades. For more details about experimental and novel methods, the original papers should be consulted. General aspects of amine analysis in various samples by GC have been detailed in several books and reviews [2–4, 14, 15].

2. Derivatization reactions for amines

GC analysis of free amines, without column modification, is unsatisfactory, due to the adsorption and decomposition of the solute, and the resulting peak tailing and loss. Amines are derivatized not only to reduce their polarity but also to improve their volatility, selectivity, sensitivity and separation during chromatography. In addition, derivatization can serve to enhance mass spectrometric properties. These mass spectra are easy to interpret, and the characteristic high-mass ions, derived from the molecular ions, can be used for trace analysis in the SIM mode. The most commonly used amine derivatization reactions for GC analysis are listed in Table 1, and the reaction schemes are shown in Fig. 1. Derivatization reactions, which are often selective for amine type (*i.e.* primary, secondary, or tertiary amines), have also been used to improve their detection and separation. The derivatization of chiral amines to form diastereomeric derivatives has been widely researched, and an equally large number of chiral reagents are available. In contrast, the combination of derivatization and SPME has also been used effectively for the selective and sensitive analysis of aliphatic and aromatic amines.

2.1. Silylation

Silylation is probably the simplest, quickest and most versatile technique available for enhancing GC performance. It involves the blocking of protic sites, thereby reducing dipole-dipole interactions and increasing volatility and GC properties, yielding rather narrow and symmetric peaks. However, silyl derivatives are less used for the GC analysis of amines, owing to their hydrolytic instability and high silyl-donor activity. Commonly, the silyl derivatives of amines can be prepared by using stronger silylating reagents and catalysts. As shown in Fig. 1A, *N,O*-bis(trimethylsilyl)acetamide (BSA), *N,O*-bis(trimethylsilyl)-trifluoroacetamide (BSTFA), *N*-methyl-*N*-(trimethylsilyl)trifluoroacetamide (MSTFA), *N*-methyl-*N*-(*tert*- butyldimethylsilyl)trifluoroacetamide (MTBSTFA) and pentafluorophenyl-dimethylsilyl (flophemesyl) reagents have been used as silylating reagents. The addition of trimethylchlorosilane (TMCS) and/or trimethylsilylimidazole (TMSIM) as catalysts generally ensures the effective derivatization of amino groups. These reagents, however, react not only with amino groups but also with hydroxyl and carboxyl groups under anhydrous conditions [7,12]. The ease of reaction with these reagents is generally in the order alcohols > phenols > carboxylic acids > amines > amides, and is higher for primary than for secondary amines [12].

Table 1 Derivatizating reagents for gas chromatography of amines

Reagents	Abbreviation	Amine type[a]	Detector[b]	Reference
1. Silylation				
N-Methyl-*N*-trimethylsilyltrifluoroacetamide	MSTFA	P, S	M	16
N,O-Bis(trimethylsilyl)trifluoroacetamide	BSTFA	P, S	M	17
N-Methyl-*N*-(*tert*-butyldimethylsilyl)-acetamide	MTBSTFA	P, S, H	M	18,19
2. Acylation				
Acetic anhydride		P, B	E, M	20
Trifluoroacetic anhydride	TFAA	P, S, B	F, E, M	21-25
Chloro or dichloro or trichloroacetic anhydride		P, S	E, M	16,21
Pentafluoropropionic anhydride	PFPA	P, S	E, N, M	16,21,26-33
Heptafluorobutyric anhydride	HFBA	P, S, B	E, N, M	21,34-41
Trichloroacetyl chloride	TCA-Cl	P, S, B	E	42
Heptafluorobutyryl chloride	HFB-Cl	P, S	M	43-45
Perfluorooctanoyl chloride	PFO-Cl	P, S	M	46,47
4-Carbethoxyhexafluorobutyryl chloride	CHFB-Cl	P, S	M	48-50
Benzoyl chloride		A, P, S	M	51
Ditrifluoromethylbenzoly chloride	DTFMB-Cl	B, H	M	52,53
Pentafluorobenzoyl chloride	PFB-Cl	P, S	M	54
N-Methyl-bis(trifluoroacetamide)	MBTFA	P, S	M	55
N-Succinimide benzoate		P	F	56
N-Hydroxysuccinimide tetrafluorobenzoate		P	F	57
N-Hydroxysuccinimide phenylacetate		P	F, M	58
3. Alkylation				
Formaldehyde/sodium borohydride		B	N	59
Propionaldehyde/sodium borohydride		P, S	F	60
Pentafluorobenzyl bromide	PFBz-Br	P, S, H	M	61,62
3,5-Bistrifluoromethylbenzyl chloride	BTFMBz-Cl	H	M	63-66
2,4-Dinitrofluorobenzene	DNFB	P, S	M	67
4. Halogenation				
Bromine		P, S	E	68,69
Iodine		P, S	E, M	70,71

5. Formation of carbamates				
Methyl chloroformate	MCF	P, S	N	72
Ethyl chloroformate	ECF	P, S, T, B	F, N, M	73-77
n-Propyl chloroformate	n-PCF	P, S	N, M	78-80
n-Butyl chloroformate	n-BCF	P, S	N, M	76,78
Isobutyl chloroformate	isoBCF	P, S	N, S, M	73,81,83
2,2,2-Trifluoroethyl chloroformate	TFECF	P, S	N, M	84
2,2,2-Trichloroethyl chloroformate	TCECF	P, S	M	85-88
Menthyl chloroformate		P, S	M	76
6. Formation of sulfonamides				
Benzenesulfonyl chloride	BS-Cl	A, P, S	E, S, M	23,67,89-95
p-Toluenesulfonyl chloride	TS-Cl	S, T	M	96
Pentafluorobenzenesulfonyl chloride	PFBS-Cl	P, B	E	97
7. Formation of phosphonamides				
Dimethylthiophosphoryl chloride	DMTP-Cl	P, S	P	98,99
Diethylthiophosphoryl chloride	DETP-Cl	S	P	100-102
8. Formation of Shiff base				
Trifluoroacetylacetone		B	F	103
Furfural		P	F	104
Cyclohexanone		P, S	M	105
Benzaldehyde		P	M	106
Pentafluorobenzaldehyde	PFBA	P	E, M	107-109
Dimethylformamide dimethyl acetal	DMF-DMA	A, P, H	N, P, S	95,110-114
9. Formation of thiourea and isothiocyanate				
Carbon disulfide		P	M	115
Allyl isothiocyanate		P	M	116
10. Formation of chiral derivative				
l-N-Trifluoroacetyl-1-propyl chloride	TFAP-Cl	P, S	M	117-120

[a] P: primary amine; S: secondary amine; T: tertiary amine; B: biogenic amine; H: heterocyclic amine.
[b] F: FID; E: ECD; N: NPD; S: FPD (S-mode); P: FPD (P-mode); M: MS.

Primary and secondary amines can be trimethylsilylated with BSTFA (or BSA) + TMCS (60°C), MSTFA + TMSIM in acetonitrile (60°C) or BSA + TMCS + TMSIM (20°C). BSTFA is a more powerful reagent than BSA, and its by-product is volatile, so it does not interfere in

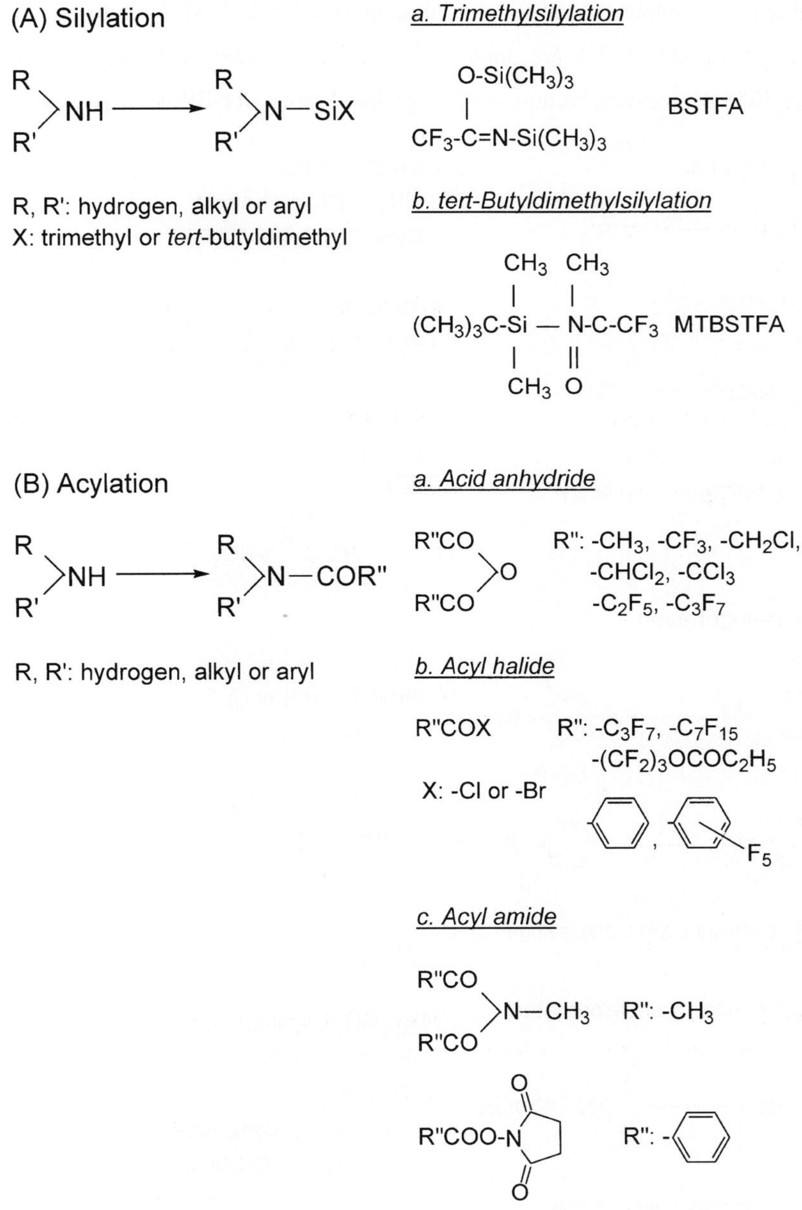

Figure 1-1. Typical derivatization reactions and derivatizing reagents for amines.

the subsequent analysis. However, the *N*-trimethylsilyl (TMS) derivatives produced by these reactions are unstable to moisture. An additional problem in the silylation of primary amines arises from the possibility of replacing both protones, resulting in the formation of mono- and di-TMS products. In contrast, the *N-tert*-butyldimethylsilyl (*t*-BDMS) derivatives produced by the reaction with MTBSTFA are about 10^4 times more stable to hydrolysis than the corresponding TMS derivatives, because the bulky *t*-butyl group of *t*-BDMS derivatives

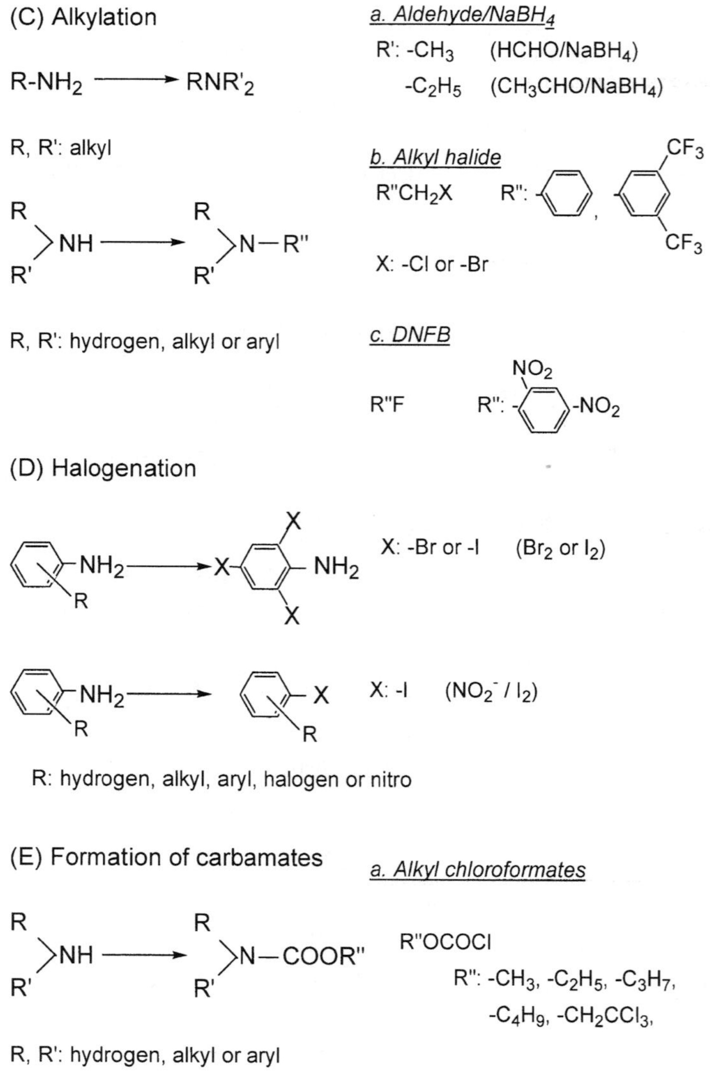

Figure 1-2 Typical derivatization reactions and derivatizing reagents for amines.

protect silyl groups from moisture. Compared with TMS derivatives, mono-*t*-BDMS and di-*t*-BDMS derivatives are 10 and 100 times more sensitive, respectively [17]. Flophemesyl reagents also react with alcohols, phenols, carboxylic acids and amines, and these derivatives are sensitive to ECD. The sequence of silyl donor power of reagents in pyridine solution is flophemesylamine > flophemesyl chloride > flophemesyldiethylamide.

2.2. Acylation

Acylation is one of the most widely used derivatization procedures for GC analysis of primary and secondary amines. The introduction of acyl protective groups improves the volatility, chromatographic mobility and chemical stability of these amines. Acylation is occasionally preferable to silylation, because acylated amines are more stable than the corresponding

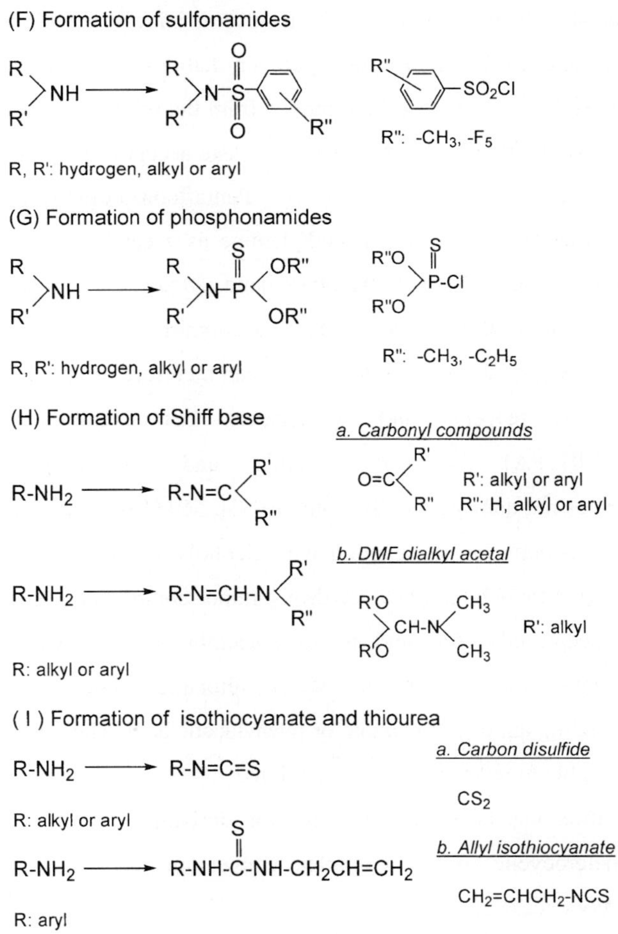

Figure 1-3 Typical derivatization reactions and derivatizing reagents for amines

N-silyl derivatives and are readily analyzed by GC. Acid anhydrides [16,20-41], acyl halides [42-54] and acyl amides [55-58] have been used as acylating reagents (Fig. 1B). Acylation with perfluoro-acylating agents strongly increases the volatility and GC mobility of derivatives. In addition, derivatization with fluoro (or chloro)-containing acyl groups, which possess electron-capturing properties, increases the electron affinity of amines and are therefore particularly useful for detection by ECD and GC/MS due to negative-ion chemical ionization (NICI). These reagents easily react with amino groups under mild reaction conditions, and many of them also react with hydroxyl, phenol and thiol groups [8,12].

In the reactions of amines with acid anhydrides and acyl halides, it is usually necessary to remove excess reagents and acid by-products, because these agents have damaging effects on GC columns. These reagents are often used with bases, such as pyridine, trimethyl- or triethyl-amine, which are not only excellent solvents but also act as catalysts to promote smooth reactions and as acceptors for hydrogen halides by-products (mainly hydrogen chloride). Hydrogen halides can also be removed from the mixture by heating. The reaction with acid anhydrides, at times in the presence of a basic acceptor, may be preferred because the excess reagent is easier to remove. Pentafluoropropionylation with pentafluoropropionic anhydride (PFPA), using trimethylamine as a catalyst, has also been used to derivatize aromatic amines [32]. On the other hand, some acyl amides, such as fluorinated acylimidazoles, trifluoroacetamides and acyl succinimides, are especially active and do not form undesirable by-products. These derivatives can therefore usually be analyzed without removal of excess reagents and by-products. For example, *N*-methyl-bis(trifluoroacetamide) (MBTFA) is very volatile and its by-product *N*-methyltrifluoroacetamide does not damage the GC column [55]. MBTFA is very active in its reaction with amines, but less active in its reaction with alcohols, making it useful for *N*-selective acylation after trimethylsilylation of hydroxyamino compounds. Several new reagents, such as acyl succinimide [56] and hydroxysuccinimides [57,58], have been suggested for the derivatization of low molecular weight aliphatic amines. The resulting derivatives are the corresponding amides of benzoic or phenylacetic acid. These reagents also have the advantage of being able to react in aqueous solutions.

On-column acylation has been described for the derivatization of primary and secondary aliphatic and heterocyclic amines. Complete conversion into *N*-acyl derivatives was observed when a 3-5 v/v excess of acyl anhydride or acyl halide was introduced into the flash-heater of a GC 3-5 s after sample injection. Extractive acylation of amines has been also described using a variety of reagents and a solid-phase extraction cartridge column

[25,31,43,51]. Although amines are usually derivatized after distillation or solvent extraction, extraction and derivatization can be performed simultaneously in the extractive acylation procedure. After adsorption of amines onto the column, they can be eluted with solvents containing acylating reagents. Extractive heptafluorobutylation of amphetamines using heptafluorobutyryl chloride (HFB-Cl) and an Extrelut® column has been described [43]. Recently, the combination of acylation and SPME has been proposed for GC analysis of amines [24,39,40,54,56,57], as described in section 2.11.

Among acylating reagents, fluorinated compounds have been widely used to introduce electron-capturing properties. The ECD gives a greatly enhanced response to halogenated derivatives, with the sensitivity of detection increasing in the order F < Cl < Br < I. However, volatility and stability tend to decline in this order, so that, in practice, fluoro derivatives are the most widely used. Although fluorinated acyl derivatives are more volatile than their corresponding nonfluorinated derivatives, there is generally little difference in retention times between trifluoroacetyl (TFA) and pentafluoropropionyl (PFP) or HFB derivatives. The ECD response generally increases with increasing fluorine content. Mono-, di- or tri-chloroacetyl derivatives also possess higher ECD responses, which are surpassed by those of pentafluorobenzoyl (PFB) derivatives. In comparing the ECD responses of chloro and fluoro derivatives of anilines, it was shown that chloroacetyl and PFB derivatives tend to be less volatile than the smaller perfluoroacyl derivatives, and the reactions of lower aliphatic amines with PFB-Cl give white, curdy precipitates, which may be the hydrochlorides of amines, and these reactions are not quantitative [21]. Although the derivatives containing nitro groups are also sensitive to ECD, they are generally less volatile and tend to give tailing peaks on chromatography, limiting their use for GC analysis. In contrast, nitrogen-phosphorus detector (NPD), which is very specific to nitrogen and phosphorus compounds, has been widely used to detect picogram levels of amines. In using NPD, the limits of detection of acyl derivatives are dependent essentially on the nitrogen content of the derivatives. Although GC-MS-SIM is also specific and sensitive to each amine, the sensitivity of this detection is generally lower than that of ECD and NPD (Table 1).

2.3. Alkylation

Replacement of H atom(s) in amino groups by an alkyl group leads to a reduction in polarity and improved GC properties. Alkylation of primary amines gives rise to secondary or tertiary amines, and alkylation of secondary amines gives rise to tertiary amines. Among the compounds used as alkylating reagents are aldehyde/sodium borohydride [59,60], pentafluorobenzyl bromide (PFBz-Br) [61,62], 3,5-bistrifluoromethylbenzyl chloride (BTFMBz-Cl) and 2,4-dinitrofluorobenzene (DNFB) [67], all of which react easily with amino groups under mild reaction conditions (Fig. 1C). Some of these compounds also react with carboxylic acid, phenol and thiol groups [9,12].

The permethylation reaction, which converts primary and secondary amines to tertiary amines, can be performed in aqueous solutions using formaldehyde/sodium borohydride and has been applied to the analysis of polyamines [59]. Permethyl derivatives eliminate the polar N-H groups but retain the troublesome adsorptive properties of tertiary amines. Derivatization with propionaldehyde/sodium borohydride to form *N*-propyl derivatives has been used to analyze amphetamines [60].

PFBz-Br has been used for the extractive alkylation of amphetamines [61] and heterocyclic amines [62]. In this method, extraction into the organic phase and pentafluorobenzylation can be performed simultaneously by stirring and heating the mixture on hot plate stirrer. BTFMBz-Cl has also been used as a derivatizing reagent for GC-MS analysis of heterocyclic amines [63-66]. Although these amines give a mixture of mono- and di-BTFMBz derivatives, the spectra of the latter show high mass fragment ions suitable for SIM work.

Dinitrophenylation, which can be performed in aqueous media, is used for the derivatization of primary and secondary amines. Among the dinitrophenylating reagents are DNFB and 2,4-dinitrobenzenesulfonic acid (DNBS), and dinitrophenyl (DNP) derivatives are sensitive to ECD, and making them particularly suitable for low-molecular mass amines that have inconveniently short retention times. Although the DNFB reaction procedure is simple, this reagent irritates the skin and occasionally causes allergic dermatosis. DNBS is water soluble and its resulting DNP derivatives are soluble in organic solvents, making them easy to separate from an excess of reagent by solvent extraction after derivatization. Another important advantage of DNBS is its greater specificity for amino groups compared with DNFB, which also reacts with thiol, imidazole and hydroxyl groups. However, DNBS generally reacts more slowly than DNFB, so that longer reaction times or stronger alkaline conditions may be required to complete the reaction. DNFB has been used for the analysis of primary and secondary aliphatic amines [67].

2.4. Halogenation

Halogenation with bromine and iodine has been used to derivatize aromatic amines and for subsequent GC-ECD analysis. Bromination or iodination of aromatic rings is performed in an acetic acid medium (Fig. 1D). In general, all free ortho- and para-positions to the amino group are attacked via electrophilic substitution. This derivatization method is easy to perform and the derivatives can be easily extract and separated by GC. Using this method, 56 aromatic amines have been analyzed [68,70]. Recently, a method for the in situ derivatization and SPME of polar aromatic amines was developed [71]. In this method, aromatic amines were directly derivatized in aqueous solution by diazotation and subsequent iodination in a one-pot reaction, and then extracted by SPME.

2.5. Carbamate formation

Carbamate derivatives, which are generally made by Schotten-Baumann type procedures, are useful for the analysis of primary, secondary and tertiary amines. Various alkyl chloroformates [72-88] have been used as reagents for this purpose (Fig. 1E). These reagents are less reactive to water than other acylating and silylating reagents. The reaction of amines with alkyl chloroformates is rapid and can be easily accomplished in a few minutes at room temperature in aqueous alkaline media, and the resulting carbamate derivatives have good GC properties. Alkyl chloroformates can also react with alcohol, phenol, thiol and imidazole groups, but anhydrous conditions may be necessary. The general aspects of chloroformates as derivatizing reagents in GC have also been reviewed by Husek [6].

Although alkoxycarbonylation of amines has been directly performed in alkaline aqueous solution with alkyl chloroformates, direct derivatization requires relatively large amounts of reagents because of their instability in alkaline solution [73,74]. With two-phase alkoxycarbonylation procedure [73,74,82,83], the derivatization takes place in the organic phase, thus allowing for more efficient and more rapid reactions with smaller amounts of reagents. The major advantage of the two-phase method is the simultaneous solvent extraction and derivatization of amines present in aqueous solution. The reaction rate and derivatization yield of alkoxycarbonylation strongly depend on the pH of the aqueous medium, and the appropriate pH is selected depending on the alkalinicities of the amines. Most alkoxycarbonylation reactions have been performed under strongly alkaline conditions (pH 10-12), but high pH is not suitable for alkaline-labile amines such as phenolic amines and catecholamines, which are prone to oxidative degradation.

As shown in Table 1, many different alkyl chloroformates are commercially available, and the carbamate derivatives of primary and secondary amines can be selectively and sensitively determined by GC-NPD [72-74,78,84], GC-FPD [81] and GC-MS [73,76,77,80,82-88]. Chloroformates with relatively short chain alkyl groups, such as ethyl, propyl and butyl compounds, have prove useful in the derivatization of amino groups. Some halogen-containing reagents such as 2,2,2-trichloroethyl (TCE) and PFBz chloroformates were developed for selective detection with ECD and GC-MS. TCE derivatives of amines are suitable for identification and quantitation by GC-MS, because molecular ion isotope clusters based on the isotope effect of three chlorine atoms in the derivative molecule have been shown [85-88]. On the other hand, tertiary amines, which tend to give tailing peaks on chromatograms owing to their polar nature, can be also converted into carbamate derivatives with alkyl chloroformate after dealkylation. Treatment with an alkylchloroformate can displace the smallest alkyl group to the nitrogen, particularly if it is a methyl group, to form a carbamate (Fig. 1E). This reaction proceeds in the presence of the alkaline catalyst sodium carbonate, but its yield is generally low. The carbamate derivatives of tertiary amines, formed by reaction with PFBz chloroformates, can be sensitively analyzed by GC-ECD.

Recently, alkoxycarbonylation procedures in combination with SPME have been proposed for amphetamine analysis [76-79]. Amphetamines can be derivatized with n-propyl or n-butyl chloroformates and the derivatives analyzed by headspace SPME/GC or GC-MS [77-79]. Amphetamines can be also directly derivatized on SPME fibers by exposing the latter to the headspace above the alkyl chloroformates in vials for 30 s, followed by desorption in the injection port of the GC-MS [76]. This on-fiber derivatization using alkyl chloroformates is simple and rapid, and can be conducted simultaneously with headspace SPME.

2.6. Sulfonamide formation

As shown in Fig. 2, the reaction of amines with sulfonyl chloride is useful for the separation and identification of amine types; because sulfonamides derived from primary amines are soluble in alkaline solution and those from secondary amines are insoluble, and tertiary amines do not react with sulfonamides. This reaction easily proceeds in aqueous alkaline media at room temperature. Benzenesulfonyl chloride (BS-Cl) [67,89-95], *p*-toluenesulfonyl chloride (TS-Cl) [96] and pentafluorobenzenesulfonyl chloride (PFBS-Cl) [97] have been used for the selective determination of low-molecular mass primary and secondary amines (Table 1). After benzenesulfonylation of amines, the BS derivatives of secondary amines can be completely extracted into n-hexane in 50% potassium hydroxide-22.5% methanol solution, and any remaining BS derivatives of primary amines in the aqueous layer can be extracted into diethyl ether under acidic conditions [92]. These BS derivatives can be selectively and sensitively analyzed by GC-FPD [92,93], and GC-chemiluminesence detector with a modified thermal energy analyzer [90,91]. The BS derivatives of primary amines can be also analyzed by GC-ECD, following conversion of these compounds into their *N*-TFA-*N*-BS derivatives by reaction with TFAA [23]. This method is selective and sensitive to primary amines, because the BS derivatives of secondary amines do not react with TFAA. PFBS-Cl is also a useful derivatizing reagent for GC-ECD of primary amines [96]. Furthermore, ammonia is converted into its BS-dimethylaminomethylene derivative by a convenient procedure involving benzenesulfonylation with BS-Cl and subsequent reaction with dimethylformamide dimethyl acetal (DMF-DMA), and the derivative analyzed by GC-FPD [95]. This reaction is based on the combination of benzenesulfonylation and Schiff base type condensation.

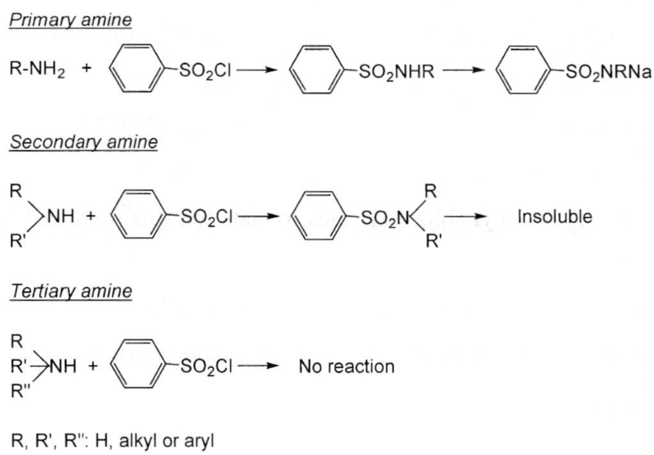

Figure 2 Separation and identification of primary, secondary and tertiary amines by Hinsberg method

2.7. Phosphonamide formation

The reaction of an amino group with dialkylthiophosphoryl chlorides can be performed rapidly in aqueous alkaline media to yield the corresponding *N*-dialkylthiophosphonamides [98-102]. Although these reagents react easily with both primary and secondary amines, the N-P bond of phosphorus amide is not more stable than the N-C and N-S bonds of the above derivatives. Using these reagents, a selective and sensitive method for the determination of aliphatic [100-102] and aromatic [99] amines by GC-FPD has been developed. In particular, secondary amines can be selectively converted into their *N*-diethylthiophosphoryl (DETP) derivatives with diethylchlorothiophosphate (DECTP) after treatment with *o*-

phthaldialdehyde (OPA), because OPA reacts only with primary amino groups (Fig. 3). On the other hand, aromatic amines can be detected as single and symmetrical peaks by capillary-GC following the formation of their *N*-dimethylphosphoryl (DMTP) derivatives. In these methods, excess reagents, which can interfere with the analysis of low-molecular mass amines, are removed by reaction with cysteic acid prior to the solvent extraction of the derivatives, because dialkylthiophosphoryl derivatives of cysteic acid are not extracted. Furthermore, these reactions can be applied to the analysis of *N*-nitrosamines, in which these amines are denitrosated with hydrobromic acid to produce the corresponding secondary amines, following by derivatization with DECTP [102].

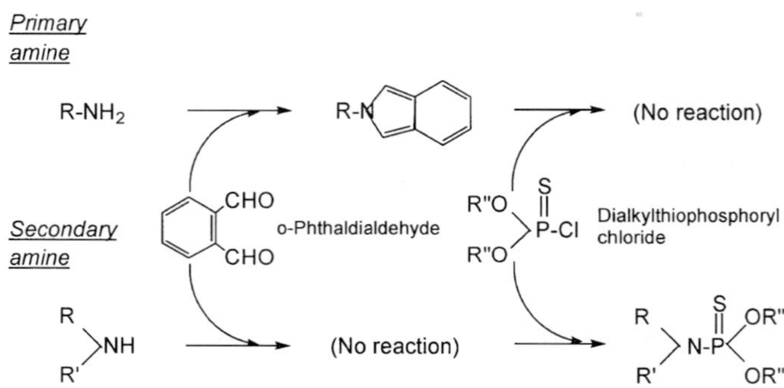

Figure 3 Selective derivatization of primary and secondary amines with *o*-phthaldialdehyde and dialkylthiophosphoryl chloride.

2.8. Shiff base formation

Primary amines react with a carbonyl compound to yield imines, whereas secondary amines react to yield eneamines (Fig. 1H). Acetone, 1,1,1-trifluoroacetone, trifluoroacetylacetone, methylisobutylketone, cyclopentanone, cyclohexanone and cycloheptanone have been used for a series of long chain and alicyclic monoamines, aliphatic diamines and aromatic amines. As shown in Fig. 4, basic drugs gave various condensation products with cyclohexanone [105]. For example, ephedrine yields a cyclic derivative, oxazolidine, due to the presence of a β-hydroxy (OH) group in the ephedrine molecule. Benzaldehyde, furfural, 2-thiophenealdehyde and pentafluorobenzaldehyde (PFBA) have also been used for derivatization of low-molecular mass amines. Biogenic primary alkylamines in wines have been derivatized with PFBA to from their corresponding pentafluorobenzylimines and analyzed by GC-MS [109]. The condensation reactions with these reagents proceed rapidly in aqueous medium (usually alcohol or acetic acid) at room temperature or on warming, and Schiff bases are obtainable in good yield. These carbonyl reagents are selective for primary amines, and their by-product, water, does not undergo further reactions in the systems used. However, excess reagents, which often interfere with the amine analysis, have to be removed in a separate clean-up step. Selective and sensitive methods based on the formation of sulfur- and fluorine-containing Schiff bases of primary amines have been developed for GC combined with FPD and ECD, respectively. Although the sensitivity of FPD is higher than that of FID for derivatives of C_1-C_4 amines, their sensitivities for derivatives of C_5-C_7 amines are similar. Derivatization of primary amines with PFBA to the corresponding pentafluorobenzylimines is compatible with GC-MS and affords picogram level sensitivity by SIM of the characteristic α-cleavage products [109]. Comparable sensitivity is achieved by GC-ECD, but unreacted PFBA saturates the ECD unless additional cleanup is applied.

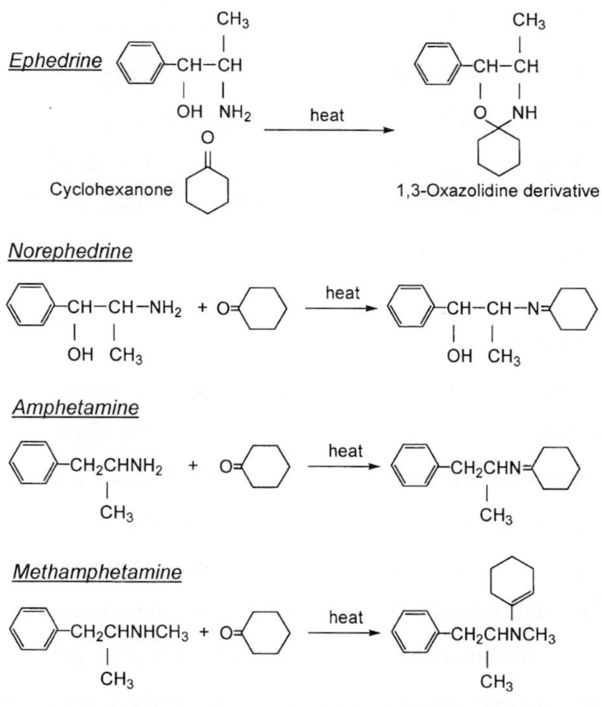

Figure 4 Reactions of some basic drugs with cyclohexanone at high temperature

Although, *N,N*-dimethylformamide (DMF) dialkyl acetal forms Schiff base type derivatives with primary amines (Fig. 1H), this reagent also reacts with carboxyl groups of fatty acids and amino acids to yield the corresponding esters. These reagents are sensitive to moisture, making it necessary to perform the reaction under dry conditions. Derivatization of primary amines with DMF-DMA is used to analyze heterocyclic amines [110-114]. These amines can be easily converted to their corresponding *N*-dimethyaminomethylene derivatives in methanol solution by heating, and they can be selectively and sensitively analyzed by GC-NPD. The excess DMF-DMA can be removed by the addition of water after the reaction.

2.9. Isothiocyanate and thiourea formation

Aromatic primary amines can be derivatized to *N*-allyl-*N'*-arylthioureas by reaction with allyl isothiocyanate (Fig. 1I). Pyrolysis of their derivatives always occurs in the heated GC injector, yielding the corresponding aryl isothiocyanates [116]. These derivatization reactions can be performed in aqueous medium at ambient temperature. Although the reaction is faster and more quantitative in alkaline solution since free amines are more nucleophilic than their protonated species, two broad peaks identified as *N,N'*-diallylurea and *N,N'*-diallylthiourea are observed due to the hydrolysis and subsequent side reactions of allyl isothiocyanate in solutions of pH≧8. The optimum pH is range 5 to 6.5. During pyrrolysis, the allyl group decompose leaving aryl isothiocyanate, and thus the amino group of aromatic amines is exchanged for the thiocyanate group of allyl isothiocyanate. The method is rapid, and together with the SPE clean-up, can be automated and used in combination with mass spectrometric and sulfur-sensitive detection methods.

2.10. Chiral derivatization

To selectively separate enantiomeric pairs, they are converted into diastereomers by reaction with an optically pure reagent, and the products are subsequently separated on achiral chromatographic matrices. The derivatization of chiral amines to form diastereomeric derivatives has been widely researched, and an equally large number of chiral reagents is available [117-120]. The major types of reaction for chiral amines are based on the formation of amide, carbamate, and thiourea (Fig. 1). These diastereomeric amide formation reactions are widely used for the resolution of various primary amines. The reactions with acid chloride and chloroformate reagents proceed rapidly to the produce corresponding amides and carbamates, respectively. The acid halogenides, such as *N*-substituted propyl chloride, are good labels because the amides are produced from acylation under mild conditions at room temperature or in an ice bath. Since hydrolysis of the reagents with water proceeds easily in sample solutions, contamination by water in the medium should be definitely avoided. Durden et al. [117] has applied this technique to the measurement of enantiomeric purities of 2-

alkylamines using *N*-(trifluoroacetyl)propyl derivatives and a chiral capillary column. Wang et al. [120] developed a method to simultaneously determine amphetamine and methamphetamine enantiomers in urine, consisting of diastereomeric derivatization with *l*-*N*-trifluoroacetyl-1-propyl chloride (TFAP-Cl) followed by GC-isotope dilution mass spectrometry. Although labeling with isocyanates and isothiocyanates is slower than with the above halogen reagents, the former are suitable for the resolution of racemate due to the introduction of polar functional groups of urea and isourea. Derivatization reactions using isocyanate reagents need rather drastic conditions, including high temperatures and long time periods.

2.11. Combination of derivatization and solid-phase microextraction

There are three techniques for derivatization-SPME. In the first, sequential derivatization, the amine is mixed with the derivatizing reagent and, after an appropriate time, the derivatized amine is recovered by SPME. The other two methods are on-fiber techniques, in which the amine and the derivatizing reagent are mixed within the SPME fiber. We decided to pursue on-fiber derivatization-SPME, we hypothesized that this approach would consume less derivatizing reagent and have greater recovery and time-efficiency. On-fiber derivatization-SPME involves exposing the fiber to the amine, allowing derivatization to occur on the fiber or within the injection port. It was also possible to rearrange the sequence of events, such that the fiber is exposed to the derivatizing reagent prior to exposure to the amine. If derivatization takes place within the fibers, and if the fiber is loaded with derivatizing reagent before exposure to the amine, then the amine may become "trapped" on the fiber. Essentially, the equilibrium between the amine external to the fiber and the amine within the fiber will be perturbed towards the latter because, in the fiber, the amine is in its derivatized form and will have a lower vapor pressure.

For derivatization in the sample matrix, Liu et al. [44] added HFB-Cl to an alkaline digest of hair and used a PDMS-coated fiber to perform HS-SPME extraction of methamphetamine and amphetamine derivatives. Cai et al. [57] determined aliphatic amines in lake water and human urine by SPME/GC-FID. The amines were derivatized with *N*-hydroxysuccinimide tetrafluorobenzoate and extracted with sol-gel dibenzo-18-crown-6 SPME fiber. A sol-gel technique was also used to fabricate silica fibers coated with poly(phenylmethylsilicon), which were used in the HS-SPME sampling of amphetamine derivatives. Zimmermann et al. [71] used a method consisting of diazotation with sodium nitrite and iodination with hydroiodic acid to decrease substantially the polarity of 18 aromatic amines in aqueous samples. The subsequent 30-min immersion of a PDMS/DVB-coated SPME fiber permitted the extraction of the derivatives, which were analyzed by GC-MS. The volume of each sample was 10 mL, which were acidified with hydroiodic acid and mixed for 20 min with sodium nitrite at room temperature to form the iodinated derivatives. SPME was performed after clean-up steps to remove excess iodine. Namera et al. [77] developed a method for the determination of amphetamine-related drugs in urine samples, consisting of in-matrix derivatization with ethyl chloroformate (ECF) and subsequent HS-SPME/GC-MS.

Of methods involving derivatization on fiber after sampling, Jurado et al. [24] reported a technique for the determination of amphetamines in urine using derivatization with trifluoroacetic anhydride (TFAA). After a PDMS-coated fiber was exposed for 10 min to the vapors of the sample solution and transferred to another vial, it was exposed for 20 min to TFAA, and the derivatives were analyzed by GC-MS. Another technique is headspace post-derivatization in the GC injection port [39,40]. In this method, heptafluorobutyric anhydride (HFBA) is injected into the injection port, and the amphetamines extracted into SPME fiber were simultaneously desorbed and derivatized by exposing the fiber in the injection port. Huang et al. [45] developed a one-step method for the determination of methamphetamine and amphetamine. The sample and additives were placed in a vial, which was closed tightly after the introduction of a glass insert; the vial was provided with 2-mm orifices in its upper part and contained HFB-Cl as the derivatizing reagent. A PDMS–coated fiber was inserted into the vial and exposed for 20 min to the HS of the solution.

3. Selective detectors of amine derivatives

A wide variety of detectors can be used for amine analysis, including NPD, FPD and ECD, which offer increased selectivity for specific amines. Furthermore, the GC-MS can provide structural information for the unequivocal identification of amines, which can be determined by MSD based on SIM. Using these detectors, sub-nanogram quantities can be measured. Although most detectors respond directly to amines, some detectors, such as FPD and ECD, require conversion of amines to suitable derivative prior to detection. Derivatization reactions, which are often selective for amine type (primary, secondary, or tertiary amines), have also been used to improve the detection and separation of these amines. For example, introducing sulfur-, phosphorus- and halogen-containing groups into a molecule enhances the response of FPD and ECD. The selection of a detector is decided by instrument availability, the required accuracy and precision, sensitivity, sample preparation needed and cost. To function properly, selective detectors must be operated under optimum conditions. These derivatization reactions and selective detectors are described in detail in previous reviews [2–4], and the general aspects of GC detectors have been reviewed in several books [121,122].

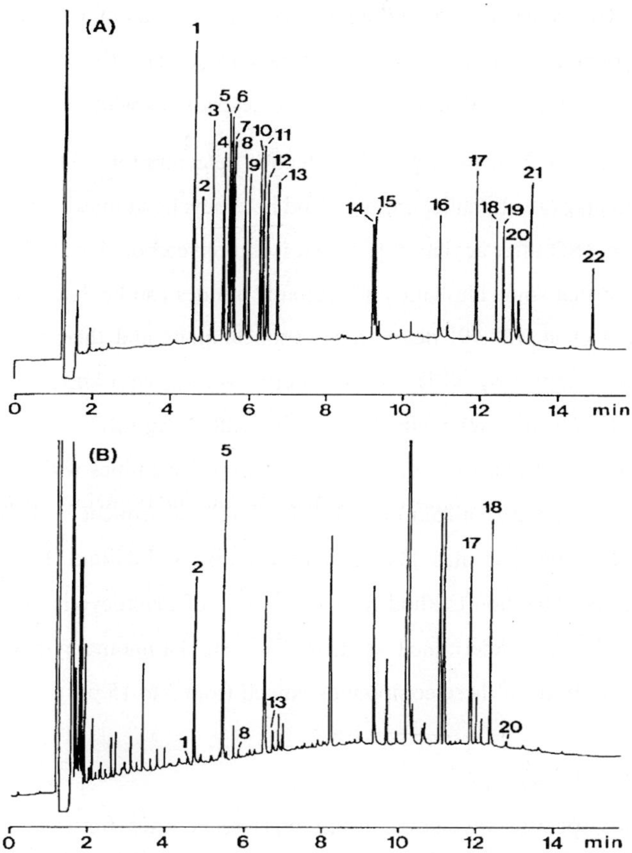

Figure 5 Gas chromatograms obtained from n-propoxycarbonyl derivatives of aromatic amines and cigarette smoke sample. (A) Standard amine derivatives (each peak corresponding to 1-5 ng of amine); (B) cigarette smoke sample. GC conditions: column, HP-5 (30 m x 0.32 mm I.D.); oven temperature, programmed from 120 to 220°C at 10°C min^{-1} and then programmed from 220 to 320°C at 20°C min^{-1} and isothermal at 320°C for 1 min; injection and detector temperatures, 330°C; carrier gas, helium, programmed from 80 to 120 kPa at 4 kPa min^{-1} and then programmed from 120 to 270 kPa at 30 kPa min^{-1} and held at 270 kPa for 1 min; detector, NPD. Peaks: 1=aniline, 2=internal standard (*N*-methyl-*m*-toluidine), 3=2-toluidine, 4=2,6-xylidine, 5=3-toluidine, 6=4-toluidine, 7=2-ethylaniline, 8=2,5-xylidine, 9=2,4-xylidine, 10=2,3-xylidine, 11=3,5-xylidine, 12=4-ethylaniline, 13=3,4-xylidine, 14=α-naphthylamine, 15=*o*-aminobiphenyl, 16=*N*-phenylnaphthylamine, 17=internal standard (2,4-diaminotoluene), 18=2-aminofluorene, 19=*p*-aminoazobenzene, 20=benzidine, 21=*p,p'*-diaminodiphenylmethane, 22=2-aminoanthracene.

3.1. Thermionic detector

The thermionic detector is selective for nitrogen- and phosphorus-containing compounds, which are efficiently ionized by a thermionic source of alkali salt. There are two types of flame type thermionic detectors, an FTD and an alkali flame-ionization detector (AFID), and a flameless type detector, a thermionic specific detector (TSD) or a nitrogen-phosphorus detector (NPD). The sensitivity of thermionic detectors is about 0.4-10 pg for nitrogen-containing compounds, which is about 50 times more sensitive than FID. The main disadvantage of this detector is that its performance deteriorates with time.

The thermionic detector is popular selective detectors for the GC analysis of amines and can be used to analyze aliphatic, aromatic and heterocyclic amines. Aliphatic amines have been detected with TSD after acylation [73] and isobutoxycarbonylation [82]. Skarping et al. [26,59,74] reported that some aliphatic and aromatic amines can be detected at the fmol level with TSD. Ugland et al. [78,79] determined amphetamines and their methylenedioxylated analogues in urine samples by NPD-GC after derivatization with alkyl chloroformates with detection limits of 5 ng mL^{-1} for methamphetamine and 15 ng mL^{-1} for other amines. Using HP-5 capillary columns, Kataoka et al. separated 22 aromatic amines within 15 min (Fig. 5A). This method is based on derivatization with n-PCF and subsequent NPD analysis, and can detect pg levels of aromatic amines. Furthermore, Kataoka and Kijima [110-114] developed a simple and rapid derivatization method for GC analysis of heterocyclic amines, in which 10 of these compounds were determined as their *N*-dimethylaminomethylene derivatives with NPD. The detection limits of these compounds ranged from 2 to 15 pg.

3.2. Flame photometric detector

FPD measures only specific wavelength, using interference filter, from the light emitted when a sample is burned in a hydrogen-rich flame, and is therefore highly selective for phosphorus- or sulfur-containing compounds. By choosing a filter having a transmission window of wavelength at 390-410 nm for sulfur-containing compounds or at 520-560 nm for phosphorus-containing compounds, the maximum emission from sulfur- or phosphorus-containing compounds can be sensed by the photomultiplier. In contrast, the light emitted by the burning of hydrocarbons alone is between 390 and 520 nm, which will be negligible using the filter described above. The selectivity ratio for detecting sulfur- or phosphorus-containing compounds versus carbon compounds is 10^4:1. The sensitivity of FPD is about 20 pg for sulfur-containing compounds and about 1 pg for phosphorus-containing compounds.

Although FPD is widely used for sulfur- and phosphorus-containing compounds, such as agricultural chemicals in the environment, it can also detect amino compounds by converting them to their sulfur- or phosphorus-containing derivatives (Fig. 1F) [23,92,93,99-102,123]. Kataoka et al. [92,93] reported that 14 primary and 9 secondary aliphatic amines could be detected by FPD with sulfur mode as their *N*-BS derivatives, with detection limits ranging from 10 to 100 pg. Furthermore, Kataoka et al. reported that aliphatic [100,101] and aromatic [99] amines and *N*-nitrosamines [102,123] could be analyzed as their *N*-dialkylthiophosphoryl derivatives by FPD with phosphorus mode.

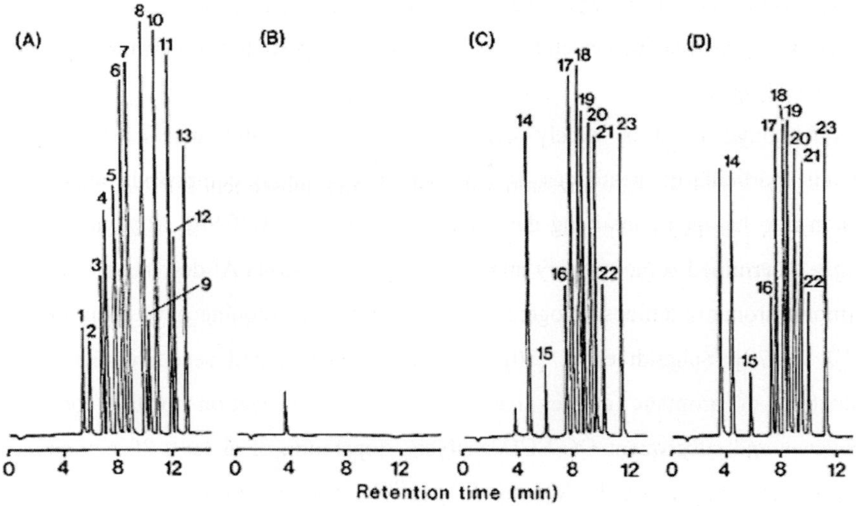

Figure 6. Gas chromatograms obtained from *N*-diethylthiophosphoryl derivatives of primary and secondary amines. (A) Primary amines; (B) primary amines pretreated with OPA; (C) secondary amines; secondary amines pretreated with OPA. GC conditions: column, DB-1701 (15 m x 0.53 mm I.D.), oven temperature, programmed from 100 to 260°C at 10°C min^{-1}; injection and detector temperatures, 280°C; nitrogen flow-rate, 10 mL min^{-1}; detector, PFD (P mode). Each peak corresponds to 20 pmol of amine. Peaks: 1=methylamine, 2=ethylamine, 3=isopropylamine, 4=*n*-propylamine, 5=isobutylamine, 6=*n*-butylamine, 7=isoamylamine, 8=*n*-amylamine, 9=hexylamine, 10=cyclohexylamine, 11=heptylamine, 12=benzylamine, 13=octylamine, 14=β-phenylethylamine, 15=dimethylamine, 16=diethylamine, 17=dipropylamine, 18=pyrrolidine, 19=morpholine, 20=piperidine, 21=dibutylamine, 22=hexamethyleneimine, 23=*N*-methylbenzylamine. (From H. Kataoka et al. [101]. Reproduced with permission of Elsevier Science)

In particular, secondary amines can be selectively converted into their *N*-DETP derivatives by treatment with OPA, which reacts only with primary amino group. As shown in

Fig. 6, primary amines are also derivatized without OPA treatment and detected by FPD, but they are not detected when pretreatment with OPA. In contrast, secondary amines are detected irrespective of OPA pretreatment, because these amines do not react with OPA. This technique has also been applied to the determination of *N*-nitrosamines, which are denitrosated with hydrobromic acid to produce the corresponding secondary amines and then derivatized.

3.3. Electron-capture detector

ECD, which consists of a radioactive source such as Ni^{63}, is a selective detector of halogenated or nitrated compounds, organometallics and conjugated carbonyls. All of these compounds are capable of capturing low energy electrons to form negatively charged ions. Although the selectivity depends on the affinity of these compounds for the electrode, the selectivity ratio for detection of poly halogen compounds versus carbon is 10^5-10^6 : 1. ECD is one of the most sensitive GC detectors available, with a sensitivity for chlorinated compounds of about 0.05-1 pg.

Although ECD is widely used in the detection and analysis of halogenated compounds, particularly, pesticides, it can also detect amino compounds following their conversion into halogen-containing derivatives (Fig. 1) [21,64-70,96]. Miyamoto et al. [96] selectively determined some primary amines as their *N*-BS-*N*-TFAl derivatives by GC-ECD. Furthermore, aromatic amines halogenated with bromine and iodine have been analyzed by GC-ECD [68-70]. Schmidt et al. [70] reported a selective and sensitive method for the determination of aromatic amines based on their derivatization to the corresponding iodobenzenes and subsequent GC-ECD analysis. Separation of at least 30 compounds in a single chromatographic run in 30 min is possible, and 52 aromatic amines have been investigated.

3.4. Mass selective detector

MSD has been recognized as an excellent detector for GC, because the spectral data provides the qualitative information lacking from other GC detection methods. GC-MS can provide structural information for the unequivocal identification of amines and can be applied to the quantitative determinations of aliphatic [67,85,89,107] and aromatic [21,28,73,88,90] and heterocyclic amines [18,61-66]. The development of quantitative GC-MS analysis based on SIM sometimes demands chemical derivatization to obtain derivatives whose mass spectra contain abundant ions in the less contaminated higher-mass region. Involvement of such ions greatly increases the sensitivity and specificity of the analysis. Introduction of mass-deficient atoms (for example halogens) can lower the chemical noise level and, hence, facilitate high-resolution MS determination. In negative-ion chemical ionization (NICI)-MS the introduction of groups with electron capture properties (mainly halogen-containing) into the analyte molecule can greatly improve the sensitivity and specificity of the analysis.

Aliphatic amines have been analyzed by GC-MS after derivatization with benzoyl-Cl [57], DNFB [67], BS-Cl [67] and PFBA [109]. Sacher et al. [67] reported that aliphatic primary and secondary amines could be determined as their *N*-BS and DNP derivatives, and that this method was selective and sensitive. In contrast, aromatic amines have been analyzed by GC-MS after derivatization with TFAA [25], PFPA [30,31,33], HFBA [36,41], CHFB-Cl [50], benzoyl-Cl [57], iodine [71] and TCECF [86]. Brede et al. [25] developed a method for the solid-phase trifluoroacetylation of primary aromatic amines, followed by GC-MS analysis. Dasgupta and Jagannath [50] reported a GC-MS method for the determination of aniline in serum samples after derivatization with CHFB-Cl. Longo and Cavallaro [36] developed a method for the simultaneous identification of 73 primary and secondary aromatic amines as their heptafluorobutyramides. The electrophoretic derivatives were analyzed by GC combined with electron-capture NICI-MS, with detection limits in the range of 0.3-66.3 pg injected in the full-scan mode and 0.01-0.57 pg injected in the SIM mode. Furthermore, Murray et al. [63,64] developed a sensitive and selective method for the determination of several heterocyclic amines as their BTFMBz derivatives, followed by NICI-MS in the SIM mode. The detection limits were about 1 pg. Amphetamines and related drugs were also determined by GC-MS after acylation [16,24,39,40,43,54], alkylation [61], carbamate formation [76,77,80,87] and Shiff base formation [105].

4. Recent applications of amine analysis in various samples

Amines are present in environmental and biological samples at low parts per billion or less. When the samples are analyzed by GC with nonselective detector such as FID, many peaks with the same retention times as those of amines are often present. Although GC analysis with selective detectors can save sample preparation, certain clean-up procedures for each complex sample matrix are necessary for reliable and accurate analyses. To remove co-eluting interfering compounds and to preconcentrate amines, the extraction and clean-up of the sample has been performed using a number of different purification techniques, including distillation, liquid-liquid extraction, adsorption with various adsorbent tube or column, preparative high performance liquid chromatography (HPLC), solid-phase extraction (SPE) with various cartridges and SPME. Some amines can be separated by acid-base partition, because these compounds can be extracted with organic solvents at high pH (pH>10). Adsorption with adsorbent tube or column is effective for the separation of amines, but it is time-consuming. Blue-Cotton or Blue-Rayon [124], consisting of cotton or rayon bearing covalently linked copper phthalocyanine trisulfonate, can selectively adsorb heterocyclic amines and other mutagens/carcinogens having polycyclic planar molecular structures. Normal or reversed phase HPLC fractionations are also useful for the preparation of samples. SPE is simple and rapid, and good recoveries of amines are obtained. In contrast, SPME, which was recently developed by Pawliszyn and co-workers [125], is an extraction technique using fused-silica fiber coated on the outside with an appropriate stationary phase. This method has been found to save preparation time, solvent usage and disposal cost, as well as improving detection limits. SPME has been used routinely in combination with GC and GC-MS, and has been successfully applied to a wide variety of compounds including amines.

4.1. Aliphatic amines

Low-molecular mass aliphatic amines cause unpleasant odors, such as those associated with the decay of foods and fish, and are hazardous to human health. These amines are also emitted from our living environment and various industries. Secondary amines such as dimethylamine and pyrrolidine are particularly important, because they are broadly distributed in various environments and are precursors of carcinogenic *N*-nitrosamines.

GC methods for the determinations of aliphatic amines using selective detectors have been applied to various environmental samples such as air, cigarette smoke, water and soil. Zhao et al. [56] developed a method for the determination of aliphatic amines by derivatization with an aqueous *N*-succinimidyl benzoate solution, followed by SPME/GC-FID. Fig. 7 shows the chromatograms obtained from lake water sample. Detection limits were below 0.13-7.2 pmol mL^{-1}, and the recoveries of standard amines spiked into the samples were 94.1-102.7%. Sacher et al. [67] analyzed primary and secondary amines in industrial wastewater and surface water samples as their *N*-BS derivatives by GC-MS, and detected methylamine, dimethylamine, morpholine and ethanolamine in concentrations up to 100 ng mL^{-1}. Furthermore, ethylamine and diethylamine were found in concentrations up to 30 ng mL^{-1}. Pietsch et al. [85], who analyzed aliphatic and alicyclic amines in wastewater samples by GC-MS after derivatization with TFECF and liquid-liquid extraction, detected morpholine and piperidine. Terashi and et al. [89] reported the determination of 8 primary and secondary

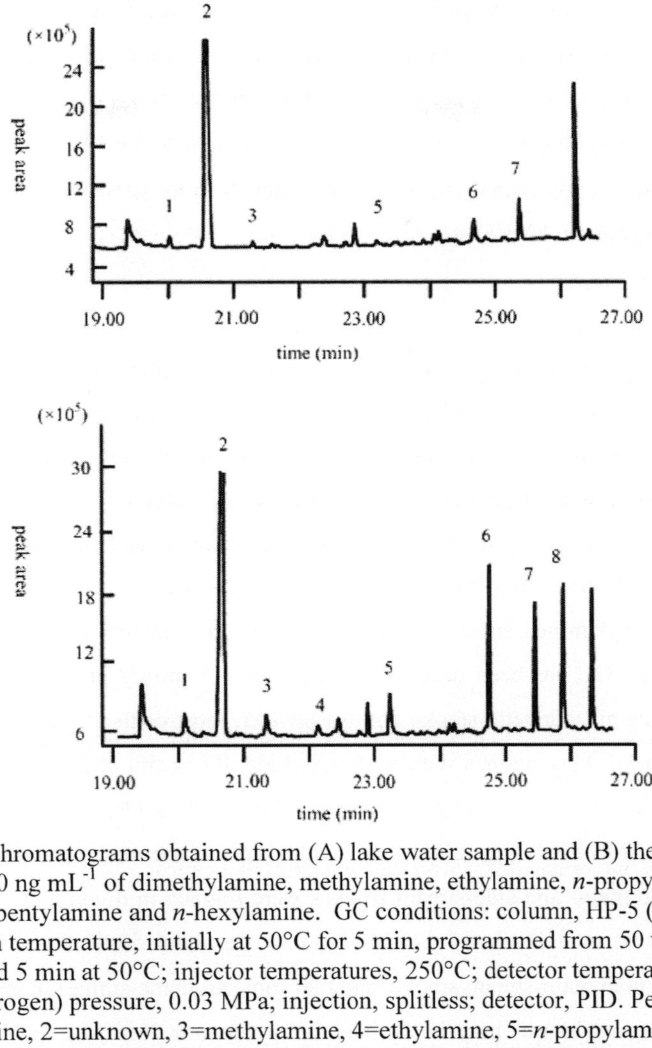

Figure 7 Gas chromatograms obtained from (A) lake water sample and (B) the same sample spiked with 100 ng mL^{-1} of dimethylamine, methylamine, ethylamine, *n*-propylamine, *n*-butylamine, *n*-pentylamine and *n*-hexylamine. GC conditions: column, HP-5 (30 m x 0.32 mm I.D.), oven temperature, initially at 50°C for 5 min, programmed from 50 to 220°C at 5°C min^{-1}, then hold 5 min at 50°C; injector temperatures, 250°C; detector temperatures, 260°C; carrier gas (nitrogen) pressure, 0.03 MPa; injection, splitless; detector, PID. Peaks: 1=dimethylamine, 2=unknown, 3=methylamine, 4=ethylamine, 5=*n*-propylamine, 6=*n*-butylamine, 7=*n*-pentylamine, 8=*n*-hexylamine. (From Zhao et al. [56]. Reproduced with permission of Elsevier Science)

---

amines in river water, seawater and sea sediment by GC-MS-SIM based on the distillation of samples under alkaline conditions and subsequent benzenesulfonylation. Using this method, the detection limits of amines in water and sediment were 0.02-2 and 0.5-50 ppm, respectively. In addition, n-butylamine, di-n-propylamine and di-n-butylamine were detected in trace amounts in a seawater sample. Kataoka et al. [101] reported the determination of secondary amines as their DETP derivatives in cigarette smoke by GC-FPD. This method is selective and sensitive for secondary amines, with limits of detection of 0.05-0.2 pmol. This method has, confirmed that dimethylamine, pyrrolidine, piperidine and morpholine occur in main- and side-stream smokes of cigarettes, and that the amounts of these amines in side-stream smoke is higher than those in main-stream smoke. Ngim et al. [109] determined biogenic primary alkylamines in wines by derivatization with PFBA followed by GC-MS analysis. Alkylamine concentrations in wines ranged from 0.048 to 91 μg mL^{-1}, and recoveries were generally 81-100%.

4.2. Aromatic amines

Aromatic amines such as aniline and its chlorinated analogues, phenylene diamine and diphenylamine are used in the manufacture of carbamate and urethane pesticides, dyestuffs, cosmetics and medicines. These amines are also employed in the rubber industry as antioxidants and antiozonants and as components of epoxy and polyurethane polymers. These compounds have been shown to be highly toxic to mammals and fish. Most are not only poisonous in themselves but may be converted into carcinogens such as *N*-nitroso compounds, naphthylamines, substituted phenylamines and benzidine analogues.

Although GC has been used to analyze aromatic amines in various environmental samples, including air, cigarette smoke and wastewater, most of these methods are based on the derivatization of these amines with acid anhydride. Pieraccini et al. [27,28] reported the determination of 17 primary aromatic amines as their PFP amides in cigarette smoke and indoor air by GC-MS-SIM. Cigarettes were smoked in a home-made smoking machine and the amines in the main- and side-stream smokes were trapped in dilute hydrochloric acid. It was confirmed that total levels of aromatic amines are about 50-60 times higher in side-stream smoke than in main-stream smoke, and that some aromatic amines in ambient air, such as that in offices and houses may be derived from contamination with side-stream smoke. Kataoka et al. also analyzed aromatic amines in cigarette smoke samples as their *N*-n-propoxycarbonyl and *N*-DMTP derivatives by GC-NPD and GC-FPD (P mode), respectively. Stabbert et al. [33] developed a method for the analysis of aromatic amines in cigarette smoke, which combined the sensitivity of PFP derivatives in the NICI mode with the selectivity of the GC-MS-MS technique. Longo and Cavallaro [36] analyzed aromatic amines in ground water samples as their HFB derivatives by GC-EC-NICI-MS after concentration with Kuderna-Danish evaporator. Schmidt et al. [70] reported a method for the selective determination of aromatic amines in water samples. This method is based on solid-phase extraction at pH 9, derivatization to the corresponding iodobenzenes, and GC-ECD analysis. This method detected aniline and nitroaniline compounds in groundwater samples.

Hsu et al. [30] reported using an isotope dilution GC-MS-NICI method to determine urine concentrations of the carcinogen benzidine and its metabolites. Urine samples were purified by SPE, reduced with $LiAlH_4$/THF, and derivatized with PFPA. The limits of detection were 0.5-1.5 pg mL^{-1}. Dasgupta determined aniline in serum sample by GC-MS after derivatization with CHFB-Cl [50] and TCECF [86]. Detection limits of these derivatives were 0.1 μg mL^{-1} serum.

4.2. Heterocyclic amines

Most heterocyclic amines are polar and less volatile, and, without derivatization, tend to elute as broad and tailing peaks due to their strong adsorption on the column and injector during GC analysis. Analysis of these amines has therefore been generally carried out by HPLC. For GC analysis of heterocyclic amines, several derivatizations using acetic anhydride, TFAA, HFBA, PFBz-Br and BTFMBz-Cl have been tested. While acylation with acid anhydrides yielded derivatives with very poor GC properties, alkylation with PFBz-Br[62], BTFMBz-Cl or -Br [63-66] heterocyclic amine derivatives with good GC properties. However, these methods gave a mixture of mono- and di-alkylated forms, which can be used for GC-MS but not for GC analysis. Vainiotalo et al. [65] analyzed 2-amino-3,8-dimethylimidazo[4,5-*f*]-quinoxaline (MeIQx) and 2-amino-3,4,8-trimethylimidazo[4,5-*f*]quinoxaline (4,8-DiMeIQx) in cooking fumes as their BTFMBz derivatives by GC-MS with SIM mode. The cooking fumes were collected through a glass fiber funnel into a sampler consisting of two glass fiber filters and an XAD-2 sorbent tube. Although these heterocyclic amines yielded a mixture of mono- and di-BTFMBz derivatives, the spectra of the di-BTFMBz derivatives possess high mass fragment ions suitable for SIM. When ions m/z 438 (MeIQx) and m/z 452 (DiMeIQx) were specially monitored, the detection limits of these compounds was 2 pg. Casal et al. [19] assayed 12 heterocyclic amines in food samples by GC-MS after derivatization with MTBSTFA. This derivatization method, however, is not applicable to the pyridoimidazoles 2-amino-6-methyldipyrido[1,2-*a* : 3'2'-*d*]imidazole (Glu-P-1) and 2-aminodipyrido[1,2-*a* : 3'2'-*d*]imidazole (Glu-P-2) and to the β-carboline Harman due to its incompleteness. The instability of the imidazolquinoline and imidazoquinoxaline derivatives, requiring their injection on the same working day, is a further drawback. Recently, Kataoka and Kijima [110] developed a simple and rapid derivatization method for GC analysis of mutagenic heterocyclic amines. Ten heterocyclic amines were converted to their *N*-dimethylaminomethylene derivatives (Fig. 8) with DMF-DMA and measured by GC-NPD. As shown in Fig. 8A, these heterocyclic amines were separated within 7 min, although 2-

amino-9*H*-pyrido[2,3-*b*]indole (AαC) and Glu-P-2 coeluted. The detection limits of these compounds ranged from 2 to 15 pg. Using this method enabled the identification of AαC, 3-amino-1,4-dimethyl-5*H*-pyrido[3,4-*b*]indole (Trp-P-1) 3-amino-1-methyl-5*H*-pyrido[3,4-*b*]-indole (Trp-P-2), Glu-P-1, IQ, 2-amino-3,4-dimethylimidazo[4,5-*f*]quinoline (MeIQ) and 2-amino-1-methyl-6-phenylimidazo[4,5-*b*]pyridine (PhIP) in combustion smoke samples [111] (Fig. 8B). In contrast, there is little GC-MS-SIM data using positive or negative electron ionization for the analysis of heterocyclic amines after derivatization. Analytical methods for the determination of heterocyclic amines have also been described in detail in a previous review [126].

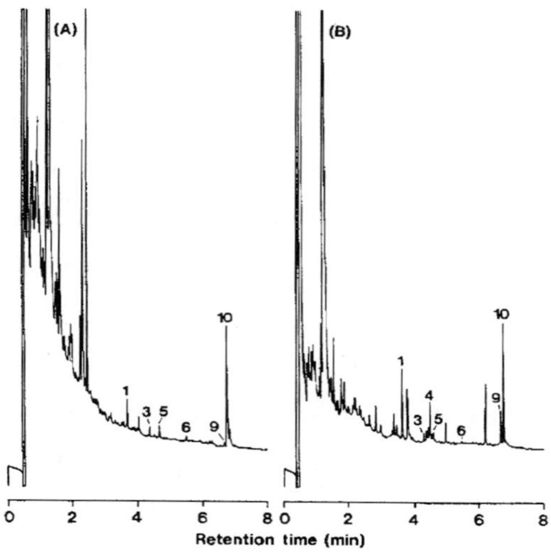

Figure 8 Gas chromatograms obtained from (A) standard heterocyclic amines and (B) cigarette sidestream smoke. GC conditions: column, connected DB-1 (10 m x 0.25 mm I.D.) and DB-17ht (10 m x 0.25 mm I.D.); oven temperature, programmed from 230 to 280°C at 10°C min^{-1}, from 280 to 330°C at 25°C min^{-1} and held at 330°C for 1 min; injection and detector temperatures, 340°C; carrier gas, helium, programmed from 180 to 230 kPa at 10 kPa min^{-1}, from 230 to 280 kPa at 25 kPa min^{-1} and held at 280 kPa for 1 min; split ratio, 10:1; detector, NPD. Peaks: 1=AαC, 2=Glu-P-1, 3=Trp-P-1, 4=Trp-P-2, 5=IQ, 6=MeIQ, 7=MeIQx, 8=DiMeIQx, 9=PhIP, 10=TriMeIQx (internal standard). (From Kataoka et al. [111]. Reproduced with permission of Springer-Verlag)

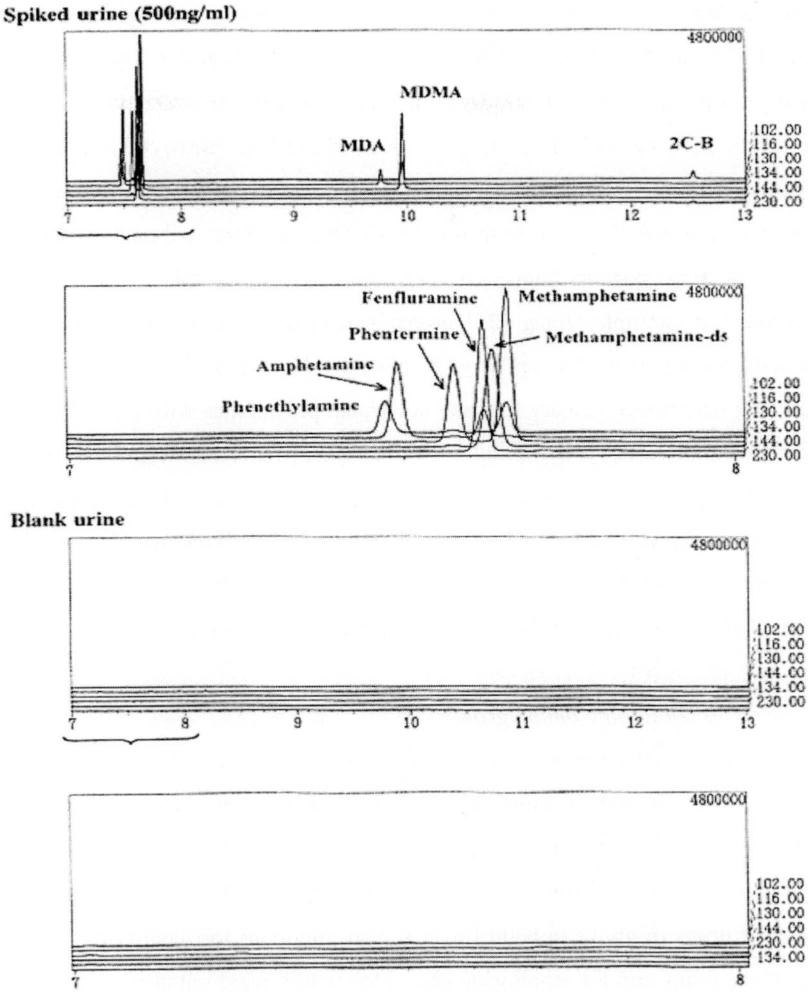

Figure 9 Typical SIM chromatograms of the ethylformate derivatives of amphetamine related drugs. (A) Spiked urine (500 ng mL^{-1}), (B) Blank urine. GC-MS conditions: GC column, Supelco PTE-5 (30 m x 0.25 mm I.D.); oven temperature, initially at 80°C for 3 min, programmed from 80 to 220°C at 40°C min^{-1}, from 220 to 280°C at 8°C min^{-1} and held at 280°C for 3 min; injector and interface temperatures, 250 and 230°C, respectively; carrier gas, helium, 0.8 mL min^{-1}. The ions used for quantitation were m/z 102 for phenethylamine, m/z 116 for amphetamine and (methylenedioxy- amphetamine: MDA), m/z 130 for methamphetamine, (methylenedioxymethamphetamine: MDMA) and phentermine, m/z 144 for fenfluramine, m/z 230 for 4-bromo-2,5-dimethoxymethamphetamine (2C-B), m/z 134 for methamphetamine-d_5 (the internal standard). (From Namera et al. [77]. Reproduced with permission of Preston Publications)

4.3. Biogenic amines

The use of GC for the analysis of biogenic amines in biological samples is gradually increasing. Although GC of biogenic amines without derivatization is possible, derivatization with suitable reagents has been shown to improve GC elution and resolution, with considerable enhancement in the sensitivity of the detection system. The enormous capabilities of high-resolution GC capillary column, together with selective detection systems, have enabled the use of GC for the quantitation of polyamines. For example, Jiang [22] determined polyamines in the urine of normal human and cancer patients by capillary GC after derivatization with TFAA. Suh and co-workers [37] determined urinary free and acetylated polyamines with capillary GC-NPD after derivatization with HFBA. Khuhawar et al. [103] reported a capillary GC method for the determination of putrescine and cadaverine in the serum of cancer patients and healthy persons by GC-FID based on derivatization with trifluoroacetylacetone, with a limit of detection of 0.5-0.6 ng. This method has been used to measure these amines in the serum and urine of cancer patients, before and after radiotherapy.

4.5. Basic drugs containing amino groups

Analysis of drugs in pharmaceutical products and biological samples is growing in importance, both for the development of more selective and effective drugs and in understanding their therapeutic and toxic effects. In contrast, drugs of abuse, illicit drugs, intoxicating drugs and poisons are analyzed in clinical and forensic toxicology. The screening of drugs of abuse in body fluids is also important for identifying and treating users of these drugs and for monitoring drug addicts following withdrawal from therapy. Derivatization and GC analysis of basic drugs have been used for these purposes.

Amphetamine and methamphetamine in urine and serum samples have been analyzed as various derivatives in combination with SPME. For example, Koster et al. [54] measured amphetamine in urine samples by on-fiber derivatization with PFB-Cl and subsequent GC-MS-NICI analysis. Extraction was performed by direct immersion of a 100 μm PDMS-coated fiber into human urine, and on-fiber derivatization was performed at the same time as or after extraction. Using this method, the limit of detection was 100 pg mL^{-1}. Lee et al. [40] reported a method for the determination of amphetamine and methamphetamine in serum samples by headspace post-derivatization with HFBA and SPME/GC-MS. This method yielded limits of detection at the pg mL^{-1} level, or one order of magnitude more sensitive than obtained without derivatization. Wang et al. [119] simultaneously determined amphetamine and methamphetamine enantiomers in urine by simultaneous liquid-liquid extraction and diastereomeric derivatization with TFAP-Cl followed by GC-isotope dilution MS. The GC-MS quantitation was carried out in the SIM mode using m/z 237 and 251 as the quantifier ions for the respective diastereomeric pairs of amphetamine and methamphetamine derivatives. The limits of detection and quantitation were as low as 40 and 45 ng mL^{-1}, respectively. Alkyl chloroformates [72,76-80,87] have often been used for the derivatization of amphetamines, and applied to the analysis of urine and blood samples. Fig. 9 shows typical chromatograms obtained from urine sample [77].

Palmer et al. [32] reported the simultaneous determination of fenfluramine and phentermine in urine samples by GC-MS after derivatization with PFPA. Dasgupta et al. developed new derivatization methods for phenmetrazine as an anorectic agent in urine using PFO-Cl [46],CHFB-Cl [48] and TCECF [88]. These derivatives were analyzed by GC-MS-NICI, with the limits of detection being 0.5 and 0.2 μg mL^{-1} for PFO and CHFB derivatives, respectively. Furthermore, Way et al. [49] reported stable isotope dilution GC-MS measurement of tricyclic antidepressant drugs. Desipramine, nortriptyline, imipramine and amitriptyline in plasma samples were derivatized with CHFB-Cl. This method had a lower limit of quantitation (25 ng mL^{-1}) and correlated well with HPLC measurements.

References

[1] H. Kataoka, S. Yamamoto, S. Narimatsu, in I.D. Wilson, T.D. Adlard, C.F. Poole and M. Cook (Editors), Encyclopedia of Separation Science, Academic Press, London, UK (2000), 5: 1982-1990.

[2] H. Kataoka, J. Chromatogr. A 733 (1996) 19-34.

[3] H. Kataoka, in R.A. Meyers (Editor) Encyclopedia of Analytical Chemistry: Instrumentation and Applications, John Wiley & Sons, Chichester, UK (2000) 3: 2587-2618.

[4] H. Kataoka, S. Yamamoto, S. Narimatsu, in W. Kleibohmer (Editor) Handbook of Analytical Separations, Elsevier Science, Amsterdam (2001) 3: 1-37.
[5] J. Segura, R. Ventura, C. Jurado, J. Chromatogr. B 713 (1998) 61-90.
[6] P. Husek, J. Chromatogr. B 717 (1998) 57-91.
[7] J.M. Halket, V.G. Zaikin, Eur. J. Mass Spectrom. 9 (2003) 1-21.
[8] V.G. Zaikin, J.M. Halket, Eur. J. Mass Spectrom. 9 (2003) 421-434.
[9] J.M. Halket, V.G. Zaikin, Eur. J. Mass Spectrom. 10 (2004) 1-19.
[10] V.G. Zaikin, J.M. Halket, Eur. J. Mass Spectrom. 10 (2004) 555-568.
[11] J. Drozd, Chemical Derivatization in Gas Chromatography, Elsevier, Amsterdam, 1981.
[12] K. Blau and J. Halket, Handbook of Derivatives for Chromatography, 2nd ed., John Willey & Sons, Chichester, 1993.
[13] M. Krogh, S.Pedersen-Bjergaard, K.E. Rasmussen, in J. Pawliszyn (Editor) Applications of Solid-Phase Microextraction, Royal Society of Chemistry, Cambridge, (1999) 461-469.
[14] H.A.H. Billiet, J. Chromatogr. Libr. 51B (1992) 583-595.
[15] F. Bruner, Gas chromatographic environmental analysis: principles, techniques, instrumentation, VCH, New York, 1993.
[16] D.-L. Lin, W.-T. Chang, T.-L. Kuo, R.H. Liu, J. Anal. Toxicol. 24 (2000) 275-280.
[17] C. Drouett-Coassolo, C. Aubert, P. Coassolo and J.P. Cano, J. Chromatogr. 487 (1989) 295-311.
[18] M.A.D. Ibanez, M. Chessebeuf-Pachiu and P. Padieu, J. Chromatogr. 488 (1989) 237-248.
[19] S. Casal, E. Mendes, J.O. Fernandes, M.B.P.P. Oliveira, M.A. Ferreira, J. Chromatogr. A 1040 (2004) 105-114.
[20] G.B. Baker, R.T. Coutts, A. Holt, J. Pharm. Toxicol. Methods 31 (1994) 141-148.
[21] H-B. Lee, J. Chromatogr, 457 (1988) 267-278.
[22] X. Jiang, Biomed. Chromatogr. 4 (1990) 73-77.
[23] Y. Miyamoto, H. Kataoka, S. Ohrui and M. Makita, *Bunseki Kagaku* 43 (1994) 1113-1118.

[24] C. Jurado, M.P. Gimenez, T. Menendez, M. Repeto, J. anal. Toxicol. 24 (2000) 11-16.

[25] C. Brede, I. Skjevrak, H. Herikstad, J. Chromatogr. A 983 (2003) 35-42.

[26] S. Skarping, L. Renman, C. Sango, L. Mathiasson and M. Dalene, J. Chromatogr. 346 (1985) 191-204.

[27] G. Pieraccini, F. Luceri and G. Moneti, Rapid Commun. Mass Spectrom. 6 (1992) 406-409.

[28] F. Luceri, G. Pieraccini, G. Moreti and P. Dolara, Toxicol. Indust. Health 9 (1993) 405-413.

[29] J.S. Thomson, J.B. Green, T.B. McWilliams, S.K.-T. Yu, J. High Res. Chromatogr. 17 (1994) 415-426.

[30] F.-F. Hsu, V. Lakshmi, N. Rothman, V.K. Bhatnager, R.B. Hayes, R. Kashyap, D.J. Parukh, S.K. Kashyap, J. Turk, T. Zenser, B. Davis, Anal. Biochem. 234 (1996) 183-189.

[31] C. Kutzer, B. Branner, W. Zwickenpflug, E. Richter, J. Chromatogr. Sci. 35 (1997) 1-6.

[32] R.B. Palmer, N.-H. Kim, A. Dasgupta, Ther. Drug Monit. 22 (2000) 418-422.

[33] R. Stabbert, K.-H. Schafer, C. Biefel, K. Rustemeier, Rapid Commun. Mass Spectrom. 17 (2003) 2125-2132.

[34] G.A. Van den Berg, A. W. Kingma, F.A.J. Muskiet, J. Chromatogr. 415 (1987) 27-34.

[35] M. Dalene, G. Skarping and T. Brorson, J. Chromatogr. A 516 (1990) 405-413.

[36] M. Longo, A. Cavallaro, J. Chromatogr. A 753 (1996) 91-100.

[37] J.W. Suh, S.H. Lee, B.C. Chung, J. Park, J. Chromatogr. B 688 (1997) 179-186.

[38] B. Dorhout, A.W. Kingma, E. de Hoog, F.A.J. Muskiet, J. Chromatogr. B 700 (1997) 23-30.

[39] A. Namera, M. Yashiki, J. Liu, K. Okajima, K. Hara, T. Imamura, T. Kojima, Forensic Sci. Int. 109 (2000) 215-223.

[40] M.-R. Lee, Y.-S. Song, B.-H. Hwang, C.-C. Chou, J. Chromatogr. A 896 (2000) 265-273.

[41] J.B. Forehand, G.L. Dooly, S.C. Moldoveanu, J. Chromatogr. A 898 (2000) 111-124.

[42] J.S. Salonen, J. Chromatogr., 527 (1990) 163-168.

[43] K. Hara, S. Kashimura, Y. Hieda, M. Kageura, J. Anal. Toxicol. 21 (1997) 54-58.

[44] J. Liu, K. Hara, S. Kashimura, M. Kashiwagi, M. Kageura, J. Chromatogr. B 758 (2001) 95-101.

[45] M.-K. Huang, C. Liu, S.-D. Huang, Analyst 127 (2002) 1203-1206.

[46] A. Dasgupta, C.E. Mahle, J. Forensic Sci. 42 (1997) 937-941.

[47] A. Dasgupta, O. Yousef, J. Chromatogr. B, 705 (1998) 283-288.

[48] A. Dasgupta, A. Hart, P. Humphrey, W. Blackwell, J. Forensic Sci. 43 (1998) 636-640.

[49] B.A. Way, D. Stickle, M.E. Mitchell, J.W. Koenig, J. Turk, J. Anal. Toxicol. 22 (1998) 374-382.

[50] A. Dasgupta, C. Jagannath, Ther. Drug Monit. 21 (1999) 238-242.

[51] S. Mishra, V. Singh, A. Jain, K. Verma, Analyst 126 (2001) 1663-1668.

[52] D.G. Watson, P. Zhou, J.M. Midgley, C.D. Milligan, K. Kaiser, J. Pharm. Biomed. Anal. 11 (1993) 1145-1149.

[53] S. Murray, N.J. Gooderham, V.F. Barner, A.R. Boobis and D.S. Davies, Carcinogenesis, 8 (1987) 937-940.

[54] E.H.M. Koster, C.H.P. Bruins, G.J. de Jong, Analyst 127 (2002) 598-602.

[55] M.S. Lelowx, E.G. De Jong and R.A.A. Maes, J. Chromatogr. 488 (1989) 357-367.

[56] Y.-Y. Zhao, L.-S. Cai, Z.-Z. Jing, H. Wang, J.-X. Yu, H.-S. Zhang, J. Chromatogr. A 1021 (2003) 175-181.

[57] L. Cai, Y. Zhao, S. Gong, L. Dong, C. Wu, Chromatographia58 (2003) 615-621.

[58] Y.-Y. Zhao, Z.-Z. Jing, H. Wang, H.-S. Zhang, J.-X. Yu, Anal. Chim. Acta 468 (2002) 255-261.

[59] M. Dalena, T. Lundh and L. Mathiasson, J. Chromatogr. 322 (1985) 169-176.

[60] P. Jacob, E.C. Tisdale, K. Pauganiban, D. Cannon, K. Zabel, J.E. Mendelson, R.T. Jones, J. Chromatogr. B 664 (1995) 449-457.

[61] M. Sato, T. Mitsui, J. Pharm. Biomed. Anal. 16 (1997) 139-145.[62] M.D. Friesen, L. Garren, J-C. Bereziat, F. Kadlubar, D. Lin, Environ. Health Perspect. 99 (1993) 179-181.

[63] S. Murray, N.J. Gooderham, A.R. Boobis and D.S. Davies, Carcinogenesis, 9 (1988) 321-325.

[64] S. Murray, A.M. Lynch, M.G. Knize and N.J. Gooderham, J. Chromatogr. 616 (1993) 211-219.

[65] S. Vainiotalo, K. Matveinen and A. Reunanen, Fresenius. J. Anal. Chem. 345 (1993) 462.

[66] L.M. Tikkanen, T.M. Sauri and K.J. Latva-Kala, Food Chem. Toxicol. 31 (1993) 717.

[67] F. Sacher, S. Lenz, H.-J. Brauch, J. Chromatogr. A 764 (1997) 85-93.

[68] T.C. Schmidt, R. Haas, E. von Low, K. Steinbach, Chromatographia, 48 (1998) 436-442.

[69] A.E. Dados, C.D. Stalikas, G.A. Pilidis, Chromatographia 59 (2004) 335-341.

[70] T.C. Schmidt, M. Less, R. Haas, E. von Low, K. Steinbach, G. Stork, J. Chromatogr. A 810 (1998) 161-172.

[71] T. Zimmermann, W.J. Ensinger, T.C. Schmidt, Anal. Chem., 76 (2004) 1028-1038.

[72] J. Jonsson, R. Kronstrand, M. Hatanpaa, J. Forensic Sci. 41 (1996) 148-151.

[73] G. Skarping, T. Bellander and L. Mathiasson, J. Chromatogr. 370 (1986) 245-258.

[74] G. Skarping, M. Dalene, T. Brorson, J.F. Sandstrom, C. Sango and A. Tiljander, J. Chromatogr. 479 (1989) 125-133.

[75] P. Husek, J. Microcol. Sep., 5 (1993) 101-103.

[76] H. Brown, K.P. Kirkbride, P.E. Pigou, G.S. Walker, J. Forensic Sci. 48 (2003) 1231-1238.

[77] A. Namera, M. Yashiki, T. Kojima, J. Chromatogr. Sci. 40 (2002) 19-25.

[78] H.G. Ugland, M. Krogh, K.E. Rasmussen, J. Chromatogr. B 701 (1997) 29-38.

[79] H.G. Ugland, M. Krogh, K.E. Rasmussen, J. Pharm. Biomed. Anal. 19 (1999) 436-475.

[80] M. Nishida, A. Namera, M. Yashiki, T. Kojima, Forensic Sci. Int. 125 (2002) 156-162.

[81] H. Kataoka, Y. Imamura, H. Tanaka and M. Makita, J. Pharm. Biomed. Anal. 11 (1993) 963-969.

[82] T. Lundh and B. Akesson, J. Chromatogr., 617 (1993) 191-196.

[83] K.R. Kim, M.J. Paik, J.H. Kim, S.W. Dong, D.H. Jeong, J. Pharm. Biomed. Anal., 15 (1997) 1309-1318.

[84] M. Dalene, G. Skarping and H. Tinnerberg, *J. Chromatogr.* 656 (1994) 319-328.

[85] J. Pietsch, S. Hampel, W. Schmidt, H. Brauch, Anal. Bioanal. Chem. 355 (1996) 164-173.

[86] A. Dasgupta, J. Chromatogr. B 716 (1998) 354-358.

[87] A. Dasgupta, J. Spies, Am. J. Clin. Pathol. 109 (1998) 527-532.

[88] A. Dasgupta, M.S. Handler, J.S. Nine, J. Forensic Sci. 43 (1998) 630-638.

[89] A. Terashi, Y. Hanada, A. Kido and R. Shinohara, J. Chromatogr. 503 (1990) 369-375.

[90] B. Pfundstein, A.R. Tricker and R. Preussmann, J. Chromatogr. 539 (1991) 141-148.

[91] B. Pfundstein, A.R. Tricker, E. Theobald, B. Spiegelhalder and R. Preussmann, Food Chem. Toxicol. 29 (1991) 733.

[92] S. Ohrui, H. Kataoka, Y. Miyamoto, K. Ohtsuka and M. Makita, Bunseki Kagaku, 40 (1991) 119-123.

[93] H. Kataoka, S. Ohrui, Y. Miyamoto and M. Makita, J. Chromatogr., 578 (1992) 120-123.

[94] H. Kataoka, S. Ohrui, Y. Miyamoto and M. Makita, Biomed. Chromatogr. 6 (1992) 251-254.

[95] H. Kataoka, S. Ohrui, A. Kanemoto and M. Makita, J. Chromatogr. 633 (1993) 311-314.

[96] K.A. da Costa, T.J. Vrbanac and S.H. Zeisel, Anal. Biochem., 187 (1990) 234-.

[97] G.B. Barker, T.S. Rao, R.T. Coutts, J. Chromatogr. 381 (1986) 211-217.

[98] H. Kataoka, K. Nagao, N. Nabeshima, M. Kiyama and M. Makita, J. Chromatogr. 626 (1992) 239-243.

[99] H. Kataoka, K. Kijima and M. Makita, J. Chromatogr. A 738 (1996) 83-90.

[100] H. Kataoka, M. Eda and M. Makita, Biomed. Chromatogr. 7 (1993) 129-133.

[101] H. Kataoka, S. Shindoh and M. Makita, J. Chromatogr. A 695 (1995) 142-148.

[102] H. Kataoka, S. Shindoh and M. Makita, J. Chromatogr. A 723 (1996) 93-99.

[103] M.Y. Khuhawar, A.A. Memon, P.D. Jaipal, M.I. Bhanger, J. Chromatogr. B 723 (1999) 17-24.

[104] L. Maros, Acta Chim. Hung. 121 (1986) 339.

[105] B.M. El-Haj, A.M. Al-Amri, M.H. Hassan, H.S. Ali, R.K. Bin Khadem, Forensic Sci. Int. 135 (2003) 16-26.

[106] M.L. Di Gioia, A. Leggio, A. Le Pera, A. Liguori, A. Napoli, F. Perri, C. Siciliano, J. Chromatogr. A 1066 (2005) 143-148.

[107] M.J. Avery and G.A. Junk, Anal. Chem. 57 (1985) 790.

[108] M.J. Avery and G.A. Junk, J. Chromatogr. 420 (1987) 379.

[109] K.K. Ngim, S.E. Ebeler, M.E. Lew, D.G. Crosby, J.W. Wong, J. Argic. Food Chem., 48 (2000) 3311-3316.

[110] H. Kataoka, K. Kijima, J. Chromatogr. A 767 (1997) 187-194.

[111] H. Kataoka, K. Kijima, G. Maruo, Bull. Environ. Contam. Toxicol. 60 (1998) 60-67.

[112] H. Kataoka, T. Hayatsu, G. Hietsch, H. Steinkellner, S. Nishioka, S. Narimatsu, S. Knasmuller, H. Hayatsu, Mutation. Res. 466 (2000) 27-35.

[113] H. Kataoka, S. Nishioka, M. Kobayashi, T. Hanaoka, S. Tsugane, Bull. Environ. Contam. Toxicol., 69 (2002) 682-689.

[114] Y. Totsuka, H. Kataoka, T. Enya-Takamura, T. Sugimura, K. Wakabayashi, Mutation. Res. 506-507 (2002) 49-54.

[115] V.S. Gaind, F. Chai, Analyst, 115 (1990) 143-145.

[116] V. Singh, M. Gupta, A. Jain, K.K. Verma, J. Chromatogr. A1010 (2001) 243-253.

[117] D.A. Durden, B.A. Davis, A.A. Boulton, J. Chromatogr. B 689 (1997) 165-173.

[118] I.F. Van Bocxlaer, W.E. Lambert, L. Thienpont, A.P. De Leenheer, J. Anal. Toxicol. 21 (1997) 5-11.

[119] Q.F. Tao, S. Zeng, J. Biochem. Biophys. Methods, 54 (2002) 103-113.

[120] S.-M. Wang, T.-C. Wang, Y.-S. Giang, J. Chromatogr. B 816 (2005) 131-143.

[121] M. Dressler, Selective gas chromatographic detectors in J. Chromatogr. Libr., Vol. 36, 1986.

[122] R.P.W. Scott, Chromatographic detectors: design, function and operation, Chromatographic Science Series Vol. 73, Marcel Dekker Inc., New York, 1996.

[123] H. Kataoka, M. Kurisu and S. Shindoh, Bull. Environment. Contam. Toxicol., 59 (1997) 570-576.

[124] H. Hayatsu, J. Chromatogr., 597 (1992) 37.

[125] J. Pawliszyn, Solid Phase Microextraction: Theory and Practice, Wiley-VCH: New York, 1997.

[126] H. Kataoka, J. Chromatogr. A 774 (1997) 121-142.

Ibolya Molnár-Perl (Editor)
Quantitation of Amino Acids and Amines by Chromatography
Journal of Chromatography Library, Vol. 70

2.2.1.

HPLC of amines as *o*-phthalaldehyde derivatives

Ibolya Molnár-Perl

Contents

Summary

1.) HPLC quantitation of amines as OPA derivatives has been evaluated and discussed.

2.) Stability and characteristics of the C_1-C_8 aliphatic monoamines, several diamines, including biogenic amines, derivatized with various SH-additives containing OPA reagents of different compositions have been studied from an analytical and theoretical point of view, equally.

3.) Stoichiometric studies have been followed as a function of the reaction time by using different SH-additives, varying the molar ratios of OPA/SH-additive from 1/0.5 to 1/50.

4.) The composition of derivatives was determined by on-line HPL/MS(ESI) measurements.

5.) As a result of an exhaustive derivatization program, performed under strictly the same practical conditions, we obtained comparable results and new knowledge: (i) in the case of the C_1-C_5 aliphatic amines it has been shown that the use of the OPA/MPA and/or the OPA/NAC=1/50 reagent resulted in two benefits: in an increased stability of the derivatives and in a lower number of species formed, consequently these reagents proved to be proper for their quantitation purposes. (ii) Derivatization studies performed with hexyl, heptyl and octyl amines revealed that applying the OPA/SH-additive=1/50 reagents, in order to inhibit the formation of the two OPA derivative-containing product, resulted in an additional, transformed OPA derivative: detected and determined by HPLC for the first time, (on the basis of on line HPLC/MS(ESI measurements), these proved to be the two SH-additive-containing OPA derivatives. The proportion of the transformed derivatives can be unambiguously influenced by the quality of SH-additive, by the composition of the OPA reagent, i.e., by the molar ratio of the OPA to the SH-additive and by the pH of derivatizations. In terms of side reaction free derivatization the OPA/ET reagents proved to be superb compared to the OPA/MPA one. (iii) In order to improve stability and to increase responses of spermidine and spermine a new principle, the two step derivatization of biogenic amines, has been introduced, applying the OPA/ET/FMOC reagent.

1. Introduction

Recently, (in addition to our study on the o-phthalaldehyde (OPA) derivatives of amino acids (AAs), detailed in chapter 1.2.3. of this book), the characteristics of the OPA derivatized amines (As) have been investigated and clarified, in parallel [1-7]: with special interest in the behavior of the C_1-C_8 aliphatic monoamines, aliphatic di and polyamines (PAs) designated also as biogenic amines (BAs).

In the frame of early structure-elucidation of the reaction product obtained from OPA and the primary amino group [8-10], — (as well as in the pioneer stability studies of this product [11-14) —n-propylamine as model compound was investigated: the isoindole character of the product has been confirmed by NMR evidence using various SH-group containing reagents [9], such as 2-mercaptoethanol (MCE), ethanethiol (ET), tert.-butylthiol, thiophenol, etc.). At that time, in the early 1980s, the product was not separated by chromatographic technique. Consequently, its transformation into the successively formed product(s) was not realized, and, to the author's knowledge, it has not been reported since either. Unfortunately, when the chromatographic technique became common and the unambiguous phenomena of the considerably low stability of OPA-As, in comparison even to the OPA-AA derivatives, became known, the intrinsic reason for and background of this fact, has not yet been investigated. However, this simple, selective and sensitive OPA-derivatization principle, also in the analytical practice of As, according to a recent compilation, has gained wide acceptance [15-66]: the HPLC of As, in particular those of BAs proved to be performed most commonly as OPA derivatives [66].

For the SH-group containing additive in the OPA derivatization of As, 2-mercaptoethanol (MCE) [3,15-47], ethanethiol (ET) [5,6,7,48,49], N-acetyl-L-cysteine (NAC) [1,2,4,18,50-58], 3-mercaptopropionic acid (MPA) [1,2,4,18,57,58], or 1-thio-D-glucose (TG) [18] have been used: applying fluorescence (Fl) [1-7,15-19,21,24-37, 39-56], photodiode array (DAD) [38] simultaneous Fl and DAD [1-7,37,53], electrochemical (El) [20,22,23] or mass selective (MS) [1,2,5-7] detections. A considerable part of the OPA/MCE-amines have been detected in the post-column mode [26,29,31,32,34,35,39-43,45], while the BAs of wines were on-column derivatized [51,52], pumping the OPA reagent through the column together with other solvents used as mobile phases. The absorption [59] and fluorescence [59,60] properties of the histamine-OPA complex proved to be suitable to histamine's quantitation of various biological matrices in the absence of any SH-additive [59,60].

The role and importance, as well the specificity of the impact of various As to several, different field of human life were compiled in review articles [61-65].

The pathway(s) of polyamine metabolism [61] explain their source and their formation-mechanism in biochemistry, microbiology, oncology and parasitology. A very illuminating compilation addressed the significance of biogenic amines in food safety and human health [62]. As in plant materials are regarded to fulfill an array of roles in cellular metabolism [63]. The multiple function of BAs in living organism, being metabolites, are considered to be involved in the process of cell multiplication and its regulation [64], serving also as cancer markers under (patho)physiological conditions [65].

On the basis of a recent monograph [66] and on a large number of papers, it is clear that, in particular in the HPLC of BAs, OPA derivatization is the method of choice. Out of the cited research articles [1-60], including basic studies [1-14,18,48,51,60], the overwhelming majority of proposals deal with the quantitation of As present in biological tissues or fluids

[15,16,19,20,23,27,33,45,49,50,53,54,57-59], in different food samples: such as in: must and wine [21,22,24,25,30,34,36,40,41,44,52], beer [46], fish [28,37,43], meat [26,31,32,35,39], cheese [29,37], vegetables [37,42,47], various agrochemical and pharmaceutical compositions [38,56] and in waste water [55].

The main dilemma of basic studies including also our own [1-7] was the low stability of the OPA-A derivatives. Tto improve the stability of the OPA-As, their extraction into organic solvents has been recommended [15-17,48], or the use of micellar surfactants proposed [18] which resulted in partly improved stability of the OPA-amine species [15-18,48].

The OPA/MCE derivatives [15-17] of histamine, norepinephrine, normetanephrine, dopamine, serotonin and tyramine obtained from plasma, urine and tissues [15], the norepinephrine, dopamine and serotonin content of plasma [16], as well as 11 As of red must and wine [17] have been extracted subsequently to their formation into ethyl acetate: in these extracts derivatives proved to be stable for 20 h [15]. The recovery of added As varied between 54.3 and 77.8 %, including sample cleanup and extraction procedures, respectively [15].

A basic research study [48] proved the increased stability of the OPA/ET derivatives of spermine (Spm), spermidine (Spd), putrescine (Put), cadaverine(Cad) and 1,6-hexanediamine in their ethyl acetate extracts: after an initial reaction time of 90 s followed by their quantitation from the ethyl acetate extracts after 2.5-, 6.5- and 23 h, in order of listing, revealed the following recovery percentages: 80-, 31- and 5 % for Spm, 93-, 83- and 67 % for Spd, 96-, 92- and 90% for Put, 97-, 93- and 85 % for Cad and 97-, 92- and 90% for 1,6-hexanediamine.

The histamine, tryptamine and tyramine content of bacterial cultures have been derivatized in 2-propanol/ethyl methyl ketone (10:90, v/v) mixture (1.3 mL) with the OPA/ET reagent (0.2 mL), in a capped tube and vortex mixed (15 s) [49]: after 30 min the organic phase, containing the OPA/ET derivatives, has been diluted with methanol and water, in the volume ratios of the organic phase/methanol/water=1/1/2 and injected into the HPLC system. The majority of amines may be estimated reliably with recoveries of added amounts ranging from 74-96%.

The addition of surfactant micelles (sodium dodecylsulfate) resulted in improved stability of the OPA/MCE derivatives of methylamine, Tyr, Put and Cad, but did not increase the stability of the corresponding OPA/MPA and OPA/NAC derivatives, respectively [18].

The histamine level from various biological matrices [19], and the histamine and 1-methyl-histamine contents from rat peritoneal mast cells [50,51] have been determined in the presence of ammonia, histidine, Spm and Spd [19] performing post column Fl- [19], El- [20,23] and precolumn Fl detections [50,51].

The mono- and diamine contents of different wines [22-24,52] were quantitated partly by Fl [24,52] and partly by El detections [22,23]. A coulometric array of sixteen electrodes increased the selectivity of the method [22]. The agmatine concentrations in brain and plasma- [27], the β-phenylethylamine content of human plasma [33] have been determined by Fl detections [27,33].The diamine (DA) content of urine and plasma samples has been measured separately from amino acids applying optimized eluent composition and the fully end-capped material containing Inertsil column [53].

On the basis of our experience [1-7], after measuring the instability of the OPA-As, in particular of those of diamines (DAs), it seemed unbelievable that the transformation of the initially formed products to the forthcoming ones, with a single exception relating to ephedrine [58], was not realized. The formation of double species of ephedrine obtained upon its pre- or postcolumn derivatization with the OPA/NAC reagent [58] was reported without any further explanation of the finding.

In this chapter the characteristics and stability of the aliphatic mono- and DAs, including the relevant Bas, have been studied, primarily, from an analytical point of view, at a basic research level. Changes in responses have been followed as a function of the reaction time and reagent's composition, applying different SH-additives in various molar ratios (OPA/SH-additive=1/0.5-50) using partly DAD and Fl detections, simultaneously, partly MS detection in the ESI positive mode.

2. Derivatization study of amines: Expectations based on experience with amino acids: the role and impact of the mole ratios of reagent's composition

With our knowledge of the reason for and background of the special behavior and characteristics of all those primary amino compounds that have in their initial structure the $-CH_2-NH_2$ moiety, including all n-aliphatic mono and diamines ([1-7], chapter 2.2.1. of this book), we assumed that changing the molar ratios of OPA to the SH- additive from 1/3 to 1/50, i.e., decreasing the ready to react, free OPA concentration responsible for the transformation of the initially formed classical isoindole, might result in two benefits, simultaneously:

(i) in a decrease of the transformation rate of the initially formed derivative, and,

(ii) in an increase of the overall stability of the total of derivatives.

In all cases investigated with the OPA/MPA, OPA/NAC and OPA/MCE reagents (as a function of the reaction time and reagent's composition) the transformation of the initially formed derivatives into the forthcoming ones has been demonstrated.

3. Studies on the behavior and characteristics of the C_1-C_5 aliphatic As and ethanolamine upon their reaction with the OPA/SH-additive=1/3 and OPA/SH-additive=1/50 reagents

Exhaustive derivatization study with selected members of aliphatic monoamines (Figures 1,2 Table 1: data for the OPA/NAC(MPA)=1/3 – 1/50 derivatives) are shown as a function of the reaction time. Responses, expressed in the total of derivatives were given in arbitrary units (Table 1: integrator unit/pM As, from 90s up to 6h).

In the cases of C_1-C_5 aliphatic As including ethanolamine, both the decreased transformation rate of the initially formed OPA derivatives and the increased stability of the total of species proved to be equal to our expectations [1]. The same tendency was observed with their OPA/MCE derivatives [3], however, a considerably lower overall stability was confirmed compared to the corresponding OPA/NAC(MPA) ones (Figure 2).

(i) The initially formed OPA derivatives, in all those cases where the neighboring group to the primary amino group was a CH_2 moiety, transformed to a second species of longer retention time containing additional OPA molecule(s), as expected: the composition of the transformed derivatives and the reaction pathway they originate from have been confirmed on the basis of their molecular masses by on line HPLC-MS measurements [2].

(ii) The exceptions of i-propyl-, s-butyl- and t-butylamines that furnish a single OPA derivative, can be attributed to their initial molecular structure: their neighboring groups to their primary amino groups, in order of listing, are the -CH-, -CH- and -C= moieties, respectively.

(iii) Stability of derivatives prepared strictly under the same condition, proved to be associated with the chain length of the aliphatic As: the longer the chain length the slower the decomposition of the total of derivatives formed. Evaluating response values from analytical point of view- (shown in the horizontal lines of Table 1, expressed as integration unit/pM As), they proved to be of importance in choosing optimum reaction time for the amine in question.

In cases of ethanolamine, members of C_1-C_4 n-As and i-amyl-A, between 90s and 7 min reaction times, maximum responses can be expected.

(iv) Unfortunately, the derivatization rate of the sterically hindered amino groups of the sec.- and tert.-aliphatic As, applying the OPA/NAC(MPA)=1/50 reagents, considerably decreased even in comparison to the OPA/NAC(MPA) =1/3 reagents. Thus, due to sterical hindrance, to

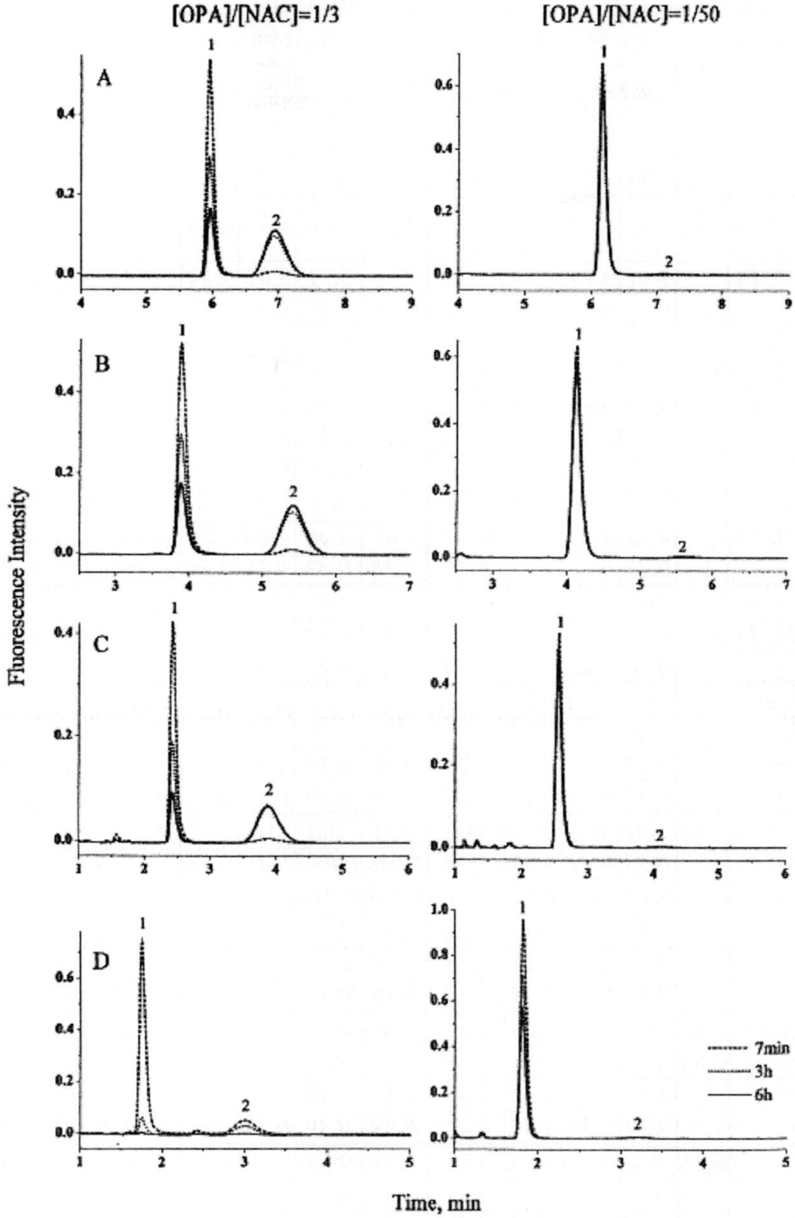

Figure 1 Fluorescence detected chromatograms of n-butyl- (A), n-propyl- (B), ethyl- (C) and methyl amine (D) obtained after 7min---, 3 h…, 6 h—, with reagents of various composition: [OPA]/[NAC]=1/3, ([OPA]/[NAC]/[A]=20/60/1), and [OPA]/[NAC] =1/50, i.e., ([OPA]/[NAC]/[A])=20/1000/1, ($1=1x10^{-9}$ M); (Data in Table 1); With permission from Reference 1

Table 1 Stability/characteristics of the OPA/NAC(MPA) derivatives of Amines and Ethanolamine as a Function of the Reaction Time and *Reagent's Composition*

Amines⇓	Ret time min ⇓	Fluorescence detection										
		[OPA]/[NAC]=1/3				[OPA]/[NAC]=1/50				[OPA]/[MPA]=1/50		
		response %*				response %*				response %*		
		90s	7m	3h	6h	90s	7m	3h	6h	7m	3h	6h
Ethanolamine1	1.35	98.9	98.6	97.7	97.9	100	100	100	100	100	100	100
Ethanolamine2	1.65	1.1	1.4	2.3	2.1	-	-	-	-	-	-	-
Int unit/pM: Fl		***4.07***	***3.79***	***2.99***	***2.20***	***3.96***	***3.71***	***3.62***	***3.54***	***3.90***	***3.36***	***2.98***
UV		***0.37***	***0.35***	***0.25***	***0.19***	***0.36***	***0.35***	***0.34***	***0.32***	***0.37***	***0.32***	***0.29***
Methylamine1	1.85	93.8	82.0	38.5	42.4	99.1	97.9	96.4	96.4	98.2	96.5	97.2
Methylamine2	3.26	6.2	18.0	61.5	57.6	0.9	2.1	3.6	3.6	1.8	3.5	2.8
Int unit/pM: Fl		***4.68***	***4.50***	***0.86***	***0.11***	***4.55***	***4.35***	***3.39***	***2.76***	***4.51***	***2.84***	***1.89***
UV		***0.42***	***0.41***	***0.08***	***0.014***	***0.40***	***0.38***	***0.28***	***0.21***	***0.39***	***0.25***	***0.17***
Ethylamine1	2.58	94.5	93.1	43.5	28.1	100	99.7	99.4	99.3	94.8	92.7	91.3
Ethylamine2	3.87	5.5	6.9	56.5	71.9	-	0.3	0.6	0.7	5.2	7.3	8.7
Int unit/pM: Fl		***4.42***	***4.44***	***4.33***	***3.37***	***4.41***	***4.40***	***4.04***	***3.76***	***4.48***	***3.68***	***3.09***
UV		***0.39***	***0.40***	***0.38***	***0.29***	***0.38***	***0.42***	***0.38***	***0.35***	***0.40***	***0.33***	***0.28***
i-Propylamine	3.83	100	100	100	100	100	100	100	100	100	100	100
Int unit/pM: Fl		***4.25***	***4.82***	***4.74***	***4.67***	***2.73***	***4.81***	***4.69***	***4.89***	***3.90***	***4.83***	***4.77***
UV		***0.37***	***0.44***	***0.43***	***0.42***	***0.25***	***0.42***	***0.42***	***0.42***	***0.34***	***0.42***	***0.42***
n-Propylamine1	4.17	98.2	95.0	53.3	37.6	100	99.5	98.2	98.1	98.5	97.7	97.7
n-Propylamine2	5.71	1.8	5.0	46.7	62.4	-	0.5	1.8	1.9	1.5	2.3	2.3
Int unit/pM: Fl		***4.69***	***4.71***	***4.82***	***4.18***	***4.57***	***4.56***	***4.34***	***4.19***	***5.04***	***4.32***	***3.72***
UV		***0.41***	***0.43***	***0.43***	***0.37***	***0.39***	***0.40***	***0.38***	***0.36***	***0.44***	***0.37***	***0.32***
tert.-Butylamine	4.85	100	100	100	100	100	100	100	100	100	100	100
Int unit/pM: Fl		***0.58***	***0.71***	***1.33***	***1.29***	***0.029***	***0.050***	***0.45***	***0.68***	***0.055***	***0.46***	***0.69***
UV		***0.17***	***0.23***	***0.42***	***0.41***	***0.007***	***0.014***	***0.14***	***0.21***	***0.016***	***0.14***	***0.21***
sec.-Butylamine	5.47	100	100	100	100	100	100	100	100	100	100	100
Int unit/pM: Fl		***4.34***	***4.23***	***4.19***	***4.15***	***1.74***	***2.63***	***4.18***	***4.22***	***2.27***	***3.34***	***3.18***
UV		***0.39***	***0.38***	***0.38***	***0.39***	***0.16***	***0.33***	***0.38***	***0.38***	***0.28***	***0.41***	-
i-Butylamine1	5.88	98.9	98.5	83.9	74.1	100	99.7	97.8	97.6	98.8	96.1	96.0
i-Butylamine2	6.88	1.1	1.5	16.1	25.9	-	0.3	2.2	2.4	1.2	3.9	4.0
Int unit/pM: Fl		***4.01***	***4.06***	***4.31***	***4.48***	***4.37***	***4.30***	***4.21***	***4.13***	***4.31***	***3.85***	***3.38***
UV		***0.34***	***0.39***	***0.40***	***0.41***	***0.38***	***0.39***	***0.38***	***0.37***	***0.39***	***0.34***	***0.30***
n-Butylamine1	6.23	98.3	94.2	50.3	33.9	100	99.4	97.7	97.6	98.7	97.1	97.1
n-Butylamine2	7.15	1.7	5.8	49.7	66.1	-	0.6	2.3	2.4	1.3	2.9	2.9
Int unit/pM: Fl		***4.07***	***4.04***	***4.18***	***3.62***	***4.02***	***4.06***	***3.88***	***3.68***	***3.96***	***3.24***	***2.71***
UV		***0.37***	***0.38***	***0.38***	***0.32***	***0.36***	***0.38***	***0.36***	***0.34***	***0.37***	***0.30***	***0.25***
i-Amylamine1	7.71	99.8	96.3	72.1	58.5	100	99.2	97.5	97.6	nd	nd	nd
i-Amylamine2	8.06	0.5	3.7	27.9	41.5	-	0.8	2.5	2.4	nd	nd	nd
Int unit/pM: Fl		***4.46***	***4.28***	***4.41***	***4.30***	***4.30***	***4.24***	***4.07***	***3.84***			
UV		***0.40***	***0.42***	***0.43***	***0.42***	***0.41***	***0.42***	***0.40***	***0.38***			

Indications: [OPA]/[MPA]([NAC])/[A]=20/60/1 or 20/1000/1 (1=1 x 10^{-9} M) correspond to mole concentrations, indicated by OPA/NAC=1/3 or 1/50; response %*= based on the total of derivatives, obtained with Fl detection; nd=not determined

apply optimum conditions for the C_1-C_4 n-monoamines and the corresponding isopropyl, isobutyl, sec. butyl- and tert.-butylamines must be a matter of compromise.
(v) Regarding the advantages and disadvantages of condition to be selected considerations depend on the compound(s) to be derivatized.
The OPA/NAC=1/50 reagent and 90s reaction time proved to be the optimum (except isopropylamine is also present): OPA/NAC derivatives of ethanolamine, ethyl, n-propyl, n- and i-butyl, as well as, i-amylamines can be determined on the basis of a single derivative and that of methylamine on the basis of its two species (methylamine1 & methylamine2). In case of i-propylamine for its quantitative reaction 7 min reaction time is needed. Except for methylamine (2.1% methylamine2) the amounts of transformed species, were below 1%. Thus, depending on their absolute and relative concentrations they should be quantitated either on the basis of their single, or, on both of their derivatives. Because of the very slow reaction rate of tert.-butyl and sec.-butyl amines their selected derivatization time with the OPA should be a matter of compromise: depending on the analyt's other components.
(v)-Comparing the percentages of the transformed OPA/NAC derivatives to the corresponding OPA/MPA ones, the advantage of the OPA/NAC derivatization was repeatedly proved (higher responses, slower transformation of the initially formed derivatives). In terms of stability OPA/MPA and OPA/NAC derivatives do not differ considerably but are much more stable that the corresponding OPA/MCE ones (Figure 2)

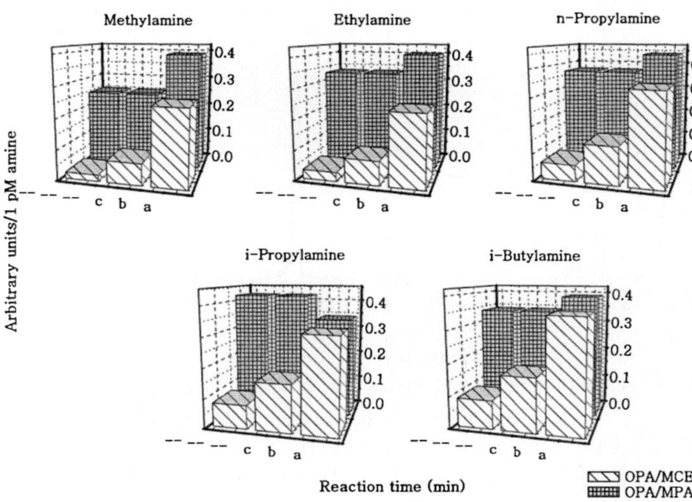

Figure 2 Comparison of responses and stabilities of the OPA/MCE =1/50 and OPA/MPA=1/50 derivatives of C_1-C_4 aliphatic As, as function of the reagent composition and reaction time; a: 7min, b: 47min, c: 87min (OPA/MCE), or, a: 7 min, b: 3 h, c: 6 h (OPA/MPA) (With permission from-Reference 3)

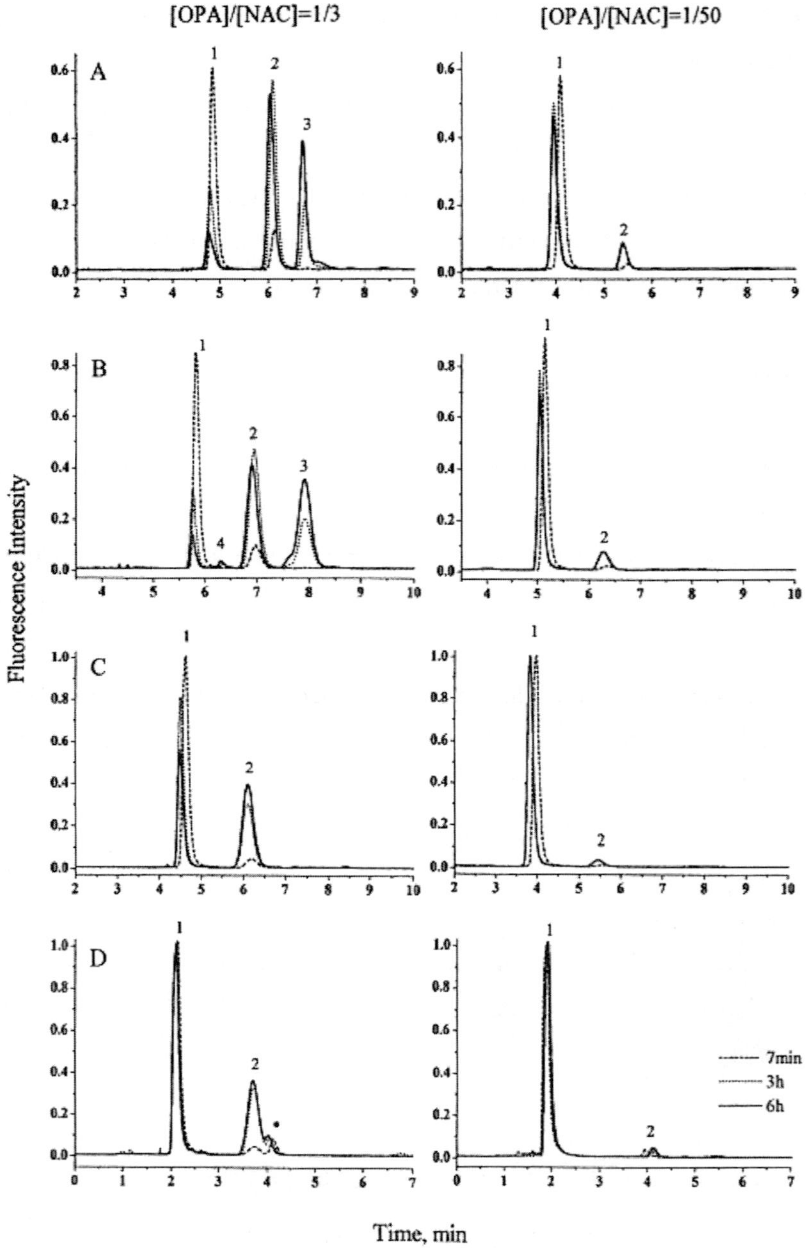

Figure 3 Fl chromatograms of putrescine (A), cadaverine (B), tyramine (C) and agmatine (D) obtained after 7min---, 3 h..., 6 h—, with reagents of various composition: [OPA]/[NAC]=1/50, ie., [OPA]/[NAC]/[A]=20/60/1, ($1=1x10^{-9}$ M) and [OPA]/[NAC]=1/50, i.e., [OPA]/[NAC] /[A]=20/1000/1, ($1=1x10^{-9}$ M); *=impurity peak (With permission from Reference1)

4. Studies on the behavior of α,ω-diamines upon reaction with the OPA/MPA and OPA/NAC=1/3 and 1/50 reagents

Introductory results obtained with the OPA/MPA=1/3 and OPA/NAC=1/3 reagents (Figure 3) provided unexpected, at the same time horrendous results: taking into account that several pro-tocols are based on the quantitation of their OPA derivatives. The only question remains, which of their derivatives were the basis of their quantitation, described in the literature?

4.1. Characteristics of the 1,2-ethylenediamine (1,2-EDA), 1,2- and 1,3 propylenediamines (1,2-PDA, 1,3-PDA)

The first three members of the homologous series of aliphatic DAs are providing more than one derivative of different type: 1,2 EDA1, 1,2-EDA2, 1,2-PDA1,2 and 1,3-PDA1 are fur-nishing UV absorbancy, exclusively.

(i) The fast forming and stable 1,2-EDA1, 1,2-EDA2 and 1,2-PDA2, as well as the continuously increased amounts of 1,2-PDA1 and 1,3-PDA1 derivatives obtained with the OPA/NAC=1/3 reagent manifested similar properties: they are not fluorescent, provide an UV maximum at 305 nm and 234 nm, (i.e., they are not of isoindole character, probably originat-ing from the special reaction of the vicinal dioxo with the vicinal DAs resulted in cyclic azine species, except for 1,3-PDA1 which might result in a seven membered ring).

(ii) Derivatives of 1,2-EDA3 and 1,2-PDA3 manifest special behavior: (i) with the OPA/NAC=1/3 reagent they furnish fluorescence responses only, unfortunately increasingly over time and with low intensities. With the OPA/NAC=1/50 reagent the fluorescence intensi-ties of the OPA derivatives of 1,2-EDA3 and 1,2-PDA3 species considerably increased (0.67-1.27 and 0.36- 0.63 integrator units/pM, in order of listing for the 1,2-EDA3 and for the 1,2-PDA3 derivatives, respectively).

4.2. Characteristics of the BAs and PAs

(i) Investigations performed as a function of the reaction time revealed [1-3] that the initially formed OPA-derivatives of the BAs, having in their initial structure the NH_2-CH_2- moiety, without exception, became transformed to further ones. As a consequence of our experience with two reagents of various composition, the use of a reagent containing the OPA/SH-additive in the 1/50 mole ratio, and, 90 s reaction time is to be preferred. In this case agmatine can be determined on the basis of a single derivative, tyramine (Tyr), Put, Cad, Spm, hista-mine and bis(hexamethylene)triamine on the basis of two, while Spd with three derivatives.

(ii) As to the composition of the transformed derivatives confirmed by on-line MS measurement (Table 2, Figures 4,5, as examples) in the case of n-propylamine (n-PA) and four diamines they have been determined as their OPA/NAC-derivatives obtained with the OPA/NAC=1/3 reagent [2]. Data proved to be in thorough accord with the more than 1 OPA derivative providing AAs (having in their initial structure the NH_2CH_2-R moiety).

Table 2 Fragmentation Patterns of the OPA/NAC Derivatives of Amines (MW_{OPA}=134.1; MW_{NAC}=163.2)

Amines, ⇓ Ions ⇒	MW	***OPA/NAC derivatives***										
		Initially Formed				***Transformed***						
		$MH^{\ddagger}$ calc.	$MH^{\ddagger}$ obt.	$MNa^{\ddagger}$	$MH^{\ddagger}$ - m/z= 129	$MH^{\ddagger}$ + OPA	$MNa^{\ddagger}$ + OPA	$MH^{\ddagger}$ + OPA - H_2O	$MNa^{\ddagger}$ + OPA +Na	$MH^{\ddagger}$ + 2OPA	$MNa^{\ddagger}$ + 2OPA	SFI*
n-Propyl	59.1	321.4	321.3	343.3	192.2	455.5	477.4	**437.4**	499.4	-	-	-
Putrescine	88.2	611.5	611.5	633.5	-	745.6	**767.6**	727.6	789.6	879.5	**901.6**	600.6
Cadaverine	102.	625.5	625.5	647.5	-	759.6	**781.6**	741.6	803.6	893.7	**915.6**	614.6
Agmatine	130.	392.5	392.5	414.4	263.3	**526.5**	-	-	-	-	-	-
Tyramine	137.	399.5	399.4	421.3	270.3	533.5	555.4	**515.4**	577.5	-	-	-
Spermidine	145.	407.5	-	-	-	-	-	**505.6**	-	-	-	-

Indications: Concentrations: [OPA]/[NAC]/[A]=20/60/1, 1=~1x10^{-9}; bold printed values=abundant ions; **505.6#=**$MH^{\ddagger}$ + OPA-2H_2O; SF*=selective fragment ions (Figures 4,5): 600.6=m/z=879.5-279.1 and 614.6=m/z=893.7-279.1;

It means that the initially formed isoindoles (abundant masses of the protonated molecular ions ($MH^{\ddagger}$) became transformed to an additional OPA molecule containing species ($MH^{\ddagger}$+OPA); The common reaction mechanism pathway of the transformation was described [2] and given also in chapter 1.2.3.(Figure 8). In cases of putrescine and cadaverine, being diamines, both of their isoindoles react with an additional molecule of OPA (Table 2: $MH^{\ddagger}$+2OPA=putrescine3=m/z=879.5, $MH^{\ddagger}$+2OPA=cadaverine3=m/z=893.7, Figures 5AB: putrescine3, cadaverine3). These very special fragment ions serve as unambiguous evidence for our common completing reaction mechanism pathway confirmed by the fragments of the putrescine and cadaverine derivatives.

Results of our derivatization studies called attention to the fact that all those data found in the literature should be treated cautiously; in terms of reliability and comparability. As to the, one by one, reported reproducibilities, they might no doubt be true, but not

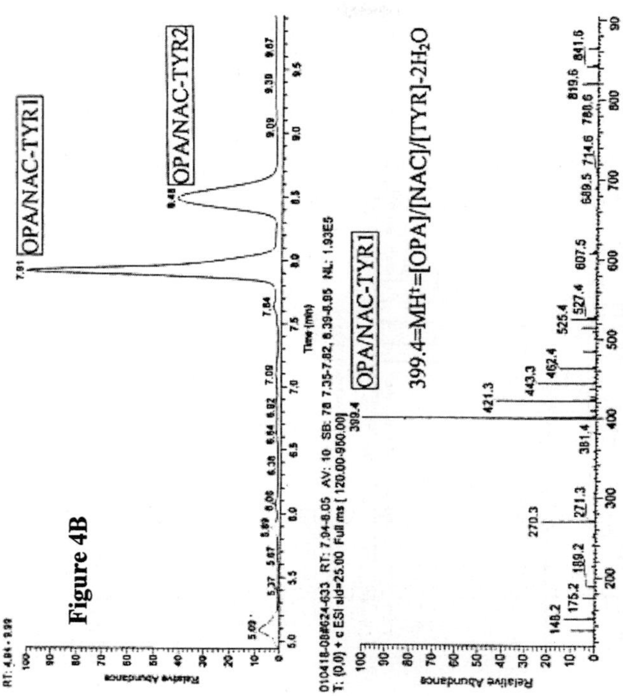

Figure 4B

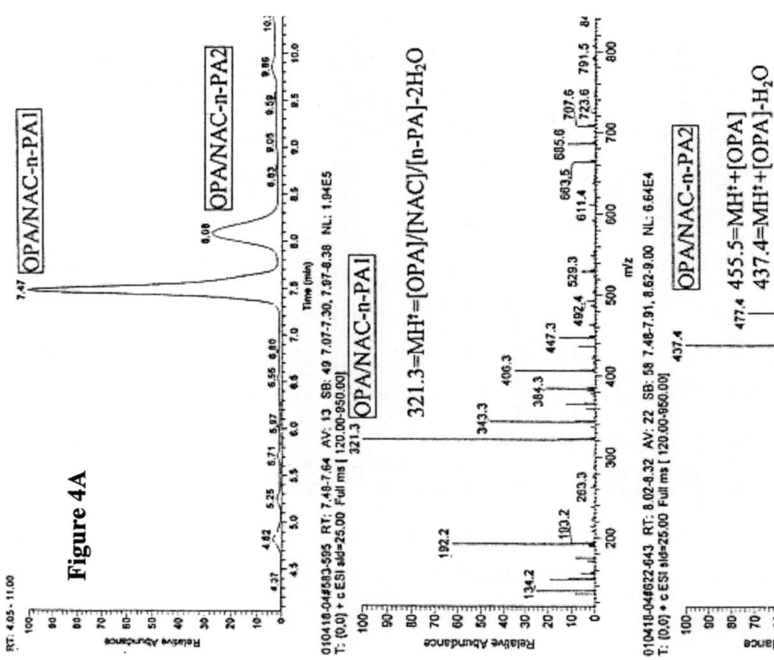

Figure 4A

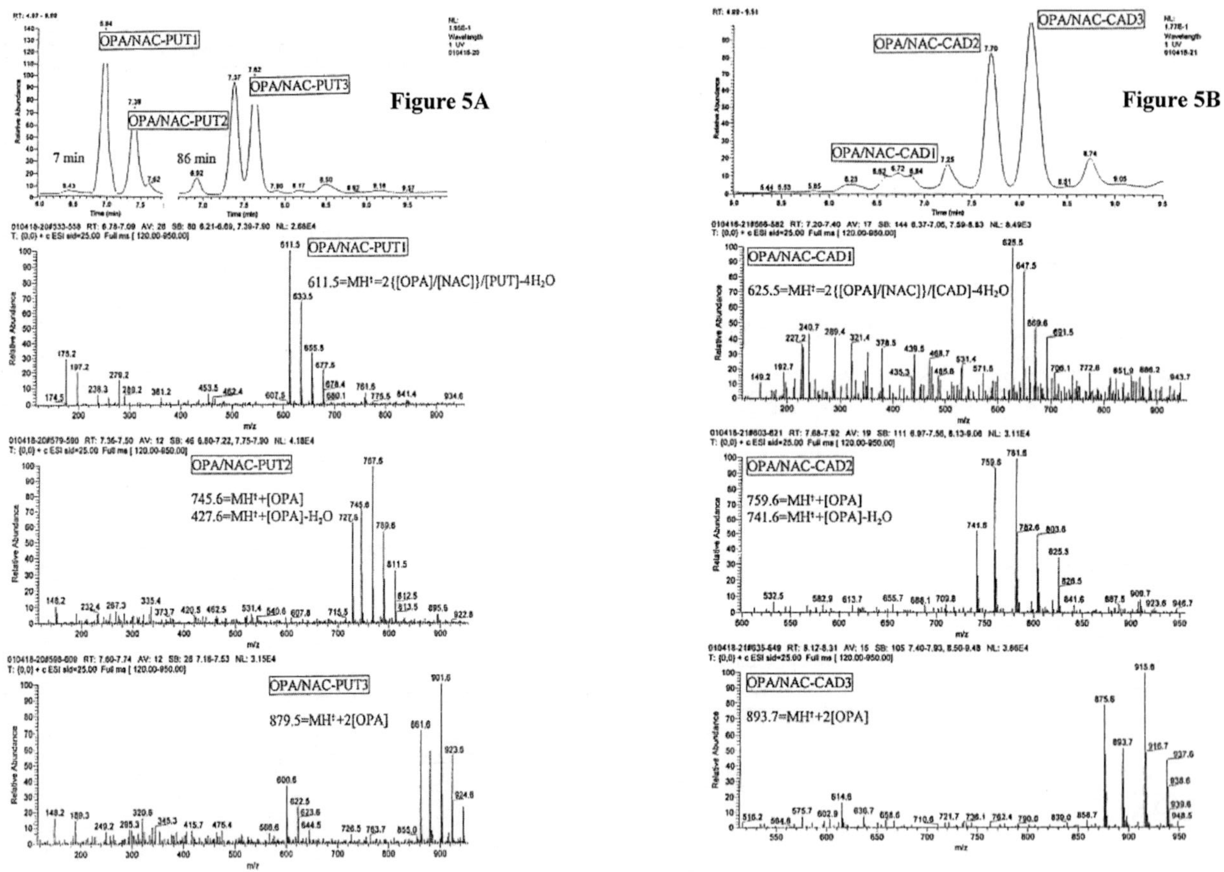

Figures 4AB, 5AB PDA/UV chromatograms (first lines), MS spectra of the initially obtained (spectra in second lines) & transformed (spectra in 3. and fourth lines) OPA/NAC derivatives of n-propyl amine (n-PA: 4A); tyramine, (TYR: 4B); putrescine, (PUT: 5A); cadaverine, (CAD: 5B); (Detailed composition of fragments in Table 2); (With permission from reference 2)

reliable and comparable: the peak, that served as the basis for the quantitation of the amine in question, remains to be defined, because they could be different from procedure to procedure.

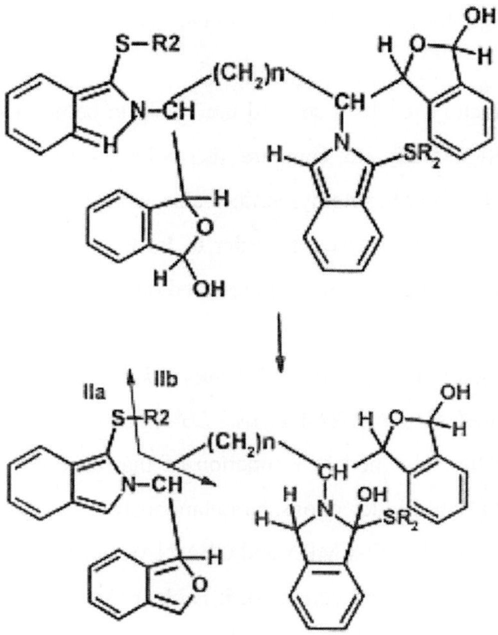

I

II

Figure 6 Origin of ions m/z=278.2-279.1, m/z=600.6 and m/z=614.6 obtained from the transformed molecular ions (MH^+) of putrescine3 (m/z=879.6-279.0) and cadaverine3 (m/z=893.7-279.1); (MH^+=2{[OPA][NAC]}[A]+2[OPA]-4H_2O); Detailed data in Figures 5AB & Table2; With permission from Reference [2]

---

5. Studies on the special behavior of the C_6-C_8 aliphatic amines and phenyl ethyl amine

Introductory investigations, performed with the OPA/MPA and with the OPA/NAC reagents applying both of them in the 1/3 and in the 1/50 molar ratios, in solutions of pH 9.3, furnished two unexpected results:

(i) Derivatizations of the C_6-C_8 As carried out with the OPA/MPA and OPA/NAC=1/3 reagents resulted in the formation of the classical isoindoles and their expected, transformed product, containing one additional OPA molecule [2]. However, these transformed species eluted before their classical isoindoles, in contrary to the transformed, corresponding species of AAs C_1-C_5 aliphatic amines (chapter 1.2.3. of this book) and diamines (3, 4 paragraphs of this chapter) that eluted after their classical isoindoles [5].

(ii) Applying the OPA/MPA(NAC)=1/50 reagent — (in order to eliminate/decrease the transformation of the initially formed isoindoles into the two OPA molecule-containing ones — a new product was detected, with considerably smaller retention times, compared both to the initially formed classical isoindoles and to the transformed derivatives, containing one additional OPA molecule.

Based on these introductory results it seemed inevitably to be necessary, in the frame of an exhaustive derivatization study, to compare the behavior of the OPA/MPA and OPA/NAC derivatives of C_6-C_8 and phenylethyl amines also with the OPA/ET, and with the most commonly used, OPA/MCE derivatives, in order to have an overview to select from analytical point of view the most advantageous, the preferred derivatives.

5.1. Studies on the OPA/MPA, OPA/NAC, OPA/MCE and OPA/ET derivatives of the C_6-C_8 and phenylethyl As obtained with the OPA/SH-additive=1/3 reagents

Results of stoichiometric studies, obtained as a function of the reaction time and reagent's composition, followed by DAD and FL detections, simultaneously are compiled in Table 3.
(i) Evaluating first the behavior of the OPA/MPA and OPA/NAC derivatives of the C_6-C_8 and phenylethyl As (Table 3: first two vertical columns), it is clear that in accord with the NH_2-CH_2- moiety-containing AAs and As [1-4], their initially formed derivatives (HexA1, HepA1, OctA1) became transformed (HexA2, HepA2, OctA2): their transformed species, elute with shorter retention times, probably due to their more polar properties. The transformed derivative of PheEtA, i.e., PheEtA2, in accord with earlier experience [1-4], elutes after its initially formed isoindole (Table 3).

The composition of the initially obtained and transformed species has been confirmed by their HPLC/MS(ESI) spectra [5], shown as example by the chromatogram and spectra of HexA (Figure 7: DAD chromatogram first line, spectra in second and third lines). As determined, without exception, the protonated molecular ions of the initially formed isoindoles (MH^+=m/z= HexA, HepA, OctA, PheEtA, m/z values of spectra in the third lines, in order of listing: MH^+=306.2, 320.2, 334.3, 326.2) do transform into their two OPA molecule containing derivatives (MH^++OPA), including their corresponding dehydrated (MH^++OPA-H_2O) and/or cationized versions (MNa^++OPA) (Figure 7, spectra in the second lines, in order of listing, MH^++OPA=440.3, 454.3, 468.4, 460.2)

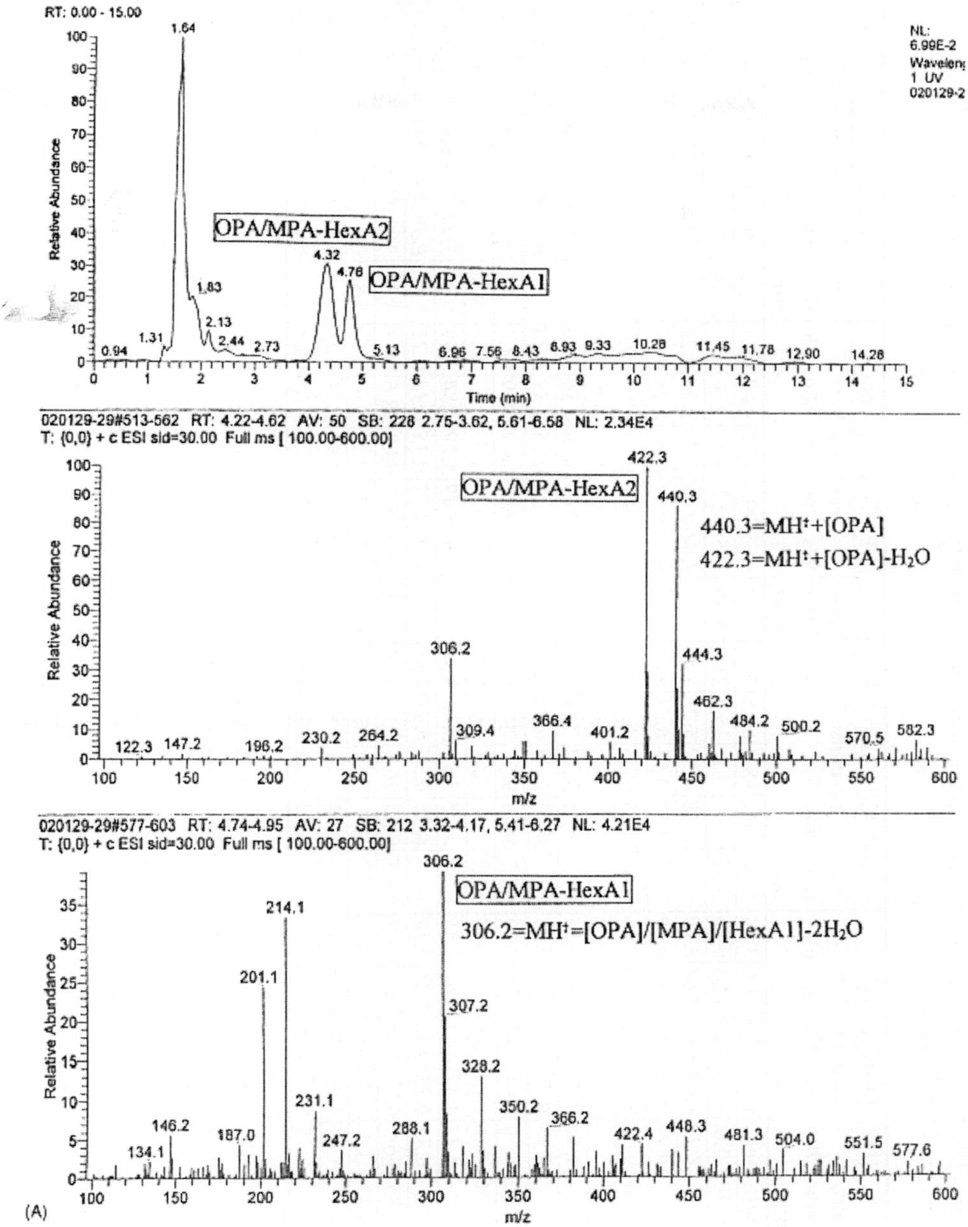

(A)

***Figure** 7* UV chromatogram (first lines) and MS spectra (second and third lines) of the initially obtained hexylamine1: HexA1, and its transformed OPA/MPA derivatives (indicated by number 2, i.e.,: hexylamine2: HexA2, (Detailed composition of fragments in the text); With permission from Reference [5]

Table 3 Stability/characteristics of OPA/MPA, OPA/NAC, OPA/MCE and OPA/ET derivatives of Hexyl (HexA), Heptyl (HepA), Octyl (OctA) and Phenylethyl (PheEtA) amines as a function of reaction time obtained with the OPA/SH-additive=1/3, at pH 9.3, based on simultaneous fluorescence (FL) and UV Detections

Amines⇓	UV	[OPA]/[MPA]=1/3						[OPA]/[NAC]=1/3					[OPA]/[MCE]=1/3						[OPA]/[ET]=1/3					
	max nm⇓	ret. time min⇓	response %* 90s	7m	1h	3h	6h	ret. time min	response %* 90s	7m	3h	6h	ret. time min	response %* 90s	7m	30m	60m	3h	ret. time min	response %* 90s	7m	30m	60m	3h
HexA1	333.9	6.18	99.4	84.2	64.4	51.6	39.4	3.66	100	98.2	84.5	78.1	4.70	99.5	99.3	99.1	98.7	98.1	3.55	100	100	100	100	100
HexA2	338.7	5.70	0.6	15.8	35.6	48.4	60.6	3.51	-	1.8	15.5	21.9	5.90	0.5	0.7	0.9	1.3	1.9	5.14					
Iu/pM: Fl			6.18	6.67	6.99	7.84	6.10		4.68	4.73	5.21	4.80		4.84	4.58	3.23	2.24	0.28		5.47	5.53	5.47	5.26	5.12
UV			0.47	0.55	0.59	0.59	0.59		0.52	0.51	0.51	0.47		0.34	0.31	0.22	0.15	0.02		0.35	0.36	0.36	0.33	0.32
HepA1	333.9	8.84	97.7	89.5	66.3	40.4	25.6	4.67	99.3	97.6	76.7	67.2	6.01	99.5	99.4	99.4	99.4	99.0	6.43	100	100	100	100	100
HepA2	338.7	7.93	2.3	10.5	33.7	59.6	74.4	4.30	0.7	2.4	23.3	32.8	7.79	0.5	0.6	0.6	0.6	1.0						
Iu/pM: Fl			4.96	4.72	5.29	4.55	3.20		3.30	3.32	3.89	3.60		5.46	5.49	5.19	4.80	3.59		6.08	6.03	6.05	5.94	5.27
UV			0.38	0.39	0.39	0.32	0.22		0.36	0.39	0.38	0.35		0.38	0.37	0.35	0.33	0.25		0.39	0.39	0.39	0.38	0.34
OctA1	333.9	13.13	98.0	89.1	69.2	44.0	30.0	6.37	99.6	97.8	70.6	57.2	7.61	99.0	99.1	98.9	98.9	98.7	7.86	99.6	99.5	99.5	99.5	99.8
OctA2	338.7	11.50	2.0	10.9	30.8	56.0	70.0	5.82	0.4	2.2	29.4	42.8	9.98	1.0	0.9	1.1	1.1	1.3	10.16	0.4	0.5	0.5	0.5	0.2
Iu/pM: Fl			4.37	4.37	4.91	3.67	2.58		3.27	3.32	3.81	3.50		5.46	5.49	5.19	4.80	3.59		5.90	5.80	6.07	5.85	5.47
UV			0.35	0.36	0.36	0.36	0.18		0.36	0.36	0.38	0.35		0.38	0.37	0.35	0.33	0.25		0.39	0.40	0.40	0.38	0.37
PheEtA1	333.9	7.85	96.5	92.3	59.6	25.5	11.7	5.03	98.0	96.3	70.0	55.0	3.41	99.0	99.2	98.8	98.6	98.3	3.31	100	99.9	99.9	99.9	99.6
PheEtA2	338.7	8.43	3.5	7.7	40.4	74.5	88.3	6.27	2.0	3.7	30.0	45.0	4.00	1.0	0.8	1.2	1.4	1.7	5.85	-	0.1	0.1	0.1	0.4
Iu/pM: Fl			5.69	5.93	6.51	5.39	2.40		3.73	3.78	3.94	3.78		5.19	4.33	3.59	2.58	0.43		6.03	6.07	5.98	5.95	5.66
UV			0.35	0.35	0.36	0.29	0.13		0.38	0.40	0.38	0.38		0.37	0.34	0.26	0.18	0.04		0.41	0.41	0.40	0.40	0.38

Indications: [OPA]/[SH-additive])/[A]=20/60/1 ($1=1 \cdot 10^{-9}$ M) correspond to molar concentrations; response %*= based on the total of derivatives, obtained with FL detection; -=no data available; Iu/pM inegrator units/picomole

(ii) As to the characteristics of the OPA/NAC derivatives of C_6-C_8 and phenylethyl amines (Table 3: data in second vertical column) it has been repeatedly proven [1-4] that their responses are smaller and the stability of their initially formed isoindoles are more stable compared to their corresponding OPA/MPA ones.
(iii) In order to compare the characteristics of the OPA/MCE and OPA/ET derivatives of C_6-C_8 and phenylethyl As (Table 3, third and fourth vertical columns), they have been investigated, strictly, under the same conditions.
Results revealed that,
–the OPA/MCE derivatives proved to be the less stable ones. Decomposition of the classical isoindoles, especially in the cases of HexA and PheEtA, takes place already after 7 min reaction time. Responses, expressed in integration units/picomole amine (Iu/pM), decrease in order of listing HexA, 90 s: 4.84, 7 min: 4.58; PheEtA, 90 s: 5.19, 7 min: 4.33.
–The stability of the OPA/ET derivatives is as excellent as of those of their OPA/MPA or OPA/NAC counterparts, with the accompanied, favored advantage of the almost side-reaction free derivatization. In the cases of the HexA and HepA transformation could not be detected, while transformation of OctA1 and PheEtA1 to OctA2 and PheEtA2 proved to be also quite negligible (less than 0.5% expressed in the total).

5.2. Studies on the OPA/MPA, OPA/NAC, OPA/MCE and OPA/ET derivatives of the C_6-C_8 and phenylethyl As obtained with the OPA/SH-additive=1/50 reagents; Reactions with the OPA/MPA=1/50 reagents, as a function of the pH

Based on our earlier experience, associated with the increased stability of the initially formed isoindoles of C_1-C_5 aliphatic As [1], our first aim was to define the characteristics of the C_6-C_8 and phenylethyl As, strictly under the same conditions (OPA/MPA=1/50 reagent, pH 9.3; data in Table 4, second vertical column).
(i) As expected, the formation of the two OPA molecule containing species (derivatives designated by indices 2) could be decreased to negligible concentrations, but, simultaneously, a new species appeared, in considerable amounts: proportions of the new product (derivatives designated by indices 0) proved to be dependent on the pH of derivatzations.
(ii) Evaluating the impact of the pH of derivatizations (Table 4: proportions of transformed products at pH 8.8, pH 9.3, pH 9.75 and pH 10.25) it is clear that the higher the pH the smaller the extent of side reactions. After 7 min reaction, important from an analytical point of view, at pH 10.25, products of both side reactions could be unambiguously influenced: either

Table 4 Stability/Characteristics of OPA Derivatives of Hexyl (HexA), Heptyl (HepA), Octyl (OctA) and Phenylethyl (PheEtA) Glycine and Alanine as a Function of Reaction Time and the pH of Reactions, Obtained with the OPA/MPA=1/50 Reagents, Based on Fluorescence (Fl) and UV Detections, Applying Isocratic Elutions

Amines⇓	UV max nm⇓	ret. time min⇓	pH 8.80 response %*				pH 9.30 response %*				pH 9.75 response %*				pH 10.25 response %*			
			7m	1h	3h	6h	7m	1h	3h	6h	7m	1h	3h	6h	7m	1h	3h	6h
HexA0	343.5	2.09	4.16	15.5	31.2	97.6	4.1	10.6	25.2	39.0	3.8	7.5	11.4	22.7	1.1	2.1	4.1	6.6
HexA1	333.9	4.56	95.8	83.5	66.9	2.0	95.9	88.6	73.4	59.7	96.1	91.5	87.5	76.2	98.9	97.9	95.9	93.4
HexA2	338.9	4.17	0.04	1.0	1.9	0.4	-	0.8	1.4	1.3	0.1	1.0	1.1	1.1	-	-	-	-
Iu/pM: Fl			***2.05***	***4.38***	***4.46***	***4.95***	***3.98***	***5.03***	***5.19***	***5.36***	***5.02***	***5.00***	***4.92***	***4.86***	***4.77***	***4.89***	***4.86***	***4.79***
UV			***0.19***	***0.29***	***0.32***	***0.34***	***0.28***	***0.38***	***0.39***	***0.44***	***0.7***	***0.37***	***0.37***	***0.37***	***0.37***	***0.36***	***0.36***	***0.36***
HepA0	343.5	2.56	3.5	13.0	25.7	94.1	4.1	7.9	20.2	32.5	3.0	8.2	17.9	21.5	2.9	4.4	6.1	8.8
HepA1	333.9	6.71	96.4	85.6	72.0	5.1	95.8	91.1	78.1	65.9	96.8	90.3	80.4	76.9	97.1	94.8	92.7	90.2
HepA2	338.7	6.10	0.1	1.4	2.3	0.8	0.1	1.0	1.7	1.6	0.2	1.5	1.7	1.6	-	0.8	1.2	1.0
Iu/pM: Fl			***2.63***	***4.61***	***4.79***	***4.92***	***4.48***	***5.20***	***5.16***	***5.15***	***5.40***	***5.08***	***4.72***	***4.62***	***5.29***	***5.32***	***5.16***	***4.26***
UV			***0.19***	***0.33***	***0.35***	***0.35***	***0.33***	***0.39***	***0.38***	***0.38***	***0.42***	***0.38***	***0.36***	***0.37***	***0.40***	***0.39***	***0.39***	***0.40***
OctA0	343.5	3.25	2.1	11.5	28.8	95.3	2.4	6.8	21.8	37.5	1.5	6.5	16.2	22.1	0.8	1.9	3.9	6.9
OctA1	333.9	9.95	97.8	87.0	68.9	4.1	97.5	92.2	76.3	60.8	98.4	92.0	82.0	76.2	99.2	97.5	95.2	92.1
OctA2	338.7	9.11	0.1	1.5	2.3	0.6	0.1	1.0	1.9	1.7	0.1	1.5	1.8	1.7	-	0.6	0.9	1.0
Iu/pM: Fl			***2.54***	***4.51***	***4.67***	***5.65***	***4.27***	***4.96***	***4.58***	***4.33***	***5.16***	***4.28***	***4.14***	***3.93***	***5.31***	***5.09***	***4.88***	***3.98***
UV			***0.19***	***0.35***	***0.35***	***0.39***	***0.32***	***0.37***	***0.34***	***0.32***	***0.39***	***0.36***	***0.31***	***0.29***	***0.9***	***0.38***	***0.37***	***0.30***

Indications as in Table 3, as well as: [OPA]/[MPA]/[A]=20/1000/1 (1=1 · 10^{-9} M) correspond to molar concentrations;

quantitatively inhibited (HexA2, HepA2, OctA2) or considerably decreased (HexA:1.1 %, HepA0: 2.9 %, OctA0: 0.8 %).

(iii) As to the spectral characteristics of derivatives (Figure 7), in the cases of the initially formed products they proved to be the classical isoindoles manifesting an UV maximum at 333.9 nm (Figure 8: HexA1; HepA1, OctA1, Tables 3-5 derivatives designated by indices 1.

Their transformed versions, already identified [6,7] with the OPA/MPA and OPA/NAC 1/3 derivatives of AAs and As, providing UV maximum values shifted to 338.7 nm, indicating the different structure of species (Figure 7: HexA2, HepA2, OctA2, Tables 3-5 derivatives designated by indices 2). Note: Chromatograms and Spectra of HepA, OctA and PhEtA are shown in the original paper, only [5]).

The products identified at the first time, formed in considerable amounts, obtained with the OPA/SH-additive=1/50 reagents manifested a maximum value at 343.5 nm; (Figure 8: HexA0, HepA0, OctA0, Tables 3-5 derivatives designated by indices 0).

The composition of these, at the first time identified compounds by HPLC, was assumed to correspond to the two SH-additive containing "dithio" OPA derivatives. This assumption is based on considerations as follows: The only literature data [14] relating to the identification of the OPA/di-tert.butylthiol derivative of n-propyl amine, has been reported a compound with an UV maximum of 344 nm. The formation of this type of compounds, in our practice, is unambiguously associated with the extremely high SH-group containing OPA reagents. The dithio-OPA derivatives of C_3-C_8 As obtained with the OPA/ethanethiol (ET) reagents have been identified recently in enormously high concentrations, by GC/MS, [6].

The above detailed assumptions have been confirmed by on-line HPLC/MS(ESI) studies (Figure 9: UV chromatograms and spectra of the OPA/MPA derivatives of C_6-As, obtained with the OPA/MPA=1/50 reagent, at pH 8.80).

The first line shows the UV detected chromatogram of the OPAMPA derivatives of As at 343 nm; The second lines furnish the spectra of the one additional SH-additive-containing derivative (HexA0); The third lines represent the spectra of the protonated molecular ion (HexA1). Results proved in all three cases unambiguously, that the one additional SH-additive-containing OPA derivatives were identified (MH^{+}+MPA, indicated as HexA0), formed according to Scheme 1. These two SH-additive-containing species (Figure 9, spectra in second lines, henceforth: dithio-derivatives) were obtained together with their very informative crude molecular ions

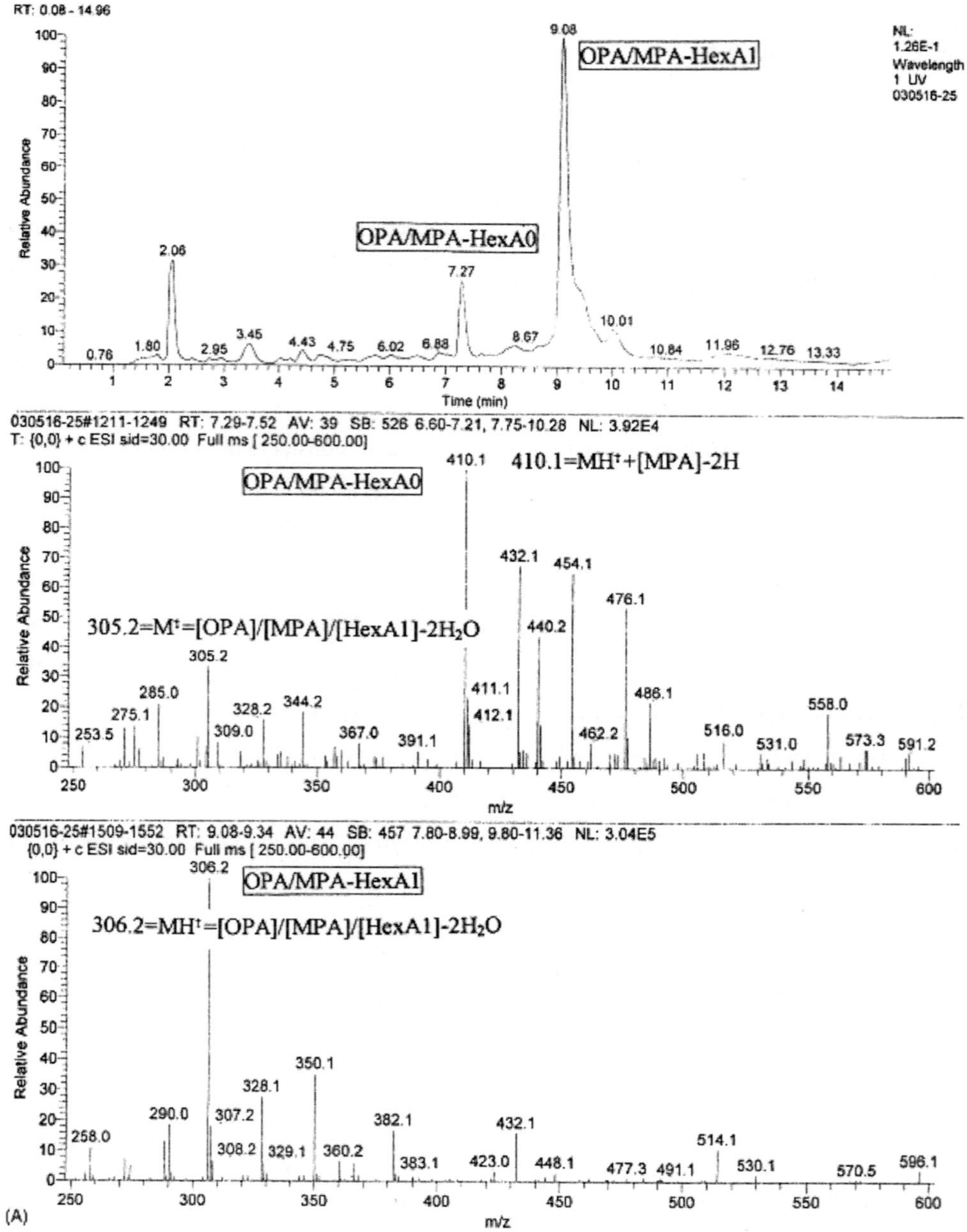

Figure 9 UV chromatograms (first lines) and MS spectra (second and third lines) of the initially obtained (indicated by number 1, i.e., hexylamine1: HexA1, and its transformed OPA/MPA derivatives (indicated by number 0, i.e., hexylamine0: HexA0 (Detailed composition of fragments in the text); With permission from Reference [5].

Table 5 Stability/Characteristics of OPA Derivatives of Hexyl (HexA), Heptyl (HepA), Octyl (OctA) and Phenylethyl (PheEtA) Amines as a Function of Reaction Time Obtained with the OPA/SH-additive=1/50 Reagents, Based on Fluorescence (FL) and UV Detections, Applying Isocratic (OPA/MCE, OPA/ET) and Gradient (OPA/NAC) Elutions (pH 8.80, pH 9.30)

Amines⇓	UV	[OPA]/[NAC]=1/50					[OPA]/[MCE]=1/50											[OPA]/[ET]=1/50										
	max nm⇓	ret. time min	response %* pH 8.80				ret. time min	response %* pH 8.80					response %* pH 9.30					ret. time min	response %* pH 8.80					response %* pH 9.30				
			7m	1h	3h	5h		90s	7m	30m	2h	3h	90s	7m	30m	2h	3h		90s	7m	30m	2h	3h	90s	7m	30m	2h	3h
HexA0	343.5	7.05	0.2	1.4	3.1	6.8	3.87	-	0.1	0.5	3.5	11.5	-	-	-	-	-	3.67	-	-	-	0.1	0.4					
HexA1	333.9	9.68	99.8	97.9	95.5	91.5	4.71	99.5	99.4	98.8	95.8	88.0	99.4	99.3	98.1	97.6	99.7	5.34	100	100	100	99.9	99.6	100	100	100	100	100
HexA2	338.7	7.53	-	0.7	1.4	1.7	5.90	0.5	0.5	0.7	0.7	0.5	0.6	0.7	1.9	2.4	0.3		-	-	-	-	-					
Iu/pM: Fl			***0.84***	***2.02***	***2.47***	***2.57***		***1.79***	***2.39***	***3.23***	***2.29***	***1.12***	***4.88***	***4.46***	***3.57***	***2.23***	***0.70***		***4.07***	***4.59***	***4.68***	***4.65***	***4.55***	***5.23***	***5.45***	***5.44***	***5.21***	***5.14***
UV			***0.12***	***0.28***	***0.35***	***0.36***		***0.12***	***1.16***	***0.23***	***0.16***	***0.08***	***0.33***	***0.32***	***0.25***	***0.15***	***0.05***		***0.30***	***0.33***	***0.33***	***0.32***	***0.32***	***0.32***	***0.34***	***0.34***	***0.33***	***0.32***
HepA0	343.5	7.30	0.1	0.8	1.4	1.3	4.56	0.3	0.4	0.8	2.1	6.0							-	-	-	-	-					
HepA1	333.9	10..35	99.9	98.1	93.8	91.7	5.83	99.3	99.0	97.6	95.7	92.6	99.5	99.6	99.2	98.6	98.3	6.68	100	100	100	100	100	100	100	100	100	100
HepA2	338.7	8.08	-	1.1	4.8	7.0	7.48	0.4	0.6	1.6	2.2	1.4	0.5	0.4	0.8	1.4	1.7	6.43	-	-	-	-	-	-	-	-	-	-
Iu/pM: Fl			***0.83***	***1.96***	***2.27***	***2.50***		***1.89***	***2.75***	***3.75***	***3.11***	***1.85***	***5.61***	***5.29***	***5.09***	***4.76***	***4.03***		***4.40***	***4.61***	***4.78***	***4.88***	***4.70***	***5.79***	***5.98***	***5.99***	***5.64***	***5.53***
UV			***0.11***	***0.28***	***0.35***	***0.35***		***0.13***	***0.19***	***0.27***	***0.22***	***0.13***	***0.37***	***0.37***	***0.35***	***0.32***	***0.28***		***0.32***	***0.33***	***0.34***	***0.33***	***0.32***	***0.36***	***0.37***	***0.37***	***0.35***	***0.34***
OctA0	343.5	8.93	0.4	2.2	4.1	7.0	5.87	0.2	0.4	0.6	1.4	3.2	1.4	1.4	1.2	1.0	0.7	4.97	-	-	-	-	-	-	-	-	-	-
OctA1	333.9	11.00	99.6	96.8	94.8	92.0	7.46	98.9	98.4	97.5	95.6	93.3	97.6	97.6	97.8	97.7	97.7	7.86	100	100	100	100	100	100	100	100	100	100
OctA2	338.7	10.83	-	1.0	1.1	1.0	9.86	0.9	1.2	1.9	3.0	3.5	1.0	1.0	1.0	1.3	1.6		-	-	-	-	-	-	-	-	-	-
Iu/pM: Fl			***0.89***	***2.36***	***2.55***	***2.60***		***1.79***	***2.94***	***3.96***	***3.89***	***2.98***	***5.65***	***5.45***	***5.28***	***5.19***	***4.86***		***4.49***	***4.76***	***4.94***	***4.92***	***4.61***	***4.82***	***5.97***	***5.87***	***5.47***	***5.29***
UV			***0.13***	***0.35***	***0.38***	***0.38***		***0.12***	***0.20***	***0.28***	***0.27***	***0.21***	***0.38***	***0.38***	***0.37***	***0.35***	***0.32***		***0.33***	***0.34***	***0.36***	***0.34***	***0.32***	***0.36***	***0.37***	***0.37***	***0.34***	***.33***
PheEtA1	333.9		-	-	-	-	3.41	-	-	-	-	-	99.5	99.0	98.1	98.0	100	3.37	100	100	99.9	99.5	99.0	100	100	99.9	99.8	99.7
PheEtA2	338.7		-	-	-	-	4.00	-	-	-	-	-	0.5	1.0	1.9	2.0		6.33			0.1	0.5	1.0			0.1	0.2	0.3
Iu/pM: Fl													***5.36***	***4.90***	***3.73***	***2.25***	***0.59***	***2.25***	***4.55***	***4.73***	***4.71***	***4.83***	***4.48***	***5.79***	***6.08***	***5.98***	***5.74***	***5.65***
UV													***0.37***	***0.35***	***0.20***	***0.16***	***0.04***		***0.35***	***0.35***	***0.35***	***0.35***	***0.33***	***0.38***	***0.39***	***0.39***	***0.38***	***0.37***

Indications as in Tables 1, 2 as well as: [OPA]/[SH-additive]/[A]=20/1000/1 ($8 \cdot 10^{-9}$ M) correspond to molar concentrations;

($M^{\ddagger}$), originated from the corresponding dithio derivatives by the loss of one MPA molecule (Figure 9: HexA0=$MH^{\ddagger}$+MPA–2H=m/z=410.1 and $M^{\ddagger}$=305.2).

Certainly, beside the dithio-derivatives also the initially formed classical, isoindoles have been repeatedly detected (Figure 9, spectrum in third line: HexA1=$MH^{\ddagger}$=m/z=306.2,),

5.3. Reactions with the OPA/NAC, OPA/MCE and OPA/ET=1/50 reagents, at pH 8.80 and 9.30

Based on the behavior of C_6-C_8 As, furnishing two different types of transformed derivatives in their reactions with the OPA/MPA =1/50 reagent, it would be of interest to clarify also the characteristics of their OPA/NAC, OPA/MCE and OPA/ET derivatives, obtained under strictly the same conditions.

(i) Selection of pH's, based on data, shown in Table 4, was expected to be important in order to define the impact of SH-additive on the proportions of transformed product(s), as well as, on the overall response values and stability of derivatives (Table 5).

–Comparing percentages of the dithio-derivatives (HexA0, HepA0, OctA0 PheEtA0 in Tables 4 and 5) they decrease from the OPA/MPA through the OPA/NAC to the OPA/MCE derivatized ones, from pH 8.80 to pH 9.30. In case of the OPA/ET derivatizations dithio-derivatives have not been found, at all.

Evaluating response values of C_6-C_8 and phenylethyl As, in general, obtained with the OPA/SH-additive=1/50 reagents (Tables 4, 5) it is clear that

–the beneficial effect of decreased, free OPA concentration [1,2], resulting in its limited reactivity, could be exploited applying the OPA/MPA reagent, at pH 10.25, exclusively.

–In the cases of the less stable OPA/MCE derivatives the decreased reactivity of the OPA reagents, at pH 8.80 resulted in decreased stability, i.e., in decreased response values.

–Derivatization of C_6-C_8 and phenylethyl As with the OPA/ET reagent(s), in particular at pH 9.30, can be regarded as a derivatization technique of choice. Practically, the amounts of transformed products are negligible, including also the products obtained with the OPA/ET=1/3 reagent (Table 3, last vertical column). Response values, obtained with the OPA/ET=1/3 and OPA/ET=1/50 reagents, equally reflect the reproducibility and stability of the process, predestinating these derivatives for analytical purposes.

Based on stoichiometric studies summarized in Tables 3-5 analytical reproducibility has been investigated with various concentrations of the C_6-C_8 As, under the most promising conditions: applying the OPA/ET=1/3 reagent at pH 9.30 (Table 6).

Table 6 Simultaneous quantitation of different amounts of hexyl (HexA), heptyl (HepA) and octyl (OctA) amines obtained with the OPA/ET=1/3 reagent, in model solutions, on the basis of their UV and fluorescence (Fl) intensities

Amine	ret. t. min	Integration units(Iu)/1 pM amino acid[a], (injected pM)										RSD %
		1800	900	450	225	112.5	62.25	31.12	15.65	7.81	Av.	
HexA, Iu/pM: Fl	5.58	-	7.52	7.53	7.49	7.53	7.53	7.42	7.70	7.54	7.53	0.97
UV		0.52	0.50	0.50	0.50	0.51	0.50	0.51	0.51	0.52	0.51	1.64
HepA, Iu/pM: Fl	7.00	5.95	5.89	5.87	5.78	5.79	5.70	5.40	5.46	5.43	5.70	3.7
UV		0.40	0.39	0.39	0.38	0.38	0.36	0.36	0.36	0.38	0.38	3.9
OctA, Iu/pM: Fl	8.95	6.94	6.82	6.72	6.58	6.61	6.43	6.14	6.16	6.14	6.50	4.7
UV		0.47	0.46	0.45	0.44	0.44	0.44	0.42	0.43	0.41	0.44	4.0

Indication: pM=picomole; [a]=obtained from three separate derivatization tests; Av.=averages, obtained with molar ratios of $[OPA]/[ET]:[A]^T$= 7:1, 14:1, 28:1, 56:1, 112:1, 224:1, 448:1, 996:1, 1982:1; [OPA]=1.86 $x10^{-6}$M

As seen, derivatization of the C_6-C_8 As is to be preferred by the OPA/ET reagent: furnishing acceptable RSD values (≤ 4.7 %) and limit of quantitation (8 pM).

6. The two step derivatization of biogenic amines with the OPA/ET/fluorenylmethyl chloroformate (FMOC) reagent

6.1. Derivatization of BAs with the OPA/ET reagents

Investigations of BAs with OPA/ET reagent was initiated on our recent promising experiences with the C_6-C_8 aliphatic As [5] - carried out with the OPA/ET reagent.

Introductory results revealed – however the peak profiles and responses of Put, Cad and the single secondary amino group-containing Spd could be accepted, in favor of the two secondary amino group-containing Spm – that derivatization protocol should be altered/improved (Table 7, Figure 10). The tiny OPA/ET-Spm's response (~0.2 integrator unit/ pM Spm) does not allow its reproducible/sensitive quantitation.

Table 7 Quantitation of Different Amounts of the OPA/ET Derivatives of Biogenic Amines in Model Solutions on the Basis of Their Fluorescence Intensities (Ex/Em=337/454); OPA/ET=1/10 Reagent's Methanol Content 20 % (v/v); (Figure 10)

Amines ⇓	Ret. time, min⇓ Injected pM ⇒	Integrator units/1pM Biogenic Amine							
		300	150	75	37.5	18.75	12.0	Aver-age*	RSD %
spermidine (Spd)	3.97-4.15	4.10	4.40	4.55	4.44	4.61	4.46	4.43	4.0
putrescine (Put)	4.23-4.43	3.89	4.27	4.38	4.31	5.49	5.23	4.60	13.5
cadaverine (Cad)	4.98-5.02	6.54	7.18	7.36	7.18	8.92	8.40	7.60	11.6
spermine (Spm)	5.88-5.90	0.37	0.27	0.29	0.23	0.24	0.20	0.27	22.3
1,7-diaminoheptane (Diah)	6.88-7.02	5.49	5.95	6.22	6.02	7.48	7.69	6.48	13.8

Indications: pM = picomole; *Average(s), obtained from three separate tests have been calculated on the basis of Integration units/1pM values

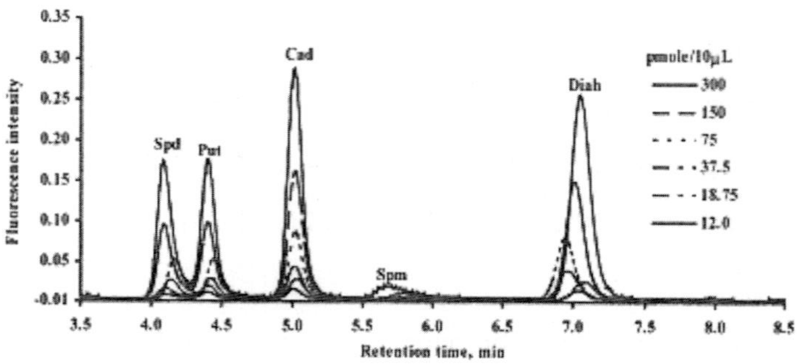

Figure 10 Fluorescence detected chromatograms of various amounts of the OPA/ET derivatives of Spd, Put, Cad, Spm and Diah (internal standard) obtained with the [OPA]/[ET] =1/10 reagent; Reaction time 90 s (detailed results in Table 7). With permission from Reference [7]

6.2. Derivatization of BAs with the OPA/ET/FMOC reagents

To increase OPA/ET-Spm's response our attention was drawn to the two step derivatization process, based on the pioneering experience with proline and hydroxyproline [67]. This principle [67], related at the beginning exclusively to the two secondary amino group-containing AAs, gained wide acceptance [66]: applying in the first step the OPA/MPA, and in the second step the FMOC reagents. Later on it was also used to amines containing matrices [68].

According to our process BAs were transformed into their OPA/ET derivatives in the first step, followed derivatization with FMOC in the second step. As a result of FMOC labeling Spd and Spm provided spectacularly excellent peak profiles, increased fluorescent responses and good reproducibility (Table 8, Figure 11).

Table 8 Quantitation of different amounts of the OPA/ET/FMOC derivatives of biogenic amines in model solutions on the basis of their fluorescence intensities (Ex/Em=337/454) and on their UV absorbance at 334 nm; OPA/ET Reagent's methanol content 20 % (v/v); (Figure 11)

Amines ⇓	Ret. time, min⇓ Injected pM ⇒	Integrator units/1pM Biogenic Amine							
		300	150	75	37.5	18.75	12.0	Average*	RSD %
on the basis of fluorescence detection									
putrescine (Put)	6.00-6.06	4.55	4.62	4.67	4.77	4.52	4.70	4.64	2.0
cadaverin (Cad)	7.08-7.16	8.64	8.61	8.62	8.35	8.21	8.26	8.45	2.3
1,7-diaminoheptane (Diah)	10.30-10.36	7.91	8.12	8.28	8.13	7.95	7.97	8.06	1.8
spermidine (Spd)	10.83-10.91	7.15	6.90	7.00	*6.06*	*5.70*	*5.56*	7.02	1.8
spermine (Spm)	12.90-13.01	*4.93*	*4.61*	*4.28*	*3.81*	*3.66*	*2.80*	-	-
on the basis of UV detection									
putrescine (Put)	5.97-6.03	0.69	0.70	0.70	0.66	0.72	0.68	0.69	3.0
cadaverin (Cad)	7.05-7.13	0.66	0.67	0.68	0.66	0.69	0.64	0.67	2.7
1,7-diaminoheptane (Diah)	10.26-10.32	0.63	0.62	0.61	0.61	0.60	0.60	0.61	1.9
spermidine (Spd)	10.80-10.88	0.56	0.50	0.54	0.53	0.52	0.52	0.53	3.9
spermine (Spm)	12.87-13.07	0.40	0.39	0.41	*0.33*	*0.30*	*0.28*	0.40	2.5

Indications: as in Table 7, as well as: *italic printed values* = have been omitted from the average values, they are read and used from the calibration curves; - = no data available

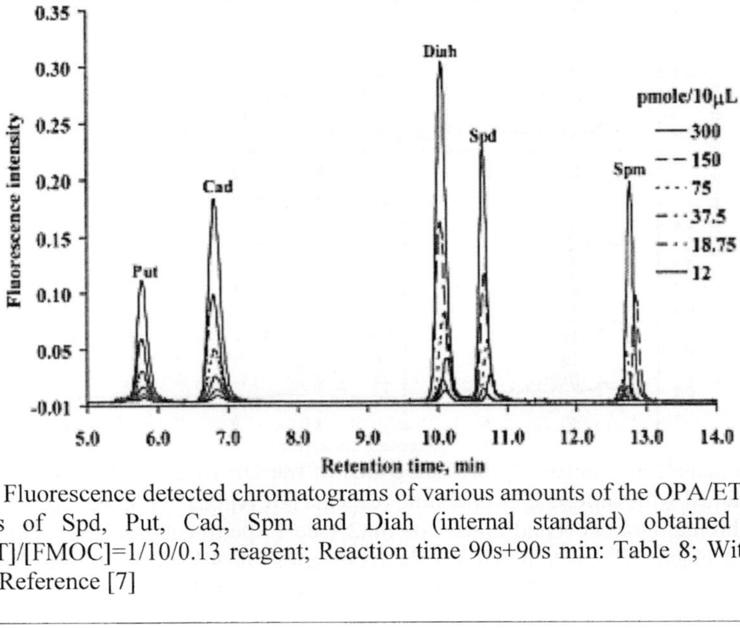

Figure 11 Fluorescence detected chromatograms of various amounts of the OPA/ET/FMOC derivatives of Spd, Put, Cad, Spm and Diah (internal standard) obtained with the [OPA]/[ET]/[FMOC]=1/10/0.13 reagent; Reaction time 90s+90s min: Table 8; With permission from Reference [7]

6.3. Extension of the OPA/ET/FMOC derivatization protocol to Orn and Lys

This special derivatization and chromatographic elution method was developed in the presence of the rest of free protein amino acids, in order to quantitate the free Orn and Lys contents of biological tissues, together with BAs, being primarily their precursors (Orn ⇒Put, Lys⇒Cad) [7].

Introductory data, obtained with the OPA/ET/FMOC reagent, revealed that both the OPA/ET-Orn and the OPA/ET-Lys derivatives do transform into various products. These unexpected derivatives, in addition to their isoindole structure (UV maximum at 334 nm), do contain also the characteristic UV maximum values of the FMOC derivatives at 262 nm. On the basis of this experience, in order to inhibit/decrease this undesired process to the possible minimum extent, detailed stoichiometric investigations were needed.

Varying the molar ratios of the OPA/ET from 1/1 to 1/50, it turned out, that on the impact of the OPA/ET/FMOC reagent, the initially formed OPA/ET derivatives (Figure 12: Orn1, Lys1) transform into various species (Figure 12: Orn2-Orn5, Lys2-Lys5). The amounts of different products could be considerably influenced by the OPA/ET molar ratios, though not quantitatively eliminated: the higher the SH-additive concentration in the reagent the *lower the amount of transformed species.*

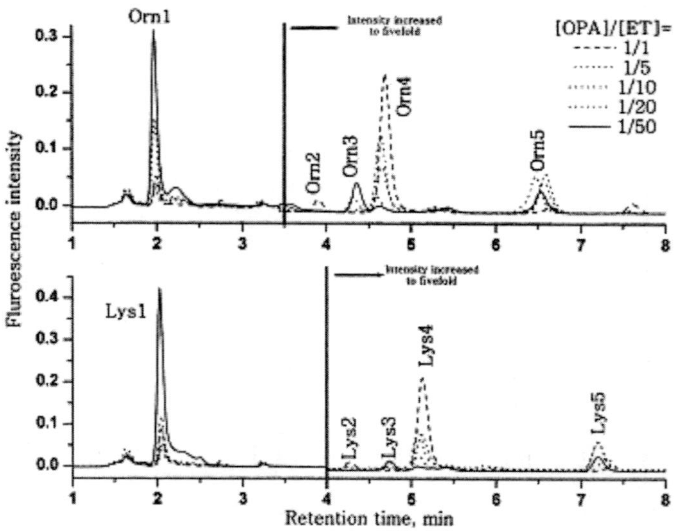

Figure 12 Fluorescence detected chromatograms of the OPA/ET/FMOC derivatives of Orn and Lys obtained with various ET containing reagents [OPA]/[ET]/[FMOC]=1/1/0.3 – 1/50/0.3; Chromatographic conditions: column2 (see Experimental secion); elution: 0min: MET 70%, A eluent 30%, 4 min: MET 75% , A eluent 25%, 4.1 min – 10 min: 100% methanol, 10.1–18 min: MET 70%, A eluent 30%; flow rate 1.5 mL·10^{-1}; temperature 50 °C; With permission from Reference7

Evaluating these data from an analytical point of view, it turned out that minimum transformation takes place when applying the reagent of OPA/ET=1/50 molar ratio: unfortunately, also this slowed down reaction provided transformation products (Figure 12: Orn2-Orn5, Lys2-Lys5). Completing these data with the behavior of BAs, on impact of OPA/ET ratios in the OPA/ET/FMOC reagent, as compromise for further studies the OPA/ET=1/10 molar ratio was selected (fast reaction, maximum response values for Spm). However to optimize the two step protocol, a further approach seemed to be necessary.

The next change in the derivatization conditions related to the MET content of the medium: this alteration proved to be of primary importance: the higher the alcohol content the lower the number and the amounts of the transformation species (Figure 13). Increasing the reagent's MET content (38%-80%, v/v) resulted in an additional advantage: the medium remains clear also after addition of the FMOC reagent.

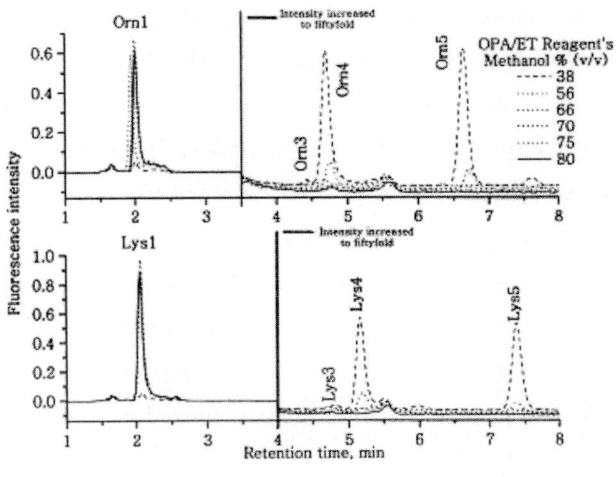

Figure 13 Fluorescence detected chromatograms of the OPA/ET/FMOC derivatives of Orn and Lys obtained with various MET (38%-80%, v/v) containing reagents [OPA]/[ET]/[FMOC]=1/10/0.3; Chromatographic conditions as in Figure 12; With permission from Reference 7

6.4. Stability studies on Put, Cad, Diah, Spd, Spm, Orn and Lys derivatives obtained with the OPA/ET/FMOC reagents

Based on derivatizations summarized in 6.1.-6.3. sections (Tables 7,8, Figures 10-13) it seemed to be necessary to characterize stability of derivatives under the most promising, 80% (v/v) MET containing conditions: depending on the reaction time in thorough connection with the OPA/ET/FMOC reagent's composition and with the optimum pH values of the two-step derivatization process.

In the first step the stability of the OPA/ET derivatives of Orn, Lys, Put, Cad and Diah have been separately tested in 80% (v/v) MET containing solution: in order to be able to follow also the tiny amounts of their transformed species. Results (not shown) [7] revealed that less than 1% transformation of the initially formed products can be expected after 1-3 min reaction times, only.

In the second approach the optimum conditions of the second step process was to be defined (reagent's FMOC content and its pH value).

The reagent's FMOC concentration at pH 9.3 were varied, in the range of the molar ratios as follows: [OPA]/[ET]/[FMOC]=1/10/0.06-1/10/0.6. It turned out that for spermine's maximum responses the molar ratios of FMOC/OPA ≥0.3/1 is to be followed.

As to the optimum pH values of the simultaneous derivatizations of the Put, Cad, Spd and Spm derivatives — carried out with the OPA/ET reagents of pH 8.6, 9.2, 9.6 and 10.6 — it has been proved that the isoindole formation is fast and quantitative in the pH range tested (pH 8.6–10.6) compared to the FMOC labeling step (pH 9.2–9.6). Thus, the pH 9.3–9.4 has been chosen as optimum reaction medium.

Prior to characterizing overall stability properties of the [OPA]/[ET]/[FMOC] derivatives of Spd and Spm, reactions were carried out after 1+1 min and 3+1 min (OPA+FMOC) derivatization steps. Since differences between the 1 and 3 min lasting OPA/ET reactions have not been found, the stability of the OPA/ET/FMOC derivatives have been followed and compared after 1+1, 1+3, 1+7, 1+16, 1+32 and 1+64 min reaction times of the mixed species (Table 4). Results (Not shown) proved [7] that from an analytical point of view the stability of derivatives is excellent. Transformation of the initially formed species do not exceed 1%.

6.5. Studies on the composition of the Put, Cad, DiaH, Spd, Spm, Orn and Lys derivatives obtained with the OPA/ET/FMOC reagents

6.5.1. UV characteristics of the OPA/ET/FMOC derivatives of Spd and Spm

As a result of DAD detection, following the UV maximum values of the OPA/ET/FMOC-Spd- and -Spm derivatives in the 190-400 nm range, it turned out that they are providing two maximum values: at 262 nm and at 334 nm. Evaluating the extent of these UV absorbency values in various methanol-containing medium (38-80%, v/v) they proved to be unambiguously consistent (Table 9). At 262 nm (characteristic of FMOC labeling), the OPA/ET/FMOC-Spd furnishes two times higher, the OPA/ET/FMOC-Spm derivatives three times higher absorbencies compared to their at 334 nm evaluated ones (characteristic to isoindole derivatives). This finding indicates that the molar absorbency of the two isoindole-functions seems to be approximately identical with the molar absorbency of one FMOC labeled function.

Table 9 UV Maximum values of the OPA/ET/FMOC-Spd and the OPA/ET/FMOC-Spm derivatives at 262nm and at 334nm; reagent: OPA/ET/FMOC=1/10/0.5

Methanol % (v/v)	Spermidine			Spermine		
	$A_{262\ nm}$	$A_{334\ nm}$	A_{262}/A_{334}	$A_{262\ nm}$	$A_{334\ nm}$	A_{262}/A_{334}
38	519	251	**2.06**	672	207	**3.25**
56	570	275	**2.07**	782	241	**3.25**
66	564	275	**2.05**	758	244	**3.11**
70	664	317	**2.09**	920	289	**3.18**
75	554	266	**2.08**	789	244	**3.23**
80	637	305	**2.09**	971	299	**3.25**

Indications: A= absorbency

6.5.2. On line HPLC/DAD/MS(ESI) studies on the composition of the OPA/ET/FMOC derivatives of Put, Cad, Diah, Spd, Spm, Orn and Lys

These investigations have been carried out separately with BAs, including Diah as internal standard (Figures 14A-F), as well as with Orn (Figures 15A-G) and Lys (Figures not shown [7]): in order to have reliable possibility to evaluate all transformed species obtained from the reactions of Orn and Lys with the OPA/ET/FMOC reagent.

6.5.2.1. HPLC/DAD/MS(ESI) study on the composition of the Put, Cad, Diah, Spd and, Spm derivatives

In the cases of BAs and Diah HPLC/DAD/MS(ESI) data confirmed their assumed, theoretical composition (Figures 14A-F). As seen the first line (Figure 14A) shows the DAD profile of Put, Cad, Diah, Spd and Spm., while the forthcoming ones (Figures 14B-F) represent their MS spectra, in order of their retention times. The protonated molecular ions and their by K cationized versions {Figure 14B, Put: MH^+=409.3=[Put]([OPA]/[ET])$_2$, MK^+=447.2; Figure 14C, Cad: MH^+=423.3, MK^+=461.2; Figure 14D, Diah: MH^+=451.3, MK^+=489.3; Figure 14E, Spd: MH^+=688.3, MK^+=726.3; Figure 14F, Spm: MH^+=967.5, MK^+=1005.5}.

6.5.2.2. HPLC/DAD/MS(ESI) study on the composition of Orn and Lys

As for derivatives obtained from the interaction of the OPA/ET/FMOC reagent with Orn (Figures 15A-G) and Lys (Figures not shown [7]) they originated from identical reactions: without exception: masses of the characteristic Orn and Lys derivatives are different in 14

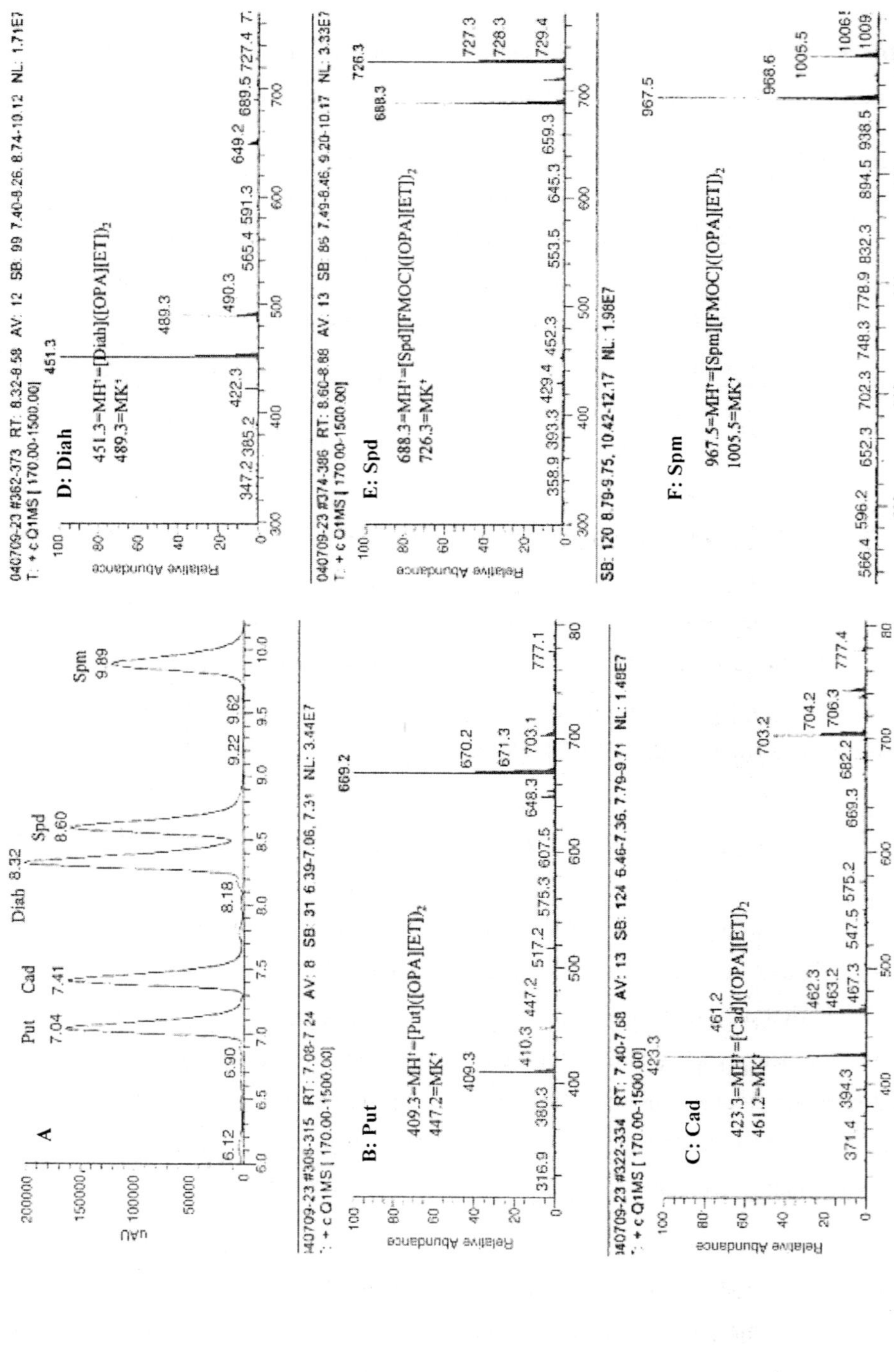

Figure 14 DAD chromatogram of the OPA/ET/FMOC derivatives at 334nm (A) and MS spectra (B-F) of Put, Cad, Diah, Spd and Spm With permission from Reference7

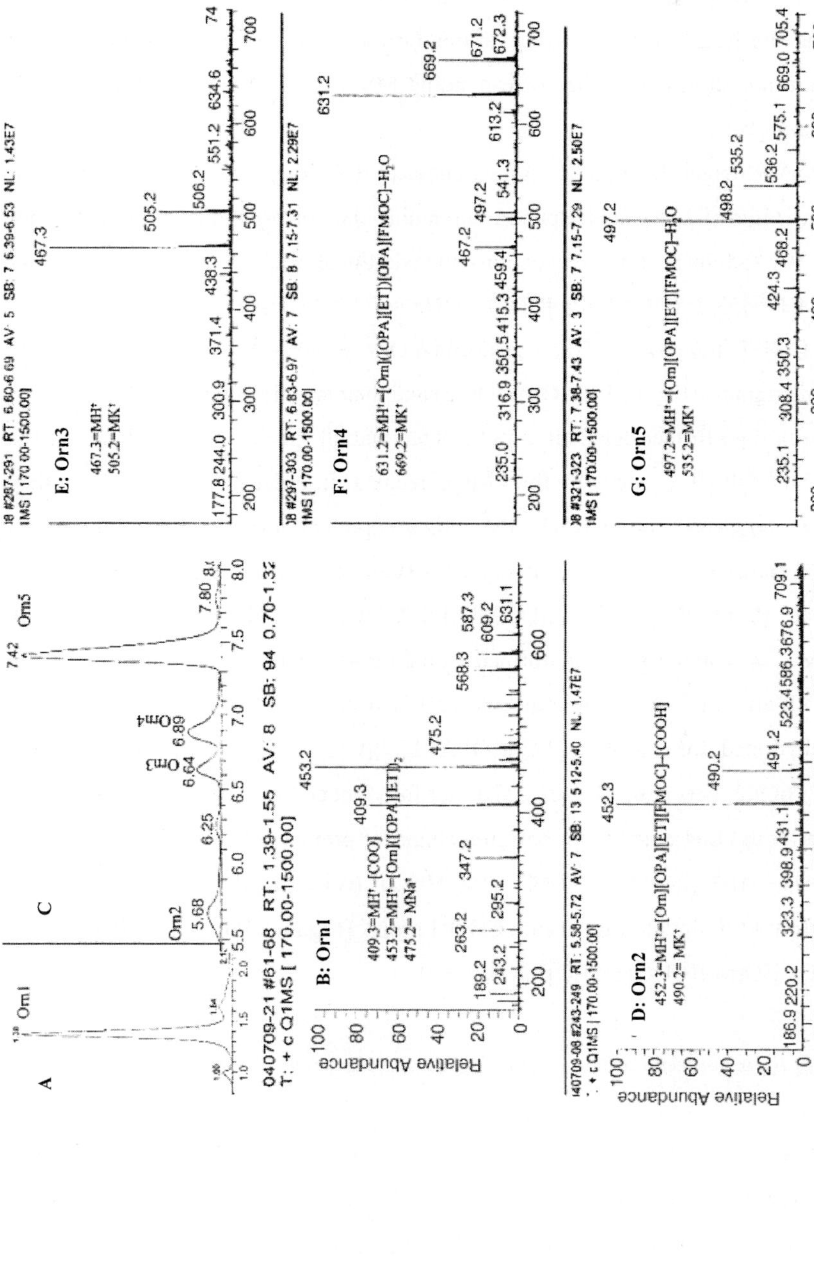

Figure 15 DAD chromatogram shown at 334 nm (A, C) and MS spectra (B, D-G) of the OPA/ET/FMOC derivatives of Orn1–Orn5; From Reference 7

mass units only (m/z=-CH_2-). (Note: Due to the different HPLC systems, to the prolonged reaction times and to the ten times diluted ammonium ion containing eluent, retention times and ratios of species — compared to those shown at Figures 12,13 — became modified: however, the number of species (Orn1-5, Lys1-5) and their retention orders were the same [7].

Evaluating the DAD profiles of the various Orn and Lys derivatives (Orn: Figures 15A,C Lysin's are not shown) and the corresponding MS spectra of species (Figures 15B, 15D-G,) it turns out that:

Figures 15A,B depict the initially formed classical isoindoles, i.e., the DAD profile of the OPA/ET-Orn (Figures 15A), their spectra containing the corresponding protonated molecular ion, their (by sodium) cationized versions and additional selective fragment ions {Figure 15B, Orn1: MH^+=453.2=[Orn]([OPA][ET])$_2$, MNa^+=475.2, $[MH\text{-}COO]^+$=409.3).

The HPLC/MS(ESI) profile of the transformed Orn species (Figures 15C-G) represent their DAD chromatograms (Figure 15C: Orn2-Orn5) and their spectra (Figure 15D-G).

Before going into further details it is to be noted that all species are mixed derivatives labeled partly by the OPA/ET, partly by the FMOC reagent. In order of their formation they are the followings: (i) Orn5 (Figure 15G), the main component, are representing the protonated, dehydrated molecular ions and their (by potassium) cationized version, the simplest, mixed derivatives {Orn5: MH^+=497.2=[OPA][ET][FMOC][Orn])-H_2O, MK^+=m/z=535.2); (ii) Orn4 (Figure 15F) contain one transformed isoindole (the -CH_2-NH_2- moieties of these AAs do react with an additional OPA molecule [2]), and one by FMOC derivatized amino group, in its dehydrated form MH^+=631.2=([OPA][ET])[OPA][FMOC][Orn]-H_2O, {Figure 15F, Orn4: MK^+=669.2; (iii) Orn3 (Figure 15E), not fully understood species, might be thoroughly correlated with Orn4 containing the same abundant protonated and cationized masses (Figure 15E, Orn3: MH^+=m/z=467.3, MK^+=m/z=505.2); (iv) Orn2 (Figure 15D) meet the masses formed from Orn5 by the loss of one COOH group {Figure 15D, Orn2: MH^+=452.2= [OPA][ET][FMOC][Orn]-([COOH+H_2O]), MK^+=490.2).

6.5.3. Behavior of the α,ω-amino group containing carboxylic acids in the two step derivatization procedures

The reaction between the FMOC reagent and the OPA/ET-derivatives of Orn and Lys called attention to the possibility of reactions also with other SH-group containing OPA reagents and with additional α,ω-amino group containing carboxylic acids. Thus, remaining on the safe side all protein AAs have been tested, one by one, in the two step process. FMOC labeling has been carried out not only subsequently to the OPA/ET derivatization but also after reactions with the OPA/MPA and with the OPA/MCE reagents. In all cases investigated, with the only exceptions of the OPA/ET derivatized Orn and Lys, they remained intact. On this basis we assumed that in this special process the crucial role is associated with the two amino groups containing AAs and, with the neutral end-group of the OPA reagent's SH-additive.

To confirm this assumption we reacted other α,ω-diamino carboxylic acids, such as the 1,3-diaminopropionic acid (Dapa), the 2,4-diaminobutyric acid (Daba) and the 2,6-diaminopimelic acid (Dpia), one by one, in the first step with the OPA/ET and in the second one with the OPA/ET/FMOC reagent.

Data obtained both with the simultaneous DAD/Fl and with the HPLC/DAD/MS(ESI) detected derivatizations proved that, in addition to the classical double isoindole, in all three above detailed reactions, mixed derivatives are formed. In cases of Dapa and Daba several derivatives, in the case of Dpia three, mixed species were obtained, characteristic to its diamino/dicarboxylic functions [7].

6.5.4. Practical utility of the simultaneous quantitation of Orn, Lys together with Put, Cad, Diah, Spd and Spm in biological tissues

In this context our main concern was associated with the elimination of the matrix effect (high salt and hydroxide ion concentrations). Preliminary studies confirmed, that without any extraction/isolation steps, Orn, Lys, Put, Cad, Spd and Spm content of mouse tissues could be quantitated in the presence of the matrix containing considerable amount of potassium perchlorate, originated from the deproteinization by perchloric acid, followed by neutralization with potassium hydroxide ([7] and manuscript in preparation).

7. Experimental

(Note: in order to save space, general considerations are shown, only: for detailed chromatographic procedures the original papers should be examined [1-7])

Materials

OPA, MPA, NAC, ET, FMOC, AAs (Orn·HCL, Lys,) and As, such as mono-(methyl-, ethyl-, n-/i-propyl, n-/i-butyl-, *tert.*-butyl-, *sec.*-butyl-, i-amyl, hexyl, heptyl, octyl amines and ethanolamine), di- (ethylenediamine, 1,2-propylenediamine, 1,3-propylenediamine, cadaverine: $NH_2(CH_2)_5NH_2$, putrescine: $NH_2(CH_2)_4NH_2$), and different polyamines {histamine: 4-(imidazolyl)ethylamine: $C_3H_3N_2CH_2CH_2NH_2$, tyramine: 4-(2-aminoethyl)phenol: $HOC_6H_4(CH_2)_2NH_2$, agmatine: 4-(aminobutyl)guanidine: $NH_2(CH_2)_4NHC(=NH)NH_2$, , Diah (as internal standard, IS), Spd·3HCl and Spm·4HCl were obtained from Sigma (St. Louis,

MO, USA) and from Serva (Heidelberg, Germany). HPLC-grade methanol (MET) and acetonitrile (ACN) were product of Riedel-deHaën purchased from Sigma-Aldrich Laborchemicalien GmbH (D-30926 Seelze, Germany). All other reagents were of the highest purity available.

Standard solutions

Standard solutions of free As have been prepared with distilled water in the concentrations of ~1-2 x 10^{-2} M and further diluted before use. Stock solution of OPA contained 0.25 — 0.75g OPA (weighed with analytical precision) in 50 mL methanol (further on: methanolic OPA solution).

Buffer solution

Borate buffer was mixed in 50/50 volume ratios from 0.2 M – 0.8 M boric acid (dissolved in 0.2 M – 0.8 M potassium chloride) /0.2 M– 0.8 M sodium hydroxide (pH 9.3 ± 0.05), if not otherwise stated.

Reagent solutions

OPA/MPA reagent was obtained by mixing, in order of listing, 5.0 mL methanolic OPA, 20.0 mL borate buffer and various amounts of MPA, MCE or ET solutions or NAC: finally, if necessary it was adjusted by 1 M sodium hydroxide, to pH 9.3 ± 0.05; The molar ratios of OPA to the SH-additives were varied from1/0.5, to 1/50 as detailed in the corresponding sections.

Derivatization

Characterization of the Reagent Solutions.

Blank elutions were performed with freshly prepared (reagent's age ≥90 min [1-7]) reagent solutions, saved in the refrigerator, (~4 ^{o}C) and injected by the robotic Autosampler, every day at least twice (Waters 717, thermostatted for ~4 ^{o}C).

Chromatography

Stability and stoichiometric studies: simultaneous DAD and Fl detection

The system was a Waters HPLC instrument (Waters Pharmaceutical Division, Milford, MA, U.S.A.), consisted of a Waters 996 DAD and a Waters 474 Fl Detectors, a Waters 600 Controller quaternary pump with a thermostattable column area and a Waters 717 Autosampler, operating with the Millennium Software (version 2010, 1992-95, validated by ISO 9002). The columns were Hypersil ODS bonded phase (5μm), 200mm x 4.6mm+ 30mm x 4.6mm guard column (column1), or 150mm x 4.6 mm + 20mm x 4.6mm guard column (column2).

Detections have been performed simultaneously: DAD and Fl detectors were connected in order of listing. Blank tests, stoichiometric investigations have been recorded between 190 and 400 nm (DAD) and evaluated at 334/262 nm, as well, as at the optimum fluorescence wavelengths of isoindoles (Ex/Em=337/454 nm).

On-line HPLC/MS(ESI) studies simultaneous UV and MS detection was carried with a Thermo Finnigan TSQ Quantum AH apparatus (Thermo Finnigan, LC-MS Division, San Jose, CA, USA), consisted of a Surveyor DAD detector a TSQ Quantum AH detector, a Surveyor Autosampler, operating with the Xcalibur^{TM} software 1.4 SRI.

Detections have been performed simultaneously, applying the Surveyor DAD and the TSQ Quantum AH detectors, connected in order of listing. Blank tests, concentration dependence have been recorded between 190 and 900 nm (UV), evaluated at 334 nm (OPA/ET-AAs and As), MS detections were performed with ESI in the positive mode (Mass range: 50-1600 mass units; Gas temperature: 200 ^{o}C (flow rate 200μL/min) or 380 ^{o}C (flow rate 1mL/min); Vcapillary: 3.5kV;

References

[1] D. Kutlán, I. Molnár-Perl J. Chromatogr. A 949 (2002) 235-248.

[2] Y. Mengerink, D. Kutlán, F. Tóth, A. Csámpai, I. Molnár-Perl, J. Chromatogr. A 949 (2002) 99-124.

[3] R. Hanckó, I. Molnár-Perl, Chromatographia 57 (2003) S-103–S-123.

[4] D. Kutlán, I. Molnár-Perl J. Chromatogr. A 987 (2003) 311-322.

[5] R. Hanckó, D. Kutlán, F. Tóth, I. Molnár-Perl, J. Chromatogr. A 1031 (2004) 51-66.

[6] T. Törő, Cs. Ágoston, I. Molnár-Perl, Chromatographia 60 (2004) S-153–S-159.

[7] R. Hanckó, Á. Kőrős, F. Tóth, I. Molnár-Perl, J. Chromatogr. A (2005) in press.

[8] S.S. Simons Jr., D.F. Johnson, J. Chem. Soc. Chem. Commun. (1977) 374-375.

[9] S.S. Simons Jr., D.F. Johnson, J. Org. Chem. 43 (1978) 2886-2891.

[10] S.S. Simons Jr., D.F. Johnson, Anal. Biochem. 82 (1977) 250-254.

[11] J.F. Stobaugh, A.J. Repta, L.A. Sternson, K.W. Garren, Anal Biochem. 135 (1983) 495-504.

[12] J.F. Stobaugh, A.J. Repta, L.A. Sternson, J. Org. Chem. 49 (1984) 4306-4309.

[13] J.F. Stobaugh, A.J. Repta, L.A. Sternson, J. Pharm. Biomed Anal. 4 (1986) 341-351.

[14] S.S. Simons Jr., D.F. Johnson, Anal. Biochem. 90 (1978) 705-725.

[15] T.P. Davis, C.W. Gehrke, C.W. Gehrke Jr., T.D. Cunningham, K.C. Kuo, K.O. Gerhardt, H.D Johnson, C.H. Williams, Clin. Chem. 24 (1978) 1317-1324.

[16] T.P. Davis, C.W. Gehrke Jr., C.H. Williams, C.W. Gehrke, K.O. Gerhardt, J.Chromatogr. 228 (1982) 113-122.

[17] C. Buteau, C.L. Duitschaever, G.C. Ashton, J. Chromatogr. 284 (1984) 201-210.

[18] F. Dai, V. Prelevic-Burkert, H.N. Singh, W.L. Hinze: Microchem. J. 57 (1997) 166-198.

[19] A. Yamatodani, H. Fukuda, H. Wada, T. Iwaeda, T. Watanabe, J. Chromatogr. 344 (1985) 115-123.

[20] L.G. Harsing Jr., H. Nagashima, E.S. Vizi, D. Duncalf, J. Chromatogr. 383 (1986) 19-26.

[21] P. Lehtonen, M. Saarinen, M. Vesanto, M.L. Riekkola, Z. Lebensm. Unters. Forsch. 194 (1992) 434-437.

[22] G. Achilli, G. P. Cellerino, G.M. d'Eril, J. Chromatogr. 661, (1994) 201-205.

[23] T.B. Jensen, P.D. Marely, J. Chromatogr. B, 670 (1995) 199-207.

[24] O. Busto, J. Guasch, F. Borrull, Chromatographia 40 (1995) 404-410.

[25] O. Busto, J. Guasch, F. Borrull, J. Chromatogr. A 718 (1995) 309-317.

[26] T. Hernandez-Jover, M. Izquirdo-Pulido, M.T. Veciana-Nogues, M.C. Vidal-Carou, J.Agric. Food Chem. 44 (1996) 2710-2715.

[27] Y. Feng, A.E. Halaris, J.E. Piletz, J. Chromatogr. B 691 (1997) 277-286.

[28] I.M. Mackie, L. Pirie, H. Yamanaka, Food Chem. 60 (1997) 57-59.

[29] S. Vale, M.B.A. Glória, Food Chem. 63 (1998) 343-348.

[30] G.J. Soleas, M. Carey, D.M. Goldberg, Food Chem. 64 (1999) 49-58.

[31] S. Bover-Cid, S. Schoppen, M. Izquierdo-Pulido, M.C. Vidal-Carou, Meat Science 51 (1999) 305-311.

[32] S. Bover-Cid, M. Izquierdo-Pulido, M.C. Vidal-Carou, Int.J. Food Microbiol. 46 (1999) 95-104.

[33] K. Kawamura, T. Matsumoto, T. Nahakara, M. Hirano, H. Uchimura, H. Maeda, J. Liq. Chrom.Rel Technol. 23 (2000) 1981-1993.

[34] A. Sass-Kiss, E. Szerdahelyi, G. Hajós, Chromatographia 51 (2000) 316-320.

[35] S. Bover-Cid, M. Izquierdo-Pulido, M.C. Vidal-Carou, Meat Science 57 (2001) 215-221.

[36] M. Venza, M. Visalli, D. Ciccui, D. Teti, J. Chromatogr. B 757 (2001) 111-117.

[37] J. Lange, K. Thomas, C. Wittmann, J. Chromatogr. B 779 (2002) 229-239.

[38] E. Wang, E. Struble, P. Liu, A.P. Cheung, J. Pharm. Biomed. Anal. 30 (2002) 415-427.

[39] C.M.G. Silva, M.B.A. Glória, Food. Chem. 78 (2002) 241-248.

[40] M. V. Moreno-Arribas, M.C. Polo, F. Jorganes, R. Muňoz, Int. J. Food Microbiol. 84

(2003) 117-123.
[41] M.C. Vidal-Carou, F. Lahoz-Portolés, S. Bover-Cid, A. Mariné-Font, J. Chromatogr. A 998 (2003) 235-241.
[42] M.P.G. Cirilo, A.F.S. Coelho, C.M. Araújo, F.R.B. Gonçalves, F.D. Nogueira, M.B.A. Glória, Food Chem. 82 (2003) 397-402.
[43] M. Kanki, T. Yoda, M. Ishibashi, T. Tsukamoto, Int. J. Food Microbiol. 92 (2004) 79-87.
[44] A. Marcobal, M.C. Polo, P.J. Martín-Álvarez, M.V. Moreno-Arribas, Food Res. Int. in press
[45] Y. Miyamoto, R. Yoshimoto, M. Yumoto, A. Ishihara, K. Takahashi, H. Kotani, A. Kanatani, S. Tokita, Anal, Biochem. 334 (2004) 89-96.
[46] S. Loret, P. Deloyer, G. Dandrifosse, Food Chem. 89 (2005) 519-525.
[47] S. Moret, D. Smela, T. Populin, L.S. Conte, Food Chem. 89 (2005) 355-361.
[48] T. Skaaden, T. Greibrokk, J. Chromatogr. 247 (1982) 111-122.
[49] N. Bilic, J. Chromatogr. A 719 (1996) 321-326.
[50] K. Saito, M. Horie, N. Nose, K. Nakagomi, H. Nakazawa, J. Chromatogr. 595 (1992) 163-168.
[51] K. Saito, M. Horie, H. Nakazawa, Anal. Chem. 66 (1994) 134-138.
[52] O. Busto, M. Miracle, J. Guasch, F. Borrull, J. Chromatogr. A 757 (1997) 311-318.
[53] R.Herráez-Hernández, P. Campíns-Falcó, J. Chromatogr. A 893 (2000) 68-80.
[54] P. Campíns-Falcó, C. Molins-Legua, A. Sevillano-Cabeza, L.A.T. Genaro, J. Chromatogr. B 759 (2001) 285-297.
[55] S. M. Lloret, J.V. Adrés, C. Molins Legua, P. Campíns- Falcó, Talanta 65 (2005) 869-875.
[56] D. Lochmann, S. Stadlhofer, J. Weyermann, A. Zimmer, Int. J. Pharm. 283 (2004) 11-17.
[57] J.F. Bowyer, P. Clausing, G.D. Newport, J. Cromatogr. B 666 (1995) 241-250.
[58] H.M.H. van Eijk, D.R. Rooyakkers, N.E.P. Deutz, J. Chromatogr. A 730 (1996) 115-120.
[59] Y. Arakawa, S. Tachibana, Anal. Biochem. 158 (1986) 20-27.
[60] S.E. Douabalé, M. Dione, A. Coly, A. Tine, Talanta 60 (2003) 581-590.
[61] F.A.J. Muskiet, B. Dorhout, G.A. van den Berg, J. Hessels, J. Chromatogr. B 667 (1995) 189-198.
[62] A.R. Shalaby, Food Res. Int. 29 (1996) 675-690.
[63] A. Bouchereau, P. Guénot, F. Larher, J. Chromatogr. B 747 (2000) 49-67.
[64] M.Y. Khuhawar, G.A. Qereshi, J. Chromatogr. B 764 (2001) 385-407.
[65] D. Teti, M. Visalli, H. McNair, J.Chromatogr. B 781 (2002) 107-149.

[66] Handbook of Derivatization Reactions for HPLC, G. Lunn, L.C. Hellwig, Editors, Amines, Wiley (1998), p. 253-624.
[67] R. Schuster, J. Chromatogr. 431 (1988) 271-284.
[68] P. Herbert, L. Santos, A. Alves, J. Food Sci. 66 (2001) 1319-1325.

Ibolya Molnár-Perl (Editor)
Quantitation of Amino Acids and Amines by Chromatography
Journal of Chromatography Library, Vol. 70

2.2.2.

Quantitation by HPLC of Amines as Dansyl Derivatives

Manuel Silva

Contents

Summary

The potential of dansyl chloride as labelling reagent for the derivatization of amines prior their determination by high performance liquid chromatography is overviewed. Different approaches for carrying out the dansylation reaction, such as bath derivatization, the use of solid supports and automated on-line procedures are discussed. Dansylation, separation and detection of low-molecular-mass aliphatic amines, biogenic amines including polyamines and aromatic amines of clinical significance are summarized. From a critical study, the most favourable conditions for the determination of these amine compounds in real samples such as waters, foods, beverages and clinical samples are reported.

1. Introduction

High performance liquid chromatography (HPLC) is one of the most widely used separation techniques for the selective and sensitive determination of a wide range of species in many areas of interest. Although simplicity and robustness are also two important features of the ensuing HPLC methods, sometimes, these analytical properties should be forgone in order to achieve the appropriate sensitivity for the determination of the analytes, which requires an additional experimental step. Thus, although most HPLC separations are carried out using UV detection due to its ease of use, owing to its poor limit of detection (LOD), the resulting HPLC method can be unsuitable for the determination of the analytes in the addressed sample and therefore, labelling becomes a necessary procedure. This is by no means uncommon, and thus labelling techniques are widely used in HPLC and have long been accepted as a method of choice for enhancing the sensitivity and/or selectivity for the overall process.

Compounds containing amine functional groups such as amines and amino acids are a kind of analyte that requires derivatization for their determination by HPLC [1]. Previous chapters of this book deal with the HPLC determination of amino acids based on the use of different derivatization reagents, and the former is related to the HPLC determination of amines using o- phthalaldehyde (OPA) as a fluorescence probe. OPA reacts with primary amines at alkaline pH in the presence of certain thiols such as mercaptoethanol and ethanethiol, the reaction being essentially complete within one minute. The derivatives, which have an average quantum yield of 0.4, are unstable and decompose to non-fluorescent products.

In this chapter, the use of another popular derivatizing agent for the HPLC determination of amines such as dansyl chloride (5-dimethylaminonaphthalene-1-sulfonyl chloride, DNS-Cl) is reviewed. Since its development by Weber in 1951, this labelling reagent has been used extensively to determine the N-terminal amino acid residue of proteins and to prepare derivatives of amino acids, amines and amine compounds.

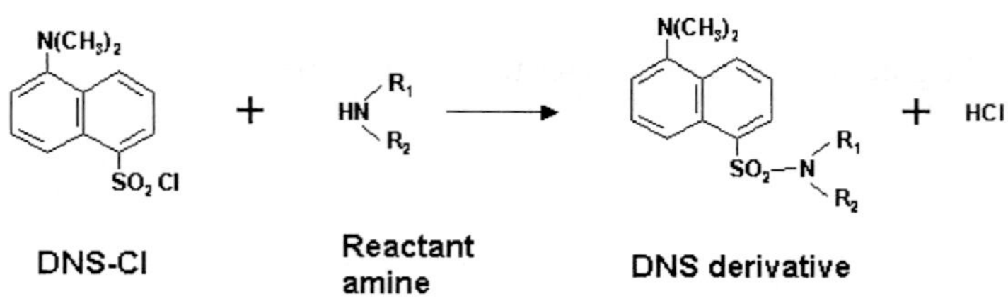

Figure 1 Scheme of the labelling reaction between DNS-Cl and primary or secondary amines.

As can be seen in Figure 1, DNS-Cl is composed of a highly fluorescent aromatic moiety (5-dimethylaminonaphthalene) and a reactive group (sulfonyl chloride). The reactive group attaches to the analytes by means of a nucleophilic substitution reaction to yield blue or blue-green-fluorescent sulphonamides (dansyl amides). DNS derivatives, formed under weakly alkaline conditions, are fairly stable and have long Stokes shifts; however, in contrast to OPA, the labelling reaction follows slower kinetics and therefore higher reaction times and temperature are required, which precludes the performance of the post-column derivatization and so, pre-column labelling is mandatory. Despite the weak molar absorptivity, ca. 4000 l mol^{-1} cm^{-1}, and moderate fluorescence quantum yield of derivatives, DNS-Cl, as stated in this chapter, is widely-used as a probe for the HPLC determination of amines with UV and luminescence (fluorescence and chemiluminescence) detection. In summary, DNS-Cl is a well-established alternative to OPA for the HPLC determination of amines with important significant practical features: the possibility of determining secondary amines; the derivatives are more stable; the optimum labelling conditions are not limited by the HPLC separation conditions; and a greater diversity of detection systems can be used (only fluorescence detection is used for OPA derivatives), which is of great practical interest. Thus, UV detection is a useful alternative to fluorescence when the aqueous buffer solution used for separation quenches the fluorescence of DNS derivatives, whereas chemiluminescence detection based on peroxyoxalate exploits the increase in the analytical signal of DNS derivatives as increasing the non-polarity of the mobile phase.

To outline the analytical features of DNS-Cl as a labelling reagent for the HPLC determination of amines, this chapter will be divided into four main sections; the first, deals with aspects related to the proposal of different approaches to improve the kinetics of the labelling reaction and their influence on the subsequent separation of the DNS derivatives and/or the sensitivity of the final HPLC determination. Methods focused on its partial or total automation (the procedure is in some cases tedious and time consuming) are also considered. The following sections refer to the kind of amine to be studied, such as low-molecular-mass aliphatic amines, biogenic amines including aliphatic, alicyclic and heterocyclic amines and aromatic amines; in the last case only amines of clinical significance such as amphetamines, among others, are considered.

2. Approaches for improving the dansylation reaction

Although DNS-Cl is a well-known label for the HPLC determination of amines, there is no consensus in the literature on the derivatization conditions that should be used even for a particular type of these compounds, such as aliphatic and biogenic amines, and therefore a wide variety of experimental conditions have been suggested. Thus, many approaches have been proposed regarding the labelling reaction of amines with DNS-Cl in order to improve several analytical aspects, such as the kinetic parameters involved in the dansylation reaction, the separation and sensitivity in the subsequent HPLC determination of DNS derivatives, and also in order to reduce sample handling based on the use of flow systems coupled on-line or off-line to the chromatographic system.

Temperature and time are two important inversely related variables that affect the performance of the labelling reaction of amines with DNS-Cl; the use of higher temperatures allows the derivatization time to be decreased. Although, in principle, this option could be used to increase the reaction rate, higher temperatures decrease the stability of DNS derivatives and consequently diminish the final labelling yield. Since microwave ovens are a good way for supplying high temperatures in a short time, a microwave-assisted approach has been reported as a useful alternative to accelerate the dansylation of amines [2,3]. The optimal procedure for biogenic amines [2] involves microwave treatment, ca. 250 W for just 5 min, keeping a maximum pressure of 3.4 bars inside the reactor. The sample throughput of the method can be easily increased by using various derivatization reactors; in fact, six or ten dansylations can be simultaneously carried out by prolonging the reaction time until 10 or 25 min, respectively, keeping the energy and pressure constant. In comparison to traditional overnight dansylation, the microwave-assisted procedure is 150-times faster.

The pH level is one of the most significant factors affecting the dansylation reaction of amines because its variation provides two concomitant effects: a) it influences the amine's reactivity by altering its basicity; and 2) it provokes the appearance of DND-Cl by-products such as follows [4]:

$$H_2O + DNS\text{–}Cl \rightarrow DNS\text{–}OH + HCl$$

$$R\text{–}NH\text{–}DNS + DNS\text{–}Cl \rightarrow DNS\text{–}NH_2 + \text{other products}$$

The presence of dansyl sulfonic acid (DNS–OH) and dansyl amide (DNS–NH_2) in the labelling medium has been recently confirmed by HPLC-atmospheric pressure chemical ionisation mass spectrometry [5]. Since the dansylation reaction is accelerated at higher pH levels, a compromise situation should be assumed because both side-reactions are also favoured at these pH values. Furthermore, DNS-OH is highly fluorescent and can interfere with the subsequent chromatographic separation of amines because it can be co-eluted with DNS derivatives of amine with a greater hydrophilic character. In order to overcome these shortcomings, several approaches have been reported, in many cases involving an additional feature such as increasing the sensitivity of the final analytical determination by using a separation technique.

The most traditional option consists of adding proline to the reaction medium once the labelling reaction has been completed and separating the DNS-derivatives by liquid-liquid extraction (LLE). This procedure removes the excess of DNS-Cl by forming the ionised proline derivative, which remains in the aqueous phase together with DNS-OH [6-11]. In general, after the labelling reaction, the sample was treated with proline at 5-10 fold molar excess with respect to DNS-Cl for 15-30 min and the derivatives were extracted into an organic solvent such as toluene [8,9], chloroform [10,11], diethyl ether [6] or cyclohexane [7]. The organic extract is dried under nitrogen stream and the residue reconstituted in a small volume of mobile phase [7-9] or acetonitrile (MeCN) [6,10,11] for HPLC analysis, which provides an additional enrichment step. Sodium glutamate [12,13], ammonia [14,15] and methylamine [16] have been also used as alternatives to proline for removing the excess of DNS-Cl. After the addition of these reagents, the mixture under reaction is then centrifuged and filtered to isolate derivatives.

Although numerous LLE methods for isolating DNS derivatives from the reaction side-products and excess of DNS-Cl have been reported, even without the previous addition of proline or a related reagent [17, 18], nowadays, solid-phase extraction (SPE) procedures are preferred to LLE because of its obvious advantages in achieving higher enrichment factors and clean-up effectiveness. Several studies have been reported for the SPE of the derivatives through C_{18} cartridges [5,19,20] in which, amines can be determined at $\mu g\ l^{-1}$ levels. The experimental procedure reported for biogenic amines involves first of all the removal of acetone under nitrogen from the labelling medium, and then the introduction of the resulting solution into the SPE cartridge, which has been previously activated with methanol (MeOH) followed by water. After the sample has passed through the cartridge, it is washed with water-acetone (80:20 v/v), dried under a vacuum, and finally DNS derivatives are eluted with MeCN. The eluates are evaporated to dryness under a nitrogen stream and the residue reconstituted to MeCN. This procedure allows the determination of biogenic amines in wines and beers with LODs ranging from 10–100 $\mu g\ l^{-1}$ using UV detection at 254 nm on account of the enrichment factor provided by the SPE system [19].

Over the past few years, new alternatives have been proposed to simplify the labelling reaction of amines with DNS-Cl, "solid support assisted derivatization" being the most interesting choice. In this approach, the analyte and the label react in a conventional C_{18} SPE cartridge (off-line systems) [21-25] or in a C_{18} SPE mini-column coupled to a flow manifold (on-line systems) [21,26]. In addition to simplifying the derivatization reaction, the SPE system can be used to clean up and for the pre-concentration of the amines prior to their labelling with DNS-Cl with the related increase in the selectivity and sensitivity of the developed analytical method.

The off-line procedure involves the following steps: 1) activation of the C_{18} sorbent with MeOH followed by the derivatization buffer (carbonate buffer adjusted to the labelling pH); 2) clean up and/or pre-concentration of amines by adding the variable sample volume (1-25 ml) to the cartridge; 3) addition and retention of the label on C_{18} by flushing a small volume of the DNS-Cl solution through the cartridge. The concentration of this solution is a critical parameter because it should be high in order to obtain yields similar to those provided using batch derivatization; however, an increase in the label concentration causes two adverse effects: a) the peak responses for undesirable peaks increases; and b) the low solubility of the DNS-Cl in water forces us to use less polar solvent that can elute amines previously trapped in the column. A DNS-Cl solution (20-30 mM) prepared in an acetone-hydrogen carbonate buffer solution (pH 9.5, 2 + 3 v/v) provides good results; 4) derivatization of amines after a given reaction time at room temperature. In order to decrease the labelling time the cartridge can be incubated in an oven, even at high temperatures such as 85–100° because, in contrast to the batch procedure no degradation products are observed. A reaction time of 10 or 15 min at 100 or 85°C, respectively can be proposed, according to the kind of amine under study. After this treatment, the cartridge is dried under a vacuum with air; and 5) elution of the DNS derivatives with MeCN. In general, the derivatization efficiency of this procedure is high when compared to the values achieved by the batch alternative, ca. 90%. However, in some cases the amount of reaction products is only ca. 60–70%, which is the case with the determination of spermine and spermidine [24], which can be assigned to their polyamine structure.

The on-line mode is based on similar principles, and, as can be seen in Figure 2, minimum instrumentation is required beyond conventional HPLC equipment, namely, an isocratic (peristaltic) pump, a C_{18} mini-column and a switching valve [21,26]. This figure shows the scheme corresponding to a fully automated method for the analysis of alkylamines based on pre-column derivatization on solid support. In such a method, samples are injected in the C_{18} mini-column inserted into a flow system for sample clean up, pre-concentration and derivatization. The derivatives formed are transferred to the analytical column, separated and detected by fluorescence.

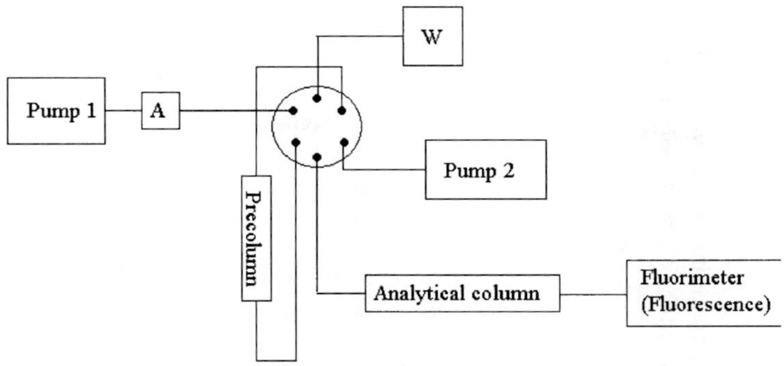

Figure 2 Scheme corresponding to on-line derivatization on solid SPE cartridges coupled to an analytical column, detector (fluorescence). (Reprinted from refetrence 21 with permission).

Table 1 lists the different parameters involved in this system and the optimum conditions established for the determination of some aliphatic amines in water samples [26], for which the full analysis is only feasible in ca. 30 min.

On the basis of the features and the results reported by both approaches, it seems that the off-line mode is the most useful and flexible alternative for the dansylation of amines on C_{18} solid-phase packing. In fact, aliphatic [22,23] and biogenic [24] amines as well as amphetamines [25] have been determined with fluorimetric and chemiluminescence detection using this approach, whereas only two references have been reported based on the on-line mode referring to the determination of aliphatic amines [21] and amphetamine and related compounds [26] taking into account that the on-line mode was the first attempt at the dansylation of amines on solid-support [26]. In general, both modes offer several advantages over other labelling procedures, namely: 1) shorter analysis time with the subsequent increase of the sample throughput; 2) lower consumption of solvent and reagents providing cleaner methods for the environment; 3) heating of the sample is not necessary; and 4) different experimental steps can be carried out sequentially on the solid support such as clean-up, labelling and sample enrichment.

Table 1 Experimental conditions and time schedule used for determination of mono amines by using a fully automated system, using DNS-Cl as derivatization reagent. (Reprinted from reference [21] with permission)

Cumulative time (min)	Trapping column	Analytical column	Valve position
0	Injection sample: 150 μl	50:50 (v/v) (acetonitrile:imidazol buffer) (flow rate of 0.1 ml min^{-1})	A
0–3.75	Injection of DNS-Cl solution 40 μl Injection buffer 60 μl Sample clean un (gradient from 0.5 ml min^{-1} to zero at 3.75 min)		
9–13.5		50:50 (v/v) (acetonitrile:imidazol buffer) (gradient flow rate: 0.1 to 1 ml min^{-1})	
13.5–16.5	Analytical separation 70:30 (v/v) acetonitrile:imidazol buffer	50:50 (v/v) (acetonitrile:imidazol buffer) mode gradient (flow rate of 1 ml min^{-1})	B
16.5–28.5	Water acetonitrile (gradient to 100% acetonitrile)	50:50 (v/v) (acetonitrile:imidazol buffer) mode gradient up to 90% acetonitrile (at a flow rate of 1 ml min^{-1})	A
28.5–29.5	Water:acetonitrile (gradient to 100% water)	50:50 (v/v) (acetonitrile:imidazol buffer) (gradient flow rate: 1 to 0.1 ml min^{-1})	
30.5	End		

In order to achieve the full automation of the HPLC process for the determination of amines via pre-column dansylation, an on-line liquid chromatographic system involving SPE, dansylation and chemiluminescence detection has been developed for the determination of aliphatic amines in water samples [27] and biogenic amines in table olives [28] (Figure 3)

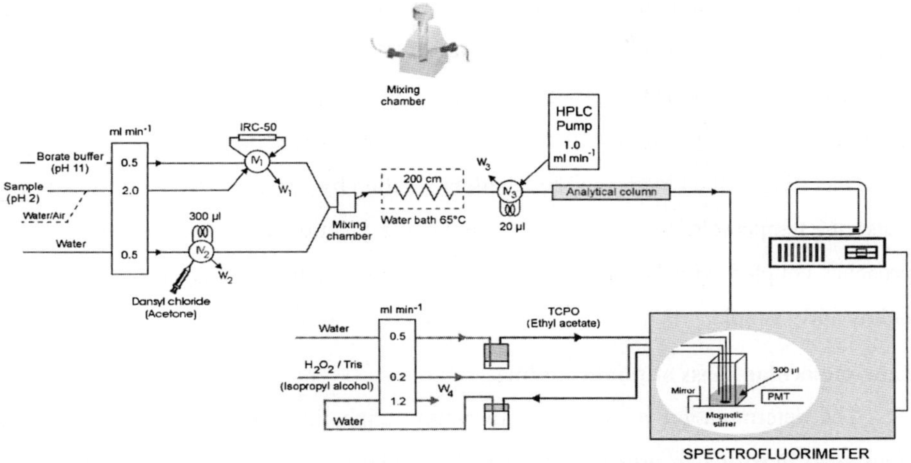

Figure 3 Experimental setup used for the continuous SPE extraction and dansylation of amines, and their determination by HPLC with chemiluminescence detection. IV, Injection valve; W, waste. (Reprinted from reference [27] with permission from Elsevier).

In order to achieve the full automation of the HPLC process for the determination of amines via pre-column dansylation, an on-line liquid chromatographic system involving SPE, dansylation and chemiluminescence detection has been developed for the determination of aliphatic amines in water samples [27] and biogenic amines in table olives [28]. Figure 3

shows the instrumental set-up required for the determination of low-molecular-mass amines in water samples [27]. As can be seen, the implementation of the continuous SPE and dansylation of amines can be carried out with minimum instrumental requirements such as a peristaltic pump, a mixing chamber and an injection valve in view of the flow system design. Thus, the water sample (up to 50 ml) is continuously aspirated through the cation-exchange column IRC-250 (previously conditioned with hydrochloric acid) located inside the loop of the injection valve for the clean up and pre-concentration of amines, and then the column is washed with water and air-dried. After switching the injection valve on, a stream of borate buffer solution for the elution of the retained amines and a water carrier stream, into which DNS-Cl solution is injected, are simultaneously propelled and mixed in the mixing chamber. Dansylation is performed in a PTFE reaction coil immersed in a thermostated water bath at

65°C for 10 min, and finally a 20-μL aliquot is injected into the HPLC. This continuous dansylation system provides two major advantages over the reported batch alternative, such as an appreciably shortened reaction time and avoidance of the tradition additional step involving the removal of any excess label reagent. Regarding on-line dansylation based on the use of C_{18} solid support, the on-line SPE-dansylation system provides higher enrichment factors and amines recovered from water samples, which can be attributed to the better performance of the cation-exchange resin IRC-250 versus C_{18} for the clean up and pre-concentration processes. In addition, the continuous dansylation system allows a more flexible handling of the kinetics of the labelling reaction by varying not only the reaction time but also the temperature. Finally, for fully automatic functioning of both on-line dansylation approaches coupled with HPLC, the pump and valves should be controlled by a computer, which is fairly simple.

3. Low-molecular-mass aliphatic amines

The determination of low-molecular-mass aliphatic amines is well documented in various areas in particular water and food analysis. Short-chain aliphatic amines are important intermediates in chemical, pharmaceutical and manufacturing industries, and therefore certain amounts of aliphatic amines in the environment are considered a serious risk. Thus, they are target analytes of US Environmental Protection Agency (EPA) Method 8260 and European legislation sets a content value below 1 mg l^{-1} for water for consumption, and between 15 and 85 mg l^{-1} for waste waters. Higher aliphatic amines (> C_{10}), which are widely used in industry as flotation agents, are fairly toxic; their maximum permissible concentration in water is 0.25μg l^{-1}. Aliphatic amines also exist in biological systems and different foods as biodegradation products of organic material such as amino acids and proteins. These amines are not only irritants to the human body, but can also react with nitrite forming N-nitrosamines, which are potential mutagenic and carcinogenic compounds.

A number of HPLC methods for the determination of low-molecular-mass amines based on the formation of their DNS derivatives have been reported. In some of them, the analytical determination of amines is not the main aim, and thus, amines are used as analytes for evaluating some of the approaches described in the above section or for a specific separation and/or detection system. However, over the past few years the proposed HPLC methods have focused on determination in water samples, the most common feature being the attempt to increase sensitivity and selectivity, because aliphatic amines occur in small quantities in certain kinds of waters.

Table 2 shows the features of the most significant procedures reported in the literature regarding HPLC determination of low-molecular-mass aliphatic amines (including primary and secondary) using pre-column dansylation. As can be seen, various spectroscopic detection systems have been used such as UV, fluorescence and chemiluminescence based on the above-mentioned spectroscopic properties of DNS-Cl; although on rare occasions, electrochemical detection has been also employed. Chemiluminescence detection is based on the chemical excitation of DNS derivatives using bis (2,4,6-trichlorophenyl oxalate) (TCPO)-hydrogen peroxide [34]. Thus, the dansyl derivatives previously separated in the analytical column, are transferred to the post-column system and mixed with the ingredients of the chemiluminescence reaction. This mixture is the key to the performance of the chemiluminescence system, and thus, when a flow system is used [22,29] the distance between the reagents confluence and the detector (PMT) should kept to a minimum to avoid loss of analytical signal due to the fast kinetics of the chemiluminescence reaction. To overcome this shortcoming, a zero-dead volume detector can be used (see Figure 3) in which the mixture of the chemiluminescence reagents and the eluate of the analytical column is carried out simultaneously [27]. Figure 4 shows a typical chromatogram obtained using this detection system. In summary, chemiluminescence detection provides higher sensitivity for the determination of amines, LODs at pg ml^{-1} level can be achieved [27]; however this kind of detection requires an additional post-column system, although it is very simple and inexpensive [22,27,29].

Table 2 Summary of different reported methods for the HPLC determination of low-molecular-mass aliphatic amines via pre-column dansylation

Amine	Labelling	Separation (C_{18})	Detection	LOD, ng ml^{-1}	Sample	Comments	Reference
$C_1 - C_6$	85°C 15 min	Gradient elution MeCN-imidazole buffer (pH 7.0)	UV (333 nm) F	3 – 340 2 – 4	Tap, irrigation ditch, lake and residual waters	SPE on C_{18} cartridges. Dansylation on solid support (off-line mode).	23
$C_1 - C_6$	25°C 10 min	Gradient elution MeCN-imidazole buffer (pH 7.0)	F	11.2 – 1500 (*)	Residual waters	SPE on C_{18} cartridges. Dansylation on solid support (on-line mode).	21
$C_1 - C_6$	110°C 10 min	Gradient elution MeCN-imidazole buffer (pH 7.0)	CL	0.15 – 0.9	Tap, irrigation ditch, lake, residual and marine waters	SPE on C_{18} cartridges. Dansylation on solid support (off-line mode).	22
$C_1 - C_7$	65°C 10 min	Gradient elution from 25-75% MeCN-water	CL	0.015 – 0.3	Tap, surface, ground and river waters	SPE on Amberlite IRC-50. On-line continuous dansylation.	27
$C_4 - C_{10}$	40°C 10 min	MeCN-Tris buffer (pH 7.7) 75:25	CL (photon counting)	0.003 – 0.1		Comparison with OPA and NBD-Cl	29
$C_{10} - C_{20}$	60°C 20 min	Gradient elution from 60-90% MeCN-water	UV	2 (C_{12})		Study of the effect of organic solvent on the amine retention	30
NH_3	55°C 90 min	MeCN-phosphate buffer (pH 7.2)	UV (340 nm)	1000 (*)	Water	In the presence of other amines	11
NH_3 methylamine	40°C 30 min	MeCN-MeOH-water (3:7:7)	F	> 8.5 > 0.75	Nitrogenase assays	Amines produced from diazomethane or CN^-	31
Primary and secondary	55°C 90 min	MeCN-phosphate buffer (pH 7.2) 62:38	UV (340 nm)	< 10 μM	Industrial waters	Comparison with NQS	10
Dimethyl and diethylamine	40°C 30 min	MeOH-phthalate buffer (pH 3.3) 70:30	E	0.012 , 0.3 pg		+1.1 V vs. Ag/AgCl	32
Dimethylamine-dipentylamine	40°C 30 min	MeCN-water (73:27)	F	0.55 – 4.75		Coefficient of variation from 2.52 to 3.24%	33

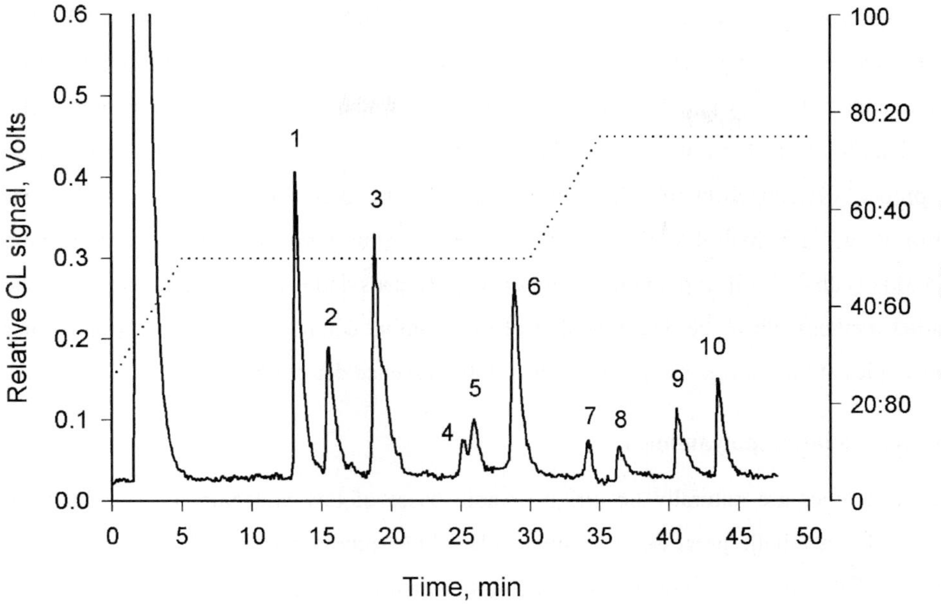

Figure 4 Optimum chromatographic separation of the dansylamine standard solution: 1 methyl-, 2 ethyl-; 3 propyl-, 4 isobutyl-, 5 butyl-, 6 diethyl-, 7 isopentyl-, 8 pentyl-, 9 hexyl- and 10 heptylamine; 20 pmol of amine injected in all cases. (Reprinted from ref.erence [27] with permission from Elsevier)

Comparing the labelling chemistry (time and temperature) and separation conditions (mobile phase composition) reported by the proposed methods, not many differences are observed even considering primary and secondary amines. Thus, a temperature of 55–65°C for 10–30 min can be recommended for the dansylation reaction, while a mobile phase composed of MeCN-water (pH 7.0–7.7) delivered in an isocratic (70:30 MeCN:water) or gradient mode can be used for the separation of the DNS derivatives. With regard to the earlier reported methods based on dansylation on a solid support, the off-line mode provides higher sensitivity due to the higher yields that can be achieved in the labelling reaction [21-13]; however, in any case their LODs are smaller than those provided by the method based on the on-line dansylation [27], which ranged from 15 to 300 pg ml^{-1}. This increased sensitivity is based on the superior features of the Amberlite IRC-50 for the clean-up (higher recoveries are achieved in the analysis of water samples) and pre-concentration of amines versus C_{18} and of the zero-dead volume detector in comparison to the traditional continuous flow system.

N-nitrosamines such as nitroso-dimethylamine (NDMA), -diethylamine (NDEA), -dipropyl (NDPA), -dibutyl (NDBA), -pyrrolidine (NPy) and –piperidine (NPip) are potential carcinogenic substances widely distributed in foods and also in environmental samples. The determination of this kind of amine has been considered in this section because it is based on the proper determination of the low-molecular-mass secondary amine formed after denitrosation with hydrobromic acid-acetic acid of the corresponding N- nitrosamine [2,35,36]. Bath [35,36] and microwave-assisted [2] dansylations have been used and the reported methods allow the determination of these amine compounds in beer [2] and water [36] samples at fmol levels using UV or chemiluminescence detection.

4. Biogenic amines (polyamines)

Biogenic amines are naturally occurring organic bases of low molecular mass, which are produced by metabolic processes in animals, plants and microorganisms. Figure 5 shows the structure of the most common biogenic amines, which can be produced in foodstuffs as a result of bacterial decarboxylation of amino acids, e.g. histamine (HI) from histidine, tryptamine (TR) from tryptophan, tyramine (TY) from tyrosine, 2-phenylethylamine (PEA) from phenylalanine, cadaverine (CAD) from lysine, putrescine (PUT) from arginine via ornithine, and it serves as a precursor to spermine (SPM) and spermidine (SPD).

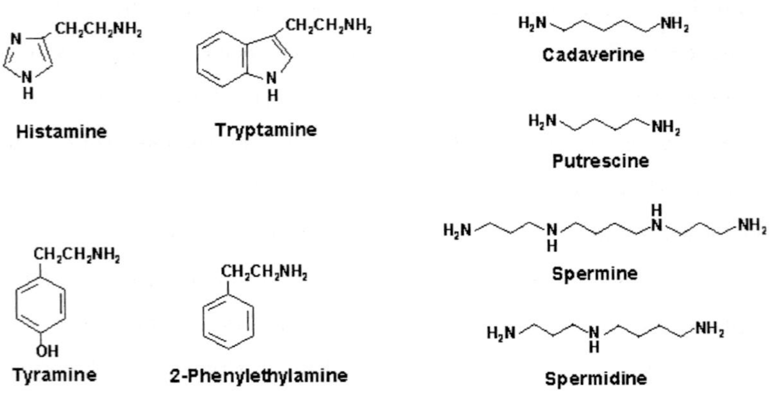

Figure 5 Structures of some common biogenic amines.

These amines are normal constituents of many foods and beverages; however, their presence in high concentrations in foods is associated with food deterioration. Biogenic amines in low concentrations are essential for many physiological functions, whereas at high concentrations they could have deleterious effects. Several symptoms occur following excessive oral intake of biogenic amines, such as headaches, hypo-or hypertension, nausea, cardiac palpitations, renal intoxication and in more severe cases intracerebral haemorrhages and death. Considering the relevance of these amines in foods and alcoholic beverages from both toxicological and technical points of view and also in clinical analysis - some biogenic amines are essential for normal growth and cellular differentiation - a wide variety of analytical methods have been reported. In order to study the HPLC determination of biogenic amines via pre-column dansylation (the scope of this section), the reported methods have been divided into subsections, according to their significance from the point of view of their application, covering the determination of biogenic amines in foods, alcoholic beverages and biological samples.

Biogenic amines are normal components in many foods such as fish, cheese, meat, wine and vegetables. However, high concentrations of these biologically active amines can be found in putrid, aged or fermented foods, which are associated with food deterioration. In contrast, biogenic amines in fruits and vegetables shape the typical and characteristic taste of mature foods and are precursors of certain aroma compounds. Regarding toxicity, HI, TY and PEA (at concentrations of 50, 100-800 and 30 mg Kg^{-1}, respectively) are the most hazardous biogenic amines for human health, whereas PUT, CAD, SPM and SPD, although they do not have a directly toxic effect, can increase the toxic effects of TY and HI.

Table 3 shows a survey of HPLC methods reported for the determination of biogenic amines in foods via pre-column dansylation. UV detection at 254 nm is widely used as a detection system on account of the fact that the concentration level of these compounds in food samples can be achieved with the sensitivity provided by this detector. In some cases, an additional LLE step is used, although for clean up rather than pre-concentration purposes. Luminescence detection has been scarcely used [40,41] even though its higher intrinsic sensitivity enables the use of small samples, minimizes handling and can even detect other biogenic amines that occur at very low levels, as is the case with the determination of biogenic amines in table olives using chemiluminescence detection [28]. As with low-molecular-mass aliphatic amines, the separation of DNS derivatives is generally based on a gradient elution with MeCN and water or a buffer solution in some cases [15, 38]. The traditional bath approach is used to carry out the dansylation reaction, yet the experimental labelling conditions are very different. In fact, derivatization times of 1 h and 5 min, both heated to 60°C, have been reported by the same authors [8,9] for the determination of biogenic amines in fish samples. In summary, and in view of the reported methods, a temperature of 50°C for ca. 1 h could be recommended for the dansylation of these amines. In contrast, there is a strong consensus in the literature regarding the sample treatment of food samples prior to their reaction with DNS-Cl. The most widespread procedure consists of a homogenisation of the sample with perchloric acid [2,8,9,15,40,41] (5% trichloroacetic acid [17,37] and 0.1 mol l^{-1} HCl [6,39,40] can be also used) followed by centrifugation. In some cases, an aliquot of the supernatant is directly derivatized or after the LLE clean up step [17,37,39,40].

The importance of biogenic amines in alcoholic beverages is due to their possible toxicological risks and therefore knowledge about where the compounds originate and how they can be controlled at the lowest levels is of great interest to the beverage industry. In brewing, the types of amines are dependent on the raw material in the beverage, as well as the method of brewing, and any microbial contamination that may have occurred during the process or during storage. More than 20 amines have been identified in wine and their total concentration has been reported to range from a few mg l^{-1} to about 50 mg l^{-1} depending on the quality of wine. In addition to the biogenic amines depicted in Figure 5, methyl-, ethyl- and isoamylamine and agmatine, among others, are commonly determined in alcoholic beverages.

Although the HPLC determination of biogenic amines in alcoholic beverages is difficult owing to their complex matrix, the treatment of the sample is not markedly complex, and thus a useful procedure for wine samples combines the use of poly(vinylpyrrolidinone) as a wine bleaching agent (phenolic compounds present in the wine, especially in red wine, are adsorbed) and a SPE of the DNS derivatives through C_{18} cartridges to remove interferences associated with matrix effects [5,19,20]. In the case of beer, a degasification followed by a LLE step, for example with 50% butanolic chloroform [43], is enough to prepare the sample for derivatization. Similar conclusions as stated above for food samples can be drawn regarding the dansylation of biogenic amines, separation and detection of their derivatives in alcoholic beverages. Thus, dansylation is usually carried out at 60°C for ca. 1 h; gradient elution with a binary system of MeCN/water is employed for the HPLC separation of DNS derivatives, which are normally detected at 254 nm, although photodiode array [12] and fluorimetric [5] detection have also been used for this purpose.

Table 3 Some principal methods for the HPLC determination of biogenic amines in food samples

Sample	Biogenic amines	Sample treatment	Labelling	Separation (C_{18})	Reference
Salted and fermented fish products	HI, TY, CAD, PUT, SPM, SPD	0.4 mol l^{-1} $HClO_4$	40°C 20 min	Gradient elution NH_4 acetate/MeCN	15
Fish (decomposition indexes)	HI, TR, TY, PEA, CAD, PUT, SPM, SPD, NH_3, NH_2CH_3	0.2 mol l^{-1} $HClO_4$ (20 min at 2°C)	60°C 5-60 min	Gradient elution water/MeCN	8, 9
Tuna, salmon and salami	HI, TY, PEA, CAD, PUT, SPM, SPD	5% trichloroacetic acid + ELL in butanol/$CHCl_3$ (1:1)	40°C 60 min	Gradient elution water/MeCN	37
Protein foods	HI, CAD, PUT	MeOH + cation-exchange with Amberlite IRC-50	60°C 60 min	Gradient elution phosphate/MeOH/MeCN	38
Cheese	HI, TR, TY, PEA, CAD, PUT, SPM, SPD	0.1 mol l^{-1} HCl + LLE with butanol at pH 11.5	40°C 60 min	Gradient elution water/MeCN	39, 40
Table olives	HI, TR, TY, PEA, CAD, PUT, SPM, SPD	5% trichloroacetic acid + LLE with butanol	35°C 120 min	Gradient elution H_3PO_4/MeOH-MeCN	17
Table olives[a]	TY, CAD, PUT, SPM, SPD	5% trichloroacetic acid	65°C,15 min (on-line)	Gradient elution water/MeCN	28
Fresh and preserved vegetables	HI, TR, TY, PEA, CAD, PUT, SPM, SPD	0.1 mol l^{-1} HCl	40°C 60 min	Gradient elution water/MeCN	6
Plant tissue	CAD, PUT, SPM, SPD	5% $HClO_4$ (ice for 60 min)	Microwave 5 min	Gradient elution water/MeCN	2
Fruits[b]	PUT, SPD	5% $HClO_4$	60°C 60 min	Gradient elution water/MeCN	40
Tomato callus[b]	CAD, PUT, SPM, SPD	5% $HClO_4$ (ice for 60 min)	60°C 25 min	$CHCl_3$/toluene/trietyhlamine (89:10:1)	41

[a and b)] fluorescence and chemiluminescence detections, respectively

Figure 6 shows the chromatograms obtained when Greek red wine and beer samples were analysed using fluorimetric detection [5]. As can be seen, biogenic amines as well as volatile low-molecular-mass amines can be detected at sub-mg l^{-1} level. The determination of TR is not feasible due to the dansyl-hydrazine interfering peak, which is present in this kind of sample.

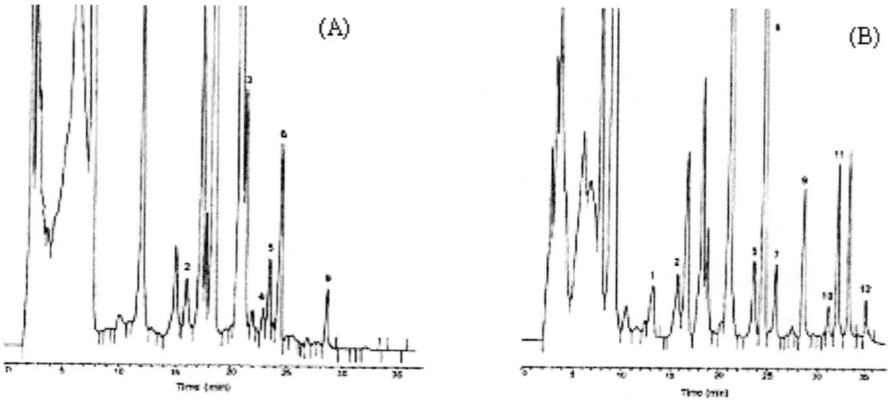

Figure 6 RP-HPLC chromatograms of (A) red wine; peaks: 2 ethylamine (0.098 mg l^{-1}); 4, 2-phenylethylamine (0.178 mg l^{-1}); 5, isoamylamine (0.174 mg l^{-1}); 6, putrescine (0.432 mg l^{-1}); 9, internal standard (0.843 mg l^{-1}); attenuation = 1024 and (B) beer; peaks: 1, methylamine (0.085 mg l^{-1}); 2, ethylamine (0.098 mg l^{-1}); 5, isoamylamine (0.127 mg l^{-1}); 6, putrescine (0.470 mg l^{-1}); 7, cadaverine (0.110 mg l^{-1}); 9, internal standard (0.843 mg l^{-1}); 10, tyramine (0.390 mg l^{-1}); 11, spermidine (0.269 mg l^{-1}); 12, spermine (0.098 mg l^{-1}); attenuation = 512. Chart speed = 0.5 cm min^{-1}; λ_{ex} = 320 nm, λ_{em} = 523 nm; column, Inersil ODS-3 (250 × 4 mm I.D., 5 μm); mobile phase: acetonitrile (40-80% from 0-25 min, 80-100% from 25-30 min, 100-40% from 30-35 min)-water; injection solvent, acetonitrile; injection volume, 20 μl. (Reprinted from reference [5] with permission from Elsevier).

The biogenic amines PUT, SPM and SPD (polyamines) are found in all living organisms and have been proved to be necessary for life. Recent studies have demonstrated that increased polyamine levels are implicated in cancer progression and apoptosis, making the study of polyamine pools and their regulation crucial in order to obtain a fuller understanding of carcinogenesis. Tumour cells contain a much higher concentration of polyamines, and an enhanced urinary excretion of polyamines is associated with patients suffering from various types of cancers.

The analytical determination of biogenic amines in clinical samples such as human tissues and urine has generated a great deal of interest in recent years because of their carcinogenic significance. In this context, HPLC with fluorimetric detection via pre-column dansylation is one of the preferred detection choices because it combined simplicity in the sample treatment with the suitable sensitivity for amine analysis in this kind of sample, at concentration levels of ng ml^{-1}. As for sample treatment, several procedures have been suggested, depending on the analysed sample, in order to achieve a protein-free amine solution. For urine samples, C_{18} [24] or cation-exchanger [44] SPE cartridges can be used to isolate the amines from the sample matrix, whereas tissues [4,18,45,46] and other samples such as faeces [14] are treated with perchloric acid [18,46] (in some cases trichloroacetic acid [4]) for de-proteinization, and the amines in the resulting supernatant are dansylated. As with the determination of biogenic amines in food and alcoholic beverage samples, the separation of DNS derivatives is achieved using a mobile phase with a gradient elution between MeCN and water; however, other alternatives including MeOH have been reported, such as MeOH/water [4,18] and water/MeCN/MeOH [7,45], which probably enhances the features of the fluorimetric detection. Finally, and with respect to the analytical sensitivity, pmol amounts of these amines as well as of their acetylated derivatives [7,44,45] can be easily determined in the analysed samples with high recoveries, over 90%.

5. Aromatic amines of clinical significance

There are a great number of HPLC methods for the determination of aromatic amines of clinical significance based on their previous derivatization with DNS-Cl, which are remarkable alternatives in comparison to other reported chromatographic methods. So, anti-tumour agents [47], serotonin reuptake inhibitors [48], broncho-relaxants [49], neurotransmitters [50], and stimulants, among others, have been determined by HPLC via pre-column dansylation. A detailed study of these methods comes under the scope of this chapter, and therefore, the determination of stimulants such as amphetamines is only considered in this section taking into account the greater research interest in these compounds.

The papers that report on the HPLC determination of methamphetamine (MA) and its main metabolite amphetamine (AP), based on their previous dansylation, focus on two aims: first, the determination of these stimulants in biological samples in order to extract clinical information regarding their addictive effects; and secondly, amphetamines are selected as model analytes to test improvements on HPLC methods based on their previous dansylation reaction, which is the case of the papers reported by Prof. Campins' team relating to the evaluation of dansylation on solid supports using off- and on-line modes and also the potential of fluorescence or chemiluminescence for the detection of resulting DNS derivatives [21,25,26]. LODs at μg l^{-1} can be achieved and in some cases, the proposed methodology is applied to the determination of amphetamines in urine samples [25].

The methods focused on the determination of amphetamines in human biological samples such as urine [51,52] and hair [53,54] as well as in rat's blood and brains [55,56] by in vivo microdialysis, which in addition to their clinical significance, also provide in some case valuable analytical information. Thus, the DNS-Cl/acetone mixture used for the dansylation of amphetamines also acts as de-proteinizing agent for the urine sample [51]; after centrifugation for protein removal, the supernatant is directly injected into the chromatograph. In addition, fluorescence [51,55,56] and chemiluminescence [52-54] detection, based on the TCPO/H_2O_2 system, are used for the detection of DNS derivatives; the latter provides LODs below ng ml^{-1}. With regards to the chromatographic conditions, an isocratic elution with a mobile phase composed of MeCN/water (70:30 v/v) allows the separation of DNS derivatives in a few minutes by using fluorescence detection; whereas a mixture of 1 mM imidazole buffer (pH 7.0) and MeCN (2:3, v/v) is recommended as a mobile phase for chemiluminescence detection. The determination of MA and AP in a single human hair sample as a control test for stimulant addicts is of particular note. In this method, the hair sample is washed, cut and sonicated for 1 h with 0.5 ml of a mixture of MeOH and 5 mol l^{-1} HCl (20:1 v/v), and allowed to stand at room temperature overnight. The residue obtained after filtration and evaporation to dryness under nitrogen is derivatized with DNS-Cl [53]. The method allows the determination of nanogram amounts of MA and AP in a single hair sample with recoveries ranging from 94 to 100%.

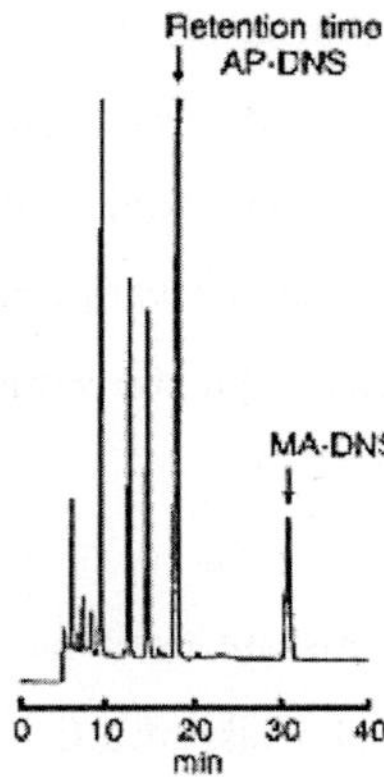

***Figure** 7* Chromatogram of stimulant user's hair simple. AP-DNS and MA-DNS correspond to dansylated amphethamine and methylamphetamine, respectively (Reproduced from reference [53] with permission from John Wiley & Sons Limited).

Figure 7 shows the typical chromatogram provided by a sample obtained from a man suspected of having used MA. As can be seen, a small amount of AP can be also detected, as a metabolite of MA.

6. Experimental

6.1 Dansylation of amines

Low-molecular-mass aliphatic amines, biogenic amines including polyamines and aromatic amines of clinical significance such as amphetamines alone and in mixtures are generally derivatized with 1% DNS-Cl in acetone. The concentrations of the analytes ranged from few μg ml^{-1} to μg l^{-1} according to the detection system used (UV or luminescence) or the SPE step utilized for the previous preconcentration of the sample (see Table 2). In general, the solution procedure involves the reaction between DNS-Cl and amine in a basic medium (saturated solution of hydrogen sodium carbonate) with heating at 50-60°C during 10-30 min or about 1 h for low-molecular-mass aliphatic or biogenic amines, respectively. In many cases, the excess of the labelling reagent is removed by addition of proline at 5-10 fold molar excess with respect to DNS-Cl. Microwave and solid-support assisted derivatization in addition to automated on-line procedures can be used as alternatives for the simplification of the dansylation process (see section 2).

6.2 Chromatography

Different commercial HPLC systems are used for the separation and determination of dansylated amines. They generally consist of a C_{18} analytical column (e.g. 250 × 4 mm I.D., 5 μm), a binary gradient pump for delivering the mobile phase at ca. 1.0 ml min^{-1} in a gradient mode (MeCN-buffer of pH ca. 7-7.5), an injection valve having a 20-μl loop, and a UV (340 nm), diode-array or fluorescence detector (see Tables 2 and 3 for other separation conditions). In the case of chemiluminescence detection, such as the case of low-molecular-mass aliphatic amines and amphetamines, a post-column device is required to develop the chemiluminescence reaction, generally based on the TCPO-H_2O_2 system. Flow systems and the so-called zero-dead-volume detector (see Fig. 3) are generally used.

6.3 Determination of amines in real samples

In spite of the great diversity of procedures described in the bibliography for the determination of these analytes in several samples, some conclusions can be extracted. Table 4 shows the typical recommended procedures for the determination of dansylated amines in water, food and clinical samples.

Table 4 Recommended procedures for sample preparation

Sample	Treatment
Water	Clean-up and preconcentration by SPE with C_{18} cartridges or cation-exchange column, such as IRC-250.
Food	Homogenisation of the sample with 0.4 $HClO_4$ (5 % TCA or 0.1 mol l^{-1} are also used) and centrifugation for removing proteins.
Wine	Poly(vinylpyrrolidinone) as a wine bleaching agent and a SPE of the DNS derivatives through C18 cartridges are used to remove matrix effect.
Beer	Degasification followed by a LLE step with 50% butanolic-chloroform is enough to prepare the sample for derivatization.
Urine	C_{18} or cation-exchanger SPE cartridges can be used to achieve a protein-free amine solution.
Tissues	Homogenisation of the sample with 0.4 $HClO_4$ (5 % TCA are also used) and centrifugation for removing proteins.

See Tables 2 and 3 for additional information

7. References

[1] I. Molnar-Perl, J. Chromatogr. A 987 (2003) 291–309.

[2] R.M. Linares, J.H. Ayala, A.M. Afonso, V.G. Diaz, J. Chromatogr. A 808 (1998) 87–93.

[3] L. Cardenes, J.H. Ayala, V. Gonzalez, A.M. Afonso, J. Liq. Chromatogr. Relat. Technol. 25 (2002) 977–984.

[4] S. Fu, X. Zou, X. Wang, X. Liu, J. Chromatogr. B 709 (1998) 297–300.

[5] Z. Loukou, A. Zotou, J. Chromatogr. A 996 (2003) 103–113.

[6] S. Moret, D. Smela, T. Populin, L.S. Conte, Food Chem. 89 (2005) 355–361.

[7] Y.A. Hassan, A.A. Ibrahim, Biomed. Chromatogr. 12 (1998) 291–293.

[8] M. Valle, P. Malle, S. Bouquelet, J. AOAC Int. 80 (1997) 49–56.

[9] M. Valle, P. Malle, S. Bouquelet, J. AOAC Int. 79 (1996) 1134–1140.

[10] J.R. Lindsay-Smith, A.U. Smart, F.E. Hancock, M.V. Twigg, J. Chromatogr. 547 (1991) 447–451.

[11] J.R.L. Smith, A.U. Smart, F.E. Hancock, M.V. Twigg, Chem. Ind. (London). 11 (1989) 353–354.

[12] R.E. Anli, N. Vural, S. Yilmaz, Y.H. Vural, J. Food Compos. Anal. 17 (2004) 53–62.

[13] N.M. Tamin, L.W. Bennett, T.A. Shellem, J.A. Doerr, J. Agric. Food Chem. 50 (2002) 5012–5015.

[14] M.T. Saarinen, Chromatographia 55 (2002) 297–300.

[15] J.H. Mah, H.K. Han, Y.J. Oh, M.G. Kim, H.J. Hwang, Food Chem. 79 (2002) 239–243.

[16] K. Nakashima, A. Kaddoumi, M.N. Nakashima, W.R.G. Baeyens, Luminescence 15 (2000) 74–75.

[17] D. Hornero-Méndez, A. Garrido-Fernández, Analyst 119 (1994) 2037–2041.

[18] C. Cann-Moisan, J. Caroff, E. Girin, J. Chromatogr. Biomed. Appl. 112 (1992) 134–136.

[19] A. Zotou, Z. Loukou, E. Soufleros, I. Stratis, Chromatographia 57 (2003) 429–439.

[20] O. Busto, Y. Valero, J. Guasch, F. Borrull, Chromatographia 38 (1994) 571–578.

[21] C. Molins-Legua, P. Camping-Falcó, S. Meseguer-Lloret, Chromatographia 58 (2003) 15–27.

[22] S. Meseguer-Lloret, C. Molins-Legua, J. Verdú-Andrés, P. Camping-Falcó, J. Chromatogr. A 1035 (2004) 75–82.

[23] S. Meseguer-Lloret, C. Molins-Legua, P. Camping-Falcó, J. Chromatogr. A 978 (2002) 59–69.

[24] C. Molins-Legua, P. Camping-Falcó, A. Sevillano-Cabeza, M. Pedrón-Pons, Analyst 124 (1999) 477–482.

[25] C. Molins-Legua, P. Camping-Falcó, A. Sevillano-Cabeza, Anal. Chim. Acta 378 (1999) 83–93.

[26] C. Molins-Legua, P. Camping-Falcó, A. Sevillano-Cabeza, Analyst 123 (1998) 2871–2876.

[27] M. Cobo, M. Silva, J Chromatogr. A 848 (1999) 105–115.

[28] M. Cobo, M. Silva, Chromatographia 51 (2000) 706–712.

[29] G. Mellbin, B.E.F. Smith, J. Chromatogr. 312 (1984) 203–210.

[30] A.V. Khryashchevskii, O.A. Shpigun, P.N. Nesterenko, J. Anal. Chem. 52 (1997) 871–876.

[31] M. Bravo, H. Eran, F.X. Zhang, C.E. McKenna, Anal. Biochem. 175 (1988) 482–491.

[32] A.Z. Zhu, Y.H. Wang, C.G. Fu, Sepu. 15(1997) 381–383.

[33] Z. Wang, H. Xu, C. Fu, Sepu. 8 (1990) 325–327.

[34] D. Pérez-Bendito, M. Silva, Kinetics in Chemiluminescence Analysis in: A.M. García-Campaña, W.R.G. Baeyens (Eds.) Chemiluminescence in Analytical Chemistry, Marcel Dekker, Inc. New York, 2001, chapter 8.

[35] C. Fu, H. Xu, Z. Wang, J. Chromatogr. 634 (1993) 221–227.

[36] C. Fu, H. Xu, Analyst 120 (1995) 1147–1151.

[37] S. Moret, L.S. Conte, F. Callegarin, Ind. Aliment. (Pinerolo, Italy) 35 (1996) 650–657.

[38] F.V. Carlucci, E. Karmas, J. Assoc. Off. Anal. Chem. 71 (1988) 564–568.

[39] S. Moret, L.S. Conte, E. Spoto, Ind. Aliment. (Pinerolo, Italy) 35 (1996) 788–792.

[40] S.Moret, R. Bortolomeazzi, G. Lercker, J. Chromatogr. 591 (1992) 175–180.

[41] S. del Rodriguez, B. López, A. R. Chaves, J. Agric. Food Chem. 47 (1999) 1431–1434.

[42] Z. Ma, R. Zang, K. Chu, Fenxi Huaxue 19 (1991) 1317–1319.

[43] S. Buiatti, O. Boschelle, M. Mozzon, F. Battistutta, Food Chem. 52 (1995) 199–202.

[44] P.M. Kabra, H.K. Lee, W.P. Lubich, L.J. Marton, J. Chromatogr. Biomed. Appl. 53 (1986) 19–32.

[45] L. Hawel, C.V. Byus, Anal. Biochem. 311 (2002) 127–132.

[46] F. Gaboriau, R. Havouis, J.P. Moulinoux, J.G. Delcros, Anal. Biochem. 318 (2003) 212–220.

[47] C.Maraschiello, E. Miranda, E. Millan, F. Floriano, J. Vilageliu, J. Chromatogr. B 791 (2003) 1–11.

[48] A. Lucca,G. Gentilini, S. Lopez-Silva, A.Soldarini, Ther. Drug Monit. 22 (2000) 271–276.

[49] J.E. Haky, W.M. Foss, B.L. Marks, J. Liq. Chromatogr. Rel. Techn. 20 (1997) 2399–2414.

[50] G. Shao, D.S. Wang, F. Wu, S.J. Chen, X. Luo, J. Liq. Chromatogr. 18 (1995) 2133–2145.

[51] T.K. Wang, M.S. Fuh, J. Chromatogr. B 686 (1996) 285–290.

[52] K. Hayakawa, Y. Miyoshi, H. Kurimoto, Y. Matsushima, N. Takayama, S. Tanaka, M. Miyazaki, Biol. Pharm. Bull. 16 (1993) 817–821.

[53] N. Takayama, S. Tanaka, K. Hayakawa, Biomed. Chromatogr. 11 (1997) 25–28.

[54] N. Takayama, S. Tanaka, R. Kizu, K. Hayakawa, Biomed. Chromatogr. 13 (1999) 257–261.

[55] M.S. Fuh, T.K. Wang, W.H.T. Pan, J. Liq. Chromatogr. Relat. Technol. 20 (1997) 1605–1615.

[56] M.R. Fuh, H.T. Lin, W.H.T. Pan, F.R. Lin, Talanta 58 (2002) 1357–1363

Ibolya Molnár-Perl (Editor)
Quantitation of Amino Acids and Amines by Chromatography
Journal of Chromatography Library, Vol. 70

2.2.3. HPLC of Amines as 9-Fluorenylmethyl Chloroformate Derivatives

Paul C. Ho

Contents

Summary

9-fluorenylmethyloxycarbonyl chloride, also known as 9-fluorenylmethyl chloroformate (FMOC-Cl), reacts with primary and secondary amines and yields highly fluorescent and stable derivatives. It has been used for the derivatization and analysis of various types of amines including biogenic amines and polyamines; short-chain aliphatic amines; neuro-transmitters, including catecholamines, and metanephrines. FMOC-Cl can also react with the amino groups of a number of drugs to yield the highly fluorescent derivatives for HPLC analysis. This approach is particularly useful to the analytes without chromophore or fluorophore for detection. This chapter reviews the experimental procedures on pre-column derivatization of these amines with FMOC-Cl for fluorescence detection after reversed phase separation. The increase in sensitivity by FMOC derivatization can be more than 3 orders of magnitude, when compared to the usual detection by UV absorbance. Table 1 summarizes the HPLC conditions for the detection of these amines. Despite the high sensitivity of these FMOC derivatives for HPLC analysis, there are some disadvantages associated with this approach, especially in the analysis of biological samples. One of the disadvantages is that many endogenous components in the biological matrix are also reactive towards the derivatizing reagents. Thus, extensive sample clean-up is usually required. Very often, solid phase extraction is employed to extract the analytes from the matrix, followed by derivatization, and in many instances, re-extraction of the derivatized analytes to remove the un-reacted reagent and concentration of the derivatives. As a result, the sample preparation can be very tedious and time-consuming. The second disadvantage of the approach is the interference to the analytes during chromatography from the un-reacted FMOC-Cl and also its hydrolysis product, FMOC-OH. To overcome this problem, an excess of glycine can be added to the reacting mixture, after derivatization of the analytes is complete. By doing this, the excess FMOC-Cl is converted to the relatively hydrophilic glycine derivative that elutes early in the chromatogram, not interfering detection of the much smaller peak of the analyte derivatives. With the advance in column technology, the problem of interference from the derivatizing agent may be further minimized by using the new generation of columns with higher resolution power.

1. Introduction

The fluorenylmethyloxy carbonyl (FMOC) group was first introduced by Carpino and Han [1] in 1972 as a base labile protecting group for primary and secondary amines. This protecting group has been widely utilized in organic chemistry and as a mild orthogonal alternative to N^{α}-*tert*-butyloxycarbonyl (t-BOC) in solid phase peptide synthesis. The FMOC group is stable toward acids and hydrolysis but readily cleaved in a variety of organic bases [2]. 9-fluorenylmethyloxycarbonyl chloride, also known as 9-fluorenylmethyl chloroformate (FMOC-Cl) was first introduced in HPLC fluorescence methods based on pre-column derivatization and reversed phase separation of FMOC-amino acids by Einarsson et al. [3,4]. The method was later commercialized by Cunnico and others [5] and has been found wide application for the analysis of amino acids and their derivatives [6-9]. Under mild conditions, FMOC-Cl reacts with primary and secondary amines and yields highly fluorescent derivatives. It has been used for the derivatization and analysis of various types of amines including biogenic amines [10], polyamines [11,12,13] and neurotransmitters such as catecholamines [14], and metanephrines [15]. FMOC-Cl can also react with the amino groups of a number of drugs [16,17,18,19] to yield the highly fluorescent derivatives for HPLC analysis.

There are other fluorescent derivatizing agents that have been employed for the analysis of various amines, e.g., ninhydrin [20], 5-dimethylaminonaphthalene-1-sulponyl chloride (Dns-Cl) [21] and *o*-phthaladehyde (OPA) [22]. Ninhydrin is highly specific towards amino compounds, and favors the use of channel rationing, thus permitting peak identification [23]. However, with the obligatory post-column derivatization system, the sensitivity is impaired due to effluent mixing, and peak widening in the reaction coil. The post-column mixing causes pump pulsations that further impair the signal-to-noise ratio. Dns-Cl [24] permits pre-column derivatization, giving high fluorescent sensitivity, but the derivative is unstable, thus storage of the derivatized sample is impossible. Similarly, OPA is highly specific and permit pre-column derivatization [25], but its fluorescent derivatives are also unstable that impair the versatility of the method. In contrast, FMOC-Cl is specific and its fluorescent derivatives are stable [3]. As a result, it is more versatile than OPA and other fluorescent derivatizing agents. In view of these advantages, FMOC-Cl is the fluorescent reagent of choice for derivatizing various amines to enhance the sensitivity of their detection for HPLC analysis. In this chapter, liquid chromatographic determination of various amines using FMOC-Cl as the fluorescent derivatizing agent is reviewed.

2. HPLC determination of biogenic amines and polyamines

Biogenic amines are natural products derived from aromatic and cationic amino acids. They are related to intracellular communication and present in specific cell types, namely amine-handling cells [26]. The common biogenic amines involved in mammalian physiology are serotonin, histamine, dopamine, and noradrenaline. Arginine/ornithine-derived polyamines might also be considered biogenic amines. They are involved in bio-signaling and possess pleiotropic effects. The polyamines are present not only in amine-handling cells, but in almost every kind of cells. These biogenic amines and polyamines appear to associate with some human pathologies, e.g., tyramine is related to migraine, hypertension, schizophrenia, Parkinson disease and depression [27]; tryptamine to depression, schizophrenia, hepatic encephalopathy [27]; and some polyamines such as spermine, spermidine, putrescine and cadaverine to cancer prognosis [28]. Though the relevance of some of these compounds to their related diseases has been known for years, the biochemistry and molecular biology of their effects are still unclear.

Biogenic amines are also found in musts and wines [29,30,31]. These biogenic amines may come from raw materials and fermentation processes. Some biogenic amines, such as histamines, tyramine [32], several volatile amines and polyamines [33] are already present in grapes. In addition, both alcoholic and malolactic fermentations can also produce amines in wine [34,35].

Since these biogenic amines (e.g., catecholamines [14], metanephrines [15]) are either present in minute quantities in biological samples or some aliphatic diamines (e.g., putrescine, cadaverine) and polyamines (e.g., sperimine, spermidine) do not contain a suitable chromophore or fluorophore group (Figure 1), hence, they cannot be determined with adequate sensitivity by spectrophotometric or fluorescence detection [28]. Therefore, FMOC-Cl has been used in many studies as a suitable derivatizing agent to increase the sensitivity for the detection of these biogenic amines and polyamines.

Tyramine, tryptamine and polyethylamine

These three biogenic amines pose considerable health hazards. It is important to make sure their concentrations in the meat products within the acceptable levels. FMOC-Cl has been found to produce specific, stable and compatible derivatives of these amines for either fluorescence or UV-absorbance detection [10].

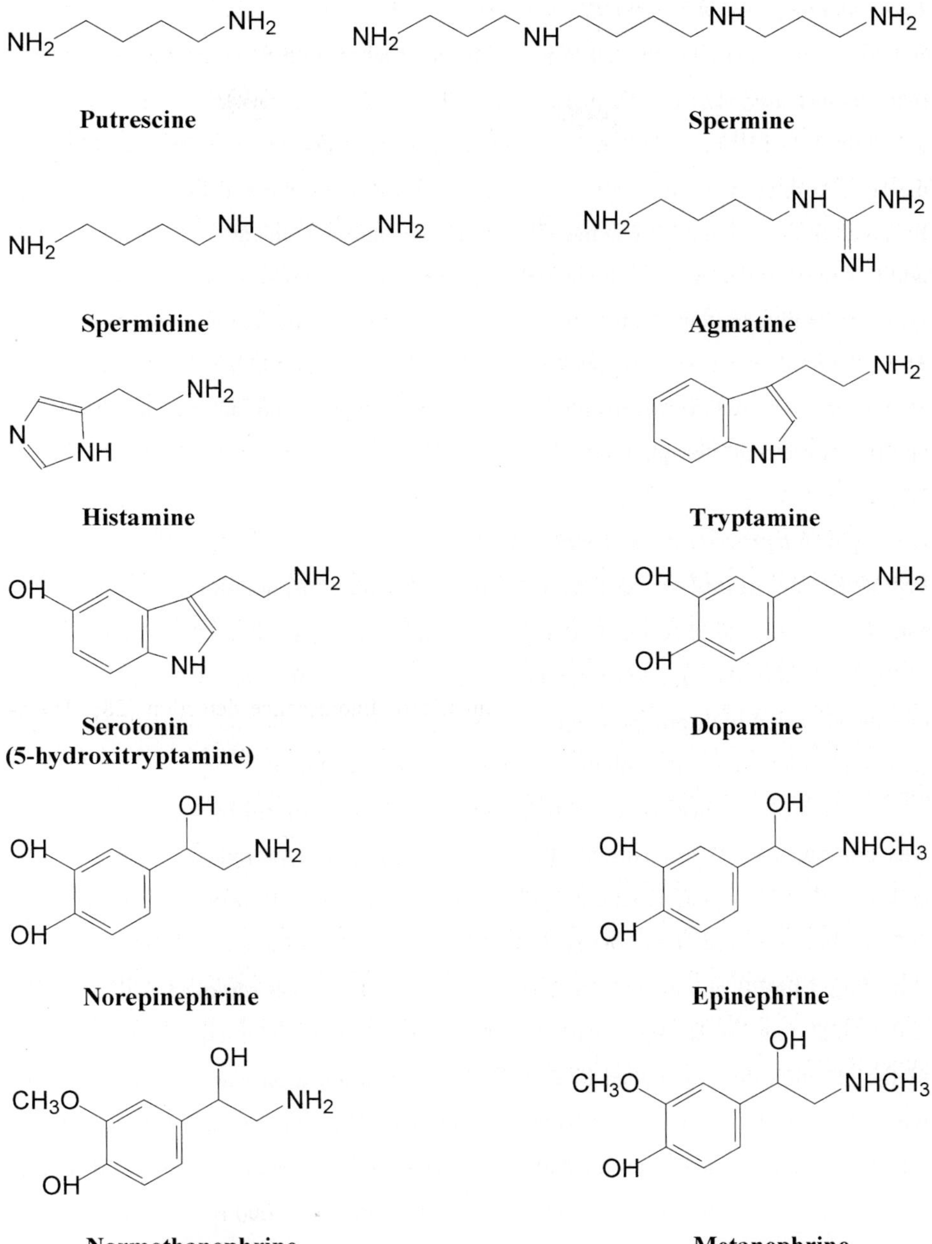

Figure 1 The main biogenic amines and polyamines

Experimental procedures for HPLC analysis

A HPLC column (Lichrochart, 125 mm x 4 mm) packed with Supersphere RP-18, 4 μm (Merck) was used. The mobile phase consisted of A, 25% methanol in aqueous sulphuric acid (pH 2.25); B, methanol. A linear gradient was performed from 20% eluent B at 0 min to 77.6% at 21.6 min, 100% at 22.5 min, 20% B at 25 min and finally equilibration for an additional 10 min. When meat samples were analysed, eluent B was pumped for an additional 10 min to prevent the late-eluting peaks of the previous analysis from appearing in the following one. The flow-rate was 1 ml/min, and the column temperature was 35°C. Two detection wavelengths were used: 263 $\pm$ 4 nm and 299 $\pm$ 2 nm. The ratio of absorbance at these two wavelengths served for peak identity and the absolute absorbance at either wavelength for quantitation.

Derivatization procedure and sample preparation

FMOC-Cl was freshly prepared as a 1mg/ml (3.87 μmol/ml) solution in methanol. The sample comprised approximately 0.6 μmol of total amines per ml in 0.2 M sodium bicarbonate. 0.5 ml of the sample or standard solution and reagent were mixed in polypropylene tubes sealed with polypropylene screw caps. After 10 min, the derivatized sample was either injected into the column or frozen at –20°C until used.

Each of six slices of 1 g deboned chicken meat, sterilized by gamma irradiation, was homogenized (Ultra-Turrax TP 189-10) in 50-ml polypropylene centrifuge tube, spiked with the appropriate amount of the amine solution and 10 ml of ice-cold 1 M perchloric acid-methanol (1:1). The homogenates were incubated in an ice batch for 2h, and centrifuged for 20 min at 4°C and 10,000 g. The fat layer was removed with a spatula, and 5-ml aliquots of the supernatant were pipetted into 10 ml volumetric flasks. One drop of the pH indicator was added to each flask and neutralization was performed by drop wise addition of 2 M potassium hydroxide solution. The volume was made with water. The flasks were cooled in an ice bath, and potassium perchlorate was removed by centrifugation of 1-ml aliquots at 2°C in 1.5 ml conical tubes, at 7000 rpm, for 15 min. Aliquots 750 μl of the supernatants were mixed with 250 μl of 0.8 M sodium bicarbonate and 500 μl were used for derivatization. With this approach, the recoveries of amines were 79.4%, 79.6% and 69.2% for tyramine, tryptamine and phenylethylamine from spiked meat samples (in the concentration range of in the concentration range of 0.125 to 2 nmol in 40 μl). The derivatized samples were found stable for 8 h at room temperature or for 7 days when stored at –17°C in a refrigerator.

2.2. Polyamines – putrescine, cadaverine, spermidine and spermine

Naturally occurring polyamines (e.g., putrescine, cadaverine, spermidine and spermine) are widely distributed in all tissues and body fluids. They are involved in the process of cell multiplication and its regulation [36,37]. Significantly increased levels of polyamines have been observed in cancer patients [28,38,39], but also in patients with benign disorders [40]. This observation is considered due to either increased secretion from the proliferating cells or release from dead cells. FMOC-Cl has been used in the pre-column derivatization of these polyamines, forming fluorescent derivatives for HPLC analysis.

Experimental procedures for HPLC analysis

The polyamines were determined by formation of FMOC derivatives that were separated by gradient elution (methanol-water, 82:18 to 100% methanol) through a NovaPak C_{18} column, 4 μm, 150 mm x 3.9 (Waters, Milford, MA) [41]. Detection was made at excitation 260 nm and emission 313 nm. Putrescine, cadaverine, spermidine and spermine were separated and eluted within 13 min with the detection limits in the range of 22.7 to 109 fmol. Similar method has reported for the determination of polyamines in milk [42]. The calibration range for that assay was linear in the range of 0.5 – 5 nmol/ml.

Derivatization procedure and sample preparation

A clean up procedure for the isolation of polyamines using a weakly acidic cation-exchange resin, Biorex-70 (Bio-rad) has been proposed [43]. Biorex-70 (Na^+ form, suspended in water) was packed to a 0.7 ml resin bed in a 12 ml syringe and washed with 10 ml of water prior to sample loading. The sample (8-10 ml) was loaded to the ion-exchange chromatography without pretreatment. Following sample loading, the resin was washed with 1.0 ml of water followed by 3.0 ml of 0.15 N NH_4OH. Polyamines were then eluted with 2.5 ml of 4.0 N NH_4OH, and the eluent was then treated with 50 μl of 2.0 N HCl to stabilize the polyamines. The solution was then lyophilized to dryness. The dried polyamine samples were then suspended in 0.1 – 1.0 ml ml of 0.1 M sodium bicarbonate (pH 7.9). An aliquot (300 μl) of the re-suspended sample was allowed to react with 300 μl of FMOC-Cl solution (4 mM in acetone). After 10 min, the mixture was extracted two times with 2 vol (0.6 ml) of pentane to remove acetone, FMOC-OH, and excess unreacted FMOC. The FMOC-derivatives in the aqueous layer were then separated by a reversed-phase HPLC.

2.3. Short-chain aliphatic amines

The determination of short-chain aliphatic amines is of great relevance, because many of these aliphatic amines are hazardous to human health. Significant levels of dimethylamine, trimethylamine, and tri-methylamine N-oxide (TMAO) are present in fish [44]. Fish consumption leads to increased dimethylamine levels in human urines and other body fluids [45.46]. By endogenous nitrosation, dimethylamine may be converted into the carcinogenic N-nitrosodimethylamine [46,47]. Therefore, it is particularly important, as these amines are increasingly used in different types of industries [48]. Considerable attention has been paid to their quantitation in water samples [49] and biological samples [50,51].

Experimental procedures for HPLC analysis

2.3.1. Methylamine

Herráez-Hernández et al. reported an assay for methylamine in water samples by solid-phase micro extraction (SPME) and HPLC after on-fiber derivatization with FMOC-Cl [49]. The mobile phase selected for the separation of methylamine-FMOC derivative was acetonitrile-water in gradient elution form. The percentage of acetonitrile was increased from 0% at 0 min to 40% at 2.5 min, to 70% at 9.5 min, and then to 100% at 17.5 min. The mobile phase flow-rate was 1.0 ml/min. Detection was made at 230 nm. In their study, a SPME assembly with replaceable extraction fibers (carbowax-templated resin, CW-TPR, 50 μm, Supelco) was selected for the extraction of methylamine and FMOC-Cl. The fiber was immersed into a glass vial containing the sample (25 ml) for an adsorption period of 15 min. The fiber was then removed from the sample vial, and immersed into a 2 ml glass vial containing the derivatization solution (1.0 ml of 1 mM FMOC-Cl in actonitrile). After the reaction of 2 min, the SPME device was placed into the SPME-HPLC interface, and the methylamine-FMOC derivative was desorbed from the fiber under the static mode: the fiber was soaked with 200 μl of acetonitrile for a desorption time of 5 min. Then the valve of the SPME-HPLC interface was rotated, so the methylamine-FMOC derivative was sent to the precolumn (Hypersil C_{18}, 30 μm, 20 mm x 2.1 mm i.d.). During this stage, the eluent was 100% water. Therefore, the methylamine-FMOC was retained in the precolumn for a short time, while the unreacted FMOC-Cl was washed out from the precolumn and sent to the waste through the additional switching valve placed between the precolumn and analytical column (LiChrospher 100 RP18, 125 mm x 4 mm i.d., Merck, Darmstadt, Germany). After elution for 0.5 min, the switching valve was rotated to connect to the analytical column, and the gradient elution was started. At the end of the assay, the valve was changed to the original position, and the fibers were cleaned by pumping through the interface 1 ml of acetonitrile at a flow of 0.95 ml/min. This method was less sensitive than the solid support-assisted derivatization method that employed reaction of the analyte with FMOC-Cl in solution followed by extraction with solid phase extraction cartridge [52]. Despite this, the SPME based method allowed the quantitation of methylamine over the range of 2.5-10.0 μg/ml with linearity, reproducibility and accuracy. The limit of detection was 0.75 μg/ml. The main advantages of the proposed SPME procedures are: sample handling involved in the extraction and derivatization steps was substantially reduced; it was free from organic solvent and non-destructive. Moreover, the employed conditions allowed the selective determination of methylamine in the presence of primary and secondary short-chain aliphatic amines.

2.3.2. Dimethylamine

A reversed-phase HPLC method was proposed for the determination of dimethylamine in serum and urine [51]. Analysis was performed on a Microspher C_{18} column, 10 cm x 4.6 mm i.d., 3 μm particles (Chrompack, Middleburg, Netherlands). The analytical column was protected by a reversed-phase guard column (1 cm x 2.0 mm i.d. from the same supplier). Potassium dihydrogen phosphate buffer (50 mM) adjusted to pH 6.5 with KOH was used for preparation of the mobile phases. The mobile phases consisted of 2 eluents:

A, acetonitrile-buffer (50:50, v/v) and B, acetonitrile-buffer (75:25, v/v). The flow-rate was 1.0 ml/min. The column was equilibrated with 100% eluent A. After injection of 5-μl volume of sample, the percentage of eluent A was reduced from 100% at 0 min to 0% in 5 min. 100% eluent B was then maintained for another 5 min, before the mobile phase was returned 100% eluent A in 1 min. Then, the column was equilibrated with 100% eluent A for 4 min before the next injection. By this gradient elution, the dimethylamine derivative was resolved from derivatives of other amines and amino acids with a total run time of 15 min. Detection was at excitation 260 nm and emission 320 nm. The lower limit of determination in serum was 0.1 μmol/l. Recoveries from spiked serum samples were 99-107%. Within- and between-run precision were better than 6%.

In this study, samples were prepared by mixing 300 μl of standard solutions or biological samples (urine, serum, or CAPD effluent) with 200 μl 5-sulfosalicylic acid solution (7.5% w/v). After centrifugation (5 min at 3000 g, 4°C), 200 μl of the clear supernatant was transferred to a glass tube containing 200 μl of 0.8 M borate buffer (pH 9.5). Derivatization was performed by adding 300 μl freshly prepared FMOC-Cl solution (10 mM in acetonitrile). After 1 min, excess of reagent was neutralized by adding 100 μl of a 100 mM glycine solution. After centrifugation (5 min at 3000 g, 4°C), 400 μl of the supernatant was transferred to a 4 ml glass vial which was sealed with a Teflon lined screw cap. The vials were heated for 10 min at 80°C in a water-bath. After cooling the vials were placed in the autosampler for injection.

A drawback of FMOC-Cl for derivatization and subsequent separation on HPLC is that both the reagent and its hydrolysis product FMOC-OH are intensely fluorescent and therefore interfere in the chromatographic separation of the amine derivatives. Excess FMOC-Cl may be removed by extraction with diethylether or pentane but this may result in partial loss of hydrophobic amine derivatives. In this study, an attempt was made to neutralize the reagent by adding an excess of glycine after derivatization of dimethylamine was complete. By doing this, the excess FMOC-Cl is converted to the relatively hydrophilic glycine derivative that elutes early in the chromatogram, not interfering detection of the much smaller peak of dimetylamine derivative.

2.4. Catecholamines and their metabolites

Catecholamines and their metabolites often serve as biomarkers for clinical diagnosis of altered function of catecholamine-synthesizing tissues, assessment of disease severity and prognosis and study of adaptive responses to stress and pharmacology [53,54]. Levels of catecholamines (norepinephrine (NE), epinephrine (E) and their metabolites metanephrine (MN) and normetanephrine (NM)) in biological fluids are monitored for the diagnosis of suspected pheochromocytoma, a tumor of chromaffin cells [55,56,57,58]. If neuroblastoma is suspected, dopamine (DA) is also measured [59]. Patients in clinical shock are often treated intravenously with pressor amine, including isoproterenol, NE, E, DA and dobutamine [53]. Clearance of these drugs is characterized by large individual variations, which are augmented in patients with hepato-renal failure. Therefore, the therapeutic levels of these amines are closely monitored to improve the therapeutic outcome.

FMOC-Cl has been successfully applied to determine NE, E, and DA in urine after solid-phase extraction with Fe(III)-loaded silica (SQ-Fe) and derivatization with FMOC-Cl [14]. The advantages of using FMOC-Cl as a derivatizing agent are the reaction is straightforward, rapid and can be performed at ambient temperature; the reaction products are stable at room temperature and more importantly, FMOC-Cl can react with both primary and secondary amines. The assay was subsequently improved for the simultaneously determination of NE, E, DA, MN and NM after pre-column derivatization with FMOC-Cl [15]. The assay was shown to be suitable for the detection of E, DA, MN and NM in human urine samples without prior solid-phase extraction or pre-concentration of these analytes. Atmospheric pressure chemical ionization mass spectrometry (APCI-MS) was also employed in that study to confirm the chemical structures of the FMOC-catecholamine and FMOC-metanephrine derivatives.

Experimental procedures for HPLC analysis

In that study [59], a Spherisorb C_8 5 μm reversed-phase column (150 x 4.6 mm, Phase Separations Inc., CT, USA) protected by a C_{18} guard column (4 x 3.0 mm, Phenomenex Inc., CA, USA) was used for separation. Mobile phase consisted of solvent A (acetic acid solution, 3.0 g/l) and solvent B (acetonitrile). The solvents were mixed on-line and run at a flow-rate of 1 ml/min on a linear gradient from 50 to 78% solvent B in 40 min for the optimal resolution of the analytes of interest in urine samples. Detection was made at excitation and emission wavelengths of 263 nm and 313 nm, respectively.

For derivatization of NE, E, DA, MN, NM and 3,4-dihydroxybenzylamine (DHBA, the internal standard) with FMOC-Cl, 100 μl of the standard solution or urine sample was added to 100 μl of 1.0 M borate buffer (pH 8.0). The mixture was further reacted with 200 μl of FMOC-Cl solution in acetonitrile. The optimal amount of FMOC-Cl for the reaction with 1000 pmol/ml of the respective analyte was 3.75 mg/ml in the final reaction mixture. Higher FMOC-Cl concentrations were found to decrease the detection signals. It was suggested that with more FMOC-Cl added, the excessive reagent could co-precipitate with the FMOC-derivatized analytes in the reaction mixture. It was also shown that the derivatization reaction was fully completed after 15 min. A 15-min reaction time was subsequently used in the experiments. The FMOC-derivatives were found to be stable at ambient temperature for at least 3 days. It was concluded that these FMOC-derivatives are sufficiently stable at ambient temperature for routine analysis in clinical laboratories.

After the reaction with FMOC-Cl, 800 μl of chloroform was added; the reaction mixture was shaken on a horizontal shaker at 200 rpm for 5 min and then centrifuged at 500 g for 2 min. The top aqueous layer was removed and further extracted with another 800 μl of chloroform. The pooled chloroform solution was blown dry under a stream of nitrogen gas. After reconstituting the sample with 400 μl of the reaction mixture (100 μl of water, 100 μl of borate buffer and 200 μl acetonitrile), 100 μl of the solution was subjected to HPLC analysis. Results of direct derivatization of spiked aqueous, blank urine and spiked urine samples with FMOC-Cl are shown in Figure 2. By comparing the three chromatograms in Figure 2A-C, it was observed that peaks of NM, MN, E and DHBA were all detectable in urine sample. However, the outsized peak of NE in the chromatogram (Figure 2B,C) was probably caused by co-elution of the analyte with other endogenous compounds in urine sample.

The structures of these FMOC derivatives were further confirmed by atmospheric pressure chemical ionization mass spectrometry (APCI-MS). The molecular ions $[M + H]^+$ of FMOE-NE, FMOC-E, FMOC-DA, FMOC-NM, FMOC-MN were *m/z* 836, 850, 820, 628 and 642, respectively (Figure 3). Based on these findings, the FMOC-derivatives of metanephrines and catecholamines were confirmed to be bi-substituted and tri-substituted respectively at the amino and catechol functional groups (Figure 4).

The developed assay was also established to be stability-indicating. In other word, the assay can resolve the parent analytes in the study from the degraded products. With this capacity, the assay was able to show that catecholamines and metanephrines are more resistant to degradation in water than in acid. These findings were consistent with observations in the previous studies [60,61], in which the analytes were measured by a stability-indicating reversed-phase ion-pair HPLC method with electrochemical detection.

In summary, this assay could be used in the clinical laboratory for routine measurements of catecholamines (E, DA) and/or metanephrines (MN, NM). The limit of quantitation (LOQ) and limit of detection (LOD) of these analytes per injection (100 μl) were approximately 1250 fmol and 375 fmol, respectively [15]. This HPLC assay for catecholamines and metanephrines using fluorimetric detection with pre-column FMOC-Cl derivatization was found to be more sensitive than the previous assay using a reversed-phase ion-pair HPLC method with amperometic detection with LOQ and LOD of 4000 and 1200 fmol, respectively [61]. Although this fluorescent assay could not measure NE in urine accurately, this shortcoming could be overcome by using mass spectrometric detection or solid-phase extraction of urine sample prior to FMOC-Cl derivatization

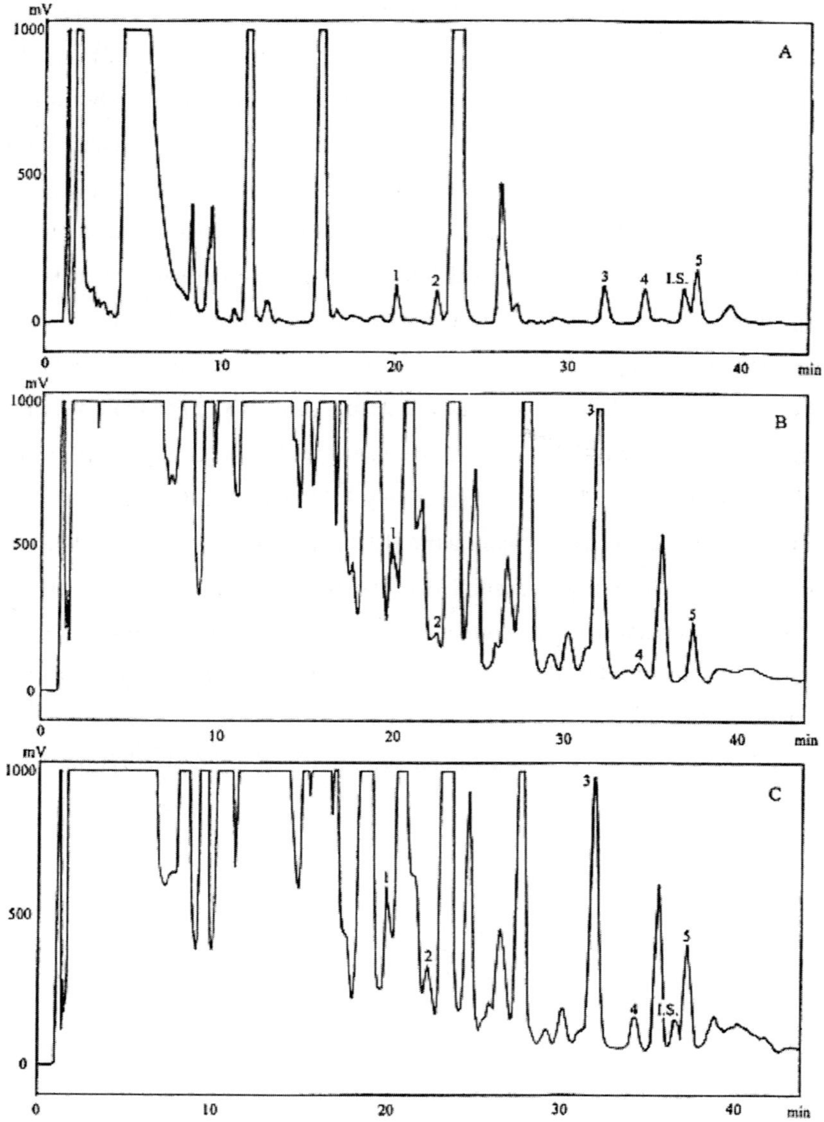

Figure 2 Chromatograms of FMOC-Cl derivatization of (A) aqueous sample, (B) urine sample and (C) spiked urine sample. Peaks: 1 = NM, 2 = MN, 3 = NE, 4 = E and I.S. = internal standard, DHBA. Peak at 32 min of (B) and (C) is co-elution of FMOC-NE and other compounds. Conditions as in Section 2.4. Reproduced with permission from Reference [15].

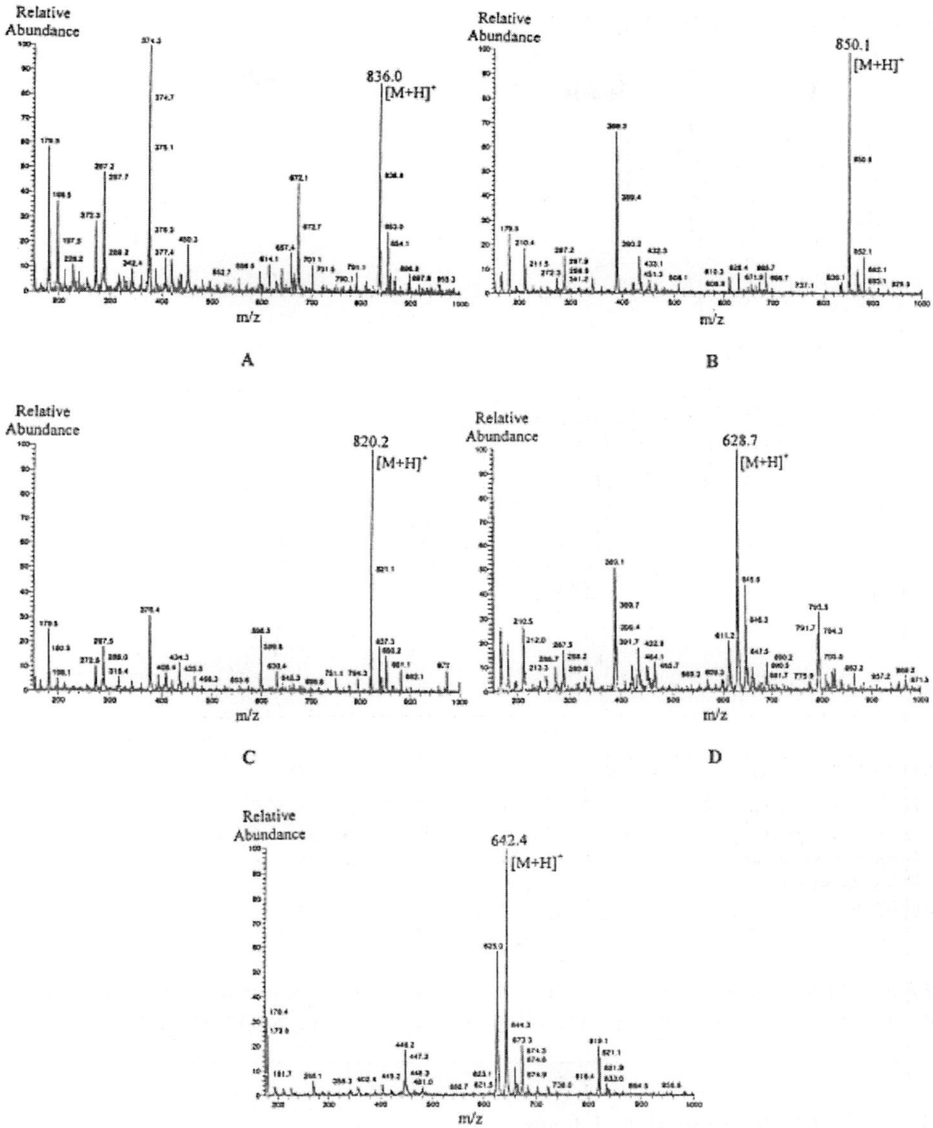

Figure 3 Mass spectra of FMOC-derivatives of (A) NE, (B) E, (C) DA, (D) NM and (E) MN. Reproduced with permission from Reference [15]

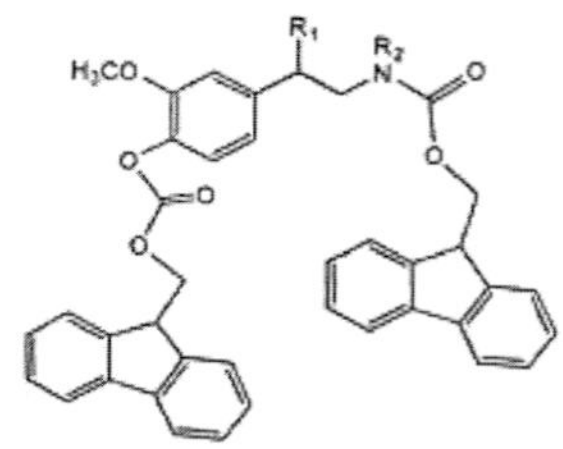

A

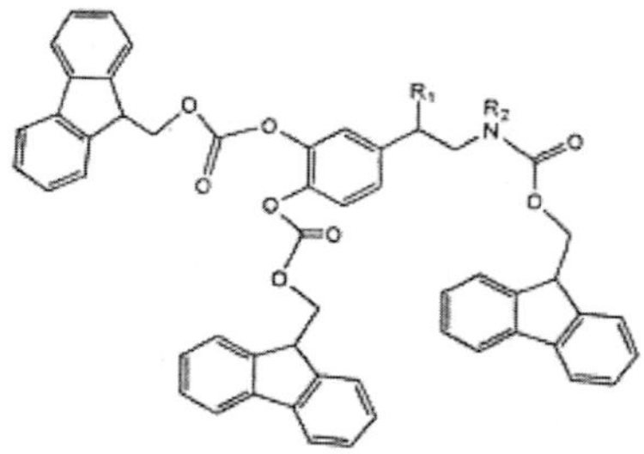

B

Derivatives	Structures	R_1	R_2	Exact mass
FMOC-normetanephrine	A	OH	H	627.68
FMOC-metanephrine	A	OH	CH_3	641.24
FMOC-norepinephrine	B	OH	H	835.28
FMOC-epinephrine	B	OH	CH_3	849.29
FMOC-dopamine	B	H	H	819.28

Figure 4 Proposed chemical structures and exact molecular masses of FMOC-derivatives of catecholamines and metanephrines. Reproduced with permission from Reference [15].

3. HPLC determination of drugs

FMOC-Cl is suitable for the pre-column derivatization of primary and secondary amines [10,62]. As opposed to *o*-phthaladehyde (OPA), the derivatives formed by FMOC-Cl are stable. Therefore, FMOC-Cl is commonly used as a derivatizing agent for a numbers of drugs containing the amino groups.

3.1 Aminoglycosides

Aminoglycosides are a class of broad spectrum antibiotics widely used in the treatment of serious infections. Because of their narrow therapeutic windows and the danger of their toxic side effects, it is important to monitor their plasma concentrations, especially in patients with in infants, elderly and patients with renal failure [63,64]. Gentamicin, an aminoglycoside produced by *Micromonospora purpurea,* is a complex mixture consisting of three major components C_1, C_{1a}, and C_2, and several minor components including C_{2a} and C_{2b}. The composition can vary considerably, but most commercial preparations have been shown by HPLC to contain 30-45% C_1, 12-30% C_{1a}, 19-43% C_2 and 7-14% C_{2a} [65]. The variable composition of gentamicin makes its quantitation difficult. The situation is further complicated, as gentamicin has no ultraviolet or visible absorption and cannot be detected by these conventional techniques without derivatization. FMOC-Cl has been shown to be a suitable reagent for the fluorescent labeling of the amino groups of gentamicin [66]. The derivatives were then analyzed by reversed-phase HPLC. This procedure allowed the fluorimetric detection of the gentamicin components C_1, C_{1a}, C_2 and tentatively C_{2a} (the identity could not be confirmed, as no authentic standard was available) that were well separated from the reagent front.

In the study, a stainless-steel column (200 x 4.6 mm i.d.) slurry packed with 3-μm ODS (Hypersil) was used for the separation of the gentamicin-FMOC derivatives. Detection was made at excitation 260 nm and emission 315 nm. The mobile phase consisted of acetonitrile-water (87:13, v/v) and the flow-rate was 1.0 ml/min.

Because of the strongly basic nature of the aminoglycosides, a carboxypropyl (CBA)-bonded solid phase extraction (SPE) column was adopted for the extraction of gentamicin from the plasma sample. The CBA-bonded SPE column was wetted with 1 ml methanol and conditioned with 1 ml phosphate buffer (20 mM, pH 7.4). An aliquot of 1 ml plasma containing the analytes was applied to the column, which was then washed with 2 ml phosphate buffer followed by 4 ml borate buffer (0.2 M, pH 9.0). The column was dried by application of a volume of air, then eluted into a 2-ml glass vial with 1 ml acetonitrile-borate buffer (0.2 M, pH 10.5) (1:1, v/v) followed by a further volume of air. The eluate (1 ml) was adjusted to pH 8.9 by the addition of 0.2 ml boric acid (0.8 M) and derivatized with 0.2 ml FMOC-Cl (2.5 mM in acetonitrile) at ambient temperature for 15 min. The reaction was stopped by adding 25 μl glycine (0.1 M) and, after a further 2 min, a volume (50 μl) of the reaction mixture was injected into the HPLC for analysis. The calibration lines for each of the major components were linear in the range of 0.2 to 5 mg/l and the limits of detection were between 10 and 50 μg/l. The developed assay was subsequently utilized for the determination of other clinically important aminoglycosides including neomycin, netilmicin and sisomicin in bacterial culture media or plasma [67].

3.2. *Amphetamines*

The stimulants such as amphetamine, methamphetamine, ephedrine, and norephedrine are common drugs of abuse. Ephedrine and norephedrine, the nasal decongestants, have been use illegally as the starting materials for amphetamine and methamphetamine synthesis [68]. Therefore, quantitative analysis of these drugs has become important in clinical and forensic sciences.

An on-line pre-column derivatization FMOC-Cl has been developed for the determination of low concentrations of amphetamines [16]. The sample solution (0.9 ml/min) was allowed to mix on-line for about 10 min in a glass coil (capacity 31.4 ml) with 5 x 10^{-4} mol/l (0.9 ml/min) FMOC-Cl in acetonitrile and the sodium hydrogen carbonate-sodium carbonate buffer solution 0.2 mol/l, pH 9.0 (0.9 ml/min). After that, the column valve was switched to connect to the analytical column (Nucleosil C_{18}, 5 μm, 200 mm x 4 mm i.d.) for the separation of the derivatized products. The mobile phase consisted of acetonitrile-water (58:42 v/v) and the flow-rate was 1.5 ml/min. The absorbance of 265 nm was chosen for the detection of these amphetamines. By using this method, the FMOC-OH (the degradation product of FMOC-Cl), and the amphetamines (amphetamine, methamphetamine, ephedrine, and norephedrine) were well separation over 45 min. Linear calibration curves were obtained over the range of 1 x 10^{-6} – 1 x 10^{-4} mol/l for each of the amphetamines. A comparison of these results with that obtained by the off-line method showed that the sensitivity of the reported method was increased by a factor of about 50.

Sulfamethazine

Sulfonamides are widely used for the prevention of infectious diseases in livestock animals. However, to the consumers, it is essential to ensure that the meats are residue-free of the antibiotics at the time when they are in the markets. Residues of sulfonamides in foods derived from treated animals could lead to the development of drug resistant strains of microorganisms [69]. Recently, the safety of sulfamethazine to consumers has further been questioned because of its potential carcinogenic effects [70]. Hence, a stringent control has been set on the levels of sulfonamides for edible animal tissues [71]. Numerous analytical methods [71,72,73,74] have been proposed for the sulfamethazine residue analysis. As FMOC-Cl can react with the amine group of a compound to form a highly fluorescent FMOC-derivative, it has been applied to derivatize sulfamethazine to form the fluorescent derivative for the HPLC analysis [75]. In the study, a TMS-capped Micro Park C18 5 μm, 15 cm x 4 mm i.d. column (Varian, Walnut Creek, CA) was used for the separation of the analytes. The standard HPLC buffer was prepared from NaH_2PO_4 and Milli-Q water and adjusted to pH 3.5 with H_3PO_4. The pH 8 reaction buffer was 0.1 M $NaHCO_3$. Using the optimized conditions for the reaction of FMOC-Cl and sulfamethazine, 1 ml of the sulfamethazine solution, 1 ml of the pH 8 buffer, and 2 ml of 10 mM FMOC-Cl in acetone were mixed and allowed to react for 15 min. After that, 20 μl of the reaction mixture was injected into the HPLC system for analysis. For the satisfactory separation of the analytes, the mobile phase was run in pattern of a medium concave gradient from 65% A (0.05 M NaH_2PO_4, pH 3.5)/35% B (methanol) to 15% A and 85% B over 4 min, followed by isocratic elution with 85% methanol. The flow rate was 1 ml/min. Baseline resolution was achieved for acetone, sulfamethazine, the FMOC-Cl hydrolysis product (FMOC-OH) formed during reaction and the FMOC-sulfamethazine derivative over a run time of 16 min. In comparison with usual UV absorbance detection of sulfamethazine, the fluorescent detection of FMOC-sulfamethazine extended the linear response range from above 100 ppb down to 100 ppt, which was equivalent to an increase in sensitivity over 3 orders of magnitude.

3.3. Macrolide antibiotics

Erythromycin, roxithromycin, clarithromycin and azithromycin are macrolide antibiotics widely used for the infections with Gram-positive bacteria such as *Streptococci*, *Staphylococcus aureus*, *Propionibacterium acnes*, *Listeria monocytogenes* and *Corynobacteria* species, and Gram-negative bacteria, such as *Legionella pneumophilia* and *Haemophilus* species. Roxithromycin, clarithromycin and azithromycin, the newer macrolides are more chemically stable and better tolerated than erythromycin, and they have a broader antimicrobial spectrum than erythromycin against *Mycobacterium avium* complex (MAC), *Haemophilus influenzae*, *nontuberculous mycobacteria*, and *Chlamydia trachomatis*. All macrolides have excellent activity against the atypical respiratory pathogens (*C. pneumoniae* and *Mycoplasma* species) and the *Legionella* species. Azithromycin and clarithromycin have pharmacokinetics that allow shorter dosing schedules because of prolonged tissue levels, Both azithromycin and clarithromycin are currently recommended for MAC prophylaxis in patients with late-stage acquired immunodeficiency syndrome (AIDS) [76,77].

Efficacy of an antibiotic can only be achieved at drug concentrations exceeding the minimal inhibitory concentration for a pathogen at the site if infection. However, due to variations in individual pharmacokinetic profiles, and drug-drug interaction, effective drug concentration might not be achieved clinically. Therefore, a suitable analytical assay for the determination of macrolide concentrations in biological specimens is needed to control and assure drug efficacy. Zierfels and Petz [78] first proposed to determine erythromycin in foods using fluorescence detection after derivatization with FMOC-Cl. The proposed assay was subsequently applied to quantitate erythromycin, azithromycin, clarithromycin and roxithromycin in human serum [79]. The mobile phase of the assay was prepared by mixing 600 ml acetonitrile with 400 ml of a buffer comprising potassium dihydrogen phosphate (0.05 mol/l) and 500 µl of triethylamine, adjusted to pH 7.5 with a 10% potassium hydroxide solution. The flow rate of the mobile phase was 2 ml/min. The separation was performed on a 125 x 4.6 mm based-deactivated Supelcosil silica C_{18} column packed with 5 µm particles and protected by a guard column (20 x 4.6 mm) packed with the same solid phase.

In the prescribed method, roxithromycin was used as an internal standard for the erythromycin, clarithromycin and azathromycin assay, and clarithromycin was used for the roxithromycin assay. For the sample preparation, a 20 µl of internal standard was added to a 1.0 ml sample containing the macrolide, in a 10 ml disposable glass tube, and 200 µl of a warmed saturated sodium carbonate solution (0.5 g/ml, with a pH of approximately 12) was added and solution was vortex-mixed for 10 s. A volume of 6 ml of diethyl ether was added to the sample solution. After shaking for 15 min at a frequency of 270/min, the solution was centrifuged for 5 min at 2700 g. The solution was allowed to freeze at -20°C in an ethanol bath. The organic layer was transferred to a dry disposable glass tube and evaporated at 40°C under a flow of nitrogen. The residue was dissolved in 200 μl of acetonitrile and vortex-mixed for 30 s. The solution was centrifuged at 2700 g and quantitatively transferred to a 1.8 ml reaction vial. A 100 μl of 2.5 mg FMOC-Cl in acetonitrile and a 75 μl of 0.1 M phosphate buffer (adjusted to pH 7.5 with 10% potassium hydroxide solution) were added. The reaction vial was incubated in a water bath at 40°C for 40 min. After derivatization, 50 μl of the solution was injected into the HPLC system for analysis. Fluorescence detection was made at an excitation wavelength of 255 nm and an emission wavelength of 315 nm.

The lower limits of quantitation of the assay were found to be 0.20, 0.065, 0.20 and 0.092 mg/l, and the higher limits of quantitation were 16.0, 19.5, 20.0 and 5.0 mg/l for erythromycin, roxithromycin, clarithromycin and azithromycin respectively.

3.4. Bisphosphonates

Bisphosphonates, such as olpadronate and alendronate are recommended for the treatment of calcium metabolic disorders such as, Paget's disease, hypercalcaemia and osteoporosis [80,81]. The molecules of bisphosphonates do not possess chromophores. Being highly potent, olpadronate and other bisphosphonates are administered in very low doses. All these pose extreme difficulties for their detection in biological samples. Numerous analytical methods have been proposed for their detection, including derivatization prior to gas chromatography [82]; ion-exchange chromatography with post-column decomposition of the bisphosphonate into orthophosphate and derivatization of the phosphate into a colored complex, followed by spectrophotometric detection [83]; reversed-phase liquid chromatography with selective UV [84], fluorescence [85] or electrochemical detection [86]; reversed-phase liquid chromatography after derivatization of the primary amino group of the bisphosphonate into a fluorescent derivative. For the last approach, 2,3-naphthalene dicarboxyaldehyde [87] and 1-naphthlisothiocyanate [88] were used as the fluorescence derivatization agents. However, all these chromatographic assays of bisphosphonates could only achieve very low detection limits; whereas other methods required expensive or complicated instruments, such as negative ion electrospray ionization mass spectrometry [89] or evaporative light-scattering detector [90]. Subsequently, Sparidan et al. proposed a complicated semi-automatic liquid chromatographic method for the analysis of olpadronate in urine and serum using FMOC-Cl as the fluorescence derivatizing agent [19]. The latter study was further simplified for the analysis of alendronate in human urine [91]. In the study, the separation was done on a 150 x 4.6 mm i.d. column filled with Nucleosil 100-3 C_{18} stationary phase, particle size 3 μm. A simple gradient method was used to elute the compounds: the starting mobile phase acetonitrile-methanol-buffer (25 mM citric acid and 25 mM sodium pyrophosphate) (20:15:65, v/v) was directly changed to acetonitrile-methanol-water (13:57:30, v/v) at 7.5 min and hold at this composition for 9 min. Then the mobile phase was restored to the starting composition. The total run time was 21 min. The flow-rate was 1 ml/min at 40oC. The excitation and emission wavelength was 260 and 310 nm, respectively and the time constant was set to 2 s.

For the preparation of sample, 10 μl of the internal standard solution (4 μg of pamidronate disodium) were added to 5 ml of urine. After shaken, 100 μl of 0.1 M KH2PO4 and the same amount of 0.1 M CaCl2 were added. The mixture was then made alkaline with 200 μl of 1 M NaOH. The sample was then centrifuged for 3 min at 2000 g and the supernatant was discarded. The precipitate was then dissolved in 0.5 ml of 0.2 M acetic acid before the addition of 5 ml of water. Subsequently, the procedure of precipitation with NaOH was repeated twice. The resulting precipitate was dissolved in 1 ml of 0.2 M acetate buffer (pH 4.5) and diluted with 2 ml of water. The sample was then loaded on DEA SPE cartridge pre-washed with water. After washing the cartridge with 2 x 0.5 ml water, the drug was eluted with 1 ml of 0.2 M sodium acetate.

For derivatization of the compound, 270 μl of the eluate was added to 100 μl of 1 M sodium carbonate (pH 11.9), followed by the addition of 100 μl of FMOC solution (1 mg in 4 ml acetonitrile). The mixture was allowed to react for 3 min, then 100 μl of 1 M citric acid was added to adjust pH and 50 μl of the sample was injected into the HPLC for analysis.

Adsorption of bisphosphonates onto analytical columns has been the severe problems, affecting their separation and detection. In the previous study [19], a bisphosphonate analog, etidronate was added in the mobile phase to minimize the adsorption of olpadronate. In this study [91], combination of structurally similar pyrophosphate anion and chelating citrate ions was used in the mobile phase as adsorption suppressors. The results showed that with this mobile phase composition, the conventional reversed-phase C_{18} column was able to give symmetrical well resolved peaks. The validated method allowed quantitation of alendronate in 5-ml urine sample in the range of 3.5 – 300 ng/ml. The sensitivity of the assay allowed its use for pharmacokinetic studies based on urinary excretion data.

Table 1 Summary of HPLC conditions suggested for the determination of various amines.

Amine	Column	Mobile phase	Isocratic/Gradient	Detector	Quantitation Limit	Reference
Tyramine, tryptamine and polyethylamine	Supersphere RP-18 column, 4 μm, 125 mm x 4 mm (Merck)	Eluent A: 25% methanol in aqueous sulphuric acid (pH 2.25); Eluent B: Methanol. Flow-rate: 1 ml/min.	Gradient: 20% eluent B at 0 min to 77.6% at 21.6 min, 100% at 22.5 min, 20% at 25 min and finally equilibration for an additional 10 min.	UV: 263 and 299 nm.	0.125 nmol in 40 μl	[10]
Polyamines – putrescine, cadaverine, spermidine, and spermine	NovaPak C_{18} column, 4 μm, 150 mm x 3.9 mm (Waters, Milford)	Eluent A: Methanol Eluent B: Water Flow-rate: 1 ml/min.	Gradient: methanol-water (82:18) to 100% methanol in about 13 min.	Fluorescence: excitation – 254 nm, emission – 313 nm.	22.7 to 109 fmol	[41,42,43]
Methylamine	LiChrospher 100 RP18, 4 μm, 125 mm x 4 mm (Merck)	Eluent A: Acetonitrile Eluent B: Water Flow-rate: 1 ml/min.	Gradient: 0% A at 0 min to 40% at 2.5 min, to 70% at 9.5 min, and then to 100% at 17.5 min.	UV: 230 nm.	2.5 μg/ml	[49]

Dimethylamine	Microspher C_{18} column, 3 µm, 100 mm x 4.6 mm (Chrompack)	Eluent A: Acetonitrile-buffer (50:50, v/v) Eluent B: Acetonitrile-buffer (75:25, v/v) Buffer: Potassium dihydrogen phosphate (50 mM) adjusted to pH 6.5 with KOH Flow-rate: 1 ml/min.	Gradient: Eluent A reduced from 100% at 0 min to 0% in 5 min, 100% eluent B was then maintained for another 5 min, before returned to 100% eluent A in 1 min. The column was equilibrated with 100% eluent A for 4 min	Fluorescence: excitation – 260 nm, emission – 320 nm.	0.1 µmol/l	[51]
Catecholamines and metabolites	Spherisorb C_8 column, 5 µm, 150 mm x 3.0 mm (Phase Separation, Inc.)	Eluent A: Acetic acid, 3.0 g/l Eluent B: Acetonitrile Flow-rate: 1 ml/min.	Gradient: 50 to 78% solvent B in 40 min.	Fluorescence: excitation – 263 nm, emission – 313 nm.	1250 fmol/100 µl	[15]
Gentamicin	ODS column, 3 µm, 200 mm x 4.6 mm (Hypersil)	Acetonitrile-water (87:13, v/v). Flow-rate: 1 ml/min.	Isocratic	Fluorescence: excitation – 260 nm, emission – 315 nm.	0.2 mg/l	[66]

Amphetamines	C_{18} column, 5 µm, 200 mm x 4.0 mm (Nucleosil)	Acetonitrile-water (58:42, v/v). Flow-rate: 1.5 ml/min.	Isocratic	UV: 265 nm.	1 x 10^{-6} mol/l	[16]
Sulfamethazine	TMS-capped Micro Park C18, 5 µm, 150 mm x 4.0 mm (Varian)	Eluent A: 0.05 M NaH_2PO_4, pH 3.5 Eluent B: Methanol Flow-rate: 1.0 ml/min.	Gradient: 65% A/35% B to 15% A/85% B over 4 min, followed by isocratic elution with 85% methanol for 14 min.	UV: 254 nm. Fluorescence: excitation – 264 nm, emission – 307 nm.	100 ppt	[75]
Macrolide antibiotics	Silica C_{18} column, 5 µm, 125 mm x 4.6 mm (Supelcosil)	Acetonitrile-buffer (600:400, v/v). Buffer: 400 ml of 0.05 mol/l potassium dihydrogen phosphate, and 500 µl of triethylamine and adjusted to pH 7.5 with a 10% potassium hydroxide solution. Flow-rate: 2.0 ml/min.	Isocratic.	Fluorescence: excitation – 255 nm, emission – 315 nm.	Erythromycin: 0.320 mg/l Roxithromycin: 3.24 mg/l Clarithromycin: 0.190 mg/l Azithromycin: 0.0988 mg/l.	[79]

Alendronate	C_{18} column, 150 mm x 4.6 mm, 3 μm (Nucleosil)	Eluent A: acetonitrile-methanol-buffer (25 mM citric acid and 25 mM sodium pyrophosphate) (20:15:65, v/v) Eluent B: acetonitrile-methanol-water (13:57:30, v/v) Eluent B: acetonitrile-methanol-water (13:57:30, v/v) Flow-rate: 1.0 ml/min at 40°C	Gradient: Eluent A was changed to eluent B at 7.5 min and held at this composition for 9 min. Then the mobile phase was restored to the starting composition. The total run time was 21 min.	Fluorescence: excitation – 260 nm, emission – 310 nm.	3.5 ng/ml in 5 ml urine sample.	[92]

References

[1] L.A. Carpino, G.Y. Han, J. Org. Chem. 37 (1972) 3404-3409.

[2] T.W. Greene, P.G.M. Wuts, Protective Groups in Organic Synthesis, Wiley, New York, 1991.

[3] S. Einarsson, B. Josefsson, S. Lagerkvist, J. Chromatogr. 282 (1983) 609-618.

[4] S. Einarsson, S. Folestad, B. Josefsson, S. Lagerkvist, Anal. Chem. 58 (1986) 1638-1643.

[5] R.L. Cunico, A.G. Mayer, P. Dimson, C. Zimmerman, BioChromatography 1 (1986) 6-14.

[6] T. Naesholm, G. Sandberg, A. Ericsson, J. Chromatogr. 396 (1985) 225-236.

[7] G.E. Kisby, D.N. Roy, P.S. Spencer, J. Neurosci. Methods 26 (1988) 45-54.

[8] D.T. Blankenship, M.A. Krivanek, B.L. Ackerman, A.D. Cardin, Anal. Biochem. 178 (1989) 227-232.

[9] T. Teerlink, D.E. Boer, J. Chromatogr. 491 (1989) 418-423.

[10] Z. Harduf, T. Nir, B.J. Juven, J. Chromatogr. 437 (1988) 379-386.

[11] J.R. Price, P.A. Metz, H. Veening, J. Chromatogr. 424 (1987) 795-799.

[12] Z. Lu, H. Veening, J.E. Coutant, J. Chromatogr. 540 (1991) 199-206.

[13] M.I. Sabri, A.I. Soiefer, G.E. Kisby, P.S. Spencer, J. Neurosci. Methods 29 (1989) 27-31.

[14] A.A. Descombes, W. Haerdi, Chromatographia 33 (1992) 83-86.

[15] E.C.Y. Chan, P.Y. Wee, P.Y. Ho, P.C. Ho, J. Chromatogr. B 749 (2000) 179-189.

[16] G. Maeder. M. Pelletier, W. Haerdi, J. Chromatogr. 593 (1992) 9-14.

[17] G.S. Liang, Z. Zhang, W.L. Baker, R.F. Cross, Anal. Chem. 68 (1996) 86-92.

[18] J.S. Toraño, H.J. Guchelaar, J. Chromatogr. B 720 (1998) 89-97.

[19] R.W. Sparidans, J den Hartigh, S. Cremers, J.H. Beijnen, P. Vermeij. J. Chromatogr, 738 (2000) 331-341.

[20] P. Vandekerckhove, H.K. Henderick, J. Chromatogr. 82 (1973) 379-381.

[21] J.Y. Hui, S.L. Taylor, J. Assoc. Off. Anal. Chem. 66 (1983) 853-857.

[22] M.T. Salazar, T.K. Smith, A. Harris, J. Agric. Food Chem. 48 (2000) 1708-1712.

[23] Z. Harduf, J. Chromatogr. 363 (1986) 261-266.

[24] N. Seiler, Methods Enzymol. 94 (1983) 10-25.

[25] W.A. Jacobs, M.W. Leburg, E.J. Madaj, Anal. Biochem. 156 (1986) 334-340.

[26] M.Á. Medina, J.L. Urdiales, C. Rodriguez-Caso, F.J. Ramirez, F. Sánchez-Jiménez, Crit. Rev. Biochem. Mol. Biol. 38 (2003) 23-59.

[27] R.T. Premont, R.R. Gainetdinov, M.G. Caron, Proc. Natl. Acad. Sci. USA 98 (2001) 9474-9475.

[28] M.Y. Khuhawar, G.A. Qureshi, J. Chromatogr. B. 764 (2001) 385-407.

[29] C.S. Ough, C.E. Daudt, E.A. Crowell, J. Agricultural Food Chem. 29 (1981) 938-941.

[30] I. Mafra, P. Herbert, L. Santos, P. Barros, A. Alves, Am. J. Enology Viticulture 50 (1999) 128-132.

[31] J.A. Zee, R.E. Simard, L. L'Heureux, J. Tremblay, Am. J. Enology Viticulture 34 (1983) 6-9.

[32] M.C. Vidal-Carou, A. Ambatlle-Espunyes, M.V. Utla-Ulla, A. Mariné-Font, Am. J. Enology Viticulture 41 (1990) 160-167.

[33] F. Radler, K.P. Fäth, in J. Rantz (Editor), Proceedings of the international symposium on nitrogen in grapes and wines, American Society for Enology and Viticulture (1991), 185-195.

[34] C. Buteau, C.L. Duitschaever, G.C. Ashton, Am. J. Enology Viticulture 35 (1984) 228-236.

[35] M.C. Vidal-Carou, R. Codony-Salcedo, A. Mariné-Font, Food Chem. 35 (1990) 217-227.

[36] A.E. Pegg, P.P. McCann, Am. J. Physiol. 243 (1982) C212-C221.

[37] A.E. Pegg, Biochem. J. 234 (1986) 249-262.

[38] J.W. Suh, S.H. Lee, B.C. Chung, J. Park, J. Chromatogr. B. 688 (1997) 179-186.

[39] M. Bandyopadhyay, A.K. Ganguly, Med. Sci. Res. 27 (1999) 645-647.

[40] C. Loser, U.R. Folsch, C. Paprontny, W. Creutzfedt, Cancer 65 (1990) 958-966.

[41] Z. Yun, R.E. Zhung, Biomed. Chromatogr. 2 (1987) 173-176.

[42] F. Bellagamba, V.M. Moretti, T. Mentasti, A. Albertini, U. Luzzana, F. Valfre, J. Chromatogr. A. 791 (1997) 79-84.

[43] R.S. Gilbert, G.G. Gonzalez, L. Hawell III, C.V. Byus, Anal. Biochem. 199 (1991) 86-92.

[44] S.H. Zeisel, K. DaCosta, Cancer Res. 46 (1986) 6136-6138.

[45] A.Q. Zhang, S.C. Mitchell, R. Ayesh, R.L. Smith, Biochem. Pharmacol. 45 (1993) 2185-2188.

[46] S.H. Zeisel, K. DaCosta, B.M. Edrise, J.G. Fox, Carcinogenesis 7 (1986) 775-778.

[47] S.H. Zeisel, K. DaCosta, J.T. LaMont, Carcinogenesis 9 (1988) 179-188.

[48] H. Gu, H.M. Ma, S.C. Liang, Fresenius J. Anal. Chem. 371 (2001) 570-573.

[49] R. Herráez-Hernández, C. Cháfer-Pericás, P. Campíns-Falcó, Anal. Chim. Acta 513 (2004) 425-433.

[50] P. Campíns-Falcó, R. Herráez-Hernández, A. Sevillano-Cabeza, I. Trümpler, Anal. Chim. Acta 344 (1997) 125-136.

[51] T. Teerlink, M.W.T. Henneskes, C. Mulder, H.F.H. Brulez, J. Chromatogr. B 691 (1997) 269-276.

[52] J. Verdú-Andrés, P. Campins-Falcó, R. Herráez-Hernández, Chromatographia 55 (2002) 129-134.

[53] B. Kagedal, D.S. Goldstein, J. Chromatogr. 429 (1988) 177-233.

[54] G. Eisenhofer, I.J. Kopin, D.S. Goldstein, Pharmacol. Rev. 56 (2004) 331-349.

[55] T.G. Rosano, T.A. Swift, L.W. Hayes, Clin. Chem. 37 (1991) 1854-1867.

[56] E.A.M. Gerlo, C. Sevens, Clin. Chem. 40 (1994) 250-256.

[57] R.T. Peaston, C. Weinkove, Annals Clin. Biochem. 41 (2004) 17-38.

[58] G. Eisenhofer, J.W. Lenders, D.S. Goldstein, M. Mannelli, G. Csako, M.M. Walther, F.M. Brouwers, K. Pacak, Clin. Chem. 51 (2005) 735-744.

[59] J. Hanai, T. Takeda, Screening 4 (1995) 91-100.

[60] E.C.Y. Chan, P.Y. Wee, P.C. Ho, Clin. Chim. Acta 288 (1999) 47-53.

[61] E.C.Y. Chan, P.Y. Wee, P.C. Ho, J. Pharm. Biomed. Anal. 22 (2000) 515-526.

[62] J.A. Shah, D.J. Weber, J. Chromatogr. 309 (1984) 95-105.

[63] M.M. Tod, C. Padoin, O. Petitjean, Clin. Pharmacokinet. 40 (2001) 803-814.

[64] E. Triggs, B. Charles, Clin. Pharmacokinet. 37 (1999) 331-341.

[65] P.J. Claes, R. Busson, H. Vanderhaeghe, J. Chromatogr. 298 (1984) 445-457.

[66] D.A. Stead, R.M.E. Richards, J. Chromatogr. B 675 (1996) 295-302.

[67] D.A. Stead, R.M.E. Richards, J. Chromatogr. B 693 (1997) 415-421.

[68] F.T. Noggle, J. DeRuiter, C.R. Clark, J. Chromatogr. Sci. 28 (1990) 529-536.

[69] D.J. Wilson, B.B. Norman, C.E. Fanti, J. Am. Vet. Med. Assoc. 199 (1991) 759-766.

[70] L.A. Poirier, D.R. Doerge, D.W. Gaylor, M.A. Miller, R.J. Lorentzen, D.A. Casciano, F.F. Kadlubar, B.A. Schwetz, Reg. Toxicol. Pharmacol. 30 (1999) 217-222.

[71] L.S.G. Vanpoucke, G.C.I. Depourcq, C.H. Vanpeteghem, J. Chromatogr. Sci. 29 (1991) 423-427.

[72] C.D.C. Salisbury, J.C. Sweet, R. Munro, J. AOAC Int. 87 (2004) 1264-1268.

[73] K.E. Maudens, G.F. Zhang, W.E. Lambert, J. Chromatogr. A 1047 (2004) 85-92.

[74] T. Korpimaki, V. Hagren, E.C. Brockmann, M. Tuomola, Anal. Chem. 76 (2004) 3091-3098.

[75] G.S. Liang, Z. Zhang, W.L. Baker, R.F. Cross, Anal. Chem. 68 (1996) 86-92.

[76] S. Alvarez-Elcoro, M.J. Enzler, Mayo Clinic Proc., 74 (1999) 613-634.

[77] N. Rastogi, K.S. Goh, P. Ruiz, M. Casal, Antimicrob. Agents Chemother., 39 (1995) 1162-1165.

[78] G. Zierfels, M. Petz, Lebensm. Unters. Forsch. 198 (1994) 307-312.

[79] J. Sastre Toraño, H.-J. Guchelaar, J. Chromatogr. B, 720 (1998) 89-97.

[80] H. Fleisch, Bisphosphonates in Bone disease. From laboratory to the patient, Parthenon Publishing Group, New York, 1995.

[81] R.W. Sparidans, I.M. Twiss, S. Talbot, Pharm. World Sci. 20 (1998) 206-213.

[82] S. Auriola, R. Kostiainen, M. Ylinen, M. Mönkkönen, P. Ylitalo, J. Pharm. Biomed. Anal. 7 (1989) 1623-1629.

[83] P.T. Daley-Yates, L.A. Gifford, C.R. Hoggarth, J. Chromatogr. 490 (1989) 329-338.

[84] J.-P. Fels, J. Guyonnet, Y. Berger, W. Cautreels, J. Chromatogr. 430 (1988) 73-79.

[85] T. Usui, R. Kawakami, T. Watanabe, S. Higuchi, J. Chromatogr. B 652 (1994) 67-72.

[86] T. Usui, T. Watanabe, S. Higuchi, J. Chromatogr. 584 (1992) 213-220.

[87] L.E. King, R. Vieth, J. Chromatogr. B 678 (1996) 325-330.

[88] R.W. Sparidans, J. den Hartigh, S. Cremers, J.H. Beijnen, P. Vermeij, J. Chromatogr. B 730 (1999) 95-99.

[89] H. Mönkkönen, P. Moilanen, J. Mönkkönen, J.C. Frith, M.J. Rogers, S. Auriola, J. Chromatogr. B 738 (2000) 395-403.

[90] R. Niemi, H. Taipale, M. Ahlmark, J. Vepsäläinen, T. Järvinen, J. Chromatogr. B 701 (1997) 97-102.

[91] P. Ptácek, J. Klíma, J. Macek, J. Chromatogr. B 767 (2002) 111-116.

Ibolya Molnár-Perl (Editor)
Quantitation of Amino Acids and Amines by Chromatography
Journal of Chromatography Library, Vol. 70

2.2.4.

HPLC of Biogenic Amines as 6-Aminoquinolyl-N-hydroxysuccinimidyl Derivatives

Thomas S. Weiss

Contents

Summary

6-aminoquinolyl-N-hydroxysuccinimidyl carbamate (AQC), a recently discovered derivatizing reagent for amines, is presented to analyze biogenic amines, in particular polyamines using reversed-phase high performance liquid chromatography (HPLC). Adducts of AQC and polyamines were characterized regarding their spectroscopic properties and their suitability for fluorescence detection. Conditions for a derivatization of amines with AQC are discussed in detail and possible fluorescent by-products within the derivatization process are described. Further, extraction procedures for amines and polyamines from various tissues and matrices prior to AQC derivatization are presented and stability of N-acetylated polyamines within acidic precipitation was investigated. An overview of HPLC methods for the analysis of polyamines and other amines is given with their respective data of validation. Additional, recovery study of

polyamines from various tissues is discussed and further, potential interferences of adducts of AQC with commonly used inhibitors of polyamine metabolism were presented.
AQC is a powerful derivatization reagent for amines forming stable and highly fluorescent derivatives. The reagent reacts with primary as well as secondary amines under mild alkaline conditions within a broad pH range. The presented HPLC methods for a chromatographic separation of AQC adducts with polyamines demonstrated high accuracy as well as reproducibility and is suitable for analysis of unconjugated and acetylated polyamines without interference of fluorescent byproduct.

1. Introduction

Biogenic amines are mostly small molecules with an amine moiety and fulfill important function in regulating cell physiology. Among them the polyamines are of major interest and are under intensive investigation. The naturally occurring polyamines such as putrescine, spermidine and spermine as well as their acetylated derivatives are aliphatic polycations which are ubiquitous in prokaryotic and eukaryotic cells. Their essential role in regulation of cell growth and differentiation is firmly established [1-3]. Furthermore their involvement in diseases like cancer and other pathological conditions, such as inflammatory bowel disease, infections, psoriasis, liver cirrhosis and neurodegenerative disorders was described (see recent reviews [4-9]). Although their metabolic pathways are well established and their intracellular concentrations are highly regulated [2,9,10], there are still many aspects of their functions remain elusive.

Versatile improved analytical techniques for polyamines are not only invaluable for diagnostic purposes and metabolic as well as pharmacokinetic studies associated with the development of new therapeutic strategies, but might also elucidate the physiological and pathophysiological functions of these essential compounds.

Among many methods for the analysis of polyamines from biological material, high-performance liquid chromatography is the most widely used [4,11]. Separation via reversed-phase chromatography and detection of the polyamines requires a labeling by postcolumn [12-14] or precolumn [15-33] derivatization for ultraviolet [15-19] or the more sensitive and selective fluorescence [12-14,20-33] detection. A common drawback using precolumn derivatization is the necessity to remove the derivatized polyamines from an excess of derivatization reagent and by-products. This is either performed by liquid-liquid (cyclohexane

[34], benzene [21], toluene [22], chloroform [18], diethyl ether [15], hexane/ethane acetate [26]) or solid phase extraction [23]. Additionally, most of the derivatized polyamines have a low stability and are light sensitive, which makes the use of automated sample preparations necessary [35]. Despite the many efforts made to improve fluorescence labelling techniques, which are also very popular in amino acid analysis, all methods as those based on 9-fluorenylmethyl chloroformate [26-28], dansyl chloride [21-25] and o-phthalaldehyde (OPA) derivatization [12-14,20] suffer from various shortcomings. Interferences with the reagent, formation of multiple derivatives, and failing derivatization of secondary amino groups, motivated researchers to quest for new labelling reagents [36].

1.1. The QAC derivatives

In the early nineties a new reagent for fluorescence labelling of amino acids was introduced by Cohen and Michaud, overcoming most of the problems described above [36]. The reagent, based on analogues of N-hydroxysuccinimide activated carbamates, was named 6-aminoquinolyl-N-hydroxysuccinimidyl carbamate (AQC). Additionally, following the recommendations of the IUPAC for nomenclature of organic compounds [37] it was also called N-hydroxysuccinimidyl 6-quinolinyl carbamate (HSQC) [31] {alternative names: succinimidyl 6-quinolyl carbamate; 2,5 dixopyrrolidine-1-yl quinolin-6-yl carbamate; quinolin-6-yl carbamic acid succinimidyl ester; 1-[[(6-quinolinylamino)carbonyl]oxy]-2,5-pyrrolidine; CA registry number 148757-94-2}. For reasons of simplification and regarding to the common use in the literature, within this chapter it was named AQC. Very rapidly not only the applicability to amino acid, peptides and protein analysis (see chapter 1.3.6) was demonstrated, but also to polyamines and other biogenic amines [30-33,38-45]. Reversed phase HPLC methods with precolumn derivatization using AQC were introduced for the separation of polyamines [30] or both polyamines and their acetylated derivatives [31,46] and were afterwards the most widely used assays [30-33,40,44,45,47-51].

N-hydroxysuccinimide activated carbamates such as AQC react with primary or secondary amines forming fluorescent derivatives and N-hydroxysuccinimide (cf. Figure 1). The reaction can be performed under room temperature for polyamines [31], but is recommended to be increased to 50°C completing derivatization of all amino acids [30].

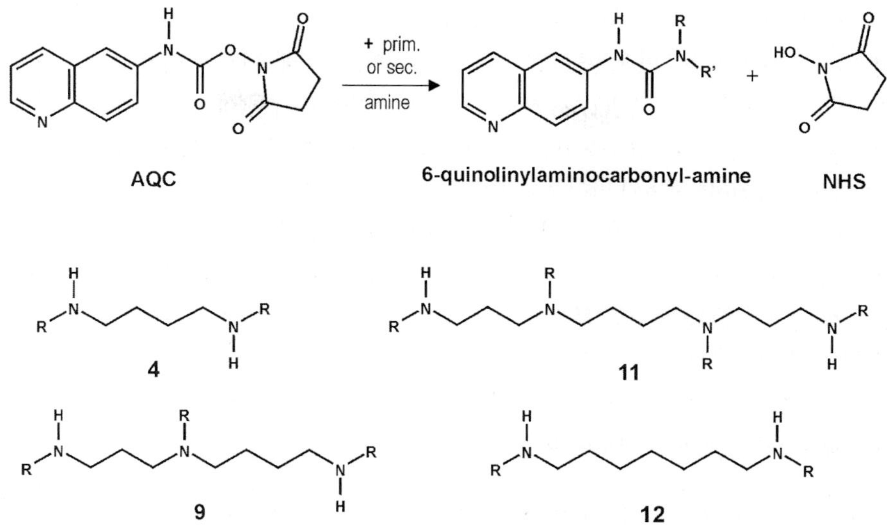

Figure 1: Primary and secondary amines react with AQC yielding 6-quinolinylaminocarbonyl amines and N-hydroxysuccinimide (NHS). Fluorescent adducts of AQC with putrescine (4), spermidine (9), spermine (11) and 1,7-daminoheptane (12) are shown (Table 1).

The reaction of AQC with polyamines leads to complete derivatized 6-quinolinylaminocarbonyl (QAC) amines, of which Figure 1 shows adducts with the most important polyamines putrescine, spermidine, spermine and 1,7-daminoheptane (internal standard). Further, Table 1 shows a list of polyamines (and ornithine) derivatized with AQC followed by a chromatographic separation within a single run [31].

Table 1 List of separated 6-quinolinylaminocarbonyl derivatives

No	Compound	Amine
1	(6-quinolinyl)urea	ammonia
2	N^2,N^5-bis(6-quinolinylaminocarbonyl)ornithine	ornithine
3	N^1-(6-quinolinylaminocarbonyl)-N^4-acetylputrescine	N-acetylputrescine
4	N^1,N^4-bis(6-quinolinylaminocarbonyl)putrescine	putrescine
5	N^1-(6-quinolinylaminocarbonyl)-N^4-acetylcadaverine	N-acetylcadaverine
6	N^1,N^5-bis(6-quinolinylaminocarbonyl)cadaverine	cadaverine
7	N^4,N^8-bis(6-quinolinylaminocarbonyl)-N^1-acetylspermidine	N^1-acetylspermidine
8	N^1,N^4-bis(6-quinolinylaminocarbonyl)-N^8-acetylspermidine	N^8-acetylspermidine
9	N^1,N^4,N^8-tris(6-quinolinylaminocarbonyl)spermidine	spermidine
10	N^1,N^4,N^9-tris(6-quinolinylaminocarbonyl)-N^{12}-acetylspermine	N-acetylspermine
11	N^1,N^4,N^9,N^{12}-tetrakis(6-quinolinylaminocarbonyl)spermine	spermine
12	N^1,N^7-bis(6-quinolinylaminocarbonyl)-1,7-diaminoheptane	1,7-diaminoheptane

1.2. The characteristics of QAC derivatives

Synthesis and purification of some selected QAC polyamines on a preparative scale enabled us to confirm the structures from Figure 1 by ^{1}H-NMR and mass spectroscopy [31] and theReferenceore the excellent reactivity of HSCQ with primary and secondary amines. Further the spectroscopic properties of the QAC derivatives were investigated and revealed comparable UV absorptivities with the maximum at 248 nm (cf. Table 2, Figure 2 A). TheReferenceore UV absorption might be an option for detection of QAC polyamines after chromatographic separation, but fluorescence detection is still more sensitive as seen by the good quantum yields (cf. Table 2, Figure 2 A). No shifts of absolute excitation (λ_{ex} = 248 nm) and emission (λ_{em} = 398 nm) maxima were observed, allowing the fluorometric detection of complex polyamine mixtures after precolumn derivatization.

Table 2 UV and fluorescence properties of selected 6-quinolinylaminocarbonyl derivatives: molar absorptivities (ε_{248}) and quantum yields (Q) were measured in methanol at 25 °C (Reproduced with permission from Reference. [31]).

compound	ε_{248}	Q (λ_{ex} = 248 nm)	analyte/internal standard
4	100572	0.203	putrescine
9	121560	0.096	spermidine
11	130684	0.092	spermine
12	117092	0.346	1,7-diaminoheptane

Obviously, fluorescence sensitivity of the QAC polyamines within a chromatographic analysis is dependent on properties of the corresponding mobile phase such as buffer pH and the volume ratio of aqueous solution and organic solvent. This correlates with the behavior of QAC amino acids [36] and by the different peak heights and peak areas of an equimolar amine mixture (cf. Figure 2 B). As described before AQC reacts in contrast with OPA [12-14] with both primary and secondary amines. Not only polyamines but also their acetylated derivatives were labeled and formed adducts, stable at room temperature for several days [31,36]. A standard chromatographic separation of a polyamine calibration mixture with 1,7-diaminoheptane as internal standard is shown in Figure 2 B.

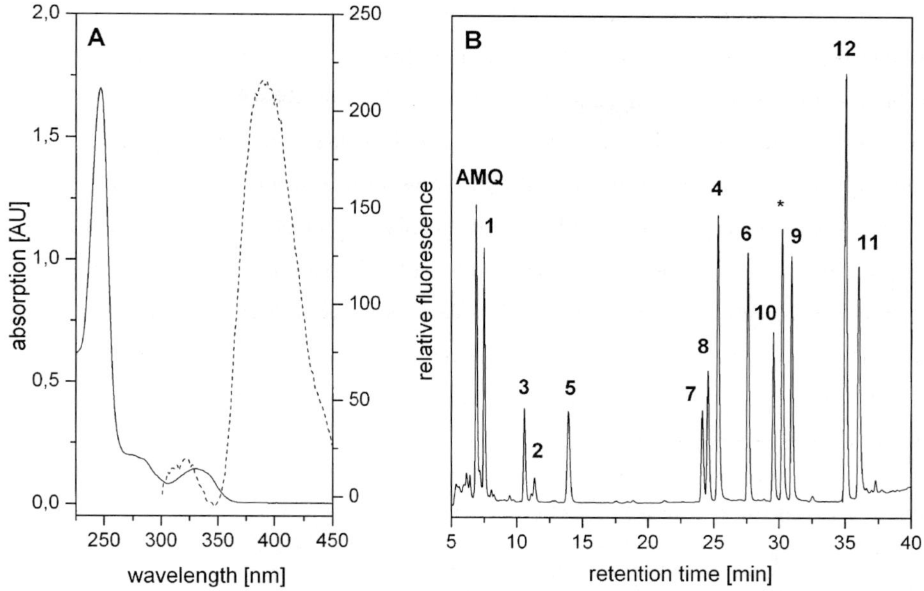

Figure 2 A) Absorption (solid line, 10μM) and fluorescence (dashed line, 0.1μM, λ_{ex} = 248nm) spectra of purified N^1,N^4,N^9,N^{12}-tetrakis(6-quinolinylaminocarbonyl)spermine in methanol are shown. B) Chromatogram of a polyamine calibration mixture and ornithine with 1,7-diaminoheptane as internal standard. Gradient elution (cf. Table 3) of a 10 μl sample containing 6.5 pmol of each component was performed at a flow rate of 1.3 ml/min at 33°C with fluorescence detection (λ_{ex} = 248 nm, λ_{em} = 398 nm). Peaks indicated see Table 1 and *, N,N'-bis(6-quinolinyl)urea, a reagent related by-product (Figure 3).

2. Sample preparation and derivatization

2.1. Extraction of amines/polyamines

Biogenic amines such as polyamines are expressed ubiquitously and are found in extra- as well as intracellular sources. Using AQC as derivatizing reagent they have been determined in food [38,45], cerebrospinal fluid [33], hippocampal slices [32], erythrocytes [30], pneumocystis carinii [47,48], tumour cell lines [31,40,44], primary hepatocytes [49], human colorectal carcinomas [50] and intestinal epithelial cells [51]. In order to make them accessible for a labelling step, firstly, cells and tissues have to be homogenized either by sonication [30,31,48-51], repeated freeze thaw cycles [44] or in a potter homogenizer [45]. After removing cells and cell debris by centrifugation, additionally, soluble proteins in the clear supernatant were separated from amines/polyamines either by boiling [47,48] or acid precipitation [31,44,45,49-51]. Particularly treatment with trichloroacetic acid [31,44,49-51] or perchloric acid [45] was thought to influence the recoveries of N-acetylated polyamines during sample preparation, since acidic hydrolysis might occur [31]. This was investigated using N,N'-hexamethylene-bis-acetamid as a model compound for diacetylated polyamines which are, e.g. excreted in human urine [52,53] and might theoretically interfere after partial hydrolysis. HPLC analysis proved that all mono- and di-acetylated conjugates were stable in presence of 0.2 N trichloroacetic acid (cf. Figure 3).

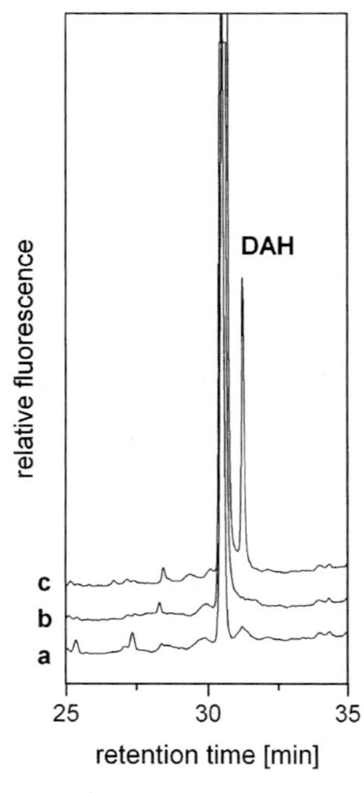

Figue 3 Analysis of diacetylated N,N'-hexamethylene-bis-acetamide a) including and b) excluding a trichloroacetic acid extracting step before derivatization procedure. Stability of acetylated compounds was demonstrated by absence of derivatized products. c) Positive control, derivatized diaminohexane (DAH).

Furthermore, high amounts of polyamine metabolizing enzymes such as polyamine oxidase (EC 1.5.3.3) were observed in pancreas and colon. This enzyme is active in breaking down N-acetylspermine to spermidine and N-acetylspermidine to putrescine. Omitting a major impact of this enzyme activity on polyamine levels it is necessary to use at the time of homogenization either specific inhibitors of polyamine oxidase (such as MDL72527, see 3.3.) or consequently keep solutions at 4°C [31].

2.2. Derivatization reaction

Basic findings concerning the conditions of how AQC reacts with primary and secondary amines was firstly shown for amino acids [36] and later intensively investigated for polyamines [31]. The reaction of AQC with amines (cf. Figure 1) competes with its slower hydrolysis yielding N-hydroxysuccinimide (NHS) and 6-aminoquinoline (AMQ) (cf. Figure 4). To favour a rapid derivatization with amines pH value is a critical value and its optimum for formation of various QAC derivatives was found in the range of pH 8.2 – 9.0 in 0.2 M borate buffer [31]. Compared to precolumn derivatization using dansyl chloride, requiring an optimal pH range around 10.5 [21,23], AQC derivatization can be performed under relative mild alkaline conditions, being of potential advantage, when deproteinization of biological samples is made by acid precipitation. Further, the proposed borate buffer did not interfere with the derivatization reaction in contrast to alternative buffer solutions such as phosphate or Tris producing troublesome fluorescent by-products (Figure 5 B).

A second task to favour reactions of AQC with amines (cf. Figure 1) over competing reactions (cf. Figure 4) is a sufficient excess of AQC reagent guaranteeing the complete derivatization of all amino groups of each analyte. However, in homogenates of biological samples, the amino groups of matrix components such as proteins, peptides, and especially amino acids far exceed the concentration of the derivatization sites of the polyamine analytes. In a model experiment this situation was taken care of, and it could be shown that a 1.5-fold molar excess of AQC with respect to potential derivatization sites, i.e., the total concentration of amino groups in the derivatization mixture, gave >90% fluorescent polyamine derivatives [31]. Considering the fact that homogenates obtained from colon and pancreatic tissue contain approximately 3 mM of free amino acids and that the polyamine concentration does not exceed 10 μM, 20 μl of a 10 mM AQC solution (see 4. Experimental) has proven to be a sufficient amount for complete polyamine derivatization [31].

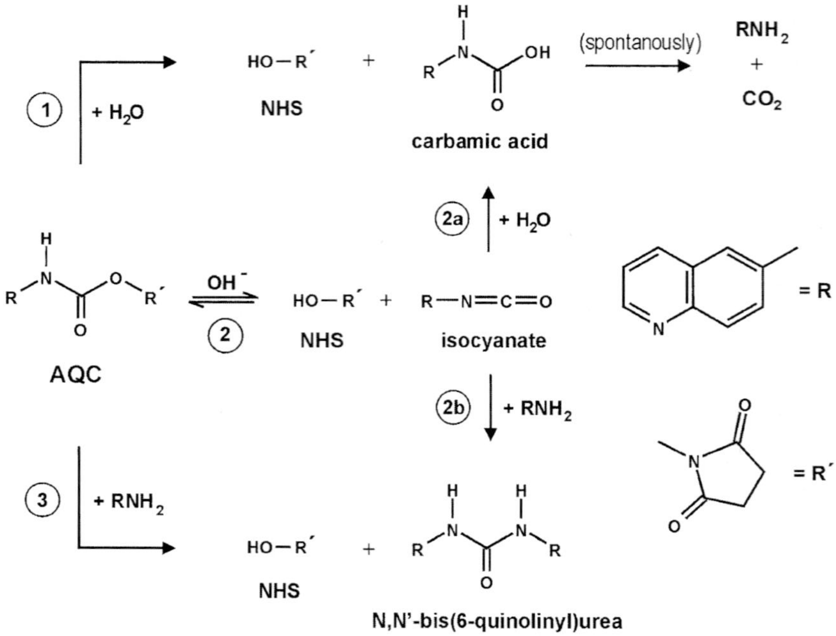

Figure 4 Reactions of AQC, which might occur in buffered aqueous solutions and/or during storage. 1) Hydrolysis of AQC (acid or base catalyzed) results in NHS and 6-quinolinyl carbamic acid. The latter is an unstable compound and breaks down to carbon dioxide and AMQ. 2) Alkaline hydrolysis of AQC eliminates NHS and gives 6-quinolinyl isocyanate. 2a) Catalyzed by acids or bases the addition of water to the carbon-nitrogen double bond gives an N-substituted carbamic acid. 2b) In absence of a basic catalyst, disubstituted urea, N,N'-bis(6-quinolinyl)urea, can be obtained by a nucleophilic addition of AMQ. 3) The primary amine AMQ forms N,N'-bis(6-quinolinyl)urea by a nucleophilic substitution reaction.

The fluorescent reagent AQC reacts very rapidly with amines forming stable QAC derivatives under the chosen conditions, but, additionally, reactions leading to by-products still occur. In aqueous solutions, such as buffers, hydrolysis of AQC leads to AMQ (cf. Figure 4, 1 and 2a), which elutes very early during chromatographic separation (cf. Figure 5 A) under conditions described for polyamine analysis (see 4. Experimental). As reported earlier excess of AQC reagent might be a critical issue if resulting AMQ or NH_3 (trace element in reagents) adducts are not discriminated from derivatized analytes [38]. Further, an additional signal identified as adduct of AMQ with AQC, N,N'-bis(6-quinolinyl)urea, was observed during polyamine analysis [31] which was related to aging of the AQC reagent (cf. Figure 4, 2b and 3; Figure 5 A). Interfering with this by-product can be omitted by adapting the chromatographic conditions (cf. 3.1) or limiting analysis to a few major polyamines using more rigid eluting conditions [30,33,40,47,48].

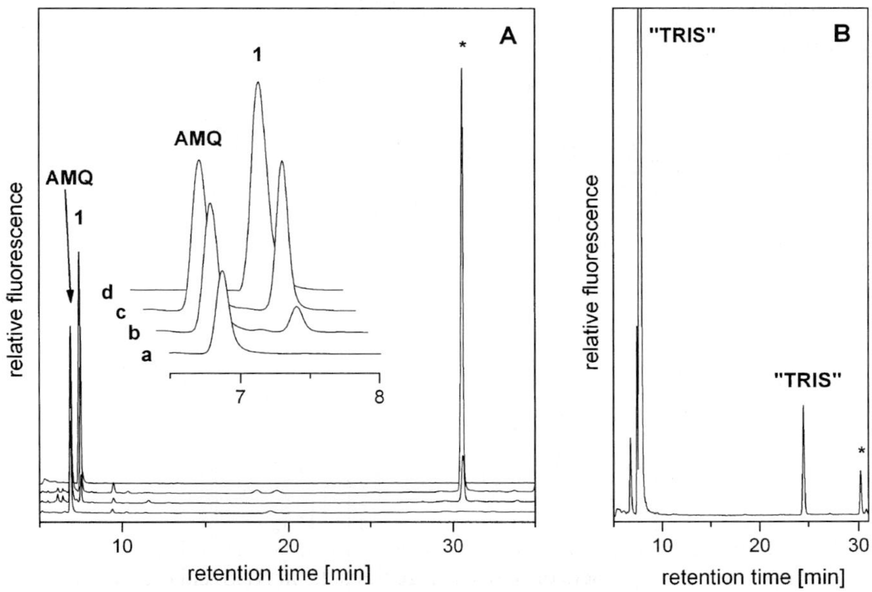

Figure 5 Chromatographic analysis of by-products resulting from AQC derivatization. A) (From bottom to top, insert a - d) a) 6-aminoquinoline (AMQ), b) hydrolysis of AQC in borate buffer pH 8.6, c) derivatization of AMQ with AQC, and d) derivatization of NH_3 (1%) with AQC. B) AQC form adducts with common buffer component tris(hydroxymethyl)aminoethane (TRIS) potentially interfering with analysis of acetylated spermidines. Peaks indicated 1 = (6-quinolinyl)urea and * N,N'-bis(6-quinolinyl)urea.

3. Chromatographic behaviour of QAC derivatives

3.1. Stationary and mobile phase: Chromatographic conditions

The chemical properties of the QAC derivatives stipulate a separation on reversed-phase columns, which are either based on C_8 [30,42,44,48] or C_{18} [31-33,39,40,45] material. Once the QAC amines loaded onto the column they have to be eluted by a gradient mixture yielding sharp and baseline separated peaks. To achieve the required separation of unconjugated [30,38-40,42,44,48] or both free and acetylated polyamines [31,49-51], a simple or a more complex gradient mixture of acidic buffer and organic solvent was used. The acidic nature of solvent A (Table 3, Table 4) and the addition of triethylamine [54,55] were necessary for the complete separation of all analytes, especially of the isomeric N^1- and N^8-acetylspermidine adducts (cf. Figure 2 B, peaks 7 and 8). Elution was accomplished by successive addition of acetonitrile [30,33,39,40,42,44,45] or methanol [32,38] or both [31,49-51].

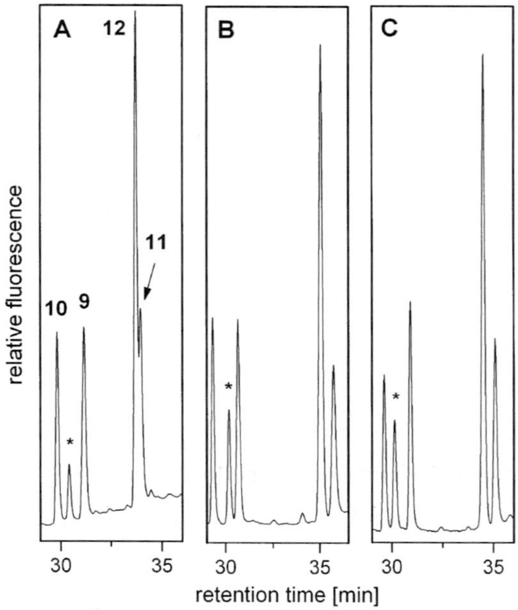

Figure 6: Influence of mobile phase and column temperature on chromatographic separation of derivatized polyamines. Standard gradient elution (Table 3) with A) 30 % and B) 70 % methanol at 34 min and 22°C column temperature. C) Additional increase of column temperature to 33°C. Peaks indicated Referenceer to derivatized polyamines 10 = N-acetylspermine, 9 = spermidine, 12 = 1,7-diaminoheptane, 11 = spermine (cf. Table 1) and * N,N'-bis(6-quinolinyl)urea.

The successful usage of acetonitrile and methanol as eluting solvents to get full separation of QAC derivatives of free and acetylated polyamines (Table 1, Figure 2 B) is demonstrated in Figure 6 A and B. Further, elevated column temperature is an additional tool for QAC peak sharpening and separation (cf. Figure 6 C) [31,38,44].

3.2. Reproducibility study and validation

For linearity determination, dilutions of single amines and amine calibration mixtures were derivatized with AQC and responses of fluorescence detector were plotted against analyte concentrations. Excellent linearity was obtained for (R)-4-oxo-5-phosphononorvaline (0.005 – 1.2 μM, r > 0.9996) [39], epinephrine, norepinephrine, dopamine (1.25 – 50 nmol, r > 0.99) [42], main polyamines (0.66 – 40 pmol, r > 0.99) [30] and both, free and acetylated polyamines (1-100 μM, r > 0.998) [31]. For all these methods 1,7 diaminoheptane, a non physiological compound, was used as internal standard. Detection limit and accuracy studies revealed good values for analysis of the derivatized amines in various matrices, such as plasma and brain dialysate [39], food [38], cell culture [42] and erythrocytes [30]. In more detail, the fluorescence detection limits of QAC derivatives listed in Table 1 were calculated from an injection of 1.83 pmol of each derivatized polyamine ranging from 30 for 1,7 diaminoheptane to 130 fmol for N-acetylcadaverine at a signal to noise ratio of 10 (n = 3) [31]. Due to their high molar absorptivities (cf. Table 2), the labelled polyamines were also amenable to UV detection at 245 nm. For injections of 9.1 pmol of each derivatized polyamine the detection limits were estimated ranging from 1.1 to 5.1 pmol at S/N = 10 (n =3). Accuracy of the method was determined by measuring the responses of 10 identical samples from a pool of derivatized calibration mixtures (cf. Table 1). Excellent reproducibility was found for both retention times and peak areas, with coefficients of variation ranging from 0.1 to 0.9% and from 1.8 to 3.8%, respectively.

Furthermore, recovery values were determined for biogenic amines in food (90 – 100%) [38], plasma and brain dialysate (102 – 114%) [39] and for polyamines in erythrocytes (putrescine 97 ± 4%, spermidine 94 ± 5%, spermine 98 ± 3%) [30]. Recovery studies with tissue homogenates of pancreas and colon revealed high intrinsic activities of polyamine metabolizing enzymes, which had some impact on recovery values (see 2.1.): N-acetylated polyamines ranging from 79 – 89%, 1,7 diaminoheptane, putrescine, spermidine, spermine, from 96 to 78% [31].

3.3. Application

As mentioned before biogenic amines, mainly polyamines were successfully analyzed using AQC as derivatizing reagent in various tissues and cell lines (cf. 2.1.). Applying the derivatization procedure and the chromatographic conditions described in the Experimental section (cf. 4.) chromatograms were obtained as demonstrated in Figure 7. Due to its extremely high sensitivity, the method has the advantage that very small samples containing as little as 5000 cells are sufficient for accurate analysis. Additionally, the method was applied to solid tumors and despite many co-eluting substances (especially amino acids, cf. Figure 7) the most important polyamines were excellently resolved [31,50].

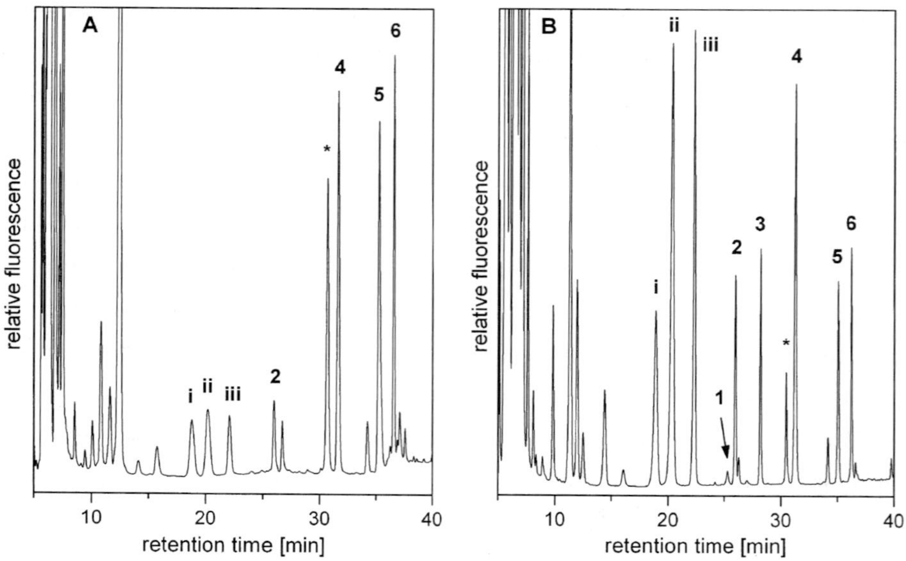

***Figure** 7* Chromatographic separation of derivatized polyamines and internal standard measured in A) cultured pancreatic tumour cells and B) colorectal carcinoma sample. Mainly non acetylated polyamines are detected with cadaverine in B) potentially originating from bacteria. Additionally, signals of amino acids are indicated by i – iii and peaks Referenceerring to 1 = N^8-acetylspermidine, 2 = putrescine, 3 = cadaverine, 4 = spermidine, 5 = 1,7 diaminoheptane, 6 = spermine and * N,N'-bis(6-quinolinyl)urea.

Good reproducibility and linearity of QAC polyamines encouraged researchers to develop assays for the determination of ornithine decarboxylase [47], polyamine oxidase [48] and spermidine/spermine acetyltransferase [48] by monitoring putrescine and N^1-acetylspermidine concentrations, respectively. Further, AQC was used to label S-adenosylmethionine, a key molecule in methylation reactions and polyamine biosynthesis, and a

HPLC method for its determination was established [41]. In addition, catecholamines in parasite *Entamoeba* were analyzed by RP-HPLC with AQC precolumn derivatization [42].

A major step forward to elucidate the metabolic pathways of polyamines was the finding and synthesis of specific inhibitors of polyamine metabolism [3,10,56-58]. Further, the identification and increased polyamine concentrations in a variety of diseases from cancer and psoriasis to parasitic infections has led to the hypothesis that manipulation of polyamine metabolism is a realistic target for therapeutic of preventative intervention in the treatment of certain diseases [5,58]. Some of these specific inhibitors (cf. Figure 8) have primary or secondary amines and theReferenceore are amenable to AQC derivatization.

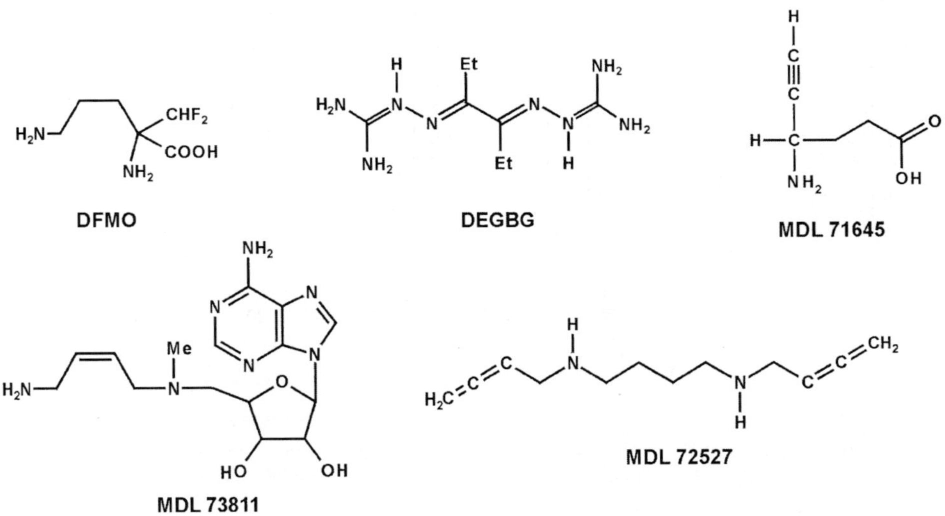

Figure 8 Specific inhibitors of polyamine metabolism: DFMO, α-difluoromethylornithine; DEGBG, diethylglyoxal bis(guanylhydrazone); MDL 71645, amino-4-hex-5-ynoic acid; MDL 72527, N,N'-bis(2,3-butadienyl)-1,4-butanediamine; MDL 73811, {[(Z)-4-amino-2-butenyl]methylamino}-5'-desoxyadenosine

DFMO, α-difluoromethylornithine, an enzyme activated irreversible inhibitor of ornithine decarboxylase effectively blocks synthesis of putrescine [5,58-62]. MDL 72527, N,N'-bis(2,3-butadienyl)-1,4-butanediamine, inhibits the FAD dependent polyamine oxidase resulting in strongly reduced degradation of N^1-acetylspermine/N^1-acetylspermidine to spermidine and putrescine, respectively [57,63-65]. MDL 71645, amino-4-hex-5-ynoic acid, blocks the gamma-aminobutyric acid transaminase (EC 2.6.1.19) disabling transamination of ornithine [66,67]. MDL 73811, {[(Z)-4-amino-2-butenyl]methylamino}-5'-desoxyadenosine, an enzyme activated irreversible inhibitor [68-72] and DEGBG, diethylglyoxal bis(guanylhydrazone) [73-75], were shown to block S-adenosylmethionine decarboxylase yielding an inhibited synthesis of spermidine and putrescine.

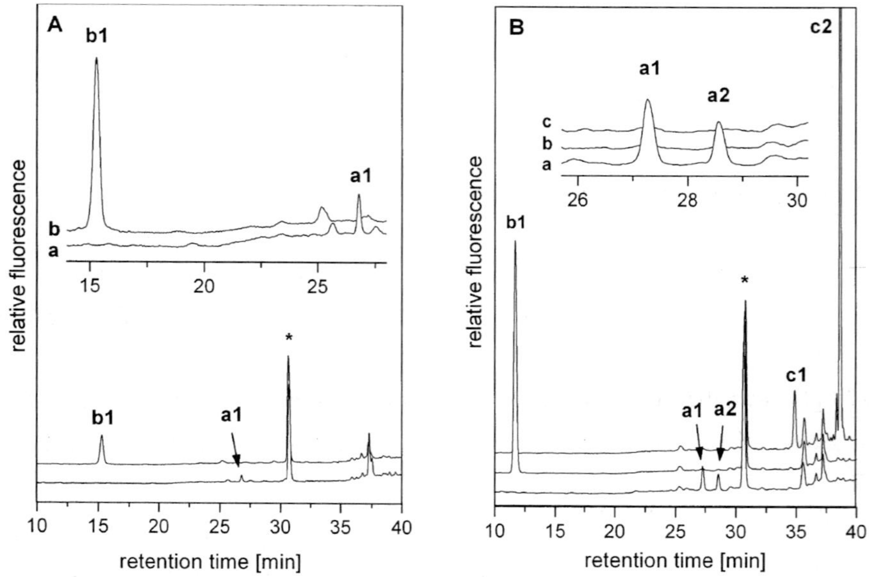

Figure 9: Chromatographic analysis of specific inhibitors of polyamine metabolism. A) Standard derivatization (for polyamines) of DEGBG (a) and DFMO (b) with AQC leads to signals (a1 and b1), which do not affect quantification of polyamines and acetylated polyamines. B) Derivatization of MDL 71645 (b) results in one signal (b1), whereas using MDL 73811 (a) and MDL 72527 (c) two signals each (a1 and a2, c1 and c2) are obtained corresponding to mono and di-derivatized compounds under standard conditions. Signal c2 might interfere with the internal standard and impair correct polyamine quantification. * N,N'-bis(6-quinolinyl)urea

Application of these inhibitors in animal or cell culture studies followed by a polyamine analysis using AQC derivatization revealed specific QAC derivatives during chromatographic analysis (cf. Figure 9). Due to their number of primary and secondary amino groups one or two signals were obtained for DFMO, DEGBG, MDL 71645 or MDL 73811 and MDL 72527, respectively. Comparing their retention times with a standard calibration mixture (cf. Figure 2 B) demonstrated that presumably the di-derivatized MDL 72527 would interfere with 1,7 diaminoheptane. It would be of great interest to adapt the derivatization procedure and the chromatographic conditions to evaluate the suitability of AQC as a derivatization reagent for the analysis of these inhibitors. Already currently available is HPLC technique for the quantification of DFMO in biological fluids or tissue/cell homogenates involving either pre- or postcolumn derivatization, with ultraviolet or fluorescence detection [76-79].

4. Experimental

4.1. Derivatization of polyamines with AQC

A polyamine stock solution containing 4.55 µmol/l of each polyamine (putrescine, N-acetylputrescine, cadaverine, N-acetylcadaverine, spermidine, N^1-acetylspermidine, N^8-acetylspermidine, spermine, N-acetylspermine, 1,7-diaminoheptane) was prepared in 0.2 M borate buffer pH 8.5. AQC solution (10 mM) consisted of 2.85 mg AQC dissolved in 1 ml anhydrous acetonitrile and was stored in glass tubes. To 10 µl of the polyamine stock solution 40 µl of 0.2 M borate buffer (pH 8.8) and 20 µl of the 10 mM AQC solution were added and gently mixed at room temperature. For standard chromatograms 10 µl of this calibration mixture (i.e. 6.5 pmol per compound) were injected onto the HPLC column.

From cell culture: Cells were trypsinized, resuspended in 50 mM borate buffer pH 7.4 containing 0.1 mM EDTA, counted in a hemocytometer, and homogenized by sonication (step 2, 3 bursts for 10 seconds). After centrifugation (13,000 g for 10 min, 4°C) and mixing of 50 µl of the supernatant with 10 µl of internal standard (60 µM 1,7 diaminoheptane in 0.2 M borate buffer pH 8.8), 30 µl of 0.6 M trichloroacetic acid were added for deproteinization. After a second centrifugation step (13,000 g for 10 min), 10 µl of the cell extract were adjusted to pH 8.6 with 30 µl of 0.2 M borate buffer, pH 9.3. This mixture was derivatized at room temperature by mixing with 20 µl of the 10 mM AQC solution and injected.

From tissue. Samples of histologically proven tumors and normal tissue, as well as specimens from animals were rapidly dissected after surgical removal and immediately snap frozen in liquid nitrogen. The tissue samples were stored at - 80°C and assayed for polyamine content within 2 weeks. 100-150 mg specimens were homogenized with a micro dismembrator (B. Braun,

Melsungen, GER), dissolved in 50 mM borate buffer pH 7.4 containing 0.1 mM EDTA and processed as described above for cell culture samples. For protein determination, the BCA assay (Pierce, Rockford, USA) was used with bovine serum albumin as Referenceerence.

4.2. Chromatography

The HPLC system consisted of a Merck-Hitachi L-6200A pump, equipped with an AS-2000A autosampler, a L-4000A UV and a F-1050 fluorescence spectrophotometer (Merck, Darmstadt, GER). D-7000 Manager software (Merck, Darmstadt, GER) was used to control system operation and data collection. All separations were performed on a LiChroCART column (250 mm x 4 mm I. D.) fitted with a guard column (4 mm x 4 mm I.D.), both packed with 5-µm LiChrospher 100 RP-18 endcapped material (Merck, Darmstadt, GER). The columns were mounted in a Hitachi Peletier column oven (Merck, Darmstadt, GER).

Complete separation of polyamines and their monoacetylated derivatives. The mobile phase consisted of solvent A (70 mM acetic acid / 25 mM triethylamine, titrated with phosphoric acid to pH 4.82), solvent B (acetonitrile / water 80:20 (v/v)) and solvent C (methanol). Optimized gradient analyses (cf. Table 3 for gradient composition) of 10 µl samples were performed at a flow rate of 1.3 ml/min at 33 °C with column temperature online UV (λ=254 nm) and fluorescence (λ_{Ex}= 248 nm, λ_{Em}=398 nm) detection.

Table 3 Gradient composition for the separation of polyamines and their acetylated derivatives (including N^1/N^8-acetylspermidine).

elution time [min]	solvent A [%]	solvent B [%]	solvent C [%]
0.0	100	0	0
1.0	78	22	0
15.0	78	22	0
27.0	55	39	6
27.5	53	33	14
34.0	20	10	70
37.0	0	100	0
57.0	0	100	0

Alternative method. Sensitivity of the chromatographic analysis is enhanced, if the mobile phase and the gradient profile were adjusted as follows. The mobile phase consisted of solvent A (70 mM acetic acid / 25 mM triethylamine, titrated with phosphoric acid to pH 5.6), solvent B (acetonitrile / water 80:20 (v/v)) and solvent C (methanol). Gradient analyses according to Table 4 were performed under conditions as described above. As a consequence of these alterations the QAC derivatives of N^1- and N^8-acetylspermidine were not separated.

Table 4. Gradient composition for the separation of polyamines and acetylated derivatives (excluding N^1/N^8-acetylspermidine).

elution time [min]	solvent A [%]	solvent B [%]	solvent C [%]
0.0	100	0	0
1.0	78	22	0
15.0	78	22	0
20.0	65	35	0
27.0	60	40	0
34.0	20	10	70
37.0	0	100	0
57.0	0	100	0

References

[1] C.W. Tabor, H. Tabor, Annu. Rev. Biochem. 53 (1984) 749-790.

[2] A.E. Pegg, P.P. McCann, Am. J. Physiol. 243 (1982) C212-21.

[3] A.E. Pegg, Cancer Res. 48 (1988) 759-774.

[4] D. Teti, M. Visalli, H. McNair, J. Chromatogr. B 781 (2002) 107-149.

[5] E.W. Gerner, F.L. Meyskens, Jr., Nat. Rev. Cancer 4 (2004) 781-792.

[6] T. Thomas, T.J. Thomas, J Cell Mol. Med. 7 (2003) 113-126.

[7] K. Stromgaard, I. Mellor, Med. Res. Rev. 24 (2004) 589-620.

[8] H.M. Wallace, A.V. Fraser, A. Hughes, Biochem J 376 (2003) 1-14.

[9] A. Gugliucci, Clin. Chim. Acta 344 (2004) 23-35.

[10] A.E. Pegg, Biochem. J. 234 (1986) 249-262.

[11] N. Seiler, J. Chromatogr. 379 (1986) 157-176.

[12] N. Seiler, B. Knödgen, J. Chromatogr. 339 (1985) 45-57.

[13] C. Loser, U. Wunderlich, U.R. Folsch, J. Chromatogr. 430 (1988) 249-262.

[14] T. Hyvonen, T.A. Keinanen, A.R. Khomutov, R.M. Khomutov, T.O. Eloranta, J. Chromatogr. 574 (1992) 17-21.

[15] S. Wongyai, P.J. Oefner, G.K. Bonn, J. Liquid Chromatogr. 12 (1989) 2249-2261.

[16] P.J. Oefner, S. Wongyai, G. Bonn, Clin. Chim. Acta 205 (1992) 11-18.

[17] S. Watanabe, T. Saito, S. Sato, S. Nagase, S. Ueda, M. Tomita, J. Chromatogr. 518 (1990) 264-267.

[18] E. Schenkel, V. Berlaimont, J. Dubois, M. Helson Cambier, M. Hanocq, J. Chromatogr. B 668 (1995) 189-197.

[19] R. Romero, D. Gazquez, M.G. Bagur, M. Sanchez-Vinas, J. Chromatogr. A 871 (2000) 75-83.

[20] M. Venza, M. Visalli, D. Cicciu, D. Teti, J. Chromatogr. B 757 (2001) 111-117.

[21] C. Stefanelli, D. Carati, C. Rossoni, J. Chromatogr. 375 (1986) 49-55.

[22] J.M. Mates, J. Marquez, M. Garcia Caballero, I. Nunez de Castro, F. Sanchez Jimenez, Agents Actions 36 (1992) 17-21.

[23] P.M. Kabra, H.K. Lee, W.P. Lubich, L.J. Marton, J. Chromatogr. 380 (1986) 19-32.

[24] S. Fu, X. Zou, X. Wang, X. Liu, J. Chromatogr. B 709 (1998) 297-300.

[25] N.M. Tamim, L.W. Bennett, T.A. Shellem, J.A. Doerr, J Agric. Food Chem. 50 (2002) 5012-5015.

[26] M.I. Sabri, A.I. Soiefer, G.E. Kisby, P.S. Spencer, J. Neurosci. Methods 29 (1989) 27-31.

[27] F. Bellagamba, V.M. Moretti, T. Mentasti, A. Albertini, U. Luzzana, F. Valfre, J. Chromatogr. A. 791 (1997) 79-84.

[28] T. Ekegren, C. Gomes-Trolin, I. Nygren, H. Askmark, J Neurol. Sci. 222 (2004) 49-53.

[29] V. Lozanov, S. Petrov, V. Mitev, J. Chromatogr. A 1025 (2004) 201-208.

[30] S. Merali, A.-B.-J. Clarkson, J. Chromatogr. B 675 (1996) 321-326.

[31] T. Weiss, G. Bernhardt, A. Buschauer, K.W. Jauch, H. Zirngibl, Anal. Biochem. 247 (1997) 294-304.

[32] D.A. Gibson, B.R. Harris, M.A. Prendergast, S.R. Hart, J.A. Blanchard, R.C. Holley, N.W. Pedigo, J.M. Littleton, Alcohol Clin. Exp. Res. 27 (2003) 1099-1106.

[33] M. Vivo, L. Camon, N. de Vera, E. Martinez, Neurosci. Lett. 330 (2002) 74-78.

[34] L. Gerbaut, Clin. Chem. 37 (1991) 2117-2120.

[35] H.M. van Eijk, D.R. Rooyakkers, N.E. Deutz, . J. Chromatogr. A 730 (1996) 115-120.

[36] S.A. Cohen, D.P. Michaud, Anal. Biochem. 211 (1993) 279-287.

[37] http://www. iupac.org/nomenclature/index.html.

[38] O. Busto, J. Guasch, F. Borull, Determination of biogenic amines in wine after precolumn derivatization with 6-aminoquinolyl-N-hydroxysuccinimidyl carbamate. J Chromatogr. A 737, 205-213. 1996.

[39] K.E. Cornelius, G.M. Fadayel, B.M. Baron, C.J. Schmidt, K.D. Haegele, T.M. Chen, J Pharm. Biomed. Anal. 14 (1995) 143-150.

[40] J. Satriano, S. Matsufuji, Y. Murakami, M.J. Lortie, D. Schwartz, C.J. Kelly, S. Hayashi, R.C. Blantz, J Biol. Chem. 273 (1998) 15313-15316.

[41] S. Merali, D. Vargas, M. Franklin, A.B. Clarkson, Jr., J Biol. Chem. 275 (2000) 14958-14963.

[42] A. Coppi, S. Merali, D. Eichinger, J Biol. Chem. 277 (2002) 8083-8090.

[43] G. Liu, J. Chen, Y. Ma, J Chromatogr. B 805 (2004) 281-288.

[44] A.L. Miller, B.H. Johnson, R.D. Medh, C.M. Townsend, E.B. Thompson, Neoplasia. 4 (2002) 68-81.

[45] G.E. Berge, H. Sveier, E. Lied, Effects of feeding Atlantic salmon (Salmo salar L.) imbalanced levels of lysine and arginine. Aquaculture Nutrition 8, 239-248. 2002.

[46] T. Weiss, E. Heinmoeller, A. Buschauer, G. Bernhardt, R. Gruber, J. Rueschoff, H. Zirngibl, A new HPLC-Method for the determination of polyamines (PA) in the subpicomolar range: PA and ornithine-decarboxylase (ODC) in human pancreatic carcinoma cell lines. Eur. J. Cell Biol. 69, 471. 1996.

[47] S. Merali, A.-B.-J. Clarkson, Antimicrob. Agents Chemother. 40 (1996) 973-978.

[48] S. Merali, J Biol. Chem. 274 (1999) 21017-21022.

[49] T.S. Weiss, B. Jahn, M. Cetto, K.W. Jauch, W.E. Thasler, Cell Prolif. 35 (2002) 257-267.

[50] T.S. Weiss, G. Bernhardt, A. Buschauer, W.E. Thasler, D. Dolgner, H. Zirngibl, K.W. Jauch, Int. J. Colorectal Dis. 17 (2002) 381-387.

[51] T.S. Weiss, H. Herfarth, F. Obermeier, J. Ouart, D. Vogl, J. Scholmerich, K.W. Jauch, G. Rogler, Inflamm. Bowel. Dis. 10 (2004) 529-535.

[52] M. Sugimoto, K. Hiramatsu, S. Kamei, K. Kinoshita, M. Hoshino, K. Iwasaki, M. Kawakita, J Cancer Res. Clin. Oncol. 121 (1995) 317-319.

[53] K. Hiramatsu, M. Sugimoto, S. Kamei, M. Hoshino, K. Kinoshita, K. Iwasaki, M. Kawakita, J Cancer Res. Clin. Oncol. 123 (1997) 539-545.

[54] A. Duchateau, M. Crombach, M. Aussems, J. Bongers, J. Chromatogr. 461 (1989) 419-428.

[55] R.W. Roos, C.A. Lau Cam, J. Chromatogr. 370 (1986) 403-418.

[56] A.E. Pegg, P.P. McCann, Pharmacol. Ther. 56 (1992) 359-377.

[57] N. Seiler, Prog. Brain Res. 106 (1995) 333-344.

[58] H.M. Wallace, A.V. Fraser, Amino. Acids 26 (2004) 353-365.

[59] B.W. Metcalf, P. Bey, C. Danzin, M.J. Jung, P. Casara, J.-P. Vevert, J Am. Chem. Soc 100 (1978) 2551-2553.

[60] P.S. Mamont, M.C. Duchesne, J. Grove, P. Bey, Biochem. Biophys. Res. Commun. 81 (1978) 58-66.

[61] H.W. Herr, E.L. Kleinert, P.S. Conti, J.H. Burchenal, W.F. Whitmore, Jr., Cancer Res. 44 (1984) 4382-4385.

[62] W.B. Harris, V.B. Grossie, D.M. Ota, K. Nishioka, J.A. Ajani, T. Chang, D. Patenia, J. Surg. Res. 38 (1985) 592-598.

[63] R.A. Casero, Jr., A.E. Pegg, FASEB J. 7 (1993) 653-661.

[64] P. Bey, F.N. Bolkenius, N. Seiler, P. Casara, J. Med. Chem. 28 (1985) 1-2.

[65] N. Seiler, B. Duranton, F. Raul, Prog. Drug Res. 59 (2002) 1-40.

[66] F.N. Bolkenius, P. Bey, N. Seiler, Biochim. Biophys. Acta 838 (1985) 69-76.

[67] N. Seiler, S. Sarhan, C. Grauffel, R. Jones, B. Knodgen, J.P. Moulinoux, Cancer Res. 50 (1990) 5077-5083.

[68] P. Casara, P. Marchal, J. Wagner, C. Danzin, J Am. Chem. Soc 111 (1989) 9111-9113.

[69] C. Danzin, P. Marchal, P. Casara, Biochem. Pharmacol. 40 (1990) 1499-1503.

[70] A.J. Bitonti, T.L. Byers, T.L. Bush, P.J. Casara, C.J. Bacchi, A.B. Clarkson, Jr., P.P. McCann, A. Sjoerdsma, Antimicrob. Agents Chemother. 34 (1990) 1485-1490.

[71] T.L. Byers, B. Ganem, A.E. Pegg, Biochem. J. 287 (1992) 717-724.

[72] T.L. Byers, R.S. Wechter, R.H. Hu, A.E. Pegg, Biochem. J. 303 (1994) 89-96.

[73] H. Elo, I. Mutikainen, L. Alhonen Hongisto, R. Laine, J. Janne, Cancer Lett. 41 (1988) 21-30.

[74] F. Svensson, I. Kockum, L. Persson, Mol. Cell Biochem. 124 (1993) 141-147.

[75] Q. Yuan, M.J. Viar, R.M. Ray, L.R. Johnson, Am. J Physiol Gastrointest. Liver Physiol 278 (2000) G49-G56.

[76] J.L. Cohen, R.J. Ko, A.T. Lo, M.D. Shields, T.M. Gilman, J Pharm. Sci. 78 (1989) 114-116.

[77] N.D. Huebert, J.J. Schwartz, K.D. Haegele, J Chromatogr. A 762 (1997) 293-298.

[78] M.L. Kilkenny, M. Slavik, C.M. Riley, J.F. Stobaugh, J Pharm. Biomed. Anal. 17 (1998) 1205-1213.

[79] W. Hanpitakpong, B. Kamanikom, V. Banmairuroi, K. Na-Bangchang, J Chromatogr. B 788 (2003) 221-231.

2.3. Quantitation of Amines by Capillary Electrophoresis/ Capillary Chromatography

Ibolya Molnár-Perl (Editor)
Quantitation of Amino Acids and Amines by Chromatography
Journal of Chromatography Library, Vol. 70

2.3.1

Determination of Underivatized Amines by Capillary Electrophoresis and Capillary Electrochromatography

Christian W. Klampfl

Contents

Summary

Electroseparation methods are perfectly applicable for the analysis of relatively polar compounds such as underivatized amines. These systems provide a series of advantages over the commonly employed chromatographic methods such as the more straightforward possibility of miniaturization and integration and low cost of analysis. If the analysis of amines is performed without a precedent derivatization step additional benefits like the avoidance of all types of pitfalls related to this procedure can be achieved. Unfortunately the analysis of amines in their underivatized form also generates some new problems such as the necessity to separate compounds with often quite similar structures and the need to find detection systems providing sufficient sensitivity for these compounds. The capillary electroseparation method most commonly employed for underivatized amines still is capillary zone electrophoresis and up to now a substantial number of reports exist, describing the application of this technique for a variety of amines in different of matrices ranging from biological fluids to environmental samples. Detection is mostly performed using spectrophotometric detectors (mainly because of their presence in most commercially available instruments) in particular UV-vis detectors either in the indirect mode for simple amines or in the direct mode for heterocyclic or aromatic amines. Some work describing the use of electrochemical detection (mostly conductivity detection) also exists. Optimum choice for the detection of underivatized amines, especially due to their easy ptonizability, is the use of a mass spectrometric detector. This can also be seen from the increasing number of publications dealing with this issue.

1. Introduction

Amines are a class of chemical substances that are of significant concern with respect to a series of chemical and biochemical processes. For this reason their determination in a wide range of matrices including biological fluids and tissues [1], natural products [2], foods and food related products [3,4] or environmental samples [5] is of particular importance. For this reason numerous publications deal with their analysis and almost all types of analytical methods have to some degree been tested with respect to their suitability regarding the determination of these compounds. Similar as in the case of amino acids, the high polarity of amines can be seen as a problem with respect to their separation whereas the lack of strong chromophore groups (except for aromatic or heterocyclic amines) complicates their detection. Therefore the majority of analytical methods for amines include a derivatization step, to transform these analytes into derivatives facilitating both their separation as well as their detection. A wide range of derivatization reagents for amine analysis can be found in the literature and research in this field is still going on [6,7]. Unfortunately derivatization is also affected with some disadvantages including derivative instability, interferences caused by the reagent or time consuming derivatization procedures. To avoid these problems, methods allowing the analysis of amines without derivatization are of significant interest.

2. Electroseparation methods for the analysis of underivatized amines

2.1 Electrophoretic methods

Electrophoretic methods are perfectly suited for the separation of charged analytes showing a specific electrophoretic mobility in the electric field, such as amines [8]. Some basic principles with respect to the separation mechanism involved can be found in the chapter about the analysis of underivatized amino acids and will therefore not be discussed further in this section. As amines are basic compounds they are usually separated in their cationic form employing acidic carrier electrolytes favoring the protonation of the amino group. If bare fused silica capillaries are used the electroosmotic flow (EOF) generated by the silanol groups of the capillary surface is directed towards the cathodic end of the separation capillary. So the mobility of the EOF points in the same direction as the electrophoretic mobility of the amines, helping to decrease analysis times. A principle drawback of electrophoretic methods is that due to their simple separation mechanism only substances with a different charge/ionic radius ratio can be distinguished. For this reason additional interactions have to be introduced if the analysis of substances with a very similar structure or even of enantiomeric compounds is required.

2.2. Electrochromatographic methods

Electrochromatographic methods combine electrophoretic with chromatographic principles, introduced by an additional pseudo-stationary or true stationary phase. Methods including pseudo-stationary phases are micellar electrokinetic chromatography (MEKC) [9] and microemulsion electrokinetic chromatography (MEEKC) [10]. Some additional information on the basic principles of operation of these techniques can be found in the chapter dealing with the analysis of amino acids. Both techniques, MEKC [11,12] and MEEKC [13] have also been applied to the separation of amines. True stationary phases can be found in capillary electrochromatography (CEC). The two major modes of CEC, namely, open tubular (OT)-CEC, where only the capillary walls are coated with a stationary phase and CEC with fully packed or monolithic columns have both been used in the analysis of underivatized amines.

3. Detection of underivatized amines

Detection is a crucial point in chemical analysis, co-determining essential parameters like sensitivity or selectivity of the method employed. High sensitivity is generally desired whereas the wish for a universal detection method suitable for a wide range of solutes even with completely different chemical properties is often surpassed by the necessity for high selectivity, allowing the determination of a single analyte in a complex sample matrix. Focusing on the detection of underivatized amines, it can be seen from the literature that a series of different detection methods has been employed in combination with electromigration methods for these analytes. Those include spectrophotometric techniques (mainly direct and indirect UV-detection), chemiluminescensce detection, electrochemical detection and mass spectrometric detection.

3.1 Spectrophotometric methods

Focusing on commercially available instrumentation for CE it can be seen that the majority of devices are equipped with spectrophotometric units in particular UV-vis detectors. Unfortunately this type of detection, similar as in the case of amino acids, is not well suited for most underivatized amines (except the aromatic species) as they do not show substantial absorbency in the UV-vis region. Whereas for amino acids direct detection at very low wavelengths is possible, due to the presence of the carboxylate group, this is not the case for simple amines. A possible way to overcome this obstacle is the use of indirect UV detection which is a very common technique in electromigration methods if it comes to the analysis of non-absorbing solutes like amines [14]. Indirect detection is based on the use of a carrier electrolyte with an UV absorbing probe, so a decrease in absorbency is monitored as the analytes pass the detection unit. A typical example for an electropherogram obtained with such a setup is shown in Figure 1 showing the separation of volatile corrosion inhibitors employing an electrolyte system typical for the indirect detection of amines. In this case detection is based on the UV-absorbency of imidazole at 214 nm which is attenuated by the non-absorbing analyte ions that are (according to the law of electroneutrality) displacing the probe ions. Fluorescence detection and in particular laser induced fluorescence (as mainly used in combination with capillary electroseparation methods) generally provides improved sensitivity and selectivity compared to methods based on absorbency [16]. Unfortunately native fluorescence can only be found in a small number of amines such as tryptamine [17,18] whereas the majority of this class of compounds (similar as in the case of UV detection) is only accessible via indirect fluorescence detection [19,20]. Chemiluminescence detection, a technique that has also been employed for underivatized amines, can be seen as the emission of radiation in the UV-vis range originating from a molecule in an electronically excited state produced by a chemical reaction at ordinary temperature without any associated generation of heat [21]. A simple and cost effective way to use chemiluminescence for the analysis of amines is electro-chemiluminescence with Tris(2,2′-bipyridyl)ruthenium [22-24].

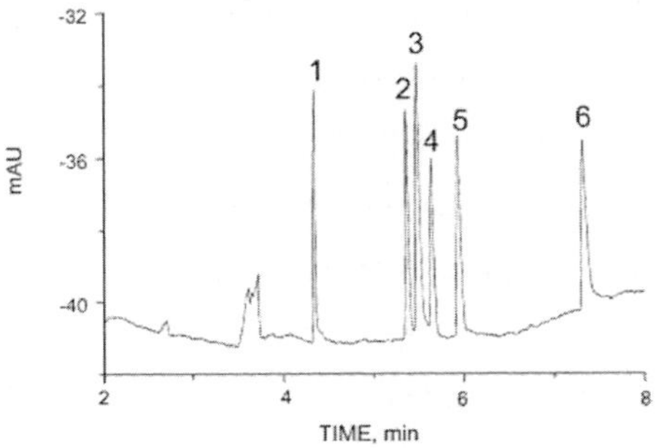

Figure 1 Optimized separation of volatile corrosion inhibitors by CZE. Conditions: electrolyte, 10 mM imidazole, 10 mM HIBA, 10 mM 18-C-6 (pH 4.3); separation voltage 18 kV; capillary, 53.5 cm x 75 μm i.d.; detection, direct UV at 214 nm; peaks: 1, monoethanolamine; 2, diethanolamine; 3, monocyclohexylamine; 4, diethylethanolamine; 5, triethanolamine; 6, dicyclohexylamine; from [15] with permission.

3.2. Electrochemical methods

Electrochemical detection methods are widely employed in the determination of non-UV-absorbing analytes, as they do not rely on the presence of chromophores, unlike spectrophotometric methods. An additional benefit, in particular in the combination with capillary or even chip separation methods, is the fact that in the case of electrochemical detectors sensitivity does not depend on the size of the detection cell. Unfortunately (as already mentioned in the chapter about amino acids) no commercially available instrumentation for CZE provides the possibility of electrochemical detection. Although this means that lab-made devices have to be designed and built, the advantages of these detection methods in combination with capillary electroseparation methods have been frequently utilized as can be seen from a variety of reports in the recent literature [25-28].

The type of electrochemical detector most widely used is the conductivity detector. Its basic principle as well as some technical aspect have already been discussed in the chapter about amino acids and will not be addressed further in this section. Due to its universality and its simplicity conductometric detection is well suited for the determination of underivatized amines after their separation with electroseparation methods [29-32], although the number of papers published on this issue is definitely not as high as in the case of amino acids. Besides conductometric detection also other modes of electrochemical detection in particular amperometric detection have been used for underivatized amines, as can be seen from several papers available on this topic [33-39].

3.3. Mass spectrometric detection

Over the last decade mass spectrometric (MS) detection has become increasingly important in combination with electroseparation methods. Particularly the development of electrospray ionization (ESI) can be seen as a great incentive when it comes to the hyphenation of MS with liquid phase separation methods. The development of commercially available interfaces for CZE-MS can be seen as a further step towards an increased use of this hyphenated method also in routine analysis. Examples for comprehensive review papers on this topic published within the last few years can be found in the literature [40-43]. Due to the easily protonizable amino group, underivatized amines are exceptionally well suited for detection with ESI-MS in particular when acidic carrier electrolytes are used. MS detection provides a series of advantages such as high selectivity, allowing the evaluation of overlapping or even co-migrating signals. Whereas in the early days most combinations of CZE and MS were restricted to simple single quadrupol MS instruments, mostly operated in the selected ion monitoring mode to achieve sufficient sensitivity, a more recent development is the increasing use of multistage MS detectors. This allows to obtain a significantly higher amount of information with respect to structure elucidation or just an increased certainty in peak assignment. An interesting example for the advantages of such a setup can be seen in Figure 2. Heterocyclic amines in an acid hydrolyzed urine sample have been separated and identified employing full scan MS and to improve the quality of peak assignment, additional product ion scan experiments [44].

4. Applications

In the following section applications of electrophoretic and electrochromatographic methods for the separation of amines are discussed according to the nature of the investigated samples as well as according to the presence of stereoisomers that have to be resolved. In addition to the information included in the text below, Tables 1 and 2 give an overview of these applications in the order of their publication date.

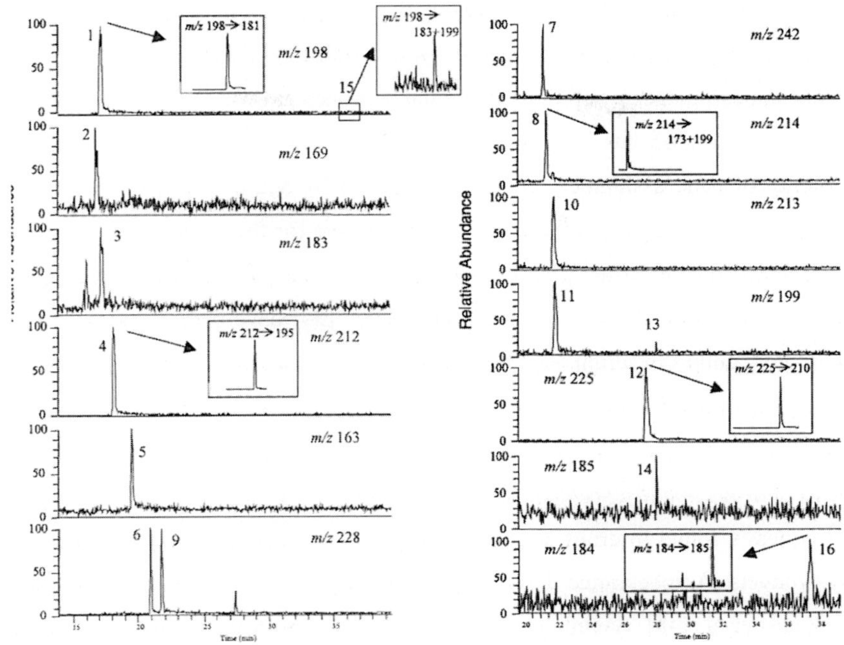

Figure 2 CE-MS (full scan) and CE-MS-MS (product ion scan) of an acid hydrolyzed spiked human urine (~ 45 ng ml^{-1}). Conditions: electrolyte, 40 mM formic acid/ammonium formate (pH 4.5) in 60 % MeOH; separation voltage 28 kV; capillary, 86 cm x 50 μm i.d.; detection, MS; peaks: 1, 3-amino-1-methyl-5H-pyrido[4,3-b]indole; 2, 9H-pyrido[4,3-b]indole; 3, 1-methyl-9H-pyrido[4,3-b]indole; 4, 3-amino-1,4-dimethyl-5H-pyrido[4,3-b]indole; 5, 2-amino-1,6-dimethylimidazo[4,5-b]pyridine; 6, 2-amino-3,7,8-trimethylimidazo[4,5-f] quinoxaline; 7, 2-amino-3,4,7,8-tetramethylimidazo[4,5-f]quinoxaline; 8, 2-amino-3,4-dimethylimidazo[4,5-f]quinoxaline; 9, 2-amino-3,4,8-trimethylimidazo[4,5-f]quinoxaline; 10, 2-amino-3,4-dimethylimidazo[4,5-f]quinoline; 11, 2-amino-3-methylimidazo[4,5-f]quinoline; 12, 2-amino-1-methyl-6-phenylimidazo[4,5-b]pyridine; 13, 2-amino-6-methyl-dipyrido[1,2-a:3′,2′-d]imidazole ; 14, 2-amino-dipyrido[1,2-a:3′,2′-d]imidazole; 15, 2-amino-3-methyl-9H-pyrido[2,3-b]indole; 16, 2-amino-9H-pyrido[2,3-b]indole. From [44] with permission

4.1. Non-Chiral separations of underivatized amines

4.1.1. Analysis of amines in biological matrices

Amines play an essential role in a series of biological processes. Therefore their analysis in biological matrices such as biological fluids or tissue is of significant importance [8]. Biologically important amines, in particular 5-hydroxytryptamine and its major metabolites have been determined in urine samples by Hsieh et al. [17]. Due to the usage of laser induced native fluorescence detection excellent selectivity and sensitivity with limits of detection in the low pM range could be obtained. Nanospray-MS detection after micro solid phase extraction has been employed by Viberg et al. for the analysis of heterocyclic aromatic amines in urine [45]. These compounds commonly are formed in protein rich food during cooking and show some mutagenic and carcinogenic activities. A similar approach allowing the determination of heterocyclic aromatic amines has been taken by Sentellas and co-workers [44,46]. In this work the certainty of peak assignment could be enhanced by additional MS^2 experiments. Two applications of electrochemiluminescence detection, one for the determination of polyamines in hydrolyzed human urine [22], the second for the analysis of urine samples with respect to the presence of analgesics and local anesthetics [24] have been published by the group of Wang. Polyamines in serum have been determined employing a simple CZE method with indirect UV detection by Zhou et al. [47].

4.1.2. Analysis of drugs and drug related substances

Amines are frequently found in drugs and related substances either as active components or unwanted by-products. Amphetamine derived designer drugs have been analyzed in human urine by G. Boatto et al. [48]. Sample enrichment and clean up in this work has been performed using solid phase extraction. A validated CZE method for the determination of some phenothiazines, a group of basic drugs commonly used as antipsychotics, neuroleptics or antihistamines, has been presented by Lara et al. [49]. Several ingredients of Chinese medicinal preparations including one amine have been analyzed by Huitao et al. using CZE with two-marker technique to improve the repeatability of the electrophoretic runs [50]. A method for the determination of Lamotrignine, an antiepileptic drug, in human plasma employing CZE-MS has been presented by Zheng et al. [51]. The results obtained with the analytical method developed in this work showed excellent correlation with those from the commonly used HPLC-UV method. Diphenylhydramine is an antihistaminic druc that is metabolized via a multistep mechanism including demethylation, oxidation and conjugation with glycine. Baldacci et al. reported a CZE-MS method allowing the separation and determination of all relevant metabolites of this drug in human urine [52]. Vanhoenackker et al. tested the performance of a dynamic coating of the capillary for fast and reproducible separations on the example of several of basic drugs [53]. A microchip for CZE together with a contactless conductivity detector has been developed by the group of Hauser []. The performance of this device has been evaluated on the example of several groups of basic drugs. CZE with MS detection is commonly restricted to the use of volatile running buffers. Unfortunately, over the years, a substantial number of CZE methods based on the non-volatile phosphate electrolytes have been developed. The possibility to combine these types of electrolytes with MS detection employing atmospheric pressure photoionization has been evaluated by Nilsson et al. using a mixture of basic drugs as test substances [54]. Only a few reports on the validation of CZE-MS methods exist. The group of Veuthey reported the development of a validated method for the determination of lidocaine and procaine in a pharmaceutical preparation [55]. A modified non-aqueous CZE method for the characterization of a series of pharmaceutical drugs has been presented by Peri-Okonny et al. [56]. Sildenafil and its metabolite originating from de-methylation of this drug has been analyzed by CZE-MS. The proposed method showed excellent figures such as a detection limit in the low ng(ml range and a wide dynamic range [57]. Adamantane derivatives with antiviral activity have been determined using a CZE method with a mixed aqueous/organic solvent electrolyte and indirect UV detection [58]. CZE combined with a commercially available conductivity detector has been used for the analysis of small amines like methylamine, propylamine and tetrabutylamine in pharmaceuticals by Bowman et al. [31] Non aqueous CE (NACE) has been used by Cherkaoui et al. for the rapid separation of basic drugs [59]. A similar approach has been taken by Bjornsdottir allowing the determination of 16 seizure drugs in less than 2.5 minutes as can be seen in Figure 3 [60]. CZE has also been used for the quantitation of residual quaternary amine in bulk drug and process streams [61].

4.1.3.Analysis of amines in foods and beverages

The presence of biogenic amines and polyamines in foods and beverages is of relevance from both toxicological and technological points of view. High amounts of these compounds can lead to health problems and accurate methods of determination as well as knowledge on where these compounds originate and how they can be controlled at the lowest levels are important in the food and beverage industries [62]. A method for the analysis of multiple constituents in beer including several amines as well as purine and pyrimidine bases, that can be regarded as fairly representative for the composition of most usual beers, has been

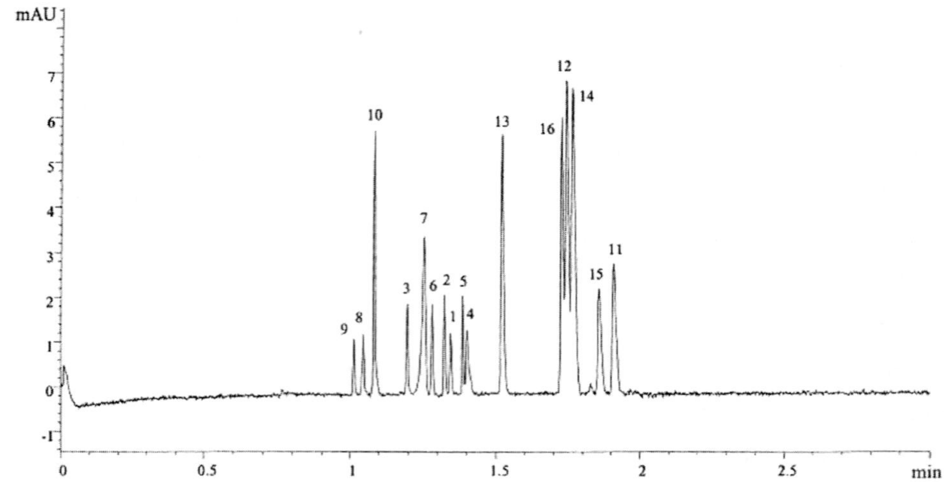

Figure 3 Separation of 16 seizure drugs by NACE. Conditions: electrolyte, 25 mM ammonium acetate, 1 M acetic acid in I/MeOH/glycerol/water (75/15/8.5/1.5); separation voltage -30 kV; capillary, 32.5 cm x 25 μm i.d.; detection, direct UV at 214 nm; peaks: 1, phenylethylamine; 2, amphetamine; 3, metamphetamine; 4, nicotine; 5, 3,4-methylenedioxyamphetamine; 6, 3,4-methylenedioxymethamphetamine; 7, lidocaine; 8, phencyclidine; 9, cocaine; 10, methadone; 11, strychnine; 12, codeine; 13, thebaine; 14, ethylmorphine; 15, heroin; 16, acetylcodeine; from [60] with permission.

---

developed by Cortacero-Ramirez et al. [63]. Two papers describing CZE methods with integrated sample pretreatment for the analysis of nine biogenic amines in wine samples have been published by the group of Valcarcel [64,65]. CZE with electrochemical detection has been used by Sun et al. for the determination of putrescine, cadaverine, spermidine and spermine, four amines that can be seen as indicators for food spoilage, in milk [34]. A method to deterimine small amines and their oxidation products in fish extracts has been published by Timm et al. [66]. Similar analytes in solid food samples have been determined by Lista et al. [67. For the enrichment of the solutes a gas extraction sampling device flow system has been employed. Histamine, a biogenic amine originating from histidine present in the food sample, has been analyzed by Gardana et al. [68]. The intake of heterocyclic aromatic amines poses a potential dietary health hazard. They are commonly formed during the heating of protein rich foods such as meat and fish. Two CZE methods for the determination of these analytes in beefsteak, meat extracts and fish have been published by Mardones et al. [69] and Puignou et al. [70]. Pan residues have been analyzed with respect to these mutagenic substances by Olson et al. [71]. A number of intoxications has been reported following the consumption of sea snails. The primary causative agent for such incidents is the presence of tetramine in this seafood. Zhao et al. developed a CZE-MS method for the determination of this substance in different types of sea snails [72]. The use of tandem MS helped to increase the selectivity and sensitivity of the developed method. A simple method for the determination of a variety of food relevant amines has been presented by Arce et al. [73]. Mendonsa et al. studied the influence of different organic solvents as well as CD´s added to the carrier electrolyte on the quality of separation obtained for several carcinogenic heterocyclic aromatic amines, important in food analysis [74].

4.1.4. Analysis of amines in environmental samples

Aliphatic and aromatic amines are popular as important environmental pollutants as it is well known that these compounds have been widely employed as synthetic intermediates for the manufacturing of dyes, pesticides, rubbers, plastics, pharmaceuticals and a series of other substantial industrial products. Therefore a series of reports exist describing the analysis of environmentally relevant amines in particular in water samples. Alkylamines have been separated analyzed in tap and river water employing pressure assisted CEC and MS detection [75]. Microchip CE with a newly developed electrochemical detector has been employed for the determination of aromatic amines in river water by Shin et al. [76]. Microporous membrane liquid-liquid extraction followed by a CZE separation has been used for the analysis of aromatic amines in tap water, river water and water from a wastewater treatment factory [77]. MEKC has been employed for the analysis of polluted water samples with respect to their contamination with arylamines by Ganesh et al. [78]. The group of Terabe demonstrated the advantages of MEKC with sweeping for the in capillary enrichment of the selected analytes and UV detection [79] or atmospheric pressure chemical ionization MS detection [80] for the trace determination of environmental pollutants. Anilines in waste water have been determined by CZE with electrochemical detection and for comparison also with fluorescence detection after their preceding derivatization [81]. Aliphatic amines have been analyzed in wastewater samples using a simple CZE method based on the commonly employed imidazole electrolyte, after pre-concentration using solid phase extraction [82]. A similar method has been used for the determination of various solutes including some amines in Space Shuttle and MIR Space Station drinking water [83]. Ground water has been analyzed with respect to its content in aromatic amines by Cavallaro et al. [84]. Dabek-Zlotorzynska used a simple CZE method with indirect UV detection for the separation and quantitation of cations including a few aliphatic amines in atmospheric aerosol [85].

4.1.5. Miscellaneous applications

A CZE method for the determination of volatile corrosion inhibitors in particular alkanolamines and cyclohexylamines has been presented by Pereira et al. [15]. Jeffamines, bifunctional spacers that are used for the cross-linking of collagen-based material in the manufacturing of artificial tissue valves, have been characterized by capillary isotachophoresis and CZE with UV detection [86]. The formation of carcinogenic aryl amines in azo dyes has been studied by Borros et al. using a CZE method with as simple phosphate buffer as carrier electrolyte [87]. A CZE method for the quality control of histamine and metacholine in diagnostic solutions has been developed by van der Schans et al. [88].

4.1.6. Separation of standard mixtures and method development

Besides the applications of electrophoretic and electrochromatographic methods for the determination of underivatized amines in real samples, a substantial number of separations of standard mixtures employing different systems and conditions for separation as well as detection have been published. The group of Fritz developed a CZE method for the separation of organic cations using electrolytes with very high ionic strengths[89]. NACE or the use of electrolytes with the addition of organic modifiers provides interesting possibilities to manipulate separation selectivities. This strategy has also been applied to the analysis of a variety of different amines [90-94]. Studies on the migration behavior of amines including strategies for method optimization can also be found in the literature [95-97]. Stacking of cationic analytes using acetonitrile[98] and three phase microextraction[99] have both used to enhance detection limits for amines in CZE analysis.

A series of investigations have been conducted dealing with the performance of CEC systems (either OT or fully packed or monolithic capillaries respectively) as well as capillary coatings for CZE for the separation of underivatized amines. Enlund et al. described the analysis of hydrophobic amines on a continuous bed CEC column[100]. Multimodal OT-CEC has been employed by Pesek et al. for the separation of amines and peptides. They tested different stationary phases including fluorosurfactants adsorbed to the capillary wall, physically adsorbed polymers and chemically modified capillaries [101]. CZE and OT-CEC with magnesia-zirconia coated capillaries has been evaluated with respect to the separation of different classes of analytes including aromatic amines [102]. The performance of capillaries coated with metal complex substituted polysiloxanes in CZE and CEC has been evaluated by Wu et al. [103]. The set of solutes employed for this task consisted of anionic and cationic analytes including several aromatic amines. Different capillary coatings have been tested for their applicability in the analysis of amines. The group of Haddad published a paper describing the separation of anilines on capillaries with a biopolymer coating[104]. In a further paper the same group investigated the migration behavior of aromatic bases in mixed mode electrokinetic chromatography[105]. The interaction of the solutes with ß-CD and polyvinylsulfonic acid has been described by a model developed in this work. An anionic polymer has been employed by Musial et al. for the CZE analysis of cationic molecules in particular inorganic cations and small amines[106]. Capillaries with a hydrophobic coating have been evaluated for the separation of some aromatic amines. The coated capillaries provided good stability (also against strong bases) and the separation performance achieved showed excellent reproducibility[107]. A further work comparing the performance of coarted and uncoated capillaries with respect to the separation of small amines has been published by Burt et al.[108]. An interesting approach to enhance CZE separations of aromatic amines has been Published by the group of Grushka [109]. They added gold nanoparticles together with a polymer t o introduce additional interactions between the capillary and the solutes.

Electrophoretic methods are also well suited for the determination of dissociation constants (pKa values) for acidic as well as basic compounds. The ionization behavior of chemical substances is a key parameter because of different reasons: firstly, the dissociation behavior of a compound is closely related with other key physico-chemical properties playing a crucial role in drug discovery, such as lipophilicity or solubility[110]; secondly, coming to separation sciences, the ionization of solutes under a wide range of different conditions is important for the calculation of their electrophoretic mobilities and consequently for the development and optimization of electrophoretic methods for their analysis. For this reason a series of reports dealing with this issue exist [111-118]. These works include investigations of the dissociation of primary, secondary and tertiary amines in aqueous, mixed aqueous/organic and purely organic media.

As the detection of underivatized amines is not a trivial task, these analytes have also been employed in the development and optimization of detection units. A simple and low cost electrochemiluminescence detector has been presented by Chiang et al. [23]. The performance of the detector has been evaluated on the example of underivatized amino acids and amines. The group of Sweedler reported a NeCu Laser based fluorescence detector for CZE [18]. Sertonine related analytes have been used as test substances for this new setup. Amperometric detection combined with UV detection has been employed for the determination of some aromatic amines by Huang et al. [119]. The same group presented a newly designed end-column electrochemical detector for the analysis of aromatic amines by CZE[120]. Electrochemical detection has been the subject in two more papers. Gallagher et al. described the fabrication of spliced capillaries for a (formerly) commercially available conductivity detector[121] and the group of Hauser presented a detection device for amines based on potentiometric detection using a coated wire ion-selective electrode[122]. MS detection with a conventional ESI interface [123] and with a laser vaporization/ionization interface has also been employed for the detection of heterocyclic amines and serotonine[124].

4.2. Chiral CE for enantiomeric separation of underivatized amines

Chirality is an important principle with respect to natural products but also to substances produced synthetically e.g. for their use as drugs which is to a great extent determining the biological properties of these compounds. This also applies to a substantial part of the amines showing pharmacologically relevant activity. For this reason chiral separations play a dominant role in amine analysis. Due to its simple separation mechanism which is mainly based on the differences in migration velocities for substances with a different charge/ionic radius ratio, CZE alone is not suitable for the separation of enantiomers. For this reason additional interactions between the analytes and a so called chiral selector have to be introduced. To achieve enantiorecognition a three point interaction between the analyte and the selector is required. All these interactions have to take place simultaneously and at least one interaction has to be stereoselective[125]. Some common selectors employed in the electrophoretic separation of amines are cyclodextrines [126,127,128,[126-128], crown ethers[129] or sometimes a mixture of both[130]. Table 2 provides an overview of the applications of chiral CE for the determination of underivatized amines since 1995.

Table 1 Non-Chiral separations of underivatized amines using electrophoretic and electrochromatographic methods (since 1995)

Samples and Analytes	Mode	Electrolyte / Column (for CEC)	Detection	Ref.
Amphetamine derived drugs	CZE	50 mM Ammonium acetate (pH 4.5);	MS	448
Thiazinamium, promazine, promethazine	CZE	100 mM Tris (pH 8.0), with 15 % ACN;	Direct UV at 254 nm	49
Biologically active amines	CZE	1.5 Tris-borate (pH 10.0) and 50 mM propanoic acid-Tris (pH 9.0) in 10% ACN (anodic)	LIF, 266 nm (ex)	17
Chinese medicinical preparation	CZE	10 mM Borate; pH 9.2;	Direct UV at 254 nm	50
Line in plasma	CZE	20 mM Ammonium acetate (pH 3.0);	MS	51
Diphenhydramine metabolites	CZE	20 mM Ammonium acetate (pH 5.6);	MS	52
Basic drugs	CZE	100 mM Formic acid, 1 mM TFA;	MS	53
Basic drugs	CZE on chip	10 mM Mes-His (pH 6.2) or 10 mM citric acid (pH 3.0) or 20 mM lactic acid (pH 2.83) or 100 mM HAc, 1 mM His;	Contactless conductivity	29
Amines in beer	CZE	70 mM Na-borate (pH 10.25);	Direct UV at 210 nm	63
Biogenic amines in wine	CZE	25 mM Citric acid (pH 2.0);	MS	64
Aliphatic amines	CEC	25 mM Citrate with 10 % MeOH / monolithic ODS column;	MS	75
Alkyl- and alkanolamines	CZE	10 mM Imidazole, 10 mM HIBA (pH),10 mM 18-C-6;	Indirect UV at 214 nm	15
Aromatic amines	CZE	100 mM Na/K/Li chloride (pH 3.0); in some cases addition of citrate and phosphate;	Direct UV at 214 nm	89
Aromatic amines	CZE on chip	30 mM Acetate (pH4.5);	Amperometry	76
Heterocyclic amines in urine	CZE	5 mM Ammonium formate (pH 2.0);	MS	45
Heterocyclic amines in urine	CZE	40 mM Formic acid-ammonium formate (pH 4.5) with 60% MeOH;	Direct UV at 213 nm and MS	44

Anilines	CZE	30 mM Tris-H_3PO_4 (pH 3.7);	Direct UV at 254 nm	99
Benzylamines, chloroanilines	CZE	10 mM Na-phosphate (pH 2.0-3.0); ChA-DS coated capillary;	Direct UV at 200 nm	10c 4
Aliphatic amines	CZE on chip	5 mM 2,6-Dihydroxybenzoic acid, 5 mM tetraethylammonium salt in DMF, DMA, DMSO;	Conductivity	90
Biogenic amines	CZE	30 mM Citrate (pH 3.5) with 20 mM NaOH in detection cell;	Pulsed amperometry	34
Aromatic amines in water	CZE	60 mM H_3PO_4, 65 mM NaH_2PO_4, 7 mM ethylene diamine in 12 % MeOH;	Direct UV at 210 nm	77
Polyamnines	CZE	200 mM Phosphate (pH 2.0), 1 M H_3PO_4 (9:1); 5 mM $Ru(bpy)_3^{2+}$ in detection cell;	Electrochemiluminescence	22
Pharmaceutical formulation	CZE	100 mM Formic acid (pH 2.4);	MS	55
Pharmaceutical drugs	CZE	80 mM Ammonium formate (pH 8.7) in MeOH/ACN (8:2);	UV at 195 nm and MS	56
Sildenafil	CZE	10 mM HAc (pH 4.5);	MS	57
Heterocyclic aromatic amines	CZE	16 mM Formic acid-ammonium formate (pH 4.5) with 60% MeOH;	MS^2	46
Trialkylamines	CZE	15 mM Na-borate (pH 9.5); 3,5 mM $Ru(bpy)_3^{2+}$ in detection cell;	Electrochemiluminescence	23
Aromatic bases	CEC	20 mM Citrate (pH 3.5) with polyvinylsulfonic acid and ß-CD;	Direct UV 200 - 211 nm	105
11 Amine compounds	CZE	50 mM Formic acid, 0.88 mM ammonium formate (pH 2.6); 50 Potassium phosphate (pH 2.5) ;	ESI-MS, APPI-MS	54
Tramadol, lidocaine	CZE	10 mM Phosphate (pH 9.0); $Ru(bpy)_3^{2+}$ in detection cell;	Electrochemiluminescence	24
Adamantane derivatives	CZE	5mM 4-Methylbenzylamine (pH 9.0), with 20 mM ß-CD or 2 mM α-CD;	Indirect UV at 210nm	58
Small amines in fish extract	CZE	5 mM $CuSO_4$, 4 mM formic acid. 3m M 18-C-6;	Indirect UV at 214 nm	66
Aromatic amines	CEC	10 mM Tris (pH 4.5) in 80 % ACN; / silica coated with Co(1,4,7-triazacyclononane)$_2^{3+}$	Direct UV at 215 nm	103
Arylamine isomers	MEKC	50 mM Phosphate (pH 8.5), 300 mM SDS, 100 mM borate;	Direct UV at 214 nm	78
Heterocyclic aromatic amines	OT-CEC	30 mM Phosphate-19 mM Tris (pH 2.1) or 30 mM Citric acid-25 mM ß-alanine (pH 3.2) or 30 mM HAc-37.5 mM GABA (pH 4.4) or 30 mM MOPS-20.7 mM imidazole (pH 6.7);	Direct UV at 263 nm	101

Alkylamines	CZE	50 mM CHES or 30 mM MES, 0.03% Triton X-100 and mixtures of different quaternary amines, cholines and polyamines;	Conductivity	106
Biologically important amines	CZE	Borate (pH 8.7);	LIF 248.6 nm (ex)	18
Small amines in fish	CZE	5 mM $CuSO_4$, 4 mM formic acid. 3m M 18-C-6;	Indirect UV	67
Aromatic amines	CZE	20 mM Acetate (pH 5.0), PDAMAC coated capillary with Au-particles;	Direct UV at 254 nm	109
Benzylamines, Amidoamines	CZE	75 mM Phosphate (pH 5.2), 40 % ACN;	Direct UV at 200 nm	91
Aromatic amines	MEKC	10 mM Phosphate-20 mM borate (pH 7.0), 50 mM SDS; 30 mM H_3PO_4, 5 mM phosphate (pH 2.2), 50 mM SDS, 2% MeOH;	APCI-MS	80
Hydrophobic amines	CEC	10 % Phosphate buffer (I=0.115; pH 2.75), 50 % CAN/acrylamide based monolith;	Direct UV at 211 nm	100
Aromatic amines	OT-CEC	16 mM Tris (pH 2.5) in 20 % MeOH or 16 mM Tris (pH 2.5), 8 mM Na_2SO_4 in 20 % MeOH;	Direct UV at 254 nm	102
Aromatic amines in water	CZE	10 mM Na acetate, 5mM HAc (pH 5.0), 1 mM NaCl;	Amperometry	81
Amines in pharmaceuticals	CZE	30 mM His-30 mM glycylglycidine;	Conductivity	31
Benzenediamines and aminophenols	CZE	50 mM Phosphate (pH 12.0);	Direct UV at 200 nm	
Pharmacologically relevant amines	CZE	30-40 mM Phosphate (pH 2.5);	Direct UV at 200 nm	107
Environmentally relevant aromatic amines	CZE/ MEKC	50-100 mM H_3PO_4, 5 mM triethanolamine (pH 2.0), 50 mM urea, 0-50 mM SDS in 20 % ACN;	Direct UV at 214 nm	79
Carcinogenic aryl amines in azo dyes	CZE	50 mM Phosphate (pH 3.1) in 10 % MeOH;	Direct UV at 254 nm	87
Aromatic amines	CZE	10 mM Phosphate;	UV at 254 nm, electrochemical	119
Aromatic amines	CZE	10 mM Phosphate (pH 6.07);	Electrochemical	120
Heterocyclic amines	CZE	Phosphate (pH 2.05), MeOH and CD´s;	Direct UV at 254 nm	74
Basic drugs	CZE	25 mM Ammonium acetate, 1 M HAc in ACN/MeOH ;	UV at 200 nm	59
Seizure drugs	CZE	25 mM Ammonium acetate, 1 M HAc in ACN/MeOH/glycerol/water (75/15/8.5/1.5);	Direct UV at 214 nm	60

Jeffamines	CZE	10 mM His-MES (pH 6.0);	In direct UV at 220 nm	86
Cationic drugs and endogenous substances	CZE	160 mM Triethanolamine, 50 mM tricine (pH 8.6) in 10 % ACN;	Direct UV at 254 nm	98
Small amines	CZE	5 mM Imidazol in MeOH/HAc (99/1);	Indirect UV at 214 nm	92
Heterocyclic amines	CZE	20 mM Ammonium acetate (pH3.0), 20 % MeOH;	MS	123
Histamine in fish	CZE	50 mM NaH_2PO_4, 0.5 gL^{-1} octanesulfonate, (pH3.0), 12% ACN;	Direct UV at 210 nm	68
Aromatic amines in fried food	CZE	30 mM NaH_2PO_4, 20 mM NaCl (pH2.0), 30% MeOH;	Direct UV at 254,263,314 nm	69
Alkylamines	CZE	2.5 mM Tartaric acid;	Potentiometry	122
Small amines in Space Shuttle and MIR water samples	CZE	5 mM Imidazol, 10 mM HIBA;	Indirect UV at 214 nm	83
Biogenic amines in wine	CZE	5 mM $CuSO_4$, 4 mM formic acid. 3m M 18-C-6;	Indirect UV at 214 nm	65
Biogenic amines	CZE	5 mM $CuSO_4$, 4 mM formic acid (pH 3.0), 4 mM 18-C-6;	Indirect UV at 210 nm	73
Heterocyclic aromatic amines	MEKC	15 mM borate (pH 9.1), 5 mM CTAB;	Amperometry	71
Quaternary amine in drug	CZE	3 mM Na acetate, 12 mmM quinine in 25% THF (pH 4.5);	Indirect UV at 235 nm	61
Heterocyclic amines in meat	CZE	10 mM KCl-HCl (pH 2.2);	Direct UV at 210 nm	70
Serotonine	CZE	0.5 mM $CuCl_2$;	TOF-MS	124
Waste water samples	CZE	5 mM Imidazole (pH 5.0);	Indirect UV at 214nm	82
Tetramine in sea snails	CZE	100 mM Formic acid;	MS^2	72
Aliphatic amines	CZE	20 mM NaF, 1 mM TEA in 10 % ACN;	Conductivity	121
Amine neurotransmitters	CZE	10 mM $N(Et)_4^+$ (pH 5.5), 0-100 mM18-C-6 in 50 % MeOH; 10 mM N-Methylimidazol (pH 4.5), 0-200 mM 18-C-6 in 30 % MeOH; 50 mM Octanesulfonate (pH 2.8), 0-10 mM 18-C-6 in 40 % MeOH;	Indirect UV at 210 and 214 nm	96
Biogenic amines	CZE	50 mM Ammonium acetate (pH 7.5) with 40 % MeOH;	Direct UV at 200 nm	93
Histamine and methacholine in Diagnostic solutions	CZE	100 mM Tris-borate (pH 8.3), 0.005 % PVA, 0.05mM CTAB (for histamine); 10 mM Creatinine-chloride (pH 4.85) or 10 mM His/phthalic acid (pH 5.3), 0.05 mM CTAB (for methacholine);	Direct UV at 214 nm (histamine) and indirect UV at 230 (methacholine)	88
Small organic amines	CZE	5 mM Imidazole (pH 3.5);	Indirect UV at 220 nm	108

Small aliphatic amines (together with inorganic cations)	CZE	5 mM Di-n-heptyl-4,4′-bipyridinium hydroxide, 6 mM glycine (pH 6.5), 2 mM 18-C-6, 2 % MeOH;	Indirect UV at 280nm	85
Aromatic amines	CZE	50 mM NaH_2PO_4, 7 mM 1,3-diaminopropane (pH 2.35);	Direct UV at 280 nm	84
Polyamines in serum	CZE	10 mM Quinine sulfate (pH 3.0) in 20 % EtOH;	Indirect UV at 236 nm	47
Drugs	CZE	25 mM Ammonium acetate-0-1M HAc, in MeOH/ACN mixtures;	Direct UV at 214 nm	94

Abbreviations:

Tris	Trishydroxymethylaminomethane	ChA-DS	α-Chymotrypsinogene-A	CHES	Cyclohexylaminoethanesulfonic acid
ACN	Acetonitrile	SDS	Sodiumdodecylsulfate	PDAMA	Polydiallyldimethyl ammonium-chloride
LIF	Laser induced fluorescence	α,β,γ-CD	a,β,γ-Cyclodextrine	Tricine	N-Tris(hydroxymethyl)methylglycine
MES	Morpholinoethanesulfonic acid	ESI	Electrospray ionization	CTAB	Cetyltrimethylammonium bromide
His	Histidine	APCI	Atmospheric pressure chemical ionization	PVA	Polyvinylalcohol
HIBA	Hydoxyisobutyric acid	APPI	Atmospheric pressure photo ionization	MOPS	Morpholinopropane sulfonic acid
18-C-6	18-Crown-6	GABA	γ-Aminobyturic acid	TFA	Trifluoroacetic acid

4.2.1. Separations employing cyclodextrines as chiral selector

Cyclodextrines (CD) and their derivatives can be regarded as the most commonly used chiral selectors in the electrophoretic separation of enantiomers. These compounds show interesting properties due to their unique structure. Their good solubility in aqueous media is advantageous in electrophoretic separation systems. Basic principle of operation of these electrolyte additives is their ability to form inclusion complexes with the analytes and thereby discriminate between different shapes and size [126,127]. This property is also beneficial in the analysis of amines and therefore a series of papers reporting the use of CD´s for this purpose can be found in the literature [131,132,133,134,135,136,137,138,139,140,141,142, 143,144,145,146,147,148,[131-148]. Besides native α, ß and γ CD´s derivatized species (either neutral or charged ones) have been employed. Neutral derivatized CD´s are commonly obtained by derivatization of the hydroxyl groups at the upper and/or lower rim of the depth of the cavity, altering cavity dimensions and also binding properties. Charged CD´s, in the case of amine analysis mainly sulfated species, can serve as a carrier uncharged analytes but also provide additional ionic interactions between the CD and the solute [126].

Stettler and Schwarz reported the use of microchip CE with a sulfated ß-CD for the separation of a model mixture including several neurotransmitters [131]. An interesting study investigating the in vitro metabolism of amphetamines and related substances has been published by Szökö et al. [132]. As can be seen from Figure 4, they managed to separate all compounds of interest in less than 16 minutes employing an acidic electrolyte including a neutral derivatized CD. Lurie et al. tested a series of dynamically coated capillaries together with electrolytes containing neutral derivatized CD´s for the routine analysis of a wide range of seized drugs [133]. Highly sulfated α and γ CD´s have been employed for the quantitation of sertraline stereoisomers. The two chiral selectors provided different selectivities by this resulting in two unequal separation profiles [134]. Rivastigmine, a non-competitive acetylcholinesterase inhibitor, has been analyzed by CE with neutral derivatized CD´s by Kavalirova et al. [135]. An interesting combination of chiral CE with highly sulfated ß-CD and MS detection has been reported by Kindt et al. for the separation of drug candidates [136]. The authors observed that the minimum concentrations of the chiral selector needed for an acceptable separation of the enantiomers did not negatively interfere with the MS detection. A similar setup has been used by Iio et al. for the determination of metamphetamine and its major metabolites in urine[149]. Amphetamine type stimulants have been analyzed using reversed polarity CZE with ESI-MS detection and a highly sulfated γ-CD as electrolyte[150]. The group of Veuthey published two papers on the enantioselective separation of drugs and their metabolites followed by MS detection [151,152]. HPLC and CE have been compared with respect to their performance in the chiral separation of pheniramine like compounds by Schuster et al. [137].CD modified CE has been also employed for the analysis of a variety of chiral drugs [138-140]. Liu et al. published a systematic screening approach for chiral separations of basic analytes employing CE with modified CD´s [141]. Their screening strategy used a bare fused silica capillary, an acidic pH electrolyte and different derivatized neutral and charged CD´s. Cai and Vigh published two papers on the use of the single enantiomer heptakis-(2,3-dimethyl-6-sulfato)-ß-CD as a resolving agent in chiral CE with purely non-aqueous and mixed organic/aqueous electrolytes [142,143]. Cationic CD´s have been employed by Wang et al. for the chiral separation of several basic drugs [144]. Illicit drugs related to ecstasy have been analyzed by Tagliaro et al. using CE with an uncoated capillary and native ß-CD as the chiral selector [145]. Different ß-CD´s, derivatized and native ones, have been compared with respect to their suitability for the resolution of basic drugs by Wang et al. [146]. Chiral separation of biogenic amines at subzero temperatures has been reported by Ma and Horvath [147]. It has been found that the number of phenolic hydroxyl groups in the analyte molecule determined the effect of temperature on the enantioselectivity. Chiral separations using NACE and various native and derivatized CD´s has been published by Wang and Khaledi [148]. The effect of non-chiral crown ethers combined with CD´s in chiral CE has been investigated by Huang et al. [153] and Koide et al. [153].

4.2.2. Separations employing chiral crown ethers as chiral selector

Crown ethers are macrocyclic polyethers which can form stable inclusion complexes with alkali, alkaline-earth and primary ammonium cations [129]. The first crown ethers synthesized by Pedersen [154] could not be used for the resolution of enantiomers but a few years later the first chiral crown ethers suitable for this purpose have been presented [155,156]. Consequently chiral crown ethers have found their way into enantioselective CE and have been employed for the separation of a wide range of chiral analytes including amines. Important factors influencing chiral resolution are the crown ether concentration, the electrolyte pH as well as the electrolyte composition [129]. In the case of the first parameter studies have shown that a higher concentration of the crown ether also enhances complex

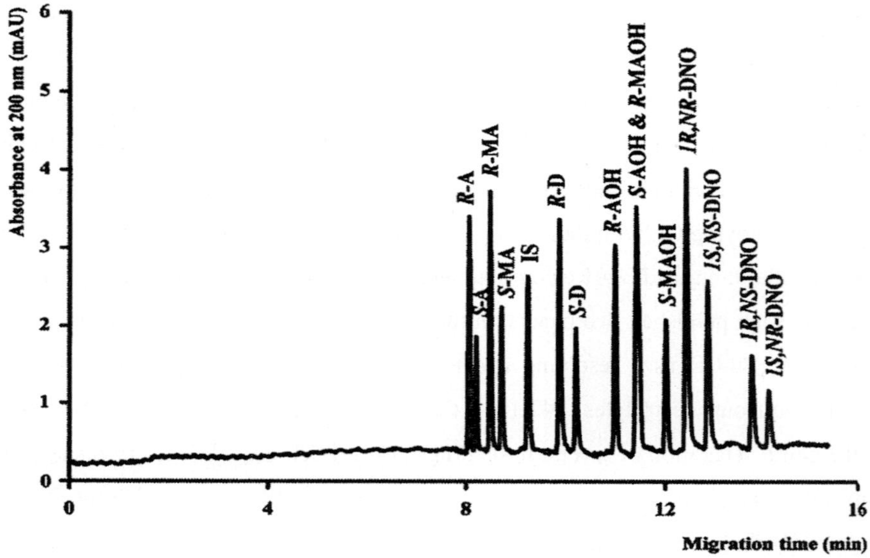

Figure 4 Simultaneous separation of substances from the FMO catalyzed metabolic pathway of deprenyl, methamphetamine and amphetamine. Conditions: electrolyte 20 mM Tris-phosphate, 10 mM hydroxypropylcyclodextrin (pH 2.7); separation voltage 25 kV; capillary, 50.2 cm x 75 μm i.d.; detection, direct UV at 200 n; peaks: R/S-A, R/S-amphetamine; R/S-D, R/S-deprenyl; R/S-AOH, R/S-amphetamine hydroxylamine; R/S-MAOH, R/S-methamphetamine-hydroxylamine; 1 R/S,N R/S-DNO, 1 R/S, N R/S deprenyl-N-oxide; IS, internal standard (parafluoro-desmethyldeprenyl). From [132] with permission.

formation and thereby increases resolution [157-159]. A chiral crown ether that has been employed for the CE analysis of amines is 18-crown-6-tetracarboxylic acid ($18C6H_4$). It has been used for the determination of gemifloxacin [19,160], primary aromatic and amino compounds [161] and drugs containing an amine functional group [162].

4.2.3. Separations with other chiral selectors

Besides the most commonly employed CD′s and the chiral crown ethers several other chiral selectors have been used for the separation of amines by CE. Those include amongst others linear polysaccharides [163], amino acid derivatives [164] and a gluconic acid derivative [165,166].

Table 2 Chiral separations of underivatized amines using electrophoretic and electrochromatographic methods (since 1995)

Samples and Analytes	Mode / Chiral selector	Electrolyte	Detection	Ref.
Epinephrine	Microchip CE / Sulfated CD´s	10 mM Tris pH (6-9 with citric, boric or phosphoric acid); 10 mM MES (pH 6 with NaOH);	Amperometric detection and direct UV at 210 nm	131
Deprenyl, amphetamine and metamphetamine metabolites	CE / hydroxypropyl-ß-CD	20 mM Tris-phosphate (pH 2.7); 10 mM Phosphate-10mM acetate-Tris (pH 3.6);	Direct UV at 200nm	132
Seized drugs	CE / different CD´s	CElixir Reagent B (pH 2.5);	Direct UV at 195, 205, 222 nm	133
Sertraline	CE / Sulfated CD´s	25 mM TEA-phosphate (pH 2.5);	Direct UV at 200, 215, 230 nm	134
Gemifloxacin in urine	Microchip CE / 18-C-6-TCA	50 mM Bis-Tris, 10 mM quinine (pH 4.0 with citric acid);	Indirect LIF at 325 nm (ex), 405 nm (em)	160
Rivastigmine in pharmaceuticals	CE / various CD´s	100 mM Phosphate,Tris or TEA (pH 2.5);	Direct UV at 200 nm	135
Drug test substances	CE / Sulfated CD´s	5 mM NH_4acetate (pH 6.1), MeOH (3:1), 1 % HAc;	MS	
Methamphetamine and metabolites	CE / ß-CD and 2,6-di-O-methyl-ß-CD	1 mM Formic acid (pH 2.2):	MS	149
Amphetamines	CE / Sulfated γ-CD	2.5 mM Highly sulfated–γ-CD (pH 3.1);	MS	150
Anaesthetic drugs	CE / Sulfated ß-CD	50 mM Tris-phosphate (pH 2.5);	MS	151
Drugs and metabolites	CE / HP-ß-CD	40 mM Ammonium formate (pH 3.0);	MS	152
Pharmacologically active amines	NACE / N-Benzoylcarbonylglycyl-L-proline	50 M NH_4-acetate with various % of 1,2 dichloroethane, dichloromethane and 2-propanol in MeOH;	Direct UV at 214, 254, 272 nm	164
Test mixtures including pharma-cologically active amines	NACE / DIKGA	20-40 M NH_4-acetate/0-30 mM HAc with 0-25 % of 2-propanol in MeOH;	Direct UV at 214 nm and MS	165
Gemifloxacin in urine	Microchip CE / 18-C-6-TCA	50 mM Bis-Tris, 0-10mM EDTA (pH 4.0 with	Indirect LIF at 325 nm	19

		citric acid) ;	(ex), 405 nm (em)	
3-Phenyl-3-heteroarylamines	CE / α,β,γ-CD´s and heparin	CD´s: 125 mM Na_3PO_4, 5 M urea (pH 2.5); Heparin: 10-100 mM Na_3PO_4 (pH 5-6);	Direct UV at 210 nm	137
Primary amines	CE /18-C-6 and 2,6-di-O-methyl-ß-CD	200 mM Tris, 300 mM boric acid (pH 7.0);	Direct UV at 214 nm	130
Bis(phenetyl)amines	CE / ß-CD, DIAC/HDMS/HDAS/HS-ß-CD	50 mM phosphate (pH 3.0) ;		138
Substituted imidazoles	CE / various CD´s	100-150 mM H_3PO_4 (pH 2.5-3.0);	Direct UV at 200 nm	139
Pharmacologically active amines	NACE / DIKGA	40 M NaOH in MeOH;	Direct UV at 214 nm	166
Pharmacologically active amines	CE / Sulfated ß-CD	50 mM Tris-phosphate, 0.1 % PVA (pH 2.5); 10 mM $Na_2B_4O_7$, 0.1 % PVA (pH 11.6) ;	Direct UV at 214 nm	140
Pharmacologically active amines	CE / HP-α,β,γ-CD´s, dimethyl-ß-CD and sulphated γ-CD	30 mM Phosphate, TBA (pH 2.5); 25 mM Phosphate, TEA; (pH2.5);	Direct UV at 214 nm	141
Amines and amino acids	CE / Pentosan polysulfate	10 mM Phosphate (pH 2.0) with 20-50 % MeOH;	Direct UV at 214 nm	163
Weak bases	NACE / HDMS-ß-CD	25 mM H_3PO_4, 12.5 mM NaOH in MeOH;	Direct UV at 214 nm	142
Some weak bases	NACE / HDMS-ß-CD	25 mM H_3PO_4, 12.5 mM TEA with 0-50 % MeOH;	Direct UV at 214 nm	143
Basic compounds	CE / QA-ß-CD	50 mM Tris, 25 mM acetate (pH 8.2); or 10 mM Na-borate, 0.1 % PVA (pH 11.6);	Direct UV	144
Amphetamine related substances	CE / ß-CD	100 mM K-phosphate (pH 2.5);	Direct UV at 200 nm	145
Basic drugs	CE / ß-CD and CM-ß-CD	75 mM H_3PO_4-Tris, 8 mM TBAB (pH 3.0); 75 mM H_3PO_4-Tris (pH 5.0);	Direct UV at 212 nm	
Primary amines	CE /18-C-6 and HP-ß-CD, Dimethyl-ß-CD, Hydroxyethyl-ß-CD and γ-CD	50-100 mM Phosphate (pH 2.0);	Direct UV at 214 nm	153
Primary amines	NACE / 18-C-6-TCA	0-100 mM TBAP in formamide;	Direct UV at 260, 300 nm	161
Primary amines	CE / 18-C-6, DCH-18-C-6, 18-C-6-TCA and Crownpack CR	20 mM Tris- H_3PO_4 (pH 1.9);	Direct UV at 210, 220, 235 nm	162

Biogenic amines	CE / HDMS-CD	100-150 mM Na-phosphate, 5 M urea (pH2.5) in 10% MeOH	Direct UV at 214 nm	147
Basic drugs	NACE / ß and γ CD´s	50-150 mM Citric acid, 25-100 mM Tris in formamide, methylformamide or dimethylformamide	Direct UV at 254 nm	148

Abbreviations:

α,β.γ CD	α,β,γ Cyclodextrin	DIKGA	2,3,4,6-Di-O-isopropylidene-2-keto-L-glukonic acid	HS	Highly sulfated
Tris	Trishydroxyaminomethane	18-C-6		HP	Hydroxypropyl
MES	Morpholinoethanesulfonic acid	EDTA	18-Crown-6	PVA	Polyvinylalcohol
TEA	Triethylamine	TBA/TBAB/TB	Etylenediaminetetraacetic acid	QA	Quarternary Ammo-
Bis-Tris	Bis(2-hydroxyethyl)imino tris (hydroxymethyl)methane	AP	Tetrabutylammonium hydroxide/bromide/perchlorate	CM	nium
		DIAC			
LIF	Laser induced fluorescence	HDMS			
		HDAS		NACE	Non aqueous capillary electrophoresis

5. *EXPERIMENTAL*

5.1 CZE analysis of underivatized small amines using indirect UV detection

Instrumental parameters:	capillary: 60-70 cm x 50 or 75 μm i.d.; (smaller i.d. leads to improved resolution but also to a decrease in sensitivity); eventually 50 μm i.d. capillaries with extended light path e.g. from Agilent can be used. indirect UV detection at ~210 nm nm separation voltage: +30 kV
Electrolyte:	5 mM imidazole; 6.5 mM hyxroxyisobutyric acid, pH adjusted to 4.0 with H_3PO_4;

This system well suited for the separation and detection of small aliphatic amines; As an alternative system an electrolyte consisting of 5mM $CuSO_4$ and 4 mM formic acid can be employed under similar conditions.

5.2 CZE analysis of underivatized amines using MS detection

Instrumental parameters:	capillary: 70-100 cm (minimum length depending on the type of CE-MS interface used) x 50 μm i.d.; MS detection in SIM or scan mode separation voltage: +25 kV
Electrolyte:	Ammoniumacatae/formic acid pH 2.0;
Sheath liquid:	either 0.6% formic acid in i-propanol/water (8/2) or 10 mM NH_4formate in methanol/water (1/1) at 2-4 μl min^{-1}

This system allows the separation of underivatized amines even in problematic matrices like food samples or biological fluids.

References

[1] M. Y. Khuhawar, G. A. Qureshi, J. Chromatogr. B 764 (2001) 385-407.

[2] A. Bouchereau, P. Guénot , F. Larher, J. Chromatogr. B 747 (2000) 49-67.

[3] L.M.L. Nollet (Ed.) , Handbook of food analysis, 2nd Ed., Marcel Dekker, New York, 2004.

[4] P. Pais , M.G. Knize, J. Chromatogr. B 747 (2000) 139-169.

[5] H. Kataoka, J. Chromatogr. A 774 (1997) 121-142.

[6] D. Kutlán, P. Presits I. Molnár-Perl, J. Chromatogr. A 949 (2002) 235-248.

[7] J.C.M. Waterval, H. Lingeman, A. Bult , W.J.M. Underberg, Electrophoresis 21 (2000) 4029-4045.

[8] S. Oguri, J. Chromatogr. B 747 (2000) 1-19.

[9] S. Terabe, K. Otsuka, K. Ichikawa, A. Tsuchiya , T. Ando. Anal. Chem. 56 (1984) 111-113.

[10] H. Watarai, Chem. Lett. 231 (1991) 391-394.

[11] J.C. Olsson, A. Dyremark, B. Karlberg, J. Chromatogr. A 765 (1997) 329-335.

[12] Y.-M. Liu, J.-K. Cheng, J. Chromatogr. A 1003 (2003) 211-216.

[13] H. Sirén, A. Karttunen, J. Chromatogr. B 783 (2003)113-124.

[14] P. Doble, M. Macka ,P. R. Haddad, TrAC Trends Anal. Chem. 19 (2000) 10-17.

[15] E.A. Pereira, M.F.M. Tavares, J. Chromatogr. A 1051 (2004) 303-308.

[16] B. M. Simonet, A. Ríos, M. Valcárcel, TRAC Trends Anal. Chem. 22 (2003) 605-614.

[17] M.-M. Hsieh, H.-T. Chang, Electrophoresis 26 (2005) 187-195.

[18] X. Zhang, J.V. Sweedler, Anal. Chem. 73 (2001) 5620-5624.

[19] S.I. Cho, K.-N. Lee, Y.-K. Kim, J. Jang, D.S. Chung, Electrophoresis 23 (2002) 972-977.

[20] N.P. Beard, A.J. de Mello, Electrophoresis 23 (2002) 1722-1730.

[21] F. Li, C. Zhang, X. Guo, W. Feng, Biomed. Chromatogr. 17 (2003) 96-105.

[22] J. Liu, X. Yang, E. Wang, Electrophoresis 24 (2003) 3131-3138.

[23] M.-T. Chiang, M.-C. Lu, C.-W. Whang, Electrophoresis 24 (2003) 3033-3039.

[24] W. Cao, J. Liu, H. Qui, X. Yang, E. Wang, Electroanalysis 14 (2002) 1571-1576.

[25] L.A. Holland, A.M. Leigh, Electrophoresis 23 (2002) 3649-3658.

[26] J. Tanyaniwa, S. Leuthardt, P.C. Hauser, Electrophoresis 23 (2002) 3659-3666.

[27] A.J. Zemann, Trends Anal. Chem. 20 (2001) 346-354.

[28] S. Polesello, S.M. Valsecchi, J. Chromatogr. A 834 (1999) 103-116.

[29] J. Tanyaniwa, P.C. Hauser, Electrophoresis 25 (2004) 3010-3016.

[30] B.A. Musial, M.N. Martin, N.D. Danielson, J. Sep. Sci. 25 (2002) 311-318.

[31] J. Bowman, L. Tang, C.E. Silverman, J. Pharm. Biomed. Anal. 23 (2000) 663-669.

[32] P.A. Gallagher, C.M. Oertel, N.D. Danielson, J. Chromatogr. A 817 (1998) 31-39.

[33] D. Shin, D.A. Tryk, A. Fujishima, A. Muck, G. Chen, J. Wang, Electrophoresis 25 (2004) 3017-3023.

[34] X. Sun, X. Yang and E. Wang, J. Chromatogr. A, 1005 (2003) 189-195.

[35] A. Asthana, D. Bose, A. Durgbanshi, S.K. Sanghi and W. Th. Kok, J. Chromatogr. A, 895 (2000) 197-203.

[36] X. Huang, T. You, X. Yang, E. Wang, Talanta 49 (1999) 425-431.

[37] X. Huang, T. You, X. Yang, E. Wang, Electroanalysis 11 (1999) 969-972.

[38] P. Schnierle, T. Kappes, P.C. Hauser, Anal. Chem. 70 (1998) 3585-3589.

[39] J.C. Olsson, A. Dyremark, B. Karlberg, J. Chromatogr. A 765 (1997) 329.

[40] C.W. Klampfl, J. Chromatogr. A 1044 (2004) 131-144.

[41] P. Schmitt-Kopplin, M. Frommberger, Electrophoresis 24 (2003) 3837-3867.

[42] C.G. Huber, G. Hölzl, Z. Deyl, F. Svec (Ed.) Capillary Electrochromatography, Elsevier, Amsterdam 2001.

[43] A. V. Brocke, G. Nicholson, E. Bayer, Electrophoresis 22 (2001) 1251-1266.

[44] S. Sentellas, E. Moyano, L. Puignou, M. T Galceran, J. Chromatogr. A 1032 (2004) 193-201.

[45] P. Viberg, S. Nilsson, K. Skog, Anal. Bioanal. Chem. 378 (2004) 1729-1734.

[46] S. Sentellas, E. Moyano, L. Puignou, M. T Galceran, Electrophoresis (2003) 3075-3082.

[47] G. Zhou, Q. Yu, Y. Ma, J. Xue, Y. Zg, B. Lin, J. Chromatogr. A 717 (1995) 345-349.

[48] G. Boatto, M. Nieddu, A. Carta, A. Ppalomba, B. Asproni, R. Cerri, J. Chromatogr. B 814 (2005) 93-98.

[49] F.J. Lara, A.M. Garcia-Campana, F. Ales-Barreo, J.M. Bosque-Sendra, Anal. Chim. Acta XXX(2005)XXX

[50] L. Huitao, W. Ketai, C. Xingguo, H. Zhide, Biomed. Chromatogr. 18 (2004) 288-292.

[51] J. Zheng, M.W. Jann, Y.Y. Hon, S.A. Shamsi, Electrophoresis 25 (2004) 2033-2043.

[52] A. Baldacci, F. Prost, W. Thormann, Electrophoresis 25 (2004) 1607-1614.

[53] G. Vanhoenacker, F. de l'Escaille, D. De Keukeleire , P. Sandra, J. Chromatogr. B 799 (2004) 323-330.

[54] S. Nilsson, C. Andersson, P.J.R. Sjöberg, D. Bylund, P. Petersson, M. Jörnten-Karlsson, K.E. Markides, Rapid Commun. Mass Spectrom. 17 (2003) 2267-2272.

[55] L. Geiser, S. Rudaz, J.L. Veuthey, Electrophoresis 24 (2003) 3049-3056.

[56] U.L. Peri-Okonny, E. Kenndler, R.J. Stubbs, N.A. Guzman, Electrophoresis 24 (2003) 139-150.

[57] W. Qin, S.F.Y. Li, Electrophoresis 23 (2002) 4110-4116.

[58] N. Reichova, J. Pazourek, P. Polaskova, J. Havel, Electrophoresis 23 (2002) 259-262.

[59] S. Cherkaoui, L. Geiser, J.-L. Veuthey, Chromatographia, 52 (2000) 403-407.

[60] I. Bjornsdottir, S.H. Hansen, J. Biochem. Biophys. Methods 38 (1999) 155-161.

[61] B.D. Johnson, N. Grinberg, G. Bicker, D. Ellison, J. Liq. Chrom.& Rel. Technol. 20 (1997) 257-272.

[62] P. Kalac, M. Krizek, J. Inst. Brew. 109 (2003) 123-128.

[63] S. Cortacero-Ramirez, A. Segura-Carretero, C. Cruces-Blanco, M.L. Romero-Romero A. Fernandez-Gutierrez, Anal. Bioanal. Chem. 380 (2004) 831-837.

[64] B. Santos, B.M. Simonet, A. Rios, M. Valcarcel, Electrophoresis 25 (2004) 3427-3433.

[65] L. Arce, A. Rios, M. Valcarcel, J. Chromatogr. A 803 (1998) 249-260.

[66] M. Timm, B.M. Jorgensen, Food Chemistry 76 (2002) 509-518.

[67] A.G. Lista, L. Arce, A. Rios, M. Valcarcel, Anal. Chim. Acta 438 (2001) 315-322.

[68] C. Gardana, P. Pietta, S. Ciappellano, G. Testolin, J. Sci. Food Agric. 79 (1999) 91-94.

[69] C. Mardones, L. Arce, A. Rios, M. Valcarcel, Chromatographia 48 (1998) 700-706.

[70] L. Puignou, J. Casal, F.J. Santos, M.T. Galceran, J. Chromatogr. A 769 (1997) 293-299.

[71] J.C. Olsson, A. Dyremark, B. Karlberg, J. Chromatogr. A 765 (1997) 329-335.

[72] J.-Y. Zhao, P. Thibault, T. Tazawa, M.A. Quilliam, J. Chromatogr. A 781 (1997) 555-564.

[73] L. Arce, A. Rios, M. Valcarcel, Chromatographia 46 (1997) 170-176.

[74] S.D. Mendonsa , R.J. Hurtubise, J. Liq. Chromatogr. & Rel. Technol. 22 (1999) 1027-1040.

[75] B. Santos, B.M. Simonet, A. Rios, M. Valcarcel, Electrophoresis 25 (2004) 3231-3236.

[76] D. Shin, D.A. Tryk, A. Fujishima, A. Muck Jr., G. Chen, J. Wang, Electrophoresis 25 (2004) 3017-3023.

[77] Q. Zhou, G. Jiang, J. Liu. Cai, Anal. Chim. Acta 509 (2004) 55-62.

[78] R.J.G. Jeevan, M. Bhaskar, R. Chandrasekar, G. Radhakrishnan, Electrophoresis 23 (2002) 584-590.

[79] J.P. Quirino, Y. Iwai, K. Otsuka, S. Terabe, Electrophoresis 21 (2000) 2899-2903.

[80] K. Isoo, K. Otsuka, S. Terabe, Electrophoresis 22 (2001) 3426-3432.

[81] A. Asthana, D. Bose, A. Durgbanshi, S.K. Sanghi, W.Th. Kok, J. Chromatogr. A 895 (2000) 197-203.

[82] W.H. Matchett, W.C. Brumley, J. Liq. Chrom. & Rel. Technol. 20 (1997) 79-100.

[83] D. Orta, P.D. Mudgett, L. Ding, M. Drybread, J.R. Schultz, R.L. Sauer, J. Chromatogr. A 804 (1998) 295-304.

[84] A. Cavallaro, V. Piangerelli, F. Nerini, S. Cavalli, C. Reschiotto, J. Chromatogr. A 709 (1995) 361-366.

[85] E. Dabek-Zlotorzynska, J.F. Dlouhy, J. Chromatogr. A 706 (1995) 527-534.

[86] J.L. Beckers, Th.P.E.M. Verheggen, H.J.P. Triepels, F.M. Everaerts, J. Chromatogr. A 838 (1999) 149-155.

[87] S. Borros, G. Barbera, J. Biada, N. Agullo, Dyes and Pigments 43 (1999) 189-196.

[88] M.J. van der Schans, J.C. Reijenga, F.M. Everaerts, J. Chromatogr. A 735 (1996) 387-393.

[89] S.A. Steiner, J. Hooker, J. Dederich, B. Scott, J.S. Fritz, J. Chromatogr. A 1039 (2004) 227-233.

[90] J. Wang, M. Pumera, Anal. Chem. 75 (2003) 341-345.

[91] I. Turnes-Carou, C. Prieto-Blanco, P. Lopez-Mahia, S. Muniategui-Lorenzo, D. Prada-Rodriguez, Chromatographia 56 (2002) 605-609.

[92] K.D. Altria, M. Wallberg, D. Westerlund, J. Chromatogr. B 714 (1998) 99-104.

[93] W.-C. Lin, C.-E. Lin, E.C. Lin, J. Chromatogr. A 755 (1996) 142-146.

[94] I. Bjornsdottir, S.H. Hansen, J. Chromatogr. A 711 (1995) 313-322.

[95] C.-E. Lin, Y.-T. Chen, J. Chromatogr. A 871 (2000) 357-366.

[96] C.-S. Chiou, J.-S. Shih, Anal. Chim. Acta 360 (1998) 69-76.

[97] J. Wu, M.K. Wong, S.F.Y. Li, H.K. Lee, C.N. Ong, J. Chromatogr. A 709 (1995) 351-359.

[98] Z.K. Shihabi, J. Chromatogr. A 817 (1998) 25-30.

[99] H.K. Lee, Anal. Chem. 75 (2003) 2784-2789.

[100] A.M. Enlund, C. Ericson, S. Hjerten, D. Westerlund, Electrophoresis 22 (2001) 511-517.

[101] J.J. Pesek, M.T. Matyska, S. Sentellas, M.T. Galceran, M. Chiari, G. Pirri, Electrophoresis 23 (2002) 2982-2989.

[102] M.-J. Xie, Y.-Q. Feng, S.-L. Da, D.-Y. Meng, L.-W. Ren, Anal. Chim. Acta 428 (2001) 255-263.

[103] Q. Wu, M.L: Lee, R.G. Harrison, J. Chromatogr. A 967 (2002) 289-301.

[104] M.C. Yang, M. Macka, P.R. Haddad, Chromatographia 57 (2003) S187-S193.

[105] P. Zakaria, M. Macka, P.R. Haddad, J. Chromatogr. A 997 (2003) 207-218.

[106] B.A. Musial, M.N. Martin, N.D. Danielson, J. Sep. Sci. 25 (2002) 311-318.

[107] M.V. Russo, G. Goretti, J. Chromatogr. A 871 (2000) 279-287.

[108] H. Burt, D.M. Lewis, K.N. Tapley, J. Chromatogr. A 736 (1996) 265-272.

[109] B. Neiman, E. Grushka, O. Lev, Anal. Chem.73 (2001) 5220-5227.

[110] H. Wan, A.G. Holmen, Y. Wang, W. Lindberg, M. Englund, M.B. Nagard and R.A. Thompson, Rapid Communn. Mass Spectrom. 17 (2003) 2639-2648.

[111] M. D. Cantu, S. Hillebrand, E. Carrilho, J. Chromatogr. A 1068 (2005) 99-105.

[112] S.M.C. Buckenmaier, D.V. McCalley, M.R. Euerby, J. Chromatogr. A 1026 (2004) 251-259.

[113] A.L. Simplicio, J.F. Gilmer, N. Frankish, H. Sheridan, J.J. Walsh, J.M. Clancy, J. Chromatogr. A 1054 (2004) 233-238.

[114] J. Muzikar, T. van de Goor, B. Gas, E. Kenndler, Anal. Chem. 74 (2002) 428-433.

[115] S.P. Porras, M.-L. Riekkola, E. Kenndler, J. Chromatogr. A 905 (2001) 259-268.

[116] A. Jouyban, A. Batish, S.J. Rumbelow, B.J. Clark, Analyst 126 (2001) 1958-1962.

[117] S.D. Mendonsa, R.J. Hrtubise, J. Chromatogr. A 841 (1999) 239-247.

[118] S. Fu, D. Li, C.A. Lucy, Analyst 123 (1998) 1487-1492.

[119] X. Huang, T. You, X. Yang, E. Wang, Talanta 49 (1999) 425-431.

[120] X. Huang, T. You, T. Li, X. Yang, E. Wang, Electroanalysis 11 (1999) 969-972.

[121] P.A. Gallagher, C.M: Oertel, N.D. Danielson, J. Chromatogr. A 817 (1998) 31-39.

[122] P. Schnierle, T. Kappes, P.C. Hauser, Anal. Chem. 70 (1998) 3585-3589.

[123] Y. Zhao, M. Schelfaut, P. Sandra, F. Banks, Elcetrophoresis 19 (1998) 2213-2219.

[124] S.Y. Chang, E.S. Yeung, Anal. Chem. 69 (1997) 2251-2257.

[125] C. Dalgliesh, J. Chem. Soc. 137 (1952) 3940-3942.

[126] M. Lämmerhofer, J. Chromatogr. A 1068 (2005) 1-30.

[127] M. Lämmerhofer, J. Chromatogr. A 1068 (2005) 31-57.

[128] E. Schneiderman A.M. Stalcup, J. Chromatogr. B 745 (2000) 83-102.

[129] R. Kuhn, Electrophoresis 20 (1999) 2605-2613.

[130] T. Koide K. Ueno, J. Chromatogr. A 923 (2001) 229-239.

[131] A.R. Stettler M.A. Schwarz, J. Chromatogr. A 1063 (2005) 217-225.

[132] E. Szökö, T. Tabi, T. Borbas, B. Dalmadi, K. Tihanyi, K. Magyar, Electrophoresis 25 (2004) 2866-2875.

[133] I.S. Lurie, P.A. Hays, K. Parker, Electrophoresis 25 (2004) 1580-1591.

[134] M.X. Zhou, J.P. Foley, J. Chromatogr. A 1052 (2004) 13-23.

[135] A. Kavalirova, M. Pospisilova, R. Karlicek, Anal. Chim. Acta 525 (2004) 43-51.

[136] E.K. Kindt, S. Kurzyniec, S.-C. Wang, G. Kilby, D.T. Rossi, J. Pharm. Biomed. Anal. 31 (2003) 893-904.

[137] A. Schuster, C. Götte, G. Bernhardt, A. Buschauer, Chirality 13 (2001) 285-293.

[138] M. Wedig, M. Thunhorst, S. Laug, M. Decker, J. Lehmann, U. Holzgrabe, Fresenius J. Anal. Chem. 371 (2001) 212-217.

[139] A. Sasse, W. Schunack, H. Stark, Biomed. Chromatogr.15 (2001) 25-30.

[140] F. Wang, M.G. Khaledi, J. Microcolumn Sep. 11 (1999) 11-21.

[141] L. Liu, M.A. Nussbaum, J. Pharm. Biomed. Anal. 19 (1999) 679-694.

[142] H. Cai, G. Vigh, J. Pharm. Biomed. Anal. 18 (1998) 615-621.

[143] H. Cai, G. Vigh, J. Chromatogr. A 827 (1998) 121-132.

[144] F. Wang, M.G. Khaledi, Electrophoresis 19 (1998) 2095-2100.

[145] F. Tagliaro, G. Manetto, S. Bellini, D. Scarcella, F.P. Smith, M. Marigo, Electrophoresis 19 (1998) 42-50.

[146] H. Wang, J.L. Gu, H.F. Hu, R.J. Dai, T.H. Ding, R.N. Fu, Anal. Chim. Acta, 359 (1998) 39-46.

[147] S. Ma, C. Horvath, Electrophoresis 18 (1997) 873-883.

[148] F. Wang, M.G. Khaledi, Anal. Chem. 68 (1996) 3460-3467.

[149] R. Iio, S. Chinaka, S. Tanaka, N. Takayama, K. Hayakawa, Analyst 128 (2003) 646-650.

[150] Y.T. Iwata, T. Kanamori, Y. Ohmae, K. Tsujikawa, H. Inoue T. Kishi, Electrophoresis, 24 (2003) 1770-1776.

[151] S. Cherkaoui, J.L. Veuthey, J. Pharm. Biomed. Anal. 27 (2002) 615-626

[152] S. Cherkaoui S. Rudaz, E. Varesio, J.L. Veuthey, Electrophoresis 22 (2001) 3308-3315.

[153] W.X. Huang, S.D. Fazio, R.V. Vivilecchia, J. Chromatogr. A 781 (1997) 129-137.

[154] C.J. Pedersen, J. Am. Chem. Soc. 89 (1967) 2495-2496.

[155] J.M. Dietrich, J.M. Lehn, J.P. Sauvage, Tetrahedron Lett. 10 (1969) 2885-2888.

[156] E.P. Kyba, M.G. Siegel, L.R. Sousa, G.D.Y. Sogah, D.J. Cram, J. Am.Chem. Soc. 95 (1973) 2691-2962.

[157] R. Kuhn, D. Riester, B. Fleckenstein, K.H. Wiesmüller, J. Chromatogr. A 716 (1995) 371-379.

[158] M.G. Schmid, G. Gübitz, J. Chromatogr. A 709 (1995) 81-88.

[159] P. Castelnovo, C. Albeanesi, J. Chromatogr. A 715 (1995) 143-149.

[160] S.I. Cho, J. Shim, M.-S. Kim,Y.-K. Kim, D.S. Chung, J. Chromatogr. A 1055 (2004) 241-245.

[161] Y. Mori, K. Ueno, T. Umeda, J. Chromatogrc. A. 757 (1997) 328-332.

[162] H. Nishi, K. Nakamura, H. Nakai, T. Sato, J. Chromatogr. A 757 (1997) 225-235.

[163] X. Wang, J.-T. Lee, D.W. Armstrong, Electrophoresis, 20 (1999) 162-170.

[164] Y. Hedeland, M. Hedeland, U. Bondesson, C. Petterson, J. Chromatogr. A, 984 (2003) 261-271.

[165] H. Loden, Y. Hedeland, M. Hedeland, U. Bondesson, C. Pettersson, J. Chromatogr. A 986 (2003) 143-152.

[166] Y. Carlsson, M. Hedeland, U. Bondesson, C. Petterson, J. Chromatogr. A 922 (2001) 303-311.

Ibolya Molnár-Perl (Editor)
Quantitation of Amino Acids and Amines by Chromatography
Journal of Chromatography Library, Vol. 70

2.3.2

Quantitation of Amines by Oncolumn Derivatization with *o*-Phthalaldehyde by Capillary Electrochromatography

Shigeyuki Oguri

Contents

Summary

Taking to heart the meaning of the proverb "Simple is best", we have been aiming to develop a method for the determinations of biogenic amines that is so simple that almost anyone could perform it. The present platform, oncolumn derivatization CEC, has the following advantages: 1) No additional apparatus or devices are needed. 2) Biogenic amines can be separated and detected, regardless of the presence of other substances in the sample, such as amino acids. This selectivity makes unnecessary the need for any sort of pre-treatment of the sample to remove unwanted components, thereby resulting in a streamlining of the analytical procedure. 3) A high-sensitivity detection can be easily obtained with an on-line preconcentration technique. 4) Downsizing the format onto a μ-chip is possible. Although oncolumn derivatization CEC still has some problems requiring solution or improvement, I believe this will be one of the best platforms for the high-throughput determinations of biogenic amines in food or biological sample in the near future

1. Introduction

About 10 years ago, an oncolumn derivatization technique came out of the traditional derivatization methods, pre- and post-capillary column formats as a third derivatization method [1-4]. To date, these traditional derivatization techniques have been applied in various fields of analytical separation science, such as Gas-liquid chromatography (GC), high-performance liquid chromatography (HPLC), capillary electrophoresis (CE) and capillary electrochromatography (CEC) using a large set of samples of naturally occurring substances, in order to enhance both the selectivity and sensitivity of detection [5]. We are sometimes confronted with enormous difficulties when such samples have no chromophore or fluorophore, making analysis by CE or CEC equipped with conventional UV or fluorescence detector impractical. Although traditional derivatization methods can sometimes work effectively, they have some problems, as mentioned above in Chapter 2. . Regarding the post-column derivatization method for CEC, as well as for CE, one unavoidable problem is the mechanical difficulty of installing a micro-scale derivatization reaction chamber on the CEC or CE. Therefore, post-column derivatization CEC has not been put into general practice. The pre-column method, on the other hand, requires several batchwise operations off line just before performing CE or CEC. In addition to lengthening the overall process, these operations increase the likelihood of human error occurring. Furthermore, obtaining high-precision results requires a high-degree of skillful technique. Although oncolumn derivatization CEC also has some problems, we believe that it is one of the best ways to detect amino compounds because it does not require the installation of additional devices such as a reactor chamber, batchwise operations for tagging amines during assay, etc. In this chapter, I focus mainly on the detecting biogenic amines using an oncolumn derivatization CEC method that has been developed by our group over the past three years [6-8].

2. Biogenic amines

Mono and/or polyamines involved in biological activities and which exist in nature are frequently referred to as "biogenic amines". When scientists first employed the terms "biogenic amines", whether about one hundred years ago [9], they were referring only to those amines that are synthesized by and exist in the living body. For example, dopamine, noradrenaline and adrenaline, all of which are called "chatechol amines", are the most representative biogenic amines existing in the brain and other organs. Histamine, serotonin and melatonin, which are imidazole or indol derivatives, are also called biogenic amines. These biogenic amines are synthesized from amino acids in the mammalian body and also play a role as neurotransmitters or hormones. Recently, the use of the terminology "biogenic amines" has expanded to include even those amino compounds which are neither present in nor generated by the living body. For example, heterocyclic amines [10] are also called biogenic amines, but are generated by the thermal decomposition of certain amino acids or proteins which occur during the process of cooking food. Some of these are known to be carcinogenic substances. Polyamines, which are synthesized from amino acids, are likewise referred to as biogenic amines. They are ubiquitous in all living cells and play an important role in cell growth and differentiation. Therefore, several diagnostic assays based on polyamine detection have been developed to screen for cancer. In our daily life, allergic food poisoning occurs when people accidentally ingest food contaminated with histamine, tyramine or other polyamines. Most regrettably, convenient analytical tools for detection of these harmful substances are not readily available at present. In other words, a useful biogenic amine analyzer for such amines is much needed. The term "biogenic amine" in this chapter is restricted to the following limited list of amines: histamine, cadaveline, putrescine, serotonin and spermidine. This list, for the sake of convenience, does not include either the chatechol nor the heterocyclic amines.

3. Theory

The mechanism of oncolumn derivatization CEC is the same as that of in-capillary derivatization, as previously discussed in Chapter.2, namely, free biogenic amine are automatically derivatized, separated and detected during migration in a capillary column filled with a run buffer containing a derivatizing reagent. To perform this in-capillary derivatization, the derivatization reagent needs to meet the following requirements: the reagent itself should be non-reactive to photometric detectors, such as a UV or fluorescence detector. The analyte-reagent derivative, however, should be detectable by the photometric detector. (See Figure 11 in Chapter.4) Lastly, the reaction between analyte and reagent should progress quickly at room temperature. To meet these requirements, we chose to use OPA/thiol as the reagent for use with this method.

4. Fabrication of the capillary column

The ODS particles, that are generally used for this column, should be tolerant of an alkaline environment, because the pH of the run buffer used for this oncolumn derivatization should be at pH10. Therefore, HPLC ODS column [7], Capcell Pack C18 (Shiseido, Tokyo, Japan) was chosen, which can be used under conditions of pH10. By removing the contents of the Capcell Pack, we obtained the particles used as the packing material for our study.

Preparation of the ODS-packed capillary column: Fabrication of the capillary column was carried out by using a sol-gel reaction [8] as follows:

Step 1: A temporary frit was made at an appropriate position, e.g., a few cm from one end of the capillary tube (30-cm length x 100 μm i.d., Polymicro Technologies LLC, Arizona, U.S.A) using a sodium silicate solution, by heating the capillary with an EK 1.2 CE Capillary Burner (Electro-Kinetic Technologies, West Lothian, UK).

Step 2:After flushing out, a slurry of the packing material (20 mg) in 2-propanol (1 ml) followed by sonication for several minutes was packed into the capillary with an LC-5A high-performance liquid chromatographic pump (Shimadzu, Kyoto, Japan), using acetonitrile as a pressurize solvent at 480 bar.

Step 3: After completing the packing, the capillary was flushed with acetonitrile for 1 h and then with a 1 % (v.v.) solution of sodium chloride at 480 bar for 20 min. Then, *a retaining frit* was made at a point close to the temporary frit, followed by making an *end frit* at an appropriate position under applying the high pressure with the sodium chloride solution.

Step 4: The outside portion of the retaining frit was cut off, then the capillary was flushed with a mobile phase used at 60 bar under the reverse direction for flushing out the rest of the packing material and conditioning.

Step5: A detection window was made near the end frit by removing a section of the polyimide coating with the capillary burner. These column-fabrication steps are illustrated in Figure 1.

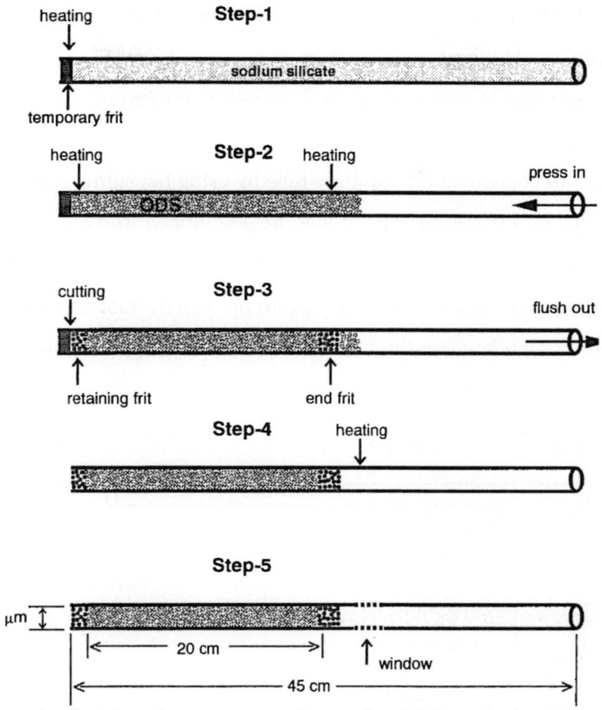

Figure 1. *Schematic showing the sequence of steps for ODS-packed column fabrication: (Step1) making a temporary frit with sodium silicate, (Step2) packing ODS particles, (Step3) making both retaining and end frit, (Step4) flushing out the rest of the packing material and conditioning, (Step5)*

Preparation of a monolithic ODS capillary column:

A monolithic ODS capillary column of the same dimension as the ODS packed capillary column (as described above) is shown in Fig 2.

Step 1: A retaining frit was made by dipping one end of the capillary tube in the sol-gel solution (a mixture of 0.20 ml of TMOS, 0.73 ml of ethanol. and 0.10 ml of 0.12 M hydrochloric acid) for a split second. Then, the capillary tube was left at room temperature overnight.

Step 2: ODS particles were added at a concentration of 300 mg/ml to the sol-gel solution (a mixture of 0.20 ml of TMOS, 0.73 ml of ethanol. and 0.10 ml of 0.12 M hydrochloric acid). This solution was sonicated for several minutes and then introduced into the opposite side of the retaining frit by a syringe pressurized with a hand-held vise. During packing, the particles stacked together at one point inside the capillary.

Step3: Packing was continued until a column 20-cm long was made and the region beyond the initial stacking point was used to fabricate a window.

Step 4: The column was kept at either 120°C for 1 h or 40°C for several days to facilitate ethanol evaporation.

Step 5: The window was created on the capillary tube by using hot sulfuric acid (> 100 °C).

Column conditioning First, each column was preconditioned with a separation buffer that had been degassed by sonication (by pressurizing the column inlet to approximately 200 *psi* with a hand-held vise). Finally, the column was further conditioned electro-kinetically in the CEC instrument by driving the buffer mobile phase through the capillary at an applied voltage of 145 V/cm until a stable baseline was achieved. This procedure typically takes 2-3 h.

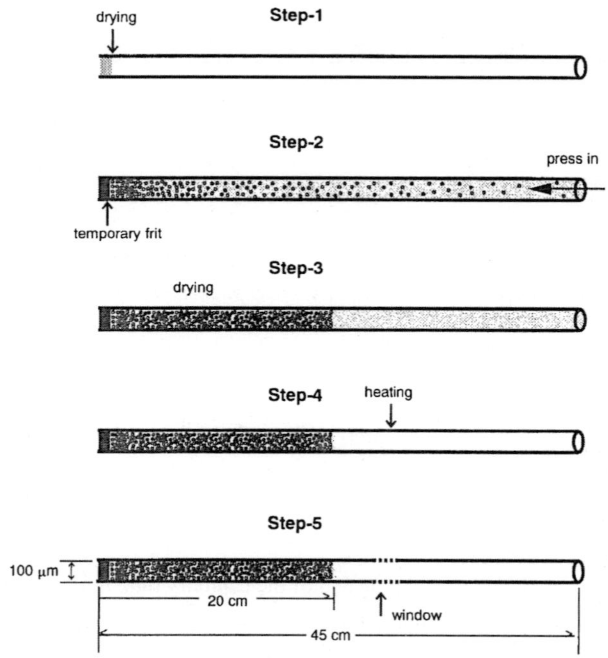

Figure 2 Schematic showing the sequence of steps for monolithic ODS column fabrication: (Step1) making a temporary frit with a sol-gel solution, (Step2) loading ODS particles in the sol-gel solution, (Step3) packing ODS particles, (Step4) pushing out the rest of the sol-gel solution, (Step5) making a detection window.

5. Oncolumn derivatization CEC

After electrokinetic injection of a test sample mixture of 5 biogenic amines (histamine, serotonin, tyramine, putrescine and cadaverine) into the ODS-packed capillary column, the free amines entered into the anodic site of the capillary column and started to travel along the column, during which time the analytes reacted with the derivatization reagent, separated out and were detected with an absorbance at 340 nm when high voltage was applied to the column. The typical chromatograms of oncolumn derivatization CEC, pre-column derivatization CEC and throughout in-capillary derivatization CE are shown in Figure 3-A, Figure 3-B and Figure 3-C, respectively. Figures, A and B show almost identical eluting profiles with each other. However, histamine, serotonin, putrescine and cadaverine are shown all fused together in one peak when in-capillary derivatization CE was applied to the same sample solution, as shown in Figure3-C.

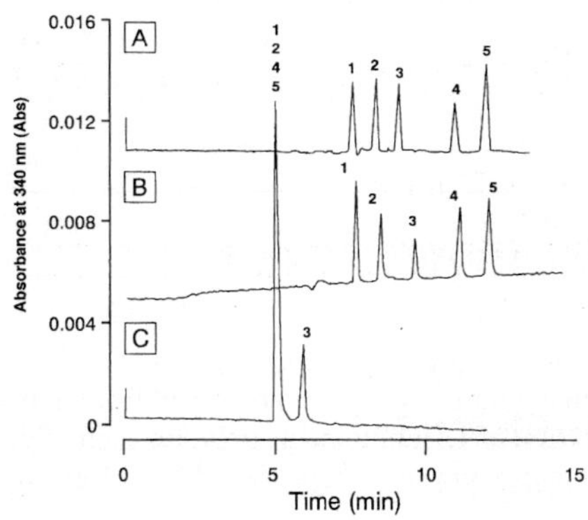

Figure 3 Typical electrochromatograms (A) and (B) and electropherogram (C) obtained from on-column, pre-column and throughout in-capillary derivatization methods, respectively. Reproduced from Reference [6] with permission.

6. Selective detection of biogenic amines

When this system was employed to analyze a mixture containing 5 biogenic amines and 17 amino acids, the 5 biogenic amines plus arginine were selectively detected, as shown in Figure 4-A. To confirm this selectivity, when a mixture containing only 17 amino acids was injected into the same system, only arginie appeared, as shown in Figure 4-B. As reference, Figure 4-C shows the electromigration profiles of the same sample solution created by in-capillary derivatization CE as mentioned in Chapter.4.

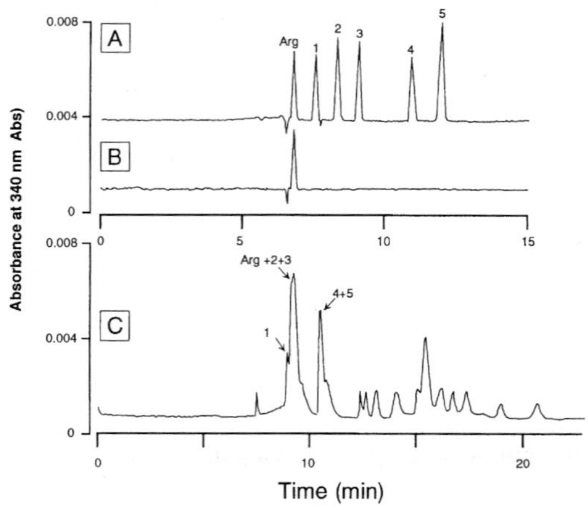

Figure 4 Typical electrochromatograms (A) and (B) obtained from a mixture of 17 amino acids plus 5 biogenic amines and the mixture of 17 amino acids only, respectively. The electropherogram of (C) was obtained from the same mixture (A). Reproduced from Ref. [6] with permission.

This selectivity can be understood as follows: This oncolumn derivatization CEC was done at pH 10. In this environment, all the amino acids, except for arginine were negatively charged, due to the dissociation of the carboxy group into anions. Consequently, all the amino acids, except for arginine, moved toward the anodic site, due to the electrophoretic driving force. The biogenic amines, along with arginine, on the other hand, having neutral charge in pH10, passed though the column by means of EOF alone and were, thus, detected. This mechanism is illustrated in Figure 5. This selectivity means that using the present method, it is possible to make determinations of biogenic amines without first having to remove undesirable compounds, such as amino acids, just prior to CEC analysis.

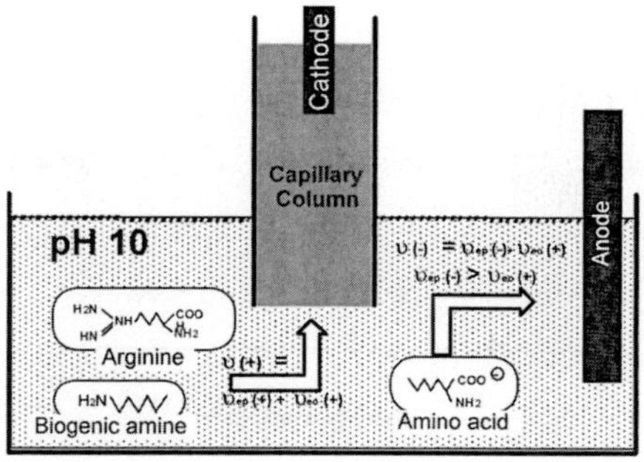

Figure 5 *Mechanism of electrokinetic injection of the analyte into the capillary column. υ (+) and υ (-) express the direction of motion of the analyte toward cathodic site from anodic site and toward the anodic site from cathodic site, respectively. When motion does not occur, it is defined as υ (0). Reproduced from Reference [6] with permission.*

There is one example of oncolumn derivatization CEC technique used to advantage for the determination of biogenic amines in food. In general, the determination of mono- and polyamines in food samples requires relatively complicated "batchwise" operations for the extraction of amines. For example, for the standard method of analysis of histamine and other amines in food, Hygiene Chemist in Japan adopts the following method: 1) Sample is extracted by homogenizing with hydrochloric acid. 2) The extract is filtered by means of an ion-exchange column (Amberlite CG-50) in order to remove unwanted compounds. 3) The solution eluted from the column with hydrochloric acid is treated with dansyl chloride as the derivatization reagent. 4) The derivatives are analyzed using optimized, reversed-phase, high-performance liquid chromatography. Although this method is an authorized method in Japan and has high reliability, it is technically difficult to perform, which increases the probability of human error. By the present method under discussion, a crude extract sample obtained from hydrochloric acid extraction can be used, and does not need any further purification, such as the ion-chromatographic purification mentioned above. After neutralization of the hydrochloric acid extraction, aliquots of the extracts are directly injected into the oncolumn derivatization CEC apparatus. Figures 6-A and B show the chromatograms obtained from samples of raw mackerel and tuna meat that were allowed to age for 1 day at room temperature, respectively. Histamine, tyramine, putrescine and cadaverine were identified in the former case, but no amines of any kind were identified in the latter case.

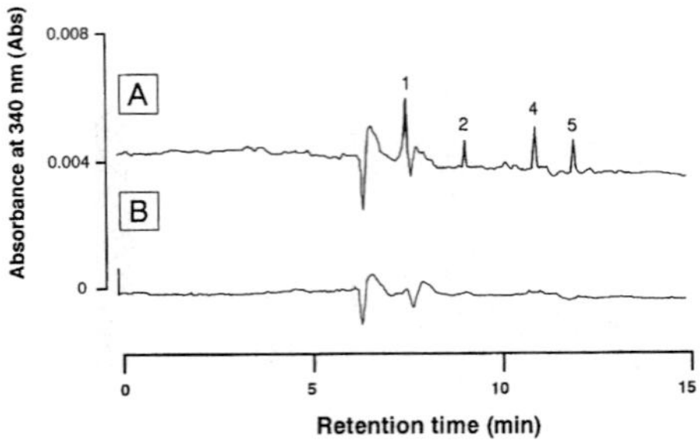

Figure 6 Electrochromatograms obtained from (A) mackerel and (B) tuna, respectively, after keeping them at room temperature for 1 day. Reproduced from Reference [6] with permission.

7. On-line concentration prior to oncolumn derivatization CEC

One of the biggest disadvantages of CE or CEC compared with HPLC is reduced detection sensitivity because of the shortness of the light-path length at the detection window. In the past decade, the technique called on-line preconcentration has enjoyed some degree of success in CE, such as stacking, sweeping and use of the pH junction [11, 12]. Normally, a sample is injected into a capillary column, and the "plug" of this sample moves past the detection point (as shown in Figure 7-A). In such a case, a sharp peak can be seen. Injection of a greater amount of sample, in order to expand the sample, does not improve the detection sensitivity (Figure 7-B). On the contrary, in the case of CE, larger amounts of sample solution, that is, a plug having greater width, has a higher resistance value (R_1) than that of the run buffer (R_2).

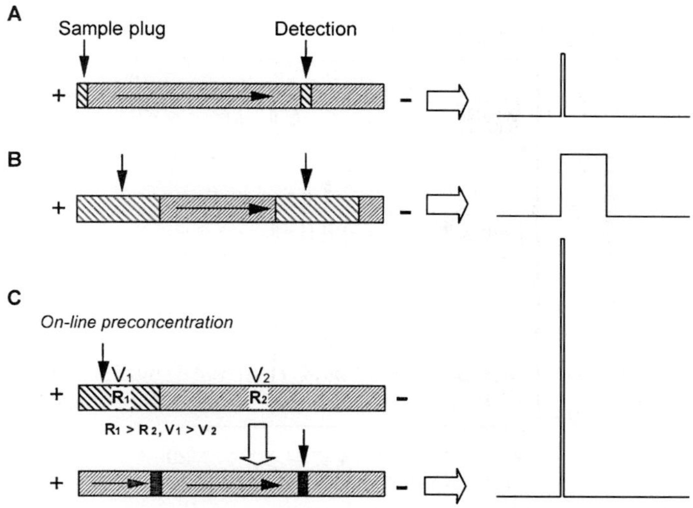

Figure 7 Diagram illustrating the expected peak shapes for the sample solution injected as (A) short or (B) long plug when sample solution was normally injected and traveling in a column. Diagram (C) expresses the principle of on-line preconcentration CE and its expected peak. R_1 and R_2 express the resistance value of the sample plug and the run buffer plug, respectively. V_1 and V_2 express the potential voltage of sample plug and run buffer plug, respectively, when a potential voltage value (V) was applied between the capillary tube.

When electrophoresis begins, sample stacking occurs at the border between the sample and run buffer plugs, because the voltage potential (V_1) of the sample plug is much higher than that (V_2) of the run buffer plug. Subsequently, the sample becomes concentrated and give a sharp peak as shown in Figure 7-C, resulting in vastly improved detection sensitivity. Actually, a degree of sample concentration equal to 1,000 times the norm has been reported with this technique. However, when a sample has no functional chromophore or fluorophore group, high-sensitivity detection of the analyte cannot be accomplished by means of conventional UV or fluorescence detectors, even if on-line preconcentration occurs. To overcome this difficulty, on-line preconcentration prior to oncolumn derivatization CEC was employed in order to expand the method to samples lacking chromophore or fluorophore groups. It is very simple and easy to do on-line preconcentration. It can be done simply by making the injection time much longer than usual.

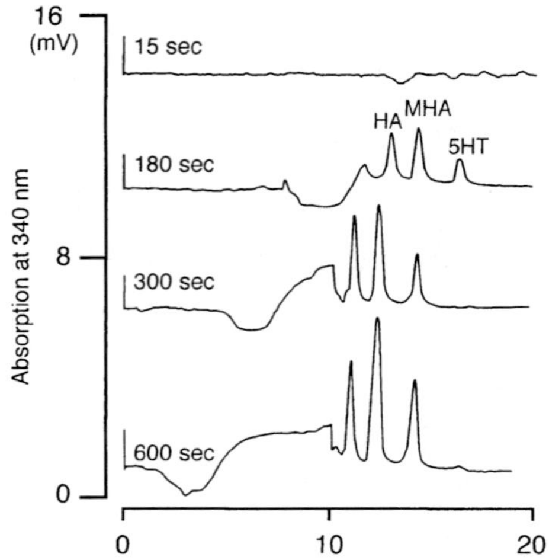

Figure 8 Typical electrochromatograms of a mixture of HA (histamine), MHA (methylhistamine) and 5HT (serotonin) with a monolithic ODS capillary column when the mixture was electrokinetically injected for 15, 180, 300 and 600 sec. Reproduced from Reference [7] with permission.

A mixture of histamine, methylhistamine and serotonin in saline at 0.1 μM each was injected with a 5-kV injection voltage for different lengths of time, e.g., 15, 180, 300 and 600 sec. Even though three peaks corresponding to the amines could hardly be seen at 15 seconds, these peak areas increased remarkably with increased injection time as shown in Figure 8.

Although on-line preconcentration CE can only be applied to sample solutions having less conductivity than that of the run buffer, the present method can be applied not only to low-conductivity sample solutions, but also to high-conductivity solutions, such as occur with food or biological samples. Figure 9 show that this on-line preconcentration/on-derivatization CEC proves effective in cases when the samples are presented in a saline solution (Figure 9-A), a water solution (Figure 9-B) and a run buffer solution (Figure 9-C). It may be considered that on-line preconcentration occurred due to both gradient elution and stacking effects. Using the present system, equipped with a fluorescence detector instead of a UV/Visible detector, the detection sensitivity for amines reached the 0.1 μM level, which increased sensitivity by a factor of 10^3 times greater than of normal oncolumn derivatization CEC

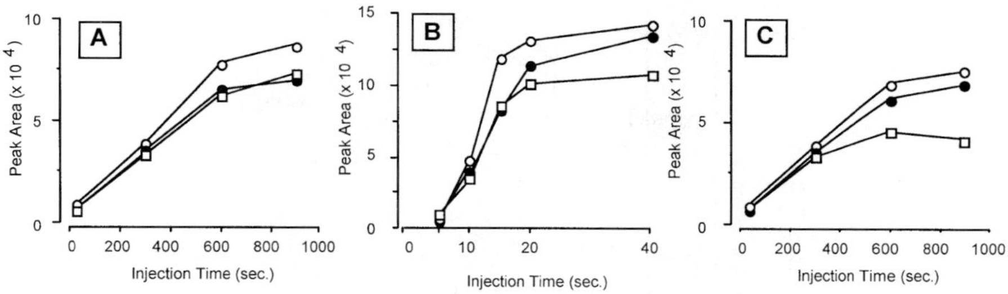

Figure 9 Relationship between the peak area response of histamine (•), methylhistamine (○), and serotonin (□) as a function of 5 kV-injection time and sample matrix, e.g., 0.9 % sodium chloride (A), water (B), or 60 % acetonitrile-10 mM borate (pH 10) (C). Reproduced from Reference [7] with permission.

8. Modification of the elution profile

There is another problem with oncolumn derivatization CEC that concerns the run buffer, namely, that the run buffer used here plays a double role as both the chemical reaction buffer for derivatization and the separation buffer. Therefore, the pH of the run buffer should be around 10, and the percentage of acetonitrile or organic solvent in the run buffer should be more than 50 %, because, with a run buffer having less than 50 % of organic solvent, hydrolysis of the linkage between ODS and the silicate or monolithic glass matrix occurs. This makes changing the eluting profile by replacing the composition of the run buffer very hard to do with oncolumn derivatization CEC.

The chemical property of OPA-amine derivative can be easily changed by replacing the thiol compound with another compound. The retention of the analyte on the ODS column becomes prolonged if the hydrophobisity of the OPA derivative is increased by using an alkyl thiol, such as propane or butane thiol instead of 2-merucaputo-ethanol. This eluting mechanism is almost the same as occurs with reversed-phase HPLC. Figure 10 shows the electrochromatograms of separations of histamine and methylhistamine when 2-merucaputo-ethanol (Figure 10-A), ethanthiol (Figure 10-B), 1-propanethol (Figure 10-C), 2-methyl-1-propanethiol (Figure 10-D) and 1-butanethiol (Figure 10-E) were employed as a thiol compound. The elutions of both histamine and methylhistame derivatives were delayed by using alkyl thiols in the order as shown in Figure10.

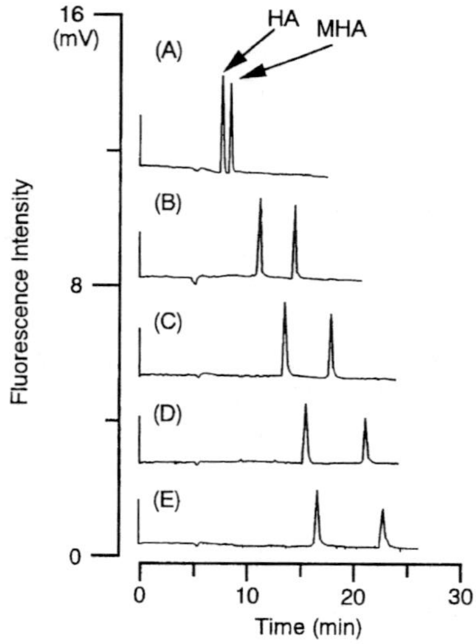

Figure 10 Electrochromatograms of the separation of HA (histamine) and MHA (methylhistamine) by using (A) 2-mercaptoethnol, (B) ethanethiol, (C) 1-propanethiol, (D) 2-methyl-1-propanethiol, and 1-butanethiol as thiol compounds used together with OPA for oncolumn derivatization CEC. Reproduced from Reference [8] with permission.

The following example shows this technique. Figure 11 shows the electrochromatograms of a natural sample human urine and a sample spiked with histamine and methylhistamine at 0.1 μM each after solid-phase extraction with "Bond Elute" cation exchange resin when 2-ME as a thiol compound was used together with OPA for this system. It is clear that the two amines cannot be identified due to interference by the urine matrix components. If, however, the elution of the two analytes from the column could be delayed, they would become separated from the others. But it is difficult to modify their elution profile by changing the pH or concentration of acetonitrile of the run buffer for the reason mentioned above. So, 1-butane thiol was used instead of 2-mercapto ethanol for this trail. The result of the separation of two amines was remarkably improved, as shown in Figure 12.

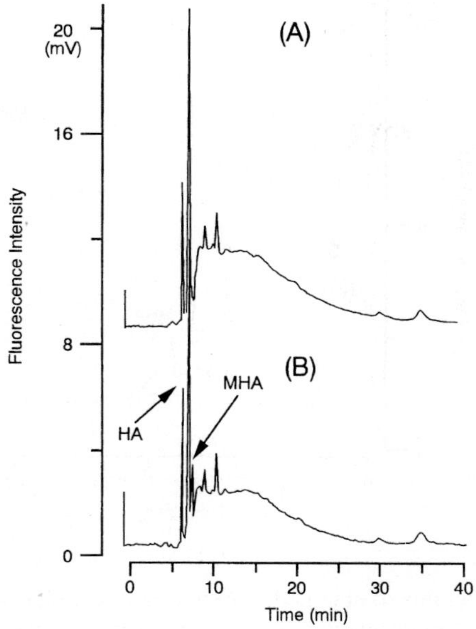

Figure 11 Electrochromatograms of : (A) non-spiked human urine and (B) urine sample spiked with HA (histamine) and MHA (methylhistamine) extracted with BondElute® cartridge. Both electrochromatograms were obtained by using 2-mercaptoethanol in the run buffer. Reproduced from Reference [8] with permission.

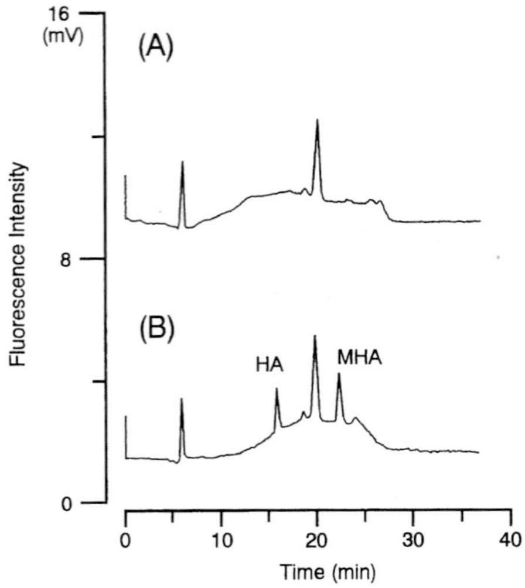

Figure 12 Electrochromatograms of obtained from the same samples of Figure 11, except for 1-butanethiol used instead of 2-mercaptoethnol. Reproduced from Reference. [8] with permission.

References

[1] S. Oguri, T. Fujiyoshi, Y. Miki, Analyst 121 (1995) 1683-1688.

[2] H. A. Bardelmeijer, J. C. M. Waterval, H. Lingeman, R. van 't Hof, A. Bult, W. J. M. Underberg, Electrophoresis 18 (1994) 2214-2227.

[3] A. Taga, M. Sugimura, S. Honda, J. Chromatogr. A 802 (1999) 243-248.

[4] Y. Zhang, F. A. Gomez, Electrophoresis 21 (2000) 3305-3310.

[5] J. Cazes (Ed.), Encyclopedia of Chromatogrphy, Marcel Dekker, Inc., New York, 2001.

[6] S. Oguri, Y. Yoneya, M. Mizunuma, Y. Fujiki, K. Otsuka, S. Terabe, Anal. Chem. 74 (2002) 3463-3469.

[7] S. Oguri, H. Tanagaki, M. Hamaya, M. Kato, T. Toyo'oka, Anal. Chem. 75 (2003) 5240-5245.

[8] S. Oguri, Y. Maeda, A. Mizunuma, J. Chromatogr. A. 1044 (2004) 271-276.

[9] R. D. Slocum, R. Kaur-Sawhney, A. W. Galston, Arch. Biochem. Biophys. 235 (1984) 283-303

[10] P. Jakszyn, A. Agudo, R. Ibáñez, R. García-Closas, G.m Pera, P. Amiano, C. A. González, J. Nutr. 134 (2004) 2011-2014

[11] E. Bonneil, K. C. Waldron, J. Capillary Electrophor. 6 (1999) 61-73.

[12] P. Britz-McKibbin, S. Terabe, J. Chromatogr. A. 1000 (2003) 917-934.

3. Quantitation of Amino Acids and Amines, Simultaneously

Ibolya Molnár-Perl (Editor)
Quantitation of Amino Acids and Amines by Chromatography
Journal of Chromatography Library, Vol. 70

3.

Quantitation of Amino Acids and Amines, Simultaneously

Ibolya Molnár-Perl

Contents

Summary

An overview is presented of chromatographic methods currently in use to determine AAs and As (i) simultaneously in a single run, (ii) in the presence of each others by separate methods, or (iii) amines alone subsequently to their isolation from AAs. Separation, derivatization and chromatographic conditions are summarized.

The advantages and drawbacks of all three possibilities were characterized on the basis of recovery, reproducibility values, time and cost phenomena of methods. (i) From the point of view of the analytical chemist the simultaneous quantitation of AAs and As seems to be the best choice: however to find optimum conditions, with the time not being considered, for AAs and As simultaneously occurring in a matrix, is an unambiguous challenge. In order to solve the problem quickly it helps to have practice in the derivatization and chromatographic conditions of the two groups (AAs, As), separately. (ii) In cases when knowledge of the A content of the sample only is needed, the best choice could be the fast chromatographic elution of AAs without separation: prior to the slow, well-resolved separation of As, such as putrescine, cadaverine, spermine and spermidine, etc. Certainly, As of low molecular weight, inserted into the AA derivatives and are eluting together with them. (iii) On the basis of the unfortunately low recoveries of As, (determined subsequently to their isolation/extraction from AAs), this possibility can be regarded as the worst solution of the task.

1. Introduction

The author's literature [1,2] and research experiences [6-8] lead to the recognition that there is a need for a simultaneous technique of reliable and reproducible quantitation of AAs and As, present in the same natural matrix, in considerably different concentrations. Based on author's derivatization and chromatographic practice in GC [1] and HPLC [2-8] equally, it seemed to be most likely that in this case HPLC is the method of choice.

After an exhaustive literature overview [9-78] it turned out that in the analyses of all those matrices, (wine, cheese, fish, meat, other foods, plant tissues, human biological matrices), that are consisting of AAs and As, (in different, changing concentrations being very often in a continuous transformation into each other), less attention has been paid to a proper, simultaneous technique. The qualitative and quantitative knowledge of this group of compounds, as well as how to determine them without loss and in a fast manner, are of primary importance from a theoretical and practical point of view, equally.

It was surprising to read [9] that the first paper dealing with a special emphasis on the quantitation of several AAs and As from a single run by HPLC, according to its authors was published ten years ago [9] (i.e., "a simultaneous determination of biogenic amines and amino acids, including the precursors of biogenic amines, has not yet been published." In contrary to this remark [9], a number of additional papers [10-78] can be found in the literature, published already in 1978 [18].

In addition, instead of the simultaneous assay of AAs and As, even recently, in order to define the transformation process of AAs into the corresponding As [45-58], or determine As only [59-78], prior to their chromatographic analyses, time consuming and tedious separation techniques were selected, resulting in two main drawbacks: (i) in considerable loss, i.e. in poor and different recovery both of AAs and As, and, (ii) in the additional cost and time consume of separate methods carried out from the two different part/extract of samples.

2. Quantitation of AAs and As from a single run

Simultaneous determination of AAs and As were performed as dabsyl [9-10], as dansyl [11], as 9-fluorenylmethylchloroformyl (FMOC) [[12,13], as OPA/3-mercaptopropionic acid (MPA) [14,15], as OPA/MPA/FMOC [16,17], as OPA/2-mercaptoethanol (MCE) [18-26], as OPA/*tert*.butyl thiol [27], as phenylthicarbamyl (PTC) [28] as fluresceinethiocarbamyl (FTC) [29-31] and 3,5-dinitrobenzoyl derivatives [32]. In addition, for the simultaneous analysis also recently introduced agents [33-36], such as carbazol-9-yl-acetyl [33], or carbazol-9-yl-propionyl chlorides [34], 5-furoylquinoline-3-carboxaldehyde [35] and N-(9-fluorenylmethoxy-carbonyloxy)-succinimide [36] were implemented.. For the histamine/histidine determination capillary zone electrophoresis, pre-column derivatization with naphthalene-2,3-dicarboxaldehyde and fluorescence detection was used [37].

Some proposals were associated with the simultaneous analysis of the underivatized AAs and As [38-43], however, quantitation in underivatized form resulted in the analysis of limited number of constituents, only. Separation techniques were HPLC [38-42], ion chromatography (IC) [43] and capillary electrophoresis (CE) [43]. Detection methods were indirect photometric [38] UV(DAD) [40-42], condensation nucleation light scattering (CNLS) [39], integrated pulse amperometric (IPAD) [43] and electrometric [44] ones.

Sample preparation and derivatization procedures, chromatographic conditions, as well as reproducibility values and affectivity of the proposal, for selected papers, have been given in detail (Tables 1-3: [9-11,18,21,22,16,7,28-31,33,35,39,

40,43]) and/or shown by original chromatograms (Figures 1-12, taken from the following references: [9,11-13,16,29,33-36,40,57]).

2.1. Simultaneous analysis of AAs and As as dabsyl and dansyl derivatives

The proteinogenic and physiological AA and biogenic A contents of complex food matrices [9] have been determined as dabsyl derivatives (Table 1, Figure 1) with a good recovery (98-104%), reproducibility (1.3-3.1%), sensitivity (1-1250 pM) and low detection limit (0.12-0.52 pM). A practical application [10] of this method relates to the evaluation of the effect of temperature on evolution of the major free AA and A contents of Azeitao cheese during storage. In both papers [9,10] derivatization was performed in deproteinized extracts at 70 °C, for 15 min, applying UV detection.

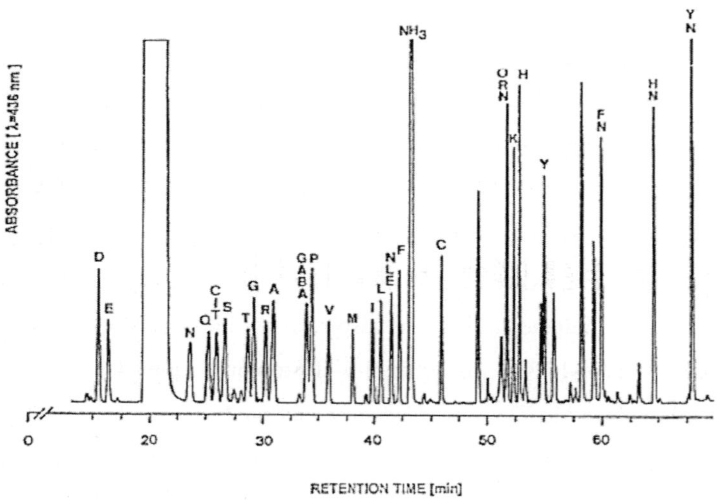

Figure 1 RP-HPLC analysis of the dabsyl derivatives of proteinogenic and physiological amino acids and amines: D: asp, E: glu, N: asn, Q: gln, CIT: citr, S: ser, T: thre, G: gly, R: arg, A: ala, GABA: GABA, P: prol V: val, M: methi, I: ileu, L: leu, NLE: nleu, F: phenala, C: cyst, ORN: orn, K: lys, H: his, Y: tyr, FN: pheta, HN: hisn, YN: tyrn; With permission from Reference [9]

The AA and A content of forest tree tissues were determined as dansyl derivatives, simultaneously and separately, as well [11]. Applying a short column the total run time for 19 AAs and 3 As, including equilibration and column washing, needed 56.6 min (Figures 2A,B)

2.2. Simultaneous analysis of AAs and As as FMOC derivatives

The drawback (wide reagent peaks) in the analysis of AAs and As, as FMOC derivatives, in wine, fruit, vegetable and cheese samples was intended to be eliminated on

Table 1
Simultaneous Analysis of Amino Acids (AAs) and Amines (As) from a Single Injection [7,9-11,13,16,1821,22] as Dabsyl [9-10], as dansyl [11] as FMOC [13] and as OPA Derivatives [7,16,18,21,22], by High Performance Liquid Chromatography [7,9-11,13,16,18,21,22]

[Ref] date	Sample Preparation (P), Derivatization (De)	Column (C), Elution (E), Flow rate (F),	Detection,	RSD %	*Compounds/min
[9] 1995	P: 0.1-1 g cheese, meat, sausage, fish, plasma, tissue+10mL 0.1M HCl cont. 0.2% TDPA: homogenization, centrifugation, ultrafiltration (0.45-µm); De: 20µL AAs+As (.2-50 nM/component)+ 180 µL buffer+ 200µL dabsyl-Cl reagent, 70 °C, 15 min	C: 150 x 4.6 mm (Spherisorb ODS 3µm); E: A, 9mM NaH_2PO_4 cont. 4% DMFA+ 1-2% TEA; B, 80% aqueous ACN; F: 1mL/min, 50°C;	UV 436nm	≤1.3-3.1	26/90: asp, glu, asn, gln, cit, ser, thr, gly, arg, ala, GABA, prol, val, methi, ile, leu, nleu, phenala, NH_3, cyst, orn, lys, his, tyr, pheta, hisn, tyrn
[10] 2001	P: 4g cheese+15 mL 0.2M $HClO_4$ homogenization, centrifugation, ultrafiltration (0.45-µm); De: dabsyl-Cl reagent, 70 °C, 15 min, ice bath 5 min, dilution with ACN+ ET, centrifugation	C: 150 x 4.6 mm (Spherisorb ODS 3µm); E: as in paper [9]	UV 436nm	≤4	18/90: gly, ala, prol, val, methi, arg, ile, leu, lys, his, tyr, put, cad, hisn, spm
[11] 2004	P: 200mg (fresh mass of cells or leaf discs or chopped needle)+1 mL 5% (v/v) ice coold $HClO_4$; (-20 °C) thawed and frozen three times; centrifugation (10m, 13500g) ; 20 µL+100µL saturated sodium carbonate+100µL dansyl-Cl reagent (20mg/mL), 60 °C, 1h, water bath; + 50 µL L-alanine, etc.	C: 100 x 4.6 mm (Luna C18, 3µm or Phenomenex Synergi Hydro RP, 4µm); E: A, ACN; B, 0.025M Na.Ac. cont. 3% 1-propanol+10% ACN (pH 5.94); F: 1mL/min;	Fl, Ex/Em 340/515	≤2.5	22/56.5: 19 AAs+put, spd, spm
[13] 1995	P/De: 20µL wine + 50µL borate buffer (pH 8.5) + 50µL FMOC reagent (8 mg FMOC/1mL ACN)After 3 min+ 300µL quenching solution (ACN/acetic acid/water=20/2/30, v/v)	C: 200 x 2.1 mm (Hypersil ODS, 5µm); E: A, ACN cont. 2-octanol 1%, v/v; B, ACN 150mL+H_3PO_4 8.8 mL+DCA 10mL/1L H_2O, pH 2.7; F: 0.3mL/min,	Fl, Ex/Em 263/313	≤1.6-13.6	13/110: his, arg, agm, phenala, pheta, orn, put, cad, tyr, tyrn, hisn, spd, spm
[18] 1985	P: tissue homogenized, urine diluted with 10mL 0.2M $HClO_4$ (1:10), centrifugation, supernatant analyzed. De: OPA/MCE post column	C: 250 x4.6 mm, ODS+70 x2.1 mm guard, 5µm); E: A, 0.1M Na.Ac (pH 4.5, B, 0.2M Na.Ac/ ACN=10/3 (A&B cont. 10mM octanesulphonic acid; C, Met: 1mL/min	UV & Fl	-	24/115: tyr, anserine, phenala, carnosine, homocarnosine, acetylput, putreanine, 3-NH_2-propionic acid, S-adenosylmet, tryp, 5'-CH_3thio-adenosine, isoputreanine lact., isoputreanine, put, cad, decarboxylated S-adenosylmethi, hisn, spd, spm
[21] 1996	P: Urine/fruit juices cation exchanged for basic and acidic fractions [15]; De: OPA/MCE, post column, 70°C	C: 150 x 4.6 mm (Inertsil ODS 5µm); E: A, 5mM SDS and 10mM $HClO_4$ cont. water: B, ACN; F: 1-2 mL/min, progr., 30 °C	Fl Ex/Em 340/430	≤5.0	20/80: 18 AAs+ creatine and creatinine
[22] 1997	P: ~3g muscle tissues of scallop or herring or mackerel + 15mL 0.2M $HClO_4$ homogenization, filtration (1. Whatman 2V filter paper, 2. 0.2µm filter): De: OPA/MCE, post column,	C: 150 x 4.6 mm (Hypersil ODS 5µm); E: A, 0.1M Na.Ac (pH 4.5, B, 0.2M Na.Ac/ ACN=10/3 (A&B cont. 10mM octanesulphonic acid; C, MET: F:1mL/min	Fl Ex/Em 300/950	-	11/30: arg, orn, lys, hisn, tyr, put, cad, his agm, spd, spm
[16] 2001	P: musts after centrifugation, wines directly diluted for 4x with nor-valine internal standard, filtered (0.45 µm; De: Two injections 1. OPA/MPA 2 min; 2. (112 min after the first) FMOC-Cl 2 min;	C: 250 x 4.6 mm (Merck LichroCART, 100 RP-18 5µm); E: A, 0.02M Na.Ac. Cont. 0.018% TEA, 0.3% THF, EDTA~ $4x10^{-4}$%; B, 20% Na.Ac. 0.01M, 40% ACN,40% MET, 0.018%TEA; F: 0.7-1.2 mL/min	Fl Ex/Em 340/450 237/340	0.5-19.2	32/138: asp, glu, aspn, ser, gln, his, gly, thre, cit, arg, ala, GABA, tyr, eta, val, methi, nval (I.S.), hisn, trp, meta, phenala, ileu, leu, orn, lys, eta, tyrn, put, cad, tyra, isam, proline
[7] 2003	P: Wine, beer, vinegar diluted for twice; De: OPA/MPA=1/50, 15-20 min, +4 °C	C: 200 x 4.6 mm + guard 20 x 4.6 mm (Hypersil ODS 5µm); E: A, 0.05M Na.Ac (pH 7.2); B, A el/ ACN/MET=46/44/10; C=MET; D= ACN; F: 1.3-1.7 mL/min, progr., 50 °C	UV & Fl	≤3.4	37/53: asp, glu, aspn, ser, gln, thre, β-ala, ala, arg, GABA, tyr, eta, val, methi, try, his, phenala, meta, ileu, orn, leu, agm, lys, etam, spm, ipa, tyrn, cad, iba, nba, put, pheta, ima, hexa, hepta, octa

Indications: Compounds/min=number of compounds/elution time, min (column equilibration time included); o-phthaldiadehyde=OPA; 2-mercaptoethanol=MCE; 3-mercap-topropionic acid=MPA; TDPA=3,3'-thiodipropionic acid; DMFA=dimethylformamide; cont.=containing; ACN=acetonitrile; ET=ethyl alcohol; FMOC=fluorenylmethyl chloroformate; DCA=dimethylcyclohexyl-amine; -=no data available; THF=tetrahydrofurane; MET=methyl alcohol;

Table 2
Simultaneous Analysis of Amino Acids (AAs) and Amines (As) from a Single Injection [28-31,33,35,39,40,43], as Phenylthicarbamyl (PTC) [28], as Fluoresceinethiocarbamyl (FTC) [29-31] as Carbazol-9-yl-acetyl [33], as Derivatives and Without Derivatization [39,40,43] by High Performance Liquid Chromatography [28,33,39,40,43] and by Capillary Electrophoresis [29-31,35]

[Ref] date	Sample Preparation (P), Derivatization (De)	Column (C), Elution (E), Flow rate (F),	Detection,	RSD %	*Compounds/min
[28] 1991	P: wine filtered through micropartition system (Centifree MOS-1, Amicon)De: PITC/water/ET/TEA=1/1/7/1, prepared freshly	C: 250 x 4.6 mm + 30 x 3.9 mm precolumn (Spherisorb ODS2, 5μm); E: A, 0.05m Na.Ac, cont. 0.25% ACN+0.05% TEA+1 ppm EDTA (pH 6.8); B, ACN; F: 1mL/min; 52 °C	PDA	2-7	25/40: asp, glu, OH-prol, ser, asn, gly, gln, his, GABA, thre, ala, arg, prol, ABA, tyr, hisn, val, methi, cystine, ileu, leu, pnenala, trp, orn, lys
[29-31] 1995-1997	P: 50μL cerebrospinal fluid [22,23], 0.23g cheese-paste extracted by 5mL 0.1M HCl, 3x, centrifuged [23] and 50μL wine [19]diluted with 0.2M carbonate buffer; De: 50 μLsamples+50μL 2.1 x 10^{-4}M FITC in acetone 2h, in dark	C:75 cm x 50μm fused silica (effective length 42cm); e: buffer 100mM SDS+100 mM boric acid +NaOH until pH 9.0; Separation=20kV, electrophoretic current 42μA	LIF	-	28/30: put, arg, lys, hisn, orn, his, eta, tyrn, β-pheta, ammonia, spd, spm, thre, ileu, leu, nleu, prol, phenala+val, try, asn, tyr, ser, ala cyst, taurine, gly, glu, asp, cysteic acid
[35] 2001	P: 4μL brain microdialysate from different tissues prepared in paper [22]; De: 4μLsample+100nM dry FQ, diluted with 0.2M carbonate buffer +2μL 25mM KCN (pH 9.2), reaction in dark, at 65 °C, for 16 min;	C:40 cm x 50μm fused silica; E: 60mM SDS+20 borate (pH 9.0), or 45mM SDS+ 15mM borate+5mM β-CD (pH 8.5); Separation=200V /cm	LIF	CC= 0.96-1.0	16/18(without flushing by NaOH, water, buffer): ser, thre, gln, glu, dopamine, his, asn, ala, taurine, asp, GABA, gly, ser, norepi, methi, β-ala
[33] 1999	P: Foods:20g cheese suspended with 40mL 0.1M HCl, stirred for 30 min (0-4°C), centrifugation, filtration of supernatant. Residue extracted 2x, by 100mL 0.1M HCl; Plant materials: 0.1-0.5g barley seedling chopped, homogenized by 5mL 5% $HClO_4$, extr. for 60 min. Centrifugation, filtration. Residue extracted 2x, with 2mL 5% $HClO_4$ De: 40-400L solution (pH 8.8) + 50μL CRA-Cl reagent 3-5min	C: 200 x 4.6 mm (Spherisorb ODS 5μm); E: A, 20mM NH_4H_2PO4 cont. 9mM TEA (pH 6.5)/Met =9/1 (v/v); B, ACN; F: 1mL/min, 35 °C	Fl Ex/Em 335/339	≤5	24/95: asn, gln, ser, his, gly, thre, ala, prol, tyr, arg, val, methi, ileu, leu, phenala, cyst, lys, pheta, put, hisn, cad, hepta, spm, spd
[43] 1998	P: 10g homogenized kiwi/meat/fish/cheese+30mL $HClO_4$ (0.375M), completed to 50-100mL, extracted with hexane (2:1, v/v), centrifuged, filtrated 0.2 μm); De: -	C: cation-exchange anal. (IonPac CS10, Dionex); E: isocratic, 1M $NaClO_4$/0.375M $HClO_4$/water =81/5/14, v/v[#]; F: 1mL/min	IPAD	3.4 - 8.8	7/30: put, his, tyr, cad, hisn, tyrn, spd
[40] 1999	P: 1g cheese homogenized for 2 min by 5 mL 5% TCA two times+with10 mL. Centrifuged supernatants combined, extracted 3 x by 10 mL diethylether. Ether removed by N_2. Residue + 25 mL water; injected 20 μL; De: -	C: 150 x 4mm + 20 x 4 mm pre-column (Spherisorb ODS2, 5μm); E: A, 0.05M KH_2PO_4 cont. 0.005M heptanesulphonic acid, acidified by H_3PO_4 (pH 3.4); B, 60%A/40% ACN; F: 1 mL/min	UV 240	-	8/40: his, tyr, phenala, hisn, tyrn, try, β-phenylethylamine, trypn
[39] 1999	P: 6.5 g tuna fish /mackerel sardines washed, ground, homogenized (magnetic stirring 45 min) +20 mLwater+10 mL 6M HCl; 100μL aliquot syringe filtered, injected	C:100 x4.6 mm (Alltech cation exchange, filled with polybutadiene-maleic acid coated silica) E: 0.01M HNO_3 cont. 10% ACN; F: 1 mL/min;	CNLSD	3.4 - 8.8	7/30: put, his, tyr, cad, hisn, tyrn, spd

Indications: as in Table 1, as well as: ABA= α-aminobutyric acid; CRA-Cl reagent=10mM carbazole-9-yl-acetyl chloride (Note: derivatization should be performed in 50% ACN and 0.2M borate buffer, pH 8.8); SDS=sodiumdodecylsulphate; IPAD=integrated pulsed amperometric detection; [#]=for detection the eluent was mixed by 0.25M NaOH (0.8mL/min); LIF=laser induced florescence; FITC=fluoresceine isothiocyanate; FQ=5-furoylquinoline-3-carboxaldehyde; CC=correlation coefficient; &=N^1-Acetylspd, N^8-Acetylspd, N^1-Acetylspm; CNLSD=condensation nucleation light scattering detection

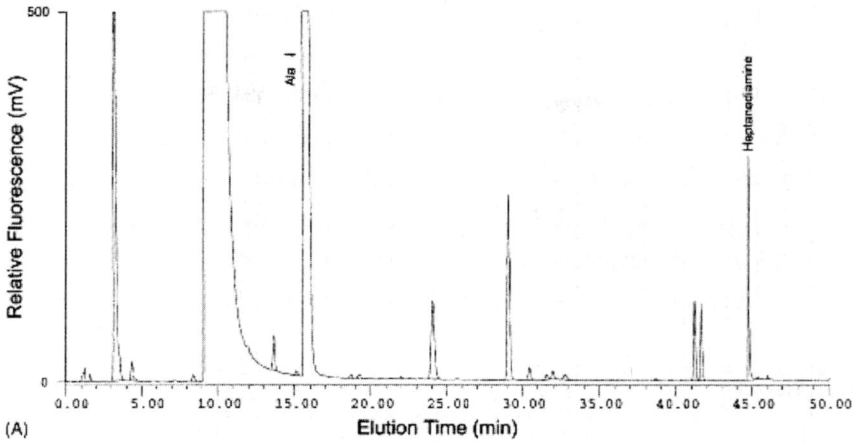

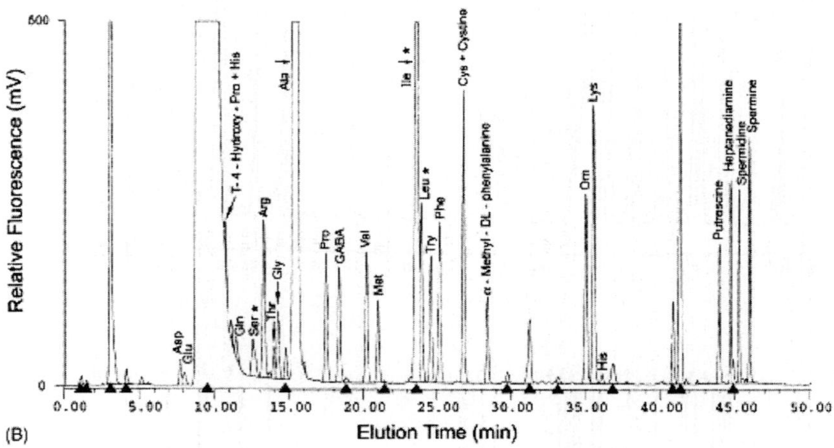

Figures 2A,B Separation of reagent blank (A), as well as dansyl amino acids and dansyl polyamines (B) obtained from mature red spruce (Picea rubens) trees; Column: Phenomenex Synergi Hydro-RP; Injected 10 μL; ▲=indicates the presence of blank peaks; With permission from Reference [11]

the basis of an exhaustive study [12] regarding: (i) derivatization conditions, such as the excess of the reagent and quenching compound, (ii) the optimum reaction time necessary for quantitative reactions, and, (iii) the optimum amount of quenching compound. Authors proved [12] that in order to use FMOC reagent in 'high concentration', (5 mM was needed for quantitative derivatization, particularly in cases of As), avoiding reagent's precipitation in concentrated buffer solutions (0.2 M), instead of ACN, as solvent acetone is to be preferred. To eliminate the high excess of FMOC-Cl requires the equivalent excess of the quenching compound, which also had to be held in solution. Thus, the amounts of the reagent's and quenching compound's excesses proved to be also a matter of compromise. The optimum reaction time in the case of

AAs 3 min (more than 3 min led to peaks which had a strong pronounced shoulder, or double peaks), while in case of As of lower reaction rate (spm, spd, pheta) 5 min reaction time proved to be the optimum: consequently, as compromise 3 min reaction time was chosen. As seen in spite of all these precautions reagent peaks could not be quantitatively eliminated (unsigned peaks at Figures 3A,B) taken from reference [12]).
Five AAs and eight As were measured simultaneously in 15 rose and 15 white wine [13]. The well-known disadvantage of FMOC derivatization, i.e. the disturbing effect of the wide reagent peaks (FMOC-Cl and/or FMOC-OH) can be seen at Figure 4 [13].

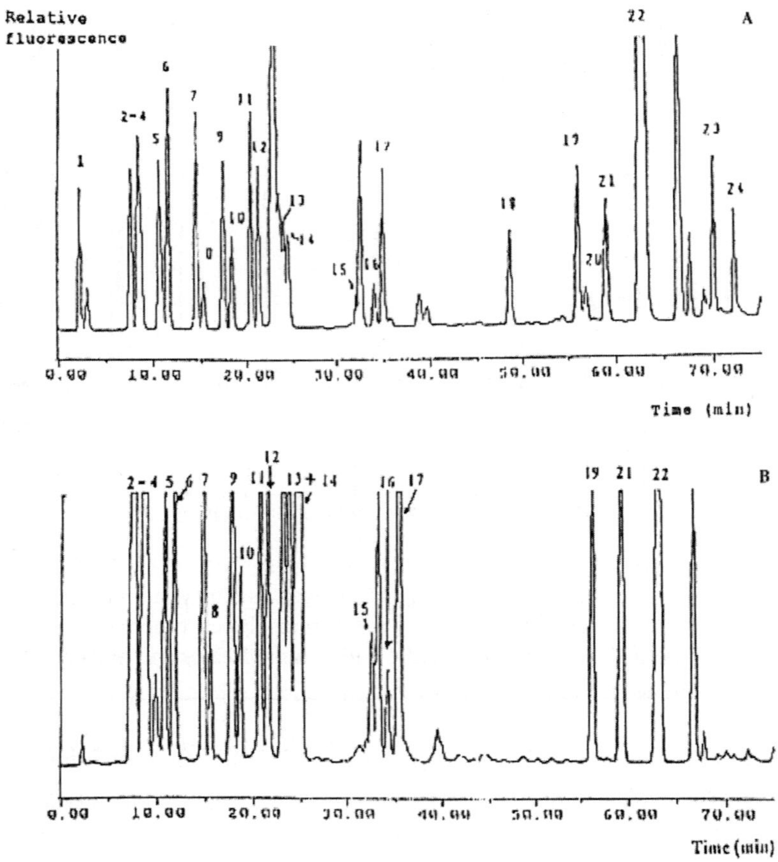

Figures 3A,B HPLC chromatogram of biogenic amines and amino acids after derivatization with FMOC-Cl and removing the FMOC reagent's excess with heptylamine (hepta), from standard mixture (3A, concentrations 5 μg/μL) and from cheese extract (3B). Peaks, 1: cysteic acid, 2: asp, 3: ser, 4: glu, 5: thre, 6: gly, 7: ala, 8; tyr, 9: prol, 10: methi, 11: val, 12: phenala, 13: ileu, 14: leu, 15: his, 16: tyr, 17: lys, 18: pheta, 19: put, 20: hisn, 21: cad, 22: hepta, 23: spd, 24: spm; With permission from Reference [12]

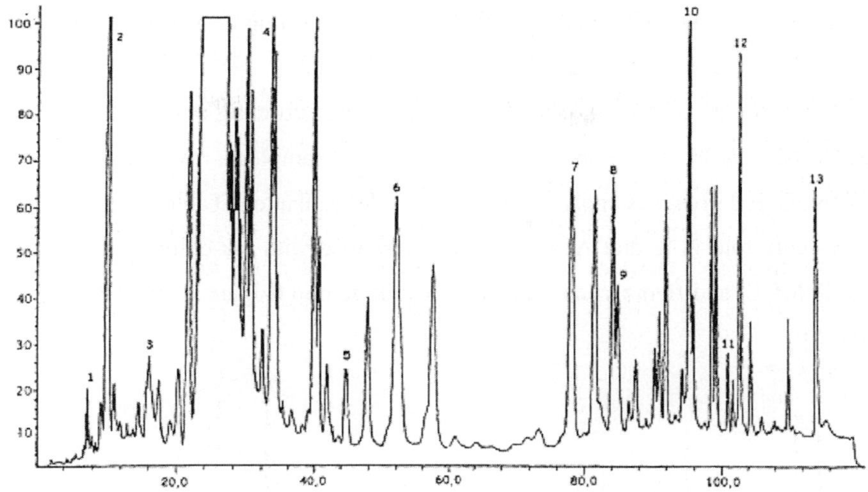

Figure 4 Chromatogram of FMOC derivatives obtained from a wine sample overloaded with 1: his 4.6 mg/L, 2: arg 5.15 mg/L, 3: agm 4.1 mg/L, 4: phenala 4.7 mg/L, pheta: 5.2 mg/L, 6: orn 3.6 mg/L, 7: put 4.0 mg/L, 8: cad 4.5 mg/L, 9: tyr 4.7 mg/L, 10: tyrn 6.2 mg/L, 11: hisn 4.6 mg/L, 12: spd 3.9 mg/L, 13: spm 3.9 mg/L; With permission from Reference [13]

2.3. Simultaneous analysis of AAs and As as OPA derivatives

For the simultaneous derivatization of AAs and As mainly the OPA/MCE reagent was selected [18-25], exceptions were the use of the OPA/MPA [7] and the two step derivatization process carried out by the OPA/MPA/FMOC reagents [16].

The first simultaneous analysis of AAs and As was performed in the last seventies [18] as OPA/MCE derivatives: separation and quantitation of 24 compounds, performed by the classical ion-exchange chromatography (IEC) required in total 115 min. Three different buffer eluents were applied with pH 5.55 for 37 min, pH 5.63 for 40 min and pH 5.73 for 93 min. Elution's temperatures were 66 °C for 55 min and thereafter 78 °C until to the end of the run. Washing (0.2 M NaOH containing 250 mg/L EDTA, for 15 min) and equilibration (first buffer for 30 min) processes needed an additional 45 min. The first HPLC of terner gradient elution, in order to quantitate AAs and As simultaneously, present in urines and tissue extracts, (in total of 24 compounds), was performed as their OPA/MCE derivatives, applying fluorescence detection [19].

A dual mode (UV and Fluorescence detections), gradient HPLC method was developed in order to determine urinary bases (creatine, creatinine) and diagnostic

AAs in a single run as their OPA/MCE derivatives, performing post-column derivatization and flow-rate gradient (1-2 mL/min), (Table 1) [21].

The amounts of four AAs and seven As — transforming under postmortem storage in the muscle of scallop, herring and mackerel samples — were followed as their OPA/MCE derivatives applying post-column derivatization (Table 1, [22]).

Twenty-four AAs and As have been determined with pre-column derivatization (OPA/MCE) and fluorescence detection within 50 min (Figure 5, taken from reference [25]).

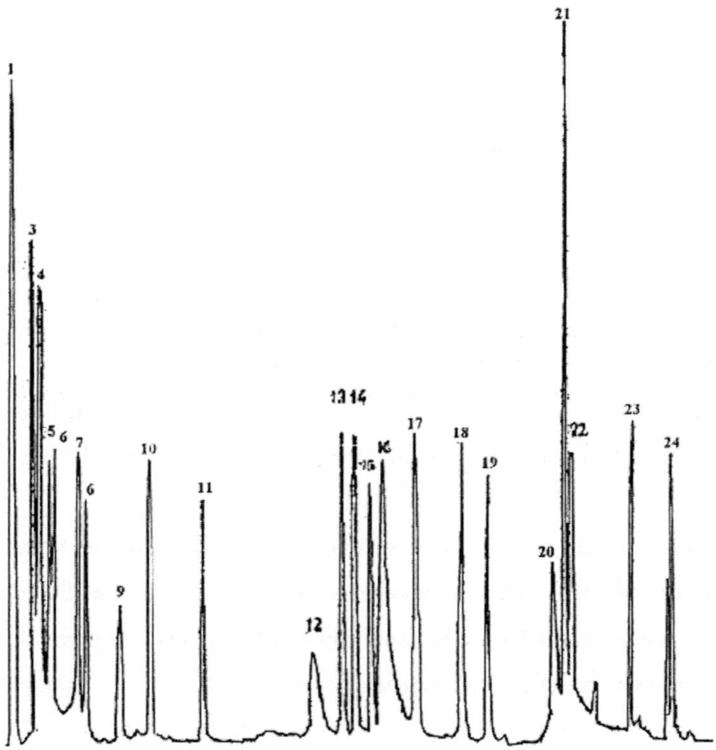

Figure 5 HPLC profile of amines and amino acids, measured as OPA/MCE derivatives. 1: asp, 2: glu, 3: asn, 4: ser, 5: gln, 6: his, 7: gly, 8: thre, 9: arg, 10: ala, 11:tyr, 12: agm, 13: methi, 14: val, 15: tryp, 16: hisn, 17: phenala, 18: ileu, 19: leu, 20: orn, 21:tyrn, 22: lys, 23: put, 24: cad; With permission from Reference [25]

The formation/transformation of free AA and A constituents of 39 musts and 33 wines were followed during the fermentation process, as their OPA/MPA/FMOC derivatives, performing fluorescence detection (Table 1) [16]. The quantitative analysis of 32 AAs and As required 138 min (Figures 6A,B).

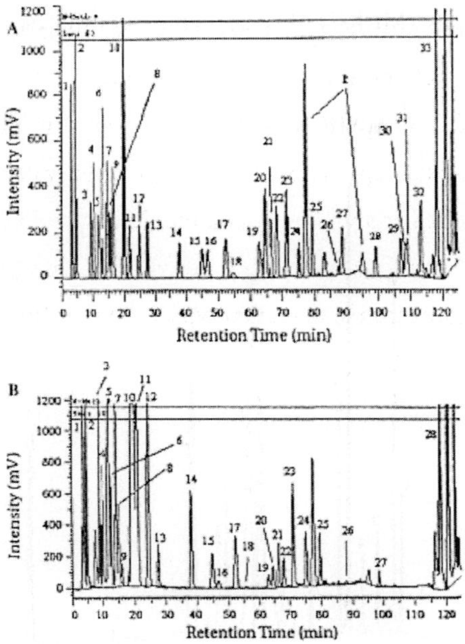

***Figures 6A*,B** Chromatogram of a standard solution (5a) and a must sample at the beginning of alcoholic fermentation (5b) obtained after derivatization with the OPA/MPA/FMOC reagent. Peaks, 1:asp, 2:glu, 3: asn, 4: ser, 5: gln, 6: his, 7: gly, 8: thre, 9: cit, 10: arg, 11: ala, 12: GABA, 13: tyr, 14: eta, 15: val, 16: methi, 17: norval, 18: hisn, 19: tryp, 20: meta, 21: phenala, 22: ileu, 23: leu, 24: orn, 25: lys, 26 & 27: tyrn, 28: put, 29: cad, 30: pheneta, 31: trypt, 32: iama, 33: prol, R: reagent peak; With permission from Reference [16]

The simultaneous determination of AAs and As, in total 37 compounds, as their OPA/MPA derivatives has been developed recently (Table 1) [7]. Detailed description of the chromatographic method and its application to the analysis of beer, wine and vinegar samples can be found in this issue.

2.4. Simultaneous analysis of AAs and As as fluoresceineisothiocarbamyl (FTC) and phenylthiocarbamyl (PTC) derivatives

Histamine together with twenty-four AAs in wine has been quantitated as their PTC derivatives by HPLC applying UV detection (Table 2) [28]. Free AAs and primary As (50 compounds/22 min: 21 quantitated, 19 identified) from child-cerebrospinal fluid (Figures 7AB) [29]) and 28 AAs and BAs from red French wines over a 14 year period [31], were determined as their FTC derivatives by CE [29] and by micellar electrokinetic chromatography (MEKC) [31], applying in both cases [29,31] LIF detection (Table 2).

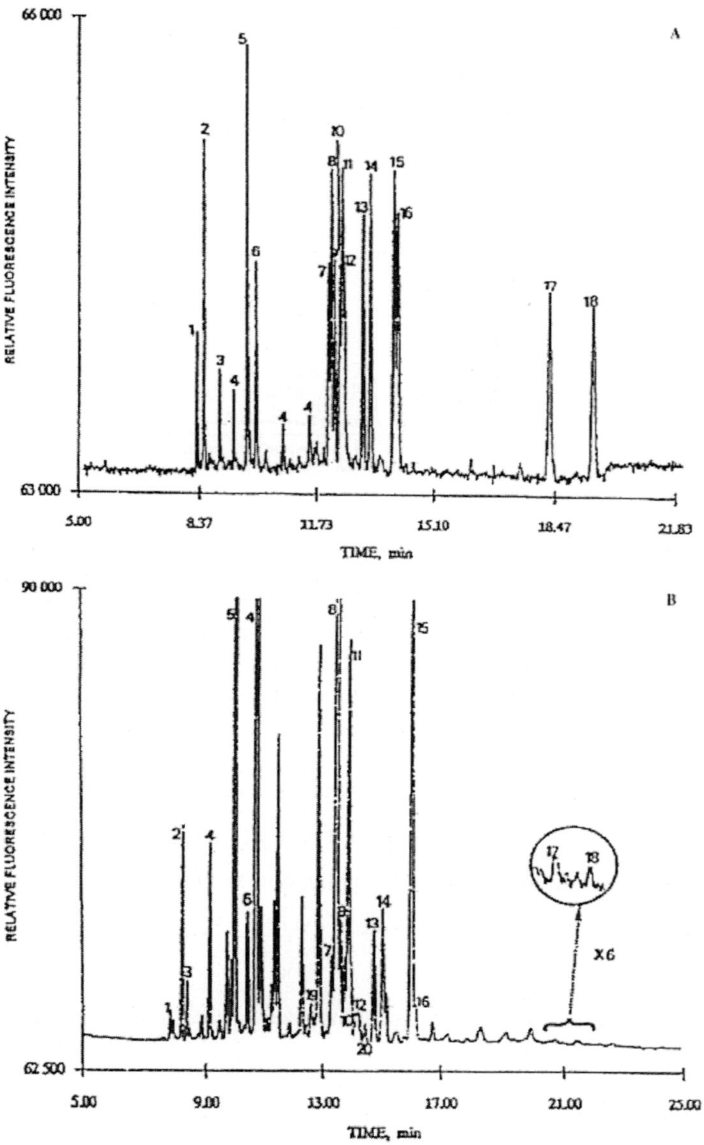

Figures 7A,B Electropherogram of seventeen FTC-standards (Figure 7A, 10^{-9} M) and AAs and As from the cerebrospinal fluids of acute lymphoblastic leucemic patient (Figure 7B), 1: lys, 2: arg, 3: orn, 4: blank, 5: ammonia, 6: tyrn, 7: leu, 8: gln, 9:tyr, 10: val, 11: thre, 12: phenala, 13: ser, 14: ala, 15:taurine, 16: gly, 17: glu, 18: asp, 19: citr, 20: asn; With permission from Reference [29]

2.5. Simultaneous analysis of AAs and As as furoilcarboxyl, carbazo(5-furoylquinoline)-3-carboxaldehyde and N-(9-fluorenylmethoxycarbonyloxy)succinimidyl derivatives

Rapid and sensitive pre-column derivatization methods [33,34] have been proposed using carbazol-9-yl-acetyl chloride (CRA-Cl) (Table 2) [33] and/or carbazol-9-yl-propionyl chloride (CRP-Cl) [34] as derivatizing reagents, for the simultaneous determination of AAs and biogenic As, via HPLC with fluorescence detection. Optimum derivatization and chromatographic conditions were given in detail [33,34]. The elution of the CRA derivatives needed 95 min (Table 2; Figures 8AB [33]) for 25 compounds, while of those of the CRP derivatives were somewhat faster (Figure 8C,D [34]): it lasted 60 min.

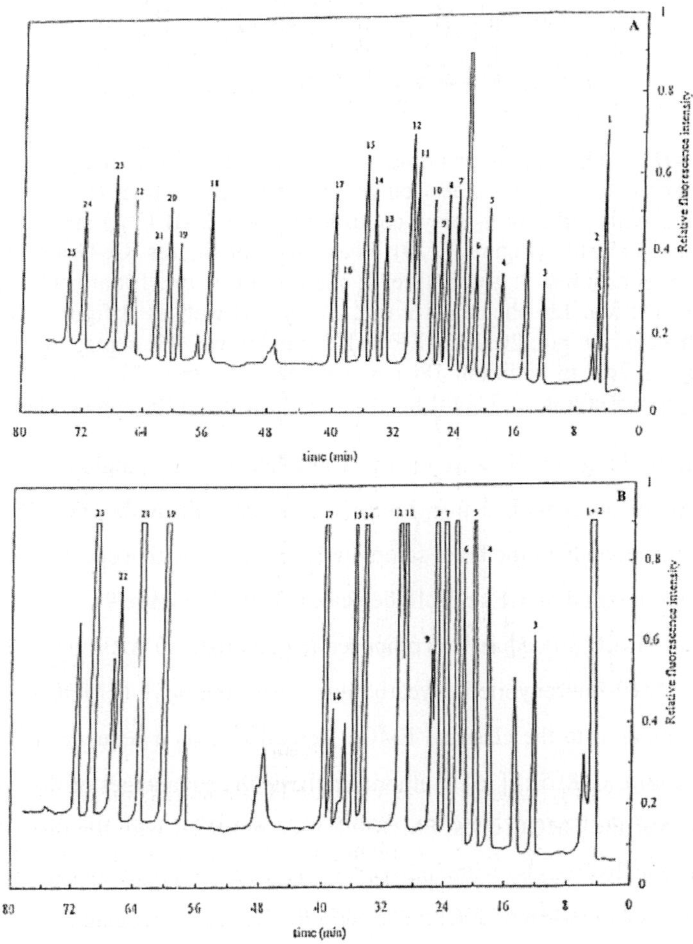

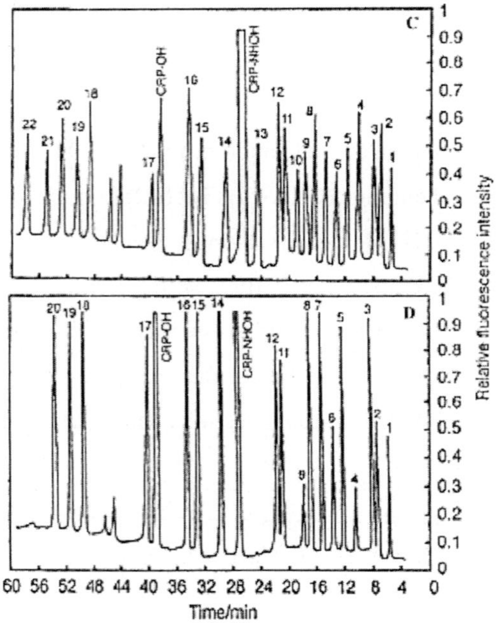

Figures 8A-D LC chromatograms obtained with a standard mixture (Figure 8A,C, 50 pM) of biogenic As and AAs, as well as with a cheese extract (Figure 8B,D) after derivatization with carbazol-9-acetyl chloride (Figures 8AB [33]) and with carbazol-9-yl-propionyl chloride (Figure 8C,D); Peaks 1-17 at figures 8A-D are the same, 1: asp, 2: glu, 3: ser, 4: his, 5: gly, 6: thre, 7: ala, 8: prol, 9: tyr, 10: arg, 11: val, 12: methi, 13: ileu, 14: leu, 15: phenala, 16: cys, 17: lys, as well as at figures 7a,b in addition: 18: pheneta, 19: put, 20: hisn, 21: cad, 22: unknown, 23: hepta, 24: spd, 25: spm and at Figures 7c,d in addition: 18: put, 19: cad, 20: hepta, 21: spd, 22: spm. With permission from Reference [33] for Figures 8A,B and from Reference [34] for Figures 8C,D]

AAs and biogenic As were resolved and detected in a single run subsequently to their derivatization with 5-furoylquinoline-3-carboxaldehyde (FQ) [35]. Sixteen FQ-tagged compounds in the brain sample were resolved with a spectacular profile in less than 16 min based on CE and LIF detection (Table 2, Figure 9).

A new HPLC procedure was proposed to quantitate 20 AAs and 4 BAs, within 32 min, as N-(9-fluorenylmethoxycarbonyloxy)succinimidyl (FMOC-OSu) derivatives and compared to the classical FMOC derivatives, on a sequence of three short columns (50 mm, C18, 5μ) [36]. Authors emphasis the advantages of the FMOC-OSu derivatives over the FMOC ones: FMOC-OSu do not react with the phenolic and selected alcoholic hydroxyls, with imidazoles, his and tyr forme single products and extraction steps can be omitted. Unfortunately the disturbing reagent peaks are present

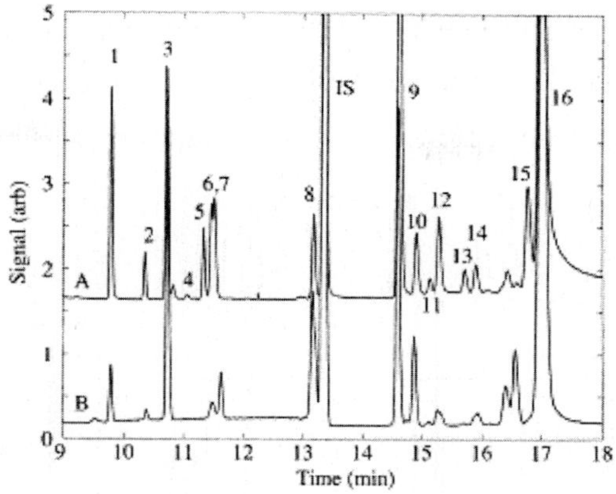

Figures 9A,B CE separation of a standard solution of 16 biogenic amines or amino acids (A) and microdialysate sample spiked with internal standard (IS=*O*-methyl-L-threonine); Peaks 1-16, 1: ser, 2: thre, 3: gln, 4: glu, 5: dopamine, 6: his, 7: aspn, 8: ala, 9: taurine, 10: asp, 11: GABA, 12: gly, 13: serotonine, 14: norepinephrine, 15: methi, 16: β-ala; With permission from Reference 35

on both chromatograms (Figures10A,B). The excellent repeatability of the method (1.1-6.7 RSD%) was shown by the analysis of beer, wine and cell culture samples.

2.6. Simultaneous analysis of AAs and As without derivatization

Six compounds (put, his, spd, spm, lys and arg) extracted from the tumor cells of brain samples were detected from nanoliter injection volumes by high performance capillary electrophoresis (HPCE). As background electrolyte quinine sulfate and indirect photometric detection was used [38].

Biogenic As (put, cad, spm, hisn, spd) and their precursor AAs (his, tyr) were determined simultaneously, in their underivatized form, in food samples (vegetables, cheese, fish) on a cation exchange column performing integrated pulsed amperometric detection (IPAD) (Table 2) [43]. The analytes were extracted from the food samples with perchloric acid and the extracts were purified by liquid-liquid extraction using hexane. Good recoveries (85.5-97.4%) and RSD percentages (3.4-8.8) were reported (Table 2) [43].

Ion pair HPLC (heptanesulfonic acid) and UV detection (215 nm) were used for the quantitation of the most relevant AAs (his, tyr, phenala, try) and As (hisn, tyrn, pheta, trypn) in cheese samples, within 40 min, with perfect resolution (Figures 8AB (Table 2) [40].

Four constituents of fish samples (his, hisn, put, cad) were determined within 8 min applying IEC and CLNS detection [39]. The high acid concentration of the eluent (10 mM nitric acid containing 10% ACN) ensures the short retention times (Table 2).

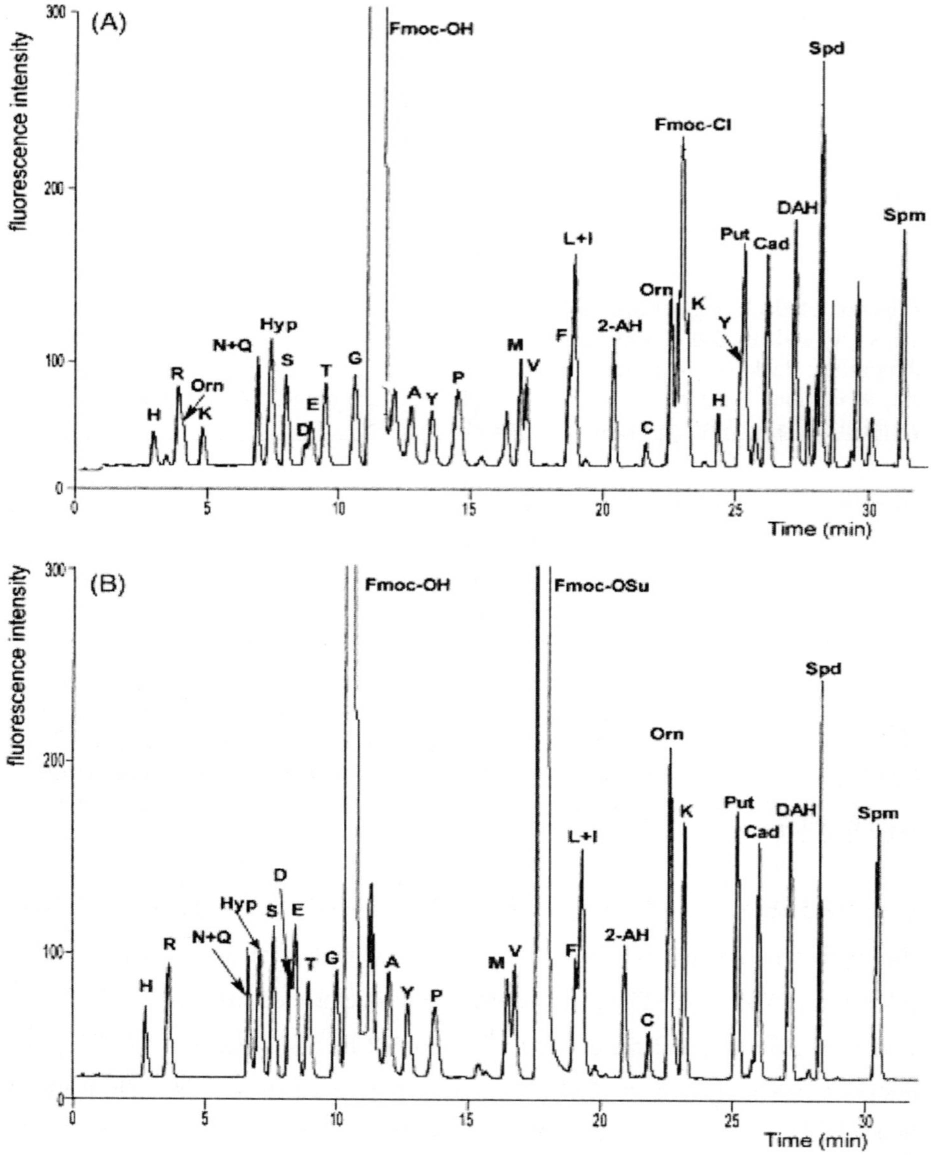

Figures 10A,B HPLC separation of 5 μM standard solution of AAs and PAs, derivatized with FMOC (A) and with FMOC-OSu (B); AA peaks are labeled with one letter abbreviations, as well as: Hyp: hydroxyproline, Orn: ornithine, 2-AH: 2-aminoheptanoic acid, DAH: 1,7-diaminoheptane. With permission from Reference 36]

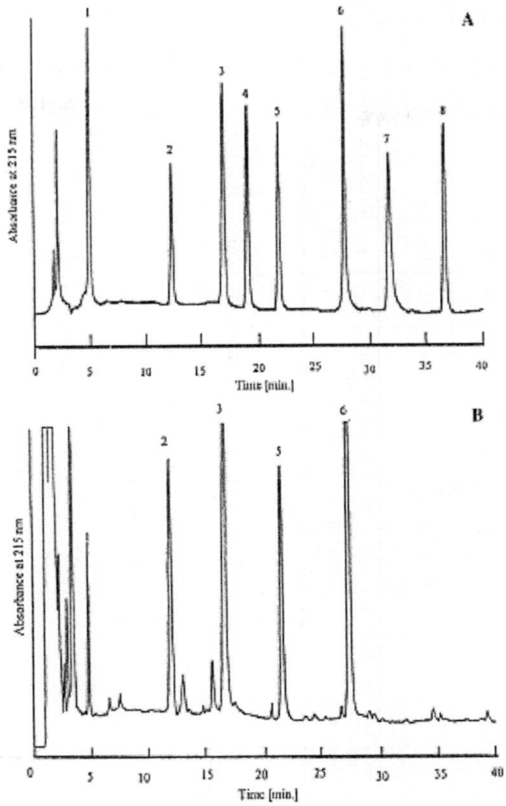

Figures 11A,B *HPLC chromatograms of standard amines and their precursor AAs (Figure 11A) as well as that of a sample of bacterial starters used for cheese production (Figure 11B) measured with ion-pair chromatography and UV detection. Peaks, 1: his, 2: tyr, 3: pheneta, 4: hisn, 5: tyrn, 6: tryp, 7: pheneta, 8: trypt. With permission from Reference [40]*

3. Determination of AAs and As from the same matrix by separate chromatographic techniques

3.1. Applying high performance liquid chromatography (HPLC)

An earlier review to this field gave a short summary [45] of common methods with particular attention to the analysis of wines: all of these proposals follow the quantitation of AAs and As from a previously, differently isolated part of the matrix. AAs were quantitated mainly with the classical ion exchange (IEC) chromatography, while amines either by their OPA or via their dansyl derivatives.

According to recent proposals [46-48], AAs and As were derivatized with the common OPA/MCE reagent in wine samples [46] and in microdialysates of biological species [47,48]. Derivatization of AAs was performed in diluted wine [46]

Table 3
Determination of AAs and As from the same matrix by separate chromatographic techniques

[Ref] date	Sample Preparation (P), Derivatization (De)		Column (C), Elution (E), Flow rate (F),		Detection,		*Compounds/min	
	AAs	As	AAs	As	AAs	As	AAs	As
[47] 2001	P: brain microdialysate 5 μL; De: OPA/MCE	P: brain microdialysate 25 μL; De: -	C: 250 x 4.6mm (Spherisorb ODS2, 5μm); E: A, 0.08M NaH_2PO_4/MET/ACN /THF=90/24/2/1(v/v); B, 0.04M NaH_2PO_4/ MET/THF=67/55.5/3 (v/v), pH 6.2; 30°C	C: 150 x 2.1 mm, Waters Symmetry Shield RP8+Waters Symmetry Guard C18; E: In 920 mL water 6.6g NaH_2PO_4+197.2g SOS+ 137.7g EDTA+80mL ACN (pH 2.9); 30°C	Fl Ex/Em 260/450	Elect r. 0.85 V	9/30: asp, glu, asn, ser, gly, taur, ala, GABA	5/30: NA, DA, DOPAC, HVA, 5-HIIA, 5-HT
[49] 1994	P: 10g ham sample + 5% SA homogenized (2°C, 17 h) centrifuged, filtrated (Whatman No.54 paper); De: PITC	P: 10g ham extracted (2/1=chloroform/ /butanol); De: dansyl-Cl [38]	C: 250 x 4.6mm (Supelcosil LC18, 5μm); E: A, 0.03M Na.Ac (pH 6.8); B, ACN/water =90/10; F: 1 mL/min; 35 °C	C: -; E: A, 0.01M Na.Ac. cont. 5% THF (pH 4.2); B, ACN/THF=9/1; F: 1 mL/min	UV 254	UV 254	22/56.5: protein AAs,	7/47: put, trypn, cad, histn, tyrn, sperm, spermd
[54] 2002	P: wine dilution 2x filtr. (0.45μm membrane), De: postcolum with ninhydrine, 121 °C	P: 15-20mL wine evapor. To 1 mL+ 2 mL Na.Citr. (pH 2.2), filtr. (0.45μm membrane), De: postcolumn ninhydrine, 121 °C	C: 36 x 0.37 cm; E: Na.Citr. buffers, A, (pH 2.6); B, (pH 3.0); C, (pH 4.25); D, (pH 7.9); F: 0.3 mL/min; 60 °C	C: 6 x 0.37 cm; E: A, (pH 6.0); B, (pH 5.7); C, (pH 5.45); F; 0.3 mL/min; 60 °C	UV 570/440	UV 570	All stand.AAs /92:	7/101: histn, tyrn, put, cad, agm, spermd, sperm
[55] 1999	P: 50 mg leaf powder extraction by ET. Insoluble residue.+1 mL water. De: 20μL extract +AQC reagent (10 min, 50 °C)	P: detailed in the text; De: supernatant $+Na_2CO_3$ + dansyl-Cl (60 °C, 1h); Excess removed (proline), extracted by toluene	C: 150 x 3.9mm (Waters Accq-Tag, 4μm); E: A, 140mM Na.ac+17mM TEA (pH 5.05); B, ACN/water=60/40 (v/v), F: 1mL/min; 37 °C	C: 250 x 4.6 mm, Beckman C18 Ultrasphere 5μm); E: Water/MET gradient from 40/60 to 5/95 (v/v); F: 1 mL/min; 27 °C	UV 254	Fl Ex/Em 336/486	22/135: asp,asn, ser, glu, gly, his+ +gln, β-ala, arg+ thre, **and	11/40: agm, tryp, dap, put, cad, octopamine ***and

Indications as in Tables 1, 2, as well as: NA=noradrenaline; DOPAC=3,4-dihydroxyphenylacetic acid; DA=dopamine; 5-HIIA=5-hydroxyindole-3-acetic acid; HVA=homovanillic acid; 5-HT=5-hydroxytryptamine (serotonine); SA=sulfosalicylic acid; Et.Ac.=ethyl acetate; dap=1,3-diaminopropane; **and=prol, GABA, cys, tyr, val, lys, methi, orn, ileu, leu, phenala; ***and=methoxy-tyramine, spermd, sperm, (elution time: 85 min/21 AAs). The chromatography of AAs and As obtained from the brain areas of rat using in vivo microdialysis [47] was detailed in Table 3.

Glutamic acid, glycine, taurine and γ-aminobutyric acid (GABA) content of microdialysis samples were measured in two runs as their OPA/MCE derivatives [48]. Both elutions were really fast: glutamic acid, glycine and taurine were eluted together with additional seven AAs, within 5 min. The separation of GABA needed also an extra 5 min. The quantitation in a single run would be easily achievable (author's remark).

Evolution of free AA and A content of cured ham during ripening were determined as PTC-amino acid and dansyl-amine derivatives [49]. AAs were derivatized in a deproteinized (by sulfosalicylic acid) homogenate after its centrifugation and filtration, while As in a chloroform/butanol (2/1, v/v) extract obtained from an other deproteinized (by $HClO_4$) homogenate subsequently to its centrifugation and filtration (Table 3).

Derivatization of AAs with the phenylisothiocyanate (PITC) and As with the 6-aminoquinolyl-N-hydroxysuccinimidyl carbamate (AQC) reagents, present in wines [50], resulted in the quantitation of six precursor AAs and six As: detailed conditions were not given.

The formation of biogenic As and liberation of free AAs during cheese ripening [51,52] were followed from the same sample by different techniques: AAs were determined in both cases by the classical IEC method {in paper [51]: 37 compounds, in paper [52]: 41 constituents}, As in acidic extracts as their dansyl derivatives, while tryptophan alone from a basic extract.

Basic AAs (hisidine, arginine, ornitine, lysine) and polyamines (putrescine, spmidine and spmine) from complex biological samples, prior to their HPLC quantitation were isolated by various ion exchange clean-up processes [53]. The AA and A contents of dried ion-exchanged residues were quantitated by HPLC as their FMOC (AAs) and dansyl (As) derivatives.

The AA and A content of wine were determined recently [54] by the classical IEC applying different columns and elution processes [53] (Table 3).

The AA, polyamine and betaine contents of freeze dried and ground plant tissues [55] (Table 3) were determined from two different extracts: (i) polyamines and betaine from the supernatant of a centrifuged homogenate obtained with 1 M ice cold

HCl, as their dansyl derivatives (derivatization: 60 min, 60 °C + extraction by toluene; chromatography: 40 min), while AAs from an ethanol extract obtained at 95 °C, prior to their precolumn derivatization with AQC (derivatization: 10 min, 50 °C; chromatography 135 min).

The AA, monoamines and their metabolites from biological tissue [56] were determined by two different HPLC-electrochemical detected (ED) process using microbore columns. AAs were quantitated as their tert.-butylthio derivatives (at 700 mV versus Ag/AgCl reference electrode, derivatizaion time: 2 min, chromatography of 14 AAs: 16 min), while As were measured at 700 mV using the Unijet microbore electrochemical detector (Bioanalytical System, chromatography of 7 As: 24 min).

The changes in the free AA and BA content during ripening of frozen sardine samples have been determined by using the classical ninhydrine method (by means of a amino acid analysator) and OPA/MCE derivatization [57].

3.2. Performing capillary electrophoresis (CE)

N-alkane diamines, lysine and diamino pimelic acid have been determined by capillary zone electrophoresis subsequently to their derivatization by salicylaldehyde-5-sulfonate [58]. Primary amino groups provided stable aldimines, at 25 °C in 10mM phosphate buffer (pH 7.8) and 60% (v/v) ethanol containing solution. Unfortunately migration solutions for the separation of As (Figure 12A) and amino acids (Figure 12B) could not be performed in a single run.

4. Quantitation of As only, followed their separation from AAs

Above methods relate mainly to the quantitation of As [59-78] accompanied by AAs, present in almost all matrices, in overwhelming concentrations compared to As. Chromatographic analyses of As were carried out as their OPA [59-64], as their dansyl derivatives [65-75], as well as without derivatization by CE [76-78]: performed subsequently to their separation from AAs by solvent extraction, ion exchange chromatography, solid phase extraction, etc.

4.1. Determination of amines as their OPA derivatives by HPLC

Derivatization of As subsequently to various separation processes were performed with the OPA/MCE [59-63] and with the OPA/NAC [64] reagents.

The most important wine amines {hisn, tyr, put, pheta, cad, isoamylamine (iama), heptylamine (hepta)} have been quantitated in the presence of AAs of ~20

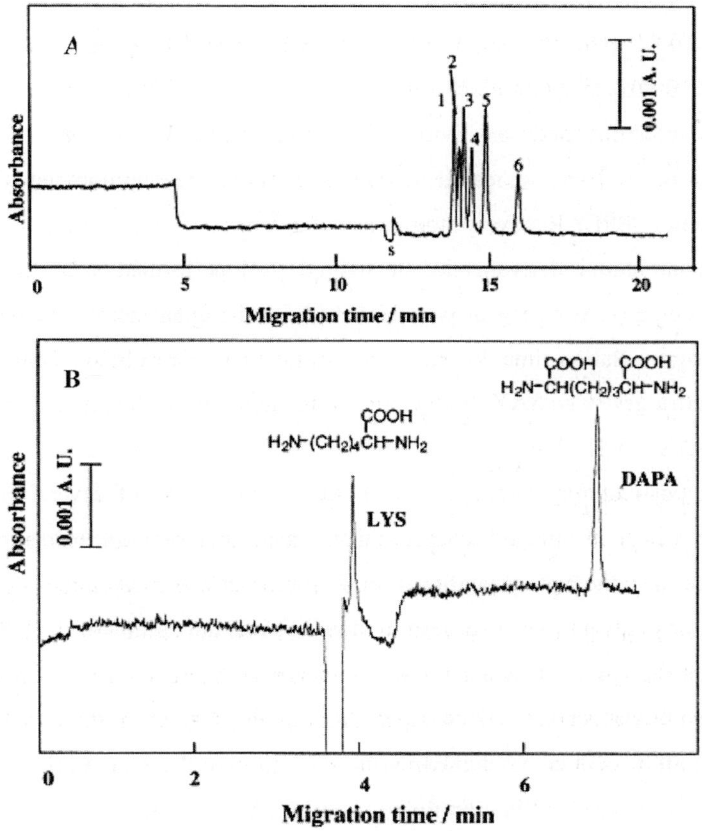

Figures 12A,B Electropherograms after derivatization with salicylaldehyde-5-sulfonate: (A) for alkane diamines and (B) for lysine (LYS) and diaminopimelic acid (DAPA); Sample solutions: 5 10^{-5}M of each; Migration solutions: (A) 10mM phosphate buffer (pH 7.8)+60%(v/v) ethanol, (B) 10mM phosphate buffer (pH 7.0+10mM Na_2SO_4+5%(v/v) ethanol; Peaks: (A)1, 1,9-diaminononane, n2, 1,8-diamonoctane, 3, 1,7-diaminoheptane, 4, 1,6-diaminohexane, 5, 1,5-diaminopentane, 6, 1,4-diaminobutane; With permission from Reference [57]

times excess [59]. According to authors AAs "cause problems only when the histamine peak was very small." However the presented chromatogram is not convincing at all: containing not identified compounds in enormous high concentrations beside the tiny amounts of As.

Exhaustive extraction studies of fifteen [60], later seventeen [61] As from wine were performed prior to their chromatographic analysis as OPA/MCE derivatives [60,61. Treatments with PVP, SAX [60] and acidic cation-exchanger catridges, [61] were tested and compared. Recoveries varied, in order of listing between 93-100% (PVP), 78-100% (C_{18} cartridge), 91-112% (SAX), 31-104% (Amberlite CG-50, Sigma) and 10-90% (SCX, Varian), respectively.

The reliability and sensitivity of the chromatographic analysis of biogenic amines in wine [62] was increased by cation exchange cleanup completed by solid phase extraction (SPE). Recovery study was not given.

An on-column derivatization technique [63] performed with the OPA/N-acetyl-L-cysteine (NAC) reagent was described for the quantitation of eight amines (derivatization + elution time 40 min) subsequently to their isolation from wines by two SPE cartridges (C_{18}, SAX, both fromVarian) reporting different recoveries (C_{18}: 65-105%, SAX: 75-103%).

The post-column derivatized OPA/MCE-biogenic As of several Hungarian wines were measured after perchloric acid treatment, homogenization/centrifugation and filtration [64]: the poor separation of eight compounds needed 73 min.

The most advantageous separation idea [65] was the elution of the OPA/MPA-AAs without the aim of their quantitation — prior to the assay of spm, spd, put and cad as their OPA/MPA derivatives. Applying a biner gradient: in the first 10 min of the gradient all AAs elute, continued by the separation of the four As (in the next 10 min) and the column equilibration time (11 min). (A eluent: 0.125 M sodium citrate containing 50 mL/L THF; B eluent: citrate buffer/ACN/THF=45/50/15, v/v; pH 6.4).

4.2. Determination of As as their dansyl, dabsyl and benzoylated derivatives by HPLC and HPTLC

Biogenic A content of different plant samples [66-68] have been determined in differently prepared acidic extracts.

The put, spd and spm content of lichens [66] were determined after the hydrolysis of samples. Free As were extracted with 5% (w/v) trichloroacetic acid (TCA) then centrifuged. Dansylation was performed at room temperature for 16 hr followed by a post-derivatization cleanup with toluene. RP-HPLC was performed from the toluene extract at 40 °C.

Five biogenic As (1,3-diaminopropane, put, cad, spm, spm) from plant and animal tissues [67] were extracted with 10% (w/v) $HClO_4$ (15 mg dry tissue/ml), dansylated at room temperature overnight, or, at 70 °C for 10 min and separated within 15 min, including column equilibration time. Recoveries varied between 89.5 and 97.1%.

Five polyamines from plant-seedlings [68] were determined after a clean-up of two steps: Tissue homogenates were extracted with 0.7 M $HClO_4$, and, the combined supernatants were further isolated by SPE (micropreparative Dowex 50-X8 column). Dansylation was performed in the dark, at 56 °C for 1 h. Dansyl-amines were extracted into toluene and separated within 40 min by HPLC. Content of As in wine samples were quantitated as their dansyl [69,70] and dabsyl [71] derivatives.

As a result of two different exhaustive isolation/enrichment procedures [69] ($method_1$: continuous liquid/liquid extraction for three days with Freon11/dichloromethane=90/10, (v/v); $method_2$: butanol/dichloromethane=70/60, v/v) twenty-five amines were identified and measured in five red vines and in four white wines. Dansylation was followed at 65 °C for 20 min, elution time was 50 min. Recovery percentages for five As, in order of listing, for method1 and method2 proved to be 9 % and 72 % (ethyl), 67 % and 74 % (butyl), 78 % and 88 % (diethyl), 85 % 40 % (phenylethyl), 93 % and 25 % (hexyl), as well as 96 % and 10 % (heptyl), respectively.

Nineteen As from wines [70] were determined after a four-step isolation/enrichment process. First phenolic compounds were removed by PVP clean-up: the mixture of 1 g PVP + 25 mL wine sample was stirred for 15 min and filtered (step1); The phenolic- free wine samples were dansylated in dark at 65 °C for 25 min (step2). Dansyl-As were subjected to SPE (step3). Finally, the acetonitrile eluatum of SPE were concentrated by a rotary evaporator (step4). Temperature of gradient elution was optimized (60 °C, 60 min). Recoveries varied in the range of 62-95 %.

Eight biogenic amines [71] of red wines were derivatized by 4-(diamino)-azobenzene-4'-sulfonyl (dabsyl) chloride (70 °C, 21 min) subsequently to their filtration and dilution (tenfold) by 0.1 M HCl containing 0.2% 3,3'-dithiopropionic acid. Gradient elution was performed at pH 5, for 50 min. Recoveries varied between 92.5 and 101.0 %.

Five biogenic As of fish meal prior to their HPTLC [62=72] were treated with 6 % TCA and centrifuged. The supernatant was dansylated (37 °C, 2 h) and derivatives extracted with ethyl acetate. Separations needed 50 min, recoveries were not given.

Nine biogenic As of various food samples (cheese, fish, meat) [73] were separated from AAs in two steps: cheese with 0.1 M HCl, fish and meat with 5 % TCA were extracted. The organic material containing extracts were saturated by NaCl and further extracted with butanol (cheese), or with butanol/chloroform (fish, meat). Dansylation of the organic extracts was followed at 40 °C for 1 h; Elution on a 15 x 1.6 mm column (particle size: 3 µm) was carried out within 6 min, with good resolutions. Recoveries varied between 47-88 % (salmon), 52-99 % (tuna), 3-91 % (salami), 3-80 % (grana cheese) and 2-92 % (gorgonzola cheese), respectively.

Changes in the polyamine contents of carbohydrates of various degree of polymerization were determined as their dansyl derivatives [74]. Prior to their derivatization first column chromatographic clean-up (Sephadex G-50), thereafter $HClO_4$ acid (5 %, w/v) treatment was performed. An aliquot of the supernatant was also hydrolyzed. Free and acid liberated As were dansylated overnight, in sealed vials. The chromatography of the toluene extracted dansyl-As was performed within 30 min. Recoveries of 85 % were reported.

Ten members of the C_1-C_7 aliphatic As, present in environmental waters at trace level, were quantitated as their dansyl derivatives applying peroxylate based chemiluminescence (PO-Cl) detection [75]. A special set-up of instruments was developed for the continuous SPE, dansylation (65 °C, 10 min) and HPLC-PO-Cl detection of samples. Elution time was 45 min, recovery values were not given.

4.3. Quantitation of As by CE

Amine contents of beer [76] and wines [77,78], were derivatized by 7-chloro-4-nitrobenzo-2-oxa-1,3-diazole [76], by AQC [77] and determined without derivatization [76]: applying simultaneous UV and LIF [76], as well as UV [77] and indirect UV [78] detections.

Ethyl-, isoamyl- and dibutylamines of beer 76] were eluted without any pretreatment, within 17 min (including washing and equilibration time), together with several unknown compounds of overwhelming concentrations (remark of present paper's author: unknown derivatives were very likely AAs). Reproducibility values varied between 10 and 19 percentages (RSD %).

Separation of seven As [77] from food samples was started by their extraction with 6% $HClO_4$ (shaking for 1h) followed by centrifugation and filtration (0.45 μm membrane). Selective enrichment of As present in food and wine samples was performed by column chromatography (Dowex 50 WX 8). Derivatization needed 10 min (55 °C) while separation 30 min. Migration time and peak area values proved to be highly reproducible (< 5 %), recovery data were not given.

Direct determination of biogenic As, developing continuous flow clean-up in wine, was optimized [78] regarding: (i) the type of SPE columns (out of the commonly used CBA, SCX and C18 columns the weak cation exchanger CBA was selected), (ii) the pH of wine subjected to clean-up (increased pH by use of the NH_4OH/NH_4Cl buffer ensuring the complete retention of As), and, (iii) the eluent (0.1 M HNO_3 in methanol). This process resulted in 96-98 % recovery of amines from the SPE column. Indirect UV detection was carried out by using 4 mom Copper II sulfate, formic acid and 18-crown-6 as running buffer.

References

[1] I. Molnár-Perl: Amino Acids, In Advanced Chromatographic and Electromigration Methods in BioSciences, (1998) (Eds. Z. Deyl, F. Tagliaro, E. Terasová) Journal of Chromatography Library, Vol. 60, Elsevier, Chapter 11, p. 415-464

[2] I. Molnár-Perl: Liquid Chromatography, In Encyclopedia of Separation Science, Academic Press, (2000), Vol III/Amino Acids, p.1999-2012

[3] Molnár-Perl: Liquid Chromatography, In Encyclopedia of Separation Science, Academic Press, (2005), Amino Acids, 2nd edition, Vol 5, Liquid Chromatography, Amino Acids, p. 239-252.

[4] I. Molnár-Perl (*Review*), J. Chromatogr. A, 913 (2001) 284-304

[5] D. Kutlán, P. Presits, I. Molnár-Perl, J. Chromatogr. A 949 (2002) 235-248

[6] I. Molnár-Perl (*Review*), J. Chromatogr. A 987 (2003) 291-309

[7] D. Kutlán, I. Molnár-Perl, J. Chromatogr. A 987 (2003) 311-322

[8] R. Hanczkó, Á Kőrös, F. Tóth[2], I. Molnár-Perl[1], J. Chromatogr. A (2005) on line available

[9] I. Krause, A. Bockhardt, H. Neckemann, T. Henle, H. Klostermeyer J. Chromatogr. A 715 (1995) 67-79

[10] O. Pinho, I.M.P.L.V.O. Ferreira, E. Mendes, B.M. Oliveira, M. Ferreira, Food

Chem. 75 (2001) 287-291

[11] R. Minocha, S. Long, J. Chromatogr. A 1035 (2004) 63-73.

[12] J. Kirschbaum, B. Luckas, W.D. Beinert, J. Chromatogr. A 661 (1994) 193-199

[13] T. Bausa, A. Blaise, F. Daumas, J.C. Cabanis, J. Chromatogr. A 707 (1995) 373-379

[14] B. Straub, M. Schollenberger,M. Kicherer, W.P. Hammes, B.Lucas, Z.Lebensm. Unt. Forsch. 197 (1993) 230-232

[15] W.Z. Zhang, D.M. Kaye, Anal. Biochem. 326 (2004) 87-92.

[16] P. Herbert, L. Santos, A. Alves, J. Food Sci. 666 (2001) 1319-1325.

[17] P. Helbert, M. J. Cabrita, N. Ratola, O. Laureano, A. Alves, J. Food Eng. 66 (2005) 315-322.

[18] V.R. Villanueva, R.C. Adlakha, Anal. Biochem. 91 (1978) 264-275

[19] N. Seiler, B. Knodgen, J. Chromatogr. 339 (1985) 45-57

[20] P. Gamach, E. Ryan, C.Svendsen, K. Murayama, I.N. Acworth, J. Chromatogr. 614 (1993), 213-220.

[21] Y. Yokojama, O. Ozaki, H. Sato, J. Chromtogr. A 739 (1996) 333-342

[22] I.M. Mackie, L. Pirie, A.H. Ritchie, H. Yamanaka, Food Chem. 60 (1997) 291-295

[23] V. Frattini, C. Lionetti, J. Chromatogr. A 809 (1998) 241-235.

[24] E.M. Siaghy, Y. Devaux, H. Schroeder, N. Sfaksi, D. Ungureanu-Longrois, F. Zannad, J.P. Villemot, P. Nabet, P.M. Mertes, J. Chromatogr. B 745 (2000) 279-286.

[25] M.R. Alberto, M.E. Arena, M.C. Manca de Nadra, Food Control 13 (2002) 125-129

[26] M.V. Moreno-Arribas, M.C. Polo, F. Jorgensen, R. Munoz, Int. J. Food Micro biol. 84 (2003) 117-123.

[27] J.A.M. Mckenzie, C.J. Watson, R.D. Rostand, I.German, S.R. Witowski, R.T. Kennedy, J. Chromatogr. A 962 (2002) 105-115.

[28] M.Callul, R.M. Marcé, J. Fabregas, F. Borrull, Chromatographia 31 (1991) 133-136.

[29] G. Nouadje, H. Rubie, E. Chatelut, P. Canal, M. Nertz, Ph. Puig, F. Couderc, J. Chromatogr. A 717 (1995) 293-298.

[30] G. Nouadje, M. Nertz, Ph. Puig, F. Couderc, J. Chromatogr. A 717 (1997) 335-343.

[31] G. Nouadje, N. Siméon, F. Dedieu, M. Nertz, Ph. Puig, F. Couderc, J. Chromatogr. A 765 (1997) 337-343.

[32] J. Kirschbaum, K. Rebscher, H. Brückner, J. Chromatogr. A 881 (2000) 517-530.

[33] J. You, H. Sun, W. Lao, Q. Ou, Anal. Chim. Acta 382 (1999) 51-65.

[34] J. You, J. You, W. Lao, G. Wang, X. Jia, Analyst 124 (1999) 281-288.

[35] Z. Chen, J. Wu, G.B. Baker, M. Parent & N.J. Dovichi, J. Chromatogr. A 914 (2001) 293-296.

[36] V. Lozanov, S. Petrov, V. Mitev, J. Chromatogr. A 1025 (2004) 201-208.

[37] L.Z. Zhang, M.X. Sun, J. Chromatogr. A 1040 (2004) 133-140.

[38] Y. Ma, R. Zhang, C.L. Cooper, J. Chromatogr. 608 (1992) 93-96.

[39] S.K. Sadain, J.A. Koropchak, J. Chromatogr. A 844 (1999) 111-118.

[40] F. Antolini, S. Franciosini, A. Florid, A.L. Floridi, Ital. J. Food Sci. 11 (1999) 335-346

[41] M. Arlorio, J.D.Coisson, A. Martelli, Chromatographia, 48(1998) 763-769.

[42] D. Barile, J.D. Coisson, M. Arlorio, M. Rinaldi, Food Control (2005) in press

[43] J.R. Draisci, L. Gianetti, P. Boria, L. Lucentini, L. Palleschi, S. Cavalli, J. Chromatogr.A 798 (1998) 109-116.

[44] Q. Wang, H. Yu, F. Ding, P. He, Y. Fang, Food Chem. 83 (2003) 311-317.

[45] P. Lehtonen, Am. J. Enol. Vitic. 47 (1996) 127-133.

[46] E. Soufleros, M.L. Barrios & A. Bertrand, Am. J.Enol. Vitic 49 (1998) 266-278.

[47] M. Parent, D. Bush, G. Rauw, S. Master, F. Vaccarino & G. Baker, Methods 23 (2001) 11-20.

[48] P.P. Piepponen & A. Skujins, J. Cheomatogr. B, 757 (2001) 277-283.

[49] J.J. Cordoba, T. Antequera Rojas, C. Garcia Gonzales, J. Ventenas Barroso, C.

[50] D.T. Goni & C.A. Azpilicueta, Am. J. enol. Vitic. 52 (2001)185-190.

[51] K.M.K. Kebary, A.H. El-Sonbaty & R.M. Badavi, Food Chem. 64 (1999) 67-75

[52] F. Galgano, G. Suzzi, F. Favati, M. Caruso, M. Martuscelli, F. Gardini, G. Salzano, Int. J. Food Sci. Technol. 36 (2001) 153-160.

[53] R.S. Gilbert, G.G. Gonzales, L. Havel III, C.V. Byus, Anal. Biochem. 199 (1991) 86-92.

[54] E. Csomós & L. Simon-Sarkadi, Chroamtographia 56 (2002) S-185 - S-188

[55] A. Bouchereau, C. Duhaz, J. Martin-Tanguy, J.P. Guégan, F. Larher, J. Chromatogr. A 836 (1999) 209-221.

[56] J.S. Soblosky, L.L. Colgin, C.M. Parrish, J.F. Davidson, M.E. Carey, J. Chro-

matogr. B, 712 (1998) 31-41.

[57] R. Mendes, A. Conçalves, M.L. Nunes, J. Food Biochem. 23 (1999) 295-306.

[58] R. Driouich, T. Takayanagi, Mitsuko, Oshima, S. Motomizu, J. Pharm. Biomed. Anal. 30 (2003) 1523-1530.

[59] P. Lehtonen, M. Saarinen, M. Vesanto, M.L. Riekkola, Z. Lebens. Unter. Forsch. 194 (1992) 434-437.

[60] O. Busto, J. Guasch & F. Borrull, J. Chromatogr. A, 718 (1995) 309-317.

[61] O. Busto, M. Mestres, J. Guasch, F. Borrull, Chromatographia 40 (1995) 404-410.

[62] P. Pfeiffer, Dtsch. Lebensm.-Rundsch. 92 (1996) 39-42.

[63] O. Busto, M. Miracle, J. Guasch, F. Borrull, J. Chromatogr. A 757, (1997) 311-318.

[64] A. Sass-Kiss, E. Szerdahelyi, G. Hajós, Chromatographia 51 (200) S-316–S-320.

[65] H.M.H. van Eijk, D.R. Rooyakkers, N.E.P. Deutz, J. Chromatogr. A 730 (1996) 115-120.

[66] M.I. Escribano, M.E. Legaz, Plant Phisyol. 87 (1988) 519-522.

[67] M. Marcé, D.S. Brown, T. Capell, X. Figuras, A.F. Tiburcio, J. Chromatogr. B 666 (1995) 329-335.

[68] K. Bencsik, T. Kremmer, M. Boldizsár, J. Tamás, M. Mák, E. Páldi, J. Chromatogr. A, 824 (1998) 175-180.

[69] P. Lehtonen, Z. Lebens. Unter. Forsch. 186 (1986) 177-181.

[71] R. Romero, D. Gázquez, M.G. Bagur & M. Sánchez-Viñas, J.Chromatogr. A 871 (2000) 75-83.

[72] M.H. Vega, R.F. Saelzer, C.E. Figueroea, G.G. Rios, V.H.M. Jaramillo, J. Planar Chromatogr. 12 (1999) 72-75.

[73] S. Moret, L.S. Conte, J. Chromatogr. A, 729 (1996) 363-369.

[74] C.W. Rodríguez, R. de Armas, C. Vicente, M.-E. Legaz, J. Chromatogr. A 881 (2000) 531-541.

[75] M.Cobo, M. Silva, J. Chromatogr. A 848 (1999) 105-115.

[76] L.M. Preston, M.L. Weber & G.M. Murray, J. Chromatogr. B 695 (1997) 175-180.

[77] Á. Kovács, L. Simon-Sarkadi, K. Ganzler, J. Chromatogr. A 836 (1999) 305-313.

[78] L.Arce, A. Ríos, M. Valcárel, J. Chromatogr. A 803 (1998) 249-260.

4. Quantitation of Polyamines by Chromatography

Ibolya Molnár-Perl (Editor)
Quantitation of Amino Acids and Amines by Chromatography
Journal of Chromatography Library, Vol. 70

4.

Chromatography of polyamines

Ynze Mengerink

Contents

Summary

This chapter reviews the most recent developments in polyamine analysis. A substantial part is devoted to the most important biogenic polyamines (putrescine, cadaverine, spermidine and spermine); some attention is also given to synthetic polyamine polymers. The chapter is subdivided into non-derivatization and derivatization methods. Besides an overview of older derivatization techniques, also recently described highly interesting procedures (e.g. PBC (4-(1-pyrene)butanoyl chloride) for selective detection of diamines beside monoamines are discussed. Special attention is given to sample preparation, recovery, repeatability and detection limit of the method. Sample preparation of different sample matrices is discussed throughout this review. The feasibility of applying comprehensive two-dimensional gas chromatography, coupled with time of flight mass spectrometry is demonstrated. Different selective approaches have been discussed. Derivatization approaches, including sample preparation of different samples were reviewed. Selective approaches, like Ni- inhibition of monoamines and pyrene labeling for selective polyamine determination are described. Ideas are discussed from synthetic polyamines (non aqueous derivatization, critical chromatography, high pH mobile phase), which could be applied to the separation for biogenic polyamines. It was demonstrated that polyamines can be highly selectively determined in urine by means of GC*GC-TOF MS.

1. Introduction

In this introductory section, some general aspects of biogenic and synthetic polyamines are discussed.

1.1. Nomenclature

Putrescine (PUT), cadaverine (CAD), spermidine (SPD) and spermine (SPM) are probably the most often cited polyamines in scientific literature. Lately, synthetic polyamines are becoming more and more popular. Polymers with two amine endgroups relate to biogenic diamines. Other important classes are synthetic polyamines like dendrimers (Figure 1) or hyperbranched polymers.

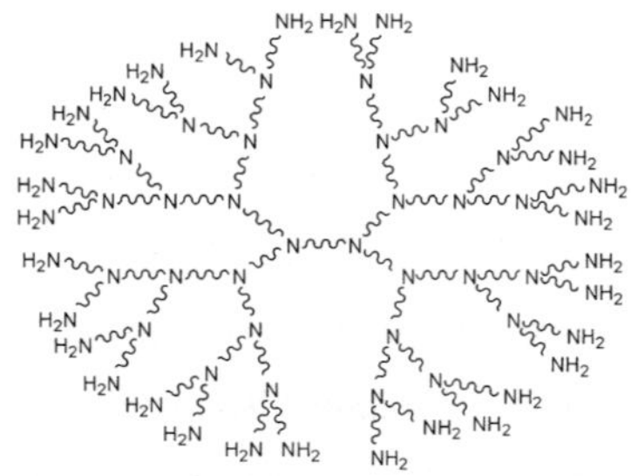

Figure 1 Typical example of a polyamine dendrimer DAB-PA_{32}, containing 32 primary amine groups

---

1.2. Background of Biogenic Polyamines

Biogenic polyamines are an important class of natural occurring compounds, which were discovered by Antoni van Leeuwenhoek in 1678. These polyamines are present in tissues and cells of living organisms and play an important role in biological and pathological processes. They are involved in the growth and differentiation of cells, the synthesis and stabilization of proteins and in nucleic acids, brain development, stabilization of lipids and nerve growth and regeneration [1-5]. Increased concentrations of polyamines have been correlated with the growth of tumor cells [6]. The biogenic diamines are both common bacterial degradation products and useful markers of spoilage or aging of fish, meat and cheese [7,8].

Recently, excellent reviews have been written with respect to the formation, the biologic function, the roll as cancer markers, and dietary applications of these biogenic polyamines as well as to their determination in all kind of different matrices [9-15]. These reviews have been used gratefully throughout this chapter.

1.3. Background of Synthetic Polymers

Linear synthetic polymers with amine endgroups are not very common. A nice example of these polymers is polyamide-6,6. The synthesis of polyamide-6,6 fibers was patented in 1931. The syn thesis is based on the polycondenzation between the two monomers hexamethylenediamine and adipic acid [16]. Polyamide chains are terminated (on both sides) with either an acid or an amine. Therefore, basically three different functionality distributions can be distinguished (diacids, acid-amine and diamines). Synthetic polyamides can be applied in an enormous amount of different applications (e.g. textile, carpets etc) [17].

Linear polyamine polymers are used in e.g. membrane technology [18], but were also applied within separation science for the investigation to deactivate electrophoretic capillaries [19].

Dendrimers [20,21] and hyperbranched polymers [22,23] are other interesting classes of polymers. Dendrimers are highly regulated and can be applied for all kind of different purposes. Recently, the interaction with dyes has been reviewed [24]. Dendrimers and hyperbranched polymers can also be used in vitro for drug delivery, for instance to form drug complexations [25]. They were also used for curing prion infections in living cells [26] and for nonviral gene delivery [27]. In addition, specific and strong interactions of amino terminated dendrimers were observed with bovine serum albumin [28]. In addition, dendrimers found their way into analytical separation science as a pseudo stationary phase in capillary electrophoresis [29,30]. It can be anticipated that polyamine terminated dendrimers would have been perfect silanols blockers in liquid chromatography.

1.4. Background on Derivatization

Biogenic polyamines do not contain chromophores and they are not volatile as ions (salts). Moreover, amines and especially polyamines are notorious for their chromatographic behavior as they create unwanted (secondary) adsorption in the chromatographic system. Derivatization of polyamines to enhance their chromatographic behavior, their volatility or their detectability is not straightforward. Different derivatives can be formed if polyamine derivatization is not complete, therefore, it is very important that polyamines are exhaustively derivatized. For this reason, synthetically polymeric molecules with a large numbers of amines are not easily uniformly derivatized before the actual separation and direct detection or post-column derivatization is preferable. Biogenic polyamines can also be determined without derivatization, but most methods utilize pre-column or post-column reactions. An important draw back of pre-column derivatization is the sample matrix, which can significantly influence the reaction characteristics.

1.5. General Considerations of Sample Preparation of Biogenic Polyamines

Different sample preparation procedures have been applied for different (but also for identical) sample matrices, which are discussed in some detail in this section and more specifically in the next sections. The exact choice of the sample preparation procedure is closely related with the analytical technique used, but also to historical procedures in a specific laboratory. A brief overview of the strategy for the sample preparation is given here. The scientist has to investigate which step is necessary and which step can be omitted. Centrifugation can be necessary after each individual sample preparation step to remove macromolecular structures or (other) non-dissolved particles, to remove salts from (cooled) organic phases or to speedup phase separations. The addition of an internal standard can correct biased results and allows easy sample concentration by e.g. partly solvent evaporation. Biogenic polyamines occur in free form, bounded to low-molecular mass compounds or bounded to macromolecules. Therefore, the choice of performing an acid extraction and hydrolysis need to be considered. Samples are homogenized (ground by mortar, electrically mixed, magnetically stirred, ultrasonically agitated etc.) and polyamines can be extracted from the sample matrix with all kind of different procedures. Samples can be denaturized with an organic solvent (e.g. acetone) or an acidified aqueous solution (e.g. 2-10% sulfosalicylic acid, trichloroacetic acid or perchloric acid). It is plausible that the polyamines stay in the solid fraction of the sample (e.g. after centrifugation), if a non-polar organic solvent is used. Some of these steps can be combined (e.g. homogenization, extraction and denaturization). Extraction can be performed with an organic solvent and an acid or alkalized aqueous phase. If desired, an ion-pairing agent can be added (e.g.bis-2-ethylhexylphosphate). Distribution will depend on all these factors, e.g. at low pH polyamines will stay in the aqueous phase, while at alkalized conditions with an ion pairing agent, polyamines will probably move towards the organic phase. In addition or as an alternative, C_{18} or cation exchange solid-phase extraction is possible to clean up the sample. The scientist should always be aware of biased results, as a consequence of his/her sample preparation.

To conclude, a typical example is discussed concerning extraction of botanic material [31,32]. Long-lasting hydrolysis (16 hr) at high concentrations HCl (6-12 N) and high temperatures (100-120°C) can be applied to release polyamines from an organic matrix. However, better results were obtained when extractions were performed at ambient temperatures [33]. This is probably caused by formation of secondary cyclic amines due to internal condensation reactions at higher temperature [34]. As an example, putrescine was converted into pyrrolidine at elevated temperatures. The resulting secondary amines, including spermine and spermidine, may undergo nitrosation. Nitrosamines are considered highly carcinogenic [35].

1.6. Some Considerations on the Determination of Polyamines

This section contains some easy and direct analyzing techniques to identify polyamines. Direct enzymatic assay tests are easy to use, but there is not always enough sensitivity and selectivity for specific amines in comparison with chromatographic separation techniques [36]. Direct spectroscopic techniques are very rarely applied. However, it can be useful for identification purposes, e.g. of reference materials. The solid-state spectra of α-Ω polyamines (n=2-12) including the biogenic polyamines of spermidine and spermine were assigned by combining inelastic neutron scattering and Raman spectroscopy [37]. Obviously, a separation with on-line detection is necessary to enhance selectivity and sensitivity. Especially in complex matrices, like blood, serum, urine and tissue, highly efficient separations are desired.

Thin layer chromatography is not a very precise and specific technique and it is not so easy to automate. However, it is simple, rapid and can be used without sophisticated equipment. Moreover, a number of samples and standards can be analyzed simultaneously and sample preparation is often straightforward, as pollution of the chromatographic injector is not an issue as it is with gas chromatography [38]. Dansyl chloride (5-dimethyl aminonaphthalene-1-sulfonyl chloride) is often applied to derivatize the (separated) polyamines, but ninhydrine derivatization is described also. TLC was applied as an easy way to determine polyamines in tissue extracts and cell extracts [39]. Tissue samples were washed with phosphate buffer and homogenized with 2% perchloric acid. Cells were washed/extracted with phosphate buffer and solicited in perchloric acid. Tissue and cell homogenates were centrifuged. The supernatant was cooled until dansylation and separation. Detection limits were <10 pmol. Non-linear calibration curves were studied with dansylated polyamines as model compounds [40]. It was shown that a quadratic fit described best this relationship between the response of a fiber optic sensor and the amount of derivatives retained on a surface of the silica gel.

2. Non Derivatization

This chapter is divided into the analysis of biogenic polyamines and synthetic polyamines. The polyamines are not derivatized before analysis. However, post-column derivatization procedures are discussed in this chapter.

2.1. Biogenic Polyamines (Including Post-Column Reaction Detection)

Sample preparation and non-derivatized biogenic polyamines with different separations techniques (HPLC, electrophoresis and GC) are discussed in this section.

2.1.1. High Performance Liquid chromatography (HPLC)

Chromatographic separations of polyamines often result in tailing peaks, as silanols will give unwanted secondary retention mechanisms. However, the latest generation reversed-phase high performance stationary phases show very often very low "silanol activity" [41]. Retention of amines is currently not based on ion interaction, but on hydrophobic interactions of the residual part of the molecule. Polyamines do not show these kinds of interactions. Therefore, retention of non-derivatized polyamines is not easy on standard reversed-phase chromatographic columns. However, a method was described based on this principle for easy samples [42]. The method was not very sensitive and only applicable to high concentrations. Reversed phase columns (ODS-2 and PLRP-S) were applied. Polyamines eluted just after t_0 and were almost not retained. Also, standard UV-detection was not applicable and indirect detection was performed by adding 2-naphthalenesulphonate to the mobile phase.

Polar embedded groups can be used to interact with highly polar compounds and to obtain maximum retention without unwanted silanol interactions. All self-respecting column manufactures have one or more different stationary phases based on this principle. For example, Waters recently introduced the Atlantis columns and performed a highly efficient separation of all kind of highly polar amine-containing compounds (cytosine, 5-fluorocytosine, uracil, 5-fluorouracil, guanine, thymine and adenine). It can be anticipated that also polyamines could be analyzed with these kinds of stationary phase, although no literature was found based on this mechanism. The separation of non-derivatized polyamines is easily performed with ion chromatography. There are three modes of operation; (cat)ion-exchange, ion-pair and ion-exclusion chromatography. Ion-exchange chromatography is straightforward and often applied. Ion-pair chromatography is almost as popular as ion-exchange chromatography, however, ion-pair chromatography is known to be less stable and long equilibrium times are necessary. Ion-exclusion chromatography is seldom used for the separation of biogenic polyamines. Some attention was given to the specific detection of non-derivatized polyamines after ion-exchange chromatography with a condensation nucleation light scattering detector (CNLSD) [43]. This detection technique is based on the evaporative light scattering principle. After on-line evaporation of the solvent, the dry particles are introduced to a cooled and saturated environment causing a condensation onto the particles. This will increase the particle size and improve the scattering. The method was applied to tuna fish, which was homogenized with high concentrations of hydrochloric acid. The sample was injected into the chromatographic column after centrifugation. Detection limits were approximately 10 $ng.ml^{-1}$, which corresponds to 0.7 μM (100 μl injection) putrescine and cadaverine. LC-MS would be another interesting approach to directly detect non-derivatized polyamines. It is both sensitive towards amine containing compounds and selective due to the possibility of reconstructing mass chromatograms. However, an LC-MS with ion exchange systems is expensive, not easy for quantification purposes and not used very often.

Electrochemical Detection

Electrochemical detection is the most straightforward way to determine non-derivatized polyamines [44,45]. Dobberpuhl et al. separated diamines with a low-capacity cation-exchange column and aliphatic monoamines with a high-capacity cation-exchange column [46]. Pulsed amperometric detection (PAD) was applied with post-column addition of an alkaline environment, which was necessary for the detection of the amines at the gold electrodes. In 0.1 M NaOH, pulsed voltametric data showed optimum response at a detection potential of 0.2 V. However, this optimum increased to 0.4-0.6 V for diamines, if acetonitrile is present in the mobile phase. Limits of detection for 25 μl injections of aqueous samples were approximately 200 nM for diamines. The relative standard deviation (R.S.D.) for eight successive injections of 100 μM ethylamine was 0.3%.

Some interesting studies with regard to the waveform of the pulse were reported [1]. PAD is often applied for the determination of e.g. hydroxyl containing compounds. However, the formation of gold oxide due to oxidation of amines increased the background. The amperometric waveform ($E_{detection}$= 0.45 V at 0-300 ms (last 50 ms for integration), $E_{oxidation}$ = 0.80 V at 310-410 ms and $E_{reduction}$ = -0.40 V at 420-780 ms) was compared with integrated voltametric detection (IVD: E_{min} = -0.1 V at 0 ms to E_{max} = 0.60 V in 1000 ms and back to E_{min} = -0.1 V at 2000 ms) and integrated square wave detection (ISWD: E_{min} = -0.1 V at 0-40 ms, E_{max} = 0.60 V at 50-450 ms and E_{min} = -0.1 V at 460-1000 ms). Detection limits found for 1,3-diaminopropane with the latter waveform (ISWD) was 0.5 pmol (injection volume 25 μl). Detection limits with the 2nd waveform IVD were only slightly worse. However, ISWD was 8 times more sensitive compared to the first described waveform PAD. The IVD and ISWD cathodic charge compensated the anodic charge better than PAD.

Another alternative of optimizing signal to noise was performed with integrated pulsed amperometric detection, where the corrected current was integrated during the total cycle [7]. With this approach, biogenic polyamines were determined in fish with (cat)ion-exchange chromatography. Fish, meat or cheese was homogenized in a mixture of aqueous trichloroacetic acid, hydrochloric acid, and n-heptane. After homogenization and centrifugation, the residue was re-extracted with 5% trichloroacetic acid and the aqueous phases were combined, diluted and cleaned before analysis with a C_{18} solid phase extraction or with a liquid – liquid extraction with n-hexane. Detection was performed at +0.35 V and optimized by post-column addition of 0.4 M sodium hydroxide solution. Cleaning of the electrode was performed at +0.9 V and –0.9 V. Detection limits were approximately 5-25 ng (25 μl injection). The average recoveries were >92% and the RSD was less than 2% (for 125 ng).

Post-Column Addition

Another possibility to determine polyamines was performed with the post-column addition of a cupper (II)-L-tryptophan complex with indirect fluorescence detection after a strong cation-exchange column [47]. The observed sensitivity of the diamine putrescine was less compared to higher polyamines spermine and spermidine. It was assumed that a substantial structural disorder is required to coordinate with the cupper complex. Detection limits for spermine and spermidine were 5-10 pmol (10 μl injection) and the repeatability was found to be <5% RSD. The system was linear for only two decades.

Post-Column Derivatization

Ion chromatography with selective post-column reaction detection is a very popular method to separate and detect polyamines [48]. Fluorescence detection in combination with an OPA (*o*-phthalic dicarboxaldehyde) post-column reactor is most often applied. This derivatization reaction is known to be very fast at room temperature. In addition, OPA is stable in water and not native fluorescence. Interferences of the sample matrix are minimized, because the polyamines are separated from other constituents. These kinds of systems are also very suitable if samples needed to be analyzed at high buffer concentrations or at extreme pH-levels. In these latter cases, sample reaction during pre-column derivatization can cause biased results due to an increase or decrease of the optimal reaction pH, which will lead to a decrease of the reaction efficiency.

As an example, polyamines in herring muscle were determined with such a post-column derivatization [49]. Dried herring muscle was homogenized/extracted in a mortar with trichloroacetic acid. After centrifugation the sample was extracted again. The supernatants was pooled and diluted to 50 ml. The extracts were filtered and injected (5 μl). Detection limits were <5pmol. Recovery and repeatability were claimed to be good, but no data were given.

Besides OPA, other post column derivatizations are possible (e.g. polyamine-ninhydrine absorbance [50,51]. In addition, post-column addition of oxidase enzymes and subsequent electrochemical detection were described. Detection limits in the pmol range can be obtained [52-54].

Chemiluminescence

A post-column derivatization with H_2O_2 and Bis-(2,4,6-trichlorophenyl)oxalate and chemiluminescence detection was developed to improve detectability of polyamines [55]. Standard or sample was preconcentrated on a C_{18} SPE column. Reagent (dansyl chloride in acetone / buffer) was flushed through the cartridge, which was subsequently incubated. The derivatives were desorbed from the packing material and injected (20 μl) into the reversed-phase system. Oxalate derivatives turn out to be very efficient chemiluminescence emitters. 2.5 mM Bis-(2,4,6-trichlorophenyl oxalate) and 10 mM H_2O_2 were separately added into a post-column setup with minimal tubing between the reagent confluence and the detector. The detection limits were between 7-60 $\mu g.l^{-1}$ in real industrial wastewater. These samples were already 50 times diluted before analysis due to co-eluting problems. The sensitivity of the chemiluminescence was 3 to 75 times better with respect to UV detection and 2 to 10 times better with respect to fluorescence detection.

2.1.2. Electrophoresis

High voltage paper electrophoresis was probably the first electrophoretic method to separate polyamines [56]. Russell published a famous article in nature, showing elevated concentrations of spermine and spermidine in urine of cancer patients [57]. The method was based on an alkalized butanol extraction and electrophoresis. Urine samples of cancer patients were compared with samples of normal patients. Urine was hydrolyzed with 6 M HCl. The pH of the samples was adjusted to pH 9 and the amines were extracted into n-butanol. Butanol was evaporated to dryness and the residue was dissolved in 0.1 M HCl. Polyamines in 10-50 μl aliquot were electrophoretically separated and stained with ninhydrin. Recovery was determined with 14C-amines. Levels of 1-200 mg excreted polyamines per day were reported.

Capillary electrophoresis with pulsed amperometric detection was applied for the determination of polyamines in milk [58]. Milk was extracted with 5% perchloric acid. After centrifugation, the pellet was mixed with 5% perchloric acid and centrifuged. Both supernatants were combined, concentrated and injected. Detection limits of 0.1-0.4 μM were obtained and the system was linear within 5 decades! Repeatability of the response was better than 3% RSD. The sensitivity of this method was claimed to be comparable or superior to laser induced fluorescence detection. Optimum separation conditions were found at pH 3.5 with a citrate buffer of 30 mM, however, 20 mM NaOH was present in the detector cell to improve sensitivity. Additives like ethanol, acetonitrile, β-cyclodextrin, 18-crown-6, urea and Cu^{2+} could not improve the separation.

Indirect UV detection is also possible, but linearity and sensitivity is often not so good [59]. However, this approach was used to determination intracellular polyamines in PC-12 tumor cells [60,61]. The cells were centrifuged and the residue was washed with 0.3 M sucrose and 10% ice-cold trichloroacetic acid. By applying this trichloroacetic acid extraction, the sensitivity of the capillary electrophoretic system towards pH and ion strength of the injected sample could be minimized. Also, trichloroacetic acid released polyamines from denaturized biogenic polymers like proteins, phospholipids, DNA and RNA. The sample was centrifuged and the supernatant was extracted three times with ether prior to analysis. Quinine sulfate was used for indirect detection (λ=236 nm). Recovery of the method was between 94-109%. Repeatability was approximately 10% RSD. Concentrations of 0.5-1.5 nmol putrescine, spermidine and spermine were determined in 10^6 cells.

Capillary electrophoresis and indirect chemiluminescence with cobalt (II) is another interesting approach to detect polyamines [62]. Sensitivity was claimed to be comparable to laser induced fluorescence and three times better than UV detection. An intense background signal was generated with a luminol-hydrogen peroxide reaction catalyzed by cobalt (II) ions in the run buffer. Displacement of these ions, by migrating polyamine cations, resulted in a decrease of the background signal. Putrescine, spermidine, and spermine were separated within 4.5 min. The detection limits were approximately 5-10 nM. Repeatability of the area was <10%RSD and repeatability for migration time was ≤1%.

Electrophoresis and staining with ninhydrine looks a straightforward and easy separation technique. However, Gehrke et al. described severe problems with specificity, accuracy and sensitivity [51]

2.1.3 Gas Chromatography (GC)

GC of non-derivatized polyamines is not very popular. It is not very sensitive and salt free extracts of biological fluids are necessary. However, a GC method was developed to separate and detect these non-derivatized polyamines in different organs [63]. Brain, liver, pancreas, ventral prostate or testis was homogenized in water. 6% Perchloric acid was added to the homogenate and the precipitate was washed twice with 3% perchloric acid. The aqueous parts and the supernatant were mixed and alkalinized with 4 N KOH. After centrifugation, the precipitate (potassium perchlorate) was washed twice with 2 N KOH. Sodium sulfate was added to the supernatants and washings (pH>13) to obtain a saturated solution. The saturated aqueous phase was three times extracted with 1-butanol. The total organic phase was extracted with saturated sodium sulfate. The aqueous phase was removed after centrifugation. Diethyl ether and internal standard were added to the organic phase, which was dried at 60°C (vacuum). 1-butanol was added and residual salts were removed by centrifugation. The residual organic phase was concentrated before injection into the gas chromatographic system to separate and quantify the non-derivatized putrescine, cadaverine, spermidine and spermine. Recoveries of the polyamines from aqueous solutions were between 80-100%, with RSD<3%, probably because internal standard corrections were applied. Recoveries of polyamines from rat liver were even better and more precise. This method demonstrated very nicely that direct determination of polyamines is possible with acceptable recoveries, but laborious sample preparation is necessary.

2.2 Synthetic Polyamines

The elution behavior of the ethylene diamine oligomers diethyleentriamine, triethylenetetramine, tetraethylenepentamine and pentaethylenenhexamine on large pore reversed-phase silica plates was studied [64]. Staining with ninhydrine was performed after development of the plates. Retention was governed to a large extent to the salt concentration of the mobile phase. It was stated that in salt free eluent, both the dissociated and undissociated form of these polyamines are very strongly bound to the free silanols. Salts with high activity coefficients (e.g. LiCl) and organic modifiers with high dielectric constants are highly recommended to decrease the retention. Also, low pH of the mobile phase decreases retention. De Robertis et al. found that the influence of the ion strength on the protonation constants of amines is very strong. They also concluded that this is significantly affected by the N-alkyl substitution [65].

Polyamide-6,6 has a molecular mass distribution. Moreover, there is also a distribution with respect to endgroup functionality. LCCC (liquid chromatography at critical conditions) is a relatively new separation technique for polymers. Entelis et al. described the basic principles of so-called critical conditions, to perform separations independent on the number of identical backbone units, but solely on different end-group functionalities [66,67]. At very specific conditions, where the pore width is larger than the radius of gyration, conditions can be found where for a non, mono and bi functional polymer elute at conditions independent of there molecular mass as given in equation 1-3

$$K_d^{(0)} = 1 \qquad \langle 1 \rangle$$

$$K_d^{(1)} = 1 + \frac{2a}{D}\left(e^{\theta_f - \theta_c} - 1\right) \qquad \langle 2 \rangle$$

$$K_d^{(2)} = \left\{K_d^{(1)}\right\}\left\{K_d^{(1')}\right\} \qquad \langle 3 \rangle$$

where K_d is the chromatographic distribution constant of a non, mono or bifunctional polymer, a is the segment size of one backbone unit, D is the pore size, θ_f is the interaction energy of the functional group with the pore wall. θ_c is the interaction energy at critical conditions of one segment of the backbone unit with the pore wall. For a non-functional linear polymer, the critical conditions can be obtained at zero Gibbs free energy (see equation 4)

$$\ln K = \ln(1) = 0 = -\frac{\Delta G}{RT} = -\frac{\Delta H}{RT} + \frac{\Delta S}{R}$$

<4>

Thus, for non-functional homologous linear polymers, the critical conditions are reached if the entropy term TΔS of a certain backbone equals the enthalpy ΔH of the same backbone. Separation is solely based on differences in interaction energy of the endgroup or deviating backbone units. We applied this technique to separate these polyamides solely based on their endgroup functionality [68]. A typical example is given in Figure 2.

It was demonstrated that polyamides, with a molecular mass of 20-30 kD could be separated based on their endgroup functionality. Recoveries of 90-100% were obtained, relatively to the calculated recoveries. It was observed that the pressure applied across the system had a significant impact on the critical conditions. As recently discussed by Guiochon, pressure has a significant influence on the distribution in liquid chromatography as it influences compressibility of the mobile phase, the viscosity and its melting point [69].

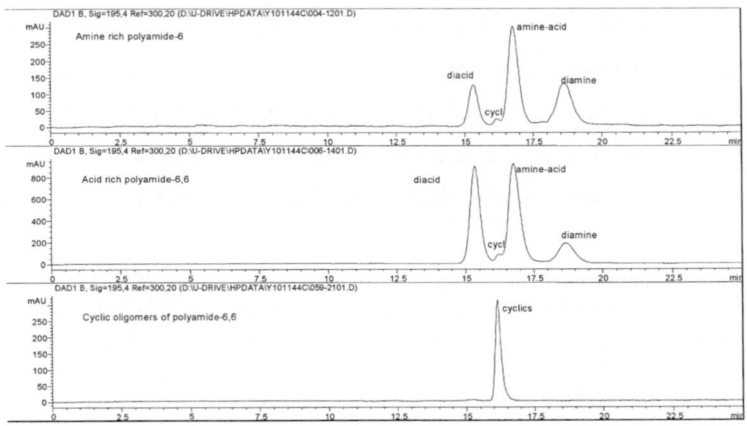

Figure 2 Some special polyamide-6,6 samples separated with LCCC: Upper trace primary amine rich polyamide-6,6 (Mw=10 kD), mid trace carboxylic acid rich polyamide-6,6 (Mw=3 kD), lower trace cyclic oligomers of polyamide-6,6 (Mw = 700 D). Column 2*(250*4) mm Nucleosil 300-5C_{18}. Mobile phase 89.5% m/m HFIP and 10.5% m/m 10 mM H3PO4, flow 0.3 ml.min^{-1}, column temperature 39°C. With permission from Reference [68].

These effects can significantly influence the separation, especially at ultra high pressures or for the separations of macromolecular compounds. It can be anticipated that a separation solely based on amine-content is also possible for biogenic polyamines. To do so, critical conditions for the aliphatic backbone have to be found

Size exclusion chromatography with evaporative light scattering detection was used to separate and detect different generations of amine terminated astramol dendrimers (4th generation DAB-PA32: Figure 1, chromatogram: Figure 3) [70].

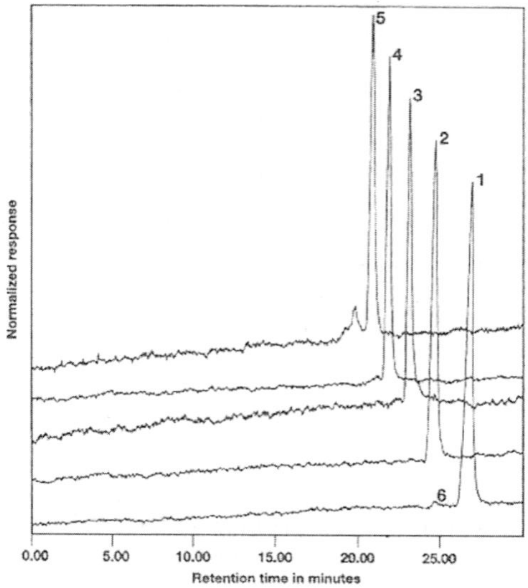

Figure 3 SEC of dendrimers: Analysis of successive generations of polypropylene amine dendrimers. Columns: eight 250*4mm Nucleosil 120-5C18 columns in series thermostated at 60°C. Injection 20 µl of 0.5 mg/ml amino-terminated dendrimers in eluent. Eluent: 0.5 ml/min 1% formic acid in water, ΔP 210 bar. 1=$DAB(PA)_4$ 2=$DAB(PA)_8$ 3=$DAB(PA)_{16}$ 4=$DAB(PA)_{32}$ 5 =$DAB(PA)_{64}$ 6=dimer of $DAB(PA)_4$. With permission from Reference [70]

With this method, also some side products like dimers were separated, which could be formed by condensation of two dendrimers. Recently, polyamidoamine dendrimers were also separated with electrophoresis on a polyacrylamide gel at acidic and alkalized conditions [71].

Nitril terminated polyamine dendrimers were determined at high pH [72]. A very nice separation was accomplished for different generations and their side products of nitril terminated polyamine dendrimers with 0.1 N NaOH as the aqueous part of the mobile phase and a polymer based PLRP-S column (Figure 4). Detection was performed both with UV detection and ESI-MS.

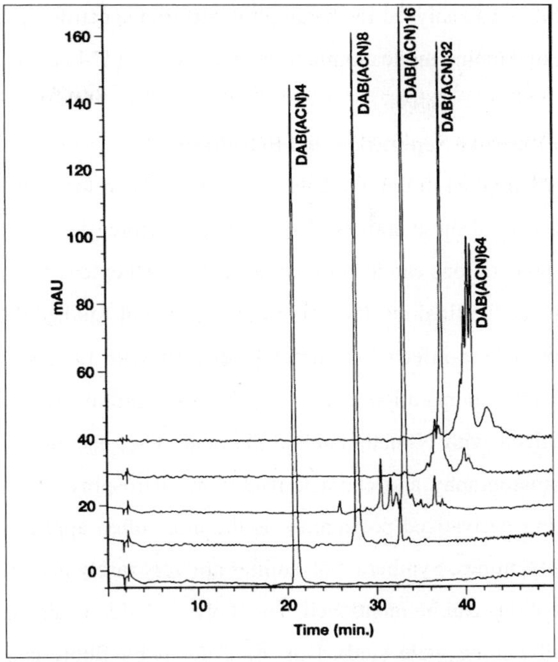

Figure 4 Separation of different generation nitril terminated poly(propylene imine) dendrimers at alkaline pH (Gradient of 95% 50mM NaOH to 95% acetonitrile in 55 min, 1 ml.min^{-1}, 125*4 mm PLRP-S at 40°C, λ=220nm. With permission from Reference [72].

This is another very interesting, but not often applied separation mode for polyamines in general. The amine is not protonated above its pKa, therefore retention will increase and it can be anticipated that UV absorbance (λ<220 nm) will increase, due to the lone electron pair on the nitrogen atom. Silica-based columns do not allow high pH operations. However, alternative columns like zirconia [73,74] or polymer stationary phases based on polystyrene divinylbenzene (as the PLRP-S column mentioned above) are stable at higher pH.

The characterization of hyperbranched polymers has been described for the unsaturated carbamate ring-opening polymerization of hyperbranched polyamine grafts on amine functionalized silicon wafers [75]. FTIR followed the decrease of the carbonyl stretch at 1738 cm^{-1}. Proton, carbon and proton-proton correlation spectroscopy NMR (^{1}H-NMR, ^{13}C NMR, ^{1}H-^{1}H-COSY NMR) were applied for the characterization and the estimation of the degree of polymerization. The surface of the final product was analyzed by X-ray photoelectron spectroscopy. The surface roughness was investigated by tapping mode atomic force microscopy (TM-AFM).

3. Derivatization to Improve Separation and Selectivity

Derivatization is mainly applied to the biogenic polyamines. Therefore, this part is primarily devoted to the determination of these amines. As mentioned before, derivatization in combination with difficult sample preparations can lead to biased results. All external factors applied to a system can directly influence the final results. Therefore, Geall et al. studied the interaction of polyamine conjugates with malaria-infected erythrocytes directly with fluorescence microscopy, circumventing off-line sample preparation errors [76]. The polyamines putrescine, spermidine and spermine were derivatized with N-methylanthranilic acid. The reaction products were isolated with preparative chromatography and characterized with elctrospray MS and NMR. However, separating and detecting derivatized polyamines is the most often applied analytical method to determine biogenic polyamines. Synthetic polyamines are very rarely determined in their derivatized form. Some exceptions can be mentioned. The motion of side chains and short segments of polylysine and polyallylamine were studied in more detail by fluorescein labeling and time-resolved fluorescence anisotropy (TRFA) [77]. Good agreement was found of the side-chain motion compared to NMR, however considerable lower values were reported for the segmental motion. With this interesting approach, polyamines entrapped into sol-gel silica could be investigated also. It was found that, although polyamines adsorb irreversible to the surface of the silica, this not necessarily reduced the segmental mobility. Versamid (Mw=1-10 kD) was separated with size exclusion chromatography. This model resin contains di and tri ethylene amines dimerized with diacids [78]. It was claimed that the primary amines needed to be derivatize to enhance sensitivity and selectivity and to improve chromatographic behavior of the primary amines. The polymers were not very well soluble in water and therefore a non-aqueous OPA derivatization system was developed. Sample, OPA and mercaptoethanol were separately dissolved in THF. Reaction times were long (60 min, 50°C) but relatively stable reaction products were obtained. The main disadvantage of this approach is that translation of the observed chromatogram to polymer characteristic parameters is not straightforward. An equivalent translation of molecular mass (based on retention time) and molar concentration at a specific time (based on intensity) can cause biased results as the assumption is probably needed that the yield of the reaction is independent of the molecular mass. However, it have been demonstrated that the OPA reaction yield different intensities for different sized oligomers [79].

3.1. High Performance Liquid Chromatography (HPLC)

Pre-column derivatization for polyamines is not straightforward. For an α-Ω diamine the reaction product of a mono derivatized diamine is identical, but for a lot of higher polyamines this is obviously not the case. Derivatization can result in more products of one compound as discussed in section 1.4 of this chapter. This is not necessarily a problem for post-column derivatization, but an important factor during pre-column derivatization. Moreover, secondary products formed during derivatization will enhance this problem. Different pre-column derivatization agents are available. An overview is given.

3.1.1. Ortho-Phthalic Dicarboxaldehyde (OPA)

OPA produces so-called isoindole derivatives of polyamines. it is a very popular and attractive derivatization agent. It reacts fast with amines at ambient conditions. OPA itself does not show any fluorescence, but the isoindol formed after reaction with a primary amine and a thiol is highly fluorescent. Recently, the characteristics of OPA derivatized polyamines were reviewed and very interesting results were presented with respect to this derivatization agent [80]. Different peaks were observed for the derivatized diamine 1,2-ethyleendiamine, 1,2-propyleendiamine and 1,3-propylenediamine (1,2-EDA, 1,2-PDA and 1,3-PDA). Besides the mono and di-derivative, (for 1,2-PDA even two different monoderivatives are possible) also the degradation products (all primary amines with -CH2-NH2 moiety can react twice with OPA) are possible. At standard reaction conditions ([OPA]:[Mercaptan]=1:3) 1,2-EDA, 1,2-PDA and 1,3-PDA produced more than two peaks i.e. secondary derivatives. The first eluting derivative of all three components showed an UV- λ_{max} at 305 nm, but did not show any fluorescence. Also, the second eluting derivative of 1,2-DEA and 1,2-PDA did not show any fluorescence (both UV λ_{max}=234 nm). The second eluting component for 1,3 PDA and also the third eluting component for all three diamines showed fluorescence. In addition, the third eluting derivative of 1,2-DEA and 1,2- PDA had a similar UV λ_{max}, while the UV λ_{max} for second and third eluting peak for 1,3-PDA deviated. With a higher [OPA]:[Mercaptan] ratio the formation of secundary products could be suppressed. It can be concluded that for 1,2-diamines quenching of the fluorescence will take place. Kutlan et al. investigated also tyramine, histamine, putrescine, cadaverine, spermine and bishexamethylene triamine. All derivatives showed fluorescence. Again, more than two derivatives were formed. Spermidine and cadaverine showed even four derivatives. At higher [OPA]:[Mercaptan] ratios (e.g. 1:50), the formation of the secondary products (containing 2 OPA groups on one primary amino-group) is probably suppressed due to minimum free OPA present in the reaction mixture. Camins-Falco et al. studied the stability of the OPA derivatives of polyamines and observed that the spermine was the most unstable isoindol [81]. Spermine was 100 times more stable on a C_{18} support compare than in solution. This is in good agreement with other observations, as the instability is caused by the reaction with a second OPA molecule on the already reacted primary amine [82].

Van Eijk et al. studied the OPA derivatization of polyamines in plasma, tissue and urine trying to minimize operator interference [83]. Blood samples were collected on ice and immediately centrifuged (5 min, 4°C). The obtained plasma was denaturized with 5% 5-sulfosalicyl acid up till a concentration of 40 $\mu g.\mu l^{-1}$. Immediately after denaturization, samples were frozen with liquid nitrogen and stored at –80°C. Before analysis, samples were thawed at 4°C, vigorously vortex mixed and centrifuged (11000 g, 10 min, 4°C). The supernatant (40 µl) was mixed with water (80 µl) and OPA-buffer reagent (80 µl, 1 M borate buffer pH=10.4 containing 3.75 $mg.ml^{-1}$ OPA, 3.75 $\mu l.ml^{-1}$ 3-mercaptopropionic acid and 10% methanol). The samples were stored in a cooled compartment of the autosampler until analysis. Tissue samples were immediately frozen with liquid nitrogen and stored at -80°C. Before analysis, they were transported to a 1.5 ml ependorf vial, containing 0.1 g, 1 mm ID glass beads and 100 µl ice cold 5% 5-sulfosalicyl acid. Tissue was homogenized in a mini-bead beater at high speed for 1 min. Next, the vials were centrifuged and the supernatant was mixed with the OPA and water identically as described for the plasma samples. Urine samples were collected after an overnight fast. Aliquots were centrifuged for 5 min at 4°C and processed the same way as the plasma samples. To optimize this method, on-column degradation products were minimized and response was optimized by cooling down the column to 10°C. Detection limits of 0.1 µM in plasma samples were obtained. Repeatability for standards was determined as RSD $\leq$4%. A typical chromatogram is given in Figure 5.

The pre-column derivatization reaction of polyamines is an interesting approach, but not that often applied. Reaction products are not stable, as discussed above. Therefore, direct on-line derivatization is necessary. However, this is actually very beneficial, as less off-line sample preparation is necessary, fewer errors can be made and less time is needed to perform the pre-column derivatization reaction.

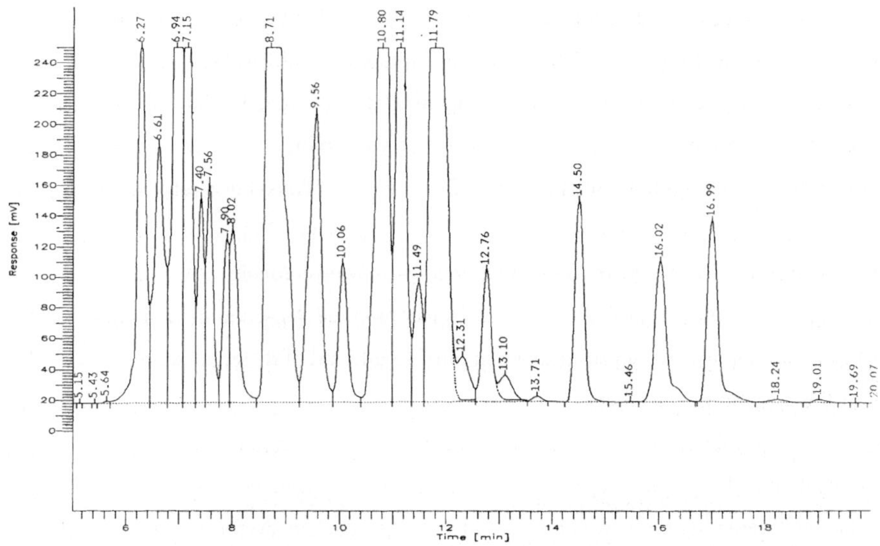

Figure 5 250 μM physiological amino acid standard (contains high concentrations of 20 *amino acids, typical present in plasma) spiked with a 250 μM standard of the polyamines (spermine, spermidine, putrescine en cadaverine).. Rt=12.76 is SPM, Rt=14.50 is SPD Rt=16.02 is PUT and Rt=16.99 is CAD, chromatographic conditions see reference). With Permission from the author and with special thanks to Dr. H.M.H. van Eijk for giving us the possibility to publish this chromatogram.*

3.1.2. Fluorescamine

Fluorescamine is an often-applied derivatization agent for amines. However, it is not that often used for the derivatization of polyamines. Rat liver samples were analyzed with a solid phase extraction prior to fluorescamine derivatization [84]. Rat liver was homogenized with 0.3 N perchloric acid. The homogenate was centrifuged and the supernatant was used as a sample. Human serum was denaturized with 1.5 N perchloric acid and vortex mixed. The sample was centrifuged and the supernatant was used as a sample. Human urine was mixed with HCl and hydrolyzed. The hydrolyzed urine was filtered and the HCl was removed in vacuum. The residue was used as a sample, which was purified on a carboxymethyl cellulose stationary phase. The residue of the evaporated fractions was dissolved in 0.1 M sodium borate and reagent (2 $mg.ml^{-1}$ fluorescamine in acetone). Sodium acetate buffer was added to the reaction mixture and after 10 min, the sample was injected into a revered phase system with fluorescence detection. Detection limits of 100 pmol in biogenic samples were reported. Repeatability was about 5% RSD. Standard addition experiments implied good recovery. The detection limits with both fluorescamine and OPA are generally better than those obtained with other reagents [9]. Also reaction is fast and elution procedures are easy. Acid chlorides are less sensitive, but yield more stable derivatives. However, better results were claimed in the low pmol range for dansyl and dabsyl derivatives. [85,86].

3.1.3 Dansyl Chloride (5-Dimethylaminonaphthalene-1-sulfonyl chloride)

Dansylation of polyamines is probably the most often applied pre-column derivatization technique for the determination of polyamines and applied to all kind of different samples [87,88]. An overview of the possibilities is given.

First, an example of the determination of polyamines in erythrocytes [89] is described. Blood was centrifuged, plasma was removed and the erythrocytes were resuspended in saline. After mixing and centrifugation, water was added to the packed erythrocytes. The cells were hemolyzed by vortex mixing. The hemolyzate was denaturized with ice-cold perchloric acid. After shaking and centrifugation the supernatant was removed from the sample and frozen. Internal standard (1,7-diaminoheptane), saturated sodium carbonate and dansyl chloride (5 $g.l^{-1}$ in acetone) were added to the supernatant or standard solutions. After vortex mixing, the samples were derivatized overnight. The dansylamides were extracted with cyclohexane and evaporated to dryness. The residue was dissolved in acetonitrile/water and 50 μl was injected into the RP-HPLC system. Fluorescence detection was performed. Recoveries ≥94% were obtained. The detection limit was 15 pmol, which corresponds to <1 μM in erythrocytes. The RSD of the method itself was approximately 3%. Reference values, evaluated in 10 healthy patients were 7.9 μM spermidine (standard deviation s=2.1) and 5.2 μM spermine (s=1.6).

The dansyl derivatization was also utilized for the determined of polyamines in Greek beer and different wines [90], after a solid phase extraction with subsequent RP-HPLC-DAD-

APCI-MS (reversed-phase high performance liquid chromatograph coupled with diode array detection and atmospheric pressure chemical ionization mass spectrometry). UV was applied for the quantification, while identification was performed with the mass spectrometer. Polyvinylpyrrolidone and samples (beer is degassed by ultrasonic agitation) were mixed and centrifuged. The polymer removes interfering contaminations for both reaction and quantification. The supernatant was filtered before use. After dansylation, samples were cleaned over a C_{18} SPE column. The repeatability of the method was better than 9% RSD (with-in and between days). Recoveries between 75 and 115% were obtained. Detection limit were approximately 0.1 ng (10μl injected).

A third example is the determination of polyamines and amino acids in forest tree tissue and cell cultures [91]. Fresh cells, leaf discs and chopped needles were brought in contact with ice-cold perchloric acid (5%). Prior to analysis, the samples were thawed and frozen (-20°C) three times, centrifuged. The supernatant was used for dansylation; internal standard (heptane diamine) and sample extract were added in a polypropylene tubes and vortex mixed. Saturated sodium carbonate solution and dansyl chloride were added, vortex mixed and incubated. Alanine was added to remove excess reagent. After 30 min, acetone was removed, toluene was added, and the sample was vortex mixed and centrifuged. 200 μl of the toluene phase (containing the derivatized polyamines) was transferred. Toluene was evaporated and 730 μl methanol was added and vortex mixed (2 min) to dissolve the dansylated polyamines. 135 μl of the aqueous fraction of the toluene extraction (containing dansylated amino acids) was added along with 135 μl of 2.9 M acetic acid. This procedure turned out to be very important to obtain full recovery of the polyamines. The polyamines stuck to the wall of the polypropylene tubes if the toluene extraction was not performed. Amino acids derivatives stayed in the aqueous phase and did not adsorb to the wall of the tube. The reaction vial was kept open for 4 min, so residual CO_2 could escape. The final solution was mixed and filtered (0.45 μm). Recoveries were 100±4% for all biogenic polyamines (10 pmol per injection of 10 μl). Detection limits were <1 pmol. The chromatographic column was cleaned after every run with 10 ml 100% acetonitrile and 5 ml water. After each sequence of 50-80 injections, the column was reversed and cleaned with 20 ml 10% acetonitrile, 20 ml 100% acetonitrile and 20 ml 75% acetonitrile. The column was only reversed towards its original direction after another sequence of injections and subsequent cleaning program. Column lifetime was increased four times with this procedure and over 1000 samples could be analyzed.

The impact of soil salinity at among other compounds polyamines produced by Limonium species was investigated in more detail [92]. Concentrations of 1,3-diaminopropane were relatively high (0.3-2.3 $\mu mol.g^{-1}$ dry material), while the concentrations of putrescine and cadaverine were much lower (<0.1 – 0.5 $\mu mol.g^{-1}$). Spermine and spermidine concentrations were in between (<0.1-1.5 $\mu mol.g^{-1}$).

Saeki et al. found that polyamines present in blood are located mainly in the erythrocytes [93]. Blood was centrifuged and the erythrocytes were suspended in sodium chloride (0.9%). After centrifugation, the sediment was washed five times with sodium chloride (0.9%) and the pellet was mixed with water. The cells were hemolysed by vortex mixing and extracted with 10% perchloric acid. After centrifugation, sodium carbonate and reagent (10 $mg.ml^{-1}$ dansyl chloride in acetone) were mixed with the supernatant and allowed to react overnight at ambient conditions by excluding light. L-proline was added to remove excess dansyl chloride. After acetone evaporation, a double extraction with benzene was performed. Benzene was evaporated and the residue was dissolved in benzene (50 μl) and injected (5 μl) into the liquid chromatographic system. It was found that the content spermidine and spermine was 14±3 and 8±3 $nmol.ml^{-1}$ packed erythrocytes respectively. Recoveries were better than 90%

Another interesting example of the applicability of the dansylation derivatization is the developed of a method for the determination of polyamines and conjugated polyamines in (hydrolyzed) cerebrospinal fluid, urine and tissue with a cationic exchange clean up procedure [94]. Cerebrospinal fluid and urine were treated identical. Sample, internal standard solution and potassium citrate buffer were mixed and brought onto a strong preconditioned cationic exchange SPE column. After washing of the SPE column, free and conjugated polyamines were eluted with 20% acetone in potassium citrate buffer and dansylated. Hydrolyzed urine samples were treated with concentrated hydrochloric acid (14-16 hr, 110°C) and freeze-dried. The residue was denaturized in 5-sulfosalicylic acid (4%) and subsequent centrifuged. 50 μl of the supernatant was used for dansylation. Hydrolyzed cerebrospinal fluid was received by adding one drop hydrochloric acid (3 M) to 1 ml sample. After freeze-drying, hydrochloric acid (0.5 ml, 6 M) was added and the sample was hydrolyzed (14-16 hr, 110°C). After renewed freeze-drying, the sample was denaturized identical to the hydrolyzed urine sample. The pH of the sample was brought to 3-4 by adding potassium hydroxide, potassium citrate and (if necessary) hydrochloric acid or again potassium hydroxide. The total sample is cleaned with the SPE pretreatment as described above.

Tissue cultures were received as followed; "Wash cells free of tissue culture medium or trypsin solution with phosphate-buffered saline". Cells were centrifuged into a pellet and saline was removed carefully without disturbing the pellet. Denaturization was performed by adding 8% 5-sulfosalicylic acid to 10^6 pellets. After ultrasonic agitation, samples were place in ice, mixed and placed back in ice. After centrifugation, the supernatant was dansylated, which was performed as follow: standard solution (50 μl), 200 μl saturated sodium carbonate and dansyl chloride (50 μl, 10 $mg.ml^{-1}$) were vortex mixed with sample (15 s) and the solution was allowed to react. Samples were cleaned on a C_{18} SPE column, eluted from this column with methanol and injected into the chromatographic system. The effluent could be stored in darkness (4°C) and was stable for at least a month. The applied reaction conditions were optimized with respect to temperature, reaction time and pH. Derivatization was completed after 10 min at 70°C. After longer reaction times and higher reaction temperatures decomposition of the dansylated polyamines and an increase of "background" peaks was observed. Optimum pH was found to be between 9.5 and 11. Detection limits were approximately 0.05-0.1 pmol. Repeatability (within day and in between days) was found to be approximately 2-6% RSD for different concentrations and different polyamines. Recovery, measured at different concentrations and for different polyamines, was between 92-117%.

Fu et al. developed an RP-HPLC method for the determination of polyamines in human prostate without extractions with organic solvents and subsequent evaporation or other condensation procedures [95]. It was claimed that this procedure could be applied to nearly all biological tissues and body fluids. For sample preparation, tissues were homogenized with phosphate buffer. Trichloroacetic acid (10%) was added to denaturize the supernatant, which was neutralized with saturated sodium carbonate. For derivatization, dansyl chloride was dissolved in acetone and mixed with supernatant, hexamethylenediamine and saturated sodium carbonate. The reactant solution was incubated in a water bath and analyzed. The within day and between days precision of the dansylated polyamines was RSD 1-3%. The recovery was >97%. The detection limits were 0.05 $nmol.ml^{-1}$.

3.1.4. Dabsyl Chloride (4-dimethylaminoazobenzene-4'sulfonyl chloride)

Dabsylation is much less popular compare to dansylation derivatization. Krause et al. developed a method for biogenic polyamines in different complex matrices [86]. The RP-HPLC method covered more than 40 components (including the polyamines). Detection was performed with UV (λ=436 nm). 0.1-1 g of homogenized solid samples of food (cheese, meat, sausages and fish) tissue or plant material, or freeze-dried samples (plasma, tissue) were dispersed with 10 ml (0.1 M HCl solution containing 0.2% 3,3'-thiodipropionic acid and 250 nM norleucine or norvaline as internal standard) and centrifuged. The tube was kept in an ultrasonic bath and was again centrifuged and any fat topping layer was removed. 3 ml of a filtrate was denaturized with ultra filtration. 20 μl standard or denaturized sample extract (if necessary extra diluted with 0.1 M HCl solution, containing 0.2% 3,3'thiodipropionic acid) was diluted with 180 μl reaction buffer (0.15 M NaHCO3, pH=8.6 with NaOH). After vortex mixing, 200 μl of dabsyl chloride reagent (12.4 mM in acetone) was added. After mixing and incubating (15 min, 70°C), the reaction was stopped in ice. After a short centrifugation step, 400 μl of dilution buffer was added and the mixture was centrifuged. The clear supernatant could be injected directly or stored at –20°C. The obtained recoveries were between 98-104%. Detection limits of 0.1-0.5 pmol (10 μl injected) were received and the obtained repeatability was 1-3% RSD

3.1.5. Benzoylchloride (quinoline-8-sulfonyl chloride)

The derivatization of polyamines with quinoline-8-sulfonyl chloride is relatively often used. As an example, the determination of biogenic polyamines in the retina of chicken embryo is discussed [96]. Retinas were stored in saline and centrifuged. Samples were homogenized (ultrasonic agitation) in perchloric acid (1 M) and centrifuged. Sample, sodium hydroxide (2 M) and 10 μl benzoyl chloride in methanol (1:1) were mixed and after 20 min extracted with chloroform. After centrifugation, the organic phase was re-extracted twice with sodium hydroxide to transfer reagent into benzoic acid. The organic phase was evaporated to dryness and re-dissolved in methanol (60% in water) before it was injected and separated on a reversed-phase system and quantified with UV detection. Repeatability was 1-2% RSD (within day) and 3-5% RSD (between days). Detection limit were <0.02 nmol (20 μl injected). Recovery was proven by an almost unit correlation coefficient for standard addition experiments.

3.1.6. Others

Fmoc (9-Fluorenylmethyl-chloroformate)

Fmoc (9-Fluorenylmethyl-chloroformate) is a well know derivatization agent for amines, but not so often applied for polyamines, however, typical examples are known. LC-MS studies with ethylene diamine, putrescine and spermidine, identified the blank peaks (Fmoc itself and its hydrolysis product Fmoh) and the polyamine derivatives [97]. As expected, it was found that both diamines reacted twice with Fmoc. The presence of a small m/z of 829.6 in the LC-MS chromatogram of the derivatized spermidine suggested also that FMOC reacts three times with this triamine. Bellagamba determined Fmoc derivatives of polyamines in milk [98]. Milk was mixed with perchloric acid (10%) and internal standard. After vortex mixing and centrifugation, KOH (1.5 M) was added to the supernatant to adjust the pH to approximately 7. After centrifugation, fat was removed with chloroform-methanol (2:1) and the sample was centrifuged again. Potassium borate (1 M) was added to the aqueous phase and mixed to adjust the pH to 8.0. Fmoc_Cl (1 ml, 5 mM in acetonitrile) was added to the alkalized extract and this solution was mixed (20 sec). After 1 min, 200 μl aspartic acid (10 mM in 10mM HCl) was added. The derivatized polyamines were extracted with n-pentane. The organic phase was transferred after centrifugation and the n-pentane extraction was repeated. The combined organic phase was evaporated to dryness. The residue was dissolved in mobile phase and injected into a column switching system (to remove matrix interferences). Detection limits of 50 pmol.ml^{-1} were obtained. Recoveries for 1-2 nmol.ml^{-1} solutions were 50-65% with a standard deviation of 2-5 %

Fmoc_OSu (N-(9-fluorenylmethoxycarbonyloxy)succinimide)

Fmoc_OSu was used for the determination of polyamines instead of the above-described Fmoc_Cl [99]. The main advantage over Fmoc_Cl was that the residual Fmoc_Osu did not interfere with the elution of the natural amines and polyamines. Also, the Fmoc_Cl derivatives of natural polyamines and the diamino acids His, Tyr, Orn and Lys were represented by more than one peak. Fmoc_OSu reacted rapidly and no double peaks were observed for the polyamines and the diamino acids. The method was applied to both beer and wine (only filtered through a 0.22μm membrane) and cell cultures. The cultured human keratinocytes and tumor cell lines (A431, Nb2 and MCF-7) were washed twice with ice-cold phosphate buffer saline and homogenized with perchloric acid (1.0 ml, 0.2 M, containing 15 nM of internal standard), in an ultrasonic bath for

30 min. The supernatants were used for analysis after centrifugation and filtration. The detection limit was 50 pM (50μl injection). Repeatability was for a 10 μM solutions of polyamines and for cell extracts approximately 1-3% RSD.

AccQ (6-aminoquinolyl-N-hydroxysuccinimidyl carbamat)

AccQ is a not very often applied derivatization agent. It was applied to e.g. the determination of polyamines in several wines from the Tarragona region [100]. AccQ derivatives of polyamines were synthesized. Only 150 analyses could be performed on an ODS-2 column, as a gradual hydrolysis of the stationary phase was observed. Wine samples were filtered before injection. With automated sample preparation, 10 μl of standard/sample was mixed with 6μl borate buffer (pH=8.8) and 0.5μl AccQ reagent (10 mM in acetonitrile) before injection. Repeatability was better than 5% RSD. Detection limits were 50 $\mu g.l^{-1}$ for putrescine and cadaverine.

Pyrene Derivatives

2-(1-Pyrenyl)ethyl chloroformate was applied for the determination of polyamines in serum [101]. Yamaguchi introduced a similar but highly interested new derivatization to determine biogenic polyamines [102-105]. PSE (4-(1-pyrene)butyric acid N-hydroxysuccinimide) and PBC (4-(1-pyrene)butanoyl chloride) reacted both with primary and secondary amines. As a result, an amine could be converted into a pyrene-labeled derivative. Monoamines were converted into a monopyrene labeled derivative, showing emission at λ=360-420nm. Due to intramolecular excimer fluorescence, the polypyrene labeled polyamines will show emission at higher wavelengths (λ=440-520 nm). At first, PBE was used, but better reactivity's were obtained for secondary amino groups with PBC. Detection limits of 1-2 fmol (20 μl injected) were received for different diamines and polyamines (including spermidine and spermine). Repeatability was ≤8% RSD. It is interesting to state that this same principle also works for bisphenols [106].

Others

There are a lot of different (exotic) agents studied for the derivatization of polyamines. Shimada et al. evaluated six different ferrocene possessing agents for pre-column derivatization and electrochemical detection [107]. Airborne polyamines were quantified with 1-naphthyl acetic anhydride [108] and 2-(9-Anthryl)ethylchloroformate was used for the pre-column derivatization of the biogenic polyamines in urine [109]. 4-Fluoro-3-nitrobenzotrifluoride was applied by Asotra et al. [110]. Sugiura used p-toluenesulfonic chloride to produce polyamine tosyl derivatives [111,112] and Vandenabeele et al. determined the polyamines with 2-chloroethylnitrosourea [113].

3.2 Capillary Electrophoresis (CE)

CE is a highly efficient separation technique. Standard UV-detection is not very sensitive as the detector path length is often relatively small. From a theoretical point of view, it could be expected that after a pre-column derivatization the benefits of separating polyamines on their charged could be argued. For electrophoretic separation, without run buffer additives to enhance migration, it is therefore important that derivatized polyamines are still charged. Linearity of the CE systems is often limited and corrections for migration times have been described. Polyamines and catecholamines were determined in PC-12 tumor cell with fluorescein isothiocyanate (FITC) and laser-induced fluorescence (LIF) detection [114]. Detection limits were reported to be between 0.03 to 2.5 $nmol.l^{-1}$. AccQ derivatives of polyamines were also separated with capillary electrophoresis and detected with LIF detection [115]. A disadvantage of both AccQ and FITC is their native fluorescence.

The inlet of the separation capillary can be used as a reaction chamber for capillary electro chromatography (CEC) [116]. This idea was applied to the OPA derivatization of polyamines [36]. The borate runbuffer contained OPA and mercaptoethanol. On-column derivatization and CEC was studied besides pre-column derivatization with CEC and throughout in-capillary derivatization with CE. Raw fish samples were homogenized with 0.5 M HCl and neutralized with NaOH, diluted with water to 50 ml. The extracts were filtered prior to analysis. With a biogenic standard solution, area repeatability was <2% RSD, retention/migration time repeatability was in-between 0.2-0.8% RSD and the detection limit was 0.05-0.1 mM.

Other derivatizations with capillary electrophoresis have been described also, like OPA derivatization [117], tosyl derivatization [118], and anhydride derivatization [119]. As an alternative, Liu et al. used a specially designed tris(2,2'-bipyridyl)ruthenium (II) electrochemiluminescence detection cell [120].

Micellar electrokinetic chromatography (MEKC) can be used to improve resolution. Although linearity is often limited, some different studies have been performed. Natural polyamines were determined in 21 different food samples (both from animal origin and plant products) with MEKC [121]. Samples were extracted with 0.6 M perchloric acid and subsequently centrifuged and filtrated. The samples were derivatized with benzoyl chloride before analysis. Detection limits of 1-2 ppm were obtained and RSD (on 5-20 ppm level) was approximately 10%.

A MEKC-method was also developed to determine polyamines and catecholamines simultaneously in pheochromocytoma (PC-12) tumor cells [6]. Some anti-cancer drugs have also a strong impact on the levels of catecholamines. After extraction, polyamines and catecholamines were derivatized with AccQ (6-Aminoquinolyl-N-hydroxysuccinimidyl carbamate). pH, SDS-concentration and 18-crown-6 ether were used to optimize the separation. Polyamines were selectively retained by the crown-ether. The authors assume that the polyamine derivatives had still a free amino acid to interact with the crown ether. Detection limits of 0.1 mM were obtained. In all cases the RSD of the relative peak areas were 4% for 6 consecutive injections.

1-Pyrenebuttanoic acid succinimidyl ester (PBE, see also section 3.1.6 of this chapter) was also used for MEKC separations [122]. Acetonitrile is needed to solvate the pyrene labeled polyamines and cholate was used as a surfactant, as it has a greater ability to separate the highly hydrophobic derivatives at higher acetonitrile concentrations. Detection limits of 10 nM (corresponding to 100-300 nM in the original samples) were reported. RSD of the repeatability was $\leq$1%, while repeatability (between days) was $\leq$8%.

3.3. Gas Chromatography (GC)

Gas chromatography is not as often used for the determination of polyamines as liquid chromatography. One of the main problems with gas chromatography is the sample introduction. Serum and urine can hardly be introduced directly into a gas chromatograph. Moreover, an aqueous phase is not the first choice for GC as the formation of non-volatile amino-salts should be circumvented. Also, the presence of a lot of non-polar contaminations in all kind of biogenic samples makes GC probably not the first separation technique to be chosen. However, GC systems are enormously efficient and methods were described for different samples like urine [123], erythrocytes [124], plasma [125], and meat [126].

Derivatizations are almost inevitable and different methods are discussed in this section. Gas chromatography also has the possibility of an almost universal detection with the flame ionization detector [127]. However, gas chromatographic systems can also be coupled with more selective detectors, like an electron capture detector (ECD), a nitrogen phosphorus detector (NPD) or a mass spectrometer (MS). In contradiction to the liquid chromatographic part, this paragraph is subdivided into the detector used.

3.3.1. ECD (Electron Capture Detection)

Electron capture detection is a selective and sensitive detection, based on the radioactive ^{63}Nickel. The emitted β particles collide with nitrogen and produce electrons, which can be captured by electronegative analytes. Specific detection can be obtained after selective reaction of the target functional group with e.g. a highly halogenated derivatization agent. Pentafluoropropionyl or even heptafluorobutyryl derivatives can be produced to obtain optimize sensitivity and selectivity [128]. GC coupled with ECD was also applied to the determined of polyamines in hepatic tissue samples of rats [129]. Tissue samples were homogenized with water and the resulting homogenates were alkalized with trisodium phosphate (10%) and the polyamines were derivatized with pentafluorobenzoyl chloride in toluene. After mixing (10 min) and centrifugation, the organic phase was separated and dried. Each residue was redissolved in toluene and analyzed. Concentrations of dietary rats were 2034±292 $ng.g^{-1}$ putrescine, 262±95 $ng.g^{-1}$ cadaverine, 179±26 $ng.g^{-1}$ spermidine.

3.3.2. NPD (Nitrogen Phosphor Detector)

The NPD detector is another interesting detection approach. The detector is sensitive towards nitrogen containing compounds (like polyamines). Heptafluorobutyryl derivatization is most popular and also compatible with ECD and GC-MS but other derivatizations have been applied also. No nitrogen rich derivatization agents have been applied for the derivatization of polyamines. A method has been described to determine free and acetylated polyamines in urine as their heptafluorobutyryl derivatives with NPD (for quantification) and MS (for identification) [130]. Urine, spiked with 1,6-diaminohexane was extracted with ethyl acetate. Both borate buffer and sodium hydroxide were added and mixed .The supernatant was introduced onto a silica SPE-cartridge. Polyamines were eluted with 0.1 M methanolic HCl. The effluent was evaporated to dryness and the polyamines were derivatized with heptafluorobutyric anhydride (1 hr, 80°C) in ethyl acetate. After cooling and evaporation of the organic solvent, the residue was dissolved in 50 μl ethyl acetate and 2 μl was injected. It was observed that all primary and secondary amines reacted once with the derivatization agent. Recoveries >80% were reported for the diamines (putrescine, cadaverine and 1,3-diaminopropane) and also for N-acetylcadaverine and N-acetylputrescine. Recoveries for spermidine, spermine, and the N_1 and N_8 acetylated form of spermidine were approximately 50-70%. For all concentration levels (0.1-10 $\mu g.ml^{-1}$) RSD was better than 10%.

3.3.3. MS (Mass Spectrometry)

Gas chromatography coupled with mass spectrometry is powerful separation and identification technique. Polyamines can be determined more selectively in complex matrices and quantification can be improved by applying isotopic labeled standards [131]. Improved sensitivity of polyamines and some metabolites in cerebrospinal fluids was by applying GC-SIM-MS (Gas chromatography single ion monitoring and mass spectrometry) with deuterated standards [132]. Cerebrospinal fluids were collected from patients by lumber puncture or a lumber external drain. Supernatant was frozen after centrifugation. A mixture of deuterated internal standards was added to 1 ml sample. Denaturization was performed with sulfosalicylic acid (12%). After vortex mixing, freezing, thawing and centrifugation the supernatant was hydrolyzed (12 N HCl, 120°C, 18 hr). The polyamines were concentrated on a SPE column containing silica and derivatized into n-heptafluorobutyryl-methylesters before analysis. Recoveries between 93-111% were reported. Repeatability's within days were determined for diaminopropane (38 nM) 13%, putrescine (159 nM) 9%, cadaverine (4 nM) 86%, spermidine (66 nM) 4% and spermine (5 nM) 82%. Alkoxycarbonylation allows derivatization and extraction simultaneously [133]. Other derivatization procedures have been reported, like trifluoroacetylatyion [134], alkylsilylation [135], perfluoroacetylacetylation [136,137], acylation [138], and perfluoroacetylacetylation [139]. The latter derivatization was applied to determine putrescine in rat brains [140]. Brains of decapitated rats were frozen and 20 specific sites were isolated from "serial frozen frontal sections". The samples were homogenized with ultrasonic agitation in HCl (0.1 N). Trichloroacetic acid (10%) extraction was performed and internal standard was added to the supernatant after centrifugation. The trichloroacetic acid extraction was repeated twice and all supernatants were extracted as one sample with diethyl ether. The sample was diluted and cleaned on a cationic SPE column. The effluent containing putrescine was dried and the residue was re-dissolved in water and freeze-dried. Recovery of this extraction procedure for ^3H-putrescine was approximately 99.5% (±1).

The freeze-dried sample was dissolved in acetonitrile and derivatized with pentafluoropropionic anhydride. After mixing, the reaction product was dried (under N_2). The residue was dissolved in acetonitrile (20 μl) and injected (0.5 μl) on a GC-MS system. The concentration of putrescine was determined in the before mentioned 20 specific sites. The concentration was found to be between 15-55 $pg.\mu g^{-1}$ of protein. The standard deviation (within the 8 different rats) was <20% RSD.

Polyamines were determined in hair by applying a double derivatization of all primary amines to obtain highly efficient separations [141]. Hair was washed with ethanol, water and acetone and cut into small pieces. Samples in HCl (0.6 M) were stored (16 hr, 40°C) and washed with n-pentane. Sodium hydroxide solution was added (2 M, pH=11-12). The sample was extracted with diethyl ether and ethylchloroformate to prepare N-EOC derivatives. After mixing and centrifugation, the water phase was extracted again with ethyl ether (2 ml). Both organic phases were combined and dried to dryness. Pentafluoropropionyl anhydride (50 μl) and ethyl acetate (100 μl) were added and the reaction mixture was heated (50°C, 30 min) and dried in vacuum. The N-EOC-N-PFP derivatives were re-dissolved in ethyl acetate (40 μl) and analyzed with GC-MS. Detection limits of polyamines of 1-10 $ng.g^{-1}$ were achieved with single ion monitoring. Recoveries were reported between 75 and 95%. Repeatability was between 5-15% RSD. Fernandes et al. developed a method to determine biogenic polyamines including 1,3-diaminopropane in port wine and corresponding grape juices [142]. They used bis-2-ethylhexylphosphate dissolved in chloroform as in ion-pair extraction reagent. After back extraction with a hydrochloric acid solution, the polyamines were derivatized with heptafluorobutyric anhydride. Recoveries between 90 and 120% were obtained. Detection was performed with electron impact mass spectrometry in the selection ion monitoring. Deuterated (internal) standards were used to optimize quantification. Linearity was obtained up till 15 $mg.l^{-1}$. Repeatability for the polyamines was between 1 and 10% RSD in different beverages with polyamine concentrations of 0.02-12 $mg.l^{-1}$. Detection limits were approximately 0.01 $mg.l^{-1}$.

4. Experimental Results

Recently, comprehensive 2 Dimensional GC*GC-TOF-MS is becoming extremely popular due to its enormous potential with respect to efficiency and selectivity. This technique has already been applied to the determination of complex flavor mixtures in food [143], urine [144] and tissue

[145]. Quantification is not straightforward as recently discussed [146]. Here we demonstrate the feasibility to analyze polyamines in spiked human urine samples. Urine sample (10 ml) is spiked with cadaverine (1.65 mg, home made), putrescine (1.87 mg, home made), spermidine (1.92 mg Fluka, Steinheim, Germany) and spermine (1.35 mg, Fluka). Samples were alkalized with sodium hydroxide (pH=11.5). Samples were extracted with 1-butanol (20 ml, Fluka). A hydrochloric acid solution (0.5 ml 2 M) was added to the organic phase, which was evaporated under nitrogen. 4 ml Dichloromethane and 400 μl trifluoroacetic anhydride (Merck, Hohenbrun, Germany) were added to the residue. 1 μl sample was injected on 2D GC*GC TOF MS system (LECO, Pegasus 4D, software chromatof 2.22). Chromatographic columns were 1st dimension column RTX-5 (10 m* 0.18 mm* 0.2 μm), 2nd dimension column DB-17 (1 m * 0.1 mm * 0.1 μm). Conditions were: constant helium flow 1.0 ml.min^{-1}, temp. program Oven 1: 50°C(1 min) – 5°C.min^{-1} – 280°C (20 min) and oven 2: 70°C (3 min) -5.75°C.min^{-1} - 280^{0}C (30 min). Modulator temp. offset was 35°C, the 2nd dimension separation time 4 s and the hot pulse time 0.4 s. TOF MS source temperature was 220°C, temp. transfer line was 250°C. Scan range was 5-600 D, (200 scan.s^{-1}). Retention times are given in Table 1.

Figure 6 Spiked urine sample analyzed with comprehensive 2D GC*GC-TOF-MS. TIC chromatogram is given. See table 1 for peak labeling, conditions see text. Not published, with permission of DSM Research.

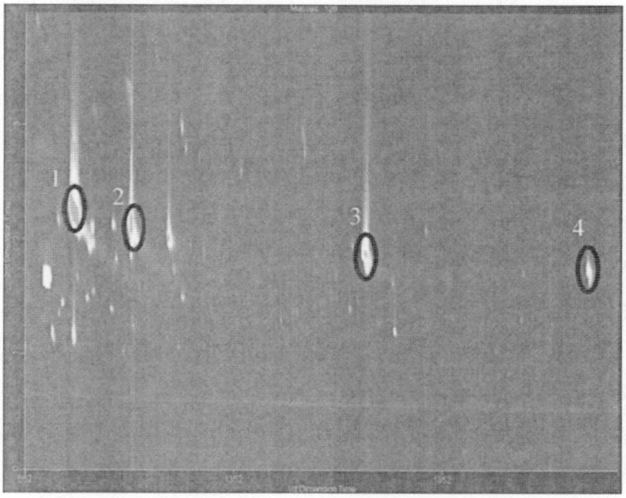

***Figure** 7* Spiked urine sample analyzed with comprehensive 2D GC*GC-TOF-MS. Reconstructed 2D chromatogram of m/z=126 chromatogram is given. See table 1 for peak labeling, conditions see text. Not published, with permission of DSM Research.

Table 1 2D retention times of the TFA derivatized biogenic polyamines

	tr: 1st dimension (s)	tr: 2nd dimension (s)
Putrescine (circle 1)	960	2.30
Cadaverine (circle 2)	1104	2.15
Spermidine (circle 3)	1668	1.88
Spermine (circle 4)	2200	1.75

Figure 6 depicts a typical comprehensive 2D GC*GC TOF-MS (TIC) of a urine sample. All kind of contaminations (including other derivatized amines) can be observed. It can nicely be observed that a lot of constituents are present. By recording the TIC, polyamines are hardly visible. Figure 7 depicts a part of the chromatogram with the reconstructed chromatogram of m/z=126, which originates from the trifluoroacetyl group (CF_3-CO-NH-CH2- fragment). Peak tailing of amino compounds on a polar (2nd dimension) colum,n can be expected, but due to the grayscaled chromatograms, peak tailing looks much worse than it actually was.

6. References

[1] J.C. Hoekstra, D.C. Johnson, Anal. Chem. 70(1998)83-88

[2] L.D. Morrison, L. Becker, L.C. Ang, S.J. Kish, J. Neuro. Chem. 65(1995)636-642

[3] I. Flink, D.E. Pettijhon, Nature 253(1975)62-63

[4] L.C. Gosule, J.A. Scheliman, Nature 259(1976)333-335

[5] C.W. Porter, R.J. Bergeron, Science 219(1983)1083-1085

[6] G. Lui, J. Chen, Y.Ma, J. Chromatogr. B 805(2004)281-288

[7] R. Draisei, L. Giannetti, P. Boria, L. Lucentini, L. Palleschi., S. Cavelli, J. Chromatogr. A 798(1998)109-116

[8] S. Tombelli, M. Mascini, Anal. Chim. Acta 258(1998)277-284

[9] D. Teti, M. Visalli, H. McNair, J. Chromatogr. B 781(2002)107-149

[10] M.Y. Khuhawar, G.A. Qureshi, J. Chromatogr. B 764(2001)385-407

[11] A. Gugliucci, Clin. Chem. Acta 344(2004)23-35

[12] P. Kalec, P. Krausove, Feed Chem. 90(2000)219-230

[13] O. Busto, J. Guasch, F. Borrul, J. Int. Scienc. Vigne Vin 30(1996)85-101

[14] F.A.J. Muskiet, B. Dorhout, G.A. van den Berg, J. Hessels, J. Chromatogr. B 667(1995)189-198

[15] P.Lehtonen Am. J. Enol. Vitic 47(1996)127-133

[16] W.H. Carothers patented (U.S.P. 2.071.2501/1) in H.J. Koslowski (editor), "Dictionary of Man-Made Fibers", International Business Press Publishers, Frankfurt an Main, Germany (1998)143-144

[17] H.J. Koslowski, "Dictionary of Man-Made Fibers", Int. Business Press Publishers, Frankfurt an Main, Germany (1998)p.292

[18] Patent, membrane technology 129(2001)p.13

[19] E. Hardenborg, A. Zurberovic, S. Ullsten, L. Soderberg, E. Heldin, K.E. Markides, J. Chromatogr. A 1003(2003)217-221

[20] A. Mecke, I. Lee, J.R. Baker, M.M. Banaszak Holl, B.G. Orr, Eur. Phys. J. E 14(2004)7-16

[21] E.M.M. de Brabander, J. Brackman, M. Mure-Mak, H. de Man, M. Hogeweg, J. Keulen, R. Scherrenberg, B. Coussens, Y. Mengerink, Sj. van der Wal, Macromol. Symp. 102(1996)6-17

[22] R.A.T.M. van Benthem, Progress in Organic Coatings 40(2000)203-214

[23] B.I. Voit, Comptes Rendus Chimie 6(2003)821-832

[24] P.E. Froehling, Dyes and Pigments 48(2001)187-195

[25] P. Kolhe, E. Misra, R.M. Kannan, S. Kannan, M Leih-Lai, Int. J. Pharmaceutics 259(2003)143-160

[26] S. Supattapone, H. Wille, L. Uyechi, J. Safar, P. Tremblay, F.S. Szoka, F.E. Cohen, S.B. Prusiner, M.R. Scott, J. Virology 75(2001)3453-3461

[27] M. Kramer, J-F. Stumbe, G. Grimm, b. Kaufmann, U. Kruger, M.. Weber, R. Haag, Chem. Biochem. 5(2004)1081-1087

[28] B. Klajnert, L. Stanislawska, M. Bryszewska, B. Palecz, Biochimica et Biophysica Acta 1648(2003)115-126

[29] Stathakis, E.A. Arriaga, N.J. Dovichi , J. of Chromatogr. A 817(1998)233-232

[30] P.G. Muijselaar, H.A. Claessens, C.A. Cramers, J.F.G.A. Jansen, E.W. Meijer, E.M.M. de Brabander-Van den Berg, Sj.v.d. Wal, J. High Resol. Chromatogr. 18(1995)121-123

[31] K.A. Nielsen, J. exp. Bot. 41(1990)849-859

[32] O. Faure, M. Mengoli, A. Nongarede, N. Bagni, J. Plant Physiol. 138(1991)5455-5465

[33] B. Fontianiella, J.L. Mateos, C. Vicente, M-E Legaz, J. Chromatogr. A 919(2001)283-288

[34] S.Yamamoto, H. Itano, H. Kataoka, M. Makita, J. Agric. Food Chem. 30(1982)435-439

[35] A. Halasz, A. Barath, L. Simon-Sarkadi, W. Holzapfel, Trends Food Sci. Technol. 5(1994)42-49

[36] S. Oguri,Y. Yoneya, M. Mizunuma, Y. Fujiki, K. Otsuka, S. Terabe, Anal. Chem. 784(2002)3463-3469

[37] M. Paula, M. Marques, L.A.E. Batista de Carvalho, J. Tomkinsons J. Phys. Chem. A 106(2002)2473-2482

[38] L. Simon-Sarkadi, A. Kovacs, E. Mincsovics J. Planar Chromatogr. 10(1997)59-60

[39] R. Madhubala, Methods Mol. Biol. 79(1998)131-136

[40] R.M Linares, J.H. Ayala, A. M. Alfonso, V. Gonzalez, J. Liq. Chromatogr. 23(2000)2653-2668

[41] H. Claessens, Thesis, "Characterization of stationry Phases for Reversed-Phase Liquid Chromatography", Eindhoven, The Netherlands (1999)7-31

[42] B. Fransson, M.L.S. Almeide, L. Grehn, U. Ragnarsson, J. Chromatogr. 508(1990)229-231

[43] S.K. Sadain, J.A. Koropchak, J. Chromatogr. A 844(1999)111-118

[44] J.C. Hoekstra, D.C Johnson Anal. Chem Acta 390(1999)45-54

[45] R. Pineda, A.D. Knappa, J.C. Hoekstra, D.C. Johnson, Anal Chem Acta 449(2001)111-117

[46] D.A. Dobberpuhl, J.C. Hoekstra, D.C. Johnson Anal. Chim Acta 322(1996)55-62

[47] M. Yang, S.A. Tomellini, J. Chromatogr. A 841(1999)9-18

[48] C. Vidal-Carou, F. Lahoz-Portoles, S. Bover-Cid, A. Marine-Font, J. Chromatogr. A, 998(2003)235-241

[49] S. Suzuki, K. Kobayashi, J. Noda, T. Suzuki, K. Takama, J. Chromatogr. 508(1990)225-228

[50] H.J. Bremer, E. Kohne Clin. Chem. Acta 32(1971)407-418

[51] C.W. Gehrke, K.C. Kun R.W Zumwalt, T.P. Waalkes, J. Chromatogr. 89(1974)231-238

[52] K. Maruta, R. Teradaira, N. Watanabe, T. Nagatsu, M. Asano, K. Yamamoto, T. Matsumoto, Y. Shionoya, K. Fujita, Clin. Chem. 35(1989)1694-1696

[53] N. Watanabe, Biomed. Chromatogr. 6(1992)1-3

[54] K. Hiramatsu, S. Kamei, M. Sugimoto, K. Kinoshita, K. Iwasaki, M. Kawakita, J. Biochem. 115(1994)584-589

[55] Meseguer Lloret, C. Molins Legua, J. Verdu Andres, P. Camp´yns Falco, J. Chromatogr. A 1035(2004)75-82

[56] D. H. Russel, C.C. Levy, S.C. Schimpf, I.A. Hawk, Cancer Res. 31(1971)1555-1558

[57] D.H. Russell. Nature 233(1971)144-145

[58] X. Sun, X. Yang, E. Wang, J. Chromatogr. A 1005(2003)189-195

[59] L. Arce, A. Rios, M. Valcarcel, J. Chromatogr. A 803(1998)249-260

[60] R. Zhang, C.L. Cooper, Y. Ma, Anal. Chem. 65(1993)704-**706**

[61] Y. Ma, R.Zhang, C.L. Cooper, J. Chromatogr. 608(1992)93-96

[62] X. Huang, J. Ren, J. Liq. Chromatogr. 26(2003)355-367

[63] S. Beninati, S. Sartori, M.P. Argento-Ceru Anal. Biochem. 80(1977)101-107

[64] G. Banghegyi, T. Cserhati, K. Valko, J. Chromatogr. 553(1991)407-414

[65] A. De Robertis, C. Foti, O. Giuffre, S. Sammartano, J. Chem Eng. Data 46(2001)1425-1435

[66] S.G. Entilis, V.V. Evreinov, A.V. Gorshkov, Adv. Polym. Sci. 76(1986)129-175

[67] Y. Mengerink, "non-exclusion separation techniques for polyamindes", thesis Eindhoven, The Netherlands, (2001)29-31

[68] Y. Mengerink, R. Peters, Sj. van der Wal, H.A. Claessens, C.A. Cramers, J. Chromatogr. A 949(2002)337-349

[69] G. Guiochon, "Is pressure in Chromatography like Esope's tongue", Abstracts of 25th Internal Symposium on Chromatography p24, ISC 04, Paris, France, 2004

[70] Y.Mengerink , M. Mure, E.M.M de Brabander, Sj. van der Wal, J. Chromatogr. A 730(1996)59-66

[71] A. Sharma, D.K. Mohanty, A. Desai, R. Ali, Electrophoresis 24(2003)2733

[72] Sj. Van der Wal, Y. Mengerink, E.M.M de Brabander, C.M. Jeronimus-Stratingh, A.P. Bruins, J. Chromatogr. A 825(1998)135-147.

[73] J. Nawrocki, C. Dunlap, A. McCormick, P. W. Carr, J. Chromatogr. A, 1028(2004)1-30

[74] J. Nawrocki, C. Dunlap, J. Li, J. Zhao, C. V. McNeff, A. McCormick, P. W. Carr J. Chromatogr. A 1028(2004)31-62

[75] D.E. Bergbreiter, A.M. Kippenberger, W.M. Lackowski, Macromolecules 38(2005)47-52

[76] A.J. Geall, J.A. Baugh, M. Loyevsky, V.R. Gordeuk, Y Al-Abed, R. Bucala Bioconjugate Chem. 15(2004)1161-1165

[77] D. Tleugabulova, W. Czardybon, J.D. Brennan, J. Phys. Chem. B 108(2004)10692-10699

[78] J.J. Broersen, H. Jansen, C. de Ruiter, U. A. Th. Brinkman, R.W. Frei, F.A. Buijtenhuis, F.P.B. Van der Maeden, J. Chromatogr. 436(1988)39-46

[79] Y. Mengerink, R. Peters, M. Kerkhoff, J. Hellenbrand, H. Omloo, J. Andrien, M. Vestjens, Sj. van der Wal, J. Chromatogr. A 878(2000)45-55

[80] D. Kutlan, P. Presits, I.Molnar-Perl, J. Chromatogr. A 949(2002)235-248

[81] P. Camins-Falco, C. Molins-Legua, A. Sevillano-Cabeza, L.A. Tortajada Genaro, J. Chromatogr. B 759(2001)285-297

[82] Y. Mengerink, F. Toth, D. Kutlan, I. Molnar Perl, J. Chromatogr. A 949(2002)99-124

[83] H.M.H. van Eijk, D.R. Rooyakkers, N.E.P. Deutz, J. Chromatogr. A 730(1996)115-120

[84] K. Samejima, M. Kawase, S. Sakamoto, M. Okada, Y. Endo, Anal. Biochem. 76(1976)392-406

[85] C. de Jong, G.J. Hughes, E. van Wieringen, J. Chromatogr. 24(1982)345-359

[86] I. Krause, A. Bockhardt, H. Neckermann, T. Henle, H. Klostermeyer, J. Chromatogr. A 715(1995)67-79

[87] K.J. Hunter, Methods Mol. Biol. 79(1998)119-123

[88] H.Y. Aboul-Enein, I.A. Al-Duraibi, Biomed. Chromatogr. 12(1998)291-293

[89] L. Gerbaut, Clin. Chem. 37(1991)2117-2120

[90] Z. Loukou, A. Zotou, J. Chromatogr. A 996(2003)103-113

[91] R. Minocha, S. Long, J. Chromatogr. A 1035(2004)63-73

[92] A. Bouchereau, C. Duhaze, J. Martin-Tanguy, J-P Guegan, F. Larher, J. Chromatogr. A 836(1999)209-221

[93] Y. Saeki, N. Uehara, S. Shirakawa, J. Chromatogr. 145(1978)221-229

[94] P.M. Kabra, H.K. Lee, W.P. Lubich, L.J. Marton, J. Chromatogr. 380(1986)19-32

[95] S. Fu, X. Zou, X. Wang, X. Liu, J. Chromatogr. B 709(1998)297-300

[96] G. Taibi , M.R. Schiavo, M.C. Gueli, P. Calanni Rindina, R. Muratore, C.M.A. Nicotra J. Chromatogr. B 745(2000)431-437

[97] Z. Lu, H. Veenig, J.E. Coutant, J. Chromatogr. 540(1991)199-206

[98] F. Bellagamba, V.M. Moretti, T. Mentasti, A. Albertini, U. Luzanna, F. Valfre, J. Chromatogr. A 791(1997)79-94

[99] V. Luzanov, S. Petrov, V. Mitev, J. Chromatogr. A 1025(2004)201-208

[100] O. Busto, J. Guasch, F. Borrull, J. Chromatogr. A 737(1996)205-213

[101] M.A. Cichy, D.L. Stegmaier, H. Veening, H.D. Becker, J. Chromatogr. B 613(1993)15-21

[102] H. Nohta, H. Satozono, K. Koiso, H. Yoshida, J. Ishida, M Yamaguchi, Anal. Chem. 72(2000)4199-4204

[103] Y. Nokano, H. Nohta, H. Yoshida, H. Saita, T. Fujito, H. Mori, M Yamaguchi, J. Chromatogr. B 2002(774)165-172

[104] H. Yoshida, Y. Nakano, K. Koiso, H. Nohta, J. Ishida, M. Yamaguchi, Anal. Science 17(2001)107-112

[105] H. Yoshida, H. Harada, Y. Nakano, H. Nohta, J. Ishida, M Yamaguchi, Biomed Chromatogr. 18(2004)687-693

[106] H. Yoshida, H. Harada, H. Nohta, M. Yamaguchi, Anal. Chem. Acta 488(2003)211-221

[107] K. Shimada, T. Oe, M. Tanaka, T. Nambara, J. Chromatogr. 487(1989)247-255

[108] W.S. Wu, V.S. Gaind, J. Liq. Chromatogr. 15(1992)267-282

[109] A.J. Faulkner, H.Veening, H-D. Becker, Anal. Chem. 63(1991)292-296

[110] S. Asotra, PV. Mladenov, R.D. Burke, J. Chromatogr. 408(1987)227-233

[111] T. Sugiura, T. Hayashi, S. Kawai, T.J. Ohno, J. Chromatogr. 110(1975)385-388

[112] T. Hayashi, T. Sugiura, S. Kawai, T. Ohno, J. Chromatogr. 145(1978)141-146

[113] O. Vandenabeele, L. Garrelly, M. Ghelfenstein, A. Commeyras, L. Mion, J. Chromatogr. 795(1998)239-250

[114] M. Du, V. Flanigan, Y. Ma, Electrophoresis 25(2004)1496-1502

[115] A. Kovacs, L. Simon Sarkadi, K. Ganzler, J. Chromatogr. A 836(1999)305-313

[116] H.A. Fischman, J.B. Shear, L.A. Colon, R.N. Zare, U.S. Patent 5 318 680 1994

[117] K.B.Male, J.H.T. Luong, J. Chromatogr. A 926(2001)309-317

[118] M.E. Legaz, C. Vicente, M.M. Pedrosa, J. Chromatogr. A 823(1998)511-512

[119] Y. Zhang, F.A. Gomez, Electrophoresis 21(2000)3305-3310

[120] J. Liu J, X. Yang. E. Wang, Electrophoresis 24(2003)3131-3138

[121] P. Kalec, M. Krizek, T. Pelikanova, O. Veskrna, Food Chemistry 90(2005)561-564

[122] R.E. Paproski, K.I. Roy, C.A. Lucy, J. Chromatogr. A 946(2002)265-273

[123] G.A. van der Berg, F.A.J. Muskiet, A.W. Kingma, W. van der Silk, M.R. Halie, Clin. Chem. 31(1986)1930-1937

[124] G.A. van der Berg, A.W. Kingma, F.A.J. Muskiet, J. Chromatogr. 415(1987)27-34

[125] M.T. Bakowski, P.A. Toseland, J.F.C. Wicks, J.R. Trounce, Clin. Chem. Acta 110(1981)273-286

[126] J. Slemr, K. Beyermann, J. Chromatogr. A 283(1984)241-250

[127] M.Y. Khuhawar, A.A. Memon, M.I. Bhanger J. Chem. Soc. Pak. 22 (2000)115-125

[128] J.M. Rattenbury, P.M. Lax, K. Blau, M. Sandler Clin. Chim. Acta 95(1979)61-67

[129] R.L.H. Clements, A. Holt, S. Gordon, K.G. Todd, G.B. Baker, J. Pharmacological and Toxicological methods 50(2004)35-39

[130] J.W. Suh, S.H. Lee, B.C. Chung, J. Park, J. Chromatogr. B 688(1997)179-186

[131] R.G. Smith, G.D. Daves, Jr., Biomed. Mass Spectrom. 4 (1977)146-151

[132] G.A. van den Berg, J.M. Schaaf, G.T. Nagel, A.W. Teelken, F.A.J. Muskiet, Clin. Chem. Acta 165(1987)147-154

[133] K.R. Kim, M.J. Paik, J.H. Kim, S.W. Dong, D.H. Jeong, J. Pharm. Biomed. Anal. 15(1997)1309-1318

[134] T. Walle, In: D.H. Russell, Editor, Polyamines in Normal and Neoplastic Growth, Raven Press, New York (1973)355-356

[135] R.D. Velasquez, G. Brunner, M. Varrentrapp, D. Tsikas, J.C. Frolich, Z. Gastroenterol. 34 (1996)116-126

[136] T. Ohki, A. Saito, K. Ohta, T. Niwa, K. Maeda, J. Sakakibara, J. Chromatogr. 233 (1982)1-8

[137] Y. Ikuina, C. Bando, M. Yoshida, H. Yano, Y. Saitoh, J. Antibiot. (Tokyo) (1997)998-1006

[138] T. Okhi, A. Saito, N. Yamanaka, K. Ohta, J. Sakakibara, T. Niwa, K. Maeda, J. Chromatogr. 228(1982)51-57

[139] S.H. Lee, S.O. Kim, H. Lee, B.C. Chung, Cancer Lett. 133(1998)47-56

[140] T. Noto, T. Hasegawa, H. Kamimura, J. Nakao, H. Hashimoto, T. Nakajima, Anal. Biochem. 160(1987)371-375

[141] M.H. Choi, K. Kim, B.C. Chung, J. Chromatogr. A 897(2000)295-305

[142] J.O. Fernandes, M.A. Ferreira, J. Chromatogr. A 886(2000)183-195

[143] M. Adahchour M, L.L. van Stee, J. Beens J, R.J. Vreuls, M.A. Batenburg, U.A. Th. Brinkman, J. Chromatogr. 1019(2003)157-172

[144] E. Sinha, J.L. Hope, B.J. Prazen, E. J. Nilsson, R.M. Jack, R.E. Synovec, J. Chromatogr. A, 1058(2004)209-215

[145] M. Harju, A. Bergman, M. Olsson, A. Roos, P. Haglund, J. Chromatogr.A 1019(2003)127-142

[146] V.G. van Mispelaar, A.C. Tas, A.K. Smilde, P.J. Schoenmakers, A.C. van Asten, J. Chromatogr. A 1019(2003)15-29

Index

D

E

P

Q

R

S

T

U

W

X

JOURNAL OF CHROMATOGRAPHY LIBRARY

A Series of Books Devoted to Chromatographic and Electrophoretic Techniques and their Applications.
Although complementary to the Journal of Chromatography, each volume in the library series is an important and independent contribution in the field of chromatography and electrophoresis. The library contains no material reprinted from the journal itself.

Recent volumes in this series

JOURNAL OF CHROMATOGRAPHY LIBRARY

A Series of Books Devoted to Chromatographic and Electrophoretic Techniques and their Applications. Although complementary to the Journal of Chromatography, each volume in the library series is an important and independent contribution in the field of chromatography and electrophoresis. The library contains no material reprinted from the journal itself.

Recent volumes in this series

Volume 60 Advanced Chromatographic and Electromigration Methods in BioSciences
edited by Z. Deyl, I. Mikšík, F. Tagliaro and E. Tesařová

Volume 61 Protein Liquid Chromatography
edited by M. Kastner

Volume 62 Capillary Electrochromatography
edited by Z. Deyl and F. Svec

Volume 63 The Proteome Revisited: theory and practice of all relevant electrophoretic steps
by P.G. Righetti, A. Stoyanov and M.Y. Zhukov

Volume 64 Chromatography—a Century of Discovery 1900–2000. The bridge to the sciences/technology
edited by C.W. Gehrke, R.L. Wixom and E. Bayer

Volume 65 Sample Preparation in Chromatography
by S.C. Moldoveanu and V. David

Volume 66 Carbohydrate Analysis by Modern Chromatography and Electrophoresis
edited by Z. El Rassi

Volume 67 Monolithic Materials: Preparation, Properties and Applications
edited by F. Svec, T.B. Tennikova and Z. Deyl

Volume 68 [illegible] Techniques in [illegible] Analysis
by [illegible]

Volume 69A Chromatography, 6th edition. Fundamentals and applications of chromatography and related differential migration methods
Part A: Fundamentals and techniques
edited by E. Heftmann

Volume 69B Chromatography, 6th edition. Fundamentals and applications of chromatography and related differential migration methods
Part B: Applications
edited by E. Heftmann